The Continuum Companion to Philosophical Logic

The *Continuum Companions* series is a major series of single volume companions to key research fields in the humanities aimed at postgraduate students, scholars and libraries. Each companion offers a comprehensive reference resource giving an overview of key topics, research areas, new directions and a manageable guide to beginning or developing research in the field. A distinctive feature of the series is that each companion provides practical guidance on advanced study and research in the field, including research methods and subject-specific resources.

The Continuum Companion to Continental Philosophy, edited by John Mullarkey and Beth Lord
The Continuum Companion to Locke, edited by S.-J. Savonious-Wroth, Paul Schuurman and Jonathan Walmsley
The Continuum Companion to Philosophy of Mind, edited by James Garvey
The Continuum Companion to the Philosophy of Science, edited by Steven French and Juha Saatsi

Forthcoming in Philosophy:

The Continuum Companion to Aesthetics, edited by Anna Christina Ribeiro
The Continuum Companion to Berkeley, edited by Bertil Belfrage and Richard Brook
The Continuum Companion to Epistemology, edited by Andrew Cullison
The Continuum Companion to Ethics, edited by Christian Miller
The Continuum Companion to Existentialism, edited by Jack Reynolds, Felicity Joseph and Ashley Woodward
The Continuum Companion to Hegel, edited by Allegra de Laurentiis and Jeffrey Edwards
The Continuum Companion to Hobbes, edited by S.A. Lloyd
The Continuum Companion to Hume, edited by Alan Bailey and Dan O'Brien
The Continuum Companion to Kant, edited by Gary Banham, Nigel Hems and Dennis Schulting
The Continuum Companion to Leibniz, edited by Brendan Look
The Continuum Companion to Metaphysics, edited by Robert Barnard and Neil A. Manson
The Continuum Companion to Political Philosophy, edited by Andrew Fiala and Matt Matravers
The Continuum Companion to Plato, edited by Gerald A. Press
The Continuum Companion to Pragmatism, edited by Sami Pihlström
The Continuum Companion to Socrates, edited by John Bussanich and Nicholas D. Smith
The Continuum Companion to Spinoza, edited by Wiep van Bunge
The Continuum Companion to Philosophy of Language, edited by Manuel Garcia-Carpintero and Max Kolbel

The Continuum Companion to Philosophical Logic

Edited by

Richard Pettigrew

and

Leon Horsten

continuum

Continuum International Publishing Group
The Tower Building
11 York Road
London SE1 7NX

80 Maiden Lane
Suite 704
New York, NY 10038

British Library Cataloguing-in-Publication Data
A catalogue record for this book is available from the British Library.

ISBN: 978-1-4411-5423-1

Library of Congress Cataloging-in-Publication Data
Library of Congress Cataloging-in-Publication Data
The Continuum companion to philosophical logic / edited by Leon Horsten and Richard Pettigrew.
p. cm.
Includes bibliographical references (p.) and index.
ISBN 978-1-4411-5423-1
1. Logic, Symbolic and mathematical. I. Horsten, Leon. II. Pettigrew, Richard.
BC135.C57 2011
160–dc22

2010052876

Typeset by Newgen Imaging Systems Pvt Ltd, Chennai, India
Printed and bound in Great Britain

Contents

List of Illustrations vii

1 Introduction 1
Leon Horsten and Richard Pettigrew

2 Mathematical Methods in Philosophy 14
Leon Horsten and Richard Pettigrew

How to Use This Book 27
Leon Horsten and Richard Pettigrew

3 Logical Consequence 29
Vann McGee

4 Identity and Existence in Logic 54
C. Anthony Anderson

5 Quantification and Descriptions 77
Bernard Linsky

6 Higher-Order Logic 105
Øystein Linnebo

7 The Paradox of Vagueness 128
Richard Dietz

8 Negation 180
Edwin Mares

9 Game-Theoretical Semantics 216
Gabriel Sandu

10 Mereology 271
Karl-Georg Niebergall

11 The Logic of Necessity 299
John Burgess

12 Tense or Temporal Logic 324
Thomas Müller

13 Truth and Paradox 351
Leon Horsten and Volker Halbach

14 Indicative Conditionals 383
Igor Douven

15 Probability 406
Richard Pettigrew

16 Pure Inductive Logic 428
J. B. Paris

17 Belief Revision 450
Horacio Arló Costa and Arthur Paul Pedersen

18 Epistemic Logic 503
Paul Égré

19 Logic of Decision 543
Paul Weirich

20 Further Reading 575
Leon Horsten and Richard Pettigrew

Bibliography 582

General Index 629

Author Index 636

List of Illustrations

Figures

Figure 17.1	*Sphere-Based Revision* (the case in which $\phi \in K \backslash \mathrm{Cn}(\emptyset)$). The grey region represents $f_{\mathscr{S}}([\![\phi]\!])$, which generates the revision of K by ϕ, $K * \phi = \mathrm{Th}(f_{\mathscr{S}}([\![\phi]\!]))$.	467
Figure 17.2	*Maxichoice Contraction* (the case in which $\phi \in K \backslash \mathrm{Cn}(\emptyset)$). The small grey disc represents the singleton proposition $\{w\}$ selected by $f_{\mathscr{S}}([\![\neg\phi]\!])$, generating the contraction of K by ϕ, $K \dot{-} \phi = K \cap \mathrm{Th}(f_{\mathscr{S}}([\![\neg\phi]\!])) = \mathrm{Th}([\![K]\!] \cup \{w\})$	469
Figure 17.3	*Full Meet Contraction* (the case in which $\phi \in K \backslash \mathrm{Cn}(\emptyset)$). The large grey region in the upper right corner represents the proposition $[\![\neg\phi]\!]$ selected by $f_{\mathscr{S}}([\![\neg\phi]\!])$, generating the contraction of K by ϕ, $K \dot{-} \phi = K \cap \mathrm{Th}(f_{\mathscr{S}}([\![\neg\phi]\!])) = \mathrm{Th}([\![K]\!] \cup [\![\neg\phi]\!])$	469
Figure 17.4	*Partial Meet Contraction* (the case in which $\phi \in K \backslash \mathrm{Cn}(\emptyset)$). The grey lens represents the proposition given by $f_{\mathscr{S}}([\![\neg\phi]\!])$, generating the contraction of K by ϕ, $K \dot{-} \phi = K \cap \mathrm{Th}(f_{\mathscr{S}}([\![\neg\phi]\!])) = \mathrm{Th}([\![K]\!] \cup f_{\mathscr{S}}([\![\neg\phi]\!]))$	470
Figure 17.5	*Levi Contraction* (the case in which $\phi \in K \backslash \mathrm{Cn}(\emptyset)$). The grey region represents $[\![K \dot{-} \phi]\!]$	481
Figure 17.6	*Severe Withdrawal* (the case in which $\phi \in K \backslash \mathrm{Cn}(\emptyset)$). The grey disc represents $\min_{\subseteq}(\mathscr{C}_{\neg\phi})$, which generates the contraction of K by ϕ, $K \dot{-} \phi = \mathrm{Th}(\min_{\subseteq}(\mathscr{C}_{\neg\phi}))$	483
Figure 18.1	A model of Ann's uncertainty	507
Figure 18.2	A model for the uncertainties of Ann and Bob	511
Figure 18.3	Epistemic structure of the Email Game	514
Figure 18.4	Updating with p	517
Figure 18.5	A doxastic epistemic model	519
Figure 18.6	An update on plausibility	520
Figure 18.7	A different revision policy	520
Figure 18.8	Epistemic model and Action model	521
Figure 18.9	The effect of Ann privately learning p	522
Figure 18.10	An impossible world structure	524
Figure 18.11	Moore's formula: a case of unsuccessful update	540

Tables

Table 9.1 The payoff matrix of the *Prisoner's Dilemma* game 252
Table 9.2 The payoff matrix of *Matching Pennies* 253
Table 9.3 The payoff matrix of Γ in Example 9.5.3 266
Table 9.4 The payoff matrix of *Eloise* in the inverted *Matching Pennies* game 268
Table 17.1 If f satisfies a condition in column I and the adjoining constraint in column II, then $*$ satisfies the adjacent postulate in column III 476
Table 17.2 If $*$ satisfies a postulate in column I, then f satisfies the adjacent condition in column II 477

1 Introduction

Leon Horsten and Richard Pettigrew

Chapter Overview

1. A Brief History of Mathematics in Philosophy 2
2. Modelling with Formal Systems 4
   2.1 Classical First-Order Logic 6
   2.2 Other Logics 7
      2.2.1 Retaining FOL 7
      2.2.2 Revising FOL 8
      2.2.3 Extending FOL 9
3. Modelling Rationality 11

What is philosophical logic? In this volume, we take an unusual view. We say that philosophical logic covers all significant uses of *mathematical modelling* in philosophy. So what is mathematical modelling, and how can it be used in philosophy? The first is a vexed question, but roughly speaking, in science, a mathematical model is a mathematical structure that is taken by scientists to represent certain features of some part of physical reality; scientists then investigate that part of reality by investigating the mathematical model. Similarly, we claim, in philosophical logic, we take a particular philosophical subject of interest, such as a part of natural language whose logical structure we wish to understand better, or the metaphysical relation of part to whole, or the norms governing beliefs and actions; then we describe a mathematical structure that we take to represent the important features of this subject; and we investigate the subject by investigating that mathematical structure.

Thus, on our definition, we are doing philosophical logic when we model parts of our natural spoken and written language using a mathematical idealization of that language, which we call a *formal language*; and we are doing philosophical logic when we ask whether certain natural language inferences are valid by asking whether their counterparts in the formal language are formally valid within a particular axiomatic theory stated in that language. Now, most philosophers would grant this. Indeed, they would probably take this to be the archetypal activity of the philosophical logician. However, we also take to be doing philosophical logic the *decision theorist*, who models an agent by a probability function that measures the strengths of her beliefs and a utility function that measures the strengths of her desires, and who states norms governing how this agent ought to choose to act in terms of this model (Chapter 19). Now some philosophers may be slower to grant this categorization. They might complain, for instance, that this is not philosophical logic because no axiomatic theory is presented, nor any model of any part of our natural language given. These they might take to be the characteristic features of philosophical logic.

Nonetheless, there are strong reasons for delimiting the subject more widely than this philosopher would wish. After all, increasingly, we see mathematical techniques from outside traditional logic being applied even to the traditional parts of philosophical logic. For instance, in Chapter 9, Gabriel Sandu applies the techniques of *game theory* – an extension of decision theory – to the very traditional question in philosophical logic that asks what determines the truth or falsity of those sentences in natural language that we model using a first-order formal theory. Such cross-fertilization has proven enormously successful; indeed, it has often revitalized parts of the subject. This suggests that we would be wise not to circumscribe philosophical logic too narrowly, especially in a handbook that seeks to introduce the subject to students and researchers. So we will not.

1. A Brief History of Mathematics in Philosophy

Aristotle is one of the most important philosophers in the Western tradition. He is also the first logician. In Aristotle's view, logic is an integral and central part of philosophy. By contrast, for Aristotle, mathematics does not play a role in philosophy. The scientific revolution of the sixteenth and seventeenth centuries established the central role of mathematics in the physical sciences. And, for a while, it seemed that the idea of *mathesis universalis* would create a similar dominance of mathematics in philosophy, undermining Aristotle's view of their relationship. But, by the end of the seventeenth century, it became clear that the Aristotelian viewpoint would prevail for some time.

For a while, famous philosophers insisted that philosophy should be done *more geometrico*. But what they meant was not that geometrical methods ought to be applied to philosophical questions. Instead, they meant that philosophical theories should, like Euclidean geometry, be formulated in an axiomatic way. A philosophical theory should be expressed as a list of evident basic principles, from which all the claims of the theory can be derived. That this programme did not succeed was due to two reasons. First, the basic principles of philosophical theories did not appear to possess the self-evidence that the axioms of geometry apparently did have. Second, it remained unclear how the derived claims of a philosophical theory follow logically from its basic principles. Since the laws of propositional and predicate logic were not yet explicated, this was a defect that philosophy shared with geometry. But geometrical constructions seemed to provide the required certainty where explicit logical derivations were lacking. Philosophy did not have a counterpart to constructions with ruler and compass.

In the nineteenth century, the laws of propositional and predicate logic were uncovered by Boole, Frege, and others ([Boole, 1854a], [Frege, 1879]). On the one hand, the new logic was more mathematical in nature than Aristotle's syllogistics. Boole, for instance, explicated the connection between logic and algebra. And Frege famously sought to bring results concerning the logic of predicates to bear on problems in the foundations of mathematics. In sum, the relations between logic and mathematics became tighter than they previously were. And this was bound eventually to have an impact upon the relation between logic and philosophy.

From the beginning of the twentieth century onwards, the new mathematised logic started to exercise an influence on philosophical practice. According to certain forms of traditional empiricism, the world is a mental construction from sense experiences. But precise hypotheses about the way in which objects and properties are constructed out of experiences were lacking. The new machinery of modern logic was taken as a tool to make the construction hypotheses precise. The slogan became that the world is a *logical* construction based on sense experience. Russell and Carnap developed construction systems in which they sought to define such notions as that of a physical object and that of a physical property ([Russell, 1914], [Carnap, 1928]). But these construction programmes became somewhat discredited due to the critiques of Quine and Goodman ([Quine, 1951b], [Goodman, 1951]). Their objections focussed on the details of the construction systems that were proposed. Even though locally, it might sometimes be possible to 'logically construct' objects, properties, or relations from more basic ingredients, it seemed that the ultimate goal of constructing *everything* out of sense experiences was overly ambitious.

Since then, and in spite of the apparent failure of that project, logical methods, and mathematical methods more generally, have crept into all areas of philosophy. They are used in metaphysics in the study of modality, vagueness, and Leibniz's Principle of the Identity of Indiscernibles, among other things. In epistemology, such methods are used in the study of epistemic norms, and they have been used to explore the principles that govern such terms as *knowledge, belief, justification*, and *evidence*. In ethics, mathematical methods are used extensively by utilitarians, as well as those studying the theory of rational decision making. And in political philosophy, the mathematical study of voting systems is highly developed. This list provides the briefest glimpse into the vast and varied discipline of philosophical logic. In this volume, we aim to provide an introduction to the central topics in that discipline.

2. Modelling with Formal Systems

To which philosophical topics have the techniques of mathematical modelling been applied? They were first applied to a topic that lies partly in the philosophy of language, partly in epistemology, and partly in metaphysics. This is the *formalization* of parts of our language. This work was begun by Aristotle in the books known collectively as the *Organon*. But it was only made explicitly mathematical by George Boole, Gottlob Frege, and the extraordinarily innovative mathematical logicians of the early twentieth century, such as Bertrand Russell, Alfred Tarski, and Kurt Gödel.

Traditionally, there are four stages in the formalization of a particular part of our language. We describe them in a little detail here, since such formalization is exactly the topic of many of the chapters of the book. We illustrate each using the example of first-order logic, which we assume to be familiar.

(I) We call the first stage the *formal language stage*. In this stage, we present a mathematical structure known as a *formal language*. This abstracts from and idealizes the part of natural language that interests us, be it the part in which we talk of objects, or the part in which we talk of properties, or duties, or knowledge, or the modal part of our language. That is, the formal language represents what we take to be the important features of that part of our language, but leaves out unnecessary complications; and it removes certain complexities of that part of our language by approximating it rather than representing it accurately. Similarly, in science, a mathematical model of a ball rolling down an incline might represent certain important features of the physical situation, such as the effect of gravity, but leave out others, such as the effect of friction – in this sense, it *abstracts* from the physical reality. And it might approximate that physical reality rather than represent it accurately by, for instance, treating the ball as perfectly spherical, when in fact it is rather irregular – in this sense, it *idealizes* physical reality. Moreover, in science, we often construct and investigate a whole class of physical models of an aspect of reality. If we are interested in collisions, we might look at the class of all infinitely extended billiard tables on which perfectly spherical and rigid balls collide with each other in accordance with certain principles.

For instance, in first-order logic, the formal language consists of a set of sequences of symbols. The symbols, such as '$\forall$', '$\wedge$', '$\neg$', 'x_i', 'P_1^1',

'R_1^2', etc. belong to a set of symbols called the *alphabet* of the formal language. And the sequences are all and only those finite sequences of those symbols built up according to a certain set of grammatical rules.

(II) We call the second stage the *semantic stage*. In this stage, we provide what is called a *semantics* for the formal language. The part of our natural language that we modelled using a formal language in the formal language stage must speak about a particular part of reality. When we give a semantics for that formal language, we do two things: first, we provide a class of mathematical models of ways that part of reality might be; second, we say what features of such a model determine the semantic status of sentences in our formal language. For instance, we might say what features of a model belonging to the relevant class determine whether a particular sentence is true or false or some other truth value; or whether it has some other semantic status, such as provability or assertability.

For instance, in first-order logic, each of the various ways the part of reality in question might be is modelled by a *Tarskian model*, which consists of a set of objects – called the *domain* of the model – together with certain subsets of the domain, which interpret the predicate symbols such as 'P_1^1', and certain sets of tuples of the domain, which interpret the relation symbols such as 'R_1^2'. Tarski's celebrated definition of *satisfaction in a model* says how the truth value of a sentence is determined.

(III) We call the third stage the *axiomatization stage*. In this stage, we provide an axiomatic theory of our reasoning in the part of our language that we are modelling. That is, we describe what we claim are *axioms* and *rules of inference* in terms of the formal language. The resulting combination of formal language and axiomatic theory is called a *formal system*. We claim that the axioms are basic truths that may be taken for granted in reasoning; and we claim that the rules of inference are valid. We call such an axiomatic theory *sound*.

For instance, in first-order logic, there are various axiomatizations. In some, such as Hilbert's axiomatization, there are many axioms, but only one rule of inference, which is usually *modus ponens*. In others, such as natural deduction, or Gentzen's sequent calculus, there are very few axioms, but many rules of inference.

(IV) We call the fourth stage the *justification stage*. In this stage, we appeal to the semantics for the formal language (introduced in the semantic stage) to justify our claim that the axiomatic theory presented in the axiomatic stage is sound relative to that semantics – that is, the axioms are justified in terms of it, and the rules of inference are valid according to it. For instance, we might show that the axioms are true in all situations, and the rules of inference preserve truth from premises to conclusion.

In first-order logic, as in most formal systems, the soundness of an axiomatization is proved by mathematical induction on the length of sentences.

(V) We call the fifth stage the *completeness stage*. This is an optional stage in formalization, and indeed sometimes it cannot be carried out. In the justification stage, we show that our axiomatization is justified, and we conclude that any inference that can be derived in it is valid. In the completeness stage, we argue for the converse. That is, we argue that every valid inference can be derived in our axiomatization. We call such an axiomatic theory *complete*. Moreover, we might ask whether the set of valid inferences is *decidable*: that is, we might wish to know whether there is a purely mechanical procedure that can determine whether a given inference is valid or not in a finite number of steps.

In first-order logic, as in most formal systems, we usually prove completeness by proving its contrapositive version, which says that any inference that is not derivable is not valid; though it turns out that the theory is not decidable. As Øystein Linnebo explains in Chapter 6, in second-order logic with the most natural semantics, it is not possible to give a complete axiomatization. Thus, the completeness stage cannot always be carried out; it is thus an optional part of formalization.

Having seen how the formalization of a part of language proceeds, we turn to particular examples.

2.1 Classical First-Order Logic

Though he did not proceed with the degree of mathematical rigour we now expect, Aristotle made the first discoveries in philosophical logic when he began the formal language and axiomatization stages for that part of our language that talks of individuals and ascribes properties to them. We call this part of language the *first-order* part, because its subject matter is individuals, which are considered first-order entities. Second-order entities are then properties of first-order entities, or concepts under which first-order entities may or may not fall. And so on.

In particular, Aristotle was interested in a certain type of inference known as a *syllogism*, which is carried out in the first-order part of our language. An example:

All philosophical logicians know about mathematics.
Frege is a philosophical logician.
___
Therefore, Frege knows about mathematics.

Frege noticed that, in mathematics, there are species of first-order inference that are not formally valid according to Aristotle's axiomatization of syllogisms. Moreover, many of these inferences concern first-order statements that cannot even be represented in Aristotle's formal language: a famous example is Euclid's celebrated theorem that there are infinitely many prime numbers; its quantifier structure is too complex to be represented in Aristotle's formalization. Further still, no second-order statements can be represented in Aristotle's purely first-order formal language. Thus, Frege extended the ambition of Aristotle's formalization. He developed the formal language stage of Aristotle's theory so that it could now represent the first- and second-order statements that it had originally omitted; and he developed the axiomatization stages to cover the greater range of inferences that could be stated using this extended language. Indeed, after the semantic and justificatory stages for first-order logic had been carried out by Tarski in 1928, it was possible for Gödel to prove in 1930 that a modification of the first-order part of Frege's axiomatization in fact captures all and only valid first-order inferences: this is called the *completeness theorem for first-order logic*. In what follows, we will call the classical version of first-order logic that results *FOL*. For a more detailed history of FOL, see Chapter 3. For the history of the second-order case, see Øystein Linnebo's Chapter 6.

2.2 Other Logics

Much subsequent philosophical logic has sought to do for other parts of language what classical first-order logic (FOL) did for first-order language. Many of the chapters of this volume are devoted to surveying these varied attempts. Typically, there have been three approaches ([Gabbay and Guenthner, 1989]).

2.2.1 Retaining FOL

On the first, we take the formalization of first-order language in FOL to have succeeded, and we try to widen its scope by accommodating within that formalization the further part of natural language that interests us. Thus, for instance, in Chapter 5, Bernard Linsky describes the various attempts to accommodate in FOL formal versions of the *definite* and *indefinite descriptions*, such as 'a big brown dog' and 'the tallest man in the room', that occur in natural language. Can they be accommodated without changing the formalization or the axiomatization of FOL, as Russell thought? Or must we introduce new features of our formal language to model them, and new axioms and rules of inference to capture the inferences that we make involving them?

Similarly, in Chapter 4, Anthony Anderson considers whether we are right to formalize our talk of *existence* as we do in FOL. There, we follow Kant and Frege in taking existence not to be a property of first-order entities. Rather, we take it to be a logical operation that acts on a formula with a distinguished variable to give back a new formula, which is considered true if the extension of the original formula is non-empty. Does this formalization succeed in modelling accurately our existence claims and the reasoning about existence in which we engage?

Soon after Gödel discovered that FOL represents all and only the valid first-order inferences, Tarksi showed that one natural and intuitively plausible way of formalizing our talk of truth will not work, because it is inconsistent: that is, it leads to contradiction. His idea was based on the ancient *paradox of the liar*. Suppose we state, as minimal assumptions in our account of truth, the following sentence, for each formula φ of our formal language:

$$\text{`}\varphi\text{' is true if, and only if, } \varphi.$$

Then this will lead to a contradiction when applied to the *liar sentence L*, which says of itself that it is not true. (Try it!) Using techniques that Gödel developed in his famous *incompleteness theorems for arithmetic*, Tarski showed that the liar sentence is a grammatically well-formed sentence that we can formalize in a first-order language. With this, Leon Horsten and Volker Halbach begin chapter 13, which goes on to survey how we might proceed to give a formal account of truth in the face of this paradox. Thus, by giving a first-order formalization of quotation of expressions, reasoning involving the truth predicate can be studied in FOL. This can be done coherently if certain sentences of the form 'φ is true if, and only if, φ' are rejected. Nonetheless, some philosophical logicians feel that even when quotation is formulated correctly, a satisfactory logical treatment calls for more drastic measures: revising FOL itself. To this stratagem we now turn.

2.2.2 Revising FOL

On the second approach to formalization, it is contended that the formalization of first-order language in FOL is not completely successful. That is, it is claimed that this formalization has omitted to represent certain important features even of first-order language, or that it has misrepresented those features. Thus, for instance, Anthony Anderson considers whether we might drop the FOL assumption that names always succeed in naming something, and then model existence as a predicate that applies to names exactly when they do so succeed. The motivation is that, in FOL, the formal inference that models the following inference is valid:

Pegasus is a winged horse.
___
Therefore, Pegasus exists.

But this seems absurd.

In Chapter 7, Richard Dietz considers another apparent shortcoming of FOL. In that formalization, we model the adjectives of our natural language as predicates and we assume that the application of a predicate is a determinate matter: that is, we assume that, for any first-order entity, and any predicate, either the predicate applies to the entity or it does not. But it seems that this idealization ignores an important feature of our language: our adjectives are often *vague*. For instance, the application of the adjective 'bald' to men seems not to be a determinate matter. If it were, it seems, there would have to be a particular number n, such that a man with n hairs on his head is bald, while a man with $n + 1$ hairs is not. Dietz describes the proposed solutions to the apparent paradox.

The problems with the FOL representation of conditional statements are notorious. These are often known as the *paradoxes of material implication*. In FOL, conditional statements of the form 'If A, then B' are represented in such a way that they are equivalent to 'B, or it is not the case that A, or both'. Two apparently worrisome consequences: a conditional 'If A, then B' is true if its antecedent A is false, or its consequent B is true, or both. Thus, on this interpretation, the statement 'If grass is blue, then the sky is green' is true, as is the statement 'If grass is green, then the sky is blue'. In Chapter 14, Igor Douven describes the putative solutions to these problems that have been proposed. Some involve changes to FOL, while some require us to extend the formalization of first-order inferences and use the new machinery available in the extension to give a better representation of conditional statements.

As mentioned above, in Chapter 9, Gabriel Sandu discusses a proposed change to FOL that does not involve the way in which it represents parts of natural language. Rather, the game-theoretic semantics he presents offers an alternative understanding of the way in which first-order sentences get their truth value. It is not directly in terms of their correspondence to reality; rather it is in terms of the sort of games that might be played between two people, one of whom wishes to verify the sentence, while the other wishes to refute it. This speaks to a particular philosophical view of how language works, and how meaning attaches to it.

2.2.3 Extending FOL

On the third approach to formalization, we do not attempt to accommodate the particular part of natural language within FOL. Rather, we accept that we must create a new formalization to model the part of natural language that interests us. In this case, we extend the logical vocabulary of the language of first-order logic by one or more new logical symbols. Treating these new symbols as logical symbols means that their interpretation will be kept invariant in the class of models that we are interested in. Now almost all symbols of first-order logic are usually treated *syncategorematically*. What this means is that the semantics explicates how these logical symbols contribute to the interpretation of larger wholes without assigning extensions to these symbols. In FOL, the only exception to this is the identity symbol: its extension consists of all ordered pairs of the form $\langle a, a \rangle$ where a belongs to the domain. Logical symbols that are added to FOL are usually treated syncategorematically.

To take an example of a notion that is treated syncategorematically, in higher-order logics, we attempt to formalize the following sort of statement: 'Every collection of natural numbers has a least element'. And the following sort of inference:

> Cicero and Tully share all the same properties.
> Cicero is Roman.
> ———————————————
> Therefore, Tully is Roman.

That is, we formalize statements and inferences that concern second-order entities. Due to a mathematical result by Georg Cantor – which was later exploited by Russell, and came to be known as *Russell's paradox* – we are not able to represent all such statements and inferences adequately in a first-order formal system. Contrary to what we might expect, we cannot accommodate properties as a special type of first-order entity without severely restricting the number of properties we can countenance. In Chapter 6, Øystein Linnebo describes how we might alter our formal language, our semantics, and our axiomatization in order to model this part of our language. This raises interesting questions in the philosophy of mathematics, where such second-order reasoning seems to be required: Which ontological commitments of our mathematical language are revealed by this formalization? How does our use of this part of natural language fix its semantics? In particular, what is it about our use of quantifiers that range over properties that determines which properties fall in that range?

Other parts of our language that are simply not captured by FOL include our reasoning about the relation of part to whole, about possibility, necessity and other modal notions, our reasoning about the truth of propositions at different times, and our reasoning about truth simpliciter. These are the subject matter of chapters 10–13.

In Chapter 10, Karl-Georg Niebergall asks what assumptions we should make about the relation of part to whole. Is it true that, whenever there is a collection of objects sharing a particular property, there is a smallest object of which each of those original objects is a part? And what is the ontological status of this object? Is it genuinely a *further* object? If it is not, can we exploit these so-called *mereological fusions* of objects to provide a nominalistically acceptable foundation for mathematics?

In Chapter 11, John Burgess surveys the many ways in which we might formalize our modal inferences. Is necessity an operator that acts on a sentence to produce a new sentence, or is it a property that sentences may or may not have? That is, should we treat it syncategorematically, as we treat the quantifiers and logical connectives of FOL, or is it a predicate with a definition given independent of its context in a formula? This is part of the formalization stage for modal logic. Whichever of these two options we choose, we must then say what assumptions we ought to make about it in our reasoning. Should we assume, for instance, that any sentence that is necessary is necessarily necessary? This is part of the axiomatization stage.

The same questions arise for our reasoning about time and the truth of propositions at times. Indeed, there is a close connection between the logic of time and the logic of necessity. This was recognized by Aristotle, for whom their philosophy was also closely linked. In Chapter 12, Thomas Müller surveys many of the techniques developed in modal logic to formalize our reasoning about time and tense.

Again, it is not always clear whether the logical treatment of a philosophical notion calls for an extension or a revision of first-order logic. In Chapter 8, Edwin Mares surveys the proposed alternatives to the representation of negation in FOL. Some philosophical logicians think that FOL does not give an accurate treatment of negation. Others think that the situation is rather more complicated. They believe that beside the classical concept of negation, we also have a concept of negation the meaning of which is not classical. These logicians would argue that a correct treatment of negation calls for an extension rather than for a revision of classical logic.

In short, the project of formalizing our reasoning about various parts of reality is rich, varied, and complicated. Different problems arise when we attempt to formalize different parts of our language. And we must use different strategies to overcome these problems. But, while this is certainly the traditional project of philosophical logic, it is by no means the only area of philosophy in which mathematical modelling has yielded insight, arguments, and results. Another important area of philosophical logic might be called the *theory of rationality*, whether that rationality is *theoretical* (or *epistemic*) and concerns beliefs, knowledge, and evidence, or whether it is *practical* and concerns actions, preferences, and choices. The chapters in the final section of the volume describe the main topics in this area.

3. Modelling Rationality

In Chapter 18, Paul Egré treats the topic of theoretical (or epistemic) rationality using the techniques of formalization employed in the previous chapters. In particular, he uses the apparatus of formal systems to model our talk of those belief states that count as knowledge, and the inferences we draw concerning them. In this topic, it is the axiomatization stage that proves most problematic. The *paradoxes of knowability* reveal that seemingly plausible universal assumptions about knowledge have strong and apparently unwelcome consequences. For instance, Fitch's notorious paradox appeals to seemingly innocuous assumptions to derive, from the premise that there is an *unknown* truth, the conclusion that there is an *unknowable* truth. But this seems too strong at first blush. Which assumptions are responsible?

Chapters 16 and 17, and the second half of Chapter 15, also treat theoretical rationality, but they do not begin by formalizing our natural language talk about such rationality. Rather, they present detailed models of epistemic states, and use these to state and justify norms that are taken to govern these states.

In Chapter 17, Horacio Arló Costa and Paul Pedersen treat belief rather than knowledge. And, in particular, they treat the problem of how we ought to update our beliefs in the face of new evidence that is presented to us as a proposition that we come to believe. In the theory of *belief revision*, which they describe, we model an agent's epistemic state at a particular time in terms of the set of propositions that the agent believes at that time, as well as a measure of the degree to which the agent is unwilling to give up her particular beliefs in the face of new evidence; that is, the degree to which the beliefs are *entrenched* for the agent at that time, where that degree is given an ordinal rather than quantitative measure.

On the other hand, in Chapter 16 and the second half of Chapter 15, Jeff Paris and Richard Pettigrew take an epistemic state at a particular time to be modelled by a mathematical function called a *belief function*, which takes each proposition about which the agent has an opinion at that time, and returns a real number that measures the degree to which the agent believes that proposition.

In Chapter 16, Paris exploits considerations of symmetry to explore the norms that govern the epistemic state that an agent ought to have at the beginning of her epistemic life, prior to learning any evidence whatsoever. And, in the second half of Chapter 15, Pettigrew surveys the justifications that have been given for the norm that demands of an agent that her belief function at any time in her epistemic life be a probability function.

Finally, in Chapter 19, we turn to practical rationality. In particular, Paul Weirich extends the degrees of belief model of an agent, which consists only of her belief functions at various times, by adding also a *utility function*, which might be thought of as a measure of the strength of the agent's desires for various outcomes of possible actions she might perform. He then describes how we might combine these two aspects of an agent's state in order to determine norms that govern how the agent ought to decide to act in given situations.

As the seventeen chapters of this volume attest, the use of formal methods and mathematical modelling is widespread, powerful, and fruitful. They are employed with significant gain in almost every discipline and subdiscipline of philosophy. Enormous progress has been made in the last century. But each chapter also contains a sketch of work that remains to be done in the future, as well as ongoing research efforts whose outcomes we await. Philosophical logic is a live discipline that holds much promise for the future. We hope that this volume will encourage young researchers to enter it, as well as more established philosophers who have previously been wary of formal methods. In Chapter 2, we try to give some advice for those using formal methods for the first time: we describe strategies that might prove fruitful, and methodology that guards against some of the more common pitfalls.

2 Mathematical Methods in Philosophy

Leon Horsten and Richard Pettigrew

Chapter Overview

1. Introduction 14
2. Logical and Conceptual Analysis 15
3. Logical Models and Possible Worlds 17
4. Mathematical Models in Philosophy 19
5. The Art of Mathematical Modelling 22

1. Introduction

Despite the fact that philosophical logic is a fairly mature discipline, since the demise of logical empiricism there has not been much critical reflection on the research methods that are used in it. Such an investigation is now long overdue. Research in philosophical logic over the previous decades has made it abundantly clear that philosophical logicians neglect methodological questions at their own peril.

Because of the growing acceptance of the use of formal methods in philosophy, the interest in the methodology of philosophical logic is fortunately increasing at present. This is witnessed, for instance, by the fact that, in recent years, workshops or sessions on the use of formal methods in philosophy appear in large international conferences on analytic philosophy. The literature on this subject goes mostly under the heading of 'formal methods in philosophy', or 'formalisation'. Some relevant recent articles on the role of logic and formal methods in philosophy are [Engel, ta], [Hansson, 2000], [Horsten and Douven, 2008], [Leitgeb, ta], [Müller, taa], [Löwe and Müller, ta], [Suppes, 1968], [van Benthem, 1982], [Wang, 1955].

In this chapter, we stand above the topics in philosophical logic discussed in this volume, and look down on them in order to investigate the methodology of formal methods that they employ. We hope to reveal the great power that these methods hold, but also to delimit what they can hope to achieve.

To a considerable extent, we will track the changes in the methodology of philosophical logic since the emergence of the discipline at around the turn of the twentieth century. Very roughly, three periods can be distinguished. The first period can be seen as a syntactic stage. It begins with Russell's investigation of the logical structure of definite descriptions, and ends in the 1950s. The second stage is characterized by a dominance of possible worlds semantics. It begins in the late 1950s, and comes to a close somewhere in the 1980s. Presently we find ourselves in a period where models drawn from an increasing variety of branches of mathematics are used to investigate philosophical problems. This widening of the methods used has resulted in a re-definition of the discipline of philosophical logic.

2. Logical and Conceptual Analysis

From the beginning of the twentieth century onwards, the new logical methods developed by Boole and Frege came to be used to analyse the logical structure of language and the conceptual relations between concepts. The idea was that when confronted with a philosophical problem, one should address it as follows. As a first step, the problem must be formalized: that is, it must be at least roughly expressed in the language of first-order logic – in the Introduction, we called this the *formal language stage*.

Showing that formalization is possible in a uniform way is in effect showing that our natural language, English, has a precise syntax. This is a highly non-trivial undertaking, as the chapter on definite descriptions by Bernard Linsky attests. As a matter of fact, it appears that one needs the syntax of intensional operators, or something like it, and even a formalized pragmatic theory to handle most of natural language. Even in this first stage of philosophical logic, there are great discoveries yet to be made.

But even so, formalization is not enough. If formalization is restricted to translation to first-order logic, then the following criticism made by Hao Wang cannot be altogether dismissed:

> [W]e can compare many of the attempts to formalise with the use of an airplane to visit a friend who lives in the same town. Unless you simply love the airplane ride and want to use the visit as an excuse for having a good time in the air, the procedure would be quite pointless and extremely inconvenient. ([Wang, 1955, p. 233])

So, more needs to be done. The key philosophical concepts in the formalization must be identified. As a next step, basic principles that express how these philosophical notions are related to other philosophical notions must be articulated, and again must be (at least roughly) formulated in first-order logic. Also, pre-theoretical convictions of truths involving these philosophical concepts must be spelled out – in the Introduction, we called this the *axiomatization stage*. Then, a precise hypothesis concerning the philosophical problem is put forward. Subsequently, a determined attempt is made to logically derive the hypothesis from the basic principles and the pre-theoretical data. If this attempt is successful, then an answer to the philosophical question has been obtained. (This answer is then, of course, not immune to criticism.) If the attempt is not successful, then the exercise has to be repeated. Perhaps more or different basic principles concerning the key philosophical notions are required – that is, we might need to repeat the axiomatization stage.

It is also possible that more facts are needed. These are traditionally taken to be supplied by our philosophical intuitions, but our intuitions are not sacrosanct either; they may be overridden by theoretical considerations ([Williamson, 2007a]). It is an integral part of philosophical research to distill the stable phenomena that must be accounted for from the raw and variable data ([Löwe and Müller, ta]).

In some instances, logical analysis can show that what seem to be genuine philosophical problems are in fact ill-conceived, or in some cases even outright senseless ([Carnap, 1935]). To take an example, traditional philosophy poses the question of the nature of being. But, according to some philosophical logicians, logical analysis shows that existence is not a predicate that expresses a property that some entities have and others lack. If there is indeed no property of existence that is expressed by the word 'exists', then it makes no sense to ask for its essence. In other cases, what appears to be a philosophical question might in the end turn out to be an empirical question. Consider, in this respect, the question about the meaning of life. According to one analysis, this might be taken to be the question: what causes a man or woman to continue living and not to commit suicide? If this analysis is correct, then the answer might not only contain a hidden parameter (which man? which woman?) but it might also be that it cannot be answered on *a priori* grounds. Instead, one would have to conduct an empirical investigation into the matter.

This approach came to be known as *conceptual analysis*, and the *locus classicus* here is Russell's logical analysis of definite descriptions ([Russell, 1905b]) (see Chapter 5). The role of logic in the traditional sense in this methodology is clear. Logical formalization forces the investigator to make the central philosophical concepts precise. It can also show how some philosophical concepts and objects can be defined in terms of others. If it emerges that certain objects are 'constructed' as classes of other objects, ontological clarification is achieved. Insisting on logically valid derivation, moreover, forces the investigator to make all assumptions that are needed fully explicit. As a result of this procedure, precise answers to philosophical questions are obtained. And if a conjectured hypothesis cannot be derived from known basic principles and data, then there must be hidden assumptions that need to be explicitly articulated. For instance, in his formal investigation of Euclidean geometry, Hilbert uncovered congruence axioms that implicitly played a role in Euclid's proofs but were not explicitly recognized ([Hilbert, 1899]). Naturally, such results are often obtained by model-theoretic techniques. That is, in order to show that a given conjecture does not follow from a collection of premises, a model is constructed in which the premises hold but the conclusion fails. By the completeness theorem for first-order logic, such a countermodel can always be found to demonstrate the invalidity of an inference.

Russell expresses the function of the method of logical analysis as follows:

> Although . . . comprehensive construction is part of the business of philosophy, I do not believe it is the most important part. The most important part, to my mind, consists in criticising and clarifying notions which are apt to be regarded as fundamental and accepted uncritically. As instances I might mention: mind, matter, consciousness, knowledge, experience, causality, will, time. I believe all these notions to be inexact and approximate, essentially infected with vagueness, incapable of forming part of any exact science. Out of the original manifold of events, logical structures can be built which have properties sufficiently like those of the above common notions to account for their prevalence, but sufficiently unlike to allow a great deal of error to creep in through their acceptance as fundamental. ([Russell, 1956, p. 341])

After World War II, this view also came under pressure. Wittgenstein in his later work became sceptical about the philosophical usefulness of formalization in logical languages. He still thought that analysis of philosophical positions and arguments is the key to the dissolution of philosophical perplexities. But he and his followers in Oxford thought that such an analysis can be carried out perfectly well in ordinary English. What we should be after is the *grammatical structure* of philosophical problems, not the first-order logical structure of such problems ([Wittgenstein, 1953]). This counter-movement to the Russellian programme, which is called *Ordinary Language Philosophy*, was very influential from the 1950s until the 1970s.

3. Logical Models and Possible Worlds

Models have been used on a daily basis in the physical and social sciences. One speaks of the Bohr model of the atom, techniques from fluid dynamics are used to model the flow of traffic, and so on. But until the 1930s, models were not used to study philosophical problems, theses, theories, and arguments.

In a monumental achievement, Tarski articulated the logical concept of a model and the notion of truth in a model ([Tarski, 1983b]) (see Chapter 3). A (logical) model is a set with functions and relations defined on it that specify the denotations of the non-logical vocabulary. A series of recursive clauses explicate how the truth values of complex sentences are compositionally determined on the basis of the truth values of their parts. This allowed an explication of the informal notion of logical consequence. A sentence ϕ follows logically from a collection of sentences Γ if and only if every model that makes every sentence in Γ true, also makes ϕ true. Similarly, a sentence is logically true if and only if it is true in all models. By Gödel's completeness theorem, the notion of logical consequence extensionally coincides with that of logical derivability, and the notion of logical truth extensionally coincides with that of logical provability.

Tarski's work yielded a new way of investigating the logical relations between philosophical concepts and between sentences that express philosophical theses. The model-theoretic approach in philosophy was at first closely associated with the investigation of the philosophical concept of truth. The liar paradox plagued attempts to provide coherent formal theories of truth. The model-theoretic or semantical perspective has proved to be very important in this area. By constructing models for certain formal truth theories, these theories were shown at least to be coherent or consistent (see Chapter 13). In a similar way, the model-theoretic perspective was very useful in evaluating proposed theories of the mereological notions of part and whole (see Chapter 10).

After the Second World War, the Tarskian notion of model was extended. A Tarskian model can in a sense be seen as a possible state of affairs. The logical properties of intensional notions such as 'it is possible that' and 'it is morally obligatory that' do not depend merely on how matters stand in one state of affairs; they depend on what is true in many possible states of affairs. In other words, we need a concept of model in which many possible states of affairs, or possible worlds, are represented. Such models are known as *possible worlds models*. They were developed in the late 1950s by Kripke and others ([Goldblatt, 2007]).

Philosophical interests have shaped the notion of a possible worlds model. Possible worlds models came to be used in philosophy from the 1960s onwards. They have been and still are used as a framework for debates in metaphysics, epistemology, and in the philosophy of language. Around this time, the term 'philosophical logic' came to be widely used. It was almost synonymous with the investigation of possible worlds interpretations of intensional notions. These intensional notions (possibility, moral obligation, temporal notions, epistemic notions, ...) were regarded as notions that are dear to philosophers' hearts. They were contrasted with extensional notions (set, number, ...) that do not require the extended notion of model. This development has led to the flourishing sub-disciplines of philosophical logic such as modal logic, epistemic logic, deontic logic, tense logic, and so on.

One feature of intensional logic in the possible worlds style is that it is philosophically not as neutral as first-order logic. Each possible worlds model contains a set of possible worlds. For this reason, possible worlds semantics is often charged with smuggling in heavy metaphysical commitments.

4. Mathematical Models in Philosophy

For a long time, it was thought that possible worlds models are the appropriate models for philosophical logic. The notion of a possible worlds model was further extended (resulting in the concept of a 'spheres model') in order to obtain a satisfactory logical treatment of counterfactual conditional sentences ([Lewis, 1973]). And in epistemic logic, even 'impossible worlds' were introduced. In this way, possible worlds semantics has proved to be an incredibly versatile modelling tool.

But from the 1960s onwards, different kinds of models made their entrance in formal approaches to philosophical problems. In the heydays of logical empiricism the method of logical analysis, as described in Section 2, was used to elucidate the confirmation relation between theory and empirical evidence. But in the early 1950s, arguments were constructed that purported to show that a satisfactory syntactic analysis of the confirmation relation can never be found ([Goodman, 1954]). In response to this impasse, philosophers of science began to try to model the confirmation relation in probabilistic terms. In a parallel development, probabilistic models came to be used in order to describe the logic of conditionals. In the first decades of the twentieth century, logicians held that the logic of indicative conditionals was adequately explicated by the truth conditions of the material implication. In the second half of the twentieth century, by contrast, logical theories of conditionals were constructed using methods from intensional logic and from probability theory. These approaches proved to be more faithful to the inferential relations that are actually operative in our conditional reasoning ([Adams, 1998]).

Probabilistic models are a different kind of model than possible worlds models or Tarskian models. Probabilistic models are *mathematical* models. Some would say that they are not *logical* models properly speaking, and that therefore logic is not of much help in investigating the confirmation relation or indicative conditionals. Others counter that logic should adopt a less austere and more pluralistic stance. The idea is that for different philosophical problems, different mathematical modelling techniques may be required. Every area of the mathematical sciences may in principle be drawn upon in philosophy for finding suitable models. In each instance, the challenge consists in finding the right tool for the job at hand. Perhaps this dispute is little more than terminological. But if logicians want to remain as relevant as possible to philosophy, then they are well advised not to spurn all classes of models other than first-order models or possible worlds models.

Probability theory is sometimes seen as a generalization of classical logic. This would make even probabilistic models logical in an extended sense. But the same cannot be said for the models that are used in disciplines such as game theory and decision theory, graph theory, algebra, or functional analysis. Yet today these mathematical disciplines are called upon to model philosophical problems. Game theory and decision theory are increasingly used to model problems in practical philosophy ([Binmore, 2009]), graph theory is used to model philosophical problems about secondary qualities and perceptual indiscriminability ([De Clercq and Horsten, 2005]), algebra is used to model mereological principles ([Niebergall, 2009b]), functional analysis is used to explore questions about epistemic norms ([Joyce, 2009]) (see Chapter 15).

Tarskian models are often taken to be static: they describe an existing state of affairs. The models that are used in contemporary philosophical logic often have a more dynamic character. For instance, the models studied in belief revision theory attempt to describe how belief states of cognitive agents change over time in response to new information (see Chapter 17). Game-theoretic models describe how players react to the 'moves' that are made by the other players (see Chapter 9). Thus contemporary modelling techniques allow us to obtain a deeper insight into dynamic phenomena.

As a result of these developments, the formal toolbox of the philosopher is now greatly expanded. And as a side effect of this, the distinction between philosophical logic and mathematical modelling in philosophy seems slowly to be evaporating. This should not be taken as a cause of concern. It is just that the term 'philosophical logic' should be given a wider scope than before. This will be reflected in the structure and content of this *Companion*. Unlike previous handbooks, guides, and companions to philosophical logic, this *Companion* will give due attention to the role that new mathematical modelling techniques play in philosophy.

In making use of these new mathematical methods philosophers are merely treading in the footsteps of the great philosophers of the first half of the twentieth century (such as Russell and Carnap) who were keen to make use of any of the (then) newest formal methods. It was merely a historical contingency that in this period axiomatic formal logic emerged as a new and exciting discipline full of promise for the future. If a *Wunderkind* of the calibre of Frank Ramsey were to turn his attention to the technical aspects of philosophy today, it is doubtful if he would accord as much attention to classical first-order logic as he in fact did. He would undoubtedly embrace the new methods enthusiastically.

Use of formal models in philosophy is sometimes described as *conceptual modelling* ([Müller, taa], [Löwe and Müller, ta]). Formal modelling can open conceptual possibilities. By isolating a class of models for a sub-discipline in philosophy, a space of possibilities is circumscribed. Such spaces are often far richer than one would expect. They provide a fertile and yet rigorous framework for exploring creative ideas:

> [. . .] as we unravel our concepts, their real wealth is unveiled, and hitherto unexplored possibilities are opened up. Hence formal precision is a stimulus to creative phantasy [. . .] The fashionable opposition of 'creative freedom' and 'logical armour' does not do justice to logic (nor, one fears, to creativity). ([van Benthem, 1982, p. 459])

Classes of mathematical models thus function as 'conceptual laboratories' in which theories about philosophical concepts can be tested 'in idealised circumstances' ([van Benthem, 1982, p. 459]).

Techniques from mathematical disciplines can also play a critical role in philosophy. For instance, according to an influential view, the human mind is structured like an algorithmic computing device such as a Turing machine. Within the computational paradigm of the mind, detailed hypotheses are sometimes proposed about the way in which humans solve certain mental tasks that they are in practice able to routinely solve quickly even in moderately complex situations. Results in complexity theory may show that the algorithms that are attributed to humans by the hypothesis are in fact intractable, i.e., that in moderately complex situations the relevant computations cannot be solved in a short time. This would cast doubt on the detailed hypothesis in the paradigm of the computational mind.

Computer simulations first came to be widely used in physics ([Galison, 1997], Chapter 8). In recent decades, computer simulations are also widespread in the rest of the physical and social sciences. Today, we see the first instances of computer simulations in philosophy of science ([Hegselmann and Krause, 2006]). It is to be expected that in the future, computer simulations will play a significant role in philosophy.

Despite these new developments, the 'traditional' methods of philosophical logic remain very powerful, and the traditional aspirations of philosophical logic remain very much alive. Even though the probabilistic approach in confirmation theory has undoubtedly shed new light on the problem of induction, substantial progress is being made on the project of finding the logic of induction, as Jeff Paris' chapter in this *Companion* shows (see Chapter 16). Even the project of the logical construction of the world from experience has in recent years seen a remarkable revival ([Leitgeb, 2007]).

5. The Art of Mathematical Modelling

Mathematical modelling in philosophy tends to take the focus away from formalization in the traditional sense of the word. Here is an example of how this can happen. Many probabilistic theories of indicative conditionals hold that indicative conditional sentences do not have truth values. Thus it cannot be a goal of such probabilistic theories to classify arguments in which conditional sentences function as premises in the traditional sense of the word. The goal rather becomes to explain how accepting conditional statements influences the personal probabilities that cognitive agents assign to other statements. As a second example, in decision theory the aim is not to formalize arguments that reasoners go through when contemplating which course of action to take or which strategies to adopt. Decision theory tries to classify actions or strategies as rational, without having to commit itself to any hypothesis as to how and why rational agents adopt a given strategy. The strategy in question might even be hard-wired as a result of evolutionary pressures, as far as decision theory is concerned.

Logicians engaged in attempts to model philosophical problems mathematically are often looking for representation theorems. These are theorems that state that if certain conditions are met, a representation of a class of phenomena in a given class of mathematical models exists. The hope is that the conditions of the representation theorem are somehow viewed as intuitively reasonable. Savage's representation theorem in the foundations of decision theory provides a good example ([Savage, 1954]). In this, Savage lays down what he hopes are intuitively plausible conditions that an agent's preferences must satisfy in order to be rational. He then shows that, for any agent who satisfies these conditions, there is a subjective probability function and a utility function that together give rise to the agent's preferences when combined using the theory of expected utility. He concludes that we should model rational agents using probability functions, utility functions, and the theory of expected utility. This, then, is one way in which a mathematical approach to philosophical problems tries to maintain the contact with our pre-theoretical intuitions. Of course these pre-theoretical intuitions are in no way sacrosanct. It is the business of philosophy to think very hard about their rational acceptability.

Also important is translations between classes of models ([Leitgeb, ta]). For instance, one might have on the one hand a class of probabilistic models for describing a class of phenomena, and on the other hand a class of possible worlds models for describing the same class of phenomena. Then it may be that for every probabilistic model, there is a possible worlds model that makes the same collection of statements true, and, conversely, for every possible worlds model, there exists a probabilistic model that makes the same collection of statements true. If our pre-theoretical intuitions are exhaustively expressed by a collection of sentences, then these two ways of modelling our pre-theoretical intuitions may be said to be 'equivalent' in a certain sense, namely, 'intuitionally' equivalent.

Mathematical models play a similar role in philosophy as they do in the sciences. They function as spectacles through which philosophical problems can be viewed. It is well known from the literature in philosophy of science that the use of models in science is far from theory-neutral. The same holds for the use of models in philosophy. Mathematical models are always impregnated with theoretical assumptions. Therefore the way in which a philosophical problem is mathematically modelled will necessarily involve non-trivial *philosophical* commitments. It is debatable whether the traditional method of logical formalization is completely philosophically neutral. But it seems that it is philosophically more neutral than the use of mathematical models.

Scientific realists argue that some models that are used in our successful empirical sciences can reasonably be taken to be approximately true. An argument of inference to the best explanation, or an argument from the likelihood principle, is invoked to argue that the models in question would most probably not have been so empirically successful if they had not approximated the truth ([Psillos, 1999]). In most other areas of philosophy, it is much harder to argue for the thesis that any mathematical models that are used represent the true state of affairs. The reason is simply that mathematical models in philosophy typically do not entail empirical predictions ([Hansson, 2000, p. 166]). The touchstone of success of models in philosophy seems to be agreement with our pre-theoretical intuitions. But this is more problematic than agreement with experiment. Even though observation and experiment are also theory-laden, they are certainly less so than our intuitions are [Engel, ta]. Indeed, it is typical of philosophical questions that they are centred around areas where we have conflicting or unclear intuitions. All this leads us to the conclusion that one should be extremely cautious when tempted to take mathematical models used in philosophy literally. Mathematical models and interpretations might shed light on a philosophical problem, but it will rarely if ever solve the philosophical problem once and for all.

Nonetheless, there are cases where mathematical models can bring order and perspective in our intuitions. Consider again the case of indicative conditionals. Sentences of the form $(\phi \rightarrow \psi) \vee (\psi \rightarrow \phi)$ are logical truths, when $\rightarrow$ is the material implication. But intuitively, we are inclined to think that such sentences are not generally logical truths. And a sentence such as 'If $2 = 3$, then plums are poisonous' just seems difficult to evaluate, even though, again, when 'if ... then' is interpreted as material implication, it is true. The probabilistic interpretation of conditionals agrees that sentences of the form $(\phi \rightarrow \psi) \vee (\psi \rightarrow \phi)$ do not generally have a high probability, so we should not accept them as true. Since conditional probabilities with antecedents that are impossible are (on standard treatments of conditional probability) ill-defined, probabilistic theories of conditionals explain why we find conditionals with impossible antecedents hard to evaluate. So perhaps the semantics of conditionals really does contain a probabilistic element.

In cases where mathematical models can make sense of our intuitions in such a way, there may be an element of truth in the models. In such cases, the mathematical models can be philosophically very fruitful. In the case of conditionals, one might think that even if the semantics of conditionals contains a probabilistic element, conditionals nevertheless also have truth values. After all, some indicative conditionals are believed, and the objects of doxastic attitudes typically are truth-evaluable. However, a famous impossibility result of Lewis shows that if the probabilities assigned to conditionals satisfy some minimal and seemingly reasonable conditions, then conditionals *cannot* also have truth values. This is a genuinely new and unexpected prediction of probabilistic treatments of conditionals.

In the sciences, models are often not taken literally at all, but often play a purely instrumental role. For instance, when engineers model the flow of traffic through a network of roads and highways by means of fluid mechanics, there is no implication that traffic really is a fluid. In philosophy, the stakes are always higher. We are not usually interested in merely 'saving the phenomena': we want to know how things really are. For this reason, a purely instrumental role is rarely played by models in philosophy. But this does not exclude subtle positions concerning the ontological significance of classes of models in philosophy. Consider possible worlds semantics for modelling modal logic. Kripke has always advocated it – indeed, he started the whole industry. But he is sceptical about Lewis' thesis that counterfactual worlds really exist. So for him, possible worlds models are illuminating, but they should not be taken to be real objects.

Mathematical modelling leads to rigour and precision in philosophical argumentation. It is the credo of analytical philosophy that precision is of the essence in philosophical theorizing; Williamson gives an eloquent defence of it:

> Precision is often regarded as a hyper-cautious characteristic. It is importantly the opposite. Vague statements are the hardest to convict of error. Obscurity is the oracle's self-defense. To be precise is to make it as easy as possible for others to prove one wrong. That is what requires courage. But the community can lower the cost of precision by keeping in mind that precise errors often do more than vague truths for scientific progress.
>
> Would it be a good bargain to sacrifice depth for rigor? That bargain is not on offer in philosophy, any more than it is in mathematics. No doubt, if we aim to be rigorous, we cannot expect to sound like Heraclitus, or even Kant: we have to sacrifice the stereotype of depth. Still, it is rigor, not its absence, that prevents one from sliding over the deepest difficulties, in an

> agonized rhetoric of profundity. Rigor and depth both matter: but while the continual deliberate pursuit of rigor is a good way of achieving it, the continual deliberate pursuit of depth (as of happiness) is far more likely to be self-defeating. Better to concentrate on trying to say something true and leave depth to look after itself. ([Williamson, 2007b])

Of course there are also dangers that are associated with mathematical modelling. One danger is that a model which is 'merely' mathematical is too easily taken literally. Perhaps this has happened in some instances with possible worlds semantics. Lewis has taken the possible worlds semantics to be literally true: there literally exist concrete possible worlds other than ours, they just aren't spatiotemporally connected to our world ([Lewis, 1986a]). Many philosophers argue that, in this case, the possible worlds theorist is led to make metaphysical assumptions for which she has no adequate justification. Of course in concrete instances it is very difficult to say whether a model can be taken literally. It depends on the connections with our intuitions. But it is nigh impossible to specify exactly when a model has shed enough light on our intuitions and has structured them sufficiently for it to be rationally allowed for us to take them at least in part literally. (Of course such judgements are always defeasible.)

Another danger of mathematical modelling is over-simplification ([Hansson, 2000, p. 168]). A model is intended to be a simplified representation of the real situation ([Leitgeb, ta]). But if a model fails to capture central features of the phenomenon under investigation, then the model should be regarded as defective. A case in point is the possible worlds semantics for epistemic logic. The objects of knowledge are propositions. In classical epistemic logic, propositions are identified with sets of possible worlds ([Hintikka, 1962]). This means that the sentence '$2 + 2 = 4$' expresses the same proposition as the sentence expressing Fermat's Last Theorem. So if a person knows that $2+4 = 4$, then it should follow logically that she knows that Fermat's Last Theorem is true. This is absurd. A reply that was given to this problem early on was to say that epistemic logic studies the notion of *implicit* knowledge. Now there may be such a notion of implicit knowledge according to which when one knows a sentence, one knows every sentence mathematically equivalent to it. But this is not the notion that epistemologists are typically interested in.

This contains a general lesson. In philosophical logic (broadly construed) mathematical modelling techniques should always remain servants of philosophy instead of the other way round. It should not be expected of the philosopher that she changes the concepts and problems she is interested in so as to make them fit the models; the models should fit the philosophical problems and concepts. If the models do not fit, then better models need to be sought.

Of course a philosophical logician may become interested in the mathematical properties of a class of models independently of the importance of this for applications to philosophy. For instance, one might want to spend a decade or so on investigating the algebraic properties of the lattice of normal propositional modal logics. But the investigator engaged in such an enterprise should have the intellectual honesty to admit that she is not working as a *philosophical* logician. It may be that the results that she obtains will, in some future time, yet turn out to play a role in philosophical applications. But that holds equally for any results in any branches of the mathematical sciences.

Yet another feature of mathematical modelling that one should be aware of is the law of diminishing returns ([Horsten and Douven, 2008, p. 159]). When a formal method has been applied in one area of philosophy, it is very natural to try to apply the same technique to problems in other areas of philosophy. But at some point the new applications will begin to look forced and somehow unnatural: the formal method does not succeed in shedding (new) light on the conceptual problems at hand.

An example may clarify this point. Possible worlds semantics was very successful in modelling the concept of necessity and has contributed greatly to contemporary metaphysics. It was natural to extend the framework of modal logic to the logic of time, and great successes were booked in this area also. But when the possible worlds semantics were further extended to model notions of knowledge and of moral obligation, the application was beginning to look distinctly forced and artificial. Once this stage is reached, it is better to look with an open mind for a better modelling technique.

To conclude, it is of paramount importance to maintain close contact, at every stage in the mathematical modelling process, with the philosophical problem under investigation. A good computer programmer documents every significant step of her programme. A good philosophical logician explains how her every technical move is motivated by aspects of the philosophical problem that she is trying to model ([Hansson, 2000, p. 170]).

There exists no algorithm or method that teaches one how to model philosophical problems successfully. Modelling is an art that can only be learned by carefully studying the paradigmatic mathematical approaches to philosophical problems from the past, and by acquiring a broad mathematical background. Any sub-discipline of the mathematical sciences can in principle play a role in modelling problems in philosophy. Having good teachers also helps enormously. Slavish imitation of the masters can of course only generate second-rate work. Truly innovative mathematical modelling in philosophy as elsewhere requires genuine creativity.

How to Use This Book

Leon Horsten and Richard Pettigrew

Any companion to a given subject serves two purposes. First, it must be possible to use the companion as a textbook from which to teach various courses on that subject. And second, it must serve as a reference book for those unacquainted with the details of particular parts of that subject.

As a reference book, it requires little explanation. The index provided is detailed, and many of the more technical chapters provide extremely good encyclopedias of results in their particular area, while the less technical chapters provide surveys of the philosophical positions taken in their area. Furthermore, Chapter 20 provides a host of references to allow readers to explore any of the topics covered in greater depth.

As a course textbook, some helpful points might be made. Designing a course in philosophical logic is a delicate balancing act. Make it too technical, and those keen to get at the philosophical meat will be left unsatisfied; make it too philosophical, and those drawn by the impressive technical edifices that have been created over the last hundred years of the subject will feel short-changed. In this book, we've tried to provide a balance of technical exposition and philosophical discussion: sometimes both are mixed evenly in a chapter; sometimes the philosophical discussion is contained in one chapter, while the technical exposition is given in another. We hope that this will allow teachers, lecturers, and professors to create a course that is equally balanced, and which provides enough of each component to keep everybody happy. Again, students looking for further detail on a particular topic are encouraged to consult the list of further reading in Chapter 20.

In this section, we suggest some course structures (chapters to be used are listed in the table below).

- *A second-year course* This is intended to be a broad course, which gives largely non-technical introductions to core topics in philosophical logic. This will be useful for students whether or not they intend to pursue study in philosophical logic itself. Each topic covered is used in many different and diverse areas of philosophy.
- *A third-year course* This is intended to be a more focused course, which introduces fewer topics, but treats them first philosophically and then technically in order to give a deeper understanding of the issues covered.

This course will give students a flavour of how philosophical logic is currently done, and it will provide them with the basic knowledge on which to build should they wish to write undergraduate dissertations on these topics.

- *A graduate course* This course assumes that students are aware of the sort of material covered in the lectures for the second-year course. It is intended to fill in technical knowledge, and to bring students to the cutting edge of the subject.

	Second year	Third year	Graduate
1	First-order logic (3)	First-order logic (3)	Vagueness (7)
2	Identity and existence (4)	Second-order logic (6)	Negation (8)
3	Definite descriptions (5)	Modal logic (11)	Games in Logic (9)
4	Conditionals (14)	Tense logic (12)	Mereology (10)
5	Second-order logic (6)	Negation (8)	Tense Logic (12)
6	Modal logic (11)	Truth (13)	Truth (13)
7	Vagueness (7)	Probability (15)	Inductive Logic (16)
8	Truth (13)	Inductive Logic (16)	Belief Revision (17)
9	Probability (15)	Epistemic Logic (18)	Epistemic Logic (18)
10	Decision Theory (19)	Decision Theory (19)	Decision Theory (19)

3 Logical Consequence

Vann McGee

Chapter Overview

1. Syllogisms 29
2. Sentential Calculus 31
3. Predicate Calculus 33
4. Truth in a Model 35
5. The Completeness Theorem 38
6. Logical Terms 42
7. Higher-Order Logic 44
8. Non-Mathematical Logic? 48
Notes 53

1. Syllogisms

Logical consequence is a hybrid notion. In part, it is a normative, epistemic notion. Logic teaches us how to reason well, by showing us patterns of reasoning with the happy property that, if we know the premises, we can know the conclusions. It is also a descriptive notion from semantic theory. φ is a logical consequence of Γ iff (if and only if) the forms of the sentences ensure that, if all the members of Γ are true, φ is true as well. What connects the two aspects is the thesis that truth is the norm of assertion and belief, so that valid arguments – arguments in which the conclusions are logical consequences of the premises – are forms of good reasoning that enable us to make good assertions.

The science of logic was created, out of whole cloth, by Aristotle, who observed that the patterns of good reasoning are always the same, no matter what the subject matter. He proposed to make the patterns of successful reasoning common to all the sciences a subject of study in their own right, and to make this study a part of the first and most general science, which he designated 'philosophy'.

Aristotle focused his attention on simple patterns called *syllogisms*, illustrated by the following examples:

All spaniels are dogs.
All dogs are mammals.
———
Therefore, all spaniels are mammals.

All spaniels are dogs.
Some spaniels don't have fleas.
———
Therefore, not all dogs have fleas.

In the *Prior Analytics*, Aristotle gave a splendidly elegant and thorough account of the valid syllogisms.

Aristotle's theory was, in a way, too successful. It was so beautifully crafted that there was very little to add to it, with the result that the store of inference patterns recognized as valid in the mid-nineteenth century was little changed from Aristotle's time. However, the sophisticated arguments found in Euclid or Archimedes go well beyond merely stringing together syllogisms.

A major impetus that pushed logic beyond syllogistic was the development of non-Euclidean geometry. As long as people, secure in the Euclidean tradition, were confident both that Euclid's axioms were true and that their spatial intuitions were reliable, it didn't make a lot of difference to their confidence in the theorems if proofs depended on spatial intuition in addition to the axioms. Once one starts doing non-Euclidean geometry, however, spatial intuitions can no longer be counted on, and it becomes vital that proofs rely on the axioms alone. The experience of working with non-Euclidean systems led people to go back and look at Euclid's proofs with a newly critical eye, and they discovered that the proofs in Euclid's *Elements*, in spite of having been regarded for generations as the paragon of rigour, were not at all watertight. Spatial intuitions, not supported by the axioms, leaked into the proofs from the diagrams, so that Euclid's theorems were not, in fact, logical consequences of his axioms. To secure the proofs, greater stringency is required than is found in Euclid's informal expositions.

Careful attention to what follows from what not only makes mathematical results more secure; it makes them more versatile. Among the ancient Greeks, mathematical methods were little used outside geometry and sciences closely allied with geometry, like statics and optics. Since Galileo, mathematical methods have been used ever more widely, until now they are employed throughout both the natural and the social sciences. If you want to apply a technique from geometry to solve a problem in economics, you need to be exactly aware of which aspects of the original geometrical problem the technique relies on.

2. Sentential Calculus

The methods of abstract algebra grew so versatile that the idea suggested itself of applying them to logic itself, so that we can carry out logical deductions using the same techniques that we use to solve equations. This program was introduced by Leibniz, but his work on the subject was mostly unpublished until long after his death.[1] It was taken up by George Boole ([Boole, 1854b]), who used the algebraic symbols '+', '×', and '–' to correspond to the English 'or', 'and', and 'not', which we symbolize '∨', '∧', and '¬', respectively. Then he let an equation hold between two algebraic expressions iff the corresponding sentences are logically equivalent, where a sentence φ implies a sentence ψ iff ψ is a logical consequence of $\{\varphi\}$, and two sentences are logically equivalent iff each implies the other. Among the equations he obtained were the familiar distributive law from high school:

$$x \times (y + z) = (x \times y) + (x \times z),$$

and a different distributive law that wasn't part of high school algebra:

$$x + (y \times z) = (x + y) \times (x + z).$$

Boole's algebra initiated the modern study of *sentential calculus*, which studies how compound sentences are built up out of simple ones.[2] (These efforts were anticipated by the ancient Stoics, but their results had largely been forgotten.) In addition to '∨', '∧', and '¬', standard sentential calculus symbols include '→' and '↔', which correspond, albeit roughly, to English 'if..., then' and 'if and only if'. What is special about these connectives is that they are *truth functional*: Whether a compound sentence is true or false only depends on whether its components are. Natural languages include connectives that are not truth functional – 'because', for example – but the sentential calculus does not. In order for 'She hit him because he insulted her' to be true, 'She hit him' and 'He insulted her' both have to be true, but knowing that the simpler sentences are both true doesn't determine whether the larger sentence is true.

The practice of translating ordinary language into an artificial language, in which '∨', '∧', and '¬' replace 'or', 'and', and 'not', is typical of logical theories, which all either employ artificial languages or restrict their attention to restricted, highly regimented fragments of natural languages. One can long for a logical theory that works with natural languages directly, but natural languages are so complicated that any such theory is well beyond our present reach.

Semantic theory for sentential calculus describes the dependence in truth values of compound sentences on simple ones. A *valuation* is a function that assigns each sentence a value, either *true* or *false*, subject to the conditions that

$(\varphi \vee \psi)$ is assigned *true* iff one or both of its components are; $(\varphi \wedge \psi)$ is assigned *true* iff both its components are; $(\varphi \rightarrow \psi)$ is assigned *true* iff either its antecedent φ is assigned *false* or its consequent ψ is assigned *true*; $(\varphi \leftrightarrow \psi)$ is assigned *true* iff both or neither of its components are assigned *true*; and $\neg\varphi$ is assigned *true* iff φ is assigned false. Why the simple sentences are true or false is a question outside the jurisdiction of sentential calculus. Because of truth functionality, we can test whether an argument is valid by examining all the possible ways of assigning true values to its atomic sentences, and seeing whether any of them provides a valuation in which the premises are assigned *true* and the conclusion *false*. If n atomic sentences appear in the argument, there will be 2^n ways to assign them truth values. (As we use the word, an 'argument' has only finitely many premises.) Having a test to determine whether an argument is valid gives us tests for implication, sentence validity (a sentence is valid iff it's a consequence of the empty set), and logical equivalence. Thus, Boole's distributive laws allege that $(\varphi \wedge (\psi \vee \theta))$ is logically equivalent to $((\varphi \wedge \psi) \vee (\varphi \wedge \theta))$ and that $(\varphi \vee (\psi \wedge \theta))$ is logically equivalent to $((\varphi \vee \psi) \wedge (\varphi \vee \theta))$. We can verify these equivalences by observing that the following *truth tables* have 't' at every line under the main connective '$\leftrightarrow$':

φ	ψ	θ	$(\varphi \wedge$	$(\psi \vee \theta))$	$\leftrightarrow$	$((\varphi \wedge \psi)$	$\vee$	$(\varphi \wedge \theta))$
t	t	t	t	t	t	t	t	t
t	t	f	t	t	t	t	t	f
t	f	t	t	t	t	f	t	t
t	f	f	f	f	t	f	f	f
f	t	t	f	t	t	f	f	f
f	t	f	f	t	t	f	f	f
f	f	t	f	t	t	f	f	f
f	f	f	f	f	t	f	f	f

φ	ψ	θ	$(\varphi \vee$	$(\psi \wedge \theta))$	$\leftrightarrow$	$((\varphi \vee \psi)$	$\wedge$	$(\varphi \vee \theta)))$
t	t	t	t	t	t	t	t	t
t	t	f	t	f	t	t	t	t
t	f	t	t	f	t	t	t	t
t	f	f	t	f	t	t	t	t
f	t	t	t	t	t	t	t	t
f	t	f	f	f	t	t	f	f
f	f	t	f	f	t	f	f	t
f	f	f	f	f	t	f	f	f

The method of truth tables gives us a *decision procedure* – an algorithm that will always provide a 'Yes' or 'No' answer – for determining whether an argument is valid or whether two sentences are logically equivalent. This stands in contrast to

Boole's algebraic technique, which begins with a finite store of starting equations and obtains new equations by the two methods of uniformly substituting terms for variables and of substituting equals for equals. Boole's equational system is complete, so that, whenever two sentences are logically equivalent, one can derive the corresponding equation. This gives us a *proof procedure*, an algorithm by which any two logically equivalent sentences can be shown to be such. It does not, however, provide a decision procedure, for it doesn't encompass a method for showing inequivalent sentences inequivalent. Failure to derive an equation doesn't show it isn't derivable, for perhaps we just haven't tried hard enough.

Sentential calculus is compact: If φ is a logical consequence of Γ, it is already a logical consequence of some finite subset of Γ. This contrasts with the informal notion of consequence that treats φ as a consequence of Γ iff it isn't possible for all the members of Γ to be true and φ not. With this more liberal notion, 'There are infinitely many stars' is a consequence of 'There is at least one star', 'There are at least two stars', 'There are at least three stars', and so on, but not of any finite subset.

3. Predicate Calculus

The development of a logic of sentential connectives fails to address the most dramatic respect in which Aristotle's logic fails to capture the kinds of reasoning found in Euclid's *Elements*. The geometry book is full of intricate and subtle reasoning about relations – 'longer than', 'between', 'congruent', and so on – and yet Aristotle's logic finds even something as simple as the following example, due to Augustus de Morgan, beyond its reach:

All dogs are animals.

Therefore, all heads of dogs are heads of animals.

During the late nineteenth century, thinkers like Ernst Schröder, Charles Sanders Peirce, and Gottlob Frege went decisively beyond Aristotelean logic by developing a logic of relations.[3] Frege's ([Frege, 1879]) treatment starts with an analysis of complex names, like 'log 27'. The name consists of two parts, a function sign, 'log', which denotes a function, and a name, '27', which denotes a object. Functions are 'incomplete' and 'unsaturated'; they require an object for their completion. Completion of the logarithm function by the object 27 results in an object, the number 1.431. Concepts are, in Frege's rather eccentric usage, functions that take either *true* or *false* as their values, and adjectives and common nouns denote concepts. Completion of the concept sign 'perfect square' with the name '27' results in the sentence '27 is a perfect square', which denotes *false*. We can also form functions of more than one argument, like sum, product, and greatest common divisor.

If we take the sentence 'Eve is a sinner', which we symbolize '$S(e)$', and we replace the name by the variable 'x', we get the *open sentence* '$S(x)$', which expresses the concept *sinner*. Prefixing the *universal quantifier* '$(\forall x)$', we get a sentence, '$(\forall x)S(x)$', that says that everyone falls under the concept, that is, that everyone is a sinner. To say that there are sinners, prefix the *existential quantifier*, '$(\exists x)$', instead. Doing the same thing to the sentence '$P(e,a)$' 'Eve is a parent of Abel', gives us sentences '$(\forall x)P(x,a)$' and '$(\exists x)P(x,a)$' which say that everyone is a parent of Abel and that someone is. We could have done the same thing with 'Eve' instead of 'Abel', getting '$(\forall x)P(e,x)$' and '$(\exists x)P(e,x)$', which say that everyone is a child of Eve and that someone is. If we take the sentence '$(\exists x)P(e,x)$' and replace the name 'e' by the variable 'y', we get an open sentence '$(\exists x)P(y,x)$', which expresses the concept *is a parent*. Prefixing the universal quantifier '$(\forall y)$' or the existential quantifier '$(\exists y)$' will result in a sentence that says that everyone is a parent or that someone is a parent. We need the two different variables 'x' and 'y' to be able to distinguish 'Everyone is a parent' from 'Everyone has a parent'.

The universal and existential quantifiers are *second-level* concepts, which take ordinary concepts as their arguments. Second-level concepts are a species of second-level functions. Another example of a second-order function is the definite integral from the calculus. Frege developed rules of inference governing the quantifiers. His notation and his formulation of the rules were different from what we'll present here, but they sanction the same arguments. *Universal specification* tells us that from $(\forall v)\varphi(v)$ you can derive $\varphi(\kappa)$, for any variable v and constant κ. *Universal generalization* tells us that, if we have derived $\varphi(\kappa)$ from the set of premises Γ, and if κ doesn't appear in $\varphi(v)$ or in any of the members of Γ, then we can deduce $(\forall v)\varphi(v)$ from Γ. What legitimates this rule is the observation that, if you can be sure, just on the basis of Γ, without knowing anything about the object denoted by κ, that the object denoted by κ falls under the concept expressed by $\varphi(v)$, and if that concept is characterized in a way that doesn't depend on κ, then the considerations that tell us that the object named by κ falls under the concept apply to other objects just as well, so that everything falls under the concept. Similar reasoning gives us *existential specification*: If you have derived ψ with the members of $\Gamma \cup \{\varphi(\kappa)\}$ as premises, and if κ doesn't appear in $\varphi(v)$, in ψ, or in any of the members of Γ, then you can infer ψ on the basis of $\Gamma \cup \{(\exists v)\varphi(v)\}$. Filling out the rules, we have *existential generalization*: $(\exists v)\varphi(v)$ is a logical consequence of $\{\varphi(\kappa)\}$.

To illustrate, let's carry out the de Morgan inference about dogs' heads:

$$\frac{(\forall x)(D(x) \rightarrow A(x))}{\therefore\ (\forall y)((\exists x)(D(x) \wedge H(y,x)) \rightarrow (\exists x)(A(x) \wedge H(y,x))).}$$

In conducting the proof, we allow ourselves to derive φ from Γ if we can show by truth tables that φ is a consequence of Γ by Boolean truth-functional logic, and we employ the rule of *conditional proof*, which lets us derive $(\varphi \rightarrow \psi)$ from Γ if we have derived ψ from $\Gamma \cup \{\varphi\}$. From the premise, we can derive '$(D(a) \rightarrow A(a))$', by universal specification. From this, together with '$(D(a) \wedge H(b,a))$', we derive '$(A(a) \wedge H(b,a))$' by truth-functional logic, and then go on to derive '$(\exists x)(A(x) \wedge H(b,x))$', by existential generalization. Putting these together, we get a derivation of '$(\exists x)(A(x) \wedge H(b,x))$' from {'$(\forall x)(D(x) \rightarrow A(x))$', '$(D(a) \wedge H(b,a))$'}. Since '$a$' doesn't appear in '$(D(x) \wedge H(b,x))$', in '$(\exists x)(A(x) \wedge H(b,x))$', or in '$(\forall x)(D(x) \rightarrow A(x))$', existential specification gives us a derivation of '$(\exists x)(A(x) \wedge H(b,x))$' from {'$(\forall x)(D(x) \rightarrow A(x))$', '$(\exists x)(D(x) \wedge H(b,x))$'}. Conditional proof converts this into a derivation of '$((\exists x)(D(x) \wedge H(b,x)) \rightarrow (\exists x)(A(x) \wedge H(b,x)))$' from {'$(\forall x)(D(x) \rightarrow A(x))$'}. Universal generalization gives us our desired derivation of '$(\forall y)((\exists x)(D(x) \wedge H(y,x)) \rightarrow (\exists x)(A(x) \wedge H(y,x)))$' from {'$(\forall x)(D(x) \rightarrow A(x))$'}.

The system of rules we just used, which is very different from Frege's system, is adapted from Mates ([Mates, 1972]), who presented a system of natural deduction. Such systems, following Gentzen ([Gentzen, 1934]), attempt a formalization that comes reasonably close to the ways people reason informally; see ([Prawitz, 2006]). There are a great variety of natural deduction systems, and a number of other procedures for recognizing valid inferences. Boole's algebraic approach was extended to the predicate calculus by Henkin, Monk, and Tarski ([Henkin et al., 1971]). Axiomatic systems, following Hilbert ([Hilbert, 1927]), obtain valid sentences by a direct, linear deduction from a fixed system of axioms. The most streamlined system of this form was obtained by Quine ([Quine, 1951a]), whose sole rule of inference was *modus ponens*, which lets you derive ψ from $(\varphi \rightarrow \psi)$ and φ. Evert Beth's ([Beth, 1970]) method of semantic tableaux is especially elegant. For an invalid argument, it lets you see a counterexample unfold before your very eyes; see ([Jeffrey, 2006]). Despite their diversity, these systems all agree on what follows from what.

4. Truth in a Model

Frege's use of the notion of *concept* is problematic. Concepts are incomplete objects. There is nothing metaphysically peculiar about incomplete buildings. An incomplete building is a perfectly ordinary sort of object, although it's an object that isn't yet suitable for habitation. However, an incomplete object isn't an object at all; so what is it? There appear to be two kinds of things, objects and non-objects. Logic is only capable of talking about the former, so that, even though there are things that aren't objects, '$(\forall x)(x$ is an object)' will be true, and logic will fall short of its ambition of being part of a first and most general science. It isn't first, because it depends on a prior inquiry into the object/non-object distinction, and it isn't fully general, since it only talks about things of a special kind.

There is also a grammatical puzzle. Singular definite descriptions, like 'the author of *Waverley*' and 'the base-10 logarithm of 27' play the same basic role as proper names: They denote objects. Grammatically, the phrase 'the concept *horse*' behaves like other singular definite descriptions. It serves as the subject of sentences, not as the predicate, and so it ought to denote an object. And yet, 'the concept *horse*' denotes the concept *horse*, if it denotes anything. The resulting contradiction led Frege ([Frege, 1892a]) to the bewildered declaration that 'the concept *horse* is not a concept'.

Yet another difficulty is an analogue to Russell's paradox, which we discuss briefly below. Any answer to the question, 'Does the concept *concept that does not fall under itself* fall under itself?' leads to inconsistency.

We can get a less ontologically perilous presentation of the semantics of the predicate calculus by using sets instead of concepts. One of the aims of the theory is to identify the logically valid sentences. Logically valid sentences are a species of *analytic* sentences, sentences that are true in virtue of the meanings of their words. Logically valid sentences are true in virtue of the meanings of their *logical* words. 'All spaniels are dogs', for example, is analytic (or so it seems, although Quine ([Quine, 1951b]) and Putnam ([Putnam, 1962]) disagree), but its truth depends on the meanings of the nonlogical terms 'spaniel' and 'dog', so it isn't logically valid. To get at the notion of logical validity, we need to cut off the truth of a sentence from any dependence on the meanings of the non-logical terms. The notion of *truth in a model* aims to do this. We get a model of the language by assigning values of appropriate types to all the non-logical terms. If a sentence is true in every model, its truth doesn't depend on the meanings of the non-logical terms. If an argument is valid, then the fact that its conclusion is true if its premises are true is ensured just by the logical form of the argument. The logical form of an argument is the skeleton that remains after all its non-logical terms have been removed. The notion of truth in a model aims to explicate the dependence of the truth conditions of a sentence on its logical form, so that an argument is valid iff its conclusion is true in every model in which its premises are.

The non-logical terms of a language of the predicate calculus are of two kinds: *constants*, which play the role of proper names, and *predicates*, which express properties and relations; each predicate has one or more argument places. (Function signs are often allowed as well, but let's keep things simple.) A model $\mathcal{A}$ of the language specifies a non-empty set, $|\mathcal{A}|$, which is to serve as the *universe* or *domain* of the model; it assigns, to each constant κ, an element $\kappa^{\mathcal{A}}$ of $|\mathcal{A}|$ that the constant denotes; and it associates each n-place predicate A with a set $A^{\mathcal{A}}$ of n-tuples from $|\mathcal{A}|$ that are to serve as its extension.

In addition to the constants, the language contains an infinite list of variables, and in addition to the non-logical predicates, it contains the logical predicate '='. The *atomic formulas* have the form $A(\tau_1, \tau_2, \ldots, \tau_n)$, where A is an n-place predicate and where each of the τ_is is either a constant or a variable, and also the form $\tau_1 = \tau_2$. The *formulas* constitute the smallest class that contains the atomic formulas and contains $(\varphi \vee \psi)$, $(\varphi \wedge \psi)$, $(\varphi \rightarrow \psi)$, $(\varphi \leftrightarrow \psi)$, $\neg\varphi$, $(\forall v)\varphi$, and $(\exists v)\varphi$, for each variable v, whenever it contains φ and ψ. Each formula is built up from atomic formulas in a unique way. An occurrence of a variable v within a formula is *bound* if it occurs within a subformula that begins with $(\forall v)$ or $(\exists v)$; if not bound, free. A formula without free variables is a *sentence*. It is sentences that are used to make assertions that are either true or false. For sentential calculus, we could specify how the truth value of a complex sentence was determined by the truth values of its simpler components. Once we turn to predicate calculus, however, we find that complex sentences typically aren't composed of simpler sentences. Complex sentences are built from simpler formulas, but the formulas might contain free variables, so if we want to give a compositional semantics, we have to show how the truth values of complex sentences depend on the semantic values of simpler formulas. Alfred Tarski ([Tarski, 1935]) discovered how to do this, defining truth in terms of satisfaction and showing how the satisfaction conditions for a complicated formula depend on the satisfaction conditions for its simple subformulas. A *variable assignment* for a model $\mathcal{A}$ is a function that assigns an element of $|\mathcal{A}|$ to each of the variables. To determine whether a variable assignment σ satisfies an atomic formula $A(\tau_1, \tau_2, \ldots, \tau_n)$ in $\mathcal{A}$, form the n-tuple $< d_1, d_2, \ldots, d_n >$, where $d_i = \tau_i^{\mathcal{A}}$ if τ_i is a constant, and $d_i = \sigma(\tau_i)$ if τ_i is a variable. σ satisfies $A(\tau_1, \tau_2, \ldots, \tau_n)$ in $\mathcal{A}$ iff $< d_1, d_2, \ldots, d_n >$ is in $A^{\mathcal{A}}$. σ satisfies $\tau_1 = \tau_2$ in $\mathcal{A}$ iff $d_1 = d_2$. σ satisfies $(\varphi \vee \psi)$ in $\mathcal{A}$ iff it satisfies either or both of φ and ψ in $\mathcal{A}$, and it satisfies $(\varphi \wedge \psi)$ in $\mathcal{A}$ iff it satisfies both. There are similar clauses for the other sentential connectives, exactly analogous to the corresponding clauses for the sentential calculus. σ satisfies $(\forall v)\varphi$ in $\mathcal{A}$ iff σ and every variable assignment that agrees with σ except in the value it assigns to v satisfies φ in $\mathcal{A}$. σ satisfies $(\exists v)\varphi$ in $\mathcal{A}$ iff either σ or some variable assignment that is like σ except in the value it assigns to v satisfies φ in $\mathcal{A}$. If two variable assignments for $\mathcal{A}$ agree in the values they assign to all the variables that occur free in φ, then both of them satisfy φ in $\mathcal{A}$ if either of them does. In particular, a sentence is satisfied by every variable assignment for $\mathcal{A}$ if it's satisfied by any of them. Defining a sentence to be *true in* $\mathcal{A}$ iff it's satisfied by every variable assignment in $\mathcal{A}$, and *false in* $\mathcal{A}$ iff it's satisfied by none, we have the principle of *bivalence*: Every sentence is either true or false in $\mathcal{A}$, but not both. A sentence $(\forall v)\psi$ is true in $\mathcal{A}$ iff every variable assignment for $\mathcal{A}$ satisfies ψ in $\mathcal{A}$, whereas $(\exists v)\psi$ is true in $\mathcal{A}$ iff at least one variable assignment for $\mathcal{A}$ satisfies ψ in $\mathcal{A}$.

Going back to de Morgan's example, let $|\mathcal{B}|$ be the set of material objects, and let 'D', 'A', and 'H' be assigned, respectively, the set of dogs, the set of animals, and $\{< x, y > \mid x \text{ is } y\text{'s head}\}$ by $\mathcal{B}$. Take any variable assignment σ. If σ('x') isn't a dog, σ doesn't satisfy '$D(x)$' in $\mathcal{B}$. If σ('x') is a dog, it's also an animal, because all dogs are animals, and so it satisfies '$A(x)$' in $\mathcal{B}$. In either case, σ satisfies '$(D(x) \rightarrow A(x))$' in $\mathcal{B}$, and so '$(\forall x)(D(x) \rightarrow A(x))$' is true in $\mathcal{B}$.

Again, take ρ to be an arbitrary variable assignment for $\mathcal{B}$. If ρ('y') is a head of a dog, let δ be the variable assignment that is just like ρ except that δ('x') is the dog whose head is ρ('y'). Then δ satisfies '$H(y,x)$' in $\mathcal{B}$. Also, since all dogs are animals, δ satisfies '$A(y)$' in $\mathcal{B}$. It follows that δ satisfies '$(A(y) \wedge H(y,x))$' in $\mathcal{B}$, and so ρ satisfies '$(\exists x)(A(x) \wedge H(y,x))$' in $\mathcal{B}$. Now suppose instead that ρ('y') isn't a head of a dog, and take σ to be a variable assignment that agrees with ρ except in the value it assigns to 'x'. Then either ρ('y'), which is the same as σ('y'), isn't σ('x')'s head, in which case σ doesn't satisfy '$H(y,x)$' in $\mathcal{B}$; or else, if ρ('y') is σ('x')'s head, σ('x') isn't a dog, and σ doesn't satisfy '$D(x)$' in $\mathcal{B}$. So, whether or not ρ('y') is σ('x')'s head, σ doesn't satisfy '$(D(x) \wedge H(y,x))$'. Since σ was arbitrary, we see that no variable assignment that agrees with ρ except (possibly) at 'x' satisfies '$(D(x) \wedge H(y,x))$' in $\mathcal{B}$, which tells us that ρ doesn't satisfy '$(\exists x)(D(x) \wedge H(y,x))$' in $\mathcal{B}$. Thus we see that, whether or not ρ('y') is the head of a dog, ρ satisfies '$((\exists x)(D(x) \wedge H(y,x)) \rightarrow (\exists x)(A(x) \wedge H(y,x)))$' in $\mathcal{B}$. Since ρ was arbitrary, '$(\forall y)((\exists x)(D(x) \wedge H(y,x)) \rightarrow (A(x) \wedge H(y,x)))$' is true in $\mathcal{B}$.

Tarski ([Tarski, 1935]) developed his compositional theory of satisfaction as a way of showing how, if you have a language for the predicate calculus in which the non-logical terms have fixed, predetermined meanings, you can define what it is for a sentence of the language to be true. He then observed, ([Tarski, 1936]), that you could factor out the dependence on the meanings of the non-logical terms, getting the more general notion of truth in a model, and that you could apply this notion to get a definition of logical consequence: φ is a logical consequence of Γ iff φ is true in every model in which all the members of Γ are true. ψ implies φ iff ψ is true in every model in which φ is. φ is valid iff it's true in every model, and inconsistent iff it's false in every model. Γ is consistent iff there is a model in which it's members are all true.

The requirement that the domain of a model be a set excludes the possibility that the language be used to talk about absolutely everything, because there isn't any set that includes absolutely everything, on account of Russell's paradox. The requirement has no justification, apart from mathematical convenience, so it is reassuring to learn from Harvey Friedman ([Friedman, 1999]) and from Agustín Rayo and Timothy Williamson ([Rayo and Williamson, 2003]) that it has no effect on what inferences are regarded as valid.

5. The Completeness Theorem

We now have a precise semantic notion of logical consequence, from Tarski ([Tarski, 1936]), and a system of rules of deduction, adapted, with substantial changes but none that affect the bottom line, from Frege ([Frege, 1879]). Our aim is to connect the two notions.

Because the semantic theory treats '=' as a logical term, we need corresponding rules of deduction. Here they are: You may derive $\kappa = \kappa$ from the empty set of premises, for any constant κ. You may derive $\varphi(\lambda)$ from $\{\kappa = \lambda, \varphi(\kappa)\}$. The second rule can be stated more fastidiously: Given a formula φ with no free variables other than v, you can derive the sentence obtained by substituting λ for all free occurrences of v in φ from $\kappa = \lambda$, together with the sentence obtained by substituting κ for all free occurrences of v in φ.

A sentence φ is said to be a *deductive consequence* of Γ iff the pair $< \Gamma, \varphi >$ appears at the end of a sequence of pairs joining finite sets of sentences to sentences, each of which is justified by the truth-functional consequence rule, conditional proof, one of the four quantifier rules, one of the two new identity rules, or the following structural rule:

> If you have a derivation of φ from Δ, and you have derivations of each member of Δ from Γ, you may derive φ from Γ.

To ensure that universal generalization and existential specification work properly we must assume that the language has infinitely many constants. We can add them before the derivation, if the language doesn't have them natively.

The following theorem is the main result of Kurt Gödel's [Gödel, 1930] doctoral dissertation:

Theorem 3.5.1 (Gödel Completeness Theorem) *If a sentence is a logical consequence of a set of sentences* Γ*, then it is a deductive consequence of some finite subset of* Γ*.*

Proof. We prove the contrapositive. Suppose χ isn't a deductive consequence of any finite subset of Γ. Add infinitely many new constants to the language, and put the sentences that result in an infinite list, $\zeta_0, \zeta_1, \zeta_2, \zeta_3, \ldots$ Put the constants in the language, old and new, into an infinite list $\kappa_0, \kappa_1, \kappa_2, \kappa_3, \ldots$ We want to start with Γ and fill in the details, until we get a story that completely describes a model in which all the members of Γ are true and χ is false. Towards this end, we form an infinite sequence $\Gamma_0 \subseteq \Gamma_1 \subseteq \Gamma_2 \subseteq \Gamma_3 \subseteq, \ldots$ of sets of sentences, as follows:

(1) $\Gamma_0 = \Gamma$.

(2) Given Γ_n with the property that χ isn't a deductive consequence of any finite subset, we define Γ_{n+1}:
   - If χ is a deductive consequence of some finite subset of $\Gamma_n \cup \{\zeta_n\}$, then $\Gamma_{n+1} = \Gamma_n$.
   - If χ isn't a deductive consequence of any finite subset of $\Gamma \cup \{\zeta_n\}$ and ζ_n doesn't begin with an existential quantifier, $\Gamma_{n+1} = \Gamma_n \cup \{\zeta_n\}$.

- If χ isn't a deductive consequence of any finite subset of $\Gamma \cup \{\zeta_n\}$ and ζ_n has the form $(\exists v)\psi(v)$, let κ_j be the first constant that doesn't appear in χ, in $\psi(v)$ or in any of the members of Γ_n, and let $\Gamma_{n+1} = \Gamma \cup \{\zeta_n, \psi(\kappa_j)\}$

The reason we added the infinitely many constants at the outset was to make sure we could find the constant κ_j that we need in the last clause. χ won't be a deductive consequence of any finite subset of Γ_{n+1}. For the last clause, this relies on the existential specification rule.

Let Γ_∞ be the union of the Γ_ns. Then Γ_∞ is a maximal set with the property that χ isn't derivable from any finite subset. Moreover, whenever Γ_∞ contains an existential sentence, it contains a witness. Our plan is to find a model in which all the members of Γ_∞ are true. This will give us what we want: a model in which all the members of Γ are true and φ is false.

For each j, let $\kappa_j^{\mathcal{A}}$ be the least number i such that $\kappa_i = \kappa_j$ is in Γ_∞, let $|\mathcal{A}|$ be $\{\kappa_j^{\mathcal{A}} : j \geq 0\}$, and, for A an m-place predicate and $< j_1, j_2, \ldots, j_m >$ an m-tuple of members of $|\mathcal{A}|$, stipulate that $< j_1, j_2, \ldots, j_m >$ is in $A^{\mathcal{A}}$ iff $A(\kappa_{j_1}, \kappa_{j_2}, \ldots, \kappa_{j_m})$ is in Γ_∞. It is straightforward, if a bit laborious, to verify that a sentence is true in $\mathcal{A}$ iff it's in Γ_∞. ■

The theorem could have been proved without the simplifying assumption that the language is countable, that is, that its sentences can be arrayed in an infinite list $\psi_0, \psi_1, \psi_2, \ldots$

The converse to the Completeness Theorem, which is known as the Soundness Theorem, is proved by an induction on the lengths of derivations, based on a careful inspection of the rules. Soundness theorems are seldom very informative, since typically we use informally, in proving the theorem, the very same rules whose soundness we are attempting to establish; see [Quine, 1936]. Apart from exotic proof systems, soundness theorems are only helpful in verifying that formalization hasn't gone badly awry.

By definition, logically valid inferences are truth preserving, and so, assuming that truth is the norm of belief and assertion, logically valid inferences are good ones. It follows by soundness that reasoning by the rules is good reasoning. Williamson ([Williamson, 2000]) has proposed that the applicable norm is knowledge, rather than truth. The Completeness Theorem assures us that, by this standard also, the logically valid inferences are good ones. If φ is a logical consequence of premises that you are in a position to know, you are capable, by putting together an appropriate proof, of coming to know φ as well.

The Completeness Theorem has three main corollaries:

Corollary 3.5.1 (Proof procedure) *There is an effective, algorithmic procedure by which a valid argument can be shown to be valid.*

A proof procedure is the most we can hope for, since Alonzo Church ([Church, 1936]) used the Gödel Incompleteness Theorem ([Gödel, 1931]) to show that there is no decision procedure. If an argument is invalid, there is a model in which the premises are true and the conclusion false, but the model will typically be infinite, so there is no way to display it concretely.

Theorem 3.5.2 (Compactness Theorem) *If φ is a logical consequence of Γ, it is a logical consequence of a finite subset of Γ.*

If φ is a logical consequence of Γ, it is a deductive consequence of a finite subset of Γ, and so, by soundness, a logical consequence of the finite subset.

Theorem 3.5.3 (Löwenheim–Skolem Theorem) *Any consistent theory has a model whose domain consists of natural numbers.*

This theorem, which does depend on the countability of the language, wasn't originally derived from the proof of the Completeness Theorem, but the other way around. Gödel proved the Completeness Theorem by applying techniques developed in Skolem's ([Skolem, 1920]) proof of the Löwenheim–Skolem Theorem. The completeness proof presented above follows Henkin's ([Henkin, 1949]) argument, rather than Gödel's.

Quine ([Quine, 1982]) invites us to consider a different way of thinking about logical validity that links it more directly to secure inference in ordinary language. We are to think of formulas of the predicate calculus as schematic. We get a substitution instance of the schema by replacing constants by proper names or definite descriptions, and replacing predicates by English open sentences. We then replace '$\vee$' by 'or', '$\wedge$' by 'and', and so on. We may also, if we like, restrict the range of the English quantifiers. An argument is valid, in Quine's alternative sense, if no substitutions result in true premises and a false conclusion.

It is clear that, if an argument is invalid in Quine's sense, it's invalid on the standard treatment. We can get a model in which the premises are true and the conclusion false by letting the extension of a predicate be the set of n-tuples that satisfy the English open sentence that is substituted for the predicate. The converse appeals to an arithmetized version of the Completeness Theorem, given by Hilbert and Bernays ([Hilbert and Bernays, 1939]), who observed that, if we use the construction given in the completeness proof to form a model with domain a set of natural numbers in which the premises are true and the conclusion false, we can describe the model arithmetically. If $\kappa_j^{\mathcal{A}} = i$, we'll substitute the Arabic numeral for i for κ_j, and for A we'll substitute a description within the language of arithmetic of $A^{\mathcal{A}}$. This gives us a substitution instance of the original argument with true premises and false conclusion, demonstrating that the two notions of 'valid argument' are coextensive.

The proof depends on arguments having finitely many premises. If Γ is a finite set of sentences, or an infinite set that can be defined (by way of a suitable coding) within the language of arithmetic, the Hilbert-Bernays argument shows that the substitutional consequences of Γ are the logical consequences in the usual model-theoretic sense, but the argument doesn't go through if Γ isn't arithmetically definable. Substitutional consequence differs from the standard, model-theoretic notion of consequence because the former isn't compact; see [Boolos, 1975].

6. Logical Terms

The partition of analytic truths into those that are and those that are not logically valid depends on the classification of terms as logical or non-logical. What is the basis for this classification? In a posthumously published lecture from 1966, Tarski ([Tarski, 1986]) proposes to address this problem by situating it within the context of Felix Klein's ([Klein, 1893]) Erlangen programme. Klein discovered that the seemingly haphazard assemblage of different geometries could be organized rather neatly by comparing geometries in terms of their transformation groups, where the transformation of a geometry is a one-one mapping of the space onto itself that preserves the properties the geometry cares about. The more specialized a geometry – if, for example, it pays attention to sizes as well as shapes – the smaller its transformation group. Klein's idea proved useful even outside geometry. Tarski, following Mautner ([Mautner, 1946]), proposed that, since logic is the most general theory, it should have the largest possible transformation group, the full permutation group consisting of all one-one maps of the universe onto itself, and so an operation should count as logical iff it's invariant under arbitrary permutations.

The familiar operations from the predicate calculus – the connectives, the quantifiers, and '=' – all count as logical by Tarski's criterion. Thus, Lindenbaum and Tarski ([Tarski and Lindenbaum, 1934–5]) show that the only binary relations invariant under arbitrary permutations are the universal relation, the empty relation, identity, and non-identity, thereby giving us a reason for including '=' among the logical terms. Tarski's criterion allows other logical operators beyond the familiar ones. Prominent among them are Mostowski's ([Mostowski, 1957]) cardinality quantifiers, things like 'there are infinitely many', 'there are uncountably many', and 'there are at least $\aleph_{12}$'.

There are reasons to think that Tarski's criterion is too liberal, for it severs the connection between logical consequence and valid deduction. To expand standard logic to accommodate the new quantifier 'there are infinitely many', '$(\exists^{\infty}v)$', we need to add two rules, one ordinary and the other not. The ordinary rule tells us that from $\{(\exists^{\infty}v)\varphi\}$ we can infer $(\exists^{>n}x)\varphi$ for each n, where we define

'$(\exists^{>n}v)$', which is not a new symbol but an abbreviation of a combination of old symbols, as follows:

$$(\exists^{>0}v)\varphi(v) =_{df.} (\exists v)\varphi(v)$$
$$(\exists^{>n+1}v)\varphi(v) =_{df.} (\exists v)(\varphi(v) \wedge (\exists^{>n}u)(\varphi(u) \wedge \neg u = v)).$$

The extraordinary rule derives $(\exists^{\infty}v)\varphi(v)$ from $\{(\exists^{>n}v)\varphi(v) : n \geq 0\}$, where we now allow a step in a deduction to have infinitely premises. This last 'permission', while perfectly reasonable as a mathematical abstraction, counts as a rule of deduction only metaphorically. Finite beings cannot carry out deductions with infinitely many premises.

Among the cardinality quantifiers, 'there are uncountably many' is distinguished by its good behaviour. There is a proof procedure and the logic is compact over countable languages. See [Vaught, 1964] and [Keisler, 1970].

Predicate calculus with the added quantifier 'there are infinitely many' follows the plain predicate calculus in satisfying the Löwenheim–Skolem Theorem, in a different form from the one presented above: For every model, there is a countable submodel – a model obtained from the original model by paring the universe down to a countable size – that preserves the conditions of satisfaction of all the formulas of the extended language. The same doesn't hold for the added quantifier 'there are uncountably many'. Indeed, a deep theorem of Per Lindström ([Lindström, 1969]) shows that no proper extension of the predicate calculus that satisfies the Löwenheim–Skolem Theorem has a proof procedure. Moreover, no proper extension that satisfies the Löwenheim–Skolem Theorem is compact.

A different reason for thinking that Tarski's criterion of logicality may be too liberal is that, whereas the boundary between logic and mathematics (or, perhaps, between logic and the rest of mathematics) isn't sharp, there is a boundary there, and one has a intuitive sense that notions like 'uncountably many' ought to fall on the mathematical side of the border. John Etchemendy ([Etchemendy, 1999]) has sharpened this complaint. Although he doesn't discuss Tarski's permutation-invariance criterion, he gives what amounts to an argument that there has to be something wrong either with Tarski's criterion for logicality or with his test for logical validity. Let κ be an inaccessible cardinal. Then '$(\exists^{>\kappa}x)$' is, by Tarski's standard, a logical operator. The power set of κ has more than κ elements, and so '$\neg(\exists^{>\kappa}x)(x = x)$' isn't valid; it isn't even true. Yet it is compatible with the standard laws of set theory that there shouldn't be more than κ sets, and indeed, that there shouldn't have been more than κ individuals altogether. If there hadn't been more than κ individuals, then there wouldn't have been any models in which '$(\exists^{>\kappa}x)(x = x)$' obtained, and so, by Tarski's criterion, '$(\exists^{>\kappa}x)(x = x)$' would be valid. That, at least, is what one wants to say, although counterfactuals with mathematical antecedents are problematic. Whether '$(\exists^{>\kappa}x)(x = x)$' is valid by Tarski's standard depends on whether there is a strongly inaccessible cardinal, and that is a mathematical question, not a question about the meanings of logical terms. Tarski's criterion for logical validity shields off questions of logical validity from any dependence on the meanings of the non-logical terms, but it doesn't thereby ensure that their answers depend solely on the meanings of the logical terms.

There are reasons to think that Tarski's criterion of logicality is too liberal, and also reasons to think it is too restrictive. Richard Montague [Montague, 1963] tried to develop a theory of necessity that treated 'necessary' as a predicate true of the sentences that express necessary truths, and he found that such efforts were snared by a variant of the liar paradox (see Chapter 13). He proposed instead that necessity be represented by an operator, so that we write '$\Box\varphi$' to mean that $\Box\varphi$ is necessary. Deductive calculi for '$\Box$' had been developed previously by C. I. Lewis ([Lewis, 1918]), and they are referred to universally as systems of 'modal logic', even though '$\Box$' isn't permutation-invariant. There are also epistemic logic, deontic logic, provability logic, and so on. They aren't 'first science' – for instance, epistemic logic rests on a foundation of epistemology – and they aren't fully general, but they are direct extensions of the predicate calculus. Their model theory is not the same as that for the predicate calculus. Instead of assigning a set of n-tuples to an n-place predicate, one assigns it a function pairing a set of n-tuples with each possible world; see [Kripke, 1963b]. But it is unmistakably model theory. To refuse to go along with common usage in applying the epithet 'logic' to them seems needlessly cantankerous.

7. Higher-Order Logic

Frege's ([Frege, 1879]) logic went beyond the predicate calculus as we have discussed it so far, the so-called *first-order* predicate calculus, in allowing quantified variables that range over concepts (see Chapter 6). These include not only ordinary concepts of various numbers of argument places, but also second- and third-level concepts. We expressed misgivings about Frege's conception of concepts, but perhaps the origin of the problems wasn't higher-order logic itself, but rather the informal exposition of it as a calculus of concepts. One of Frege's principle motives in developing his system was to demonstrate, contrary to what Kant ([Kant, 1787]) had taught, that the laws of arithmetic are analytic. He did this by identifying the natural numbers with certain sets. The number five was to be the set of all five-element sets, which he managed to define without circularity. He thought that the basic principles of set theory were analytic, regarding 'Fido is an element of $\{x \mid x \text{ is a dog}\}$' as just another way of saying that Fido is a dog, in the same way as 'Abel is a child of Eve' is just another way of saying that Eve is a parent of Abel. When he formalized the development in [Frege, 1893], the sole principle of set theory he required was that two concepts have the same set as their extension iff the same objects fall under both. This principle is contradictory, as Russell ([Russell, 1902]) realized, for it requires there to be a one-one map from concepts to objects, whereas Cantor ([Cantor, 1895–7]), in effect, shows that there have to be more concepts than objects.

Whitehead and Russell ([Whitehead and Russell, 1925]) proposed to resuscitate Frege's proposal by eliminating sets and classes from the story. There is plenty of talk about classes in *Principia Mathematica*, but it is all to be understood as shorthand for theorems that aren't about sets or classes at all, but about concepts. Or rather, about propositional functions, which have propositions as their values, which, for reasons we needn't go into here, Whitehead and Russell prefer to concepts, which have *true* or *false* as values. The inference from '$S(e)$' to '$(\exists X)X(e)$' surely looks like a logical inference, so it appears that we can have propositional functions for free, without any extralogical ontological assumptions. Unfortunately, the propositional functions we obtain by second-order existential specification aren't enough for the purposes of mathematics. Mathematics requires extra propositional-function existence assumptions that make the contention that there has been a reduction of mathematics to logic difficult to sustain. But even if they didn't restore to mathematics its ontological innocence, they did succeed in giving a version of Frege's program that is, as far as anyone knows, free of contradiction.

Once we give up on trying to establish the analyticity of mathematics, there is no advantage to working with concepts or propositional functions, rather than sets. More important, there is no longer any advantage to maintaining the immensely complicated logical structure, in which there are variables of different sorts for propositional functions at various levels with various numbers of argument places. A simpler account, which treats sets and their elements as ontologically on a par – they are all 'objects' or 'individuals', even though Fido and $\{x \mid x \text{ is a dog}\}$ are very dissimilar individuals – is able to obtain mathematically more powerful results much more easily. This observation, due principally to Gödel ([Gödel, 1944b]), explains why Zermelo–Fraenkel set theory has nearly everywhere supplanted *Principia Mathematica* as the accepted foundation of mathematics.

First-order formalization introduces distortions into classical mathematical reasoning more naturally formulated as second-order. One of the culminating achievements of Euclidean geometry was the presentation, by Oswald Veblen ([Veblen, 1904]) and David Hilbert ([Hilbert, 1903]) of *categorical* axiomatizations, systems of axioms that described the geometric structure so completely that any two models of the axioms are isomorphic. The axiom systems they presented were second-order, and indeed, if they hadn't been allowed to use second-order axioms, their efforts would have had no hope of success. The

Löwenheim–Skolem Theorem informs us that any first-order axiomatization of Euclidean geometry will have, in addition to the expected models – the model we get by taking 'points' to be ordered triples of real numbers, and models isomorphic to it – unexpected countable models.

Richard Dedekind ([Dedekind, 1888]) helped secure the conceptual foundations of number theory by providing a categorical axiomatization of number theory (misleadingly called 'Peano Arithmetic', even though Peano ([Peano, 1891]) acknowledges that he got his axioms from Dedekind). The axioms included a second-order version of the principle of mathematical induction, '$(\forall X)(X(0) \wedge (\forall y)((N(y) \wedge X(y)) \rightarrow X(s(y))) \rightarrow (\forall y)(N(y) \rightarrow X(y)))$'. Here '$N$' symbolizes 'natural number', and 's' represents the successor function, where we now allow function signs in addition to predicates and constants. First-order Peano Arithmetic replaces the second-order axiom with the infinitely many instances of the axiom schema that we obtain by deleting the initial '$(\forall X)$'. An instance of the schema is obtained by replacing all occurrences of 'X' by a formula, and then prefixing initial universal quantifiers to bind any free individual variables other than 'y' that appear in the formula. Modulo harmless arithmetical assumptions, the second-order induction axiom is equivalent to the well-ordering principle, that every non-empty collection of natural numbers has a least element. The schematic version tells us only that there is a least element for every collection that is definable (in the language we get from the first-order language of arithmetic by adding names for individual members of the model). The first-order theory isn't categorical. To see this, consider the theory that we get from the first-order theory by adding the constant 'c' and axioms '$N(c)$' and '$(\exists^{>m}x)(N(x) \wedge x < c)$', for $m \geq 0$. Each finite subset of this enlarged theory has a model, obtained by letting 'c' denote a sufficiently large positive integer, and so, by the Compactness Theorem, the whole theory has a model, but it's a model that won't be isomorphic to the natural numbers.

Magnifying a worry raised by Skolem ([Skolem, 1923]), Putnam ([Putnam, 1980]) argues that this proliferation of models forces us to a sceptical conclusion. Real analysis is a highly developed branch of mathematics with innumerable applications throughout the sciences. But all this theory, taken together, is not enough to determine what 'real number' refers to. We know this, because we know the theory has countable models. Apart from our theory, what else is there? For names of concrete things, like 'Fido', there are direct causal connections that link our usage of the name to its bearer (although Putnam argues that these connections are less efficacious in pinning down reference than one might have thought). But for mathematical objects, there are no such direct connections, and the indirect connections, like the link between the numeral '4' and Fido's paws, do not adjudicate among the models. Skolem concludes that there is nothing that distinguishes intended from unintended models of our mathematical theories, and so no way to advance from truth in a model to mathematical truth. Notions like countability have a relative significance, so that we can ask whether a collection is countable within one or another structure, but it makes no sense simply to ask whether the collection is countable.

Advancing to second-order logic offers an easy way out of Skolem's difficulty. Second-order logic has neither compactness nor Löwenheim–Skolem, and we know from the categoricity theorems that it is able to nail down intended models of arithmetic, analysis, and geometry. Adopting second-order logic means accepting a wide gap between logical consequence and provability. Second-order Peano Arithmetic is complete (because it's categorical), and so a proof procedure for second-order logic would yield a decision procedure for second-order arithmetic, and we know from the Gödel ([Gödel, 1931]) Incompleteness Theorem that there is no decision procedure even for first-order arithmetic. But at the semantic level, it neatly dissolves a knotty problem.

The suggested way out is perhaps too easy, for we don't obtain a powerful logic just by adopting a different typeface. A lesson we should have learnt from Gödel's ([Gödel, 1944b]) discussion of Whitehead and Russell is that the benefits of using lowercase variables to range over numbers and uppercase variables to range over classes of numbers, versus giving a first-order theory with a single style of variable ranging over both numbers and their classes, are, at best, the advantages of notational convenience. To suppose anything more is, as Quine ([Quine, 1986, pp. 64–66]) puts it, to disguise the theory of classes in sheep's clothing.

To get any advantage from moving to second-order logic, we need to assign to second-order variables a role different from merely ranging over collections made up of the things the first-order variables range over. George Boolos ([Boolos, 1984; Boolos, 1985]) suggested such a role, based on an investigation of the behaviour of plural noun phrases in English. The discussion centres on the Geach-Kaplan sentence 'There are some critics who admire only one another'. The sentence can be explained as declaring that there is a non-empty class consisting of critics who admire only other members of the class, but this rendering is not quite accurate, for the original sentence didn't say anything about classes. A nominalist, who denies that there are any classes, might perfectly well assent to the Geach-Kaplan sentence, because that sentence only requires the existence of critics that have a certain collective property; it doesn't require the existence of classes. Boolos offered an alternative to the standard second-order semantics, in which a variable assignment assigns an individual to each first-order variable and a class to each second-order variable. The alternative assigns individuals to both kinds of variables. Assignments to individual variables are subject to the constraint that one and only one individual is paired with the variable. Second-order variables don't have that constraint, so that it's permissible to pair many individuals with a single second-order variable. First-order variables range over individuals one at a time, whereas second-order variables range over individuals many at a time. In terms of plural quantification, the statement that the natural numbers are well-ordered can be rendered thus: It is not the case that there are some numbers among which none is least.

Boolos' proposal is highly controversial, and for those who think it goes too far, there are logical systems intermediate in strength between first- and second-order predicate calculus. For example, introducing the quantifier 'there are infinitely many', which can be defined in second-order logic, enables us to specify the natural number system; the crucial axiom is '$\neg(\exists x)(\exists^{\infty} y)(N(x) \wedge N(y) \wedge y < x)$'. Building on a suggestion of Kreisel ([Kreisel, 1969]), Lavine ([Lavine, 1998]) and McGee ([McGee, 1997]) have recommended holding onto first-order logic, but understanding the crucial axiom schemata as 'open-ended', so that all instances of the schema will continue to hold even after the language is enriched by the introduction of new predicates. There are numerous other possibilities.

8. Non-Mathematical Logic?

In his 1923 article 'Vagueness', Russell observes that, outside of pure mathematics, vagueness is ubiquitous in human languages, and he goes on to declare, 'All traditional logic habitually assumes that precise symbols are being employed. It is therefore not applicable to this terrestrial life, but only to an imagined celestial existence' ([Russell, 1923, pp. 88f]). The principle of traditional, so-called classical, logic most in doubt is the law of the excluded middle, which permits us to assert sentences of the form $(\varphi \vee \neg\varphi)$. Ordinary English adjectives and common nouns, like 'rich', leave room for borderline cases (see Chapter 7). If Carlos is such a borderline case, then English usage doesn't determine whether someone in Carlos' financial situation ought to be classified as rich or as not rich. In such a case, it is natural, although certainly not inevitable, to declare that 'Carlos is rich' is neither true nor false. Treating falsity as truth of the negation, we conclude that neither 'Carlos is rich' nor 'Carlos is not rich' is true. But how can the disjunction, 'Carlos is rich or Carlos is not rich', be true, if neither of its components is?

The question is oversimplified, because it ignores contextual variation, and the conditions of application of vague terms are heavily dependent on context. Moreover, it presumes that there are, or could be, compatibly with the way we use 'rich', contexts and persons for which usage leaves it undetermined whether 'rich', as it's used in that context, applies to that person. Epistemicists, led by Timothy Williamson ([Williamson, 1994]) deny this, arguing that usage determines, with respect to each context in which 'rich' can be meaningfully used, an exclusive and exhaustive, down to the last penny, partition. Adjectives like 'rich' are considered vague, epistemicists say, because, in cases near its border, it is impossibly difficult to determine which of the terms, 'rich' and 'not rich', applies.

Truth-value gaps have been reported in other places than the borders of vague terms: conditionals, for some theorists, notably Adams ([Adams, 1975]); moral and aesthetic statements, for expressivists; and the culprit sentences in the semantic paradoxes. Let us focus on vague sentences, however, because vagueness is so prevalent. While noticing that scientific terms are typically more precise than those found in the daily papers, Russell observes that complete precision is almost unheard of, even in the so-called exact sciences, other than mathematics. The stakes here are enormous. Classical mathematics, both pure and applied, sits squarely on a foundation of classical logic, and the methods of classical mathematics are used continually throughout the sciences and their applications. If we aren't entitled to employ classical methods in situations in which the things we are counting or measuring are imprecisely defined, the legitimacy of modern science and engineering must be thrown into doubt.

The usual response to the problem cases is to postulate truth-value gaps, but gluts have sometimes been proposed instead. The *dialetheic* position that there are judgements that are both true and false has had a bad reputation, ever since Aristotle declared that 'an exponent of this view can neither speak nor mean anything, since at the same time he says both 'yes' and 'no'. And if he forms no judgement, but 'thinks' and 'thinks not' indifferently, what difference will there be between him and the vegetables?' [Aristotle, 1933, 1008b10] Dialetheists protest that Aristotle is assuming a principle they contest, namely, that someone who is committed to the thesis that there are some judgements that are both true and false is thereby committed to the thesis that every judgement is both true and false. See [Priest, 2006].

Intuitionists, following Brouwer ([Brouwer, 1927]), think that truth-value gaps arise even within pure mathematics. Mathematical objects are, they say, creations of the human mind, and they don't have any properties apart from those our constructions built into them. If it is impossible to answer a mathematical question, that is because our constructive activity hasn't given the question an answer, in which case there isn't an answer. Intuitionists efface the distinction between truth and provability, so that if a disjunction $(\varphi \vee \psi)$ is intuitionistically true, it must be possible to prove either φ or ψ, and if a negation $\neg\varphi$ is intuitionistically true, it must be possible to derive a contradiction from φ. If φ is a conjecture that cannot be settled, so that it isn't possible either to prove φ or to derive a contradiction from it, then neither φ nor $\neg\varphi$ nor the disjunction $(\varphi \vee \neg\varphi)$ will be intutionistically true. An existential sentence will be intuitionistically true only if one can identify a witness, so that it might be possible to derive a contradiction from a generalization $(\forall v)\varphi(v)$ without being able to specify a counterexample, in which case $\neg(\forall v)\varphi(v)$ will be true but $(\exists v)\neg\varphi(v)$ will not. See [Heyting, 1971].

Michael Dummett ([Dummett, 1991]) has recommended intuitionistic logic, even outside mathematics, as a refuge from realism for those who renounce the idea of a mind-independent reality that makes statements true that lie entirely beyond our epistemic grasp.

Donald Davidson ([Davidson, 1971]) has described two approaches to the study of language, the building block method and the holistic method. He was concerned primarily with how simple sentences get their truth conditions, but we can apply the idea in trying to understand the connection between the truth conditions of complex sentences and those of their simple components. The building block theorist embraces, and the holist shuns, the thesis that the meaning of a compound sentence is obtained as a function of the meanings of its simple parts.

It's hard to see how, unless by adopting epistemicism, a building block theorist could accept classical logic, because the disjunction, 'Either Carlos is rich or Carlos is not rich' is classically true, but it isn't made true by either of its components.

The holistic method looks more promising. The guiding idea, loosely attributed to Gentzen ([Gentzen, 1969]), is that the meanings of the logical terms are given by the rules of inference, which are imposed by stipulation. Whereas for the building block theorist, the rules are justified by the fact that they're truth preserving, for the holist, the rules don't require a justification. They are laid down as law by fiat. To keep matters as simple as possible, let us imagine the logical analogue of the state of nature, introducing logical terms into a language that previously had none. The myth is ahistorical, of course, but convenient. In the mythical history, we introduce the logical terms by adopting rules of inference. To state these rules, we would need to employ logical connectives, but one can learn how to follow a rule without being able to state it.

The building block theorist utilizes the maxim that truth is the norm of assertion to obtain assertion conditions from truth conditions. Once you've established that a sentence is true, you are entitled to assert it. The holist makes use of the maxim in the other direction. We adopt certain practices for making assertions and drawing inferences. If our linguistic conventions entitle us to assert a sentence, they thereby make it true, because the maxim ensures that we aren't entitled to assert things that aren't true.

Despite romantic notions of speaker sovereignty, we aren't entitled to introduce any rules we like, pell-mell. We can see the need for limits by considering Prior's ([Prior, 1960]) rules for the new connective 'TONK': From $\{\varphi\}$, you may deduce $(\varphi \text{ TONK } \psi)$, and from $\{(\varphi \text{ TONK } \psi)\}$ you may deduce ψ. Adopting these rules would enable us to deduce anything from anything.

A natural constraint, recommended by Belnap ([Belnap Jr., 1962]), is *conservativeness*: The new rules shouldn't enable you to produce any new inferences, not containing the new connective either in their premises or their conclusions, that you couldn't produce before. We might decide, on reflection, that a rule that isn't conservative is one that we nonetheless want to embrace, because it lets us establish new truths we weren't able to see before. But we shouldn't adopt a non-conservative rule without undertaking such an investigation, merely on a stipulative whim, because it might have the opposite effect.

The classical rules are conservative. Even though, in our logical state of nature, we don't have logical terms in the language, some assignments of values to non-logical terms might be ruled out as analytically impossible. Assignments that make 'Fido is a spaniel' true without verifying 'Fido is a dog', for instance. If there are analytically permissible models that make all the members of Γ true without making φ true, these models will also make all the sentences classically derivable from Γ true without making φ true. We know this from the Soundness Theorem, which assures us that the rules preserve truth in a model.

Belnap actually asks for something more, not merely that new rules be conservative but that they be demonstrably conservative. In order for the introduction of new rules to successfully stipulate that the sentences derivable by the rules are truth preserving, the rules have to be conservative. For us to be justified in making the introduction, we need to be able to prove that the rules are conservative. In a context in which we already have a rich supply of established rules, this requirement is sensible. But in the logical state of nature, we can prove scarcely anything, so we can't prove that the rules are conservative. Our stipulation contains an unavoidable element of cognitive risk.

To justify talking about *the* connective introduced by a system of rules, Belnap proposed a second condition, *uniqueness*. To take '$\rightarrow$' as our example, consider the language with two conditionals, '$\rightarrow_1$' and '$\rightarrow_2$', and in which the rules for '$\rightarrow$' apply to both symbols. If the uniqueness condition is met, then $(\varphi \rightarrow_2 \psi)$ is derivable from $\{(\varphi \rightarrow_1 \psi)\}$ and $(\varphi \rightarrow_1 \psi)$ from $\{(\varphi \rightarrow_2 \psi)\}$. The uniqueness condition insists that there can't be two distinct, logically inequivalent symbols that play the inferential role prescribed by the rules. J. H. Harris ([Harris, 1982]) proves uniqueness, but here's the surprising thing: He proved uniqueness for the *intuitionist* rules. Since intuitionist logic is weaker than classical logic, intuitionists and classical logicians both accept the rules of intuitionist logic, and so, according to Harris's theorem, the intuitionist connectives and the classical connectives are logically equivalent. Yet the intuitionist and the classicist mean different things by the connectives, as witnessed by the fact that they accept different rules.

We haven't discussed the natural deduction rules for the sentential connectives up till now, since for classical logic, one can employ the method of truth tables, which yields a decision procedure and not just a proof procedure, instead. But now intuitionistic logic is in the picture. The two schools have the same rules for '$\vee$' and '$\wedge$': You can infer $(\varphi \vee \psi)$ from $\{\varphi\}$ or from $\{\psi\}$. If you can infer χ from $\Gamma \cup \{\varphi\}$ and from $\Delta \cup \{\psi\}$, you can infer χ from $\Gamma \cup \Delta \cup \{(\varphi \vee \psi)\}$. You can infer $(\varphi \wedge \psi)$ from $\{\varphi, \psi\}$. You can infer both φ and ψ from $\{(\varphi \wedge \psi)\}$. For '$\rightarrow$' the intuitionistic rules are *modus ponens* and conditional proof, but these rules do not suffice for classical logic. Classical logic includes Peirce's law, which derives φ from $\{((\varphi \rightarrow \psi) \rightarrow \varphi\}$, and Peirce's law isn't derivable intutionistically; one can show this by the methods of Kripke ([Kripke, 1965]). For '$\neg$', *ex contradictione quodlibet* – From $\{\varphi, \neg\varphi\}$, you may derive anything you like – and intuitionistic *reductio ad absurdum* – If you can derive $\neg\varphi$ from $\Gamma \cup \{\varphi\}$, you can derive it from Γ alone – suffice intuitionistically, even though these don't yield classical *reductio as absurdum* – If you can derive φ from $\Gamma \cup \{\neg\varphi\}$, you can derive φ from Γ alone – or double negation elimination – From $\{\neg\neg\varphi\}$, you can derive φ. There is a similar intuitionist/classical gap for '$\leftrightarrow$'.

The argument for Harris's theorem is straightforward. We'll go through it only for '$\rightarrow$'. *Modus ponens* for '$\rightarrow_1$' lets us derive ψ from $\{(\varphi \rightarrow_1 \psi), \varphi\}$, and this lets us derive $(\varphi \rightarrow_2 \psi)$ from $\{(\varphi \rightarrow_1 \psi)\}$, by conditional proof for '$\rightarrow_2$'. A symmetric argument gets $(\varphi \rightarrow_1 \psi)$ from $\{(\varphi \rightarrow_2 \psi)\}$.

From a classical point of view, the intuitionistic conditional, '$\rightarrow_I$', implies the classical conditional, '$\rightarrow_C$', but not *vice versa*. Intuitionists regard a conditional as true if there is a proof that derives the consequent from the antecedent. If there is such a proof, the conditional is true classically, but, by classical lights, the conditional could be true without there being any proof. From the assumption that $(\varphi \rightarrow_C \psi)$ is provable, you can derive $(\varphi \rightarrow_I \psi)$, but you can't derive $(\varphi \rightarrow_I \psi)$ from the mere assumption that $(\varphi \rightarrow_C \psi)$ is true; this is the distinction that intuitionists reject. From a classical perspective, $\{(\varphi \rightarrow_C \psi)\}$ doesn't imply $(\varphi \rightarrow_I \psi)$, and so, since $\{(\varphi \rightarrow_C \psi), \varphi\}$ does imply ψ, '$\rightarrow_I$' doesn't satisfy conditional proof.

From the intuitionistic point of view, there can be no meaningful sentence that plays the inferential role the classical logician ascribes to $(\varphi \rightarrow_C \psi)$, a sentence that supposedly can be true even though we have no way of determining whether φ is true or ψ is true, or of discerning any connection between them. For the intuitionist, '$\rightarrow_C$' is not a rival candidate for what we mean by '$\rightarrow$'. To suppose there is a well-defined connective that plays the role the classical logician attributes to '$\rightarrow_C$' is to presume the sort of realism intuitionists reject.

The rules identify $(\varphi \rightarrow \psi)$ as the weakest sentence that, together with φ, entails ψ; see [Koslow, 1992]. Within the intuitionistic language, $(\varphi \rightarrow_I \psi)$ is the weakest sentence that, together with φ, entails ψ, but the classical logician's metaphysical conscience allows her to express a still weaker sentence that, together with φ, entails ψ, namely $(\varphi \rightarrow_C \psi)$.

The conclusion that I am inclined to draw – you may well draw a different conclusion – is that, whereas the rules do succeed in pinning down the meanings of the connectives, they only do so with a conception of what is required for one sentence to count as a consequence of others already present in the background. The same rules fix different meanings to the connectives for classical logicians and for intuitionists, because they are working from different background conceptions of consequence. Your mature understanding of logical consequence is not something you were born with, but something you reach as a result of metaphysical and epistemological inquiry, and that inquiry will require you to make logical inferences. Thus it can happen that the logical inferences you accept at one stage will lead you to metaphysical and epistemological conclusions that will lead you to reassess your logical methods, and therefore to reevaluate your metaphysical and epistemological conclusions. The further conclusion I am inclined to draw from this is that the laws of logic do not provide an indubitable starting point for inquiry. This is obvious if you get the laws of logic by the building block method, which makes logical norms dependent on semantic theory. But even with the holistic method, the laws of logic are subject to scrutiny and vulnerable to revision. The relation between metaphysics, epistemology, and logic is dialectical, rather than hierarchical.

Notes

1. See, for instance, various writings collected and translated in [Leibniz, 1966].
2. The sentential calculus is sometimes also known as the 'propositional calculus'.
3. This is variously called 'the predicate calculus' and 'first-order logic', which is occasionally abbreviated as 'FOL'.

4 Identity and Existence in Logic

C. Anthony Anderson

Chapter Overview

1. Identity and Logic 55
   1.1 Identity and Intensional Contexts 56
   1.2 Identity and Russell's Theory of Descriptions 57
   1.3 Direct Reference Theory of Proper Names 58
   1.4 Frege's Theory of Names 59
   1.5 Defining Identity 59
   1.6 Criteria of Identity 60
   1.7 Relative Identity 61
2. Existence and Logic 61
   2.1 Parmenidean Consequences 63
   2.2 Rejecting DE: Existence and Being 64
   2.3 Rejecting PP or DE: Versions of Free Logic 67
   2.4 Mistake about Logical Form I: Russell's Theory of Descriptions Again 69
   2.5 Mistake about Logical Form II: Frege-Church Logic of Sense and Denotation 70
   2.6 How Should Logic Treat Existence? 72
Notes 74

> It depends on what the meaning of 'is' is.
>
> William Jefferson Clinton, 42nd President of the United States.

The two concepts of *identity* and *existence* both correspond to meanings of the word 'is'. Certainly they are general enough and abstract enough to initially be counted as concepts naturally treated by logic. There are of course other criteria for what makes something a logical concept, but these may sometimes clash. On balance these two notions seem quite at home in logic.

1. Identity and Logic

Identity is one of the simplest and clearest concepts we possess and yet it has given rise to much philosophical puzzlement. It is not quite obvious that identity is properly a notion to be studied directly by logic. It is fairly common to say that logic deals with arguments that are valid in virtue of their 'form', but identity is expressed by a binary *predicate*. In spite of some ambivalence, most logicians count identity as a logical concept.

The essential properties of identity are self-evident. Pretty clearly everything is identical with itself and if one thing is identical with another and the second with a third, then the first is identical with the third. Furthermore, if one thing is identical with a second, then the second is identical with the first. Already there is a certain awkwardness in stating these. How can one thing be identical with *another* or a *second* thing? Identity here means strict identity – that there is only one thing being discussed. The awkwardness is just a difficulty in ordinary language and is easily overcome in logic by using variables. To introduce some useful technical terminology, we can sum up our description of identity so far by saying that identity is a *reflexive*, *symmetric*, and *transitive* relation. Any relation R which is such that:

1. For every x, xRx (reflexivity),
2. For every x, y, and z, if xRy and yRz, then xRz (transitivity), and
3. For every x and y, if xRy, then yRx (symmetry),

is said to be an *equivalence relation*. Identity is thus an equivalence relation. There are others, but they often seem to be derivative from some kind of identity, e.g. *being the same height as*, taken as a relation between people, is identity in height.

Even some of these apparently evident claims about identity have been questioned. The political philosopher and revolutionary Leon Trotsky ([Trotsky, 1973, p. 329]) and the semantico-psychologist Alfred Korzybski ([Korzybski, 1933, p. 194]) have denied that everything is identical with itself, but their complaints seem to be based on confusions. Alas, there is no claim, no matter how evident it may seem, that has not been disputed by some philosopher.

More provocative is another alleged property of identity, *The Indiscernibility of Identicals*, stated informally:

(IndId) For any x and y, if x is identical with y, then whatever is true of x is true of y and vice versa.

We really should distinguish two closely related, but distinct, principles:

(SubId) For any x and y, if $x = y$, then $A[x]$ if and only if $A[y]$, where $A[y]$ results from $A[x]$ by substituting, without binding, one or more occurrences of y for free occurrences of x [*The Substitutivity of Identity*].

(IndIdProp) For any x and y, if $x = y$, then every property of x is a property of y and vice versa [*The Indiscernibility of Identicals with respect to Properties*].

The first of these, in some version, will be familiar from first-order (predicate) logic with identity. Notice that it mentions particular formulas of a particular language (formalized in this case). As an axiom, it typically has some such appearance as this:

(SI) $\forall x \forall y(x = y \rightarrow (A[x] \leftrightarrow A[y]))$

Or perhaps there is a rule of inference enabling one to infer, from an identity and a sentence, the result of substituting one side of the identity for the other in the sentence. It is well known that one can derive all the properties of identity stated so far (except *IndIdProp*) if (SI) is slightly simplified and there is added an axiom stating the reflexivity of identity:

(I1) $\forall x \forall y(x = y \rightarrow (A[x] \rightarrow A[y]))$
(I2) $\forall x(x = x)$

For the usual applications of logic these two suffice. But there are arguments in ordinary language that seem to be invalid and yet seem also to be instances of (I1) as it would be applied to English or other natural languages.

1.1 Identity and Intensional Contexts

Curiously, instances of the analogue of (I1) for natural languages sometimes seem to fail:

(a) If Bruce Wayne = Batman, then if Commissioner Gordon knows *a priori* that Bruce Wayne = Bruce Wayne, then Commissioner Gordon knows *a priori* that Bruce Wayne = Batman.

Of course the example is fictional, but it is the *possibility* of counterexamples that is of interest to logic.

(b) If Samuel Clemens = Mark Twain, then if it is an important fact of literary history that Samuel Clemens = Mark Twain, then it is an important fact of literary history that Samuel Clemens = Samuel Clemens.

This does not have the ring of truth. On a list of important facts in the history of literature, the sentence 'Samuel Clemens = Samuel Clemens' would seem strikingly out of place. The following examples and variants thereof have been extensively discussed in the philosophical literature.

(c) If 9 = the number of planets, then if necessarily 9 > 7, then necessarily the number of planets > 7.[1]
(d) If the Morning Star = the Evening Star, then if it is necessary that the Morning Star = the Morning Star, then it is necessary that the Morning Star is the Evening Star.
(e) If the author of *Waverley* = Sir Walter Scott, then if King George IV wished to know whether the author of *Waverley* = Sir Walter Scott, then King George IV wished to know whether Sir Walter Scott = Sir Walter Scott.

Notice that some of the examples involve proper names and others involve also definite descriptions. These of course might be treated differently in logic. Some have argued that these examples are not really instances of (I1) as it would be extended to natural language. This is no doubt in some sense correct, but we should initially just admit that the analogue of (I1), carefully stated, does not hold for ordinary language. But this should not lead us to reject *IndIdProp*! Substitutivity of Identity may fail for natural languages, but the corresponding principle about the indiscernibility of identicals with respect to *properties* is untouched by this (see especially Cartwright ([Cartwright, 1971])). *Why* Substitutivity of Identity fails, when it does, is still much disputed. Contexts in which this law fails are often called *intensional contexts*. The failure of that principle is sometimes just used to define such contexts, but the suggestion is nearby that in at least some of the cases, the meaning of the expressions substituted, as distinguished from their denotation, is somehow responsible for the failure.

These difficulties are intimately related to fundamental questions in the philosophy of language and in particular the semantics of natural language sentences. Different approaches to semantics yield different resolutions to these puzzles.

1.2 Identity and Russell's Theory of Descriptions

According to Russell, one of the first antecedents of the natural language examples are really identities. That is, they may have the syntactical form of identity statements, but the propositions expressed are not simple identities. So, in effect, the solution is that these are not really natural language analogues of the logical principle of Substitutivity of Identity. Definite descriptions are 'analyzed away' in favour of expressions involving quantifiers. According to Russell, proper names in natural languages are disguised definite descriptions. Even 'Sir Walter Scott' is not a name in the appropriate logical sense. Perhaps it means 'the knight or baronet whose given name is 'Walter' and whose family name is 'Scott'. Let us suppose we introduce a predicate expressing these properties, 'Scottizes'. Then 'Scott is the author of *Waverley*' really expresses 'There is one and only one scottizer and one and only one author of Waverley and the former is identical with the latter.'

Whitehead and Russell ([Whitehead and Russell, 1910]) adopt conventions of abbreviation that correspond to the ideas just informally explained. The sentence 'Scott is the author of *Waverley*' would be represented as

$$(\iota x)S(x) = (\iota x)AW(x).$$

This is read: 'The scottizer is the author of *Waverley*'. But this is just an abbreviation of:

$$\exists x \forall y[(x = y \leftrightarrow S(y)) \wedge \exists z \forall w[(z = w \leftrightarrow AW(w) \wedge x = z)]]$$

'There is an individual such that for all individuals, the first mentioned individual is identical with one of them if and only if it scottizes and there is an individual such that for all individuals, the just previously mentioned individual is identical with one of them if and only if it authored Waverley and the very first mentioned individual is identical with the one lately mentioned.' The formal version is a formula that contains an identity sign, but the identity sign stands between variables. A natural language paraphrase of this is extremely awkward, but its formal version is easily mastered and manipulated. Saul Kripke ([Kripke, 1972a]) has vigorously criticized the treatment of proper names this theory involves. Some philosophers accept Russell's treatment of explicit definite descriptions, but have rejected his extension of the idea to include proper names, naturally so-called in natural language.

1.3 Direct Reference Theory of Proper Names

According to this currently popular view, the puzzling inferences above involving only proper names are in fact correct(!). The proposition that Samuel Clemens is Mark Twain just *is* the proposition that Samuel Clemens is Samuel Clemens, but the historical interest attaches not to the proposition alone but to the way it is presented by the sentences 'Samuel Clemens is Samuel Clemens' and 'Samuel Clemens is Mark Twain', respectively. In a similar vein, Commissioner Gordon knows *a priori* the proposition that Bruce Wayne is Batman under a certain 'guise'. That is, it is known *a priori* as it presented by some sentences, but not necessarily as it is presented by other sentences.

In example (d) about the Morning Star, it is maintained that if the terms are read as proper names, then the identity 'The Morning Star = the Evening

Star' is really a necessary truth. This view is a sort of compromise between the idea that the meaning of a proper name is simply what it stands for and the idea that the meaning is 'given' in a certain way – as in Frege's theory. The meaning that is associated with the sentence in the second way is relegated to psychology.

1.4 Frege's Theory of Names

Gottlob Frege held that both ordinary proper names and definite descriptions have *sense* as well as (usually) denotation. The failure of the Substitutivity of Identity in intensional contexts is due to the fact that in such contexts names and descriptions *denote* what they ordinarily *express*: their ordinary senses. Failure of the logical principle of Substitutivity of Identity is thus a case of the Fallacy of Equivocation. Frege puts one version of the general puzzle in roughly this way: How can '**A** = **B**', if true, differ in meaning from '**A** = **A**'? One can see this as of a piece with the examples given above:

If **A** = **B**, then if '**A** = **A**' means that **A** = **A**, then '**A** = **A**' means that **A** = **B**.

It is not difficult to go on to infer from this that if '**A** = **B**' is true, then it means the same as '**A** = **A**'. Frege's solution was that here we have a case of substitution in an intensional context and thus an equivocation. Again Kripke argued persuasively that proper names do not have any invariant senses for different speakers that can plausibly be represented by definite descriptions. Notice that Russell and Frege agree on one point – proper names are 'really' definite descriptions. Frege says that the definite description has a sense. Russell says that it should be analysed away. There seems to be no solution to these puzzles that is presently accepted by the majority of philosophers and logicians.

1.5 Defining Identity

It was Leibniz who first indicated how identity might be defined. If we consider *second-order logic*, then a perfectly adequate definition of identity is:

$$x = y =_{df} \forall F(F(x) \rightarrow F(y))$$

Under its now standard principal interpretation, the monadic predicate variables in second-order logic range over *subsets* of the domain of individuals. For any given individual there is a subset of the domain containing that one individual, the 'singleton set' containing that individual as sole element. If anything belongs to every subset of the domain containing that individual, then it belongs to that singleton – and hence just *is* the given individual. Using the given definition in second-order logic the principles (I1) and (I2) can be proved. About this there is no reasonable debate. But it is not so with a certain interpretation of:

(IdInd) If x and y have all their properties in common, then $x = y$ [*The Identity of Indiscernibles*]

If you contemplate the definition offered above, you might think that the present principle is an easy consequence of it. This is not correct. In the definition, the variables range over subsets of a given domain of individuals. In the Identity of Indiscernibles, one speaks about *properties* and the notion of a property is by no means clearly fixed and formalized in modern symbolic logic. Suppose we think of properties as *qualities* or as *purely qualitative*. This concept is itself far from clear but it seems clear enough to support a counterexample to the claim that (*IdInd*), understood in these terms, is a necessary truth. Note well that it is not the mere truth of that principle that is in dispute, it is its *necessary* truth. Is it appropriate as a principle of logic, perhaps a future logic of properties? If so, it could be combined with (*IndId*) to produce a necessary equivalence and hence a definition of identity within the theory of properties.

Alas, Max Black ([Black, 1962]) long ago gave an example that convinces almost everyone that the Identity of Indiscernibles, understood as concerning qualities or purely qualitative properties, is *not* a necessary truth. We are asked to imagine a possible world consisting entirely of two qualitatively identical spheres, perhaps made of steel, say. It is difficult to deny that there is a clear and distinct conception of such a situation and yet the spheres are assumed to be distinct. We are invited to conclude that this is a genuine possibility and hence that IdInd, so understood, is not a necessary truth.

At present there is no clearly motivated and clearly adequate logic of properties, purely qualitative or not, and so we must look to future developments in intensional logic to throw light on these matters.

1.6 Criteria of Identity

At one time, not so very long ago, it was taken for granted that if there is no 'criterion of identity' for a kind of entity, then such entities are automatically philosophically suspect and perhaps 'ill-defined'. It is not easy to articulate the intuition and supporting arguments lurking behind this idea. The medieval philosophers and then Leibniz were keen on finding 'principles of individuation' and the idea appears again in Frege, to be taken up in some respects by Wittgenstein. If we ask 'Under what circumstances, how is it to be determined even in principle, that there is given only one individual of a certain kind, rather than two?', we may well be at a loss to understand what is wanted and why it is needed.

One is tempted to reply that identity is just identity, being the very same thing, and it need not be supported by some kind of 'criterion.' If a definition is wanted, one that applies to just about anything, then one might use the one given above in second-order logic. This retort will satisfy no one. Nor should it. There is something behind the idea and this can be seen if one contemplates a logical or mathematical theory where ever so many questions of identity and distinctness are left open. Such theories are profoundly incomplete and something like what is called a 'criterion of identity' often settles many of these questions. How to articulate this and form it into a philosophical argument or a useful methodological maxim is still quite an open question. (see [Williamson, 1986] and [Anderson, 2001] for some meager progress).

1.7 Relative Identity

Peter Geach ([Geach, 1962]) has argued that the ideas of absolute identity and absolute distinctness are ill-conceived. If this is so, then this is a defect of the logic of identity as it is now treated. Instead of just asserting that **A** and **B** are identical *simpliciter*, Geach urges that we should really say that they are *the same F*, where F is a certain kind of concept. You and I may own the same car, the 2010 Honda LX-S, tango red, and yet not own the same physical object. My motoring machine is in my garage and yours is in your garage.

If we pursue this idea, we would write, say, '$x =_F y$' to mean that x is the same F as y. This may be independent of '$x =_G y$', meaning that x and y are the same G. One application might be to the doctrine of the *Trinity*.

John Perry ([Perry, 1970]) argued that once we distinguish exactly *what* is being said to be the same, the examples supposedly supporting the idea of relative identity just evaporate. The *kind of car*, identified by make, model, and colour, say, is the same for you and me, but the cars, just the cars, are simply distinct. Or so Perry argued.

There is one considerable argument that Geach urges against the idea of absolute identity. If one tries to explain it by saying that x and y are absolutely identical if they have *all* their properties in common, then we may approach the edge of paradox. There are supposed to be contradictions lurking around the idea that one can quantify over all the properties that there are. There are indeed deep difficulties involved in the project of formulating an adequate theory of properties, but these are beyond the scope of this article.

2. Existence and Logic

The concept of *existence* is perhaps the only concept that seems even simpler and clearer than identity. Yet it gives rise to its own conundrums. One of the oldest such is what has sometimes been called 'Parmenides's Paradox'.[2] The original text by Parmenides is apparently quite difficult to translate and so its intended meaning is controversial:

> '[T]hou couldst not know that which is-not (that is impossible) nor utter it; for the same thing exists for thinking and for being.'

Kirk and Raven ([Kirk and Raven, 1957, p. 269]) take this to mean:

> [I]t is impossible to conceive of Not-being, the non-existent. Any propositions about Not-being are necessarily meaningless; the only significant thoughts or statements concern Being. ([Kirk and Raven, 1957, p. 270])

If something is not, i.e. there is no such thing, then we cannot speak truly about *it*. Indeed, we cannot even say truly about that which is not *that* it is not.

In order to focus on this claim we present an analysis of the the implicit argument for it suggested by these passages and elicit some further paradoxical consequences. We can motivate various ideas in philosophical logic as if they were responses to this paradox about existence, although in historical fact they had a number of motivations. We formulate the reasoning as involving sentences rather than thoughts. Similar arguments can be constructed about thoughts or propositions, but the terminology would be unfamiliar and the presuppositions more controversial. Here are our three Parmenidean assumptions:

(PP)

(i) A sentence of the form *s is P* (a *subject-predicate sentence*), where '*s*' is a singular term, is true if and only if an entity is designated by '*s*' and that entity has the attribute expressed by '*P*'.

(ii) Such a sentence is false if and only if an entity is designated by '*s*' and that entity lacks the attribute expressed by '*P*'.
[*Predication Principle*]

(DE) If '*s*' does designate something, then that thing exists, i.e. has the attribute of existing. [*Designation Implies Existence*]

(NC) If '*s*' designates something, then if the sentence *s is P* is true, then the sentence *s is non-P* is false. [*Non-Contradiction*].

A *singular term* is an expression that stands for, or purports to stand for, a single thing. Proper names such as 'Aristotle', 'Homer', 'Nicholas Bourbaki', and descriptive expressions (*definite descriptions*) such as 'The president of France in 2010', 'The largest prime number', 'The War Between the States', and the like, are naturally regarded as singular terms. Various qualifications are required to accommodate the fact that singular terms may have multiple uses, e.g., 'Aristotle' is the name of both a famous Greek philosopher and a famous Greek shipping magnate. As we have stated it, the Predication Principle is rather limited in its scope. As applied to thoughts or propositions, we might say more generally that every proposition is about something[3] and attributes a property to it – and the attribution is correct if that thing has the property and incorrect if it does not.

Our third premise (**NC**) is not especially Parmenidean, but is usually considered as a law of logic or a law of thought. We include it here because some of those who maintain that we can speak and think about that-which-is-not have been led to deny that the law of non-contradiction applies those things. In **(NC)** *non-P* stands for the predicate obtained from '*P*' by forming its *complement* or *negation*. In English, this is done in various ways. We get 'non-flammable' from 'flammable'.[4] Various prefixes are used for the purpose: 'non','in', 'ab', 'a', 'un', and so on. No such prefix need be available in all cases. We can still form the complement by means of an appropriate circumlocution. The three premises seem to be relatively unproblematic, but some curious consequences follow.

2.1 Parmenidean Consequences

It seems to follow immediately from (**PP**) that one cannot truly say anything directly about the unreal:

(UT) If '*s*' does not designate something, then every sentence of the form '*s is P*' is untrue. [*The Paradox of Untruth*]

Thus one cannot speak truly about what is not. Oddly, perhaps even paradoxically, one cannot even say of them that they are not. That is, it follows from (**UT**) that:

(NE) If '*s*' does not designate something, then the sentence '*s is non-existent*' is untrue. [*The Paradox of Negative Existentials*]

One slightly subtle point: we should distinguish the consequent of (**NE**) from the claim: *It is untrue that s is existent*. These might not always have the same truth value. Still, (**NE**) is already quite odd. One naturally supposes that the singular term 'Father Christmas' does not designate anything. So it follows from this observation and (**NE**) that the sentence *Father Christmas is non-existent* is untrue, i.e. *Father Christmas does not exist* is untrue! Most adults who understand what is meant tend to think that Father Christmas *doesn't* exist, i.e. that the claim *Father Christmas is non-existent* is true.

Now from (**PP**) and (**DE**) we get:

(T1) If '*s*' designates something, then *s is existent* is true. **[PP,DE]**

To indicate the assumptions upon which a conclusion depends, we note the assumptions in square brackets. From this last and (**NC**), we may infer:

(T2) If '*s*' designates something, then *s is non-existent* is untrue. **[PP,DE,NC]**

Combining (T2) and (NE), we may conclude:

(NEG) Every sentence of the form *s is non-existent* is untrue. **[PP,DE,NC]**

It should be noticed that there are versions of these puzzles involving general terms. We might ask how '*There are no unicorns*' and '*Unicorns do not exist*' can be about unicorns and be true. Slightly different issues are involved there, but for simplicity, we consider only the singular-term version.

2.2 Rejecting DE: Existence and Being

One response to these puzzles is to reject the principle that designation implies existence. We might admit that every singular term must designate something if it is to be meaningful and occur as the subject of a true sentence, but deny that such a term must designate something that *exists*. Although the terminology is not uniform among philosophers, this response to the paradox sometimes involves introducing a distinction between *existence* and *being* – the latter being a more general kind of reality. Early Bertrand Russell ([Russell, 1903]) puts the Parmenidean argument and the proposed solution thus:

> *Being* is that which belongs to every conceivable term, to every possible object of thought – in short to everything that can possibly occur in any proposition, true or false, and to all such propositions themselves. Being belongs to whatever can be counted. If *A* be any term that can be counted as one, it is plain that *A* is something, and therefore that *A* is. '*A* is not' must always be either false or meaningless. For if *A* were nothing, it could not be said not to be; '*A* is not' implies that there is a term *A* whose being is denied, and hence that *A* is. Thus unless '*A* is not' be an empty sound, it must be false – whatever *A* may be, it certainly is. Numbers, the Homeric gods, relations, chimeras, and four-dimensional spaces all have being, for were they were not entities of a kind, we could make no propositions about them. Thus being is a general attribute of everything, and to mention anything is to show that it is.

> *Existence*, on the contrary, is the prerogative of only some amongst beings. ([Russell, 1903, p. 449])

Parmenides couldn't have said it better – in fact, he didn't say it nearly as well. Essentially Russell accepts the underlying reasoning of the argument, but wishes to allow that we can significantly deny the *existence* of things. We just can't significantly deny the *being* of anything.

Some have seen this distinction as incurably obscure, as a kind of evasion, and even as philosophically dangerous. Even now the matter is debated, some maintaining that there just is no such distinction and others insisting that there is. One might suspect here that the dispute is largely about 'semantics' in the disparaging sense. We think that this reaction is partly right – even though we consider matters of semantics in general to be quite interesting and important for the philosophical enterprise. We will return to this point in the concluding section of this entry below.

Certainly, in introductory courses in predicate logic (first-order logic, quantification theory), we are taught to symbolize

(1) Unicorns don't really exist

as

(1′) $\neg\exists x \text{Unicorn}(x)$

Indeed, we call '∃' the *existential quantifier*. Logic books typically explain the semantics of this so that a (usually, non-empty) domain is chosen as the *range* of the variables and such things as (1′) are counted as true if nothing in the domain belongs to the set assigned to the predicate '*Unicorn*'. Of course we may assign a different meaning to '∃' if we choose, as long as we select our axioms and rules of inference accordingly. But we are still left with no way of saying that a certain particular unicorn[5] does not exist or that Father Christmas does not exist. Let us assume for now that we can somehow make sense of the distinction. Logic should be as general as is sensibly possible in order to be able to express the reasoning coming from various quarters.[6]

The simplest way to respect the purported distinction between existence and being is to just add predicates, say '*E!*' and '*I!*' to express *existence* and *being*, respectively. To ameliorate certain disputes that will inevitably arise, perhaps it is better to think of the latter predicate as expressing '*is-ness*'. What's that? Well, to attribute *is-ness* to something (an object or a term) is just to say there is such a thing (or object or term).

We may then understand the semantics differently. You are to choose a domain of entities as the range of the variables – things that can be counted.

To avoid possible misunderstanding of the notation, it might also be better to simply drop the usual symbols for the quantifiers and put something in their place, e.g., 'Σ' and 'Π', to be read 'There is an . . .' and 'Every item whatever . . .'. To retain the connection between the intended meanings of the predicates, we should require that an interpretation assign the entire domain to '$I!$' and a subset of the domain, proper or not, to '$E!$' – that is, we treat existence as we do any other predicate. Following (early) Russell's suggestion, we should adopt as logical axiom:

(R1) $\Pi x I!x$ ('Everything is, or has being')

If we assume that *entities that have being and the quantifiers governing them obey the usual laws of logic*, then we will be able to prove from (R1) that:

(R2) $\Pi x(E!(x) \rightarrow I!(x))$ ('Everything that exists is, or has being.')

Indeed, if something has any property, then it has being. We then allow that an individual constant may designate a being that does not exist and so we could formalize the claim that Father Christmas does not exist straightforwardly as:

(2) $\neg E!(c)$

Of course, it follows from (R1) that he has being.

We have made very minimal changes to ordinary 'classical' logic to accommodate some of the ideas of this response to the Parmenides Paradox. Since the interpretation of the 'is-ness' predicate is to be constrained to be the entire domain in every case, we are treating it as a logical constant. If we consider predicate logic with identity, we might use ideas from *Free Logic* (discussed below) and just define:

$$I!(x) =_{df} \Sigma y(y = x)$$

Then, with the usual axioms or rules for identity, we can prove (R1) and, hence, (R2). That is, we essentially make *no* changes to classical logic, except in the understanding of its interpretations and the possible addition of a predicate for existence! The 'being quantifier' looks different from the existential quantifier, but its logic is exactly the same. If we like, we can just go back to using the old notation and no one will be the wiser. True, existence is being treated as a 'predicate', but this is not obviously a mistake (see below). 'Is-ness' is being treated as a logical notion, defined in terms of identity and quantification. 'Existence' is just an ordinary predicate to be assigned an extension as we please – as long as it is a subset of the universe of objects.

2.3 Rejecting PP or DE: Versions of Free Logic

An alternative response to the Parmenidean puzzles would be to reject **PP**. One might allow that a subject–predicate sentence could be true even if the subject term does not designate anything. Alternatively, we might retain the Principle of Predication and, as before, allow that some objects do not have to exist in order to be designated, but insist that '∃' be interpreted as a genuinely *existential* quantifier. These two alternatives correspond to versions of what is called *Free Logic*.

Free logics have been extensively developed and studied. Perhaps the most general characterization is as follows. (1) In a free logic singular terms are allowed that do not designate anything that exists. Sometimes free logics also incorporate an independent idea: (2) the domain or universe of discourse of the logic is allowed to be empty. Logics satisfying both of these conditions have been called 'universally free logics'.

It is important to emphasize that there are two distinct changes being considered for logic. One difference has to do with singular terms. One may want to have singular terms that do not designate existing entities. In some treatments of Free Logic they need not designate *at all*. In others some singular terms designate *non-existent* entities. This latter involves introducing, at least in the meta-language, something like a distinction between existence and being.

It has been seen as a defect in 'logical purity' that one can prove in the usual formulations of first-order logic such things as:

$$\exists x(F(x) \vee \neg F(x))$$

But why should we be able to prove an *existence* claim in logic? Isn't logic supposed to be neutral about such matters? Even if we interpret this quantifier as concerning *being*, it still seems curious that this is a theorem of logic. Thus arose the proposal to alter the usual axioms or rules of inference of classical logic to prevent the proofs of existence claims. Corresponding to this, the semantics is altered to allow the universe of discourse to be empty. It is true that the logic is simpler if we confine ourselves to non-empty domains, but it is thought that the postulate that the universe of discourse is non-empty should be left to the one who is using logic in a particular application. This idea is not any sort of response to the Parmenides Paradox, but is independently motivated.

We state in some detail a formulation of a free logic incorporating both of these ideas.[7] Add to ordinary first-order logic without identity, but with individual constants, our monadic existence predicate '*E!*'. In this approach this should be thought of as a logical constant since the definition of an interpretation will constrain its extension. We give an axiomatic ('Hilbert-sytle') formulation.

The axioms consist of all tautologies and all the closed (N.B.) well-formed formulas of the following forms:

(MA1) $A \rightarrow \forall x A$
(MA2) $\forall x(A \rightarrow B) \rightarrow (\forall x A \rightarrow \forall x B)$
(MA3*) $\forall x A \rightarrow (E!(a) \rightarrow A(x/a))$
(MA4*) $\forall x E!(x)$
(MA5) $\forall x A(x/a)$ where A is an axiom.

The sole rule of inference is Modus Ponens. Here a is an individual constant and $A(s/t)$ means the result of substituting the term (individual constant or variable) s everywhere in A for the term t. The first axiom just allows for vacuous universal quantification. The second axiom should be familiar.

Notice especially (MA3*) and (MA4*). The first is similar to the usual axiom (or rule) of Universal Instantiation or Universal Specification. If something is true of everything, then it is true of the particular thing a – provided that a exists. The universal quantifier means here 'everything that exists' and the corresponding existential quantifier means 'something that exists'. The axiom (MA4*) just means 'every thing that exists, exists'. Here the concept of existence is contained once in the meaning of the quantifier and then again in the meaning of the predicate.

So here is a logic with existence as a predicate, the quantifiers interpreted as ranging over existents, but with constants that need not designate things that exist. In specifying a semantics we might proceed as before: the domain is to consist of things that exist together with things that are, or have being, but the quantifiers range just over the former.[8] Or we might devise a semantics whereby some of the individual constants don't designate anything – they are vacuous. We are left with choices to make about sentences containing such constants. Presumably, we want '$E!(a)$' to be false if a is such a constant and so for '$\neg E!(a)$' to be true. But can we truly say other things 'about' a? With respect to simple ('atomic') sentences $P(a)$ – we might count them as all false or as having no truth value (except for '$E!(a)$'), or some of them as true and some of them false.[9] If we extend the underlying predicate logic to include identity, then we can define existence thus:

$$E!(x) =_{df} \exists y(y = x)$$

These different choices lead to different free logics and they have all been studied in the logical literature.

We do not attempt to discuss all these options. But it is interesting to see how the Parmenides puzzles fare in the different cases.

If we incorporate a 'super-domain' in the semantics for our free logic, containing both existents and objects with being, then we are in effect rejecting **DE**.

Individual constants can designate things that don't exist. This leads to the early Russell position. We can truly deny existence, but denials of being – if we could express them – would be logically false. In passing, we note that it seems curious that standard formulations of free logic don't allow for quantification over the additional elements of the domain in this case. If we don't mind designating them and saying true and false things about them, why can't we speak generally about all or some of them? If we do allow this, then we are led back to the logic above with general ontological quantifiers and an existence predicate – which latter will not be constrained in its interpretation.

If we instead allow interpretations according to which some of the individual constants don't designate anything and we count '$E!(a)$' as false for any such, then we seem to be rejecting **PP**, at least in part. The sentence '$E!(a)$' is false, but it isn't about anything – or, at least, it isn't about what the subject of the sentence designates, there being no such thing. Curiously, '$\neg E!(a)$' although true, isn't about anything.[10] It gets counted as true because we stipulate that the negated sentence is false.

Of course for a formalized language, we may stipulate as we please. The crunch will come when we formalize thoughts expressed in a natural language. How shall we formalize 'Pegasus is the flying horse of Greek mythology'[11] or 'Sherlock Holmes is a fictional detective'? There is a considerable literature on 'the logic of fiction', but luckily it falls outside of our purview here. Here we just note that some of the alternatives reject **PP** and some reject **DE**.[12]

2.4 Mistake about Logical Form I: Russell's Theory of Descriptions Again

Russell's theory of descriptions is briefly discussed above and is thoroughly discussed by Linsky (see Chapter 5). For the present purpose, we need only recall Russell's contextual definition:

$$E!(\iota x)\phi(x) =_{df} \exists x \forall y(x = y \leftrightarrow \phi(y))$$

This doesn't really treat existence as a predicate; it's a contextual definition of certain sentences that look like they assert existence of a subject. Assertions and denials of existence only make sense when the subject expression is a definite description. And the apparent form is misleading. The proposition expressed is really an existential quantification, not a simple subject–predicate sentence.

Natural language expressions that appear to deny existence, say,

(1) Father Christmas does not exist

can be true if understood as having a misleading grammatical form. 'Father Christmas' is treated as a disguised definite description, perhaps some such thing as 'The man who lives at the North Pole and does thus-and-such'. The immediate formal counterpart of (1) is then:

(2) $\neg E!(\iota x)C(x)$

where $C(x)$ represents '*x is a man who lives at the North Pole . . .*'. This is in turn an abbreviation for

(3) $\neg\exists y\forall x(C(y) \leftrightarrow y = x)$

This will be true because there is no one who lives at the North Pole and does such-and-such, and is unique in those respects.

What about Parmenides? Strikingly, an adherent of Russell's theory of descriptions can accept all of Parmenides premises and thus his conclusion! According to Russell's theory, denials of existence are not subject–predicate sentences in the relevant sense. Or, to put it another way, the sentences are grammatically subject–predicate, but the propositions they express are not of subject–predicate form. What are the sentences about? They are about *propositional functions* – which are Russell's substitutes for properties, but are not quite the same. We can say some true things that *seem* to be about Father Christmas, but they are really about the propositional function, *being a man who lives at the North Pole and such-and-such*. In many ways this is a very satisfactory result. General denials of existence are understood in a similar way. *Unicorns do not exist* is about the propositional function *being a unicorn*, i.e., *being a naturally one-horned equine animal*, and says of it that it is not true of anything. We are not speaking about what *is not* – we are speaking about propositional functions – which *are*, they have being.

2.5 Mistake about Logical Form II: Frege-Church Logic of Sense and Denotation

According to the account of meaning and language formulated by Gottlob Frege, every independently meaningful expression has a *sense*, or meaning *properly so-called*, and – usually – a denotation. The sense (German: *Sinn*) of an expression is what is grasped when the expression is understood. The denotation (German: *Bedeutung*) is what the expression designates. Frege constructed his logic in a formalized language so that every meaningful expression designates something, but he was well aware of the fact that this does not hold in natural languages. Expressions that would otherwise be non-denoting are just arbitrarily assigned a denotation in his formalized language. Alonzo Church attempted to formalize Frege's semantical ideas, with some alterations, in a system called 'the logic of sense and denotation' ([Church, 1951, Church, 1973, Church, 1974]). We discuss these ideas only insofar as they concern existence.

According to Church[13] a statement of the form '*s exists*' is about the *concept* expressed by the name *s*. That is, an assertion of singular existence is a claim to the effect that a certain sense determines an existing object. We can truly say that Father Christmas does not exist but we do not thereby speak of Father Christmas and deny his existence. We speak of the *Father Christmas concept* expressed by 'Father Christmas'. Let $\Delta(X, x)$ express that X is a concept of the thing x. Then

(1) Father Christmas does not exist

is formalized as

(2) $\neg\exists x_\iota \Delta(C_{\iota 1}, x_\iota)$

and this is abbreviated in turn as:

(3) $\neg e_{0\iota 1} C_{\iota 1}$

This looks like the denial of a subject–predicate sentence. The subject is the concept of Father Christmas (better: the Father Christmas concept) and the predicate expresses a property of that concept, viz. being a concept *of* something. The subscript iota corresponds to the type of individuals and iota-one to the type of concepts of individuals and thus (2) may be read as 'There does not exist anything that falls under the *Father Christmas* concept'. Again, Parmenides was correct. One cannot speak of that which is not, even to say of it that it is not. But one can speak of concepts and say of them that they do not correspond to anything real. Of course, this is not very helpful unless a theory of concepts is supplied. This Church attempted to do, but the project was never quite completed.

In general all truths 'about the non-existent' will be represented, on this view, by corresponding truths about concepts. 'Pegasus is the winged horse of Greek mythology' will be paraphrased as saying about a certain concept that it has a certain place in the system of propositions constituting Greek mythology. 'Plato speculated about the site of Atlantis' does not, on this view, assert a relation between Plato and the site of Atlantis, but between Plato and the concept of the site of Atlantis. Not that he speculated about the concept, but rather that his speculation involves a certain relation to that concept.

This view might be seen as rejecting the idea that in sentences of the form *s exists*, the predicate '*exists*' expresses existence(!). In such a context, the subject term designates a concept and the predicate expresses the property of *being non-vacuous*. Again in a sense Parmenides's argument is being accepted. Denials of existence are not about things that are not. They are about concepts. We cannot say true things about things that are not, but we can say true things that *seem* to be about non-beings. They are all about concepts.

2.6 How Should Logic Treat Existence?

Our subject is philosophical logic: Logic applied to philosophy and philosophy applied to logic. Logic can and should strive for generality and neutrality, even though there are limits to both. The concept of existence is certainly important in philosophy. How is it to be represented in logic, consistent with the goals just mentioned? It is always worth considering what is conveyed to ordinary natural language speakers by such a philosophically important term as 'exists'. Of course, this will not be definitive. We may wish to make distinctions where none are recognized, or are only infrequently recognized, by ordinary speakers. And we must of course be aware of contextual factors and even inconsistent usage in natural language.

Early Russell claims that there are two senses of 'exist':

> The meaning of *existence* which occurs in philosophy and in daily life is the meaning which can be predicated of an individual: the meaning in which we inquire whether God exists, in which we affirm that Socrates existed, and deny that Hamlet existed. The entities dealt with in mathematics do not exist in this sense: the number 2, or the principle of the syllogism, or multiplication are objects which mathematics considers, but which certainly form no part of the world of existent things. ([Russell, 1905a, p. 398])

As we observed above, others are equally confident and strongly insistent that there is only one natural sense of the word, both inside and outside philosophy. Or rather, they often claim that they do not understand any such distinction.[14] We could undertake an extensive empirical study of the occurrences of the term outside of philosophy, but that would be time-consuming, tedious, and difficult to evaluate – since in every case there will be a context that may contribute 'pragmatic' meaning or 'conversational implications.' It is clear on the most cursory examination of the writings of mathematicians that they have no aversion to saying that this-or-that mathematical entity exists. But is this a *different* sense of *existence*?

We need not decide. What needs doing is to examine the connotations associated with the term and decide which are important for philosophical and/or logical discourse. Then in our philosophical use we settle on the concept that has the best prospects for being of service, carefully distinguish it from other concepts, and always observe the distinction. For logical purposes, we seek a clear, perhaps somewhat idealized, concept that is of sufficient generality and neutrality to serve its purpose as objective arbiter of competing arguments. Of course this latter won't really be completely feasible since there are perennial disputes even about what belongs at the core of logic.

Taking our cue from Michael Slote's Theory of Important Criteria ([Slote, 1966])[15], let us consider an ideal case of existence. What would something be like that exists in the strongest possible way, that has every attribute that might go into real and substantial existence, worthy to be said to be such? We use 'worthy' here advisedly. Alan Ross Anderson ([Anderson, 1959]) has emphasized that there are sometimes honorific connotations involved in disputes about existence. (see also [Fitch, 1950]).[16] A massive physical object that exists now, the larger the better, and, for good measure, has always existed[17], would be a pretty solid case. It could and perhaps does causally interact with other objects. It would exist, we suppose, even if no one had ever thought of it, so its existence is in no sense 'subjective' or 'thought-dependent'. The thing has spatial and temporal location and a good deal of both. In fact the idea that spatio-temporal location is an important aspect of the concept of existence is clearly at the basis of some of those who make a distinction between existence and being.

Pointing in another direction, numbers and other 'abstract entities' have sometimes been thought to have *necessary* existence. Not only do they exist, some claim, but they *could not* fail to exist. This is the legacy of Plato who thought that the Forms (certain abstract entities) are *more real* than physical objects. Perhaps they are the only things, according to him, that are *truly* real (*really real*?). If there are such things and they are as described, then they do exist in a very substantial way.

But notice that the paradigm cases seem to conflict. Ordinary physical objects, no matter how solid, are liable to decay, become corrupted, and cease to exist. Not so with the alleged abstract objects. However it is also claimed by some Platonists that abstract object do not causally interact, at least not directly, with the physical world. They may be timeless, eternal, and hence do not literally have a temporal duration.

Both of these kinds of things, physical spatio-temporal things and abstract objects, are important to us in different ways. (See [Anderson, 1974]). One view, perhaps a compromise of sorts, is to say that both of these kinds of things exist in the fullest sense of the word – *if* there are any things of these two rather different kinds. If one is an anti-Platonist, you can assert, using this sense of 'exists', that there simply *are* no such things (necessarily existing things) and hence there do not *exist* any such things either. If you are a Berkelean Idealist, perhaps you should say flat-out that physical objects really do not exist in this sense – there aren't any such things. One reason is that if no one had ever thought of them, then there wouldn't be any such things (There is a bit of a difficulty about this in the case of God's thoughts.).

On this showing, there are perfectly good ways to distinguish different 'modes of being'. It may even be that fictional entities, though there are such things, do not exist in the sense of existence we have attempted to delimit. If someone protests, we respond that these things do not have spatio-temporal location, they do not directly causally interact with other existents, and they would not be if no one had ever thought of them and so do not exist of necessity. So some of us give them a lower score.[18]

What is clear is that there are sensible ways to make a distinction between different kinds of being and the one who understands the distinction (as opposed to those who claim that they don't) has the advantage. He can say things that his opponent cannot say. One need not fear that such distinctions lead to a 'bloated ontology'. We need only distinguish *ontological commitment*[19] from *existential commitment*. Both are full-blooded commitments to things of certain kinds. One certainly is not automatically drawn into thinking that there are things that are impossible in the sense of actually having incompatible properties.[20] And there is no harm in saying that there are impossible things in certain stories.[21]

What about those who say 'There are things that do not have any mode of being.'? We have not left a way for them to say this without contradiction. The infinitive 'to be' is intimately connected with the noun 'being'. And it seems natural to take a mode of being as being a mode of 'is-ness'. That is, an object has a mode of being if there is such an object in some sense. One can protest this identification, but 'mode of being' really is a technical philosophical notion that needs further explanation. Presumably we do not want to go so far as to say that there are things which are such that there are no such things.[22] It is very difficult to understand those who do want to say this.

The moral for logic seems to be that a predicate for existence should be allowed if needed for some such distinction. Happily, even if the predicate is vague, often arguments involving it can perfectly well be evaluated for validity. An is-ness predicate may be added (or defined using identity and ontological quantification) if desired. Ontological quantifiers might just as well range over all the entities needed for the semantics. This could include possible things as in modal logic, past and/or future individuals, and the like (Cf. [Cocchiarella, 1969]). The minimal way to accommodate this suggestion would be to just stop calling '∃' an *existential quantifier* and to always read it as 'there is ...' rather than 'there exists ...'. Then the change would hardly be noticed in most applications.

Notes

1. The example was given and much discussed before Pluto was demoted.
2. Also called 'Plato's Beard' by W.V. Quine ([Quine, 1948]) because of its resistance to Occam's Razor.

3. Compare Gödel's suggestion (Gödel, 1944, p. 129) for a premise for a very general version of Frege's arguments that all true sentences have the same 'signification': Every proposition is about something.
4. Curiously, a previous common usage had 'inflammable' meaning what is now using expressed as 'flammable'.
5. Perhaps *Lady Almathea* (a.k.a. *'the Unicorn'*) of Peter S. Beagle's novel *The Last Unicorn*.
6. Cf. Alonzo Church's ([Church, 1956, p. 396]) remarks '... [T]he value of logic to philosophy is not that it supports a particular system but that the process of logical organization of any system (empiricist or other) serves to test its internal consistency, to verify its logical adequacy to its declared purpose, and to isolate and clarify the assumptions on which it rests'.
7. For a general characterization and more detailed discussion of free logics, see [Lambert, 2001]. Our sample free logic is from that source.
8. This way of doing the semantics for free logic may derive from a comment in [Church, 1965].
9. There is a difficulty about treating atomic predicates differently from complex ones. In applying the logic to a natural language, we must somehow determine that the predicate expresses an 'atomic' property. Some syntactically simple predicates (in some languages) might express non-existence or some property entailing it. Formally the result is the failure of substitutivity for predicates. This in turn means that we are requiring something of the interpretation that may be difficult to determine in a particular application.
10. One might count the negation as being about the proposition expressed by the sentence negated – so that they are not about the same things. This requires some account of *propositions* as opposed to a semantics that just assigns truth-values or 'truth-conditions'.
11. This first disjunct comes from an example by Parsons ([Parsons, 1980]).
12. There are interpretations of free logic that have an 'outer domain' consisting of expressions. Ordinary (extensional) semantics doesn't require that we actually assign meanings, in the full sense, to the sentences of the language. If it did, this kind of interpretation would correspond to the idea that denials of existence are about *names* or other linguistic items. This view seems to be endorsed by (early) Frege. The more natural extension of his other views would point to the Frege–Church option discussed below. Taken literally it is subject to near refutation by way of the Church Translation Argument (Cf. [Salmon, 2001]).
13. Frege's view about these same cases was (at one time), roughly, that they are about the name involved and say in effect that is does not denote ([Frege, 1979]).
14. In this they do not always appear to be sincere, since they sometimes go on to consider ways of making such a distinction that they do admittedly understand.
15. I don't suppose that 'exists' is a 'cluster term', but Slote's general strategy for highlighting what is in question in disputes about definitions seems to be helpful here all the same.
16. Consider also uses of 'real' as in 'Michael Jordan is a *real* basketball player.' 'Santa Claus doesn't *really* exist – though he exists in the hearts and minds of those who believe in him.'
17. Of course there probably isn't any such object, but we are nevertheless trying to consider an ideal case of what would be an existent object.
18. An interesting case is Frege's ([Frege, 1980, p. 35]) example of the Equator. Do we want to say that it exists? The Celestial Equator is even more challenging.
19. Given the etymology, 'ontological' commitment really should mean the things that one is committed to there *being*. You claim *there are* such things.

20. Meinongians and Neo-Meinongians do sometimes allege such things, but it is in no way intrinsic to allowing a distinction between kinds of being.
21. See Graham Priest's story Sylvan's Box in [Priest, 2005, p. 125].
22. This saying is derived from Alexius Meinong ([Meinong, 1960]) and is endorsed in some version by (some of) his followers.

5 Quantification and Descriptions

Bernard Linsky

Chapter Overview

1. Proper Names versus Definite Descriptions	77
1.1 Differences between Names and Definite Descriptions	79
1.1.1 Analytic truths involving descriptions	79
1.1.2 Reference failure	80
1.1.3 Descriptions and intensional contexts	82
2. Russell's Theory of Descriptions	83
3. Descriptions as Singular Terms	90
3.1 The Frege–Hilbert Theory of Descriptions	90
3.2 The Frege–*Grundgesetze* Theory of Descriptions	92
3.3 The Frege–Carnap Theory of Descriptions	93
3.3.1 Syntax for Frege–Carnap	94
3.3.2 Semantics for Frege–Carnap	94
3.3.3 Deduction for Frege–Carnap	96
3.3.4 The 'Slingshot Argument'	96
4. Descriptions as Quantifiers	99
4.1 Syntax, Semantics, and Rules for Descriptions as Quantifiers	102
5. Conclusion	103
Notes	104

1. Proper Names versus Definite Descriptions

Quantifiers and singular terms are very distinct categories of expressions in logical grammar. Both supplement an open formula to produce a sentence, but in different ways. A singular term t replaces the free variable in ϕx to produce a sentence ϕt. The quantifier expressions 'there is' ($\exists$) and 'for all' ($\forall$) are completed with a variable x to produce the quantifiers $\exists x$ and $\forall x$, which are then prefixed to a formula (which is in the 'scope' of the quantifier) to produce the formulas $\exists x \phi x$ and $\forall x \phi x$. Corresponding to these different ways they complete a formula, names and quantifiers are given very different roles in the definition of truth. Singular terms are assigned an object as *denotation*, which *satisfies* the formula, whereas the quantifier produces a *true* or *false* sentence depending on which objects satisfy the formula. The singular terms in a formal language include constants (which symbolize proper names), complex terms involving function symbols, e.g., '$f(x,y)$', and definite descriptions, expressions involving the definite article 'the' and a predicate, of the form 'the ϕ'. Semantically they are like the other singular terms, having a denotation, at least ordinarily, which denotation is their contribution to the semantics of formulas in which they occur. Or at least so it seemed to Gottlob Frege in his account of referring denoting expressions in [Frege, 1892b].

This chapter will trace the history of the treatment of definite descriptions from Frege's initial inclusion of examples as proper names, through Bertrand Russell's account in 1905, to the contemporary analysis of descriptions as restricted quantifiers in LF (Logical Form). Definite descriptions are the subject of perhaps the most famous essay in twentieth-century Philosophical Logic, namely Bertrand Russell's 'On Denoting', published in *Mind* in 1905. Russell's account analyses definite descriptions as neither singular terms nor quantifiers, but instead as 'incomplete symbols' which, when properly defined, do not appear in the symbolic language at all. Moreover, on the route to their elimination, in an intermediate level of expression, they present some of the features of singular terms and one of the features of quantifiers, namely a *scope*. Russell's theory of definite descriptions is a way point in the story of the treatment of definite descriptions over the last hundred years.

Definite descriptions are also crucial to the account of proper names in Philosophical Logic. The distinction between proper names and definite descriptions is at the heart of the 'new theory of reference' introduced by Saul Kripke's *Naming and Necessity* lectures from 1970 and the debate over whether names have a 'sense', as Frege held. Thus this part of Philosophical Logic has direct consequences for philosophical issues about reference and meaning more generally in the Philosophy of Language, and so illustrates the application of Philosophical Logic to Philosophy as a whole.

In grammar, names and definite descriptions are part of the class of Noun Phrases, which includes also 'indefinite descriptions'. Another, more recent, development has been to see how to capture the logical properties of names and descriptions in a uniform fashion, while still representing the differences.

The following examples are taken from this long literature and will be used in this chapter:

Proper Names: Venus, Vulcan, Mercury, Pegasus, Zeus, Sherlock Holmes, 4, Odysseus, Aristotle, Plato, Socrates, Alexander the Great, Sir Walter Scott, George IV, *Waverley*, ...

Definite Descriptions: the least rapidly converging series, the Morning Star, the Evening Star, the present king of France, the author of *Waverley*, the teacher of Alexander, the pupil of Plato, the length of your yacht, the square root of 4, the negative square root of 4, the celestial body most distant from the Earth, the girl, . . .

Indefinite Descriptions: a man, any man, all men, no man, some man, . . .

Frege treats names and descriptions as in the same class, as can be seen from his examples in 'On Sense and Reference':

> The designation of a single object can also consist of several words or other signs. For brevity, let every such designation be called a proper name. [Frege, 1892b, p. 57]

The examples he uses, 'the least rapidly converging sequence' and 'the negative square root of 4', clearly includes what we would distinguish as definite descriptions along with familiar proper names, 'Odysseus', etc.

1.1 Differences between Names and Definite Descriptions

Names and definite descriptions, however, have different logical properties. Frege, who included both the reference (*Bedeutung*), and sense (*Sinn*), of names as constituting logical features of them says in a notorious footnote:

> In the case of an actual proper name such as 'Aristotle' opinions as to the sense may differ. It might, for instance, be taken to be the following: the pupil of Plato and teacher of Alexander the Great . . . ([Frege, 1892b, p. 58])

The quotation is problematic for several reasons. One is that Frege suggests, later on in the footnote, that individuals may vary in what sense they attach to a name, and that indeed, only a 'perfect language' would attach a unique sense to a name. The other problem raised by this footnote, and relevant for our topic, is the suggestion that the sense of an expression can be expressed accurately with a definite description, thus the sense of 'Aristotle' is expressed by 'the pupil of Plato and teacher of Alexander'.

1.1.1 Analytic truths involving descriptions

Whether or not a unique definite description captures the sense of a name or not, there is a certain logical phenomenon identified which is later used by Kripke to argue that names and descriptions are very different.

The phenomenon is simply that certain truths follow logically from a true sentence with a definite description. Thus it would seem that someone who attached the sense of 'the pupil of Plato and the teacher of Alexander' to 'Aristotle' above would say that the sentence:

Aristotle was a teacher

is an analytic truth. This is because it would seem to be a logical truth (following from the logic of definite descriptions) that:

The teacher of Alexander was a teacher.

This leads to one of the first principles of the logic of definite descriptions, namely, every instance for a predicate ϕ of:

$$\text{The } \phi \text{ is } \phi \tag{5.1}$$

or, in this example, a logical consequence of an instance for 'F and G'.

Definite descriptions seem to have logical structure in a way that proper names do not. That indeed is the thrust of Kripke's arguments in *Naming and Necessity*. There he argues, for example, that names do not have a sense, precisely because such examples as 'Aristotle was a teacher' are not analytic. While 'The teacher of Alexander is a teacher' is a logical truth, and so analytic, 'Aristotle was a teacher' is not an analytic truth. Given that we could, for example, discover that Aristotle was not a philosopher by tracing back the chain of reference to someone else, it can turn out that Aristotle was not a teacher. This is one of Kripke's arguments that names do not have a sense, and it relies on the identification of a logical feature of definite descriptions that does not hold for names.

1.1.2 Reference failure

A second way in which definite descriptions and names differ arises from the phenomenon of *reference failure*, when names and descriptions don't have a referent. Frege used as an example 'the most rapidly converging sequence'. Russell used 'The present King of France'. These descriptions fail to have a reference, since it is both the case that for any converging sequence there is another that converges more rapidly and that France was a republic long before Russell wrote 'On Denoting' in 1905. Of course there seem to be also names that have no referent: 'empty names' such as 'Vulcan' (purportedly naming a planet orbiting the sun inside of Mercury), or more arguably, 'Zeus' or 'Sherlock Holmes.' The latter two are difficult cases, because some argue that they do have abstract (mythological or fictional) objects as referents after all. Although both definite descriptions and names can be empty, the logical accounts of this phenomenon differ. It is very difficult to deny that names refer, because generally names obey certain logical principles, in particular Existential Generalization (from ϕt infer $\exists x \phi x$), and Universal Instantiation (from $\forall x \phi x$ infer ϕt). It seems obvious that if Aristotle was a Greek philosopher, then someone was a Greek philosopher. If everything is ϕ then Aristotle is ϕ. But one hesitates, precisely because the description is empty, to conclude from:

> Vulcan is a planet orbiting the Sun inside of Mercury

that

> There is a planet orbiting inside of Mercury

Similarly from:

> Nothing is a planet orbiting the Sun inside of Mercury

one should not therefore conclude that:

> Vulcan does not orbit the Sun inside of Mercury

The conclusion of the first inference, at least, is surely false, so we are reluctant to accept both inferences with such 'names'. On the other hand, Russell at least thinks that there is no problem in assigning truth values to sentences with non-denoting descriptions. That the present King of France is bald, he says, is 'plainly false' ([Russell, 1905b, p. 484]).

Russell himself, and many others following him, took one accomplishment of his theory of definite descriptions to be its avoidance of an otherwise persuasive argument for Meinongian, non-existent, objects. If a definite description 'The present King of France' in fact must have a denotation, then 'the round square' must refer to something that does not exist. Russell's theory of definite descriptions allows us to avoid being ontologically committed to objects simply by virtue of using descriptions which seemingly denote them.

Whether this was in fact Russell's main use of the theory of definite descriptions is a matter of dispute among historians of logic. What's more, *Neo-Meinongian* theories, such as that of Parsons ([Parsons, 1980]) and Zalta ([Zalta, 1983]) vary with respect to how they treat the phenomenon of 'empty descriptions'. Parsons allows for non-existent objects to be the referent of otherwise non-denoting descriptions ([Parsons, 1980, p. 119]). Zalta, on the other hand, provides an account of descriptions as singular terms in which many are non-denoting. The special Meinongian objects, such as 'the round square' will be non-existent (abstract) objects which *encode* (rather than *exemplify*) the properties expressed in empty descriptions. Thus there is no object which exemplifies the properties of being round and square, even a non-existent object, but there will be an object that encodes those properties. Neo-Meinongian theories were developed to account for non-existent objects while avoiding the logical problems for them that Russell raised. Whether they have referents for seemingly empty definite descriptions or not is incidental.

1.1.3 Descriptions and intensional contexts

A third, and somewhat complicated, difference between names and descriptions is in regard to substitution in *intensional contexts*.

George IV wished to know whether Scott was the author of *Waverley*. (5.2)

is true, but not:

George IV wished to know whether Scott was Scott. (5.3)

even though

Scott was the author of *Waverley*. (5.4)

The context '(5.2) George IV wished to know whether . . .' is *intensional* for it appears to violate standard principles characteristic of 'extensional' logic. For one thing it is not *truth-functional* for it may be true when completed by one true sentence, such as (5.2) but not another, as in (5.3), and secondly, the difference between those such two cases may be solely due to the replacement of one of two, co-referring, singular terms by the other, in this case 'Scott' and 'the author of *Waverley*'. It seems important to the failure of this difference that one of the terms is a name and the other is a definite description. Indeed Russell uses the difference between

Scott was Scott. (5.5)

and (5.4) in his 'proof' that descriptions are not names, and indeed, must be 'incomplete symbols' ([Whitehead and Russell, 1910, p. 67]).

It was Russell's characterization of names as contributing constituents to propositions which is the origin of the later characterization of names as 'directly referential'. This distinguishes names from descriptions, which seem to work with something like a sense, they refer by means of those properties which are part of them. Thus 'the *F*' refers to something that is *F*, if to anything at all.

This move, which was standard until recently, when descriptions and names are given a non-uniform treatment, was the first example of a uniform syntactic class getting a different logical analysis.

Russell saw the difference between names and descriptions even before he developed the theory of descriptions in [Russell, 1905b] for which he was famous. Even with his earlier theory of 'denoting concepts' from *Principles of Mathematics* ([Russell, 1903]) there was a difference between names and descriptions. Russell noted that descriptions seem to be involved in functions 'the R of x', called 'descriptive functions', and so 'denoting seems impossible to escape from' [Russell, 1994, p. 340].[1]

2. Russell's Theory of Descriptions

The paper that introduced Russell's theory of definite descriptions, 'On Denoting', in fact begins with an account of *indefinite descriptions* such as 'A man ...', 'Some man ...' and 'Any man ...'. Russell had earlier described them all, definite and indefinite, as introducing denoting concepts in *Principles of Mathematics*:[2]

> A concept *denotes* when, if it occurs in a proposition, the proposition is not *about* the concept, but about a term connected in a certain peculiar way with the concept. If I say 'I met a man,' the proposition is not about *a man*: this is a concept which does not walk the streets, but lives in the shadowy limbo of logic-books. What I met was a thing, not a concept, an actual man with a tailor and a bank-account or a public-house and a drunken wife. ([Russell, 1903, p. 53])

Thus the proposition *A man is mortal* contains the denoting concept *a man* as a constituent, much as the proposition *Socrates is mortal* contains Socrates, but it is not about that denoting concept. Instead, and this is the difficult part of the theory to express, it is about an 'indefinite man', some real man (with a tailor or a public-house) but no man in particular, such as Socrates. Russell motivates this difference by pointing out the difference in having a belief in the propositions, for example. One can believe the indefinite proposition without having any particular individual in mind. It is true that the existential sentence will have at least one witness, but no particular witness is a part of the proposition.

The contribution of 'On Denoting' is to show how, using the familiar existential and universal quantifiers, one can do without these denoting concepts. As Russell says, this theory can be seen as one that avoids denoting. What is proposed for the denoting phrases 'All' and 'Some' is the standard analysis of elementary logic:

All ϕ's are ψ's.

and

Any ϕ's are ψ's.

become

$$\forall x(\phi x \supset \psi x)$$

On the other hand:

A ϕ is ψ

and

Some ϕ's are ψ's

are symbolized as:

$$\exists x(\phi x \wedge \psi x)$$

These indefinite descriptions are incomplete symbols because they do not turn out to be constituents of the propositions:

Some ϕ's . . .

becomes

$$\exists x(\phi x \wedge \ldots)$$

to be filled in with the symbolization of '. . . are ψ's', namely 'ψx'. That part which represents 'Some ϕ's' is a discontinuous portion of the proposition, not representing any constituent at all, even to the extent that connectives and quantifiers represent constituents, much less as well formed formulas, like 'ψx'. It is this phenomenon that Russell invokes when he says that definite descriptions are 'incomplete symbols'.

When it comes to definite descriptions, which were represented by denoting concepts in Russell's earlier thinking, again we get a complex quantificational sentence. The expression 'the' will be represented below by the iota symbol 'ι', (which Russell originally inverted), so that:

The ϕ is ψ

when symbolized as

$$\psi(\iota x\ \phi x)$$

is defined to be:

$$\exists x \forall y((\phi y \equiv y = x) \wedge \psi x)$$

Again, definite descriptions are also incomplete symbols. Because the defined expression is not a constituent of the proposition in which it occurs, the definition does not take the form of an identity or explicit definition replacing one symbol by another of the same syntactic category. As definite descriptions appear to be singular terms, an explicit definition would take the form:

$$\iota x\ \phi x =_{df} \ldots$$

But no such definition is forthcoming. Instead we get what is called a *contextual definition*, which shows how to 'eliminate' the description from a *context*, represented by ψ. In fact there are more occasions to use definite descriptions in Russell's logical system, including the notation for the expression that says that a description is proper. 'The ϕ' is proper just in case there is exactly one ϕ. In *Principia Mathematica* the notion of being proper is indicated with the symbol '$E!$'.[3] In [Whitehead and Russell, 1910] ($*14{\cdot}02$) the definition is:

$$E!(\iota x\ \phi x) =_{df} \exists x \forall y(\phi y \equiv y = x)$$

There is a difference between the apparent form of propositions, in which definite and indefinite descriptions seem to be constituents, and in syntax are parts of the class of noun phrases, and their representation in the notation of quantifiers by Russell's theory. This is the source of the view that the deep structure, or logical form, of sentences are very different from their surface or syntactic structure. Following Ramsey's description of Russell's theory of descriptions as a 'paradigm of philosophical analysis', this came to be in fact the model for all philosophical analysis; namely finding the proper analysis of propositions, which might have a very different form from what is suggested by the surface grammar of sentences.[4] In an extreme case it was felt that some terms, such as those expressing values 'good' or 'beautiful' did not express properties at all, or at least no simple, primitive properties. Ontology was reformed when expressions such as 'the nation' were felt to be logical constructions out of people, and this supported reductivist or eliminativist metaphysical projects. Gilbert Ryle proposed that this notion of logical construction was a model of how to avoid category mistakes, in his case as big as the 'myth of the mental' which reified the Cartesian mind rather than following the right path of logical behaviourism.[5]

To return to Russell's theory of descriptions, there is one aspect, the notion of the scope of a description, which would eventually lead to the notion that this is literally the *scope* of a quantifier. One of Russell's three 'puzzles' from [Russell, 1905b] has to do with descriptions that lack a referent, and so not a *proper description*.[6] Russell discusses the example:

The present King of France is bald. (5.6)

Russell says one won't find the present King of France on the list of bald things, nor on the list of things that are not bald. It would seem that this gives rise to a violation of the law of the excluded middle. Russell's solution is to invoke the notion of the 'scope' of a description. There are two similar sentences that differ with respect to the scope of the description, and so differ in truth value. One is simply the negation of (5.6) and is false precisely when that sentence is true. The other, with the wide scope for the description, amounts to saying that there is one and only one king of France and he is not bald. This sentence is the natural reading of the sentence:

The present King of France is not bald. (5.7)

and the fact that both are false if there is no king of France is what produces the apparent violation of the law of the excluded middle.

Russell indicates the scope of the description by writing the description in square brackets right before the occurrence of the context of the description, as explained above. In the official statement of the contextual definition (∗14·01) we have:

$$[(\iota x\ \phi x)]\psi(\iota x\ \phi x) =_{df} \exists x \forall y((\phi y \equiv y = x) \wedge \psi x)$$

The symbolization of the sentence with the description having a 'primary occurrence', or we would say 'wide scope' or 'scope over the negation', is the best rendering of the meaning of (5.7). It is symbolized as:

$$[(\iota x\ Kx)]\ \neg B(\iota x\ Kx)$$

The *scope indicator*, '$[(\iota x\ Kx)]$', which is simply the description placed in square brackets, immediately precedes the beginning of the *scope* of the description, i.e., what stands in for the ψ above. Here it is '$\neg B(\ldots)$' or '... is not bald'. When the description is eliminated from this context, the claim becomes:

$$\exists x \forall y((Ky \equiv y = x) \wedge \neg Bx) \tag{5.7a}$$

or, that there is one and only one x which is king of France and x is not bald. This is false because there is not even one king of France, as the country is a republic. The other scope for (5.7) takes 'The King of France is bald' and simply negates it, and it is represented as:

$$\neg[(\iota x\ Kx)]\ B(\iota x\ Kx)$$

Here the scope indicator immediately precedes the context '$B\ldots$', and so it is the negation of the expression (5.6). The sentence (5.6) is by definition:

$$\exists x \forall y((Ky \equiv y = x)\ \wedge Bx)$$

i.e., there is one and only one x which is a king of France and that x is bald. This sentence is false, for the same reason as the last. The negation of that gives the result of negating that, thus amounting to:

It is false that there is one and only one present king of France who is bald

in symbols:

$$\neg\exists x \forall y((Ky \equiv y = x) \wedge Bx) \tag{5.7b}$$

As (5.7*b*) says that there is not one and only one x which is a present king of France and x is bald, which is true. Both the original and the occurrence with *wide scope* or 'primary occurrence', are false, thus producing the appearance of a violation of the law of excluded middle, but since in fact it is the *narrow scope*, 'secondary occurrence' which is the negation of the first, and only one of those two is true and the other false, observing the law of excluded middle after all.

In 'On Denoting' Russell introduces the notion of scope of descriptions to answer his second puzzle, but this solution then returns him to the solution to the first puzzle of Scott and the author of *Waverley*. The first solution is simply to point out that this doesn't give a violation of the inference involving identity sentences known as 'Leibniz' Law' (LL), namely the inference from $t_1 = t_2$ and a formula ϕ, to $\phi[t_1/t_2]$, the result of substituting occurrences of t_2 for t_1 in ϕ:

$$\frac{t_1 = t_2, \phi}{\phi\ [t_1/t_2]} \tag{LL}$$

This does not apply directly to cases of replacing descriptions within a context, because definite descriptions are not terms but rather 'incomplete symbols' that look like terms until analysed. The complication is that in fact an apparent substitution of descriptions is derivable even when the descriptions have been eliminated via the contextual definition. As a result the inference:

$$\begin{array}{l} \text{the } \phi = \text{the } \chi \\ \text{the } \phi \text{ is } \psi \\ \hline \therefore \text{ the } \chi \text{ is } \psi \end{array}$$

what, as Russell says is 'verbally' the substitution, is in fact valid after all. The inference is not a straightforward substitution of terms, but instead is a rather complicated inference, especially as the second premise includes two descriptions that are eliminated in terms of quantificational formulas. The first stage, with scope indicators will look like this:

$$\begin{array}{l} [(\iota x\ \phi x)]\ [(\iota y\ \chi y)]\ x = y \\ [(\iota x\ \phi x)]\ \psi(\iota x\ \phi x) \\ \hline \therefore\ [(\iota x\ \phi x)]\ \chi(\iota x\ \phi x) \end{array}$$

As Russell points out, the inference is only valid when the description has wide scope, as above. Eliminating the descriptions with the contextual descriptions according to that scope, we get a complicated, but valid, inference of first-order logic that is not of the form of Leibniz' Law:

$$\begin{array}{l} \exists x \forall y((\phi y \equiv y = x)\ \wedge\ \exists u \forall v((\chi v \equiv v = u)\ \wedge\ x = u)) \\ \exists x \forall y((\phi y \equiv y = x)\ \wedge\ \psi x) \\ \hline \therefore \exists x \forall y((\chi y \equiv y = x)\ \wedge\ \psi x) \end{array}$$

For intensional contexts such as 'George IV wished to know whether Scott is the author of *Waverley*', the two scopes are not equivalent, and so, once again, we see that in this case, the original, problematic, inference does not follow. Not only is this not a case of substituting singular terms, it is also not one of the valid cases of substituting definite descriptions in the place of singular terms. In *Principia Mathematica* ∗14, the chapter on descriptions, Whitehead and Russell propose a theorem, ∗14·3, which is intended to characterize those cases where the scopes are equivalent if the description is proper, and so the limits of the cases where the apparent substitution is valid because it is of the form above.

They claim, but feel hampered by being unable to actually prove, that so long as the context 'ψ ...' is extensional, that the narrow scope will be equivalent to the wide scope, and as a consequence we learn that the above inference will be valid in just those cases. It is at this point that one of the issues of modal logic arises, namely how to give a semantic account of the two occurrences of scopes of descriptions with intensional contexts. Russell is content to use a humorous example, the story of the touchy owner who responds to 'I thought your yacht was larger than it is' with 'No, my yacht is not larger than it is.' The joke is meant to illustrate the two scopes, relied on to make the apparently contradictory sentences in fact both true, with the two scopes for:

I thought that the size of your yacht is greater than the size your yacht is. (5.8)

One reading expresses this with the scope of the description indicated intuitively as:

The size that I thought your yacht was is greater than the size your yacht is. (5.8′)

This is represented in the notation of generalized quantifiers that will be introduced below as:

[The x : size of your yacht x]I thought that x was greater than x. (5.8a)

The other reading:

I thought the size of your yacht was greater than the size of your yacht. (5.8″)

can be symbolized as:

[The x : size of your yacht x] I thought that the size of your yacht was greater than x. (5.8b)

Russell then points out that 'George IV wished to know whether Scott is the author of *Waverley*' is in fact similarly ambiguous and with one scope for the description the problematic substitution goes through. The sense in which George IV might in fact wish to know whether Scott is Scott, is that in which he might be said to want to know, of the author of *Waverley*, i.e. Scott, whether *he* is Scott, thus:

[The x : author of *Waverley* x]George IV wished to know whether $x =$ Scott. (5.2a)

This reading attributes to George IV a wish to know *de re*, as opposed to the *de dicto* attitude we would naturally attribute to George IV, namely of wishing to know whether Scott is the one and only person who wrote *Waverley*.

3. Descriptions as Singular Terms

Frege had more to say about definite descriptions than just that they should be classed as names. He was acutely aware of the problem of reference failure for definite descriptions and also of the case of *improper* descriptions, i.e. those that apply to more than one thing or to nothing at all. In his study of Frege's views, Carnap gives four different accounts of definite descriptions, which all treat them as singular terms. They will be called 'Frege–Hilbert', 'Frege–Strawson', 'Frege–Carnap', and 'Frege–*Grundgesetze*' in what follows, to keep them distinct and to acknowledge others who have developed them independently. The theory that most directly competes with the contemporary view of descriptions as quantifiers, to be described in the next section, is the view that descriptions are simply singular terms, but which use the model-theoretic device of a 'chosen object' to in fact make all descriptions proper, yet to still represent the distinctive features of descriptions. Although Carnap's name is only associated with this final account, the very classification of suggested approaches in Frege comes from [Carnap, 1948], *Meaning and Necessity*, and so it is appropriate to credit Carnap with a theory that treats definite descriptions as singular terms.[7]

3.1 The Frege–Hilbert Theory of Descriptions

The various Fregean theories of descriptions as singular terms that Carnap found can all be traced to passages in Frege's works. Thus the first, Frege–Hilbert view can be seen in the following from 'On Sense and Reference':

> A logically perfect language (*Begriffschrift*) should satisfy the conditions, that every expression grammatically well constructed as a proper name out of signs already introduced shall in fact designate an object, and that no new sign shall be introduced as a proper name without being secured a reference. ([Frege, 1892b, p. 70])

Then in discussing the example of 'the negative square root of 4' (as contrasted with the improper description 'the square root of 4'), he says:

> We have here the case of a compound proper name constructed from the expression for a concept with the help of the singular definite article. This is at any rate permissible if the concept applies to one and only one single object. ([Frege, 1892b, pp. 71–2]

Here we have a hint of the procedure that Carnap finds in Hilbert & Bernays, the familiar requirement of proving an 'existence and uniqueness theorem' before introducing a singular term. If we are guaranteed that the description is proper, then the logical properties which distinguish names from descriptions will not be relevant. (Presumably the further properties of descriptions, such as that 'The F is F', will be provable with whatever demonstrates the existence and uniqueness of 'the F' in the first place.)

Frege is aware that in natural languages, i.e., not in the 'logically perfect' language that his *Begriffschrift* is meant to be, that there will of course be many definite descriptions which are not proper:

> It may perhaps be granted that every grammatically well-formed expression representing a proper name always has a sense. But this is not to say that to the sense there also corresponds a reference. The words 'the celestial body most distant from the Earth' have a sense, but it is very doubtful if they also have a reference. In grasping a sense, one is certainly not assured of a reference. ([Frege, 1892b, p. 58])

> Is it possible that a sentence as a whole has only a sense, but no reference? At any rate, one might expect that such sentences occur, just as there are parts of sentences having sense but no reference. And sentences which contain proper names without reference will be of this kind. The sentence 'Odysseus was set ashore at Ithaca while sound asleep' obviously has a sense. But since it is doubtful whether the name 'Odysseus', occurring therein, has a reference, it is also doubtful whether the whole sentence has one. Yet it is certain, nevertheless, that anyone who seriously took the sentence to be true or false would ascribe to the name 'Odysseus' a reference, not merely a sense; for it is of the reference of the name that the predicate is affirmed or denied. Whoever does not admit the name has reference can neither apply nor withhold the predicate. ([Frege, 1892b, p. 62])

The proposal is that a sentence with an improper description in it lacks truth value. Strawson ([Strawson, 1950]) distinguishes between the sentence and the *statement*, what is said by uttering the sentence in a given context, which is in fact what has or lacks a truth value, but when applied to sentences this becomes a 'truth-value gap' account of improper descriptions, and the general approach can still be called 'Frege–Strawson'. Free logic is aimed at presenting the logic of sentences that contain singular terms which fail to refer. Some don't allow truth-value gaps, and so, modelled on examples like 'Pegasus has wings', require that sentences all have truth values, despite the occurrence of non-referring singular terms. Others allow the failure of reference to result in truth-value gaps.[8] Notice that this approach maintains the strict analogy between descriptions and names, for both can introduce reference failure, however it is treated logically.

3.2 The Frege–*Grundgesetze* Theory of Descriptions

The next approach to descriptions that is found in Frege comes from his *Grundgesetze*, using the symbol ' \' to represent the definite article. The intended semantics for the theory is explained as follows. In the *Grundgesetze*, Frege uses the symbols $\acute{\epsilon}\ F\epsilon$ to indicate the 'course of values' (*Werthverlauf*) of F, that is, the set of things that are F. *Grundgesetze* §11 introduces the symbol '$\backslash\xi$', which he calls the 'substitute for the definite article'. It is clearly only a 'substitute', for it does not represent an operation which applies directly to *concepts* which would be the denotation of predicates like 'F', but rather to particular objects, namely the extensions of concepts. Frege distinguishes two cases:

1. If to the argument there corresponds an object Δ such that the argument is $\acute{\epsilon}\ (\Delta = \epsilon)$, then let the value of the function $\backslash\xi$ be Δ itself;
2. If to the argument there does not correspond an object Δ such that the argument is $\acute{\epsilon}\ (\Delta = \epsilon)$, then let the value of the function $\backslash\xi$ be the argument itself.

And he follows this up with the exposition:

> Accordingly $\backslash\ \acute{\epsilon}\ (\Delta = \epsilon) = \Delta$ is the True, and '$\backslash\ \acute{\epsilon}\ \Phi(\epsilon)$' refers to the object falling under the concept $\Phi(\xi)$, if $\Phi(\xi)$ is a concept under which falls one and only one object; in all other cases '$\backslash\ \acute{\epsilon}\ \Phi(\epsilon)$' has the same reference as $\acute{\epsilon}\ \Phi(\epsilon)$.

In more modern notation, replacing Frege's '$\acute{\epsilon}\ (\Delta = \epsilon)$' by '$\{\epsilon : \Delta = \epsilon\}$', we get the rule that if the extension of a predicate F is in fact a unique object Δ, then the value of the description 'the F' is Δ, otherwise it is $\{x : Fx\}$. The passage above is from the introductory sections which provide a description of the syntax and an informal motivation for what is to follow. In the formal development of *Grundgesetze* there is only one axiom that deals with descriptions at all:

Basic Law (VI): $a = \backslash\ \acute{\epsilon}\ (a = \epsilon)$

(in modern notation: $a = \backslash\{x : x = a\}$). This means (given Frege's analysis of identities as including two terms with the same reference but possibly distinct senses) that a term 'a' has the same reference as '$\backslash\{x : x = a\}$'. In other words, if a is the unique member of the course of values of the concept 'is identical with a', then a is the value of the \ operation applied to that course of values. In the case of an improper description 'the F', $\backslash\{x : x = \text{the F}\}$ is just $\{x : Fx\}$, so the identity is true in that case as well. This axiom VI, however, seems to be sufficient for what follows in *Grundgesetze*, and indeed descriptions soon fade after an initial use in the very first theorem.[9] As Frege's system is second order, and so the notion of validity will be vexed, and since it is in any case inconsistent, as shown by Russell's paradox, one hesitates to put too much stress on the adequacy of one axiom to capture this theory of descriptions.

3.3 The Frege–Carnap Theory of Descriptions

The last account of descriptions as terms which can be found among Frege's different suggestions is the one developed by Carnap, which is here referred to as the 'Frege–Carnap' theory of descriptions as names. It is inspired by this remark from 'On Sense and Reference':

> This arises from an imperfection of language, from which even the symbolic language of mathematical analysis is not altogether free; even there combinations of symbols can occur that seem to stand for something but have (at least so far) no reference, e.g., divergent infinite series. This can be avoided, e.g., by means of the special stipulation that divergent infinite series shall stand for the number 0. ([Frege, 1892b, p. 70])

This passage in fact immediately precedes that quoted above, to the effect that in a logically perfect language improper descriptions should not be introduced, which was cited before as the source for the Frege–Hilbert view. Here we have the source for what might be called 'special' or 'chosen object' theories of descriptions. The idea is just to pick an object 'a^*' for improper descriptions to refer to. Notice that it depends on what object is chosen, so the present King of France is bald if the object is Yul Brynner. (As David Kaplan points out in his [Kaplan, 1970].) There are various ways of implementing this in formal semantics. One is to have the chosen object be a regular member of the domain, as in the example of Yul Brynner. If the chosen object varies from model to model, then what follows logically as true in all models will wash this out. In some models someone with a fine head of hair will be chosen to be the interpretation of 'the present King of France'.

A formal system for the Frege–Carnap theory of descriptions is presented in Kalish and Montague's textbook, *Logic*.[10] Kalish and Montague get by with two rules, one for proper descriptions, essentially justifying the inference that 'the F is F', and one for improper descriptions which captures the decision to have some one object chosen to be the 'referent' of all improper descriptions. To explain the Frege–Carnap theory, it is first necessary to show what revisions are necessary to the notion of singular term in order to treat definite descriptions as singular terms. Then a modification of standard semantics is needed, to include the interpretation of descriptions in a model, and then it will be possible to present rules which when added to a standard system of first-order logic are complete for the revised semantics.

3.3.1 Syntax for Frege–Carnap

The principal modification to standard semantics for first-order languages which is needed to treat definite descriptions as singular terms in Carnap's fashion is due to the fact that atomic formulas, those containing only a relation symbol and a series of terms, can now be of arbitrary complexity. Thus in the atomic formula $\psi(\iota x\ \phi x)$ the predicate ϕ can be an arbitrary formula containing other descriptions. The inductive definition of a formula, then, does not follow the definition of a term, but instead is simultaneous:

Definition 5.3.1 Definition of *term* and *formula*

(i) All variables and constants are *terms*.
(ii) If f is an n-place function symbol and $t_1, \ldots, t_n$ are *terms*, then $ft_1, \ldots t_n$ is a *term*.
(iii) If R is an n-place relation symbol and $t_1, \ldots t_n$ are *terms*, then $Rt_1, \ldots t_n$ is a *formula*.
(iv) If t_1 and t_2 are *terms* then $t_1 = t_2$ is a *formula*.
(v) If ϕ and ψ are *formulas*, then so are: $\neg\phi$, $(\phi \supset \psi)$, $(\phi \vee \psi)$, $(\phi \wedge \psi)$, $(\phi \equiv \psi)$.
(vi) If ϕ is a *formula* and x is a variable, then $\forall x\phi x$ and $\exists x\phi x$ are *formulas*.
(vii) If ϕ is a *formula* and x is a variable, then $\iota x\ \phi x$ is a (descriptive) *term*.

As description operators *bind* variables in the way that quantifiers do, the corresponding notions of *free* and *bound* occurrences of variables, proper substitution of a term for a variable, etc., must be extended.[11]

3.3.2 Semantics for Frege–Carnap

An account of definite descriptions as singular terms has to be able to capture the characteristic feature of descriptions that 'the F is F', and the decision to 'arbitrarily' select some special object as the 'referent' of all improper descriptions. A standard way of representing semantics for first order logic can be modified in an analogous way to this:

The semantics is based on the notion of a *model* $\mathcal{A}$ for the language, which includes a set as its domain A, and individual $c^{\mathcal{A}}$ in A for each constant c, an n-ary function $f^{\mathcal{A}}$ for each n-ary function symbol f. The model identifies an object $a^* \in A$, which will be the designated object of the model.

Because the interpretation of some terms (namely those that include definite descriptions) will depend on what objects satisfy certain formulas, the notions of interpretation and truth of a formula cannot be defined separately. The standard practice is to define a notion of structure, containing the domain A and functions and relations, and then to define a notion of 'denotation', which consists of a function that yields an object for each constant and to each variable yields the object to which it is assigned. Instead we define the two together.[12]

A *model* $\mathcal{A}$ is a sequence $\mathcal{A} = \langle A, f_1{}^{\mathcal{A}}, \ldots f_n{}^{\mathcal{A}}, R_1, \ldots R_k, a^* \rangle$

An *assignment* β is a function from variables and constants to elements of A, such that $\beta(v) \in A$ and $\beta(c) \in A$ for each variable and constant in the language.

The *denotation* of a term t on a model $\mathcal{A}$ relative to an assignment β, $\eth^{\beta}(t)$, is the value of a function $\eth^{\beta}$, defined as follows together with the *truth in a model* $\mathcal{A}$ *on an interpretation relative to a sequence* β *of a formula* ϕ , that is: ($\mathcal{A} \models^{\eth,\beta} \phi$),

Definition 5.3.2 Definition of: $\eth^{\beta}(t)$ and $\mathcal{A} \models^{\eth,\beta} \phi$

(i) For a variable x let $\eth^{\beta}(x) = \beta(x)$. For a constant c, let $\eth^{\beta}(c) = c^{\mathcal{A}}$

(ii) If f is an n-place function symbol and $t_1, \ldots, t_n$ are terms, then: $\eth^{\beta}(ft_1, \ldots t_n) = f^{\mathcal{A}}(\eth^{\beta}(t_1), \ldots, \eth^{\beta}(t_n))$.

(iii) If R is an n-place relation symbol and $t_1, \ldots t_n$ are terms, then $\mathcal{A} \models^{\eth,\beta} Rt_1, \ldots t_n$ iff $R^{\mathcal{A}}(\eth^{\beta}(t_1), \ldots, \eth^{\beta}(t_n))$.

(iv) If t_1 and t_2 are terms then $\mathcal{A} \models^{\eth,\beta} t_1 = t_2$ iff $\eth^{\beta}(t_1) = \eth^{\beta}(t_2)$.

(v) If ϕ and ψ are *formulas*, then:
   (a) $\mathcal{A} \models^{\eth,\beta} \neg\phi$ iff $\mathcal{A} \not\models^{\eth,\beta} \phi$
   (b) $\mathcal{A} \models^{\eth,\beta} (\phi \supset \psi)$ iff $\mathcal{A} \not\models^{\eth,\beta} \phi$ or $\mathcal{A} \models^{\eth,\beta} \psi$
   (c) $\mathcal{A} \models^{\eth,\beta} (\phi \vee \psi)$ iff $\mathcal{A} \models^{\eth,\beta} \phi$ or $\mathcal{A} \models^{\eth,\beta} \psi$
   (d) $\mathcal{A} \models^{\eth,\beta} (\phi \wedge \psi)$ iff $\mathcal{A} \models^{\eth,\beta} \phi$ and $\mathcal{A} \models^{\eth,\beta} \psi$
   (e) $\mathcal{A} \models^{\eth,\beta} (\phi \equiv \psi)$ iff $\mathcal{A} \models^{\eth,\beta} \phi$ and $\mathcal{A} \models^{\eth,\beta} \psi$ or $\mathcal{A} \not\models^{\eth,\beta} \phi$ and $\mathcal{A} \not\models^{\eth,\beta} \psi$

(vi) If ϕ is a *formula* and x is a variable, then
   (a) $\mathcal{A} \models^{\eth,\beta} \forall x\phi x$ iff for all $a \in A$, $\mathcal{A} \models^{\eth,\beta[a/x]} \phi x$
   (b) $\mathcal{A} \models^{\eth,\beta} \exists x\phi x$ iff for some $a \in A$, $\mathcal{A} \models^{\eth,\beta[a/x]} \phi x$
   (where $\beta[a/x]$ is just like β except possibly in assigning a to x)

(vii) If ψ is a *formula* and $\iota x\ \phi x$ is a (descriptive) *term*, then
   (a) If there is a unique $z \in A$ such that $\mathcal{A} \models^{\eth,\beta[z/x]} \phi x$, then $\eth^{\beta}(\iota x\ \phi x) = z$
   (b) otherwise, $\eth^{\beta}(\iota x\ \phi x) = a^*$

The notion of truth in a model is the standard one, modified for models of the Frege–Carnap language:

Definition 5.3.3 $\mathcal{A} \models \phi$ iff $\mathcal{A} \models^{\eth,\beta} \phi$ for all $\eth, \beta$

and the notion of logical consequence $\Gamma \models \phi$ is similarly standard:

Definition 5.3.4 $\Gamma \models \phi$ iff for all $\mathcal{A}$, if $\mathcal{A} \models \Gamma$ then $\mathcal{A} \models \phi$

(where $\mathcal{A} \models \Gamma$ iff $\mathcal{A} \models \gamma$ for every $\gamma \in \Gamma$) A formula ϕ is *valid*, $\models \phi$, just in case $\mathcal{A} \models \phi$ for all models $\mathcal{A}$.

3.3.3 Deduction for Frege–Carnap

Two inference rules are sufficient for the system of deduction for descriptions in the Kalish & Montague system. One is PD (Proper descriptions):

$$\frac{\exists y \forall x(\phi x \equiv x = y)}{\phi(\iota x\ \phi x)} \qquad \text{(PD)}$$

(where x, y are variables, ϕx is a formula in which y is not free, and $\phi(\iota x\ \phi x)$ comes from ϕx by proper substitution of the term $(\iota x\ \phi x)$ for x.) When there is exactly one ϕ, one can conclude that the ϕ is ϕ.

The other, ID (Improper descriptions) is:

$$\frac{\neg\exists y \forall x(\phi x \equiv x = y)}{\iota y\ \phi y = \iota z\ z \neq z} \qquad \text{(ID)}$$

(where x, y and z are variables, ϕx is a formula in which y is not free.) If there is not exactly one ϕ, then the ϕ = the z such that $z \neq z$, in other words, all improper descriptions have the same denotation.

These two rules, when added to a group of other standard rules related to the other connectives and logical expressions, produces a notion of provable consequence $\Gamma \vdash \phi$ which is complete in the standard sense; for all Γ and ϕ, $\Gamma \vdash \phi$ iff $\Gamma \models \phi$. (In the special case when Γ is the empty set, we have that all and only theorems ϕ are valid formulas: $\vdash \phi$ iff $\models \phi$.)

The need for only these two rules reflects the fact that in the Frege–Carnap theory definite descriptions are introduced as singular terms, and so have the logical features of all singular terms, that 'the F is F' is a logical truth whenever 'The F' is a proper description, and finally that all improper descriptions denote the same thing. The distinctive logical features of descriptions on the Frege–Carnap account are captured by these rules, in the sense that the system is complete, a formula is provable with these rules if and only if it is valid with respect to the relevant set of models defined above.

3.3.4 The 'Slingshot Argument'

The famous argument due to Gödel [Gödel, 1944] which Barwise and Perry [Barwise and Perry, 1981] named 'the slingshot' can be formulated following Dagfinn Føllesdal, in his [Føllesdal, 1961], as an argument against the Frege–Carnap theory of descriptions. The argument relies on treating descriptions both as singular terms, while at the same time attributing to them a logical structure. As singular terms they count as legitimate instances of Universal Instantiation for Descriptions (UID):

$$\frac{\forall x \psi x}{\psi(\iota x\ \phi x)} \qquad \text{(UID)}$$

This seems to follow from their nature as singular terms which always refer, even if, in the case of 'improper' descriptions, to the selected object a^*. Another principle of modal logic that Føllesdal uses is the Necessity of Identity (NI):

$$\forall x \forall y(x = y \supset \Box(x = y)) \qquad \text{(NI)}$$

Føllesdal's argument is presented in a system where the object a^* can be named in the language. (For a version of the proof in a system of modal logic combined with the Kalish and Montague system above, consider 'a^*' below to be an abbreviation for '$\iota x(x \neq x)$'.) The argument shows that if there is some object y such that $\Box y \neq a^*$ and p is true, then it follows that $\Box p$, in other words, the modalities collapse in this situation. That $\Box(y \neq a^*)$ follows from $y \neq a^*$ in most systems, by a comparable 'Necessity of Non-Identity' principle, $\forall x \forall y(x \neq y \supset \Box(x \neq y))$. The argument requires some lemmas from modal logic, but even so takes only 22 lines for Føllesdal. Here is a sketch of how it proceeds. First assume:

$$\Box(y \neq a^*) \wedge p \qquad (5.9)$$

Then, by the principle (PD):

$$\iota x(x = y \wedge p) = y \qquad (5.10)$$

Then by the Necessity of Identity (NI), it follows that:

$$\Box(\iota x(x = y \ \wedge \ p) = y) \qquad (5.11)$$

by using Universal Instantiation of the variable x to $\iota x(x = y \wedge p)$. Now the Frege–Carnap theory of descriptions has the following consequence:

$$\iota x(x = y \wedge p) = y \ \wedge \ y \neq a^* \supset p \qquad (5.12)$$

Since (5.12) is a theorem, its necessitation:

$$\Box(\iota x(x = y \wedge p) = y \ \wedge \ y \neq a^* \supset p) \qquad (5.13)$$

is a theorem, and so by an elementary principle of modal logic, we get:

$$\Box(\iota x(x = y \ \wedge \ p) = y \wedge y \neq a^*) \supset \Box p \qquad (5.14)$$

The antecedent of (5.14) follows directly from (5.9) and (5.11) and so we derive, on the assumption of (5.9), that:

$$p \supset \Box p \qquad (5.15)$$

This sentence (5.15) was proved for an arbitrary sentence p, and so this is the resulting 'collapse' of the modality $\Box$.

However, the Slingshot argument cannot be carried out in Russell's theory of descriptions, and so the argument can be taken as an objection to the Frege–Carnap theory of descriptions, as much as the objection to quantified modal logic, as Quine and Føllesdal took it to be. The Slingshot is not valid on Russell's theory because when the scope of the descriptions are to be indicated, there is no one scope that validates the move from (5.10) to (5.11) and which fits with the interpretation of (5.11) needed to deduce the antecedent of (5.12). Line (5.11) is only well formed with the scope indicator as follows:

$$\Box[\iota x(x = y \wedge p)]\iota x(x = y \wedge p) = y \tag{5.11'}$$

Only the following would follow by NI:

$$[\iota x(x = y \wedge p)]\, \Box\iota x(x = y \wedge p) = y \tag{5.11'a}$$

However, what is needed later in the proof is:

$$\Box([\iota x(x = y \wedge p)]\iota x(x = y \wedge p) = y) \tag{5.11'b}$$

A more familiar example will make the problem clear.[13] (Let 'Nx' represent 'x is the number of the planets'). From the identity:

$$[\iota x Nx]\iota x Nx = 9 \tag{5.16}$$

the rule of necessitation can only yield the false sentence:

$$\Box[\iota x\ Nx]\iota x\ Nx = 9 \tag{5.17a}$$

for it is not necessary that there are 9 planets. All that would follow correctly using NI is:

$$[\iota x Nx]\Box\iota x Nx = 9 \tag{5.17b}$$

In other words, it may be true that there is a wide scope reading of the sentence on which it is true, of the number of planets, i.e., 9, that it is equal to 9, but that does not lead to any collapse or other objection to quantified modal logic.

That Russell's theory of descriptions allows one to block the Slingshot arguments against quantified modal logic was pointed out by Smullyan in [Smullyan, 1948]. Føllesdal's version of the slingshot, however, is directed against quantified modal logic in conjunction with a different theory of descriptions, the

Frege–Carnap theory. Gödel in his original presentation of the argument suggests that pointing out that Russell can avoid the collapse, '... there is something which is not yet completely understood ...' [Gödel, 1944, p. 130]. That is if one thinks that there must be a theory of descriptions which treats them as singular terms. The argument can also be taken as an objection to the Frege–Carnap theory that definite descriptions are singular terms. It can also be taken as an argument for the view that descriptions are quantifiers, for quantifiers also introduce scope distinctions.

4. Descriptions as Quantifiers

The view that definite descriptions just are a sort of quantifier seems to emerge from a suggestion of Arthur Prior in [Prior, 1963], who proposed that definite descriptions are a special case of a quantifier, which he defines as 'a functor which forms a sentence from a variable and an open or closed sentence or sentences' ([Prior, 1963, p. 198]). In the case of definite descriptions, he sees the *inverted iota* as the expression which applies to a variable, x, and two open sentences ϕx and ψx to produce a sentence. As above, we use the uninverted iota '[ι]' in what follows.

One can see the next step, the literal identification of descriptions as quantifiers in logical form, as coming out of what almost seems to be a trick with notation. First take a statement with a definite description in Russell's notation including the scope indicators:

$$[\iota x\ \phi x]\ \psi(\iota x\ \phi x)$$

As Richard Sharvy ([Sharvy, 1969]) put it:

> ... such an expression, particularly the second occurrence of $\iota x\ \phi x$, is needlessly long and confusing. I replace this latter occurrence with just an 'x', and view the initial '[$\iota x\ \phi x$]' as a quantifier serving to bind it. This device is particularly useful when it is necessary to distinguish various scopes of given definite descriptions; it also captures directly Russell's view that a definite description is a kind of quantifier. ([Sharvy, 1969, p. 489])

Then, finding the second occurrence of the description to include redundant material, replace it simply with the variable 'x':

$$[\iota x\ \phi x]\ \psi x$$

What before was a scope indicator, '[$\iota x\ \phi x$]', has now become a quantifier.

Sharvy presents this as a revision of Russell's theory made purely for convenience (because the original is 'needlessly long and confusing') because it captures the analogy between definite descriptions and indefinite descriptions, which are more clearly kinds of quantifiers, as well as capturing the phenomenon of 'scope' for definite descriptions is treated as literally the scope of a quantifier.

The early presentation of the view holds that definite descriptions are perhaps *like* quantifiers, or best replaced by quantifiers, in a formal system. Kaplan ([Kaplan, 1970]) points out that one way of viewing Russell's theory is by focusing on the fact that what looks like a uniform class of singular terms are in fact given a very different account in logical form. In fact definite descriptions are grouped with indefinite descriptions, and both of them look more like quantifiers than names.

In 'English as a formal language' ([Montague, 1970]) Richard Montague took a further step by insisting that all noun phrases be given a uniform treatment. As quantifiers are considered classes of properties, names are now reinterpreted so that rather than referring to an individual they now stand for the class of properties that the individual in question has. Montague, however, makes use of a syntax that does not have bound variables as the logical notation for quantifiers does. Montague says that:

> The expression 'The' turns out to play the role of a quantifier, in complete analogy with 'every' and 'a', and does not generate (in common with common noun phrases) denoting expressions. ... Further, English sentences contain no variables, and hence no locutions such as 'the v_0 such that v_0 walks'; 'the' is always accompanied by a common noun phrase. ([Montague, 1970, p. 216])

Thus the quantificational nature of definite descriptions appears only in the semantic interpretation of expressions such as 'the' and all the notions of variables and binding are in the semantics, which is, famously for 'Montague Semantics', read directly off the (surface) syntax of the sentence.

Another step was taken with Barwise and Cooper ([Barwise and Cooper, 1981]), as part of their general theory of *generalized quantifiers*. So, above we will find corresponding to:

a man, any man, all men, no man, some man ...

the expressions:

[a x: man x], [any x: man x], [all x: man x], [no x: man x], [some x: man x] ...

including also 'the man' and the corresponding:

[the x: man x]

The semantics Barwise and Cooper present is taken from Montague, who treats all noun phrases as second-order functions which are true of some predicates and not others. All of these quantifiers are interpreted as functions which yield classes of properties, intuitively those that satisfy the quantifier, i.e. are true of all men or the unique man or no man ... These all satisfy Prior's definition of a quantifier as a 'functor' that applies to variables and open formulas to produce sentences.

The final step towards the view that definite descriptions are literally quantifiers was taken by Stephen Neale ([Neale, 1990]), who says that descriptions are quantifiers in Logical Form, 'LF', a distinct level of syntactic analysis, and the level that is most directly related to semantic interpretation. In the generative grammar of Chomsky's ([Chomsky, 1981]) 'Government and Binding' style grammar, the 'SS' (read as 'surface structure') of a sentence is bifurcated into a 'PP' (i.e., 'phonological form'), and an LF (or 'logical form'). The LF will include *traces*, which are unpronounced but none the less syntactically real, and, most importantly bound by noun phrases according to the rules such as that which an anaphoric pronoun in LF is bound by a quantifier that '*c*-commands' it.[14] Simply put, the variables in:

[the x: man x]

are real. Even though, as Montague says, English only includes the two words 'the man' as the pronounced element of PP, in LF there are traces with the same role, even though it might be expressed in a 'notational variant' in LF. Thus, in Neale's example the SS:

$$[_S[_{NP}\text{ the girl}][_{VP}\text{ snores}]] \tag{5.18}$$

is turned into the LF structure:

$$[_S[_{NP}\text{ the girl}]_x\ [_S\ [_{NP}\ t]_x\ [_{VP}\text{ snores}]] \tag{5.19}$$

with its trace, t, and placement of variables as subscripts, is more recognizable as:

$$[\text{the } x : \text{girl } x](x \text{ snores}) \tag{5.20}$$

We have now reached the point where definite descriptions are treated uniformly with other indefinite descriptions, just as Russell started out in 1905. Now descriptions are literally quantifiers in LF. Not only are their semantics the same as quantifiers as in Montague, as extended by Barwise and Cooper, they even bind variables which later occur in the logical form of a sentence.

4.1 Syntax, Semantics, and Rules for Descriptions as Quantifiers

For this account of descriptions as quantifiers, the definition of *term* and *formula* will be simpler, eliminating Definition (5.3.1vi) and replacing (5.3.1vii) with:

(vii′) If ϕ and ψ are *formulas* and x is a variable, then $\forall x\phi x$, $\exists x\phi x$, and [the x: ϕx] ψx are *formulas*.

In this definition *term* and *formula* are defined separately, as in standard logic.

Similarly, in Definition (5.3.2), the definitions of the semantic notions of *denotation* and *truth in a model on an interpretation relative to a sequence* are replaced by:

(vii′) If ψ and ϕ are formulas, then:

i. $\mathcal{A} \models^{\partial,\beta}$ [the x: ϕ x] ψx if $\mathcal{A} \models^{\partial,\beta[a/x]} \phi$
where $\beta[a/x]$ differs from β in assigning a to x, where a is a unique element of A such that $\mathcal{A} \models^{\partial,\beta[a/x]} \phi$.

ii. $\mathcal{A} \not\models^{\partial,\beta}$ [the x: ϕ x] ψx, if there is no such a.

With descriptions literally quantifiers in this way, it is clear that the scope distinctions necessary to block the Slingshot argument are also easily represented as:

$$[\text{the } x : (x = y \wedge p)]\Box(x = y \wedge p) = y \qquad (5.11''a)$$

and

$$\Box([\text{the } x : (x = y \wedge p)](x = y \wedge p) = y) \qquad (5.11''b)$$

'The number of planets is 9' is symbolized as:

$$[\text{the } x : Nx](x = 9) \qquad (5.16')$$

The two readings of 'Necessarily the number of planets is 9' will be represented as the false sentence:

$$\Box[\text{the } x : Nx](x = 9) \qquad (5.17a')$$

which follows by NI, and the 'scope' on which it is true as

$$[\text{the } x : Nx]\Box(x = 9) \qquad (5.17b')$$

This is literally an issue of the relative scope of a quantifier ([the x: Nx]) and the modal operator ($\Box$).

5. Conclusion

Each chapter in this book is intended to show that the field of Philosophical Logic engages in solving philosophical problems using the techniques of logic. The topic of definite descriptions has been significant more as a model of philosophy than for its application to any specific traditional problem of philosophy.

One way in which Russell's theory was taken as a 'paradigm' of philosophy was as a model of the sort of analysis of meaning that was to be the main activity of the newly emerging analytic philosophy. Thus A. J. Ayer, in Chapter III 'The Nature of Philosophical Analysis', of his *Language, Truth and Logic* ([Ayer, 1936]), presents the contextual definitions of the theory of descriptions as a model of philosophical analysis. It is thus that philosophy can consist of discovering analytic truths without simply being a catalogue of definitions of words. The accounts of the meaning of words will consist of accounts of the meaning of entire sentences in which they occur. To the extent that philosophers engage in 'transformative analyses', they are following in the footsteps of Russell's theory of descriptions.[15]

The technique of 'contextual definitions' which Russell used in his theory also led to a more specific view about the nature of the logical analysis of ordinary language, which has been the focus of this chapter. Russell's theory of descriptions was long taken as a paradigm of a theory that relies on a gap between the real logical form of a proposition and its apparent logical form, as suggested by its syntactic structure. The syntactic category of noun phrases, for Russell, denoting phrases, listed at the beginning of this chapter, do not represent constituents of propositions, but are to be analysed instead as contributing in different ways to the logical form of the sentences in which they occur. This chapter has traced the history of this role for the activity of Philosophical Logic. While Frege proposed treating definite descriptions in a class with proper names, Russell pointed out that they differ from proper names in several respects, most distinctively in introducing something like 'scope' distinctions. At the end of the twentieth century we have come to the view that definite descriptions, and indeed all of the 'denoting phrases' with which Russell began are literally quantifiers, and so they are to be classed not with proper names but with quantifiers. More generally, the moral has been drawn that in fact a theory of logical form should closely follow the (proper) syntactic analysis of sentences.

Current research on definite descriptions and indeed much of the Philosophical Logic on noun phrases, tries to give them a uniform account which fits with the syntactic role in sentences, and with other linguistic phenomena, such as anaphora, which involve noun phrases. As well, definite descriptions have a place in the discussion of the distinction between 'speaker's reference' and 'semantic reference' in [Kripke, 1979] which has now become a more general debate about the relationship between semantics and pragmatics.[16]

Notes

1. Also, 'Descriptive Functions' is the title of $*30$ of [Whitehead and Russell, 1910].
2. Chapter V of *Principles of Mathematics* is titled 'Denoting'.
3. This symbol was later used to express existence or, that a name t denotes, in the form $E!t$.
4. This famous remark occurs in the first footnote to the paper called 'Philosophy', in [Ramsey, 1931a, p. 263].
5. See [Ryle, 1979] for the citation of the theory of logical constructions as a model for philosophical method.
6. Empty, or non-denoting, descriptions are the other sort of *improper descriptions*.
7. There is no similar attempt to treat indefinite descriptions as singular terms, however, although Hilbert's Epsilon Calculus can be seen as a way of using a language with special terms to replace the use of quantifiers, and so, in that way, to treat quantifiers as terms, just not singular terms. See [Avigad and Zach, 2009].
8. For a survey of free logic see [Bencivenga, 1986]. The syntax for a formal treatment of the Frege–Strawson view will be that of Section 3.3.1 below, in which definite descriptions are included in the class of singular terms. The distinctive features of various approaches to free logic come in how they treat the notions of logical consequence and logical truth when some sentences can lack a truth value. As well there is a difference between 'positive free logic' in which atomic sentences with non-denoting singular terms can be true, and those in which the truth-value 'gaps' even apply to atomic sentences.
9. Pavel Tichý, ([Tichý, 1988, p. 151]) however, argues for a second basic law to cover just that case in which the description is not proper:

   (VI*): $[\neg(\exists a)(a = \acute{\epsilon}\ (a = \epsilon)] \supset \backslash a = a.$

10. Chapter VI, 'The', pp. 306–345. Chapter VIII, 'The' Again: A Russellian theory of descriptions, pp. 392–410, presents a version of Russell's theory which gives rules for descriptions which doesn't require eliminating the descriptions. The first theory dates from the first edition of the book, written solely by Kalish and Montague. Chapter VIII appears in the second edition, along with Mar as a third author, and so the theory of chapter VI will be attributed to Kalish and Montague in what follows.
11. In what follows we follow the use of variables in Russell's *Principia Mathematica* notation, as in $\iota x\ \phi x$ and $\exists x \phi x$, which suggests that the variable 'x' must *occur* as a *free variable* in 'ϕ'. Kalish and Montague follow the contemporary practice of allowing for 'vacuous quantification'. Similarly, a particular variable 'x' is used in the statement of meta-linguistic rules and definitions, where a meta-linguistic variable such as the 'α' and 'β' that Kalish and Montague use, which ranges over particular variables $x, y, \ldots$.
12. This is also done by those accounts which have a notion of semantic value: $\| \ldots \|^{\beta}_{\mathcal{A}}$, which is a function which applies both to terms (returning an object as a value) and to formulas, giving a truth value.
13. Based on the example in [Quine, 1943] discussed in [Smullyan, 1948] .
14. [Neale, 1990, p. 174]. Neale credits Gareth Evans [Evans, 1977] with this observation.
15. The notion of 'transformative' as opposed to 'decompositional' analysis in the philosophy of Frege and Russell is due to Michael Beaney. See [Beaney, 2009] for an account of the distinction.
16. See the papers in [Ostertag, 1998].

6 Higher-Order Logic

Øystein Linnebo

Chapter Overview

1. Introduction	106
2. A Closer Look at Second-Order Logic	107
2.1 The Language of Second-Order Logic	107
2.2 Deductive Systems for Second-Order Logic	108
2.3 Set-Theoretic Semantics for Second-Order Logic	109
2.4 Meta-Logical Properties of Second-Order Logic	110
2.5 Plural Logic	112
3. Applications of Higher-Order Logic	113
3.1 Formalizing Natural Language	113
3.2 Increased Expressive Power	114
3.3 Categoricity	114
3.4 Set Theory	115
3.5 Absolute Generality	115
3.6 Higher-Order Semantics for Higher-Order Languages	116
4. Languages of Orders Higher than Two	117
4.1 The Technical Question	117
4.2 The Conceptual Question	118
4.3 Infinite Orders	119
5. Objections to Second-Order Logic	119
5.1 Quine's Opening Argument	119
5.2 Quine's Fall-Back Argument	120
5.3 Ontological Innocence	121
5.4 The Incompleteness of Second-Order Logic	122
5.5 Second-Order Logic has Mathematical Content	123
6. The Road Ahead	124
Notes	125

1. Introduction

Different logics allow different forms of generalization. Consider for instance the claim that Socrates thinks, which we can formalize as:

$$\text{Think}(\text{Socrates}) \tag{6.1}$$

Classical *first-order logic* allows us to generalize into the noun position occupied by 'Socrates' to conclude that there is an object x that thinks:

$$\exists x\, \text{Think}(x) \tag{6.2}$$

Although classical first-order logic is quite expressive, there are stronger logics that allow additional forms of generalization.

Plural logic allows us to generalize *plurally* into this noun position to conclude that there are one or more objects xx that think:

$$\exists xx\, \text{Think}(xx) \tag{6.3}$$

Here we make use of plural variables (which we write as double letters), each of which can be assigned one or more objects as its values, rather than just one object, as in classical singular first-order logic.

Second-order logic studies yet another form of generalization: it allows us to generalize into the predicate position occupied by 'Think' in (6.1) to conclude that there is a concept F under which Socrates falls:

$$\exists F\, F(\text{Socrates}) \tag{6.4}$$

A logic that allows one or more of these additional forms of generalization is called a *higher-order logic*. We have already seen that such logics come in different forms. For although both plural logic and second-order logic provide ways of talking about many objects simultaneously, they do so in completely different ways, namely by generalizing into different kinds of position.

Philosophers and logicians have many reasons for taking higher-order logics seriously. Since the relevant claims and inferences appear to be available in natural language, it should be permissible to introduce a logical formalism capable of representing these claims and inferences. Moreover, the increased expressive and deductive power of higher-order logics make them very useful tools to employ in the philosophy of mathematics, semantics, and set theory. However, higher-order logics are also very controversial. Quine famously argues that second-order logic is 'set theory in sheep's clothing' ([Quine, 1986, p. 66]). Many philosophers and logicians agree that higher-order logic has substantial set-theoretic content and is thus not such an innocent tool as its defenders often take it to be.

2. A Closer Look at Second-Order Logic

I first describe the language and theory of second-order logic. Then I describe two different kinds of model-theoretic semantics for this language and comment on some meta-logical properties of second-order logic.

2.1 The Language of Second-Order Logic

The language of second-order logic is a simple extension of the language of classical first-order logic. Essentially, all we do is add second-order variables and quantifiers binding them. It will nevertheless be useful to give a precise definition.

A language $\mathcal{L}$ of second-order logic has the following variables and constants:

- an *individual variable* x_i and (if desired) an *individual constant* a_i for each natural number i;
- a *predicate variable* F_i^n and (if desired) a *predicate constant* A_i^n for all natural numbers i and n. The superscript n is used to indicate that the predicate takes n arguments. (The limiting case of $n = 0$ can either be excluded or seen as involving variables and constants for propositions.)

In second-order logic, identity is often defined by letting '$x = y$' abbreviate '$\forall F(Fx \leftrightarrow Fy)$'. In the standard semantics to be described below, this defined notion of identity is easily seen to coincide with the ordinary notion. But since the two notions may otherwise come apart, it is often useful to assume that one of the predicate constants is the symbol '=' for identity, which we write in the ordinary way rather than as a doubly indexed 'A'.

The *atomic formulas* of $\mathcal{L}$ are of the form $Pt_1 \ldots t_n$, where P is an n-place predicate symbol (either constant or variable) and $t_1, \ldots, t_n$ are individual terms (either constant or variable); although where P is '=', we write $t_1 = t_2$ in the ordinary way. The *formulas* of $\mathcal{L}$ are defined in the usual recursive manner:

- every atomic formula is a formula;
- when ϕ and ψ are formulas, then so are $\neg(\phi)$, $(\phi \vee \psi)$, $\forall x_i(\phi)$, and $\forall F_i^n(\phi)$;
- nothing else is a formula.

As usual, parentheses will often be omitted. The other connectives $\wedge$, $\rightarrow$, and $\leftrightarrow$ and the existential quantifiers of first and second order will be regarded as abbreviations in the usual way. An occurrence of a variable is said to be *free* if it is not in the scope of a quantifier binding this variable; otherwise it is said to be *bound*.

Sometimes variables and constants for *functions* are added to the language of second-order logic as well. We won't do this here; for claims about functions are easily expressed by means of relations instead.

2.2 Deductive Systems for Second-Order Logic

Next we would like a deductive system for second-order logic that is at least sound. (The question of completeness will be considered below.) We use as our starting point some complete axiomatization of classical first-order logic. It will be useful to assume that the first-order quantifiers are subject to the standard introduction and elimination rules. We now want to add axioms and rules that govern the second-order variables and quantifiers.

The most obvious and least controversial addition is to extend the standard introduction and elimination rules to the second-order quantifiers. The elimination rule for the second-order universal quantifier states that from $\forall F_i^n\, \phi$ we may infer $\phi[P/F_i^n]$, where P is any n-place predicate symbol (either constant or variable) that is substitutable[1] for F_i^n, and where $\phi[P/F_i^n]$ is the result of replacing every free occurrence of F_i^n in ϕ by P. The introduction rule says that, when ϕ has been proved from premises containing no occurrences of P (if P is a predicate constant) or no free occurrences of P (if P is a predicate variable), then we may infer $\forall F_i^n\, \phi[F_i^n/P]$.

Next we add *comprehension axioms* which specify what values the second-order variables can take. Each *comprehension axiom* says that an open formula $\phi(x)$ defines a value of a second-order variable:

$$\exists F \forall x[Fx \leftrightarrow \phi(x)] \qquad \text{(Comp)}$$

where $\phi(x)$ does not contain F free.[2] For terminological reasons, it will be convenient to follow Frege and call such values *concepts*, without thereby accepting any of Frege's metaphysical claims about concepts. The *full* or *unrestricted* comprehension scheme has a comprehension axiom of this form for every formula $\phi(x)$ expressive in the language.

The comprehension axioms interact in an important way with the elimination rules for the second-order quantifiers. The elimination rules formulated above allow only second-order variables and constants as instances. For example, from $\forall F(Fa)$ the rule of universal elimination allows us to infer directly that Ga but not that $\phi(a)$ for any open formula $\phi(x)$. The latter inference must proceed via the comprehension axiom $\exists F \forall x(Fx \leftrightarrow \phi(x))$, which makes explicit the assumption that $\phi(x)$ succeeds in defining a concept that can serve as the value of the variable F. It is of course possible to modify the elimination rule for the second-order universal quantifier to allow any open formula to count as a legitimate instance. But doing so is undesirable because it runs together two very different things: the uncontroversial step from a generalization to an instance, and the controversial question of what instances there are. In many situations we wish to keep tight control on what instances are regarded as legitimate. For example, when studying weak mathematical theories or investigating set-theoretic or semantic paradoxes, we often only allow formulas $\phi(x)$ without any bound second-order variables to define concepts. The resulting comprehension scheme is said to be *predicative*.

Sometimes a second-order version of the Axiom of Choice is added as well. This axiom can be expressed as the claim that for any dyadic relation R whose domain includes all individuals (that is, $\forall x \exists y\, Rxy$), there is a sub-relation S of R that is functional (that is, $\forall x \exists y \forall z (Sxz \leftrightarrow y = z)$).

2.3 Set-Theoretic Semantics for Second-Order Logic

The traditional way to develop a semantics for second-order logic is within set theory. I now describe two kinds of set-theoretic semantics. One is very general and due to the logician Leon Henkin. The other trades generality for a unique standard interpretation and is therefore known as 'standard semantics'. Both approaches are based on set-theoretic models and a Tarski-style notion of satisfaction. (An alternative semantics using higher-order logic rather than set theory will be outlined in Section 3.6.)

A *Henkin model* for a second-order language consists of the following sets:

- a domain D_1 of individuals;
- a domain D_2^n of n-adic relations for each n, where each element of D_2^n is an n-tuple of elements of D_1;
- an *interpretation function* I that assigns to each individual constant an object in D_1 and to each n-place predicate constant an element of D_2^n.

Note that each domain D_2^n must contain all definable n-adic relations if the unrestricted comprehension scheme (Comp) is to be validated. A Henkin model is said to be *standard* just in case D_2^n consists of *all* n-tuples from D_1; that is, just in case D_2^n is the power-set of the n-fold Cartesian product of D_1 with itself. A standard model thus recognizes as many n-adic relations as can be represented within set theory.

A *variable assignment* is a function s that assigns to each individual variable an element of D_1 and to each n-place predicate variable an element of D_2^n. Together, an interpretation and an assignment secure a *denotation* for every term of the language: the interpretation assigns a denotation to every constant, and the assignment does so to every variable.

A model $\mathcal{M}$ and an assignment s *satisfy* a formula ϕ (in symbols: $\mathcal{M}, s \models \phi$) just in case one of the following holds:

- ϕ is an atomic formula of the form $Pt_1 \ldots t_n$ and the sequence of objects denoted by the terms t_i is an element of the denotation of P;
- ϕ is of the form $\neg\psi$ and it is not the case that $\mathcal{M}, s \models \psi$;
- ϕ is of the form $\psi_1 \vee \psi_2$ and either $\mathcal{M}, s \models \psi_1$ or $\mathcal{M}, s \models \psi_2$ or both;
- ϕ is of the form $\forall x_i \psi$ and for every assignment s' that differs from s at most in its assignment to x_i we have $\mathcal{M}, s' \models \psi$;
- ϕ is of the form $\forall F_i \psi$ and for every assignment s' that differs from s at most in its assignment to F_i we have $\mathcal{M}, s' \models \psi$.

A formula ϕ is said to be a Henkin (alternatively: standard) *consequence* of a set of formula Γ just in case every Henkin (alternatively: standard) model and every variable assignment that satisfy every formula in Γ also satisfy ϕ. We write this as $\Gamma \models_h \phi$ (alternatively: $\Gamma \models_s \phi$).

2.4 Meta-Logical Properties of Second-Order Logic

Recall the most important meta-logical properties of first-order logic.

> **Completeness**. There is a complete proof procedure. That is, there is a recursively axiomatized proof procedure (which we write as $\vdash$) such that, whenever ϕ is a model-theoretic consequence of Γ (which we write as $\Gamma \models \phi$), then $\Gamma \vdash \phi$.

Recall that a theory Γ is said to be *satisfiable* just in case there is a model $\mathcal{M}$ and a variable assignment s such that $\mathcal{M}, s \models \phi$ for each formula ϕ in Γ.

> **Compactness**. If every finite subset of Γ is satisfiable, then Γ too is satisfiable.

> **Löwenheim–Skolem**. If Γ has a model whose domain of individuals is infinite, then for any infinite cardinal κ that is at least as large as the cardinality of the language, Γ has a model based on κ many individuals.

Second-order logic with Henkin semantics is much like a first-order theory with many different sorts of variables and constants: one sort for individuals, one for monadic concepts, and so on. This is reflected in the following theorem.

Theorem 6.2.1 *Second-order logic with Henkin semantics is complete, compact, and has the Löwenheim–Skolem property.*

The proof is similar to that for first-order logic. See for instance [Enderton, 2001, pp. 302–3] or [Shapiro, 2000, Section 4.3].

Things change dramatically when second-order logic is equipped with the standard semantics.

Fact 6.2.1 In second-order logic there is a sentence λ_∞ that is true in a standard model iff its first-order domain is infinite. To see this, let λ_∞ state that there is a relation R that is transitive, irreflexive and without an endpoint on the right:

$$\exists R[\forall x \forall y \forall z(Rxy \wedge Ryz \rightarrow Rxz) \wedge \forall x\, \neg Rxx \wedge \forall x \exists y\, Rxy]$$

For there to be such a relation, there must be infinitely many individuals to act as relata. And conversely, in any standard model with infinitely many individuals there will be such a relation.[3]

This fact has an important consequence.

Theorem 6.2.2 *Second-order logic with standard semantics is not compact.*

Proof sketch. Let λ_n be a standard formalization, in first-order logic with identity, of the claim that there are at least n objects. Let $\Gamma = \{\neg\lambda_\infty, \lambda_2, \lambda_3, \ldots\}$. Then every finite subset Γ_0 of Γ is satisfiable. For let n_0 be the largest natural number n such that $\lambda_n \in \Gamma_0$. Then Γ_0 is satisfiable in any model with n_0 individuals. But Γ itself is not satisfiable. For in order to satisfy all the sentences λ_n, a model must contain infinitely many individuals. But then the model cannot satisfy $\neg\lambda_\infty$. ⊣

Recall that a theory is said to be *categorical* (given a certain semantics) just in case all of its models (that are available in this semantics) are isomorphic.

Fact 6.2.2 In second-order logic with standard semantics we can provide a categorical axiomatization of the natural number structure. (By the Löwenheim–Skolem theorem, this cannot be done in first-order logic.) This is achieved by means of second-order *Dedekind–Peano arithmetic*, or PA_2:

(PA1) $N0$
(PA2) $Nx \wedge Sxy \rightarrow Ny$
(PA3) $Sxy \wedge Sxy' \rightarrow y = y'$
(PA4) $Sxy \wedge Sx'y \rightarrow x = x'$
(PA5) $Nx \rightarrow \exists y\, Sxy$
(PA6) $\forall F[F0 \wedge \forall x \forall y(Fx \wedge Sxy \rightarrow Fy) \rightarrow \forall x(Nx \rightarrow Fx)]$

A proof due to [Dedekind, 1888] shows that any two models of PA_2 are isomorphic. The gist of the proof is easily explained. Consider any two models $\mathcal{M}_1$ and $\mathcal{M}_2$ of PA_2, which interpret the arithmetical expressions of PA_2 as respectively $N_1, S_1, 0_1$ and $N_2, S_2, 0_2$. The key move is to define the smallest relation R that relates the initial elements 0_1 and 0_2 and has the closure property that, whenever it relates u and v, it also relates the S_1-successor of u and the S_2-successor of v. More precisely, we use comprehension to define Rxy by the following formula:

$$\forall X[X0_1 0_2 \wedge \forall u \forall u' \forall v \forall v' (Xuv \wedge S_1 uu' \wedge S_2 vv' \rightarrow Xu'v') \rightarrow Xxy]$$

It is then straightforward to prove that R defines an isomorphism from $\mathcal{M}_1$ to $\mathcal{M}_2$. The proof uses the fact that induction holds in both models.[4]

Fact 6.2.2 has important consequences concerning other meta-logical properties of second-order logic with standard semantics.

Theorem 6.2.3 *Second-order logic with standard semantics lacks the Löwenheim–Skolem property and is incomplete (in the sense that it lacks a sound and complete proof procedure).*

Proof sketch. The lack of the Löwenheim–Skolem property is immediate from the ability to provide a categorical characterization of the natural numbers: PA_2 has standard models with countably many individuals but not with uncountably many individuals.

Assume for *reductio* that the logic was complete. Then any set of formulas Γ would be consistent iff Γ is satisfiable. Since Γ is consistent iff each of its finite subsets Γ_0 is consistent, this would ensure that Γ is satisfiable iff each of its finite subsets Γ_0 is satisfiable; that is, that the logic is compact. Since this is false by Theorem 6.2.2, we conclude that the logic is incomplete. $\dashv$

2.5 Plural Logic

The above discussion is easily adapted to plural logic. Consider the fragment of second-order logic containing only *monadic* second-order variables. The language of plural logic is identical to the language of this fragment except for two minor adjustments. Instead of variables of the form F_i^1, plural logic has variables of the form xx_i. And instead of atomic formulas of the form $F_i^1 t$, plural logic has atomic formulas of the form $t \prec xx_i$ (to be read as 't is one of xx_i'). Otherwise the language remains the same.

The deductive system for plural logic is the same as that of the monadic second-order logic except for some straightforward adjustments required by the fact that there are no empty pluralities. We add an axiom to this effect: $\forall xx \exists u(u \prec xx)$. And we formulate the plural comprehension scheme so as to allow only formulas that are instantiated to define pluralities:[5]

$$\exists u\, \phi(u) \rightarrow \exists xx \forall u[u \prec xx \leftrightarrow \phi(u)] \qquad \text{(P-Comp)}$$

Just like ordinary second-order logic, plural logic can be given two sorts of set-theoretic semantics: Henkin and standard. And just like ordinary second-order logic, plural logic maintains the three mentioned meta-logical properties on the Henkin semantics but loses all three properties on the standard semantics. The proofs are analogous but complicated somewhat by the fact that plural logic does not provide any primitive device corresponding to quantification over relations. We get around this complication by adding a first-order theory of ordered pairs, which enables us to express quantification over n-place relations as plural quantification over n-tuples.[6]

However, proponents of plural languages argue that any sort of set-theoretic semantics does violence to the intended interpretation of such languages. According to Boolos, the function of plural variables is to *range plurally* over ordinary objects, not to range singularly over sets. That is, each plural variable has one or more ordinary objects as its values, not one extraordinary object, such as a set or any other special entity one may wish to assign to plural variables. I will return to this issue in Sections 3.6 and 5.3.

3. Applications of Higher-Order Logic

Higher-order logic has a wide range of applications in philosophy, mathematics, and semantics. I now describe some of the most important ones. It should be noted that many of the applications are controversial. Some criticisms will be discussed in Section 5.

3.1 Formalizing Natural Language

Various sentences of natural language are arguably most directly and naturally formalized by means of higher-order logic. Consider for instance the following three sentences.

(1) a and b have something in common.
(2) However a and b are related, so c and d are related as well.
(3) There are some critics who only admire one another.

These sentences are arguably most naturally formalized as follows:

(1′) $\exists F(Fa \wedge Fb)$

(2′) $\forall R(Rab \rightarrow Rcd)$
(1′) $\exists xx \forall u[(u \prec xx \rightarrow \textsc{Critic}(u)) \wedge \forall v(u \prec xx \wedge \textsc{Admires}(u,v) \rightarrow v \prec xx \wedge u \neq v)]$

The first two formalizations use second-order logic, and the third, plural logic.

3.2 Increased Expressive Power

Higher-order logic with standard semantics enables us to characterize a number of important logico-mathematical concepts that cannot be characterized using classical first-order logic alone, for instance the transitive closure of a relation, the notions of equinumerosity, finitude, countability, and many infinite cardinalities. The transitive closure R^* of a relation R can (as Dedekind and Frege discovered) be defined by letting R^*xy abbreviate the claim that every R-hereditary property F that is possessed by x is also possessed by y:

$$\forall F[Fx \wedge \forall u \forall v(Fu \wedge Ruv \rightarrow Fv) \rightarrow Fy]$$

And the *F*s and the *G*s are equinumerous just in case there is a dyadic relation *R* that one-to-one correlates *F*s and the *G*s. Next, the *F*s are finite just in case there is no dyadic relation *R* that one-to-one correlates all of the *F*s with all but one of the *F*s. Further, the *F*s are countably infinite just in case they can be ordered by a dyadic relation *R* to form an isomorphic copy of the natural numbers, as characterized in Section 2.4.[7]

3.3 Categoricity

Higher-order logic is used extensively in the philosophy of mathematics in order to provide categorical axiomatizations of important mathematical structures, such as the natural number structure, the real number structure, and certain initial segments of the hierarchy of sets. The ability to provide such characterizations plays an important role in many philosophical accounts of mathematics, such as structuralism.[8] We saw in Section 2.4 how to provide a categorical characterization of the natural number structure. Various other categorical characterizations of structures are explained in [Shapiro, 2000].

What about the entire hierarchy of sets? [Zermelo, 1930] showed that second-order Zermelo–Fraenkel set theory (ZF_2) is *quasi-categorical* in the sense that, given any two models of ZF_2, one is an initial segment of the other. In this sense, ZF_2 fixes the 'width' of the hierarchy of sets, leaving only its 'height' undetermined.[9]

3.4 Set Theory

In set theory we sometimes want to talk about 'collections' that don't form sets.[10] For instance, we may want to say that any 'collection' of ordinals is well-ordered by the membership relation, regardless of whether this 'collection' forms a set. This claim can be formalized very naturally as a second-order or plural generalization over a domain whose individuals include all the ordinals. We may also want to express the set-theoretic principles of Separation and Replacement as single axioms rather than axiom schemes. For instance, Separation can be formalized as the claim that for any set x and any concept X, there is a set y whose elements are precisely the elements of x that fall under the concept X:

$$\forall x \forall X \exists y \forall z(z \in y \leftrightarrow z \in x \wedge Xz)$$

Moreover, higher-order notions play a role in some of the considerations that are used to motivate 'large cardinal axioms' in set theory. For instance, the set-theoretic *reflection principle* says, very roughly, that any property that is had by the set-theoretic universe is already had by some proper initial segment of this universe. When this talk about 'properties' is cashed out in the language of first-order set theory, the resulting principle is a theorem of standard ZF. But when we use the language of higher-order set theory, the resulting principle entails the existence of certain 'large cardinals', such as strongly inaccessible cardinals and Mahlo cardinals.[11]

3.5 Absolute Generality

Higher-order logic has recently been applied to defend the possibility of quantification over absolutely everything, or *absolute generality* for short. This important application requires some explanation.

Set theory is naturally understood as a theory of all sets. For its first-order quantifiers seem to range over all sets. But this natural view gives rise to a problem when we try to develop a semantics for the language of set theory. On the standard set-based semantics of the sort outlined in Section 2.3, the first-order domain has to be a set. So the natural interpretation would require a universal set for the first-order quantifiers to range over. But standard set theory does not allow a universal set. This means that standard set-based semantics is unable to produce a model that corresponds to the natural interpretation of the language of set theory.

How serious is this problem? The answer will depend on the goals of one's semantic theorizing. If one's goal is merely to give an extensionally correct account of logical consequence, then the problem is surmountable. For first-order languages, Kreisel's famous 'squeezing argument' shows that nothing is lost by restricting oneself to set-based models ([Kreisel, 1967]). For if ϕ is provable from a theory Γ, then ϕ is a logical consequence of Γ in an informal and intuitive sense, which in turn entails that ϕ is true in every set-based model of Γ, which (by the completeness theorem for first-order logic) entails that ϕ is provable from Γ. For higher-order languages the same effect is obtained by means of set-theoretic reflection principles, which are widely accepted in the set-theoretic community (although they go beyond standard ZF).[12] However, if one's goal is the more ambitious one of providing models that faithfully represent every permissible interpretation of the language, then the problem becomes serious. For we saw that no set-based model can faithfully represent the natural interpretation described above.

One influential response to this problem is to deny that the natural interpretation of the language of set theory is coherent.[13] What the problem teaches us, this response claims, is that it is impossible to quantify over absolutely all sets. Whenever we quantify over some sets, it is possible to consider the domain of this quantification. This results in another set, which on pain of contradiction cannot be in the original range of quantification. It is thus impossible to quantify over absolutely all sets. So absolute generality is unattainable.

Recent decades have seen the emergence of a new response to such attacks on absolute generality. The idea is to develop the requisite semantic theories in higher-order meta-languages rather than rely on first-order set theory as one's meta-theory.[14] Recall that sets are individuals (in the sense that they are values of first-order variables). So for any individuals a and b, there is another individual $\langle a, b\rangle$ that represents their ordered pair; n-tuples follow in the usual way. The first novel idea is to formalize talk about the domain by means of a second-order variable 'D' rather than a first-order variable ranging over sets: 'Dx' will mean that x is in the domain. Next, the interpretation of all non-logical constants is described using another second-order variable 'I': $I\langle t, x\rangle$ will mean that t denotes x (if t is an individual constant), or that x is one of the (n-tuples) of which t is true (if t is a predicate constant). For instance, $I\langle `\in', \langle a, b\rangle\rangle$ represents that the predicate constant '$\in$' is true of a and b (in that order). Finally, we use a second-order variable 'A' to code for variable assignments: $A\langle v, x\rangle$ will mean that x is assigned to the variable v. Given these resources, we can now proceed to formulate a standard Tarskian theory of satisfaction. The upshot is that it appears possible, after all, to develop a semantics that is compatible with the possibility of absolute generality.

3.6 Higher-Order Semantics for Higher-Order Languages

The higher-order approach to semantic theorizing can be extended to object languages of order higher than one. Although logicians have been aware of this option ever since [Tarski, 1935], its philosophical significance was fully appreciated only in [Boolos, 1985]. In this article Boolos shows how to develop a theory of satisfaction for a plural object language in a plural meta-language equipped with a satisfaction predicate (not present in the object language) that takes plural arguments. The idea is a straightforward generalization of the approach outlined above. We let A code assignments not just to singular variables but also to plural ones. If v is a plural variable, then $A\langle v, x\rangle$ means that x is one of the objects assigned to the variable v; but A may assign other objects to v as well. The objects assigned to v are thus all the objects x such that $A\langle v, x\rangle$. As before, we can now proceed to formulate a standard Tarskian definition of satisfaction of a formula ϕ by an assignment A relative to a domain and an interpretation I.

Let a *generalized semantics* be a theory of all possible interpretations that a language might take, without any artificial restrictions on the domains, interpretations, and variable assignments; in particular, it must be permissible to let the domain include all objects. A generalized semantics thus goes beyond a theory of satisfaction by allowing the interpretation of the predicates to vary. What resources are needed to develop a generalized semantics for a higher-order language? The question is answered by some recent generalizations of Boolos's work. The upshot is that a generalized semantics for a language of order n can be developed in a language of order $n + 1$ but not in any language of lower order.[15] (These languages will be defined in the next section.)

The fact that the semantics of a higher-order object language can be developed in a higher-order meta-language plays a key role in the debate about the ontological commitments of higher-order languages, as will be discussed in Section 5.3.

4. Languages of Orders Higher than Two

Are there languages and logics of orders higher than two? That is, is it legitimate to add variables and constants of orders higher than two and to bind these variables by quantifiers? Many logicians have thought so, including Frege, Russell, and Hilbert. For instance, Frege thought that the first-order quantifier should be understood as standing for a second-order concept, namely the concept that holds of a first-order concept F just in case F is instantiated. Russell went even further and argued that there are concepts (or, strictly speaking, 'propositional functions') of every finite order.

4.1 The Technical Question

The development of languages and logics of orders higher than two is straightforward from a technical point of view. To keep things simple, let's focus on the case of monadic predicates, retaining only a single dyadic predicate '=' for identity. Then we may allow variables of the form x^i_j and constants of the form c^i_j, where i and j are natural numbers. The upper index is here known as the *order* of the symbol. Terms for individuals have order 1. As atomic formulas we now accept all strings of the form $t(t')$, provided the order of t is precisely one higher than the order of t'. We also accept all identities $t = t'$, where t and t' are terms of order 1. The notion of a formula is then defined in the usual recursive manner.

Let's say that a language of this form is *of order n* just in case its variables are of order no higher than n and its constants are of order no higher than $n + 1$.[16] This generalizes the ordinary notion of a first-order language; for the predicate constants of an ordinary first-order language are constants of order 2. If we allow variables and constants of arbitrary finite order, we get the language of *simple type theory*.[17]

The deductive systems for logic of order n or simple type theory are straightforward extensions of those for second-order logic. We add the obvious introduction and elimination rules for all the higher-order quantifiers. And for each natural number n such that the language contains variables of order $n+1$, we add a comprehension scheme of the form $\exists x^{n+1} \forall u^n [x^{n+1}(u^n) \leftrightarrow \phi(u^n)]$, where x^{n+1} must not occur free in $\phi(u^n)$. We may also add principles of extensionality and choice.

4.2 The Conceptual Question

The conceptual question whether such languages are legitimate is much harder. For these languages and theories to be more than uninterpreted formal systems, there must really exist expressive resources of the sort described. But how does one establish the existence of some alleged expressive resources?

One option is to show that such expressive resources are realized in natural language. Indeed, it appears that natural language contains traces of expressive resources of order three.[18] However, it is doubtful that any natural language contains any systematic machinery for expressing quantification of order three or higher.

However, there is no reason to think that all legitimate expressive resources have to be realized in human languages. Another way to defend the legitimacy of certain expressive resources is to show that they can be obtained by iterating principles of whose legitimacy we are already convinced. If we believe that it is possible to advance from a classical first-order language to a second-order language, why should it not be possible to continue to a third-order language? It is thus not surprising that most proponents of second-order languages have also accepted languages of higher orders.[19]

4.3 Infinite Orders

In the early twentieth century, higher-order logics and simple type theory competed with set theory for the status as the canonical framework in which to develop a foundation for mathematics. The competition was eventually won by Zermelo–Fraenkel set theory. Before this happened, a number of prominent mathematicians and logicians sought to extend simple type theory to languages and logics of infinite orders.[20] Although now obsolete as a foundation for mathematics, such languages and logics raise some interesting philosophical questions.

Some of these questions are investigated in [Linnebo and Rayo, shed], where (inspired by [Gödel, 1933b]) it is argued, first, that some of the motivations offered for higher-order logics also motivate logics of transfinite orders; and secondly, that such logics take on many features characteristic of set theory, with the result that they resemble fragments of set theory in a particularly restrictive notation.

5. Objections to Second-Order Logic

I now outline the main objections that have been made to second-order logic. Some are due to its arch-enemy, Quine, who challenges the very idea of a logic of second order. Later objections have been more nuanced and tied to various attempted applications of second-order logic.

5.1 Quine's Opening Argument

Quine's opening argument against second-order logic in [Quine, 1985] can be reconstructed as follows.

> **Premise 1**. It is legitimate to quantify into a position occupied by an expression *e* only if this occurrence of *e* names something.

For instance, we cannot quantify into the position occupied by a truth-functional connective; for the connectives don't name anything but rather serve a syncategorematic role, which is explained by the associated recursion clause of a Tarskian theory of truth.

> **Premise 2**. Predicates do not name anything.

According to Quine, a predicate contributes to a sentence by being true of certain objects, but this contribution is discharged without the predicate naming anything. The two premises clearly imply Quine's conclusion that it is illegitimate to quantify into the position occupied by predicates. So the question is whether the premises are true.

In a well-known response, Boolos objects to Premise 1 ([Boolos, 1975]). In order to quantify into predicate position, it is sufficient that predicates have extensions and that the second-order quantifiers be associated with a range of such extensions. To insist on naming rather than having an extension is, according to Boolos, simply to beg the question against higher-order quantification.

Who is right? The answer depends on how the notion of 'naming' is understood. If 'naming' is understood as doing what successful singular terms do, then Boolos is clearly right that Premise 1 is question begging: the premise would then amount to an outright ban on quantification into anything other than positions occupiable by singular terms. On the other hand, if 'naming' is understood more broadly as having a semantic value (or several) of the sort appropriate for the kind of expression in question, then even Boolos's notion of 'having an extension' will count as an instance of naming, thus undermining Boolos's objection to Premise 1.

Regardless of what Quine might have intended, let's focus on the more inclusive understanding of 'naming' and so avoid begging the question. Thus understood, Premise 1 is quite plausible. The role of a variable is to be assigned a value (or several). So unless an expression has a semantic value (or several), it is hard to see what sense could be made of replacing the expression by a variable. However, this increased plausibility of Premise 1 comes at the cost of putting great pressure on Premise 2. For the more inclusive the understanding of 'naming', the harder it becomes to hold on to the claim that predicates don't 'name' anything.

5.2 Quine's Fall-Back Argument

Quine realizes that some logicians will deny Premise 2. So he outlines a fall-back argument addressed at such logicians. We may reconstruct the argument as follows.

> If predicates have semantic values, then these must have an extensional criterion of identity. For we are unable to formulate any sufficiently clear intensional criterion of identity. But the only available semantic values with an extensional criterion of identity are sets. So if predicates have semantic values, then these must be sets. This shows second-order logic to have substantial ontological commitments, which logic shouldn't have.

Extrapolating slightly, the argument can be extended as follows.

This also shows that second-order logic isn't universally applicable, as logic should be. To see this, let the first-order variables range over all sets, and consider the following axiom of second-order logic: $\exists F \forall x(Fx \leftrightarrow x \notin x)$. If the variable F ranges over sets, this commits us to a Russell set, which leads to contradiction. So if the semantic values of second-order variables are sets, then second-order logic cannot be applied to discourse about all sets.

Several steps of these arguments are controversial. Many philosophers and logicians are unconvinced by Quine's insistence on extensionality. Moreover, Boolos's plural interpretation seems to provide a way of holding on to extensionality without letting the values of the second-order variables be sets. Finally, the derivation of Russell's paradox requires the controversial assumption that it is possible to let the first-order quantifiers range over absolutely all sets. So a great deal of work would be required to make these arguments persuasive.

5.3 Ontological Innocence

One way to shore up Quine's argument would be by showing that second-order logic incurs unacceptable ontological commitments. Suppose Quine is right that quantification requires the assignment of values to the variables being bound. (This is the weak understanding of Premise 1 discussed above.) Doesn't the assignment of values to variables show that higher-order logic incurs additional ontological commitments? This would threaten at least some of its applications.

As mentioned, Boolos's plural interpretation provides a way of resisting this line of argument. On this interpretation, a plural variable ranges plurally over ordinary objects. There is no need to assign to a plural variable any single value such as a set of ordinary objects. Boolos can thus insist that plural sentences such as (3) and its formalization (3′) are ontologically committed only to critics, not to sets thereof.

Attempts have been made to argue that second-order logic too is ontologically innocent. The arguments turn on the plausible idea that, when a sentence is a logical consequence of another, then the ontological commitments of the former cannot exceed those of the latter.[21] Consider the following sentences, the former of which logically entails the latter:

(4) Roses are red.
(5) $\exists F$(roses are F).

So the plausible idea entails that (5) cannot have any ontological commitments not already had by (4). And even Quine agrees that (4) has no problematic ontological commitments.

However, this argument assumes that quantification into predicate position is legitimate in the first place. To defend this assumption, we need a semantics for languages with such quantification which is compatible with their alleged ontological innocence. Again, Boolos points the way. We saw in Section 3.6 how to develop a semantics for a second-order language in a higher-order meta-theory in a way that avoids assigning to the second-order variables any objects as their values.

Where does this leave us? A *prima facie* case has been presented for the ontological innocence of certain locutions. And this view has been shown to be stable in the sense that, if one accepts that these locutions are innocent when used in the meta-language, then this can be used to demonstrate their innocence when used in the object language. However, the *prima facie* case for ontological innocence has been disputed.[22] And the ascent to a meta-language cuts both ways: someone who denies the innocence claim as applied to the meta-language can use this to challenge the innocence claim as applied to the object language. So we appear to have reached a stand-off.

My own view is that the dispute has been transformed to one about how the notion of ontological commitment is best understood. If the notion is understood as concerned exclusively with the existence of objects, and if an object is understood as the value of a singular first-order variable, then the higher-order semantics does indeed show that higher-order logic is ontologically innocent. For this semantics does not use any singular first-order variables to ascribe values to the higher-order variables of the object language; rather, this ascription is made by means of higher-order variables. On the other hand, if the notion of ontological commitment is understood more broadly as tied to the presence of existential quantifiers *of any order* in a sentence's truth condition, then even the higher-order semantics shows that plural and predicative locutions incur additional ontological commitments.

It may be objected to the broader notion of ontological commitment that the commitments associated with higher-order quantifiers should be given a different name, for instance (following Quine) *ideological commitments*. However, I see little point in quarrelling over terminology. A more interesting question is whether ideological commitments in this sense give rise to fewer philosophical problems, or is philosophically less substantive, than ontological commitments narrowly understood. It is far from obvious that this is so.

5.4 The Incompleteness of Second-Order Logic

We know from Theorem 6.2.2 that second-order logic with standard semantics is incomplete. Many philosophers have found this objectionable.

The best reason to insist on completeness is (in my opinion) of a methodological nature. One of Frege's chief contributions to modern logic and mathematics is the requirement of explicit proof, which demands that all assumptions of a scientific argument be made perfectly explicit by listing them as axioms or rules of inference, and that the argument be spelt out in steps, each of which is either an axiom or licensed by a rule of inference. This will transform the question whether to accept the conclusion to the question whether to accept the axioms and the rules of inference.

The standard second-order consequence relation is incompatible with this goal of perfect explicitness about one's assumptions. Because of its incompleteness, this notion of consequence outstrips what can be made explicit in the form of axioms and rules. So insofar as one wishes to adhere to the ideal of explicitness, standard second-order consequence is inappropriate. Note that this objection is directed only at a certain use of second-order logic, unlike the more general objections due to Quine.[23]

Supporters of standard second-order consequence will respond that they too may choose to list all of their assumptions in the form of axioms and rules. This is certainly true. But doing so would undermine the significance of their preference for the standard semantics over the general one. For if they choose to abide by these strictures, then each of their arguments can be reproduced without loss by advocates of the general semantics – with respect to which second-order logic is complete.

5.5 Second-Order Logic has Mathematical Content

Second-order logic with standard semantics (henceforth, simply 'SOL') has substantial mathematical content. For to apply SOL to a domain of individuals is from a mathematical point of view equivalent to considering the totality of subsets of this domain.

The mathematical content of SOL surfaces in several different ways. A standard example is that there is a sentence in the language of pure SOL that is a logical truth just in case the Continuum Hypothesis (CH) is true, and likewise for its negation.[24] However, Gödel's and Cohen's celebrated results show that CH is independent of the standard axiomatization ZFC of set theory. There are thus questions about second-order logical truth whose mathematical content is beyond the reach of ZFC. Another example concerns the logical *invalidity* of arguments. An argument is invalid just in case there is a countermodel. In first-order logic, such countermodels can always be chosen to be countable. By contrast, SOL requires some very large countermodels, including ones of strongly inaccessible cardinality. But such large cardinalities are beyond the reach of standard ZFC. Claims about standard second-order invalidity can thus have very substantial mathematical content.

Why would the strong mathematical content of SOL be problematic? One reason is that it compromises the topic neutrality that logic is often required to have.[25] For instance, either CH or its negation corresponds to a logical truth of SOL. This makes SOL inappropriate as the logic to be employed in any investigation of the important mathematical question of CH. Moreover, SOL will interfere with many weak set theories where one investigates set theory in the absence of (say) the Axiom of Choice or a commitment to a determinate totality of all subsets. This interference makes SOL unsuitable as a completely general background theory.

It will be objected that no interesting logic can provide a completely neutral medium in which all other debates can be adjudicated.[26] Perhaps so. But neutrality is a matter of degree. And SOL is particularly far from the neutral end of the spectrum, having implicit content that 'answers' some of the hardest questions investigated in contemporary set theory.

The strong mathematical content of SOL also calls into question some of its applications. Consider the use of SOL in categoricity arguments (Section 3.3). Since SOL is infused with set-theoretic content, any assurance provided by these arguments comes from within mathematics, rather than from some more secure logical standpoint outside of it. In particular, the use of SOL to defend the quasi-categoricity of set theory is cast in a different light. It is true that quasi-categoricity follows when we 'freeze' the subset relation by restricting our attention to standard models of second-order Zermelo–Fraenkel set theory. But this approach helps itself to the subset relation, which is one of the main objects of study of contemporary set theory.[27]

The use of SOL to defend absolute generality is also put under pressure. This defence seeks to safeguard absolutely general quantification over an ontological hierarchy of sets and urelements by using a second-order metalanguage to develop a semantics that is compatible with such quantification. But in order to develop an appropriate semantics for this meta-language in turn, we need to invoke a third-order language (Section 3.6). And this phenomenon continues: in order to develop the appropriate semantic theories, we are forced to climb up an ideological hierarchy of expressive resources associated with logics of higher and higher orders. This is a phenomenon akin to that involved in denying absolute generality. Thus, for the mentioned defence of absolute generality to do more than simply shift the bump in the carpet, the ontological hierarchy of sets and the ideological hierarchy of expressive resources must be sufficiently different in character. But in light of the strong set-theoretic content of higher-order logic, it is unclear whether the difference between the two hierarchies is very deep.[28]

6. The Road Ahead

Many open questions remain. Let me mention some that strike me as particularly worthy of investigation.

Many of the applications of higher-order logic require further investigation (Section 3). To what extent can the use of full second-order logic in categoricity arguments be replaced by so-called schematic reasoning?[29] For instance, can the second-order induction axiom (PA6) be replaced by the schematic principle that induction holds for any meaningful predicate, without specifying ahead of time what predicates are meaningful? Next, how substantive is the apparent need for second-order logic in set theory? And does the formulation of semantic theories in higher-order meta-languages provide a stable defence of absolute generality?

A better understanding is needed of logics of orders higher than two (Section 4). Do our reasons for accepting plural and second-order logic also give us reason to accept logics of higher orders? Does the same answer hold for plural and second-order logic? If higher orders are legitimate, then how high can we go? All the way into the transfinite?

A host of interesting questions remain about the relation between higher-order logics and set theory (Sections 5.2 and 5.5). If there are logics of very high orders, what is their relation to set theory? Are they fundamentally different or just alternative perspectives on a shared subject matter? Type theory was superseded by first-order set theory as the canonical foundation for mathematics in the first half of the twentieth century. Does this development hold any lessons for today's resurgence of interest in higher-order logics? How deep is the difference between variables of different orders? Are there legitimate transitions from higher orders to lower? Frege's Basic Law V was a failed attempt to effect such a transition.[30] Are there consistent and theoretically useful ways of harnessing such transitions?[31]

The debate about the ontological innocence of higher-order logic remains open (Section 5.3). I argued that the most interesting question is whether the use of higher-order variables is philosophically less problematic or substantive than the use of singular first-order variables. An answer is needed.

A topic not even broached in this article is the interaction of modalities and higher-order logics. Here plural and second-order logic are likely to come apart. For when an object is one of several, this seems to be a matter of necessity; whereas it often seems contingent whether an object falls under a concept. The formal investigation of this terrain is still in its infancy.[32]

Notes

1. An expression e is said to be *substitutable* for a variable v in a formula ϕ iff every free occurrence of v in ϕ can be uniformly replaced by e without any variables in e thus becoming bound by quantifiers in ϕ.
2. In fact, the displayed formula is short for its universal closure; that is, the result of prefixing it by universal quantifiers binding all of its free (first- and second-order) variables. The variables bound in this way are known as *parameters*.

3. The proof of this last claim uses a very weak form of the Axiom of Choice known as *countable choice*.
4. See [Shapiro, 2000, pp. 82–3] for a complete proof. A categorical axiomatization of the real number structure is also available; see *ibid*. p. 84.
5. Or, strictly speaking, the universal closure of the displayed formula: see footnote 2.
6. The theory of ordered pairs uses a three-place predicate OP and an axiom stating that any two objects have a unique ordered pair: $\forall x \forall y \exists z \forall z' (\mathrm{OP}(x, y, z') \leftrightarrow z = z')$.
7. See [Shapiro, 2000, pp. 100–6] for details and extensions to some higher cardinalities.
8. See for instance [Hellman, 1989] and [Shapiro, 2000].
9. [McGee, 1997] shows how the 'height' too is fixed if we assume (a) that the urelements form a set, and (b) we can quantify over absolutely everything. (In fact, (a) can be weakened to the assumption that the urelements are equinumerous with the ordinals.) However, each of these assumptions is controversial.
10. See [Linnebo, 2003, pp. 80–1] for more details.
11. See [Drake, 1974] for technical details and [Burgess, 2004] and [Uzquiano, 2003] for philosophical discussion.
12. See [Shapiro, 1987].
13. See for instance [Russell, 1908], [Zermelo, 1930], [Dummett, 1981], and [Parsons, 1977].
14. See [Williamson, 2003a] for an influential example.
15. See [Rayo, 2006] for this result and a more fine-grained one, and [Linnebo and Rayo, shed] for generalizations into the transfinite. The need to ascend one order is due to the fact that a language of order n contains predicates of order $n + 1$, whose various interpretations can properly be described only by using variables of order $n + 1$.
16. This notion of 'language of order n' corresponds to Rayo's [Rayo, 2006] notion of 'full n-th order language'.
17. This is a simplification of the system of Russell and Whitehead's *Principia Mathematica* suggested by Leon Chwistek and Frank Ramsey.
18. See for instance [Oliver and Smiley, 2005] and [Linnebo and Nicolas, 2008] concerning higher-order plurals.
19. Are there 'superplural' languages that stand to ordinary plural languages the way these stand to classical first-order languages? See [Rayo, 2006] and [Linnebo and Rayo, shed] for discussion of this harder question, which won't be addressed here.
20. See for instance [Hilbert, 1926, p. 184 (p. 387 of translation)]; [Carnap, 1934, p. 186]; [Gödel, 1931, fn. 48a]; and [Tarski, 1935].
21. See for instance [Rayo and Yablo, 2001] and [Wright, 2007].
22. See for instance [Resnik, 1986] and [Parsons, 1990], as well as [Linnebo, 2003] for discussion.
23. In fact, the highly circumscribed claim of the previous sentence appears to be conceded by [Shapiro, 1999, pp. 44, 53]. However, Shapiro argues that there are other uses of second-order logic where there is no need to adhere to the ideal of deductive explicitness, for instance the characterization of mathematical structures.
24. This follows fairly directly from the ability to provide categorical characterizations of the natural numbers and the reals. See [Shapiro, 2000, pp. 104–5] for details.
25. See [Jané, 2005] for a more developed argument of this sort.
26. See for instance [Shapiro, 1999, 54].
27. See [Koellner, 2010] for a related argument.
28. See [Linnebo and Rayo, forthcoming] for an argument that it is not.
29. See for instance [McGee, 1997] and [Parsons, 2008, ch. 8].
30. This inconsistent 'law' says that two concepts F and G have the same extension just in case $\forall x(Fx \leftrightarrow Gx)$.

31. [Parsons, 1983b] and [Linnebo, ta] use a modal version of such a transition to motivate and derive much of ZFC set theory.
32. I am grateful to Salvatore Florio, Leon Horsten, Marcus Rossberg, and Richard Pettigrew for discussion and comments on earlier versions, as well as for a European Research Council Starting Grant (241098-PPP), which facilitated the completion of this article.

7 The Paradox of Vagueness

Richard Dietz

Chapter Overview

1. The Paradox 130
   - 1.1 Soriticality 130
   - 1.2 Sorites Arguments 131
   - 1.3 Approaches to the Paradox 133
2. Borderline Vagueness 134
   - 2.1 Empirical Content 134
   - 2.2 Theoretical Views 135
   - 2.3 Soriticality and Bordeline Vagueness 137
3. Higher-Order Vagueness 140
   - 3.1 What the Hypothesis Says 140
   - 3.2 Some Arguments For and Against the Hypothesis 141
4. Classical Frameworks for Vagueness 143
   - 4.1 Epistemicism 144
   - 4.2 Vagueness as a Semantic Modality 150
   - 4.3 Contextualism and Connectedness 151
5. Non-Classical Approaches to Vagueness 156
   - 5.1 Paracompleteness and Paraconsistency 156
   - 5.2 Many-Valued Logics 159
     - 5.2.1 K_3 160
     - 5.2.2 *LP* 162
     - 5.2.3 $Ł_{\aleph}$ 163
   - 5.3 Supervaluationism and Subvaluationism 165
     - 5.3.1 *SpV* 165
     - 5.3.2 *SbV* 169
   - 5.4 Transitivity of Logical Consequence Reconsidered 170

Acknowledgements 171

Notes 171

In colloquial language, *vagueness* is a generic term that is loosely used in association with all sorts of linguistic phenomena such as ambiguity, context-sensitivity, obscurity, or lack of specificity in content. In the philosophical literature, the term is used rather technically, in association with two types of features that many general terms in natural language (e.g., adjectives such as 'bald', nouns such as 'walking distance', or quantifiers such as 'most') have. For one, it is a familiar feature of many general terms that they are indefinite in extension to some extent. For example, a scalp with no hairs is definitely bald, whereas a scalp with 150,000 hairs is definitely not bald; on the other hand, for some numbers of hairs in between, it is indefinite whether they make for baldness or not – in other words, 'bald' has some *borderline cases* of application (or cases of application that are indefinite in truth value). Contrast this with general terms that lack this feature (e.g., 'is four-foot in height' has no borderline cases).

More notoriously, and this brings us to the other feature, general terms with borderline cases are typically (if not generally) *soritical*, that is, susceptible to a type of argument which is also known as *sorites* argument. Arguments of this type are paradoxical. For on the one hand, they appear to be valid, and it seems odd to deny any involved premise; on the other hand, their conclusion can be hardly accepted. In effect, it follows from such arguments that the general term involved fails to be coherent – which seems a very odd result, for it suggests that the term is of no use as a means of making distinctions. Since it is hard to overstate the pervasiveness of soriticality in natural languages, the sorites paradox poses a threat to the fundamental claim that we can represent reality coherently in natural language by means of general terms.

In this view, it is far more global in scope than other paradoxes such as the Liar or the Lottery, which rather highlight a problem with particular notions (such as truth, or belief respectively).[1] The discussion of sorites paradoxes already starts in ancient philosophy. However, the idea that there is a common feature of general terms that gives rise to such paradoxes emerges only in modern analytic philosophy.[2] According to a widely held view, vagueness is not only a broad phenomenon but also a persistent one, in the sense that any general terms in which we may describe vagueness are to be vague as well – in other words, it is held that vagueness gives rise to *higher-order vagueness*.

Rather controversial is the question of whether the vagueness of general terms is an instance of an even broader type of indeterminacy. For one, it has been suggested that vagueness is a kind of indeterminacy in extension that may affect not only general terms but also other types of linguistic expressions. Some authors have argued for an even more radical thesis to the effect that vagueness is a kind of indeterminacy that may affect not only the ways in which we represent reality in language (or other kinds of representation) but even reality itself, independently of our ways of representing it. Notwithstanding some tendencies to widen the notion of vagueness to various sorts of indeterminacy, the sorites paradox remains centre stage in the philosophical discussion of vagueness. The paradox has been one of the driving motivations in the development of various non-classical semantics and logics for natural languages; and it has met with various accounts in epistemology, the philosophy of language, philosophical logic as well as in linguistics.[3] This chapter gives a survey of influential accounts of the paradox, with the focus lying on the philosophical literature.

Sections 1–3 explore more general philosophical problems related to the paradox, which may be separated from special problems arising in particular frameworks for vagueness. To start with, the paradox (Section 1), the problem of vagueness-related indefiniteness (Section 2) and the thesis of higher-order vagueness (Section 3) are introduced. Section 4 discusses ways of modelling vagueness in a classical framework. Section 5 turns to some ways of modelling vagueness in non-classical frameworks. Without loss of generality and in accordance with the general discussion, we will focus on natural language expressions that may be formalized as unary predicates.

1. The Paradox

This section gives the condition for the existence of instances of the sorites paradox (1.1), along with some standard forms of instances of the paradox (1.2) and a survey of approaches to the paradox (1.3).

1.1 Soriticality

It is a familiar feature of many general terms in natural language that it seems odd to deny that they are insensitive to changes in the objects it is predicated of, provided these changes are sufficiently small. For instance, it seems odd to deny that a walking distance is still a walking distance if we increment it by one foot; or that a bald scalp is still a bald scalp if its number of hairs increments by one. Since small changes accumulate to big ones, tolerance gives rise to a type of paradox known as the sorites paradox. For example, starting from one foot, which is definitely a walking distance, we may expand it to a distance of 1,000 miles (i.e., 5,280,000 feet) by incrementing it successively by one foot. Since one foot more does not seem to make any difference as to whether something is a walking distance, no pair of adjacent distances in the series should mark a cut-off point between walking distances and distances which are not walking distances. But then, every distance in the series should be a walking distance, including the 1,000 miles we end up with – which contradicts common sense, according to which 1,000 miles is not a walking distance. Contrast this case with general terms that are not soritical – for instance, there is no sorites series for 'is four-foot in height'.

Generalizing from particular examples, one may say that there is an instance of the *sorites paradox* for a given predicate F whenever there is a *sorites series* for F, that is a series for which F meets the following constraints:[4] (1) a 'clear-case' constraint, to the effect that the first member of the series, i, is an element of the predicate's extension and that the last member of the series, j, is an element of its anti-extension, (2) an 'unlimited tolerance' constraint, to the effect that there is a relation R such that: (2.i) R is a tolerance relation, that is, if R applies to a pair of objects $\langle x, y\rangle$, it follows that the corresponding instance of the schema

TOLERANCE (TOL): if Fx is true then Fy is true.[5]

is true; and (2.ii) the series is R–connected, that is, R applies to each pair of adjacent members in the series. More formally, we have:

SORITES CONDITION (SOR): There is a *sorites series* of objects for F, that is, a series of objects $\langle a_0, \cdots, a_i\rangle$, with S being the union of all members of this series, such that each of the following conditions is compelling:

1. Clear Case (CC): F is true of a_0 and false of a_i (i.e., $\neg Fa_i$ is true);

2. Unlimited Tolerance (UT): there is a relation R such that

2.i R–Tolerance (R–TOL): R is a *tolerance relation* for F with respect to S, i.e.: for any $i, j \in S$: if $R(i, j)$ is true, then if Fi is true, Fj is true too;

2.ii R–Connectedness (R–CON): $a_0Ra_1, \cdots, a_{i-1}Ra_i$.

If a series of objects is a sorites series for F, we also say that F is *soritical* for that series. For any relation for which it is compelling to say that it is a tolerance relation for F (with respect to a domain D), we say that it is an *indifference relation* for F (with respect to D).[6]

1.2 Sorites Arguments

Given a sorites series for a predicate, there are different argument forms that instantiate the sorites paradox. The standard version which has received most attention in the previous discussion goes by a series of conditionals

CONDITIONAL SORITES[7] – LONG (CS–L)

$$\begin{array}{ll} (1) & Fa_0 \\ (2_1) & Fa_0 \rightarrow Fa_1 \\ & \vdots \\ (2_i) & Fa_{i-1} \rightarrow Fa_i \\ \therefore & Fa_i, \end{array}$$

where an indifference relation for F applies to every pair $\langle a_n, a_{n+1}\rangle$ (with $0 \leq n < i$). It is easy to see that Fa_i can be derived from the given premises if logical consequence ($\models$) satisfies modus ponens (i.e., the inference rule that allows us to infer from conditional sentences and the antecedent to the consequent: $\{P, P \rightarrow Q\} \models Q$) and generalized transitivity (if $\Gamma \models \varphi$ and $\Delta \models \gamma$, for all $\gamma \in \Gamma$, then $\Delta \models \varphi$). For instance, undoubtedly, one foot is a walking distance. Hence, given that if a one-foot distance is a walking distance, so is a two-foot distance, by modus ponens, it follows that a two-foot distance is a walking distance as well, which we can use as an input for the next inferential step to conclude that the same holds for a three-foot distance, and so on, with the last inferential step having the conclusion that also 1,000 miles is a walking distance. By generalized transitivity of logical consequence then, it follows from the assumption that a one-foot distance is a walking distance and the relevant instances of (TOL) that 1,000 miles is a walking distance as well.

Replacing the premises $(2_1) \cdots (2_i)$ by the universal $(\forall n \in \{0, \cdots, i-1\})(Fa_n \rightarrow Fa_{n+1})$, we obtain a shorter variant of the conditional sorites:

Conditional Sorites – Short (CS–S)

(1) Fa_0
(2) $(\forall n \in \{0, \cdots, i-1\})(Fa_n \rightarrow Fa_{n+1})$
$\therefore$ Fa_i,

where an indifference relation for F applies to every pair $\langle a_n, a_{n+1} \rangle$ (with $0 \leq n < i$). The derivation of Fa_i from (1) and (2) then runs the same as in the longer for propositional logic; we just need to employ additionally universal instantiation, in order to obtain all relevant instances of (TOL), $(2_1) \ldots (2_i)$ from (2). Since sorites series are commonly finite, the use of predicate logic is in the end always dispensable (for instead of universal quantification, we can always consider corresponding conjunctions of relevant instances of (TOL)). For convenience (to avoid discussion of long-winded conjunctions), the (CS–S) will be occasionally referred to after all.

Another version of the sorites paradox goes by mathematical induction (which allows us to infer from $P(0)$ and $(\forall n)(P(n) \rightarrow P(n+1))$ to $(\forall n)P(n)$), and has the form

Mathematical Induction Sorites

(1) Fa_0 (inductive basis)
(2) $(\forall n)(Fa_n \rightarrow Fa_{n+1})$ (inductive premise)
$\therefore$ $(\forall n)Fa_n$,

For instance, it appears that for any natural number n, if n feet is a walking distance so is $n+1$. By induction then, since zero feet is undoubtedly a walking distance, for any arbitrarily high natural number n, n feet are a walking distance.[8] There are other variants of this form.[9] And yet still other forms of the sorites paradox have been suggested.[10] The philosophical literature on the paradox has been focussed primarily on the versions (CS–S) and (CS–L). The focus of this discussion is going to be the same accordingly.

1.3 Approaches to the Paradox

According to some authors, the two types of constraints that make for soriticality are to be accepted as indispensable for an adequate account of vague predicates, and the principles of deduction that allow us to generate a contradiction from these constraints do hold. In effect, the paradox is embraced (e.g., see [Dummett, 1975], [Wheeler, 1979], [Unger, 1979] [Unger, 1980], or more recently, [Eklund, 2005] and [Gómez-Torrente, 2010]).[11] Typically, advocates of this view propose that soritical terms (such as 'walking distance', 'heap' or 'bald') are empty, and that their respective negations ('non-walking-distance', 'non-heap', or 'not bald' respectively) are trivial: according to this, it is true to say that there are no walking distances, no bald men, no heaps of sand, and so on; in other words, everything is a non-walking distance, a non-heap, not bald, and so on. This view is also known as *nihilism*. (For the most outspoken defence of this view, see [Unger, 1979]; but contrast this with his later view, in [Unger, 1990].) A problem with this view, which has been widely noted, is that it is radical to an extent that brings it close to absurdity. For, considering the pervasiveness of vagueness, it suggests that most general terms we use in natural language fail to provide a means of making distinctions – either they are empty, or they are trivial.[12]

Another problem with nihilism is that, as assessed on its own terms, it seems to be not radical enough. To wit, if soritical primitive terms such as 'walking distance' or 'bald' are subject to inconsistent constraints, then the same should hold for associated complex terms such as 'non-walking distance' or 'not bald' respectively, which are as soritical as their primitive counterparts – they seem to support clear-case constraints on the extension and anti-extension (1,000 miles should be, by any standards, a non-walking distance, whereas a zero-foot distance should not be so), as well as a converse tolerance constraint (starting from a non-walking distance, one foot less should result in a non-walking distance in turn). Nihilism rests on an asymmetric treatment of soritical primitive terms and their soritical non-primitive counterparts. For the former, it is taken that they obey all constraints that give rise to paradox, whereas for the latter, clear-case constraints on the anti-extension are rejected (e.g., it is denied that a distance of 1,000 miles is not a non-walking distance). For lack of a good rationale for this asymmetry, it seems that not only soritical primitive terms, but also their complex counterparts should fail to have an extension. One way of putting this idea would be to argue for an even more radical claim to the effect that soritical terms not only fail to have an extension but even fail to fix any truth conditions that would partition the domain of objects into an extension and anti-extension.[13] Needless to say that this comes down to an even more radical proposal.

The rather prevailing type of approach in the philosophical discussion is to reject the paradox in one way or other – the remainder of the discussion will focus on this type of approach. Although the proposals in this spirit may be diverse, it seems to be common ground in this camp not to question (CC). Starting from a classical logic for vagueness, this approach commits to the assumption of some counterinstance to (TOL) pertaining to some pair of adjacent members in a sorites series, that is, the thesis that some such pair marks a cut-off point between true and false applications in the sorites series. E.g., according to this, there is a greatest distance between zero foot and 1,000 miles that is still a walking distance, even though it would fail to be one if it were incremented by one foot. Various escape routes from a conclusion of this form offer non-classical frameworks, where one can reject instances of (TOL) without being committed to assert their negation. Other non-classical approaches that allow us to keep all instances of (TOL) pertaining to adjacent members in a sorites series involve more radical departures from classical logic.

Before having a closer look at various types of resolutions to the paradox, two related controversial issues in the theory of vagueness are introduced. Either issue bears on the account of soriticality and the resolution to the paradox.

2. Borderline Vagueness

An n-ary general term is said to be borderline vague iff some n-tuple of objects is a borderline case of the term. This section describes some pre-theoretical features of borderline vagueness (Section 2.1) and some generic views on the nature of borderline vagueness (Section 2.2). Furthermore, the controversial question as to how soriticality and borderline vagueness are related is explored to some extent (Section 2.3).

2.1 Empirical Content

As Fara [Fara, 2000, p. 76] puts it:

> We are prompted to regard a thing as a borderline case of a predicate when it elicits in us one of a variety of related verbal behaviors. When asked, for example, whether a particular man is nice, we may give what can be called a *hedging* response. Hedging responses include:'He's nice*ish*', 'It depends on how you look at it', 'I wouldn't say he's nice, I wouldn't say he's not nice', 'It could go either way', 'He's kind of in between', 'It's not that clear-cut', and even 'He's a borderline case'. If it is demanded that a 'yes' or 'no' response is required, we may feel that neither answer would be quite correct, that there is 'no fact of the matter'.

On this account, the question of what is a borderline case of a predicate may be reformulated as the question of what might prompt hedging responses of the said type. In the same spirit seems to be Gaifman's suggestion ([Gaifman, 2010, p. 9]) that borderline vagueness can be manifested in two ways in linguistic behaviour:

1. Undecidedness or hesitation on the part of the speaker, which does not derive from lack of factual knowledge.[14]
2. Divergence in usage among competent speakers (in situations where they are competent judges) including, possibly, the same speaker on different occasions.

Hedging responses may have various causes, some of which are entirely unrelated to vagueness, insofar as they may prompt also hedging responses for non-soritical general terms. For example, in giving a hedging response to the question whether John is taller than Bob, despite the fact that we believe that he is, we may want to avoid the unwanted implicature that he is *significantly* taller than Bob.[15] This still leaves the possibility that some kind of cause (or kinds of causes) for hedging responses may be characteristic of soritical terms, in the sense that only hedging responses with regard to applications of such terms may have such a cause – in this case, one could reserve the term 'borderline vague' for occasions of hedging behaviour that have the said characteristic kind of cause. But in the absence of an argument in support for this hypothesis, there is no justification for taking it for granted at the outset. In view of these considerations, when raising the issue of what kind of thing borderline cases are, one should qualify it as a hypothetical question of the form: supposing there is a common kind of cause (or a distinguished class of kinds of causes) that is characteristic of hedging responses with respect to applications of soritical terms, what might this kind of cause (or distinguished class of causes) be more exactly? For brevity, this qualification is omitted in what follows, but it will be intended implicitly throughout.

2.2 Theoretical Views

The question of what borderline vagueness is is highly controversial. One may hope that a satisfying account of borderline vagueness might provide a better basis for discussing the variety of logical options that have been suggested for languages with vague expressions. For instance, if borderline vagueness is a purely epistemic feature, that does not attach to meaningful expressions absolutely but rather only as used in certain language communities, this may be seen as a motivation for adopting a standard, classical semantics for vague languages. The same point may be made with regard to the controversial question of what the logical features of 'borderline vague' are – for instance, there is no common ground on the question as to whether it is consistent to assume a sentence to be vaguely true (i.e., to make assumptions of the form of 'it is the case that *P*, though is it vague whether P'). Roughly, one may distinguish between two main approaches in dealing with borderline vagueness in language.

For one, some authors argue that borderline vagueness may be characterized in purely epistemic terms (see [Cargile, 1969], [Campbell, 1974], [Scheffler, 1979], [Sorensen, 1988], [Sorensen, 2001], [Williamson, 1994], [Horwich, 2000], and [Fara, 2000]). According to this view, also known as the *epistemic view* of vagueness, borderline vagueness is a kind of epistemic indeterminacy, which is thought to be different in kind from mere lack of information regarding relevant facts – e.g., on this type account, any application of 'walking distance' to a number of feet is a borderline case just in case competent speakers of English are ignorant as to whether the term applies, for certain reasons (that are meant to be characteristic of borderline vagueness). Typically, the epistemic view combines with a classical framework for vague languages.[16]

Other authors have suggested that borderline vagueness is a feature that attaches to linguistic expressions as used, independently of the respective epistemic capacities of the speaker – in distinction to the epistemic view, we call this generic view of vagueness here *semantic*. According to this, borderline vagueness may be characterized as some kind of semantic indeterminacy in extension (e.g., see [Lewis, 1970a], [Lewis, 1975], [Lewis, 1979], [Lewis, 1986a], [Fine, 1975], [Burns, 1991], [McGee and McLaughlin, 1995], [Soames, 1999, Chapter 7], [Heck, 2003], [Varzi, 2007], [Rayo, 2008], and [Rayo, 2010]).[17] On this account, for instance, any application of 'walking distance' to a number of feet is a borderline case just in case the semantics of the term and the circumstances of its application do not fix uniquely a classical truth value. Typically, the semantic view associates with some non-classical semantic framework for vagueness – in this case, it is often suggested that borderline cases are *truth-value gaps* (i.e., neither true nor false), or alternatively, it is suggested that they are *truth-value gluts* (i.e., both true and false). The semantic view has also been proposed in combination with a classical semantics for vagueness though (see Section 4.2).

The distinction between epistemic and semantic views of borderline vagueness is not mutually exclusive – the two approaches may combine with each other.[18] Nor is this distinction exhaustive. On an entirely different kind of account, it has been suggested that there is no genuine borderline vagueness in language, and that all apparent instances of this type are derivative from some borderline vagueness in reality itself – where there is no common ground on the question of what it would mean for reality more specifically to be affected by instances of borderline vagueneness.[19] Since our focus is on accounts that do not drop the hypothesis of *genuine* vagueness in language though, we can feel free to put ontological views of borderline vagueness aside.

For another, it has been argued that borderline vagueness is genuinely psychological in kind. According to this, a sentence is borderline vague (relative to a relevant class of epistemic subjects) just in case distributions of rational degree of belief with respect to this sentence and other sentences that embed this sentence obey certain structural constraints that are characteristic of borderline vagueness ([Schiffer, 2003]).[20] Another kind of psychological account is offered in [Douven et al., 2009], where borderline vagueness is described in terms of some sort of indeterminacy in conceptual spaces (for a different account in terms of indeterminacy in mental representation, see [Koons, 1994]).

Yet other authors have suggested that 'borderline vague' may be better treated as a primitive notion, which can be best characterized merely in terms of its logical features.[21]

2.3 Soriticality and Bordeline Vagueness

It is not an overstatement to say that there is a high correlation between occurrences of soriticality and occurrences of borderline vagueness. Yet it may still be regarded as an open question whether these two features are in fact independent. On the other hand, even if the answer is to be given in the positive, there is still reason for hope that a unified theory of vagueness may explain why the features typically occur, and if not, why not. The following considerations are not meant to give an ultimate answer on the question of how soriticality relates to borderline vagueness. But they may help to make clear that the issue leaves room for controversy.

For convenience, some notation is first introduced. Insofar as borderline vagueness is expressible in the object-language, it is standardly symbolized by means of a sentence operator $\mathcal{D}$ for 'definite truth'. Sentences of the form '$\neg\mathcal{D}P \wedge \neg\mathcal{D}\neg P$', where P is a closed sentence, abbreviate 'P is indefinite (in truth-value)' (in other words, 'it is indefinite whether P'); accordingly, complex one-place expressions of the form '... is a borderline case of F' (or 'it is indefinite of ... whether ... is an F') can be formalized as open formulas of the form '$\neg\mathcal{D}Fx \wedge \neg\mathcal{D}\neg Fx$' where F is a unary predicate and x is a free variable.

Now, consider the following argument. It is a common idea that predicates F are soritical only if (and perhaps just in case) they satisfy a principle of the following form:[22]

Gap (GP): $(\forall n \in \{0, \ldots, i-1\})(\mathcal{D}Fx_n \rightarrow \neg\mathcal{D}\neg Fx_{n+1})$,

Indeed, starting from classical predicate logic, one may reasonably argue that a predicate satisfies an associated instance of (GP) just in case it has borderline cases. Take any finite sorites series $\langle a_0, a_i \rangle$ for a predicate F, which implies that $\mathcal{D}Fa_0$ and $\mathcal{D}\neg Fa_i$ are both true. Hence, by reductio ad absurdum, the principle $(\forall n \in \{0, \ldots, i-1\})(\mathcal{D}Fx_n \rightarrow \mathcal{D}Fx_{n+1})$ is false (note, if it were true, by soritical reasoning, it would follow that $\mathcal{D}Fa_i$ is true as well). Hence, there is a member a_k (with $0 \leq k < 1$) in the series where $\mathcal{D}Fa_k$ is true and $\neg\mathcal{D}Fa_{k+1}$ is true as well. Furthermore from this, by (GP), it follows that there is a member a_k (with $0 \leq k < 1$) in the series where $\mathcal{D}Fa_k$ is true and $\neg\mathcal{D}Fa_{k+1} \wedge \neg\mathcal{D}\neg Fa_{k+1}$ is true. Hence F has borderline cases.

There is also a safe route from borderline vagueness to (GP). Apart from classical predicate logic, we only need the assumption that we have a series of objects beginning with a definite truth and ending with a definite falsity, where preceding members in the series are always better candidates for definitely true predications than their successors, and where also conversely, succeeding members in the series are always better candidates for definitely false predications than their predecessors. More precisely, if $\langle a_0, \ldots, a_i \rangle$ is the relevant series, then it is supposed to satisfy the constraints:

MONOTONICITY$_1$ (MON$_1$): $(\forall n \in \{1, \ldots, i\})(\mathcal{D}Fx_n \rightarrow \mathcal{D}Fx_{n+1})$.

MONOTONICITY$_2$ (MON$_2$): $(\forall n \in \{1, \ldots, i\})(\mathcal{D}\neg Fx_n \rightarrow \mathcal{D}\neg Fx_{n-1})$.

The argument then runs as follows: Suppose F is borderline vague and that there is a series of objects $\langle a_0, a_i \rangle$ with respect to which F satisfies (MON$_1$), and where $\mathcal{D}Fa_0$ and $\mathcal{D}\neg Fa_i$ are both true. Assume, for reductio ad absurdum, that there is a pair of adjacent members, $\langle a_n, a_{n+1} \rangle$, that marks a cut-off point between members that are definitely F and members that are definitely not F. Then by (MON$_1$), for every number k smaller than n, $\mathcal{D}Fa_k$ is true as well. By (MON$_2$) it follows furthermore for every number m larger than $n+1$ that $\mathcal{D}\neg Fa_m$ is true as well. Consequently, there is no borderline case of F in the series – which contradicts what we assumed to be the case. Hence, by reductio ad absurdum, there is no sharp cut-off between definite truths and definite falsities with respect to F in the series. Thus, the relevant instance of (GP) is satisfied – this completes the argument.

As it stands, the argument is open to various objections. Given a predicate F that is affected by borderline vagueness, one may suggest that also the definitized counterpart predidate $\mathcal{D}Fx$ is affected by borderline vagueness (see Section 3). That is, if vagueness requires a departure from classical logic, it cannot be taken for granted that the argument from soriticality to borderline vagueness goes through also on other frameworks that have been proposed for vagueness (see Section 5). On another note, it has been argued that a generalized version of (GP) is not sustainable for any finite sorites series in certain frameworks for vagueness (see Section 5.3). Notwithstanding possible objections on the part of advocates of non-classical frameworks for vagueness, it ought to be noted as well though that apart from arguments from non-classical frameworks for vague languages, there seem to be no independent reasons for doubting that borderline vagueness is adequately captured by a gap principle.

That is, assuming at least that soriticality implies a gap principle, the above argument furthermore suggests that soriticality implies borderline vagueness, which seems indeed to conform with the received view.[23] If gap principles in general conversely implied that the relevant predicate is soritical, soriticality could be accounted for as an aspect of borderline vagueness, to the effect that: whenever a predicate *F* obeys (MON_1) and (MON_2) with respect to a given series of objects, where the series begins with definite truths and ends with definite falsities, with some borderline cases in between, we have a sorites series for the predicate – or so one might suggest. However, some authors have cast doubt on this account strategy for soriticality as a viable option.

A famous type of counterargument is due to Sainsbury [Sainsbury, 1991, p. 173] and invokes partially defined terms such as:

CHILD*:

1. If *x* has not reached her sixteenth birthday, then 'is a child*' is true of *x*.
2. If *x* has reached her eighteenth birthday, then 'is a child*' is false of *x*. (The end)

According to Sainsbury [Sainsbury, 1991, p. 173], persons who are at least 16 and not yet 18 years old are borderline cases of 'child*', even though 'intuitively, this is not a vague predicate' – where the intended sense of 'vague' seems to imply soriticality (as far as general terms are concerned).[24] It seems right indeed that predicates of this type are not soritical, but one may object that the involved use of 'borderline case' is rather a misnomer, considering that 'child*'-predications of persons whose age is in the range (16, 18) do not meet the feature of divergence of usage that was mentioned as a characteristic feature of borderline cases (Section 2.1): e.g., for anybody who is 17 of age, it does not seem legitimate, being asked whether she is a child*, to answer in the hedging way that is characteristic way of borderline cases. Considering this, instances of partiality like 'child*' do not seem to provide a good case in point against any account of soriticality in terms of borderline vagueness; rather they highlight a problem with the view that partiality is a sufficient condition for borderline vagueness.[25,26,27] This said, there is still another kind of counterexample which seems more forceful.

Take the example 'has few children for an academic' (from [Weatherson, 2010, p. 80]), which is associated with a discrete dimension (number of children). The term has borderline cases – plausibly two and three children are borderline cases; and it has both definitely true and definitely false application cases (one child and five children respectively). But one can hardly generate a compelling sorites paradox with this term. Consider a sorites argument of the form:

HAS FEW CHILDREN FOR AN ACADEMIC:

1a. An academic with one child has few children.
1b. If an academic with one child has few children, then an academic with two children has few children.

1c. If an academic with two children has few children, then an academic with three children has few children.
1d. If an academic with three children has few children, then an academic with four children has few children.
1e. If an academic with four children has few children, then an academic with five children has few children.
1f. So an academic with five children has few children.

As Weatherson ([Weatherson, 2010, p. 80f]) notes, whereas (1a) is compelling and (1f) only to be denied, the tolerance instances (1b) and (1c) can be hardly considered as compelling; indeed, one may even strengthen this point, saying that for either instance, it is both agreeable to accept it in a hedging way and agreeable to deny it in a hedging way. On either account, we have a case in point for the thesis that borderline vagueness does not always go with soriticality. Importantly, the counterevidence is pre-theoretical in kind and does not rely on any account of apparent tolerance in terms of definite truth (e.g., (GP) or alternative stronger principles one may suggest).[28]

To take stock, in a classical framework for vagueness, one can indeed reasonably argue that soriticality implies borderline vagueness. However, as far as the converse case is concerned, it seems problematic in view of pre-theoretical evidence that tells against it. This result may suggest that the notion of borderline vagueness is in the end dispensable for an account of soriticality; on the other hand, granted that there may be borderline vagueness without a compelling sorites paradox, a theory of borderline vagueness may after all supply means of describing sufficient conditions for soriticality. (For instances of either type of approach, compare Sections 4.1 and 4.3 respectively).

3. Higher-Order Vagueness

This section introduces the notion of higher-order vagueness (Section 3.1) and mentions some arguments for and against the thesis that there are instances of higher-order vagueness (Section 3.2).

3.1 What the Hypothesis Says

An expression is called *higher-order vague* just in case any expressions we may choose for describing its vagueness are themselves vague. Standardly, the term is understood more specifically in terms of borderline vagueness. For the present purposes, the following informal characterization (which generalizes a characterization given in [Williamson, 1999, p. 132] for sentences) may do: An (i-ary) predicate F (where $i \geq 0$) is first-order vague just in case it has some borderline cases (in case $i = 0$, F is a sentence, and F has a borderline case iff F is borderline vague in truth value). F is second-order vague just in case any second-order expressions – that is, any expressions (such as 'definitely F', 'definitely not F', 'either definitely F or definitely not F', or 'neither definitely F nor definitely not F') in terms of which we may classify (i-tuples of) objects as to whether F definitely holds, definitely does not hold, or neither – themselves have borderline cases. More generally, F is a first-order expression that classifies (i-tuples of) objects as to whether F holds. $(n + 1)$th-order expressions classify (i-tuples of) objects as to whether nth-order expressions definitely hold, definitely do not hold, or neither. Borderline vagueness for any nth-order expression is nth-order vagueness of F.[29,30]

Inasmuch as borderline vagueness of higher-order expressions is supposed to go with soriticality, the thesis of higher-order vagueness immediately bears on the account of the paradox of vagueness. For it should be then a desirable feature of any strategy for first-order expressions that it be reapplicable to higher-order expressions.[31] Indeed, the thesis that there is higher-order vagueness seems to reflect the received, orthodox view on vagueness. Yet, there is no common ground on the scope of higher-order vagueness, or whether higher-order vagueness may terminate. For one, it may just come to the claim that there are general terms that are n-th order vague, where $n > 1$ – which may allow for the possibility of first-order vagueness without higher-order vagueness, and also for the possibility that higher-order vagueness may be terminating (i.e., for some n, we have n-th order vagueness, without any i-th order vagueness for any $i > n$). (For arguments for the thesis that higher-order vagueness may terminate at some finite level, see [Burgess, 1990] and [Dorr, 2010]). Often, the thesis seems to be put forward in a more radical version though, to the effect that every instance of vagueness gives rise to non-terminating higher-order vagueness (see esp., [Russell, 1923, pp. 63–4], [Dummett, 1959, p. 182], and [Dummett, 1975, p. 108]). Even though the thesis that there is higher-order vagueness is often presented as something like a datum to be accommodated by any satisfactory theory of vagueness, it may be questioned whether there is evidence for higher-order vagueness that is as strong as the available evidence for vagueness. In what follows, some noteworthy statements and arguments for and against the thesis are mentioned.

3.2 Some Arguments For and Against the Hypothesis

In view of its wide acceptance, it seems no surprise that there have not been many attempts to give a non-question-begging argument in favour of the thesis of higher-order vagueness. Special mention should go to the argument that is due to Sorensen and Hyde. Sorensen ([Sorensen, 1985]) gives an argument to the effect that 'vague' is itself vague. Hyde ([Hyde, 1994]) makes use of this result for an argument for the conclusion that some vague predicates must be higher-order vague. The soundness of the argument has been questioned.[32] Even granted that the Sorensen–Hyde argument is sound though, as Varzi ([Varzi, 2003]) argues, Hyde's subargument is rejectable as question-begging; for in making use of Sorensen's subargument, it already presupposes that there are borderline cases of borderline cases for some predicates.

A natural rationale for the idea of non-terminating higher-order vagueness may be the impression that genuine instances of the sorites paradox are persistent in the sense that they are not resolvable in terms of higher-order distinctions. Even for definite walking distances (definite failures of being a walking distance, or borderline cases), one may run a sorites paradox, and the paradox will equally reemerge for expressions of even higher orders – or so one may argue. Although on the face of it this reasoning may be compelling, it seems that it leaves room for reasonable doubt. To wit, it seems questionable whether there is evidence for the soriticality of higher-order terms such as 'is definitely a walking distance' or 'is a borderline case of a walking distance'. For one, as far as pre-theoretical usages of such expressions are concerned, it seems that nested occurrences of the form 'it is borderline whether it is a borderline case' or 'it is borderline whether it is definitely' are rather outlandish. For another, in the absence of strong pre-theoretical evidence for higher-order vagueness, one may argue that there is no theoretical need for adopting the assumption of higher-order vagueness even hypothetically – insofar as a perfectly precise theoretical notion of 'borderline case' may supply sufficient means for an account of first-order vagueness. For example, Koons ([Koons, 1994]) submits that all linguistic vagueness expresses at the level of first-order vagueness of expressions that make up languages. According to his account, there is no need for introducing further indeterminacy by blurring the boundary between predications with a definite truth value and those with an indefinite truth value. (For similar considerations to the effect that there is no need for a hypothesis of higher-order vagueness, see [Sainsbury, 1991, p. 178] and [Wright, 2010, Section 8]).

Wright takes an even more radical line in [Wright, 1987] and [Wright, 1992] when advancing an argument that is supposed to pose a threat to the idea that the assumption of higher-order vagueness is consistent. Specifically, (following [Fara, 2003, p. 200]) his argument may be reconstructed as hinging on two principles governing a $\mathcal{D}$-operator for definite truth, to wit

$\mathcal{D}$–Intro: If $\Gamma \vdash P$, then $\Gamma \vdash \mathcal{D}P$

and the second-order gap principle

Gap 2nd order: $(\forall n \in \{0, \ldots, i-1\})(\mathcal{D}^2 Fx_n \rightarrow \neg\mathcal{D}\neg\mathcal{D}Fx_{n+1})$.

Starting from these principles, one can derive the following sorites sentence for 'definitely F' for any sorites series of F: for all x, if the immediate successor of x (in the series) definitely is not definitely F, then x is definitely not definitely F as well. By repeated appeal to this sentence, for instance it follows for a sorites series of 'small' (where items increase in height within the series) that also the first member of the series, which may be, say just two foot in height, is definitely not definitely small. Wright's argument essentially rests on the application of ($\mathcal{D}$–INTRO) in subproofs. Edgington ([Edgington, 1993]) and Heck ([Heck, 1993]) note that these applications are not unproblematic and in fact invalid on natural interpretations of entailment and $\mathcal{D}$ that would validate ($\mathcal{D}$–INTRO).[33]

A different argument, by Fara ([Fara, 2003]), highlights a problem with accommodating the idea of non-terminating higher-order vagueness consistently for any finite sorites series, assuming merely modus ponens, ($\mathcal{D}$–INTRO) and a generalization of (GP) for k iterations of $\mathcal{D}$ (where k is arbitrarily high)

GAP GENERALISED (GP–GEN): $(\forall n \in \{0, \ldots, i-1\})(\mathcal{D}^{k+1}Fx_n \rightarrow \neg\mathcal{D}\neg\mathcal{D}^k Fx_{n+1})$,

This argument seems to have more force, for one may provide an account of definite truth and of entailment in support of all relevant provisos. Wright ([Wright, 2010, Section 5]) interprets the argument as a challenge to the consistency claim for the assumption of higher-order vagueness. Fara, by contrast, taking it that there is higher-order vagueness, directs her argument against the supervaluationist account of definite truth and of entailment, which supports all relevant provisos (in a standard framework of supervaluationism, ($\mathcal{D}$–INTRO) is valid, and (GP–GEN) may be considered as a natural prerequisite for accommodating non-terminating higher-order vagueness). (For further details, see Section 5.3).

This short synopsis may do for highlighting the need for further argument on either side of the spectrum of opinions. In view of reasonably defensible doubts, it does not seem fair to treat higher-order vagueness as an accepted matter of fact. But in the absence of a compelling proof of inconsistency, evidence against the thesis of higher-order vagueness in the form of no-need arguments may be undermined or even rebutted by evidence to the contrary.

4. Classical Frameworks for Vagueness

One way of interpreting the sorites paradox is to say that it tells us something about the logic of natural languages. According to this, we need to reconsider some principles in play in soritical reasoning. This thesis has been put more specifically and in various ways by advocates of non-classical frameworks for vagueness (see Section 5). Proponents of classical first-order logic for vagueness give a different diagnosis of the problem revealed by the paradox. According to this, the paradox tells us only something about common sense constraints governing many general terms in natural languages. Standardly, adherents to this approach do not reject the (CC) constraint, but the (UT) constraint (see (SOR), in Section 1.1). Starting from classical logic, assuming (CC), it follows that some instances of (TOL) pertaining to adjacent members in a sorites series must be false – that is, some such pair must mark a cut-off point between true and false applications. Prima facie, this way of resolving the sorites paradox seems to be merely a make-shift solution, insofar as in effect, it seems to generate a new paradox: if we have to accept the clear-case constraint involved (a zero-foot distance is a walking distance, whereas a 1,000-miles distance is not) and to deny some instances of (TOL) pertaining to adjacent objects in a sorites series (not every walking distance between zero feet and 1,000 miles is still a walking distance, if incremented by one foot), then in every sorites series, there is a pair of adjacent members in the series that marks a cut-off point (there is a number of feet that still makes for a walking distance, and where one foot more makes for failing to be walking distance), or so one may argue. One may consider this concern as one of the most serious threats (if not the most serious one) to the generic idea that vagueness can be adequately modelled in a classical framework.

This section gives a survey of the most prominent (previous) contenders in this camp, beginning with the epistemicist account of borderline vagueness (Section 4.1), and suggestions of reinterpreting it in semantic terms (Section 4.2). Moreover, some contextualist approaches to soriticality are set out (Section 4.3). As a disclaimer, we mention here Orłowska's classical modal framework (in [Orłowska, 1985]), which applies Pawlak's theory of 'rough sets' (developed more systematically in [Pawlak, 1991]) to vagueness. While her framework has interesting features from a formal semantic point of view, it is not discussed here, not least for lack of space.

4.1 Epistemicism

Epistemicism is the name of the type of view that combines a classical framework for vagueness with an epistemic view of borderline vagueness (see Section 2.2). According to this, in borderline cases, the predication does have a truth value, which we are just ignorant of. Epistemicism seems to go back as far as ancient philosophy.[34] More recent advocates of this approach are Cargile ([Cargile, 1969]), Campbell ([Campbell, 1974]), Sorensen ([Sorensen, 1988], [Sorensen, 2001]), Horwich ([Horwich, 2000]), and in particular, Williamson (esp., [Williamson, 1994]), who will be focused on here; for his theory of vagueness represents the (to date) most elaborate and serious candidate of the epistemicists.

Williamson suggests modelling vagueness in terms of a modal operator $\mathcal{D}$ for 'definite truth', which has the intended sense of 'clarity' (see [Williamson,

1994, pp. 270–5]).[35] Formally, for a language of propositional logic[36] containing $\mathcal{D}$, models M are quadruples $\langle W, d, \alpha, v\rangle$, where W is a non-empty set (of 'worlds'), d is a metric on W (that is, d is a symmetric function mapping $W \times W$ to non-negative reals such that $d(w_1, w_2) = 0$ iff $w_1 = w_2$ and $d(w_1, w_2) + d(w_2, w_3) \leq d(w_1, w_3)$), α is a non-negative real number, and v is a mapping of atomic sentences to subsets of W. The relation $w \models_M \varphi$, reading 'φ is true in a world w in a model M', is then defined the standard inductive way for the language of propositional logic:

1. $w \models_M P$ iff $w \in v(P)$ (for any atomic sentence P).
2. $w \models_M \neg\varphi$ iff $w \not\models_M \varphi$.
3. $w \models_M \varphi \wedge \psi$ iff $w \models_M \varphi$ and $w \models_M \psi$.

The interesting valuation rule is that for $\mathcal{D}$. Williamson considers two types of models; for one, a *fixed margin model*, where the relevant clause is

4. $w \models_M \mathcal{D}(\varphi)$ iff $(\forall w' \in W)(d(w, w') \leq \alpha \rightarrow w' \models_M \varphi)$.

For another, he considers a *variable margin model*, with the clause

4′. $w \models_M \mathcal{D}(\varphi)$ iff $(\exists \delta > \alpha)(\forall w' \in W)(d(w, w') \leq \delta \rightarrow w' \models_M \varphi)$.

In either type of model, a formula is *valid* if and only if it is true at every world in every model. Fixed margin models can be thought of as standard possible worlds models with $\mathcal{D}$ in place of the necessity operator $\Box$, where a world x is accessible from a world w just in case $d(w, x) < a$. The definition of a metric implies accessibility to be symmetric and reflexive, and conversely, any reflexive symmetric relation R on W is representable by a metric d on W (where for some α, xRy iff $d(x, y) \leq \alpha$);[37] validity in fixed margin models amounts hence to validity in reflexive symmetric models. That is, we end up with the Brouwersche modal logic KTB, which can be axiomatised by the set of tautologies, the modus ponens inference rule, and

(RN) If $\vdash \varphi$ then $\vdash \mathcal{D}\varphi$.
(K) $\vdash \mathcal{D}(\varphi \rightarrow \psi) \rightarrow (\mathcal{D}\varphi \rightarrow \mathcal{D}\psi)$.
(T) $\vdash \mathcal{D}\varphi \rightarrow \varphi$.
(B) $\vdash \neg\varphi \rightarrow \mathcal{D}\neg\mathcal{D}\varphi$.[38]

The comparison between variable margin models and possible worlds models is less straightforward, since the former use rather a family of accessibility relations (one for each $\delta > \alpha$) instead of a single one. But indeed, also here, a correspondence result is provable to the effect that validity in variable margin models amounts to validity in possible world models that are reflexive, that is validity in the modal system KT, which is obtainable from the axiomatisation of KTB by dropping the Brouwersche axiom (B).[39] Both types of model make room for higher-order vagueness. Specifically, on either type of model, for any formula φ, $\varphi \rightarrow \mathcal{D}\varphi$ is valid if and only if φ or its negation is valid – that is, any formula that is logically contingent permits for a margin in which it is true but not clearly true.[40] Unlike the other mentioned axioms involving $\mathcal{D}$, the axiom B seems to have no prima facie intuitive force. However, on an epistemic interpretation of accessibility as indiscriminability, one may suggest (as Williamson [Williamson, 1999, p. 130] does) that it is symmetric. The same interpretation may also be seen as an argument for the intransitivity of accessibility, and hence for the failure of the KK principle for definite truth (i.e., the principle $\vdash \mathcal{D}\varphi \rightarrow \mathcal{D}\mathcal{D}\varphi$).[41] On another note on symmetry, unlike validity in variable margin models (KT), validity in fixed margin models (KTB) is powerful enough to ensure higher-order vagueness of any finite order, given second-order vagueness for sentences (see [Williamson, 1999, p. 136]).[42]

The intuitive rationale for Williamson's margin models may be illustrated as follows. Consider a scalp with 120,000 hairs. To know that 120,000 is the number of hairs on the scalp, we would need to be able to notice any change in the number of hairs on the scalp, however small it may be. The discriminatory capacities of human epistemic subjects with regard to numbers of hairs, however, are only limited, insofar as estimates are gained on the mere basis of looking at a scalp (without counting its number of hairs): differences in number of hairs below some margin of error are not distinguishable. Thus one may illustrate the idea of inexact knowledge by margin for errors. Williamson's basic idea is to think of borderline vagueness as a special case of inexact knowledge by margin for errors. Consider a vague sentence of the form '*k* hairs make for baldness', henceforth abbreviated as '$B(k)$'. Williamson suggests that its vagueness can be accounted for as a case of inexact knowledge on the part of ordinary speakers regarding its truth conditions. According to this, as far as vague expressions are concerned, ordinary speakers are able to notice changes in their truth conditions only if they are 'big enough'. This suggests a corresponding margin for error for definite truth: for instance, whereas the margin for error relevant to knowledge of number of hairs by mere observation may be specified as the greatest indiscriminable difference in number of hairs, the margin for error relevant to definite truth for applications of 'B' may be specified as the greatest indiscriminable distance in the threshold for B.[43]

More precisely, consider for example, a fixed margin model $M = \langle W, d, \alpha, v\rangle$, where

(i) $W = \{w_n : n \in \mathbb{N} \wedge 1 \leq n\}$
(ii) $w_i \models_M B(n)$ iff $n < i$

(iii) $w_i R w_j$ iff $|i - j| \leq 1$
(iv) $w_i \models_M \mathcal{D}\varphi$ iff $(\forall w_j)(w_i R w_j \rightarrow w_j \models_M \varphi)$.[44]

Clause (ii) says that the cut-off for B occurs between 0 and 1 at w_1, shifting by one hair upwards at each successive world in the series; clause (iii) says that the distance between worlds is taken to be the difference between the respective thresholds for B, with any pair of worlds whose thresholds for B differ by at most 1 being accessible from each other; clause (iv) expresses Williamson's idea that definite truth is is characterized by a margin of error principle pertaining to indiscriminable interpretations of the language. This model satisfies (for every world) also another kind of margin for error principle, pertaining to objects with indiscriminable features relevant to B-ness:

$$(\forall n)(\mathcal{D}B(n) \rightarrow B(n+1)).$$

That is, provided that the strongest indifference relation for B (with respect to the relevant domain) comes to an absolute difference of at most 1, from this margin for error constraint, it follows that any (GP) principle (Section 2.3) for B of the form

$$(\forall n)(\mathcal{D}B(n) \rightarrow \neg\mathcal{D}\neg B(n+1)).$$

is true for every world. In fact, as noted (Section 2.3), it seems reasonable to assume that a predicate is soritical only if it satisfies an associated gap principle. Assuming that soriticality does not stop at the first level but reemerges for definitisations of B of any finite order, it would be hence desirable to have also support for the generalized principle (GP–GEN) (Section 3.2), in the form of:

$$(\forall n)(\mathcal{D}^{i+1}B(n) \rightarrow \neg\mathcal{D}\neg\mathcal{D}^{i}B(n+1)).$$

However, there is a general problem with accommodating this constraint on either mentioned type of margin models, insofar as vague predicates involve applications that are *absolutely true*, that is, definitely^n true for any n. Consider for example, it may be seen as hardly controvertible that $B(0)$ is definitely^n true for any n. Assuming $B(k)$ is absolutely true at a world w in our model, however, it can be shown that for some sufficiently large i, for some n, $\mathcal{D}(\mathcal{D}^{i}B(n) \wedge \neg\mathcal{D}^{i}B(n+1))$ is true at w; which implies that (GP–GEN) for B is false. Generalizing a result by [Gómez-Torrente, 2002],[45] Fara ([Fara, 2002]) shows that (GP–GEN) fails for any sorites series, for every fixed margin model where the margin is positive; and furthermore, that the same type of problem arises for a distinguished class of variable margin models as well. The options that offer an escape route for either model seem to be either (a) to deny that the higher-order predicate 'is definitely$^n B'$ is soritical for every n, or (b) to deny that some applications of B are absolutely true.[46] Indeed, as Fara shows in another generalization step, the problem reemerges even if we allow margins for error to be arbitrarily small, leaving no serious escape routes other than (a) and (b).[47,48]

Even if one of these options is viable and margin models supply sufficient means of accommodating the (GP–GEN) principle, whenever it is appropriate, there is still reason for doubting that they provide a satisfactory framework for describing soriticality. Specifically, as they stand, the given models leave two crucial problems unaddressed. To formulate the problems, it is not even necessary to take into acccount the possibility of higher-order vagueness; we can stick to first-order vagueness:

(1) B is obviously soritical, and (as shown) the principle (GP) can be accommodated in an appropriate margin model for B (in the sense that it is true in every world in the model). It is easy to see that from this, it follows that any sentence that marks a 'sharp' cut-off, of the form

$$B(i) \wedge \neg B(i+1),$$

is borderline vague, if true.[49] Assuming that definite truth describes a necessary condition for being known, it follows that any true statement that marks a sharp cut-off is 'unknowable', in the sense that it fails to meet a certain necessary condition for being known. But this result alone cannot serve as an explanation for the observed fact that it is odd to agree to any sentences of this type (Section 1.1), for this account strategy would overgenerate. To wit, it would also predict that also that it is odd to agree to any negation of sentences of the said type[50] – which are classically equivalent to instances of (TOL) pertaining to adjacent members in a sorites series for B, that is sentences that *are* compelling:

$$B(i) \rightarrow B(i+1).$$

Hence, more is required, to account for the noted asymmetry between sentences that mark a cut-off point between two adjacent members in a sorites series and associated instances of (TOL).[51]

(2) It seems equally odd to agree to the existential assumption of any cut-off for any soritical predicate. On the given margin for error approach, however, (since worlds in models are associated with classical interpretations, which imply the existence of a sharp cut-off), existential assumptions of this form are definitely true – that is, on the suggested interpretation of margin models, they fulfil a necessary condition for being known to be true. Needless to say that this calls for further explanation of the contravening common sense impression.[52,53]

A possible way of confronting problem (1) in terms of margin models is offered in [Williamson, 1994, pp. 244–7]. The basic idea is that reasonable belief requires a sufficiently high subjective probability conditional on what is known. Assuming, for simplicity, that the subject knows that its situation is within the margin for error δ of its world w, the probability of a belief conditional on what is known may be thought of as the proportion of worlds within δ of s in which the belief is true. A sufficiently high probability accordingly may be informally thought of as truth in most worlds within δ of w.[54]

For example, suppose the relevant epistemically possible worlds are those in which the cut-off points for 'heap' vary, with w_k being the world in which k is the least number of grains that make a heap. Suppose w_k is the world of our subject, and that the worlds within the appropriate margin for error of w_k are five worlds, $w_{k-2}, \ldots, w_{k+2}$. Suppose the required threshold for reasonable belief is truth in at least four epistemically possible worlds. It is then easy to see that for no n is it reasonable to believe 'n grains make a heap, but $n-1$ grains do not' (note: for $n \leq (k-2)$ and $n \geq (k+3)$, this belief is true at no world within the margin, and for any other n, this belief is only true at one world within the margin). On the other hand, by parity of reasoning, it follows that for any n, it is reasonable to believe the associated instance of (TOL), 'if n grains make a heap, then so do $n-1$ grains' (note: for $n \leq (k-2)$ and $n \geq (k+3)$, this belief is true at all worlds within the margin, and for any other n, this belief is true at four worlds within the margin).

More complex versions of this explanation strategy may cope with more complex cases. However, it is easy to see that this strategy is of no avail with regard to problem (2). To wit, since for all epistemically possible worlds within the margin, the existential assumption 'there is an n such that n grains make a heap, but $n-1$ grains fail to be a heap' is true, it is hence also true at most worlds, and hence, on the suggested account, reasonably believable.

It may be suggested that people are inclined to accept statements of the form $(\forall x)\varphi(x)$ if φ is true of 'almost all' instances of x. But this account would again overgenerate, considering the example (from [Halpern, 2008, p. 541]) 'for all worlds w, if there is more than one grain of sand in the pile in w, then there is still one grain of sand after removing one grain of sand' for a case where there might be up to 1,000,000 grains in the pile, and where it is yet not to be ruled out that it consists of only one grain. Even though, given what is known, the universally closed sentence is true in almost all instances, its universal closure does not seem compelling at all, for it is clear that the possible case where the pile consists of only one grain is a counterinstance. Just to reply that in the given example, the relevant complex predicate $\varphi(x)$ is perfectly precise in extension and to qualify the suggested account as intended only for genuinely vague predicates may render adequate results, but would yet owe an explanation of why people deal with universal quantification involving vague predicates in a different way.

Alternatively, it may be suggested that people are inclined to accept $(\forall x)\varphi(x)$ if they are inclined to accept the statement $\varphi(x)$ for each instance of x (e.g., compare [Fara, 2000, p. 59]). But this account would overgenerate as well, as the following instance of the Lottery paradox shows: Let $\langle c_0, \ldots, c_{1,000,000} \rangle$ be a sequence of collections of lottery tickets, where we know that c_0 is the collection of all tickets, and that for every $0 < i \leq 1,000,000$, c_i is obtained from c_{i-1} by drawing one ticket out of c_{i-1}, without knowing for any $0 < i \leq 1,000,000$ whether c_i was obtained by drawing the winning ticket from c_{i-1}. For any $1 \leq n \leq 999,999$, 'Wc_n' reads 'collection c_n contains the winning ticket'. Then, for each $0 \leq n \leq 999,999$, the corresponding sentence of the form $Wc_n \rightarrow Wc_{n+1}$, as individually taken, is compelling; for considering the large number of drawings, it is extremely unlikely that the (n+1)th draw happened to be the very draw that picked the winning ticket. On the other hand, it is certain that the associated universal sentence, $(\forall n \in \{0, \ldots, 999,999\})(Wc_n \rightarrow Wc_{n+1})$, is false; for it is certain that at some point in the series of successive drawings, the winning ticket must have been picked.[55] Again, it should be clear that it would be wanting just to restrict the account strategy to genuinely vague predicates. Since these considerations do not hinge on any philosophical interpretation of classical probability, it highlights a general problem with classical probabilistic accounts of the sorites paradox.[56]

The further philosophical discussion of epistemicism is vast and can be only mentioned in passing here. For one, some authors target the underlying idea that knowledge is in general subject to a margin for error (e.g., see Chapter 18 in this volume), or the suggestion that speakers may have only inexact knowledge regarding the factual semantic features of the language they competently use; it has also been argued that epistemicism lacks any support in the form of a substantive account of how sharp cut-offs may emerge, or that Williamson's version of epistemicism owes an account of what makes the semantic features of vague expressions more easily susceptible to change than those of precise expressions (e.g., see [Tye, 1997], [Schiffer, 1999], [Burgess, 2001], [Wright, 2001], [Jackson, 2002], and [Heck, 2003]).

4.2 Vagueness as a Semantic Modality

Instead of combining a classical logic for vagueness with an *epistemic* view of borderline vagueness, one may combine it with a *semantic* view (see Section 2.2). This approach is sometimes referred to as a non-standard version of 'supervaluationism'[57], or alternatively, as 'pragmatism'[58] or 'plurivaluationism'[59]. The standard variant of this approach is, from a logical point of view, no different from Williamson's epistemic approach. That is, definite truth may be thought of as a notion that may be modelled like a necessity operator in normal modal logics. Standard possible worlds models are, however, not thought of as spaces of epistemically possible worlds endowed with an indiscriminability relation, but rather as spaces of 'interpretations', endowed with an 'admissibility' relation (for semantic frameworks in this spirit, see esp., [Varzi, 2007] and [Asher et al., 2010]; see also [Lewis, 1970a], [Lewis, 1975], [Przełecki, 1976], [Burns, 1991], [Eklund, 2010]. For critical discussion, see [Keefe, 2000, Chapter 6][60] and [Smith, 2008, pp. 98–133 and 197–200]). The underlying idea is that there is no unique interpretation for a language involving vagueness that may be referred to as a 'the one and only admissible' interpretation of the language. Rather, we can at best only speak of a class of 'admissible' interpretations. If vagueness stops at the first level, this idea can be accommodated by an equivalence relation of accessibility (i.e., a relation that is reflexive, symmetric, and transitive). Given second-order vagueness, the notion of 'admissibility' is to be treated as vague as well, and hence as admitting of more than one interpretation, and so on. A way of accommodating higher-order vagueness is the adoption of a reflexive, symmetric, but intransitive accessibility relation (which may be interpreted as 'being about as admissible as') (for discussion of various philosophical interpretations of 'admissibility' that accord with the semantic view of borderline vagueness, see [Varzi, 2007, Section 1]). A more informative and rigorous account of accessibility in the intended semantic sense, which might offer a serious alternative to the epistemicist margin for error account, is a desideratum for further investigation. The non-standard supervaluationist view of borderline vagueness may be of philosophical interest in its own right. It remains to be seen though whether it opens up any genuinely new perspectives on the paradox of vagueness.

4.3 Contextualism and Connectedness

Most accounts (such as epistemicism and the more common proposals that adopt a non-classical framework for vagueness) seem to take the 'connectedness' constraint (R–CON) (along with (CC)) for any sorites series for granted. The paradox is accordingly supposed to reveal a problem with the assumption (R–TOL), saying that the indifference relation in play in the sorites series is a tolerance relation. There is still another way of saving soritical predicates from contradiction, which has been explored in some contextualist frameworks for vagueness. Advocates of this approach argue that, similarly to the case of indexicals such as 'I' or 'today', the extension of vague general terms (such as 'tall') may vary with contexts of use – more specifically, it is suggested that the standards for true applications (such as a threshold for 'tall') may vary with contexts (e.g., see [Lewis, 1979], [Kamp, 1981], [Bosch, 1983], [Pinkal, 1983], [Pinkal, 1995], [Burns, 1991], [Tappenden, 1993], [Raffman, 1994] [Raffman, 1996], [van Deemter, 1996], [Soames, 1999], [Fara, 2000], [Shapiro, 2006], [Halpern, 2008], [Gaifman, 2010]). A popular rationale for a contextualism about vagueness is the idea that each instance of (TOL) pertaining to a pair of adjacent members in a sorites series may be rendered as true in contexts where it is under consideration. (For a defence of this idea, see esp. [Raffman, 1994]).[61] On the other hand, it is often suggested that there is no context at which all such instances of (TOL) are true. But there are ways of salvaging all such instances of (TOL) in a contextualist framework. According to this, we can trust our impression and say that indifference relations *are* tolerant, yet have to reconsider the associated impression that an indifference relation may provide a path connecting a clearly true application and a clearly false application (for the relevant predicate) – that is, the 'connectedness' constraint (R–CON) is in effect rejected. This kind of approach may be underpinned with different accounts of indifference, and it may be implemented in different logical frameworks. In what follows, classical frameworks will be concerned.

More generally, the case against (R–CON) may be put as a case against the condition:

R–CONNECTEDNESS′ (R–CON′): The domain S with respect to which predications are made is R-connected, that is, there is no partition of S into two non-empty subsets $\langle S_1, S_2 \rangle$ such that we have for the restriction of R to S, $R \mid S$, either $(R \mid S) \subseteq S_1 \times S_1$ or $(R \mid S) \subseteq S_2 \times S_2$ (i.e., however we split up S into two non-empty disjoint and jointly exhaustive subsets S_1 and S_2, R always applies to some pair k, l of members of S where $k \in S_1$ and $l \in S_2$).

For any sorites series for a predicate F, where S is the class of all members of the series, (R–CON′) follows from the associated instance of (R–CON). That is, to the extent to which (R–CON′) can be challenged, the paradox may be contained in scope or even fully resolved. Considering this, the following contextualist idea may suggest itself (compare [van Rooij, 2009], [Gómez-Torrente, 2010], [Pagin, 2010], and [Gaifman, 2010]): The domain with respect to which we evaluate vague predications varies with contexts; in particular, in 'normal' contexts, where we are not faced with the paradox, we consider only proper subsets of a domain of objects D (that is, in effect, predicates are analysed as relations that apply to pairs of individuals and contexts). This makes room for the idea that the domain may be so coarse-grained that for no indifference relation R for the relevant predicate F, with respect to the domain, does (R–CON′) hold; specifically, for any such R, there will be a partition of the relevant class of objects D^* into a subclass of Fs and a subclass of non-Fs, where there is no $x \in D^*$ that is an F and R-related to some non-F.[62] As a result, the assumption of (R–TOL) for any indifference relation R becomes safe.

On the other hand, as far as other contexts are concerned where the relevant domain is bigger, indifference relations R (with respect to that domain) may fail to be tolerant, by (R–CON′). For example, suppose we are in a context where only a restricted class of people is relevant, $i_1, \ldots, i_6$, say the people in the room we are in. If the number of people is sufficiently small, there is no sorites series for 'smallness'. For instance, suppose we are in a context where 'small'-predications are indifferent with respect to differences in height below 0.15 foot, that only heights below 5 foot make for 'smallness', and that $i_1, \ldots, i_6$ have the heights, 4.75, 4.85, 4.95, 6.25, 6.35, and 6.45 feet, respectively. In this case, any indifference relation for 'small' (with respect to the given class) is in fact a tolerance relation (with respect to that class): for any indifference relation for 'small' with respect to the said class will apply to all pairs of the form $\langle i_n, i_{n+1} \rangle$ (for $1 \leq n \leq 5$) except for $\langle i_3, i_4 \rangle$, and 'small'-predications are tolerant exactly with respect to these pairs. E.g., if we just add a further person, j, who is 5.05 feet in height to the class of relevant people, assuming that the standards for 'small' and the threshold for indifference are not affected thereby, any indifference relation for 'small' with respect to the expanded class violates the tolerance instance 'if i_3 is small, so is j'. (For classical frameworks in this spirit, see [van Rooij, 2009] and [Pagin, 2010]). It seems that contexts in which we consider genuine instances of the paradox are the very kind of context where the relevant space of objects is fine-grained enough to ensure that the relevant instance of (R–CON′) holds; for, otherwise, (R–CON) would have no intuitive force. That is, in effect, the proposal to consider less fine-grained domains may provide an effective strategy of avoiding the paradox, but for sure, it does not supply means of resolving it effectively.

On a different kind of approach, which targets assumptions of the form (R–CON) in general, it has been suggested that the paradox rests on an equivocational fallacy. Specifically, the impression that drives the paradox is that there is one, *dyadic* relation of indifference R (for a given predicate) that gives rise to contradiction; for, so is the impression, in instances of the paradox, it both satisfies a tolerance principle (for the relevant predicate) and allows for the construction of an R-path, beginning with a clear truth and ending with a clear falsity. Contrary to this impression, one may argue that in fact, indifference is to be analysed as a *ternary* relation, which applies to pairs of objects relative to contexts, which validates the relevant tolerance constraint, but violates the relevant connectedness constraint for every context. That is, so the suggestion goes, we are in fact safe from contradiction, and the impression to the contrary rests on the fact that in giving an account of the paradox in the way of (SOR), we in fact equivocate between different dyadic relations of indifference, which relate to different contexts. This idea can been cashed out in different ways.

Van Deemter ([van Deemter, 1996]) interprets indifference (with respect to a vague predicate) as indiscriminability (or, in his terminology, as 'indistinguishability') (in certain respects relevant to the predicate) relative to a comparison class. The idea that indiscriminability is relative to comparison classes goes back to Russell ([Russell, 1926]) and has been explored systematically in [Luce, 1956] and [Goodman, 1966]. An object i may be indiscriminable from another object j, if we compare the two objects with each other, without taking other objects into consideration, and the same for j and another object k, even though this might not hold of i and k. On the basis of considerations like this one, one may argue that *direct* discriminability is not transitive.[63] Not so for the corresponding *indirect* notion of discriminability, which depends essentially on what other objects may be taken into account in discriminating objects from each other: according to this i and j are indirectly indiscriminable (relative to a comparison class c) just in case (i) i and j are not directly discriminable, and (ii) there is no $k \in c$ such that either i is directly discriminable from k, whereas j is not, or j is discriminable from k, whereas i is not.[64] For the limiting case that the comparison class does not contain any elements other than the respective pair of objects to be compared, indirect indiscriminability collapses with the direct counterpart notion. It is a well-known fact that indirect indiscriminability is transitive.[65] As van Deemter notes, this feature may be exploited for blocking the sorites paradox. Specifically, he distinguishes between two ways of disambiguating (R–TOL) in terms of a dyadic predicate F (applying to individuals relative to comparison classes of individuals) and a ternary relation R_F^* of indirect indiscriminability for F (applying to pairs of individuals relative to comparison classes), which may be put in a more simplified way as follows:

R–TOLERANCE1 (R–TOL$_1$): $(\forall i, j \in D)(\forall c \in C)(R_F^*(i, j, \{i, j\}) \rightarrow (F(i, c) \rightarrow F(j, c)))$,

R–TOLERANCE2 (R–TOL$_2$): $(\forall i, j \in D)(\forall c \in C)(R_F^*(i, j, c) \rightarrow (F(i, c) \rightarrow F(j, c)))$,

where D is a non-empty domain of objects and C is a non-empty set of subsets of D (which may be but need not be the powerset of D).[66] (R–TOL$_2$) essentially differs from (R–TOL$_1$) in that it makes use of an indirect notion of indiscriminability, whereas in effect, (R–TOL$_1$) makes use of direct indiscriminability, R_F (i.e., $R_F(x, y)$ iff $R_F^*(x, y, \{x, y\})$). Assuming that (a) there are no constraints on comparison classes, and that (b) the pairs of adjacent members in the sorites series s (for a vague predicate F) are each directly indiscriminable (with respect to F), it follows that there is an R_F-path connecting a true and a false application case of F in D. In this case, (R–TOL$_1$) gives rise to contradiction. Yet, (R–TOL$_2$) can come to the rescue then: To wit, since the first and the last member in the series are directly discriminable (the first one is clearly F and the second one is clearly not F after all), there is a least initial segment of the sorites series, s^*, for which R_F fails to be transitive. As a consequence, there is also a least initial segment of the sorites series, s', where R_F^* fails to apply to some pair of adjacent members relative to the comparison class c, where c is the domain of all members of s. As a consequence, (R–CON) fails for our sorites series for F. By generalization, this strategy may be applied to any sorites series for any vague predicate. Or so one may argue.

Granted that under this interpretation of indifference (R–TOL) can be consistently sustained, and the assumption of (R–CON), it is yet questionable whether this interpretation captures the intended sense of (R–TOL) (which is in play in assessments of instances of the paradox). For the sorites paradox arises even in cases where we can perfectly discriminate all adjacent members of a series with respect to the features relevant to applications of our predicate – e.g., even with perfectly accurate information about distances, we may generate a sorites series for 'walking distance' with such distances. If 'indiscriminability with respect to a given predicate F' is understood otherwise, as related to the way we deal with objects in terms of F-ness, it seems that what is in play in the paradox is not the indirect but rather the direct notion of indiscriminability. However, this notion is of no use, since, as noted, it gives rise to contradiction.

Fara's 'interest-relative' account of vagueness, in [Fara, 2000], may be interpreted as a different way of saving tolerance in terms of a relation of indifference that is modelled as context-relative. Fara sets out her account for adjectives, which are typically associated with a dimension of variation (e.g., 'tall' is associated with height, 'hot' with temperature, etc.); as far as other types of general terms in natural language (such as nouns) are concerned, where it is harder to find such a dimension of variation, she suggests a generalization of her account on a case-by-case basis. Modelling adjectives as predicates in a regimented language of first-order logic, one can sketch the idea of her account by way of the following account schema

$F(a,c)$ is true iff $f_c^F(a) \;>_c^!\; norm_c(F)$,

where a ranges over elements of a domain, c ranges over contexts, F is associated with a scale, and: (i) f^F is a context-sensitive function that maps objects to degrees on the scale associated with F; (ii) $>^!$ is a context-sensitive relation of 'being significantly greater than', and (iii) *norm* is a context-senstitive function that maps predicates into degrees on the scale associated with the predicate. According to Fara, indifference with respect to a vague predicate F is a context-sensitive notion, which can be informally thought of as an relation of 'salient similarity', or of 'being the same for the present purposes', and which may be modelled as identity in the f_c^F-measures.[67,68] In particular, she suggests that every instance of (R–TOL) may be rendered true by the very act of considering it. As a further consequence of the given account of indifference, the following 'similarity constraint' is derivable

if $R_F(x,y,c)$ is true, then $F(x,c)$ is true just in case $F(y,c)$ is true.[69]

A fortiori, it follows that F is indeed tolerant with respect to the associated indifference relation R_F. To illustrate Fara's account, consider the following example of hers:[70] We are in an airport, and there are two suspicious-looking men I want to draw your attention to. You ask me, 'Are they tall?'. Since the men are not much over five feet eleven inches, there may be some leeway in choosing between 'yes' or 'no'. But if the men are pretty much the same height, the option of saying 'One of them is, the other isn't' is not available, because the similarity of their heights is 'so perceptually salient – and now that you've asked me whether they're tall, also conversationally somewhat salient'. In this case, I may not choose a standard for 'tall' that one meets but the other does not, or so she suggests. Is Fara's account of indifference relations safe from contradiction, if it implies that indifference is a tolerance relation? She submits (in [Fara, 2000, p. 75]) that there will be always a cut-off between *F*s and non-*F*s – which, if R_F is an indifference relation for *F*, entails that there will be never an R_F-path that connects an instance of *F*-ness with an instance of non-*F*-ness: according to this, the initial fragment of a sorites series for *F* that are saliently similar to the first member can never be stretched out to the end of the series.[71] As it stands, this account is only schematic insofar as the informal notions of 'salient similarity' or 'being the same for the present purposes' require further explication.[72] That said, there seems to be more than commonly thought to the idea that (R–TOL) may be salvaged – at the price of rejecting (R–CON).

5. Non-Classical Approaches to Vagueness

Starting from a classical framework for vagueness, the natural way of blocking soritical reasoning is to say that some instances of (TOL) pertaining to adjacent members in a sorites series are *false* – and hence to accept the statement that some pair of adjacent members marks a cut-off between true and false predications. The only common ground among adherents to some non-classical framework for vagueness seems to be that the classical account of the paradox is no option. However, there does not seem to be any agreement on where the classical account is supposed to go wrong. For example, some opponents to the classical account argue that the commitment to there being some false relevant instances of (TOL) is too strong: according to this, no relevant instance of (TOL) should be evaluated as false; on the other hand, some other opponents to a classical framework for vagueness argue that the said commitment is too weak: according to this, some instances of (TOL) should be evaluated as both false and true. Before going into some details, it may be helpful to give first some synopsis of some types of approaches to the paradox that have been implemented in different frameworks.

5.1 Paracompleteness and Paraconsistency

Roughly, the options that have received most attention in the philosophical literature may be subdivided into two types. For one, some authors have advocated so-called *paracomplete* logics for vagueness.[73] As far as applications to vagueness are concerned, the standard options of this type are *Strong Kleene logic* (K_3), Łukasiewicz' infinite valued logic ($Ł_{\aleph}$), and *supervaluationism* (SpV).[74] The characteristic feature of these logics is that they deny the so-called *implosion principle*, which says that for any sentences A and B (of the language of propositional logic), assuming B holds, either A or its negation holds. Formally, for any given multi-conclusion consequence relation $\models$, we say that it satisfies the implosion principle iff it has the property: $B \models \{A, \neg A\}$.[75] Accordingly, a consequence relation $\models$ is then said to be paracomplete iff it satisfies

$$B \not\models \{A, \neg A\}.$$

Some provisos that are standardly taken on paracomplete approaches to vagueness allow us to reformulate the implosion principle in a catchier way. Assuming that (i) logical consequence is modelled in terms of preservation of truth, and (ii) that truth of a negation is equated with falsity, the implosion principle says: if there are truths, then there are no truth-value gaps – in this sense, if truth-value gaps implode anywhere, then they implode everywhere. Accordingly, a logic is paracomplete iff it allows for non-trivial truth-value gaps. Standardly, proponents of a paracomplete approach to vagueness postulate that borderline cases are truth-value gaps. On the standardly discussed paracomplete frameworks for vagueness, it follows that if a sorites series involves truth-value gaps, some instances of (TOL) are gappy as well, though no instance is false.[76] In this sense, it is suggested that one can reject some instances of (TOL) without being committed to their negation. In effect, this kind of approach offers a way of blocking all standard forms of instances of the paradox as unsound.

Another prominent type of frameworks that have been adopted for vagueness fall into the group of so-called *paraconsistent* logics.[77] The standard options for vagueness here are Priest's Logic of Paradox (LP) and subvaluationism (SbV). The characteristic feature of these logics is that they deny the so-called *explosion principle* (i.e., the dual to the implosion principle), which is also known as *ex falso quodlibet* principle. This principle says that for any sentence A and B (of the language of propositional logic), assuming both A and its negation, it follows that B holds. Formally, for any given (multi-premise) consequence relation $\models$, we say that it satisfies the explosion principle iff it has the property: $\{A, \neg A\} \models B$. A consequence relation $\models$ is accordingly said to be paraconsistent iff it satisfies

$$\{A, \neg A\} \not\models B.$$

Again, some provisos that are standardly taken for granted allow us to give the principle a more intelligible interpretation. Assuming (i) that logical consequence is modelled in terms of preservation of a lack of simple falsity, and (ii) that any sentence A is both true and false just in case both A and its negation lack simple falsity, the explosion principle says: if there are truth-value gluts, they are everywhere – in this sense, if truth-value gluts explode anywhere, they explode everywhere. Accordingly, a logic is paraconsistent iff it allows for non-trivial truth-value gluts. Paraconsistent accounts of vagueness standardly postulate that borderline cases are truth-value gluts – definite truths and falsities are accordingly modelled as cases of simple truth and simple falsity respectively.

The paraconsistent strategy of resolving the paradox runs in one respect similarly to the strategy of paracomplete accounts: Some members in a sorites series are borderline cases, from which, on each of the said paraconsistent semantics, it follows that some relevant instances of (TOL) are to be borderline vague as well – with the further consequence that some premises in soritical reasoning are to be borderline vague as well, with the remaining premises being definitely true. But there is an important disanalogy: Since each instance of (TOL) is either simply true or glutty, no such instance is rejectable as untrue. That is, to be safe from contradiction, another escape route is called for. In fact, the paraconsistent notions of logical consequence that are standardly discussed for vagueness offer such an escape route, for they are weaker than the standard paracomplete alternatives: preservation of lack of simple falsity (or 'definite falsity') is a stronger constraint than preservation of truth ('or definite truth'). Since no premise in standard sorites reasoning is treated as simply false, even though the conclusion *is* simply false, it follows that soritical reasoning is not valid. Or so standard paraconsistent accounts of the paradox suggest.

*K*3, *LP*, and $Ł_{\aleph}$ may be distinguished from *SpV* and *SbV* in an important respect: *SpV* is only *weakly paracomplete*, in the sense that it is paracomplete but not furthermore satisfying

$$B \nvDash A \vee \neg A,$$

which says that there are non-trivial counterinstances to the classical

LAW OF EXCLUDED MIDDLE (LEM): $A \vee \neg A$.

*K*3 and $Ł_{\aleph}$, by contrast, are *strongly paracomplete* in the sense that they are paracomplete, but not only weakly paracomplete. Likewise, *SbV* is only *weakly paraconsistent*, in the sense that it is paraconsistent, but not furthermore satisfying

$$A \wedge \neg A \nvDash B.$$

LP, by contrast, is *strongly paraconsistent*, in the sense that it is paraconsistent, but not only weakly paraconsistent.[78] The distinction between strong and weak versions of paracompleteness and paraconsistency goes with an important distinction in the semantic frameworks for these logics. *K*3, *LP*, and

$Ł_{\aleph}$ are many-valued logics, in the technical sense of logics that are characterized by logical matrices, which generalize standard classical matrices for a wider range of semantic values. A common feature of these logics is that the semantics for logical connectives and quantifiers obeys the principle of truth-value functionality: that is, the truth value of a formula is a function of the truth value of its immediate components. In frameworks of *SpV* and *SbV*, by contrast, the principle of truth-value functionality is violated. For each type of approach, arguments have been advanced in the philosophical literature. As a disclaimer, the related controversy about whether there may be truth-value gaps (or gluts) will not be gone into here, since it concerns the theory of truth in general rather than the paradox of vagueness in particular.[79] At least, in view of the earlier mentioned pre-theoretical characterizations of borderline vagueness (Section 2.1), it seems unfair to dismiss gap or glut accounts of borderline vagueness as 'inadequate' at the outset: for, whereas truth-value gaps may seem a natural choice for modelling undecidedness in borderline cases, gluts may seem a rather natural choice for modelling divergence of usage in borderline cases.[80]

The discussion continues with applications of many-valued logics to vagueness (Section 5.2), then turning to applications of *SpV* and *SbV* (Section 5.3). Finally, another option for dealing with vagueness is mentioned (Section 5.4). For brevity, we will focus on languages of propositional logic.[81] To begin with (as for Section 5.2), also possible expressions of 'definite truth' in natural languages can be ignored. That is, we start with a standard language of propositional logic, $\mathcal{L}$, the syntax of which is given by $\langle \mathcal{A}, \mathcal{C}, \mathcal{S} \rangle$, where $\mathcal{A}$ is a set of atomic sentences, $\mathcal{C}$ the set of standard logical connectives $\{\neg, \wedge, \vee, \rightarrow\}$, and $\mathcal{S}$ is the smallest set of sentences that may be obtained inductively from $\mathcal{A}$ by means of members of $\mathcal{C}$. For short, the conditional version (CS–S) will be referred to as the 'standard form' of sorites reasoning.

5.2 Many-Valued Logics

The simplest way of defining a system of many-valued logic is to fix a characteristic logical matrix for its language.[82] A *logical matrix* for $\mathcal{L}$ is a structure $\langle \mathcal{V}, \mathbb{C}, \mathcal{D} \rangle$, where $\mathcal{V}$ is a set (of 'semantic values'), $\mathbb{C}$ is a set of operators on $\mathcal{V}$, $\mathcal{D}$ is a subset of $\mathcal{V}$ (of 'designated values'). In many-valued logics, all valuations have a common base. A valuation ν has base $\mathcal{B} = \langle \mathcal{V}, \mathbb{C}, \odot \rangle$ iff $\odot$ is a mapping $\mathcal{C} \rightarrow \mathbb{C}$, and ν is a mapping $\mathcal{S} \rightarrow \mathcal{V}$ such that for all connectives $\varphi \in \mathcal{C}$, for all sentences $P_1, \ldots, P_n \in \mathcal{S}$:

$$\nu(\varphi(P_1, \ldots, P_n)) = \odot_{\varphi}(\nu(P_1), \ldots, \nu(P_n)).$$

In words, the semantic value of logical compounds governed by a connective φ is a function of the semantic values of its immediate components, where the function is characteristic of φ. The set $\mathcal{D}$ of 'designated values' is invoked to define satisfaction: A sentence $P \in \mathcal{S}$ is satisfied by a valuation ν, in short $\models_\nu P$ iff $\nu(P) \in \mathcal{D}$. Correspondingly, the semantic notion of logical consequence is defined as follows: $\Gamma \models P$ iff for all valuations ν such that $\models_\nu A$ for all $A \in \Gamma$, $\models_\nu P$; in words, logical consequence is defined as preservation of a designated value. With this general setting in place, it is straightforward to introduce K_3, LP, and $Ł_{\aleph}$ as systems of many-valued logic.

5.2.1 K_3

The logical matrix for the strong Kleene system K_3 is $\langle \{1, 0, i\}, \{\odot_\neg, \odot_\wedge, \odot_\vee, \odot_\rightarrow\}, \{1\} \rangle$, where the logical operators are defined as follows:[83]

α	$\odot_\neg \alpha$
0	1
i	i
1	0

$\odot_\wedge$	0	i	1
0	0	0	0
i	0	i	i
1	0	i	1

$\odot_\vee$	0	i	1
0	0	i	1
i	i	i	1
1	1	1	1

$\odot_\rightarrow$	0	i	1
0	1	1	1
i	i	i	1
1	0	i	1

Some explanatory remarks are in order here: (i) The given truth-value tables for logical operators of propositional logic are generalizations of the classical truth tables – that is, with respect to the input values 0 and 1, the respective operators behave like their classical counterparts. (ii) K_3 models the conditional '$\rightarrow$' as a material conditional, i.e., $P \rightarrow Q$ and $\neg P \vee Q$ are logically equivalent. (iii) Since the designated value is 1, no formula is a tautology – for any valuation that assigns to every atomic sentence the value i assigns i to every sentence of the language. As a consequence of this, K_3 is strongly paracomplete. On the other hand, modus ponens is valid. (iv) Kleene invented K_3 with view to applications to partial functions, i.e., functions that are not defined for certain input values (e.g., division (of any number) by zero) (see [Kleene, 1952, Section 64]). According to Kleene, 1, 0, and i can be interpreted as 'true', 'false', and 'undefined' respectively, or as 'true', 'false', and 'unknown (or 'value immaterial').[84] (v) The operators for universal and existential quantification may be obtained by way of natural generalizations of the conjunction and disjunction operators.[85]

Several authors have made a case in favour of K_3 as a framework for vagueness (e.g., see [Körner, 1966, pp. 37–40], [Tappenden, 1993], [Tye, 1990], [Tye, 1994], [Soames, 1999, Chapter 7], [Richard, 2010], [Field, 2003], and [Field, 2010]).[86] The common rationale for this proposal is the idea that borderline cases may be thought of as a kind of partiality.[87] It is often suggested that i is not to be interpreted as lack of truth and falsity, but rather as a placeholder status, which leaves it open whether the truth value is truth, or falsity, or undefined. In this sense, assignments of i may be interpreted as modelling a state that does not even imply a commitment to untruth or unfalsity.[88]

On either suggested interpretation, the account of the paradox is plain. Assuming that borderline cases receive the value i, the standard sorites argument (via (CS–L)), though being valid, can be blocked as unsound (in some sense, dependent on the more specific interpretation of i). For instance, take a sorites series for 'walking distance' where the distances are non-decreasing as we go down the series: since in this series, there are only immediate transitions from 0 to i, or from i to 1, there is no relevant instance of (TOL) that will receive the value 0; but some instances will receive the value i – to wit, instances where the antecedent has value 1 (or i) and the consequent the value i (or 0, respectively). By parity of reasoning, no statement of a particular counterinstance to (TOL), of the form $Fa_n \wedge \neg Fa_{n+1}$, is true, but some are gappy. By the standard 3-valued truth tables for disjunction, from this it follows furthermore that the associated disjunction of the form $(Fa_0 \wedge \neg Fa_1) \vee \ldots \vee (Fa_{i-1} \wedge \neg Fa_i)$, which says that there is a counterinstance to (TOL), is gappy as well. That is, K_3 offers a strategy of blocking standard soritical arguments, not only without being committed to any particular cut-off point in the series, but also without being committed to the existence of such a cut-off. Though this distinction may appear to make no difference, it will turn out that on other paracomplete accounts of the paradox, it does (see Section 5.3).

Opponents to K_3 typically target it on the ground that it implies that the structural features of borderline vagueness are pretty strong.[89] To wit, K_3 makes it quite hard for compound sentences to be true or false if some of their immediate components take an intermediate value. More precisely, starting from the classical truth tables, one can show that K_3 is the strongest extension of the classical tables that satisfies the following regularity constraint: A given column (row) contains 1 (0) in the i row (column), only if the column (row) consists entirely of 1's (0's). That is, the tables take the value 1 (0) if this value is compatible with the regularity constraint. The said regularity constraint indeed has a motivation in applications Kleene has in mind,[90] but there is reason for doubt that it has a motivation, as far as applications to vagueness are concerned. For example, from the K_3 tables, it follows that if P is borderline vague, so are not only the respective instances of (LEM) or the Law of Non-Contradiction (LNC), but even $P \rightarrow P$.[91]

5.2.2 *LP*

The logical matrix for Priest's system *LP* is easily obtainable from the logical matrix for K_3, just by replacing the set of designated values – adopting $\{1, i\}$ instead of $\{1\}$. *LP* is strongly paraconsistent. In fact, it is the dual of K_3, which is strongly paracomplete. That is, we have $\varphi \models_{LP} \psi$ iff $\neg\psi \models_{K_3} \neg\varphi$, and $\varphi \models_{K_3} \psi$ iff $\neg\psi \models_{LP} \neg\varphi$; more generally, for natural generalizations $\models^*_{LP}$ and $\models^*_{K_3}$ of $\models_{LP}$ and $\models_{K_3}$ for multi-conclusion logic respectively: $\Gamma \models^*_{LP} \Delta$ iff $\Delta' \models^*_{K_3} \Gamma'$, and $\Gamma \models^*_{K_3} \Delta$ iff $\Delta' \models^*_{LP} \Gamma'$, where $\Delta' = \{\neg\delta : \delta \in \Delta\}$ and $\Gamma' = \{\neg\gamma : \gamma \in \Gamma\}$. Priest suggests interpreting the intermediate value i as a truth-value glut, i.e., as 'both true and false'.

The suggested account of borderline cases and relevant instances of (TOL) in a sorites series is exactly the account we know already from K_3: borderline cases take intermediate values, and the same for some instances of (TOL) – with the only difference being that gaps are here reinterpreted as gluts. As a consequence, by parity of the above reasoning, every instance of (TOL) can be valuated as true, though not every instance is 'simply true', for some instances are also false. By the standard 3-valued truth tables for conjunction, from this it follows furthermore also that the conjunction of all relevant instances of (TOL) is true as well. In this sense, *LP* allows us to embrace in full the (UT) constraint that underlies the sorites paradox (see Section 1.1). The obvious flip-side of these results is that the strategy of blocking standard instances of the paradox as unsound, which is available in K_3 logic, is of no avail for the *LP* theorist. *LP* offers a different escape route from the paradox though, by failure of modus ponens. Specifically, it fails when the consequent is simply false without the antecedent being simply false. Since sorites series begin with a case of lack of simple falsity but end with a case of simple falsity, it follows that some applications of modus ponens in soritical chain arguments of the form (CS–S) are not safe. For instance, in the relevant instance of (CS–S) for the above sorites series for 'walking distance' (W) (which was assumed to be non-decreasing with the ordinal numbers of members), we can safely apply modus ponens to stretch out applications of W throughout the series until we reach the first distance a_n such that $W(a_n)$ is simply false. By assumption then, $W(a_{n-1})$ is still true and false, and so is $W(a_{n-1}) \rightarrow W(a_n)$;[92] however $W(a_n)$ is simply false. Hence the inference from the former two premises to the latter sentence is invalid. That is, to some extent, *LP* lends support to soritical reasoning as safe, but it fails to supply means of accommodating the pre-theoretical idea that sorites arguments are justifiable by way of conclusive inferences. Indeed, one may turn this point into a point against the account of 'if ... then' as a material conditional and suggest an alternative account, on which modus ponens is valid.[93] Whether this kind of move would result in more plausible logical option is a question to be left open here.

5.2.3 $Ł_{\aleph}$

Łukasiewicz's system $Ł_{\aleph}$[94] is a continuum valued logic[95] that is characterized by the logical matrix $\langle[0,1],\{\odot^*_{\neg},\odot^*_{\wedge},\odot^*_{\vee},\odot^*_{\rightarrow}\},\{1\}\rangle$, with the logical operators being defined as follows:

$$\odot^*_{\neg}(x) = 1 - x$$
$$\odot^*_{\wedge}(x,y) = min\{x,y\}$$
$$\odot^*_{\vee}(x,y) = max\{x,y\},$$

where $min\{x,y\}$ and $max\{x,y\}$ give the *minimum* and *maximum* of $\{x,y\}$ respectively.[96] That is, representing i by the truth value $\frac{1}{2}$, we can interpret $\odot^*_{\neg}$, $\odot^*_{\wedge}$, and $\odot^*_{\vee}$ as generalizations of the K_3 counterpart operators $\odot_{\neg}$, $\odot_{\wedge}$, and $\odot_{\vee}$ respectively. Not so for the conditional, which unlike in K_3, receives the truth value 1 if both the antecedent and the consequent take the intermediate truth value $\frac{1}{2}$ and hence is not a material conditional:

$$\odot_{\rightarrow}(x,y) = \begin{cases} 1 & \text{if } x \leq y \\ 1-(x-y) & \text{otherwise.} \end{cases}$$

The intuitive motivation for the conditional may be put as follows: $A \rightarrow B$ should increase in truth value the less slide there is between the assumed antecedent and the concluded consequence; in other words, it should be the difference between the maximal truth value and the slide from A to B. Since the maximal truth value is the designated value, it is easy to see that modus ponens is valid: for if A has the maximal truth value and there is no slide from A to B in truth value, B must have the maximal truth value as well. On the other hand, modus ponens does not have the property of preserving positive truth values that are lower than 1, that is: if A and $A \rightarrow B$ both take a value that is not lower than δ for $0 < \delta < 1$, it does not follow in general that B also takes a value that is not lower than δ. As a consequence, if 'acceptability' amounts to having a truth value greater than δ for some $0 < \delta < 1$, it follows that modus ponens does not preserve acceptability. For instance, if A and $A \rightarrow B$ both take the value .99, then B takes the value .98. Hence, if acceptability requires a truth value that is not lower than .99, the said instance of modus ponens fails to preserve acceptability. However, there is a limit to the extent to which the truth value in modus ponens may drop down. Specifically, we have:

Fact 7.5.1 $(1 - \nu(B)) \preceq [(1 - \nu(A)) + (1 - \nu(A \rightarrow B))]$.[97]

That is, an application of modus ponens always renders a conclusion that is not more distant from the maximum truth value than the sum of the respective distances of conditional and of the antecedent.

These features of $Ł_{\aleph}$ are exploited in standard applications of $Ł_{\aleph}$ to the paradox (for approaches to vagueness that operate in an $Ł_{\aleph}$ framework, e.g., see [Lakoff, 1973], [Machina, 1976], and [Forbes, 1983].[98]) Assuming that 'truth' amounts to the designated value 1, one can in general model sorites series $\langle a_0, \ldots, a_i \rangle$ for a predicate F as cases where Fa_0 is true but for any $0 < n < i$, Fa_n is untrue. Since, by assumption, the truth value for valuations of the form Fa_n will have to drop down when we go through the series, there is a pair of adjacent members $\langle a_k, a_l \rangle$ where $Fa_k \rightarrow Fa_l$ is smaller than 1. Consequently, some premise in the standard sorites argument for our series is untrue. Furthermore, if the slide in truth value from one member to the next one in the series is always lower than a threshold $0 < \alpha \leq 1$, it follows that every instance of (TOL) of the form $Fa_n \rightarrow Fa_{n+1}$ is greater than $1-\alpha$ in truth value. Hence, if 'acceptability' amounts to having a truth value greater than $\delta \leq 1-\alpha$, it follows not only that the first premise, but also the other relevant premises in a standard sorites argument, that is, all relevant instances of (TOL), are acceptable. Conversely, if we assume all relevant instances of (TOL) for a sorites series to be greater than $0 \leq \epsilon < 1$ in truth value, Fact 7.5.1 ensures that soritical chain reasoning by way of modus ponens applications involves only slight drops in truth value: for each pair of predications Fa_{n+1} and Fa_n the difference between their truth value is to be lower than $1-\epsilon$.

On this account, the fact that instances of (TOL) are that compelling amounts to the fact that the slides in truth value when we go through the series, from one member to the next, are only very small. For example, consider a sorites series $\{0, \ldots, 100,000\}$ for 'i hairs make for baldness' (B_i). For simplicity, suppose $\nu(B_i) = \frac{100,000-i}{100,000}$; $B_0, B_1 \ldots, B_{99,999}, B_{100,000}$ take then the values $1, 0.99999, \ldots, 0.00001, 0$ respectively. Furthermore, all relevant instances of (TOL) take the value 0.99999. Hence the argument is valid but unsound. However, all premises of the argument (assuming an appropriate threshold for acceptability) are acceptable – that is, the slides in truth value for predications when we go down step by step in the series are only small. Finally, it is important to note that if each relevant instance of (TOL) is acceptable, so is the associated conjunction of all these instances: for by the continuum-valued tables for conjunction, if all conjunctions take a value above a threshold, so does the conjunction. In this weak sense, the soritical constraint (UT) can be accommodated without abandoning modus ponens. (As a parenthetical note, in view of the last result, one may suggest that $Ł_{\aleph}$ shares the respective virtues of *LP* and K_3 without sharing their limitations.)

While the $Ł_{\aleph}$-based account of the paradox has some attractive features, it is highly controversial. For one, as Edgington ([Edgington, 1997]) has noted (referring to results from Adams' work on probability logic), the very features that are exploited in this account (a continuum-valued approach, validity of modus ponens, and Fact 7.5.1) are available also on classical probabilistic accounts of the paradox.[99] And, insofar as the $Ł_{\aleph}$-based account is intended as a model of 'credence' or 'degrees of assertability', one may object (as Edgington does) that degrees of this kind should have a classical probabilistic structure: e.g., whereas on $Ł_{\aleph}$, by truth-value functionality, contradictions $P \wedge \neg P$ receive a positive degree whenever P takes a positive value lower than 1, one may argue that in general, contradictions should not be believable or assertable to any positive degree. Advocates of continuum-valued semantics though express full satisfaction with these results.[100,101]

Second, whereas philosophical proponents of Łukasiewicz's system usually treat the label 'degrees of truth' like a primitive, self-explanatory term, the idea that truth may come in degrees is received rather with caution and scepticism outside this community.[102,103]

Third, Łukasiewicz' system is faced with a common *tu quoque* objection (e.g., see [Kamp, 1981, pp. 294–5], [Beall and van Fraassen, 2003, pp. 143–4], and [Weintraub, 2004, Sections 2 and 3]). To wit, one of the main counterarguments against classical semantics is that it requires a cut-off point in a sorites series between true and false application cases. The main charge is then that there is no such point, for instance, in a sories series for 'bald', there is no highest number which makes for baldness, and where just one hair more would make for lack of baldness. But even a continuum-valued framework is committed to some type of cut-off point in sorites series – to wit a cut-off between predications which are true (i.e., receiving the value 1) and predications that are untrue (i.e., receiving a value lower than 1). At least, the proponent of a continuum-value semantics is faced with this predicament if her meta-language operates in a framework of classical logic and set theory. (Obviously, this objection can be levelled against applications of other non-classical frameworks to vagueness as well, insofar as the framework of the meta-theory is classical – which is standardly the case.)[104,105]

5.3 Supervaluationism and Subvaluationism

5.3.1 *SpV*

The application of supervaluationist logics to vagueness was first suggested by Fine ([Fine, 1975]) and more recently defended by Keefe ([Keefe, 2000]). Standardly, it is motivated by a 'semantic view' about borderline vagueness (Section 2.2) and an idea that was already mentioned in connection with semantic reinterpretations of epistemicism: according to this, a sentence is borderline vague just in case it admits of more than one bivalent interpretation – generally, a language involves vagueness just in case it admits of more than one classical interpretation. This view may come in different varieties. To ease comparison with other frameworks, supervaluationism is introduced here on the basis of a standard framework of possible-worlds semantics for a language $\mathcal{L}_{\mathcal{D}}$ of propositional logic containing an operator $\mathcal{D}$ for definite truth.[106] A *frame* for $\mathcal{L}_{\mathcal{D}}$ is an ordered pair $\langle W, R\rangle$, where W is a non-empty set (of 'sharpenings (of the language)'), R is a relation (of 'admissibility') on W. A *model* for $\mathcal{L}_{\mathcal{D}}$ is a triple $\langle W, R, v\rangle$, where $\langle W, R\rangle$ is a frame and v is a bivalent interpretation (i.e., for every $w \in W$, $v_w(\varphi) = 1$ or $v_w(\varphi) = 0$) that accords with the following valuation rules:

$$v_w(\varphi \wedge \psi) = 1 \text{ iff } v_w(\varphi) = 1 \text{ and } v_w(\psi) = 1$$

$$v_w(\varphi \vee \psi) = 1 \text{ iff } v_w(\varphi) = 1 \text{ or } v_w(\psi) = 1$$

$$v_w(\neg\varphi) = 1 \text{ iff } v_w(\varphi) = 0$$

$$v_w(\varphi \to \psi) = 1 \text{ iff } v_w(\varphi) = 0 \text{ or } v_w(\psi) = 1$$

$$v_w(\mathcal{D}\varphi) = 1 \text{ iff } v_{w'}(\varphi) = 1, \text{ for all } w' \text{ such that } wRw'$$

A common postulate in supervaluationist accounts is that borderline cases are truth-value gaps. A natural way of modelling this idea is to specify truth in a model M as follows:

SUPERTRUTH: For every model $M = \langle W, R, v\rangle$, $\models^M \varphi$ (or, φ is 'supertrue' in M) iff for *all* $w \in W$, $v_w(\varphi) = 1$.

'Superfalsity' in a model, accordingly, amounts to falsity for every sharpening in the model. Depending on how logical consequence is specified in terms of this framework, one may distinguish between two main divisions in the 'supervaluationist' camp. Some authors have made suggestions to the effect that logical consequence may be defined the way it is defined in standard possible worlds frameworks (see, [Varzi, 2007] and [Asher et al., 2010])[107]:

SPV LOCAL: $\Gamma \models_{SpV-L} \varphi$ iff for every frame $\langle W, R\rangle \in \mathcal{F}$, for every model $\langle W, R, v\rangle$ based on the frame $\langle W, R\rangle$, and for every $w \in W$, if $v_w(\alpha) = 1$, for every sentence $\alpha \in \Gamma$, then also $v_w(\varphi) = 1$,

where the class of frames $\mathcal{F}$ is standardly assumed to be at least restricted to frames with a reflexive relation R, in order to ensure that $\mathcal{D}$ is factive (i.e., $\mathcal{D}\varphi \to \varphi$ is valid); however to make room for higher-order vagueness, transitivity or symmetry should fail. According to this approach, even though the notion of 'supertruth' may be still embraced as an adequate account of truth simpliciter, logical consequence is not to be defined in terms of supertruth preservation.[108] In effect, classical logic is embraced in full, and $\mathcal{D}$ is treated like a normal modal operator of necessity. The focus here is on the more 'orthodox' version of *SpV* (proposed by Fine [Fine, 1975] and Keefe [Keefe, 2000]), which involves some departure from classical logic. According to this, logical consequence is supertruth preservation, that is, we have:

SPV GLOBAL: $\Gamma \models_{SpV-G} \varphi$ iff for every frame $\langle W, R\rangle \in \mathcal{F}$, for every model $\langle W, R, v\rangle$ based on the frame $\langle W, R\rangle$: if for *every* $w \in W$, $v_w(\alpha) = 1$, for every sentence $\alpha \in \Gamma$, then also for *every* $w \in W, v_w(\varphi) = 1$.

An important difference between these two notions of logical consequence is that only the latter one validates ($\mathcal{D}$–INTRO) (Section 3.2).[109] In what follows, for brevity, logical consequence in the sense of (SPV GLOBAL) is referred to as '*SpV*', and supertruth and superfalsity (in a model) are simply referred as 'truth' and 'falsity' (in the model) respectively.

(As a parenthetical note, the given two options do not exhaust the logical space, and one may plausibly suggest still other ways of modelling logical consequence in a standard possible-worlds setting.[110] Furthermore, it ought to be mentioned that this general setting is not general enough to cover *every* kind of framework that has been proposed under the label 'supervaluationism'. In particular, one may suggest that for 'sharpenings' to be considered as 'admissible', they should not be classical interpretations, which fix a cut-off point in every sorites series, but some type of partial interpretations, which leave some area in a sorites series undefined. Depending on the way partiality is modelled (e.g., by way of Strong Kleene, or intuitionist semantics), this approach suggests logical options that are very different from the frameworks that are standardly considered under the label 'supervaluationism' (see [Fine, 1975, p. 127] and [Shapiro, 2006, Chapter 4]).[111])

SpV has some distinctive features that, prima facie, make it appear an interesting alternative to the many-valued options discussed. For one, unlike K_3 and $Ł_{\aleph}$, *SpV* is only weakly paracomplete. That is, on the one the hand, it allows for non-trivial truth-value gaps, but on the other hand, it validates all instances of (LEM) in $\mathcal{L}_{\mathcal{D}}$. More generally, unlike the strong paracomplete alternatives K_3 and $L_{\aleph}$, supervaluationist entailment ($\models_{SpV}$) preserves classical entailment ($\models_{CL}$) for $\mathcal{L}_{\mathcal{D}}$, in the sense that: if $\Gamma \models_{CL} \varphi$, then $\Gamma \models_{SpV} \varphi$.[112] A related feature of *SpV* is that its semantics for logical constants is not truth-value functional; for, even though some disjunctive sentences of the form $\varphi \vee \psi$ should fail to be true, if φ and ψ are both gappy (e.g., instances where there is no semantic or other intelligible connection between φ and ψ), some other disjunctions with the same feature are bound to be true, to wit, instances of (LEM), of the form $\varphi \vee \neg\varphi$ (note that $\neg\varphi$ is gappy, if φ is gappy). Whereas failure of truth-value functionality is commonly perceived as a serious problem by opponents of *SpV* (e.g., see [Williamson, 1994, pp. 135–8]), proponents of this framework commonly endorse it as a useful feature.[113] Specifically, they argue that *SpV* supplies means of accommodating so-called 'penumbral connections' ([Fine, 1975, pp. 123–5]), that is, semantic connections between natural language expressions outside the domain of logical constants. For example, one might require appropriate models of a natural language to accommodate 'analytic truths' such as sentences of the form 'If patch *a* is red, *a* is not orange', where the component sentences are themselves borderline vague. Whereas on many-valued logics, due to standard truth-value functional semantics for the conditional, such 'analytic truths' fail to be true, they can be validated in a *SpV* framework, on appropriate constraints on the class of models.[114] The highlighted features of weak paracompleteness and failure of truth-functionality are also in play in the standard *SpV*-based account of the paradox of vagueness.

Again, the account of the paradox is plain: Assuming that borderline cases are gappy in truth-value, the standard sorites argument (via (CS–L)) is indeed valid, but since some premises are rejectable as untrue, the argument can be blocked as unsound. For example, take a sorites series for 'walking distance' (*WD*) where the distances are non-decreasing as we go down the series: since in this series, there are only immediate transitions from truth to gappiness and from gappiness to falsity, there is no relevant instance of (TOL) that is false (note that some remnants of truth-functionality still hold on *SpV*: for $P \rightarrow Q$ to be false in an *SpV* model, the associated conjunction $P \wedge \neg Q$ is to be true in the model, which holds just in case both conjuncts are true in the model). However, some instances of (TOL) are gappy – to wit, instances where the antecedent is true and the consequent is gappy, and instances where the antecedent is gappy and the consequent is false (note, that if P is true and Q gappy in an *SpV* model, it follows that for some 'sharpenings' in the model, $P \rightarrow Q$ is false; likewise for instances where the antecedent is gappy and the consequent is false).

So far, the *SpV*-based account sounds very similar to the K_3-based account (Section 5.2). However, in contrast to K_3, where the the disjunction $(WD(d_1) \wedge \neg WD(d_2)) \vee \ldots \vee (WD(d_{i-1}) \wedge \neg WD(d_i))$ is gappy, by truth-functionality, the disjunction is true on *SpV* models. To wit, for every appropriate *SpV* model $\langle W, R, v\rangle$ for a given sorites series, $WD(d_1)$ is true for every 'sharpening' in the model, and $WD(d_i)$ is false in every sharpening in the model. Since the sharpenings $w \in W$ are classical valuations, however, each 'sharpening' fixes a cut-off point in the sorites series – which will vary with 'sharpenings', since *WD* is supposed to be vague. Hence, the disjunction $(WD(d_1) \wedge \neg WD(d_2)) \vee \ldots \vee (WD(d_{i-1}) \wedge \neg WD(d_i))$ – which asserts the existence of a cut-off point – is true in any appropriate *SpV* model for the series. Failure of truth-value functionality comes to the rescue here though, for in contrast to many-valued logics, on *SpV*, the truth of a disjunction (in a model) does not entail that some of the disjuncts are true (in the model). That is, the supervaluationist is committed to the conclusion that there is a cut-off point in the sorites series, without being committed to any particular cut-off point.

Weak paracompleteness implies a departure from classical multi-conclusion logic; for it implies that there are non-trivial counterinstances to $\varphi \vee \neg\varphi \models \{\varphi, \neg\varphi\}$. In fact, (as observed by Machina [Machina, 1976] and discussed in detail by Williamson [Williamson, 1994, Chapter 5.3]), even the single-conclusion relation of logical consequence violates classical logic, as far as applications to the full language $\mathcal{L}_\mathcal{D}$ are concerned. To wit, for $\mathcal{L}_\mathcal{D}$, $\models_{SpV}$ fails to be closed under certain classical inference rules that involve assumptions that are eventually discharged, such as:

- $(\Gamma \cup \{\varphi_1\} \models \psi$ and ... and $\Gamma \cup \{\varphi_n\} \models \psi) \Rightarrow \Gamma \cup \{\varphi_1 \vee \ldots \vee \varphi_n\} \models \psi$ *(argument by cases)*
- $\Gamma \cup \{\varphi\} \models \psi \Rightarrow \Gamma \models \varphi \rightarrow \psi$ *(conditional proof)*
- $\Gamma \cup \{\varphi\} \models (\psi \wedge \neg\psi) \Rightarrow \Gamma \models \neg\varphi$ *(reductio ad absurdum)*
- $\Gamma \cup \{\varphi\} \models \psi \Rightarrow \Gamma \cup \{\neg\psi\} \models \neg\varphi$ *(contraposition)*

Specifically, whereas in the absence of the $\mathcal{D}$-operator, the given rules hold also for $\models_{SpV}$, they have counterinstances for the more general case involving discharged premises containing a $\mathcal{D}$-operator. For example, we have $\varphi \models_{SpV} \mathcal{D}\varphi$, however, $\not\models_{SpV} \varphi \rightarrow \mathcal{D}(\varphi)$ (note that any φ that is neither true nor false for some model is a counterinstance).[115] According to Fara ([Fara, 2003]), even for the $\mathcal{D}$-free fragment of $\mathcal{L}_{\mathcal{D}}$, classical inference rules of the said type may fail, insofar as the class of *SpV* models is to be constrained to ensure the 'analytic' validity of certain inference patterns.

Fara ([Fara, 2003]) highlights still another (potential) problem relating to ($\mathcal{D}$–INTRO). She argues that a supervaluationist can only give an adequate account of vagueness if the generalized gap-principle (GP–GEN) can be accommodated for every finite sorites series.[116] However, as she can prove, for every finite series, (GP–GEN) and the ($\mathcal{D}$–INTRO) rule are jointly inconsistent.[117] Whether this result reveals a problem with *SpV* or rather with the requirement that (GP–GEN) be valid for a full-fledged account of vagueness is a question that deserves further discussion.[118]

5.3.2 *SbV*

SbV is a logic that has been defended by Hyde and Colyvan ([Hyde, 1997], [Hyde and Colyvan, 2008]). It is obtainable from a standard possible-worlds semantics by adopting the following notion of logical consequence:

SbV: $\Gamma \models_{SbV} \varphi$ iff for every frame $\langle W, R\rangle \in \mathcal{F}$, for every model $\langle W, R, v\rangle$ based on the frame $\langle W, R\rangle$: if for every sentence $\alpha \in \Gamma$, there is a $w \in W$ such that $v_w(\alpha) = 1$, then there is also a $w \in W$ such that $v_w(\varphi) = 1$.

To bring out more clearly the difference from *SpV*, one can introduce the following counterpart notion to 'supertruth' (in a model):

Subtruth: For every model $M = \langle W, R, v\rangle$, $\models^M \varphi$ (or, φ is 'subtrue' in M) iff for *some* $w \in W$, $v_w(\varphi) = 1$.

'Subfalsity' in a model, accordingly, amounts to falsity for some sharpening in the model. With this in place, the *SbV* account tells us that logical consequence should be preserving subtruth (in models). For brevity, subtruth (in a model) will be referred to here simply as 'truth' (in a model). *SbV* is weakly paraconsistent.

In fact, it is the dual of *SpV*. That is, for natural generalizations $\models^*_{SbV}$ and $\models^*_{SpV}$ of $\models_{SbV}$ and $\models_{SpV}$ for multi-conclusion logic respectively: $\Gamma \models^*_{SbV} \Delta$ iff $\Delta' \models^*_{SpV} \Gamma'$, and $\Gamma \models^*_{SpV} \Delta$ iff $\Delta' \models^*_{SbV} \Gamma'$, where $\Delta' = \{\neg\delta : \delta \in \Delta\}$ and $\Gamma' = \{\neg\gamma : \gamma \in \Gamma\}$.[119] Consequently, whereas *SpV* is weakly paracomplete (i.e., $\varphi \not\models_{SpV^*} \{\psi, \neg\psi\}$, but $\varphi \models_{SpV*} \psi \vee \neg\psi$), *SbV* is weakly paraconsistent (i.e., $\{\varphi, \neg\varphi\} \not\models_{SbV} \psi$, but $\varphi \wedge \neg\varphi \models_{SbV} \psi$). As a consequence, we have corresponding departures from classical logic; in particular, weak paraconsistency implies that there are non-trivial counterinstances to rule of *conjunction introduction* (or *adjunction*), $\{\alpha, \beta\} \models \alpha \wedge \beta$ (note, we have $\{\varphi, \neg\varphi\} \not\models_{SbV} \varphi \wedge \neg\varphi$).

We already noted the similarities between certain paracomplete accounts of the paradox of vagueness, the one applying *SpV*, the other applying K_3. It should not be very surprising that one can make the same point with respect to their paraconsistent duals, i.e., *SbV* and *LP* respectively. Like for *LP*, the *SbV*-based account starts from the postulate that borderline cases are truth-value gluts. As a consequence, since sorites series do not contain a pair of members where one is a simply true application case and its adjacent member is a simply false application case, every relevant instance of (TOL) can be valuated as true, though not every instance is 'simply true', for some instances are also false. The strategy of blocking standard instances of the paradox as unsound, which is available in *SpV* logic, is hence of no avail for the *SbV* theorist. Instead of that, another option of blocking the paradox is available, which is not available for the *SpV* theorist; to wit, modus ponens fails to be valid on *SbV*. Specifically, it fails when the consequent is simply false without the antecedent being simply false. The further reasoning that was spelt out for the *LP*-based account simply carries over to the *SbV*-based account (for further details, see Section 5.2). To some extent, standard soritical reasoning can be accommodated as safe. But the pre-theoretic impression that it is a valid form of reasoning is not sustainable, according to *SbV*. *SbV* essentially differs from *LP* in the following respect though: whereas on *LP*, not only all relevant instances of (TOL) but also their conjunction is true, on *SbV*, conjunctions of this form are simply false. That is, the soritical (UT) constraint is accommodated only to some extent.[120]

5.4 Transitivity of Logical Consequence Reconsidered

The reasoning that is commonly invoked in support of sorites arguments involves more than one inferential step and hence hinges on the proviso that logical consequence is transitive (see Section 1.2). On standard non-classical accounts of the paradox, this proviso is taken for granted (note that in particular, the proviso holds for all frameworks that were discussed in Sections 5.2 and 5.3). According to this, the paradox reveals a problem either with some of the instances of (TOL) that serve as premises (this line is suggested in the paracomplete frameworks K_3, $Ł_{\aleph}$, and SpV) or with the inference rule of modus ponens, which is invoked in soritical chain reasoning (this line is suggested in the paraconsistent frameworks LP and SbV). This leaves still a third possibility open, to wit, to block soritical chain reasoning by abandoning the transitivity constraint for logical consequence. According to this, indeed all individual inferential steps which jointly lead us from the premises to the conclusion are valid; however, there is no valid single inference leading from the premises to the conclusion. Hence, arguments of the form (CS–L) and (CS–S) are invalid – or so one may suggest. On the face of it, this suggestion may sound odd, insofar as we think of logical consequence as a relation that preserves a particular standard (such as truth, lack of falsity, or other) – for if sentences that are validly preserved from a premise set are thought to inherit a certain standard from the premises, logical consequence can hardly be intransitive. But one may suggest otherwise and let the premises of logically valid inference meet a higher standard than the conclusions. This generic idea may be cashed out in different ways, resulting in different notions of logical consequence. For further details, see the frameworks in [Kamp, 1981], [Zardini, 2008], and [Cobreros et al., 2010], the latter of which elaborates an idea that was first suggested in [van Rooij, 2010].

Acknowledgements

For helpful discussion, many thanks to Pablo Cobreros and Leon Horsten.

Notes

1. For a survey of case studies of soritical reasoning in all sorts of practical contexts, see [Walton, 1992].
2. On the history of the philosophical discussion of sorites paradoxes and of vagueness in general, see [Williamson, 1994, Chapters 1–3] and [Hyde, 2007]. For the discussion of vagueness in early analytic philosophy, see also [Rolf, 1981, Chapters 1–3]
3. For a survey of approaches to vagueness in linguistics, see [Pinkal, 1995] and [van Rooij, 2009].
4. For similar formulations of the condition for soriticality, compare [Fara, 2000, pp. 49–50] and [Gómez-Torrente, 2010, pp. 228–9].
5. Wright ([Wright, 1976, Section 2]) coined the phrase 'tolerant' for describing predicates for which there is 'a notion of degree of change too small to make any difference' to their application.
6. The qualification 'with respect to domain D' is not redundant; e.g., see [Smith, 2008, Chapter 3.4.4]. However, insofar as we consider cases where the qualification is not essential, we will not mention it.
7. The label 'Conditional Sorites' is adopted from [Hyde, 2007].
8. The inductive premise (2) is classically equivalent to $\neg(\exists n)(Fa_n \wedge \neg Fa_{n+1})$, which says that there is no pair of adjacent members in a sorites series which marks a cut-off (or a *sharp boundary*) between F-ness and lack of F-ness. In this reformulation,

the mathematical induction sorites is also known as *No-Sharp-Boundaries (Sorites) Paradox*; see [Wright, 1987].

9. For example, zero feet is a walking distance. But not every natural number of feet is a walking distance. Thus, by the *least number principle* (saying that every set of natural numbers has a least member), which is classically equivalent to mathematical induction, there is a least number of feet that still is a walking distance, and where $n+1$ feet fails to be a walking distance – which implies that, contrary to appearance, 'walking distance' is not tolerant. This chain of reasoning has the form of:
MATHEMATICAL INDUCTION SORITES – REFORMULATED

 (1) Fa_0
 (2) $\neg(\forall n)Fa_n$
 $\therefore$ $(\exists n)(Fa_n \wedge \neg Fa_{n+1})$.

10. Priest ([Priest, 1991] and [Priest, 2008, pp. 572–3]) suggests that modulo certain reasonable assumptions, each instance of the paradox pertaining to a general term generates a corresponding instance of a paradox pertaining to identity, and vice versa.
11. Some non-classical frameworks, also known as *paraconsistent logics*, make room for the possibility that a vague predicate may apply both truly and falsely to the same object. However, standard paraconsistent frameworks for vagueness accommodate contradictory applications only for borderline cases, that is the type of application cases that are not covered by common sense clear-case constraints (on the extension and anti-extension) for vague terms. Nihilism is therefore clearly to be distinguished from paraconsistent accounts of vagueness. For further discussion of applications of paraconsistent logics to vagueness, see Section 5.
12. [Williamson, 1994, Chapter 6].
13. For a position in this spirit, see [Gómez-Torrente, 2010].
14. See also [Sainsbury, 1986, pp. 99–100], [Williamson, 1994, pp. 230–4].
15. [Fara, 2000, p. 80, n. 29]
16. E.g., see [Sorensen, 1988], [Williamson, 1994], and [Fara, 2000]. For further discussion, see Section 4.1; for an exception, see [Wright, 2001], who endorses an intuitionist framework instead.
17. 'Semantic indeterminacy' is broadly conceived and may comprise also forms of pragmatic indeterminacy. For more subtle distinctions between various types of the semantic view, see [Varzi, 2007, Section 1] and [Smith, 2008, Chapter 2.5].
18. E.g., see [Wright, 2001].
19. For further critical discussion of ontological conceptions of vagueness, see [Williamson, 2003b].
20. See also [Field, 2000], who deems the question of what it is for a sentence to be considered as borderline vague to be more promising a question for further research (rather than the traditional question of what it is for a sentence to be borderline vague).
21. [Field, 1994, Section 1].
22. See [Wright, 1987].
23. Compare [Fara, 2000, p. 48].
24. The same account is suggested by Smith in [Smith, 2008, p. 133], with reference to his example 'schort'.
25. On the other side of the spectrum of opinions seems to be Fine ([Fine, 1975, p. 120]), who introduces his notion of '(extensional) vagueness' by means of the example of a partially defined predicate, '$nice_1$'.
26. Williamson ([Williamson, 1997a]) argues that 'partially defined' predicates are false for the range of application cases left out in partial definitions. On the further

assumption that vagueness is a sort of partiality, this would suggest that applications to borderline cases should be only deniable, which again, would collide with the assumption that borderline cases allow for divergence in use.

27. In effect, Sainsbury argues against the tenet that the notion of borderline vagueness should play a central role for any theory of vagueness. According to him, the notion is a theoretical artifact and primarily motivated by the idea that apparent tolerance is representable by a gap principle (or a variant thereof): to the effect that there is some sort of tripartite division between best candidates for truth (i.e., definite truths, or something even stronger), best candidates for falsities (or something even stronger), and a union of cases in between. Dismissing this idea as misconceived, he contends that soriticality may be best characterized as 'boundarylessness' – which, he suggests, may be modelled in coherent terms in the way suggested in [Tye, 1990], which adopts a K_3 framework for vagueness (see Section 5.2). See [Sainsbury, 1990], [Sainsbury, 1991].
28. E.g., for possible options of accounting for apparent tolerance in terms of certain strengthenings of (GP), see, for instance, [Sainsbury, 1991, p. 173], who does not subscribe to any given option though.
29. For a rigorous definition of higher-order vagueness, for sentences in a language of propositional logic containing an operator of definite truth, see [Williamson, 1999, p. 132].
30. The given characterization only covers orders of *extensional* vagueness, insofar as it does not take into account more than one possible state of affairs. For brevity, we leave out here orders of *intensional* vagueness. For the distinction between extensional and intensional vagueness, see [Fine, 1975, pp. 120–1].
31. However, some authors have suggested that higher-order vagueness is different in kind; e.g, see [Simons, 1992, p. 167] and [MacFarlane, 2010].
32. For defences of the Sorensen-Hyde argument against such doubts, see [Hyde, 2003] and [Varzi, 2003], [Varzi, 2005].
33. In fact, the original version of Wright's argument involves only a weakening of ($\mathcal{D}$–INTRO): if P follows from a set of premises Γ, then if all members of Γ are sentences of the form $\mathcal{D}\varphi$, $\mathcal{D}P$ also follows from Γ. However, the criticisms levelled against Wright's argument in the reconstructed version carry over to the original version as well.
34. See [Williamson, 1994, Chapter 1] and [Hyde, 2007, Section 2].
35. Williamson uses '$\mathbb{C}$' as a definite truth operator. For the sake of uniformity, we stick here to the $\mathcal{D}$-notation.
36. For some complex issues regarding the predicate logic of clarity, which are not discussed here, see [Williamson, 1994, Section 9.3].
37. See [Williamson, 1994, p. 271].
38. For further details, see Chapter 11.
39. The suggestion that higher-order vagueness makes KT the logic for the $\mathcal{D}$ operator goes back to [Dummett, 1959, pp. 182–3]. For further discussion of logical options for $\mathcal{D}$ with view to higher-order vagueness, see [Williamson, 1999].
40. [Williamson, 1994, pp. 272–3].
41. But see [Égré and Bonnay, 2010] for a different approach.
42. For other features of KTB logic for definite truth that may make it an attractive option for modelling higher-order vagueness, see [Gaifman, 2010, pp. 38–41].
43. Indeed this still leaves open the question of how to interpret Williamson's margin models more specifically. The discussion in [Williamson, 1994, Chapter 7] suggests that 'worlds' may be thought of as metaphysically possible ways of using the object-language, where the semantic features of linguistic expressions are thought to supervene on ways of using them. However, Williamson does not seem wedded to

this idea. For example, in [Williamson, 1995, p. 181], he considers also the alternative interpretation of 'worlds' as contexts of use (that are all situated at the same possible world) as a serious option.

44. Compare Williamson's model for the same example in [Williamson, 1997b, pp. 262–3].
45. Gómez-Torrente ([Gómez-Torrente, 2002]) shows that for a distinguished class of fixed-margin models, (GP–GEN) fails for any sorites series. Gómez-Torrente's and Fara's discussions refer to the operator 'K', but since they have in mind Williamson's margin for error account of 'clarity', or 'definite truth', their results carry over to definite truth.
46. Williamson seems to consider both (a) and (b) as serious options. Compare his reply in [Williamson, 1997b] to an earlier observation made by Gómez-Torrente in [Gómez-Torrente, 1997], and his reply to Gómez-Torrente and Fara, in [Williamson, 2002].
47. Specifically, the type of model considered is a 'no-minimum' margin model, that differs from fixed and variable margin models in the following valuation rule for $\mathcal{D}$:

   4″. $w \models_M \mathcal{D}(\varphi)$ iff $(\exists r > 0)(\forall w' \in W)(d(w,w') \leq r \rightarrow w' \models_M \varphi)$.

48. For another problem with accommodating (GP–GEN) for a finite sorites series within any normal modal framework for $\mathcal{D}$, see [Cobreros, 2010].
49. Note, if true, it cannot be definitely false, by factivity of $\mathcal{D}$. And by (GP) and the standard constraint $\mathcal{D}(P \wedge Q) \rightarrow (\mathcal{D}P \wedge \mathcal{D}Q)$, it can be ruled out that any such statement is definitely true.
50. Note that instances of (GP) are classically equivalent to negations of associated statements of a 'sharp' cut-off; and for the discussed types of models, a formula is borderline vague just in case its negation is.
51. Compare Keefe's objection in [Keefe, 2000, pp. 70–2].
52. Compare [Fara, 2000, p. 50].
53. Bonini et al. [Bonini et al., 1999] provide empirical evidence to the effect that estimates of an acknowledged, but unknown boundary are generated in a manner similar to estimates of the true and false regions in a continuum associated with vague predicates. In this view, the epistemicist hypothesis of a cut-off point (between some adjacent members) in a sorites series seems to be backed by empirical data about linguistic behaviour. This said, the hypothesis would be more attractive if it were associated with an explanation of why it sounds prima facie unacceptable.
54. More generally, assuming a measure of the size of sets, the size of the subset of worlds within δ of w where the belief is true is to be 'big enough'. Compare [Williamson, 2000, Chapter 10.5].
55. Needless to say that these assessments can be perfectly accommodated in terms of classical probability, such that: for every $0 \leq n \leq 999{,}999$, '$Wc_n \rightarrow Wc_{n+1}$' should receive the value 0.999999, whereas $(\forall n \in \{0,\ldots,999{,}999\})(Wc_n \rightarrow Wc_{n+1})$ is to receive the value 0.
56. For other classical probabilistic frameworks for vagueness, see for one [Lewis, 1970a] and [Kamp, 1975], and for another, [Edgington, 1997]. On the account suggested by Lewis and Kamp, probability is interpreted as measuring the size of the subset of 'admissible' classical interpretations (of the language) in which P is true. On Edgington's account, probability is interpreted as a 'degree of closeness to clear truth', also refereed to as 'verity'.
57. [Williamson, 1999, Section 1]. For standard supervaluationism, see Section 5.3.
58. [Burns, 1991].
59. [Smith, 2008, Chapter 2.5].

60. To be clear, Keefe ([Keefe, 2000]) herself subscribes to a standard version of supervaluationism, which is not at issue here (see Section 5.3).
61. Another idea that is occasionally pronounced in support of a contextualism about vagueness is the more generic, so-called 'open texture thesis': according to this, borderline vagueness is not merely divergence in usage with respect to the same relevant circumstances but also recognition on the part of competent speakers that such divergence in usage is to be expected and legitimate. The term 'open texture' was originally coined by Waismann (in his [Waismann, 1951]), but there (it rather seems) with view to intensional vagueness in general. As a label for the said thesis, it was introduced by Shapiro, in [Shapiro, 2006, p. 10]. For other authors who subscribe to the thesis, e.g., see [Wright, 1987, p. 244], [Sainsbury, 1990, Section 9], [Soames, 1999, Chapter 7], [Halpern, 2008, pp. 538f], and [Gaifman, 2010, p. 9].
62. Note that if the relevant tolerance relation is not symmetric, it will also need to be made sure that D^* fails also to be R'-connected with respect to any counterpart tolerance relation R' that satisfies a counterpart tolerance principle for failure of F-ness. It is common to specify tolerance relations as symmetric, in which case this caveat is unnecessary.
63. Van Deemter takes this line. However, there is room for argument both in favour and against the view that direct discriminability is transitive. For the ongoing controversy on this and the related issue on what the individuation criteria for qualia are, see, for example, [Horsten, 2010].
64. Van Deemter ([van Deemter, 1996, p. 66]) does not want to prejudge the question of whether i and j are to be elements of c. For this reason, the first clause is not redundant.
65. See [van Deemter, 1996, Appendix 2].
66. Van Deemter credits Frank Veltman and Reinhart Muskens with being the first to suggest this idea.
67. Fara does not give a more exact account of indifference herself. But her discussion of indifference seems to suggest strongly the above account. For lack of space, further details have to be omitted here.
68. For a different reconstruction of Fara's account, see [van Rooij, 2009].
69. [Fara, 2000, p. 57].
70. [Fara, 2000, p. 59].
71. The same line is taken on the special case of phenomenal sorites, in [Fara, 2001].
72. Further discussion of the issues raised here would go beyond the scope of this chapter, for it would lead straight into closely related discussions in empicical psychology and choice theory.
73. For the distinction between paracomplete and paraconsistent logics, see [Hyde, 2008, Chapter 4]
74. Another paracomplete logic that has been suggested for vagueness is *intuitionist logic*. For defences of an intuitionist logic for vagueness, e.g., see [Putnam, 1983], [Putnam, 1985], [Schwartz, 1987], [Schwartz and Throop, 1991], and [Wright, 2001]. For critical discussion of intuitionism for vagueness, see [Williamson, 1996b] and [Chambers, 1998].
75. For multi-conclusion logic, conclusions, like the premises, may be an arbitrary set of formulas. Given a multi-premise logic that is characterized in terms of preservation of a certain semantic status (truth, lack of falsity, or other), there is then a natural way of generalizing this logic for conclusion sets as follows: An inference from Γ to Δ is valid just in case for every interpretation (of the kind appropriate for the logic) for which *every* premise has the relevant semantic status, *some* conclusion has the relevant semantic status too. For a systematic investigation into multi-conclusion logic, see [Shoesmith and Smiley, 1978].

76. More specifically, since the logic of the metalanguage is standardly taken to be classical, we are free to assume the least number principle. Thus, assuming assignments of truth, gappiness, or falsity to predications, for each member in a sorites series, and beginning with a true predication, there is a first instance of (TOL) where the antecedent is true and the consequent is gappy. On standard paracomplete semantics for the conditional, such instances are then gappy as well.
77. See Chapter 8.
78. Note that strong paraconsistency is not to be identified with the case where there are non-trivial counterinstances to the 'Law of Non-Contradiction' (i.e., the schema $\neg(A \wedge \neg A)$). To wit, $\models_{K_3}$ has the latter property, but it fails to be strongly (and even to be weakly) paraconsistent.
79. For Williamson's argument against truth-value gaps and gluts, see [Williamson, 1994, Chapter 7.2] and [Andjelković and Williamson, 2000]. For another argument against truth-value gaps, see [Glanzberg, 2003].
80. For the question of whether truth-value gap or glut theories match better with experimental data of linguistic behaviour, see [Alxatib and Pelletier, ta]. In effect, the study suggests a kind of pluralist approach, according to which either type of theory has its virtues and its limitations.
81. For the frameworks discussed here in more detail (in Sections 5.2 and 5.3), this proviso does not affect the generality of the points made. For the respective resolution stategies proposed for such frameworks for propositional logic can be easily generalized for predicate logic.
82. Compare [Beall and van Fraassen, 2003, Chapter 7.2].
83. On all accounts discussed here, the biconditional $\leftrightarrow$ is definable the standard way, in terms of the conditional and conjunction. That is, $P \leftrightarrow Q$ is treated as equivalent to $(P \rightarrow Q) \wedge (Q \rightarrow P)$.
84. For background information on this and Kleene's other system (aka 'Weak Kleene' logic), see [Rescher, 1969, Chapter 2.5] and [Blamey, 1986, Chapter 2.5].
85. That is, $(\exists x)\varphi$ takes the maximum value of φ for assignments to x, whereas the universal $(\forall x)\varphi$ takes the minimum value.
86. For fruitful applications of K_3 in natural language semantics, see [Landman, 1991, Chapter 3].
87. To be clear, this idea is compatible with the view that partiality does not exhaust all features of vagueness; see [Soames, 1999, Chapter 7], who argues that vagueness is a sort of partiality that combines with context-sensitivity.
88. For further discussion, e.g., see [Soames, 1999, Chapter 6].
89. See esp., [Williamson, 1994, Chapter 4.5].
90. On this, see [Blamey, 1986, Chapter 2.5].
91. Parsons ([Parsons, 2000]) proposes a closely related system, Łukasiewicz's 3-valued logic Ł3, as a logic of 'indeterminacy'. Ł3 is simply obtainable from K_3, just by redefining the conditional in terms of the following operator:

$\odot_{\rightarrow}$	0	i	1
0	1	1	1
i	i	1	1
1	0	i	1

Parsons explicitly does not intend adopting the system as a logic of vagueness. Nonetheless, it may be considered as a serious alternative.
92. Note that $W(a_{n-1}) \rightarrow W(a_n)$ is LP-equivalent to $\neg W(a_n) \rightarrow \neg W(a_{n-1})$.
93. For example, assuming a strict linear order $<$ on $\mathcal{V}$ such that $0 < i < 1$, one may suggest a non-standard conditional operator $\odot_{\rightarrowtail}$, which is defined as follows on $\mathcal{V}$: $\odot_{\rightarrowtail}(x, y)$ takes value 1 iff $\neg(y < x)$ and value 0 otherwise.

94. Sometimes, the system is also referred to as 'fuzzy logic', which is a bit misleading, since the term is otherwise used technically for a wider class of logical systems. For an overview, see [Dubois et al., 2007].
95. Instead of the unit interval [0,1], one may choose for it also the set of rationals between and including 0 and 1. That the two systems are equivalent was proved by Lindenbaum; see [Łukasiewicz and Tarski, 1930, Theorem 16].
96. In a generalization of $L_{\aleph}$ for predicate logic, universal and existential quantification can be accordingly modelled in terms of greatest lower bounds and lowest upper bounds.
97. Note that the two equations that define $\odot^{*}_{\rightarrow}$ are jointly equivalent with the intuitively less perspicuous equation $\odot^{*}_{\rightarrow}(x, y) = 1 + min\{x, y\} - x$.
98. Goguen ([Goguen, 1969]) defends a different infinite-valued logic for vagueness. Like in Ł$_{\aleph}$, sentences take truth values in the unit interval, and the designated value is 1; however, the relevant logical operators are different. Another unorthodox application of infinite-valued semantics to vagueness is defended in [Smith, 2008]. He makes a case for adopting Ł$_{\aleph}$ valuations for vague languages without adopting the associated Ł$_{\aleph}$ notion of logical consequence, according to which 1 is the designated value. Smith suggests keeping to a classical notion of logical consequence, which can be modelled as follows: $\Gamma \models \varphi$ iff for every valuation on which every $\gamma \in \Gamma$ takes a value strictly greater than .5, φ takes a value that is at least as great as .5.
99. It is to be stressed that this point holds independently of whether the probability of simple conditionals (i.e., conditionals that do not involve other conditionals) is modelled as the probability of a material conditional, or as a conditional probability of the consequent given the antecedent.
100. E.g., see [Schiffer, 2003]. For studies in the structure of credence that start from a Ł$_{\aleph}$ framework, see [Milne, 2008] and [Smith, 2010]; the former paper takes into account also other systems of many-valued logics.
101. On a related point, Ł$_{\aleph}$ implies that the degree of a conditional (A) $\varphi \rightarrow \psi$ is at least as high as the degree of the associated disjunction (B) $\neg\varphi \vee \psi$, and that the latter in turn must be equal in value to a negated conjunction of the form (C) $\neg(\varphi \wedge \neg\psi)$. Assuming that degrees should preserve orderings in plausibility, Weatherson ([Weatherson, 2005]) contends that this account of the connectives does not match with ordinary speakers' assessments, as far as instances of (TOL) and their reformulations in the form of (B) and (C) are concerned. According to him, expressions of tolerance of the form (B) are the most plausible, followed by instances of the form (A) and then (C). However, empirical experiments reported in [Serchuk et al., ta] suggest that indeed, contrary to Weatherson's claim, conditional expressions of tolerance of the form (A) are the most persuasive. Contrary to what should be expected, starting from Ł$_{\aleph}$, however, rankings of persuasiveness for expressions of tolerance of the form (B) and (C) were not exactly the same.
102. There is a common argument for the assumption of degrees of truth that invokes comparisons with respect to everyday concepts like 'tall': e.g., if x is taller than y, we can infer that the degree of truth of 'x is tall' is greater than 'y is tall' (e.g., see [Forbes, 1983, pp. 241–2]). But this seems to be a *non sequitur* (see [Keefe, 2000, Chapter 4]). On the related idea that degrees of truth may be interpreted as numerical measures of an underlying property, see the discussion in [Keefe, 1998] [Keefe, 2000] [Keefe, 2003], and [Smith, 2003].
103. Indeed, there is an ongoing serious discussion in artificial intelligence on operationalist interpretations of Ł$_{\aleph}$ and other 'fuzzy semantic' frameworks (for an introduction to this discussion, see [Lawry, 2006, Chapter 1]). That said, it is hard to see that the options that have been considered in this discussion may lend continuum-valued semantics more 'intuitive content'.

104. For replies to the *tu quoque* objection, in defence of a continuum-valued semantics, see [MacFarlane, 2010, Section 25.3.1] and [Smith, 2008, Chapter 3.5.5]. MacFarlane grants that the distinction between 'true' and 'untrue' applications of vague predicates should be vague as well, but that this type of vagueness is rather epistemic and therefore requiring a different kind of model. Smith denies that there is any conflict between the assumption of higher-order vagueness and a commitment to cut-offs of the said type. According to him, the vagueness (including higher-order vagueness) of a predicate is exhaustively described by the following 'closeness' constraint: 'If a and b are very close in F-relevant respects, then 'Fa' and 'Fb' are very close in respect of truth.' [Smith, 2008, Chapter 3.4]
105. For further critical discussion of continuum-valued semantics, see [Williamson, 1994, Chapter 4].
106. [Hughes and Cresswell, 1996, Chapters 1.2 and 1.3].
107. Also, McGee's and McLaughlin's account in [McGee and McLaughlin, 1995] may be interpreted as a proposal in this spirit.
108. The most outspoken defenders of this line are [McGee and McLaughlin, 1995]. See also [Belnap, 2009] for a defence of local validity; his argument is not related to vagueness specifically though.
109. For further comparative discussion of the two relations of logical consequence, see [Kremer and Kremer, 2003], [Varzi, 2007], and [Cobreros, tab].
110. Cobreros [Cobreros, 2008] defends a so-called 'regional' notion of logical consequence, according to which: $\Gamma \models_{SpV-R} \varphi$ iff for every frame $\langle W, R\rangle \in \mathcal{F}$, for every model $\langle W, R, v\rangle$ based on the frame $\langle W, R\rangle$: if for every $w \in W$, if $v_{w'}(\alpha) = 1$, for every w' such that wRw', for every sentence $\alpha \in \Gamma$, then also $v_{w'}(\varphi) = 1$, for every w' such that wRw'. That is, logical consequence is thought of as preservation of definite truth (or 'regional truth'). For still other interesting options in a standard possible-worlds setting, see [Bennett, 1998].
111. Another non-standard version of 'supervaluationism' is Burgess' and Humberstone's natural deduction system (in [Burgess and Humberstone, 1987, pp. 200–4]), which preserves distributivity of supertruth over disjunction.
112. For this and other technical results on supervaluationist logical consequence, see [Kremer and Kremer, 2003].
113. The question whether SpV-type counterinstances to truth-value functionality have psychological reality seems still unexplored. For a model of rational credence for supervaluationist frameworks, see [Dietz, 2008], [Dietz, 2010].
114. As far as 'analytically valid' inferences involving sentences that are borderline vague are concerned, it seems that the validity of such inferences can be accommodated in many-valued frameworks as well; for example, see Landman's adoption of a refined Strong Kleene framework in [Landman, 1991, Chapter 3.5].
115. To be clear, it is not suggested that the said rules fail *whenever* they involve discharged premises containing a $\mathcal{D}$-operator. For instance, not only the inference from $\mathcal{D}\varphi$ to $\mathcal{D}\varphi$, but also the associated conditional $\mathcal{D}\varphi \rightarrow \mathcal{D}\varphi$ is valid on SpV. For the question to what extent rules of classical natural deduction are sustainable in some restricted version, see the discussion in [Keefe, 2000, Chapter 7.4], [Varzi, 2007, Section 4], and [Cobreros, tab].
116. Fara in fact means to target truth-value gap accounts of borderline vagueness in general. But, as it is not clear how to model a $\mathcal{D}$-operator that allows for higher-order vagueness in alternative frameworks that are typically associated with a truth-value gap account (K_3, $Ł_{\aleph}$), it seems legitimate to discuss her argument as a challenge to SpV in the first instance.
117. Take a sorites series for a predicate T with m members $\langle 1, \ldots m\rangle$, where $T(1)$ is clearly true and $\neg T(m)$ is clearly true as well. By $m-1$ applications of ($\mathcal{D}$–INTRO), from $T(1)$

it follows that $\mathcal{D}^{m-1}T(1)$. But this is inconsistent with (GP–GEN) and ($\mathcal{D}$–INTRO), as can be shown by the following argument:

$$\begin{array}{rl}
\neg T(m) & \\
\mathcal{D}\neg T(m) & \mathcal{D}\text{–INTRO} \\
\neg \mathcal{D}T(m-1) & \text{Gap principle for } T(x) \\
\mathcal{D}\neg \mathcal{D}T(m-1) & \mathcal{D}\text{–INTRO} \\
\neg \mathcal{D}^2T(m-2) & \text{Gap principle for } \mathcal{D}T(x) \\
\mathcal{D}\neg \mathcal{D}^2T(m-2) & \mathcal{D}\text{–INTRO} \\
\neg \mathcal{D}^3T(m-3) & \text{Gap principle for } \mathcal{D}^2T(x) \\
\vdots & \\
\neg \mathcal{D}^{m-1}T(1) & \text{Gap principle for } \mathcal{D}^{m-2}T(x)
\end{array}$$

118. As Cobreros ([Cobreros, 2010]) observes, Fara's result does not carry over to *SpV-L*, nor to his 'regional' version of 'supervaluationism'.
119. Hyde and Colyvan ([Hyde, 1997] and [Hyde and Colyvan, 2008]) exploit the duality between the two logics as an argument for the more general claim that *SbV* is as good an option for vagueness as *SpV*.
120. For a credit point in favour of *SbV* and against *SpV*, see Cobreros' [Cobreros, taa], who shows that a strengthened version of Fara's argument (in [Fara, 2003]) threatens even the weaker *SpV LOCAL*, but that it does not carry over to *SbV*.

8 Negation

Edwin Mares

Edwin Mares

Chapter Overview

1. Introduction 181
2. Classical Negation 183
  2.1 Classical Negation and Truth Functional Semantics 183
  2.2 De Morgan's Laws, Non-Contradiction, and Excluded Middle 183
3. Negation in Many-Valued Logic 185
  3.1 Kleene and Łukaseiwicz Logics 185
  3.2 Varieties of Negation in Many-Valued Logic 188
4. Application: Paraconsistent Logic 189
  4.1 Introducing Paraconsistent Logic 189
  4.2 Many-Valued Paraconsistent Logic 190
  4.3 Modal Approaches to Paraconsistent Logic 193
5. Negation in Intuitionist Logic 195
  5.1 Introducing Intuitionism 195
  5.2 The BHK Interpretation of Intuitionist Logic 196
  5.3 Kripke's Semantics for Intuitionist Logic 197
  5.4 The Falsum and Negation 199
  5.5 Natural Deduction for Intuitionist and Classical Logic 200
  5.6 Minimal Logic 202
6. Negation and Information 203
  6.1 Language, Logic, and Situations 203
  6.2 Information Conditions and the (In)compatibility Semantics for Negation 205
7. Application: Relevant Logic 207
  7.1 Introducing Relevant Logic 207
  7.2 Natural Deduction for Relevant Logic 208
  7.3 Negation in Relevant Logic 211
8. Summing Up 213
Acknowledgements 214
Notes 214

1. Introduction

Negation is an especially interesting connective. Many non-classical logics have been constructed to avoid certain aspects of classical negation. The two most controversial principles of classical negation have been the so-called law of excluded middle, that is,

$$A \vee \neg A$$

and the rule of ex falso quodlibet, i.e.,

$$\frac{\begin{array}{c} A \\ \neg A \end{array}}{\therefore\ B}.$$

The law of excluded middle is a schema. Accepting it means that we accept all substitution instances of it, such as $p \vee \neg p$, $(p \wedge q) \vee \neg(p \wedge q)$, and so on. If we treat disjunction in the standard way and take the negation of a statement A to mean that A is false, accepting excluded middle forces us also to accept the principle of *bivalence*, which is the dictum that every statement is either true or false. Some philosophers hold that vague predicates, such as 'is bald' and 'is a heap' violate bivalence (see Chapter 7). Some other philosophers think that mathematical statements do not obey bivalence (see Section 5). If one wants to reject bivalence, one must opt for either a non-standard treatment of disjunction – such as supervaluationism (see Chapter 7) – or reject classical negation.

The rule of ex falso quodlibet has been rejected by some logicians merely because it is counterintuitive. Among these are relevant logicians. For relevant logicians the problem with ex falso is that it has instances in which its premises are completely irrelevant to its conclusion, for example,

$$\frac{\begin{array}{c} 2+2=4 \\ 2+2 \neq 4 \end{array}}{\therefore\ \text{the moon is made of green cheese.}}$$

(see Section 7). Paraconsistent logicians, on the other hand, point out that logic may be made more useful by abandoning ex falso. We all have inconsistent beliefs, we sometimes tell inconsistent stories, and scientists have even used the occasional inconsistent theory. We are able to reason about inconsistent beliefs, stories, and theories in useful and important ways. We don't attribute to them the commitment that every proposition is true. Rather, we seem to use more subtle principles. Paraconsistent logicians – at least some of them – attempt to represent the reasoning process that we use in understanding inconsistent theories, stories, beliefs, and so on, in logical systems. We will examine some of these in Section 4.

In studying the logical connectives, philosophers of logic typically adopt one of two different perspectives. The first perspective is that of *model theory*. Philosophers often hold that it is an important criterion of the success of a logical system that it can be given an intuitive model theory. A model theory, as a philosophical theory, is supposed to give truth conditions connected with the various parts of the logical language. For example, the classical truth tables give an inductive method for determining the truth value of any complex sentence (of the language of classical propositional logic) given that one knows the truth value of all of the atomic sentences involved. Moreover, on one very popular philosophy of language, the meaning of a statement is the set of possible conditions under which it is true. A model theory, by setting out a theory of truth for a logical language, also gives us a theory of meaning for the sentences of that language.

A rather different perspective on logic is that of *proof theory*. A proof theory is just what is sounds like. It is a logical theory of how to prove the valid formulas of a given logic. We will look at the natural deduction systems for several of the systems that we examine. Most readers will be familiar with some form of natural deduction system from their introductory logic courses.[1] Some philosophers think that the way in which a given connective can be used in a proof system tells us the meaning of that connective. They hold, for example, that the meaning of conjunction in most logical systems is defined by the fact that it can be used to connect two formulas that have already been proven and that, given the proof of the conjunction of two formulas we can prove either or both of those statements. But even if we do not think that meaning of a connective is defined by its role in a proof system, we can see that having a good proof system is extremely important. We have very strong intuitions about what sort of inferences are good and which are not. If a proof system makes valid the good ones and not the bad ones, this is an important virtue of the proof system and a good reason to adopt it as our theory of deductive inference.[2]

In this chapter, we will look at negation from both a model-theoretic and a proof-theoretic points of view. My own view is that by going back and forth between these two perspectives can provide a useful system of 'checks and balances' on one's choice of a logical system. For if one adopts a reasonable looking model theory, but it supports a very unintuitive proof theory, then there is a problem to be sorted out – what are our intuitions about proof telling us if they are largely wrong? Unfortunately, not all of the systems we examine have intuitive proof theories.[3] In particular, the many-valued logics that we examine do not have reasonable natural deduction systems.[4] So we examine them only from the perspective of model theory.

2. Classical Negation

2.1 Classical Negation and Truth Functional Semantics

We begin with the most familiar form of negation – negation in classical logic or 'classical negation'. The best way to motivate classical negation is by examining its model-theoretic semantics.

According to the standard semantics of classical logic,[5] there are two truth values – true (1) and false (0). All of the logical operators are treated in this semantics as truth functions. An n-place operator is a function from sequences of n truth values to a truth value. The operators only distinguish between statements in so far as they can distinguish between their truth values. Because the operators are taken to be functions and there are two truth values, we can represent them by the familiar two-valued truth tables. For example, the behaviour of conjunction can be represented as follows:

$\wedge$	1	0
1	1	0
0	0	0

We can also think of conjunction as selecting the minimum value of its arguments. More formally, $V(A \wedge B) = \min\{V(A), V(B)\}$. Similarly, disjunction is a function that selects the maximum value of its arguments, i.e., $V(A \vee B) = \max\{V(A), V(B)\}$.

Thus, we have two constraints on the way we can think about the connectives: (1) the connectives are truth functions and (2) the only truth values are true and false. Given these two constraints, there really is only one choice for what negation could be. It must be a function that takes true to false and false to true, or $V(\neg A) = 1 - V(A)$. Negation's role in classical logic is to change (or 'flip') the truth value of the statement that is negated.

2.2 De Morgan's Laws, Non-Contradiction, and Excluded Middle

Classical logic has many virtues. Among these virtues is the fact that in classical logic the connectives are related to one another in elegant ways that often involve negation. Some important examples of these relationships are the De Morgan laws, which involve negation, disjunction, and conjunction. Here are four of De Morgan's laws:

$$\text{(DM1) } (A \wedge B) \leftrightarrow \neg(\neg A \vee \neg B);$$

$$\text{(DM2) } (A \vee B) \leftrightarrow \neg(\neg A \wedge \neg B);$$

$$\text{(DM3) } \neg(A \wedge B) \leftrightarrow (\neg A \vee \neg B);$$

$$\text{(DM4) } \neg(A \vee B) \leftrightarrow (\neg A \wedge \neg B).$$

What is nice about the De Morgan laws is that they enable us to select as a primitive only one of disjunction or conjunction and define the other in terms of it and negation.

In algebraic terms we understand a logical system as being characterized by a class of algebraic structures. For classical logic, these structures are called *boolean algebras*. Many of you who have studied some computer science will be familiar with the two-element boolean algebra – which has the elements 0 and 1. But there are infinitely many boolean algebras. There is one for each power of 2. This means that for all natural numbers n, there are boolean algebras with 2^n elements. In each algebra, there is an ordering relation on elements. In the two-element boolean algebra, 0 is less than 1. The disjunction of two elements in an algebra (also known as the *join* of those two elements) is their least upper bound. This means that if we have two elements a and b, then $a \vee b$ is an element of the algebra that is greater than both a and b but less than any other element that is greater than both a and b. Similarly, $a \wedge b$ (the *meet* of a and b) is an element that is less than a and less than b but is greater than any other element that is less than both a and b. If we look at the structure of the fragment of the part of the algebra that contains only the elements, meet, and join – called the lattice of the algebra – then we have his remarkably symmetrical entity. If we 'turn it upside down' and treat meets as joins and joins as meets and replacing the ordering relation on the algebra with its complement, then we also have a lattice. In boolean algebras, adding negation allows us to maintain this lovely symmetry. The De Morgan laws express these symmetries. In algebraic terms they tell us that the meet of a and b is the negation (or 'complement') of the join of the complements of a and b. Similarly, the join of a and b is the negation of the meet of the complements of a and b. In sort turning a boolean algebra upside down produces a boolean algebra. From an aesthetic point of view at least, this is a very nice quality of boolean algebras (and hence of the logic that they characterize – classical logic).

Let's set aside the De Morgan laws briefly to consider what many philosophers, from Aristotle to the present, think is a central principle of logic, that is, the law of non-contradiction:

$$\neg(A \wedge \neg A)$$

The principle of non-contradiction, on its standard reading, tells us that, for any particular proposition, it is not both true and false. The principle that no statement is both true and false is called the *principle of consistency*. The difference between the principle of consistency and the principle of non-contradiction is that the former must be stated in a semantic metalanguage, whereas the latter is a thesis of logical systems. As we shall see in Section 3.1 there are logical systems that obey the principle of consistency but do not make valid the law of non-contradiction. And, as we shall see in Section 4, there are logics that include the law of non-contradiction but whose semantics do not obey the principle of consistency. In classical logic, however, the principle of consistency can be said to be expressed adequately by the law of non-contradiction.

If we accept the law of non-contradiction, together with DM3, then we also have to accept the following formula:

$$\neg A \vee \neg\neg A$$

If we also accept the principle of double negation, i.e.,

$$\neg\neg A \leftrightarrow A$$

Then we obtain the law of excluded middle:

$$\neg A \vee A$$

The law of excluded middle tells us, on its standard reading, that bivalence holds, i.e., that every proposition is either true or false. If we want to reject excluded middle, we must reject either the law of non-contradiction, DM3, or the principle of double negation.[6] As we shall see, each of these paths has been taken by someone.

3. Negation in Many-Valued Logic

3.1 Kleene and Łukaseiwicz Logics

One simple way of rejecting bivalence is to move to a many-valued logic. With many-valued logic, we keep the truth-functionality of classical logic, but merely add more truth values. The simplest many-valued logics are three-valued logics. We start with what is perhaps the simplest of these, Kleene's strong three-valued logic [Kleene, 1952].

One reason for wanting a three-valued logic is to act as a basis of a theory of *presupposition* [Strawson, 1950]. Consider the statement

The present king of France is bald.

On the presupposition view, the description 'the present king of France' is a singular term. This sentence is true if and only if the thing denoted by the description, i.e., the present king of France is bald. It is false if the present king of France fails to be bald. But if the present king of France does not exist, then 'he' can neither be bald or fail to be bald. So, according to the presupposition theory, the displayed sentence is neither true nor false. The sentence *presupposes* the existence of a present king of France – it requires his existence in order to be either true or false. Thus, in order to formalize the theory of presupposition we need a way of making some sentences be neither true nor false. Kleene's three-valued logic provides one basis for a formal theory of presupposition.

Kleene's logic, K3, has the truth values 0, 1, and .5. Let's start with the connectives conjunction, disjunction, and negation.[7] Here are their truth tables:

$\wedge$	1	.5	0
1	1	.5	0
.5	.5	.5	0
0	0	0	0

$\vee$	1	.5	0
1	1	1	1
.5	1	.5	.5
0	1	.5	0

$\neg$	
1	0
.5	.5
0	1

Conjunction in K3 takes the values of two formulas and returns the lesser of those values. More formally, $V(A \wedge B) = \min\{V(A), V(B)\}$. Similarly, the value of a disjunction is the greater of the values of the formulas disjoined, i.e., $V(A \vee B) = \max\{V(A), V(B)\}$. And the value of a negation is determined by $V(\neg A) = 1 - V(A)$.

The equations that we have just given are the same as those that we gave for classical logic in Section 2.1. This shows that K3 is a *generalization* of classical logic. It adapts the classical treatment of the connectives to the three valued framework. There may be more than one way, however, to generalize logical ideas. Consider implication. One way of understanding implication in classical logic is through the following definition:

$$A \rightarrow B =_{Df} \neg A \vee B$$

This is, typically, the way in which implication is understood in K3 (see, e.g., [Rescher, 1969], [Urquhart, 1986], [Priest, 2008]). This way of understanding three-valued negation has its drawbacks. Consider a case in which the truth value of p is .5. Then the value of $p \rightarrow p$ is also .5. This means that $p \rightarrow p$ is not a K3-tautology – it is not true on every assignment of values to the propositional variables. In fact, in K3 there are no tautologies. This is a strange feature of this logic. We can remedy this by adopting another generalization of the classical treatment of implication. On this approach, implication is given the following truth table:

$\rightarrow$	1	.5	0
1	1	.5	0
.5	1	1	.5
0	1	1	1

If we look at just the values that are generated by the truth values 1 and 0 we get classical implication. The full three-valued logic is the implication of *Jan Łukasiewicz's* three valued logic, Ł3 [Łukasiewicz, 1970]. His logic is just defined by the K3-truth tables for conjunction, disjunction, and negation, together this truth table for the implication.

The logic Ł3 does have tautologies. Among them are the principle of double negation and all of de Morgan's laws. But it rejects bivalence and also the law of excluded middle. This means that it also rejects the law of non-contradiction, $\neg(A \wedge \neg A)$. Let p a propositional variable with the value .5. Then $\neg(p \wedge \neg p)$ also has the value .5.

There are further many-valued generalizations of classical logic. For each natural number n, we can construct an n-valued version of K3 and Ł3, merely by taking the set of truth values to be $\{0, \frac{1}{n-1}, \ldots, \frac{n-2}{n-1}, 1\}$. For example, K4 and Ł4 have the truth values $\{0, \frac{1}{3}, \frac{2}{3}, 1\}$ and K5 and Ł5 have the truth values $\{0, \frac{1}{4}, \frac{1}{2}, \frac{3}{4}, 1\}$. As usual, we have $V(A \wedge B) = \min\{V(A), V(B)\}$, $V(A \vee B) = \max\{V(A), V(B)\}$, and $V(\neg A) = 1 - V(A)$ for both of these logics. For Ł3, the truth value of implicational formulas is given by the following equation:

$$V(A \rightarrow B) = \begin{cases} 1 \text{ if } V(B) \geq V(A) \\ 1 - (V(A) - V(B)) \text{ otherwise} \end{cases}$$

If we set n to 2, then we generate the truth table for classical implication. If we set it to 3, of course we have Ł3. And so on.

There are even infinitely valued logics. The logics $Ł_\omega$ and K_ω are just those defined by calculating truth values using the above equations on the set of rational numbers between (and including) 0 and 1.[8] We can also use as our truth values the set of real numbers $[0, 1]$ – the closed real interval between 0 and 1. The logic $K_{[0,1]}$ is also called *fuzzy logic*. One use of infinite valued logics is as a basis for a theory of vagueness (see Chapter 7). For example, let $H(n)$ mean 'n grains of sand is a heap'. Then, according to this way of treating the sorites paradox, at certain points, $V(H(n)) < V(H(n+1))$, although they will be extremely close in value. Thus, we retain the intuition that adding one grain of sand doesn't turn a (complete) non-heap into a heap, but we also can see how after adding a certain number of grains we do actually create something that we can call a heap. Thus, the use of infinite-valued logics is supposed to provide a solution to the sorites paradox.

3.2 Varieties of Negation in Many-Valued Logic

Consider again the three truth values 0, .5, and 1. The negation that we have discussed merely takes 0 to 1, and vice versa, and takes .5 to itself. But this is not the only form of negation that is definable over these values. Consider the following sentence of *loglish* (a mixture of formal logic and English):

p fails to be true.

The operator 'fails to be true' is not naturally formalized using $\neg$ as defined using the truth table given in Section 3.1. For, intuitively, 'p fails to be true' should be true when p gets the value .5, since it fails then to have the true value 1. Thus, we can define another negation connective; let us formalize it by $\sim$. This second negation has the following truth table:

$\sim$	
1	0
.5	1
0	1

If we do add $\sim$ to our logical language, we get a form of the law of excluded middle, i.e., $A \vee \sim A$. It is, however, an interesting question as to whether we have bivalence. In a sense we do not. Not every statement has the value 1 or 0, and so we can correctly say that not every statement is either true or false. But we can say that every statement is either true or fails to be true. Of course we could say this without having $\sim$ in our language, but now we can express that fact in the logical language itself.

Another form of many-valued negation is due to Emil Post ([Post, 1921]). Using the same truth values as we have been using, we can represent Post's negation, $-$, as follows:

$-$	
1	.5
.5	0
0	1

Here we have a *cyclic* negation. Post developed n-valued logics for all natural numbers n. Instead of representing the truth values as real or rational numbers, he used the natural numbers themselves. He used 1 as the true value, as usual, but the number n as the false value. So we now understand disjunction as taking two values to their minimal value and conjunction as taking two values to their maximal value, inverting the equations given in Section 3.1 above.

Post's generalized form negation is given by the following table:

$-$	
1	2
$\vdots$	$\vdots$
$n-2$	$n-1$
$n-1$	n
n	1

When $n = 2$, we have the classical table for negation (replacing 0 with 2). So, Post negation counts as a generalization of classical negation, even though in the cases in which n is greater than 2 the negation of 1 is not the false value.[9]

Focusing on Post negation raises an interesting question: what makes a connective a form of negation? This is a difficult question to answer. We will see, when we discuss sequent calculus, that we can give an answer (albeit a controversial one) in a proof-theoretic framework. But it is difficult to say what truth-conditional features are necessary or sufficient for a connective to be considered a form of negation. To most of us, Post's 'negation' does not look like a form of negation, because we do not use 'not' to mean this. But it is a generalization of classical negation, and this is a good reason to treat it as a form of negation.

4. Application: Paraconsistent Logic

4.1 Introducing Paraconsistent Logic

So far we have been concentrating on the rejection of bivalence. Many-valued logics have also been used to make sense of the rejection of the *principle of consistency*. The principle of consistency says that no statement and its negation can both be true at the same time.

It is natural to think that there is a close link between the principle of consistency and the law of non-contradiction, i.e., $\neg(A \wedge \neg A)$, just as there is between the principle of bivalence and the law of excluded middle, but the link is far more tenuous in the case of the law of non-contradiction. The principle of consistency is more closely bound up with a rule of inference – the rule of ex falso quodlibet (EFQ):

$$\frac{\begin{array}{c}A\\ \neg A\end{array}}{\therefore B}$$

In classical logic, from two contradictory premises, any proposition follows. A logic is paraconsistent if and only if it does not make this rule valid.

There are various reasons for wanting to reject EFQ. We all have inconsistent beliefs. Scientists have used inconsistent theories. We read or watch, but fully understand, inconsistent stories. To explain how we can understand and use inconsistent beliefs, stories, or theories, we need to explain how we can make deductive inferences about their contents. People rarely, if ever, infer that every proposition is true in inconsistent stories or that every proposition would be made true by one's inconsistent beliefs or an inconsistent theory. In order to understand the norms that govern our uses of theories, beliefs, and stories, we need a paraconsistent logic.

Some philosophers take a more extreme view. They believe that there are true contradictions. This view is known as *dialetheism*. One motivation for dialetheism is that it can act as the basis for a semantically closed view of language, that is, the treatment of a language as being its own metatheory. Consider for the sake of contrast a theory of truth that takes K_3 as its logical basis and which treats all liar-like sentences as being neither true nor false (see, Chapter 13). Now consider the so-called strengthened liar sentence:

This sentence fails to be true.

If this sentence is given either the values 0 or .5 then, intuitively, it is true and so it should 'also' be given the value 1. But, if it is true, then it is also false. One way of dealing with the strengthened liar is to claim that it is both true and false. Then, since it is false, it is true. But since it is true it is also false.[10]

In what follows we will examine some simple paraconsistent logics through their model theories.

4.2 Many-Valued Paraconsistent Logic

Perhaps the simplest paraconsistent logic is Graham Priest's logic LP (for 'logic of paradox') ([Priest, 1979]). The truth values for LP are the same as they are for K_3 – 0, .5, and 1. Moreover, the truth tables for the connectives for LP are the same as they are for K_3. What is different is that in LP, we consider both 1 and .5 to be 'true values'. As usual 1 is understood as *true*, but now .5 is understood as *both true and false*. We thus say that $\{1, .5\}$ is the set of *designated values* for LP.

LP has some very interesting properties. First, it has exactly the same tautologies as classical propositional logic ([Priest, 1979]). An LP tautology is a formula that gets a designated value on every row of its truth table. On one reading a logic is just the set of its tautologies, and so LP can be considered to be the same as classical logic and that the LP model theory gives a paraconsistent interpretation to classical logic.

But not every inference valid in classical logic is valid in LP. An inference is LP-valid if and only if every assignment of truth values to propositional variables which give all the premises of the inference designated values also gives its conclusion a designated value. Consider, for example, the following instance of EFQ:

$$\begin{array}{c} p \\ \neg p \\ \hline \therefore q \end{array}$$

Let $v(p) = .5$ and $v(q) = 0$. Then $v(\neg p) = .5$. So, both p and $\neg p$ have designated values on v and q has a non-designated value. So, this instance of EFQ is invalid. Somewhat less pleasing is the fact that modus ponens is also invalid. In LP, as in K3, it is usual to define $A \to B$ as $\neg A \vee B$. Now consider the following inference:

$$\begin{array}{c} p \to q \\ p \\ \hline \therefore q \end{array}$$

Let $v(p) = .5$ and $v(q) = 0$ as before. Then $v(p \to q) = v(\neg p \vee q) = \max\{(1 - .5), 0\} = \max\{.5, 0\} = .5$. So, both $v(p)$ and $v(p \to q)$ are designated, but $v(q)$ is not. Therefore this instance of modus ponens is invalid.[11]

Because LP does not make modus ponens valid, LP's implication does not really look like a true form of implication.[12] To rectify this, one might want to add an implication connective to LP that has a different truth table:

$\to$	1	.5	0
1	1	0	0
.5	1	.5	0
0	1	1	1

The resulting logic is called RM3. RM3 validates modus ponens. But RM3 makes a very poor basis for a dialethic theory of truth. One reason for this concerns its treatment of Curry's paradox. Consider the sentence

(C) If this sentence is true, then the moon is made of green cheese.

Let 'g' be short for 'the moon is made of green cheese'. Then consider the truth value of $C \to g$. If C gets the value 1, then because C has the same value as (since it is a name for) $C \to g$, $C \to g$ has the value 1. Then, by the truth table, g has the value 1. So the moon is made of green cheese. Now suppose that C has the 0. Then $C \to g$ has the value 1. But C and $C \to g$ must have the same value. So, C cannot have the value 0. Finally suppose that C has the value .5. Then $C \to g$ has the value .5. But this means that g also has the value .5, because the consequent of any implication with the value .5 also has the value .5. This means that it is both true and false that the moon is made of green cheese. But it is just plain false that the moon is made of green cheese – it is not true at all! Thus, RM3 gives us a very unsatisfactory analysis of Curry's paradox.

In fact the problem of how to construct a conditional that is appropriate for a dialethic theory of truth is an important and interesting problem but one that is very difficult. We will return to this issue in Section 4.3 below.

Perhaps a better way of thinking about the values of LP is due to J. M. Dunn.[13] On this view, formulas are given *sets* of classical truth values. For LP, only the non-empty sets, $\{1\}$, $\{0\}$, and $\{0, 1\}$ are allowed as values. Given an assignment of values to propositional variables, we then can calculate the value of complex formulas using the following clauses:

- $1 \in v(A \wedge B)$ iff $1 \in v(A)$ and $1 \in v(B)$
- $0 \in v(A \wedge B)$ iff $0 \in v(A)$ or $0 \in v(B)$
- $1 \in v(A \vee B)$ iff $1 \in v(A)$ or $1 \in v(B)$
- $0 \in v(A \vee B)$ iff $0 \in v(A)$ and $0 \in v(B)$
- $1 \in v(\neg A)$ iff $0 \in v(A)$
- $0 \in v(\neg A)$ iff $1 \in v(A)$

If we read '$1 \in v(A)$' as 'A is true according to v' and '$0 \in v(A)$' as 'A is false according to v', then we have clauses that sound very much like the standard classical truth conditions for the connectives. But the difference here is that both truth and falsity conditions are required and that a formula may have more than one truth value.

A generalization of this semantics allows formulas to be assigned the empty set, $\emptyset$. The resulting logic is the system D4.[14] As in the case of LP, the D4 designated values are $\{1\}$ and $\{1, 0\}$. In other words, a value X is designated if and only if $1 \in X$. This makes sense, because it says that a value is designated if and only if truth is in it.

One way of reading the 'set of values' semantics is of course the dialethic reading – that some formulas can have more than one truth value. Another reading is due to Nuel Belnap ([Belnap, 1977b], [Belnap, 1977a]). On Belnap's interpretation, to say that 1 is in the value of a given formula is to be told that the formula is true and for 0 to be in its value is to be told that the formula is false. Of course, we may be told that a formula is true, that it is false, that it is both, or we may have no information about its truth value at all. If we have no information about a formula, then the value we assign to it is $\emptyset$.

As we have seen, we can think of the truth values as being ordered. Until now, all the models we have examined have had values that are most intuitively understood as being *linearly ordered*. A linear order is just as it sounds – the values are ordered in a line. In a linear order each value is either greater than or less than every other value. The values of D4 values, however, are not linearly ordered. They have a partial ordering. We can represent their order by a Hasse diagram:

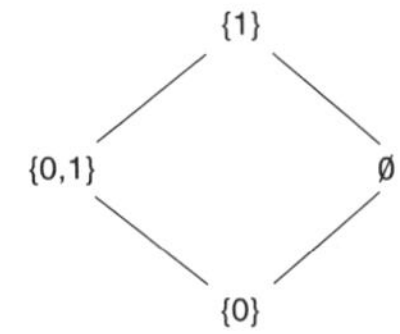

Higher values in the ordering are nearer the top of the diagram. Conjunction is understood in terms of the meet of two points (their greatest lower bound) and disjunction in terms of their join (least upper bound). The meet of $\{0, 1\}$ and $\emptyset$ is $\{0\}$ and their join is $\{1\}$. So, given the dialethic reading of the truth values, the conjunction of a formula that is both true and false and one that is neither true nor false is itself just false, and their disjunction is just true. The conjunction of formulas with the values $\{0\}$ and $\{0, 1\}$ is $\{0\}$ and their disjunction has the value $\{0, 1\}$, and so on.

Negation in D4 has two *fixed points*. The fixed point for an operator is an argument x such that $f(x) = x$. Recall Dunn's clauses for negation:

$$1 \in v(\neg A) \text{ iff } 0 \in v(A)$$

$$0 \in v(\neg A) \text{ iff } 1 \in v(A)$$

According to these clauses, if $v(A) = \emptyset$, then neither $0 \in v(\neg A)$ nor $1 \in v(\neg A)$. So, if $v(A) = \emptyset$, then $v(\neg A) = \emptyset$. Similarly, if $v(A) = \{0, 1\}$, then both $0 \in v(\neg A)$ and $1 \in v(\neg A)$, so $v(\neg A) = \{0, 1\}$. So both $\emptyset$ and $\{1, 0\}$ are fixed points for negation. The negation of $\{1\}$ is $\{0\}$ and the negation of $\{0\}$ is $\{1\}$. If we think of the values that a formula can get in D4 if its propositional variables only have either the value $\{0\}$ or the value $\{1\}$, then we just get back the classical truth tables. So D4 is (once again) a generalization of classical logic.

We say that the two-valued boolean algebra is *embedded* in the algebra for D4 (given in the Hasse diagram above). The three-point algebra that is made up of the truth values of K3 and the three membered algebra made up of the truth values of LP are also embedded in the algebra for D4. For K3, we map the values 1 to $\{1\}$, .5 to $\emptyset$, and 1 to $\{1\}$. For LP we, of course, map 1 to $\{1\}$, .5 to $\{0, 1\}$, and 1 to $\{1\}$. These translations preserve the values of conjunctions, disjunctions, and negations. This means that D4 has certain properties that LP and K3 have. Like K3, D4 has no valid formulas. Like LP, modus ponens and EFQ are invalid in D4.

4.3 Modal Approaches to Paraconsistent Logic

I call 'modal approaches' to paraconsistent logic those semantic theories that utilize worlds, like the possible worlds of Kripke's semantics for modal logic. There are two ways in which worlds are used in models for paraconsistent logic.

They are either employed to provide alternatives to the many-valued semantics or as supplements to the many-valued semantics.

Perhaps the most straightforward worlds-based alternative to many-valued semantics is due to Jean-Yves Beziau ([Beziau, 2002]). Consider a model for a modal logic, $\mathcal{M} =< W, R, v >$ (see Chapter 11). We take a standard modal language, with possibility, necessity, conjunction, disjunction, and implication. We then define a second negation, $\sim$:[15]

$$\sim A =_{Df} \diamond\neg A.$$

We now have a paraconsistent negation. For there may be in a model a world w such that wRw' and formulas A and B for which $\mathcal{M}, w \models A$, $\mathcal{M}, w \models \sim A$, and $\mathcal{M}, w \not\models B$.

A similar idea, but which requires a more sweeping reinterpretation of the semantics, is the following simplification of Stanisław Jaśkowki's discussive logic (see [Jaśkowki, 1969]). This time we drop the modal operators from our original language. We once again take a model for a modal logic $\mathcal{M} =< W, R, v >$ and define a satisfaction relation $\models'$ such that

$$\mathcal{M}, w \models' A \text{ if and only if } \exists w'(wRw' \wedge \mathcal{M}, w' \models A).$$

With this semantics we can satisfy contradictory formulas at a world without thereby satisfying every formula. We can interpret '$\mathcal{M}, w \models' A$' as saying that the formula A is accepted at w. A group of people may accept contradictory formulas in a conversation. The accessibility relation in our model connects worlds relative to a conversation in those worlds to a set of worlds that the conversation is (ambiguously) about. There are several variants that one can construct of this modelling. I leave those to the reader.

One way of supplementing many-valued paraconsistent logic is to employ worlds to provide truth conditions for a conditional. Here we look briefly at two such logics, due to Priest. The first of these logics is K_4 [Priest, 2008, pp. 163f]. A model for this logic is a pair $< W, v >$, where W is a set of worlds and v is a four-valued assignment of values to propositional variables (where the values are the subsets of $\{0, 1\}$). The value assignment treats conjunction, disjunction, and negation according to the truth and falsity clauses for D4. The clauses for implication are as follows:

$$1 \in v_w(A \rightarrow B) \text{ if and only if}$$
$$\text{for all } w' \in W \text{ if } 1 \in v_{w'}(A), \text{ then } 1 \in v_{w'}(B)$$

$$0 \in v_w(A \rightarrow B) \text{ if and only if}$$
$$\text{for some } w' \in W,\ 1 \in v_{w'}(A) \text{ and } 0 \in v_{w'}(B)$$

One problem with K_4 is that, like RM3, it cannot be used as a basis for a paraconsistent theory of truth. It also falls prey to Curry's paradox. For suppose that

w is an arbitrary world in a K4 model and that $1 \in v_w(C)$. Then, $1 \in v_w(C \rightarrow g)$. But this means that, for all $w' \in W$, if $1 \in v_{w'}(C)$ then $1 \in v_{w'}(g)$. But this means that, for every world w'' in the model, $1 \in v_{w''}(C \rightarrow g)$ and so $1 \in v_{w''}(C)$. But then $1 \in v_{w''}(g)$. So we have proven that the moon is made of green cheese (and necessarily so!).

To rectify this problem, Priest introduces another similar system, N4 ([Priest, 2008, pp. 166–8]). A model for N4 is a triple $< W, N, v >$, where $N \subseteq W$. N is the set of 'normal' worlds. At normal worlds, the truth and falsity conditions for the connectives are exactly the same as they are for K4. At non-normal worlds (the worlds in $W - N$), the truth and falsity conditions for all the connectives except for implication are the same as they are for K4 but the truth and falsity conditions for implication are different. There are no recursive truth or falsity conditions for implication at non-normal worlds. Rather, whether they are true or false (or both or neither) is determined merely by v and not by the truth or falsity of any other formulas.

5. Negation in Intuitionist Logic

5.1 Introducing Intuitionism

Intuitionist logic began as a way of formalizing *intuitionist mathematics*. Intuitionist mathematics was a form of mathematical practice that began in the early years of the twentieth century as a reaction to classical mathematics. Classical logic began (in the work of Frege, Bertrand Russell, and others) as a way of understanding the inferences made in classical mathematics. If we are to use the classical notion of validity to codify mathematical inference, then there must be a usable concept of mathematical truth. At the turn of the twentieth century, there were a few such concepts available – let us consider for the sake of contrast the *Platonist* concept of mathematical truth. According to Platonism (a view held by Gottlob Frege and the set theorist Georg Cantor among others), there are entities called 'mathematical objects'. A number is a mathematical object, so is a set, so is a function, and so on. Where are these mathematical objects? They are, according to Platonism, nowhere in space or time – they have their own 'realm'. Platonism has the virtue of giving a straightforward and rather standard theory of truth. A mathematical statement is true if and only if the things it talks about actually have the properties attributed to it by the statement. For example, the statement '$2 + 2 = 4$' is true if and only if applying the function of addition to the pair $< 2, 2 >$ has the value 4.

Platonism, however, clearly also has important difficulties. First, it seems philosophically ad hoc to postulate a special realm of objects just to explain how certain sentences can be true. Second, if these objects are nowhere in space or time, then we cannot perceive them. If we cannot perceive them, how can we know things about them? Surely there is mathematical knowledge, and this fact needs to be explained.

Intuitionism is a reaction against Platonism. We won't go over the original form of intuitionism, because although extremely interesting it is a complicated mix of nineteenth century philosophy and mysticism. Rather, we will look at a more modern form due to Michael Dummett ([Dummett, 2000]). According to this modern form of intuitionism, what is true in mathematics is what can be *constructibly proven*. The idea is that a mathematical statement is true if and only if there is a step-by-step method that will prove it. In effect, what is true is what can (ideally) be proven by a computer.[16] In this move from Platonist truth to constructive proof, we see an attempt to deal with the two problems we have stated above. First, the notion of proof is clearly central to mathematical practice – it is not ad hoc to make it central to a philosophy of mathematics. Second, the intuitionist view that takes truth to be what can be proven explains how we can know mathematical truths. Our proofs show that they are true. The Platonist has to explain why we take proofs in classical logic to show that certain statements about Platonic objects are true. For the intuitionist, mathematical truth is just provability, so no further explanation is needed.

For the intuitionist, talk of mathematical objects is rather misleading. For them, there really isn't any*thing* that we should call the natural numbers, but instead there is counting. What intuitionists study, then, are mathematical processes, such as counting (in arithmetic), collecting things (in intuitionist set theory, sometimes called the 'theory of species'), and so on. We will follow the intuitionists' practice of talking about mathematical objects, but note that this is really shorthand for talk of processes.

In classical mathematics, we talk about infinite sets. In fact, we talk about larger and larger infinite sets: the natural numbers, the real numbers, the set of functions over the real numbers, and so on. If we talk about the process of collecting things, rather than a complete collection itself, we get a rather different notion of infinity. Philosophers distinguish between a never-ending process (sometimes called a 'potential infinity') and a completed infinity. Classical mathematics deals with completed infinities, whereas intuitionists accept only never-ending processes. Given that they reject the notion that there are completed infinities, intuitionists cannot accept the notion that there are different sizes of infinity. This leads also to problems regarding the real numbers (we usually think of irrational numbers in terms of infinitely long strings of digits), and the intuitionist theory of the reals is as a result extremely complicated, as is their treatment of calculus.

5.2 The BHK Interpretation of Intuitionist Logic

In the late 1920s, Arend Heyting developed a logical system in which intuitionist mathematics could be formalized (see [Heyting, 1972]). As we have seen, intuitionism takes what can be proven to be central to its view of mathematics. The usual interpretation of intuitionist logic also takes the notion of proof to be its key notion. Whereas the standard interpretation of classical logic takes that system to formalize the preservation of truth in possible circumstances (as represented by the rows of truth tables), intuitionist logic is taken to codify what can be proven in ideal circumstances. For example, suppose that one comes to understand a property, say, the property of being red. This understanding gives her the ability to construct a set[17] – it gives her the ability to collect together the red things in the world. Let us call this set R. If this agent is a 'logically ideal' agent, then she has certain other abilities as well. She can tell that if an object a is such that $a \in R$ then $\neg\neg a \in R$, and so on.

An interpretation of the intuitionist connectives that uses the conditions under which a statement is proven rather than truth conditions is the Brouwer–Heyting–Kolmogorov (BHK) interpretation, named after L. E. J. Brouwer, Heyting, and Andrey Kolmogorov (the great Russian mathematician). These are the proof clauses for the propositional connectives (taken from [Iamhoff, 2008]):

A proof of $A \wedge B$ is a proof of both A and B
A proof of $A \vee B$ is a proof of either A or B

A proof of $A \rightarrow B$
is a proof that any proof of A can be transformed into a proof of B

A proof of $\neg A$
is a proof that any proof of A can be transformed into a proof of a contradiction.

Note that there is no general procedure given for proving atomic formulas. Our knowledge of such proofs is determined by the contents of the atomic formulas themselves. But we still have a method for understanding complex statements on the basis of our understanding of simple ones, just as in the semantics for classical logic. Thus we say that this is a *compositional* semantics for intuitionist logic.

5.3 Kripke's Semantics for Intuitionist Logic

In the late 1950s, Saul Kripke developed a model theory for intuitionist logic that is rather like his model theory for modal logic ([Kripke, 1965]). Instead of thinking of the points in the model for intuitionism as possible worlds, he thought of them as 'evidential situations'. These evidential situations are circumstances in which an agent has constructed particular mathematical objects, such as the set of red things that we discussed above. Since we will use the term 'situation' in a slightly different way in Section 6.1 below, we will use 'circumstance' for points in Kripke's models for intuitionist logic. Each circumstance is related to further situations in which more things can be constructed and more facts proven about them. Kripke's models consist of a set C of circumstances, an accessibility relation R, which relates circumstances to other circumstances that continue them in this sense. R is reflexive and transitive. The model also, as usual, has a value assignment, v. But there is an interesting added feature of value assignments for intuitionist logic – they have what is known as a hereditariness property. For any circumstances i and j, and any propositional variable p,

$$\text{if } v_i(p) = 1 \text{ and } iRj, \text{ then } v_j(p) = 1.$$

This stipulation makes sense, given the interpretation of the accessibility relation R. What is proven in one circumstance is carried over to its continuations.

A value assignment for propositional variables determines a satisfaction relation between worlds and formulas such that, where $\mathcal{M} =< C, R, v >$ is a model for intuitionist logic,

- $\mathcal{M}, i \models p$ if and only if $v_i(p) = 1$
- $\mathcal{M}, i \models A \wedge B$ if and only if $\mathcal{M}, i \models A$ and $\mathcal{M}, i \models B$
- $\mathcal{M}, i \models A \vee B$ if and only if $\mathcal{M}, i \models A$ or $\mathcal{M}, i \models B$
- $\mathcal{M}, i \models \neg A$ if and only if for all circumstances j, iRj implies $\mathcal{M}, j \not\models A$
- $\mathcal{M}, i \models A \rightarrow B$ if and only if for all circumstances j, iRj implies $j \not\models A$ or $\mathcal{M}, j \models B$.

It is easy to prove that the 'full' hereditariness property holds of this model, that is, for any formula A if $\mathcal{M}, i \models A$ and iRj, then $\mathcal{M}, j \models A$. Note that the metalanguage that we are using in which for formulate the semantics is classical. It is an interesting and very difficult question as to whether intuitionist logic is adequate for the task of formalizing its own model theory ([McCarty, 2008]).

At least with regard to conjunction, disjunction, and implication, we can see that Kripke's semantics captures the BHK interpretation, at least if the connectives used in the BHK interpretation are understood classically. Conjunction and disjunction are straightforward, so let us consider implication. Suppose that an implication $A \rightarrow B$ is proven in circumstance i. Then, on the BHK interpretation, if we are given a proof of A in any continuation of i, then we have the means to prove B. Conversely, suppose that $\mathcal{M}, i \models A \rightarrow B$. Then, if we have a proof of A in any continuation of i, according to Kripke's interpretation, we also can prove B. On the intuitionist view of proof, this is to say that we can turn a proof of A into a proof of B, since for the intuitionist it valid that

$$B \rightarrow (A \rightarrow B).$$

So, if we have a proof of B, we can turn any proof of A into a proof of B according to the BHK interpretation.

5.4 The Falsum and Negation

Relating the treatments of negation in Kripke models to that of the BHK interpretation is a little more difficult. According to the BHK interpretation to prove $\neg A$ is to prove that a contradiction follows from any proof of A. It is easier to formalize this understanding of negation if we have another logical primitive in our language. This logical primitive is a propositional constant or 'zero-place' connective, f. This connective is called a 'falsum', 'the contradiction', or sometimes merely 'the false'. We can also think of it, in intuitionist logic at least, as standing for a particular contradiction such as $0 = 1$. According to intuitionism (and classical logic), all contradictions are logically equivalent, so it does not matter which we choose in our interpretation of the falsum. When we have a falsum in our language we can think of an intuitionist negation, $\neg A$, as meaning the same thing as $A \rightarrow f$. That is, it means the same as 'from a proof of A we can prove a contradiction'.

The proof condition for f is rather simple. There are no proofs of f. Similarly, in Kripke's semantics, the set of circumstances in which f is proven is the empty set. In Kripke's semantics, $\neg A$ is equivalent to $A \rightarrow f$. Here is a brief proof. Let i be an arbitrary circumstance. Suppose first that $\mathcal{M}, i \models A \rightarrow f$. Then for all circumstances j such that iRj, either $\mathcal{M}, j \not\models A$ or $\mathcal{M}, j \models f$. But we know that $\mathcal{M}, j \not\models f$ because f is not satisfied by any circumstance. So $\mathcal{M}, j \not\models A$. Thus, by the proof condition for negation $\mathcal{M}, i \models \neg A$.

Now suppose that $\mathcal{M}, i \models \neg A$. Then, by the proof condition for negation, for all j such that iRj, $\mathcal{M}, j \not\models A$. Then, for any formula B, for all j such that iRj, either $\mathcal{M}, j \not\models A$ or $\mathcal{M}, j \models B$. So, in particular, for all j such that iRj, either $\mathcal{M}, j \not\models A$ or $\mathcal{M}, j \models f$. Hence $\mathcal{M}, j \models A \rightarrow f$. Therefore we have proven that Kripke's condition for negation and the condition using the falsum are equivalent.

We can see that the intuitionist notion of negation does not support the law of excluded middle,

$$A \vee \neg A.$$

Interpreting negation as the implication of the falsum, we obtain

$$A \vee (A \rightarrow f).$$

This schema is read, 'for any formula A, we can either prove A or find a proof that a proof of A can be transformed into a proof of a contradiction'. Clearly, we cannot prove this statement. Thus, the law of excluded middle is not valid in intuitionist logic.

There are other familiar theorems of classical logic that fail in intuitionist logic. Perhaps the most famous is double negation elimination, viz.,

$$\neg\neg A \rightarrow A.$$

On the other hand, the principle of double negation introduction is provable:

$$A \to \neg\neg A.$$

This principle is an instance of $A \to ((A \to B) \to B)$, which is also provable.

5.5 Natural Deduction for Intuitionist and Classical Logic

Intuitionist logic appears most attractive in the form of a natural deduction system. I use a Fitch-style natural deduction system in what follows, but anyone familiar with any style of natural deduction should be able to understand what is going on.

The key to natural deduction as it is understood by contemporary intuitionists (see, e.g., [Dummett, 2000] and [Prawitz, 2006]) is that the behaviour of each connective is governed by an introduction and an elimination rule. Here we are interested in two connectives: negation and the falsum. The negation introduction rule that we use appeals to both negation and the falsum:

If there is a proof of f from the hypothesis that A,

then we can discharge the hypothesis and infer $\neg A$.

The negation elimination rule is the following:

From A and $\neg A$, we may infer f.

There is no extra introduction rule for f – the negation elimination rule is a falsum introduction rule. The elimination rule for f is similar to the negation elimination rule in classical logic:

From f we may infer B.

That is, from a contradiction we may infer any formula.

We can state the introduction and elimination rules for negation in intuitionist logic without using the falsum. The falsum-free introduction rule is

If there is a proof of $\neg A$ from the hypothesis that A,

then we can discharge the hypothesis and infer $\neg A$.

and the falsum-free elimination rule is

From A and $\neg A$, we may infer B.

My reason for using the falsum will become clear when we look at minimal and relevant logic.

To see how the rules are used, consider the following proof of $\neg A \to ((B \to A) \to \neg B)$:

1.	$\neg A$	*hyp.*
2.	$B \to A$	*hyp.*
3.	B	*hyp.*
4.	$B \to A$	2, *reit.*
5.	A	$3,4, \to E$
6.	$\neg A$	1, *reit.*
7.	f	$5,6, \neg E$
8.	$\neg B$	$3-7, \neg I$
9.	$(B \to A) \to \neg B$	$2-8, \to I$
10.	$\neg A \to ((B \to A) \to \neg B)$	$1-9, \to I$

The elimination and introduction rules for negation are often used closely in sequence in this way in the system that includes the falsum. The only way in which we can introduce the falsum is through a negation elimination and we require a proof of the falsum in order to use negation introduction.

We can produce natural deduction systems for classical logic by adding a variety of rules to the system for intuitionist logic. Perhaps the most elegant of these rules is Dag Prawitz's rule [Prawitz, 2006]:

> (*Rd*) From a proof of f from the hypothesis that $\neg A$, we may discharge the hypothesis and infer A.

'*Rd*' stands for 'reductio'. Adding this rule allows an easy proof of double negation elimination ($\neg\neg A \to A$) and a somewhat more difficult proof of excluded middle:1.

1.	$\neg(\neg A \vee A)$	*hyp.*
2.	A	*hyp.*
3.	$\neg A \vee A$	$2, \vee I$
4.	$\neg(\neg A \vee A)$	1, *reit.*
5.	f	$3,4, \neg E$
6.	$\neg A$	$2-5, \neg I$
7.	$\neg A \vee A$	$6, \vee I$
8.	f	$1,7, \neg E$
9.	$\neg A \vee A$	$1-8$, *Rd*

Every inferential move in this proof is intuitionistically acceptable except the last one.

Adding the rule *Rd* spoils the lovely symmetry of the system. In intuitionist logic each connective has one introduction and one elimination rule attached to it. In the classical system we have to add an extra rule for negation. There are a variety of other ways of producing a system for classical logic, but all of them have a similar unaesthetic quality to them. Moreover, there are negation-free theorems of classical logic that, in this system, cannot be proven without negation. Perhaps the most famous of these is Peirce's law:

$$((A \to B) \to A) \to A$$

Here is a proof using R:

1.	$(A \to B) \to A$	*hyp.*
2.	$\quad \neg A$	*hyp.*
3.	$\quad\quad A$	*hyp.*
4.	$\quad\quad \neg A$	2, *reit.*
5.	$\quad\quad f$	$3, 4, \neg E$
6.	$\quad\quad B$	$5, fE$
7.	$\quad A \to B$	$3-6, \to I$
8.	$\quad (A \to B) \to A$	1, *reit.*
9.	$\quad A$	$7, 8, \to E$
10.	$\quad f$	$2, 9, \neg E$
11.	A	$2-10, Rd$
12.	$((A \to B) \to A) \to A$	$1-11, \to I$

We can add negation-free rules to the system that allow the proof of Peirce's law, but all of these look ad hoc in some way – most of them are not obviously related to the meanings of the connectives involved.

5.6 Minimal Logic

A logic slightly weaker than intuitionist logic is *minimal logic*, created by Ingebringt Johansson ([Johansson, 1937]) in the 1930s. The difference between minimal logic and intuitionist logic is that minimal logic rejects the falsum elimination rule, that we can infer any formula from f. Minimal logic is a paraconsistent logic, for in it we cannot prove the validity of EFQ.

Models for minimal logic are quite easy to construct. We take an intuitionist frame $< C, R >$ in which R is reflexive and transitive. But now we do not constrain our value assignment such that $v_i(f) = 0$ for all circumstances i. We allow that f be 'proven' in some circumstances. Thus, we allow there to be impossible (or inconsistent) circumstances. Interestingly, like LP, we can prove in minimal logic the law of non-contradiction, $\neg(A \wedge \neg A)$. Thus, once again we have an illustration of how unconnected the law of non-contradiction and the principle of consistency are.

6. Negation and Information

6.1 Language, Logic, and Situations

Logic is a normative discipline. It does not tell us how we do reason. It tells us how we should reason. The semantics for logical systems have played a key role in justifying the use of those logical systems. For example, the use of classical logic is justified because it never leads us from correct assumptions to false conclusions – an inference is valid in classical logic if it preserves truth (on the two-valued conception of truth). Paraconsistent logics have been justified, on the other hand, because either they preserve truth (on a three- or four-valued conception of truth) or because they are safe in the sense that they do not (always) allow us to infer arbitrary propositions from contradictions.

A rather different justification for certain logical systems comes to us from *situation semantics*. Situation semantics was a theory developed by Jon Barwise and John Perry in the 1980s ([Barwise and Perry, 1983]). Parts of worlds are situations. For example, consider the room that you are in right now. There is certain information available to you in that room. If it is our lecture room, then the information is available to you about whether the projector is on or off and about what the lecturer is saying right now. But there is other information not available to you that is available to people in other situations. For example, someone in Singapore will have the information available to her about whether or not it is raining there, but won't have the information about whether the projector in our lecture room is on. So, in a single possible world, there are many different situations, each containing different information. We say that each situation contains *partial information*, because it does not (necessarily) tell us about the whole world.

We often use as examples of information available in a situation facts that are perceptually present in our environments. These are good examples, but we should not be misled by them. As we shall see, situation semantics is supposed to be the basis of a theory of meaning, and human languages contain a lot of statements that are not about what can be perceived. So we have to include in situations what agents are connected to in other ways, such as by virtue of causal connections. This allows us to use situation semantics to explain how we can talk about things we cannot perceive, such as atoms and subatomic particles, laws of nature, and so on (see [Mares, tab]).

Situation semantics is an approach to the meaning, not just of the logical connectives, but of all the parts of language. The theory of meaning that is connected with situation semantics is called the 'relational theory of meaning' ([Barwise and Perry, 1983, pp. 10–13]).[18] There are two sorts of relations that are important in the relational theory of meaning. First, there are regularities between situations. We come to understand the world by noticing regularities between situations. Situations are what we confront in our experience and we abstract from them properties and even individual objects. These entities (properties, individuals, and other things such as facts and events) are then used in the semantic theory, as we shall soon see. But individuals, properties, facts, and events are treated in situation semantics as abstractions from situations.

The objects that are abstracted from real situations are used to construct *abstract situations*. An abstract situation is a representation of a part of a world. Abstract situations are constructions from individuals, properties, and so on. They may be considered as structures containing sets of states of affairs and relations to other situations ([Mares, 2004, Chapter 4]). According to ([Barwise and Perry, 1983]), a state of affairs is a structure $< P, a_1, \ldots, a_n; 1 >$ or $< P, a_1, \ldots, a_n; 0 >$, where P is an n-place property, the a_is are individual objects, and 1 and 0 are 'polarities'. The presence of $< P, a_1, \ldots, a_n; 1 >$ in a situation tells us about a particular positive fact – that $a_1, \ldots, a_n$ stand to one another in the relation P. Similarly, $< P, a_1, \ldots, a_n; 0 >$ tells us that $a_1, \ldots, a_n$ do not stand to one another in that relation. We can see that this understanding of situations and states of affairs makes a good match with the four-valued semantics discussed in Section 4.2 above. But the variant that we will look at in connection with relevant logic does away with polarities (see [Mares, 2004]). An abstract situation may be an accurate representation of some part of the real world, or it may not. It may in fact not represent any possible world at all. An abstract situation that does not accurately represent any part of any possible world is called an *impossible situation*.

The second sort of relation that is important for the relational theory of meaning is a *constraint*. According to the relational theory of meaning there are constraints between facts in situations and the information contained in those situations. We will look at the constraints that are important for understanding negation in later sections. Right now let us consider a simple constraint:

$$\text{if } s \Vdash < P, a_1, \ldots, a_n; 1 > \text{ then } s \models [P, a_1, \ldots, a_n]$$

where '$\Vdash$' means 'contains' and $[P, a_1, \ldots, a_n]$ is a proposition. So this constraint says that if a situation contains a particular state of affairs (or, rather, the fact that the state of affairs represents) then it supports the corresponding proposition. This constraint is a logical constraint that links a proposition to the state of affairs that is its content. But there are non-logical constraints. Consider the constraint that *kissing involves touching*. In any real or possible situation in which two people kiss, they touch one another ([Barwise and Perry, 1983, p. 101]).

We are interested in two distinctions between sorts of constraints. First, there is a distinction between global and local constraints. Global constraints give closure conditions for all the situations in a model. The set of formulas that are valid in a model captures the global constraints of that model. In contrast to global constraints, there are local constraints. If we have situations that do not characterize physically possible worlds, then the actual laws of nature are local constraints – they only tell us about the closure conditions for physically possible situations. Second there is a distinction between constraints that govern the behaviour of the facts in a situation and those constraints that are themselves contained as information within that situation. For example, it may be that a particular situation is physically possible but not contain as information that a particular law of nature holds.

Although I have been using laws of nature as examples of constraints, we may have constraints that are of a much more humble nature. Consider the constraint that a particular telephone connection is reliable and free of noise. This can be information available to us in a situation. If we have such information in a situation, then we can make inferences about other situations (e.g., the situation in which the person with whom we are conversing over the telephone is located) on the basis of information that is immediately available to us. As we shall see in Section 7.2, this sort of local constraint is central to my interpretation of relevant implication.[19]

In the sections that follow, we examine models that are rather like the models for modal or intuitionist logic, but contain abstract situations instead of possible worlds or circumstances as points. As we shall see, these models will typically contain both possible and impossible situations.

6.2 Information Conditions and the (In)compatibility Semantics for Negation

Consider for a moment a real situation: one that consists of the room in which you are now sitting during the time in which you are reading my chapter on negation. Certain information is present in that room – the colour of the pages in front of you, the number of chairs in the room, the presence of any other people in the room, and so on. But there are certain facts about which the information remains silent – the exact number of chairs in the universe, for example. The situation based on your room supports neither of the following statements:

There are exactly 5,493,000,000 chairs in the universe.

There are not exactly 5,493,000,000 chairs in the universe.

But it does (let us say) support the following statement:

The page on which this sentence is written is not red.

What feature of the room (or, rather a thing in the room) forces 'the page that this sentence is written on is not red' to be true? Clearly it is the fact that this page is white and black. Being white and black all over is *incompatible* with being red. We will return to the issue of negative information soon.

Situational semantics for logics consider not what is true in worlds, but what information is contained in situations. There are particular constraints that allow us to formulate information conditions – which are similar to truth conditions for classical or many-valued logic or proof conditions for intuitionist logic. For example, the following are the intuitive constraints that govern conjunction and disjunction in situations. Where φ and ψ are propositions,[20]

$$s \Vdash \varphi \wedge \psi \text{ if and only if } s \Vdash \varphi \text{ and } s \Vdash \psi$$

and

$$s \Vdash \varphi \vee \psi \text{ if and only if } s \Vdash \varphi \text{ or } s \Vdash \psi.$$

In what follows we will not be considering propositions, but only the relationship between situations and formulas. For we are interested in logic and logical *languages* here.

Let us return to the topic of negation. The example of the chairs given above illustrates our information condition for negation. We say that a negated formula $\neg A$ is supported by a situation s if and only if there is something about s that is incompatible with the truth of A. In order to formalize the notion of incompatibility, we add a compatibility relation to our model. Thus, a situated model is a triple $M =< S, C, v >$ where S is a set of situations, C is a binary relation between situations, and v is an assignment of values to propositional variables. If Cst, then we say that s and t are compatible and otherwise they are incompatible. Now we can formulate our information condition for negation:

$$s \models \neg A \text{ if and only if for all situations } t, Cst \text{ implies not-} t \models A$$

This condition says that a situation s supports not-A if and only if no situation that is compatible with s supports A. Incompatibility was first used to give a semantics for negation by Robert Goldblatt in his semantics for orthologic (a generalization of quantum logic) ([Goldblatt, 1974]).

Note the very close similarity to the condition for negation in Kripke's semantics for intuitionist logic (merely replace C with R). But there are some important differences, both conceptual and formal. The conceptual difference lies in the use of the idea that two situations can be compatible or incompatible. The standards for compatibility are applied to a whole model. Thus, for example, if we take being red an being green as incompatible, we hold that any two situations that represent the same object as being red and as being green (in the same way and at the same time) are incompatible with one another. Whether we should hold that these incompatibilities are deep metaphysical truths or part of human psychology or merely conventions is not an issue that we need to decide when doing semantics. We merely need to argue that our use (or at least a use) of negation captures a notion of incompatibility.

The formal difference comes from the logical use to which we put compatibility. The notion of a valid argument that is captured by our situated models is supposed to be one of *information preservation* or information *containment*. If $A \vdash B$ is valid over the class of these models, then we want to say that the information that A in some way contains the information that B. Now consider EFQ. According to EFQ, any formula follows from two contradictory formulas. Using the intuitive sense of 'information', it would seem that contradictions do not contain all information. On some technical understanding of 'information' it is true that contradictions are maximally informative (and classical tautologies contain no information), but this technical use of the term 'information' is contrary in this respect to our pre-theoretical understanding. In order to bring our formal treatment of information closer to our pre-theoretical understanding we invalidate EFQ in our semantics. We do so by allowing that some situations are not compatible with themselves. This makes sense in our formal framework. There is nothing to stop us from having an abstract situation contain, say, both the states of affairs $<Red, a>$ and $<Green, a>$. Thus, the situation contains two incompatible states of affairs and so is incompatible with itself. So we can have situations that support contradictory formulas but that do not satisfy every formula. Therefore, we have models that invalidate EFQ.

It is natural to make the compatibility relation symmetrical:

$$\text{If } Cst \text{ then } Cts.$$

For we say that two things are compatible with one another without placing a direction on this relationship. Making C symmetrical validates double negation introduction:

$$A \vdash \neg\neg A$$

For suppose that $s \models A$. Now consider some situation t such that Cst. By symmetry, Cts, so $t \not\models \neg A$. By the information condition for negation, then, $s \models \neg\neg A$.

7. Application: Relevant Logic

7.1 Introducing Relevant Logic

Relevant logic has its roots in the early twentieth century. It was then, after Frege, Peano, Russell, and others published work on classical logic that there were calls for a different approach to implication. There was fairly widespread dissatisfaction with the notion of material implication. C. I. Lewis ([Lewis, 1917]) and

Hugh MacColl ([MacColl, 1906]) are perhaps the best-known critics, but there are many others who thought that material implication was a form of implication in name only. The problem is that the paradoxes of material implication are valid in classical logic. Among these so-called paradoxes are the following:

- $(p \wedge \neg p) \rightarrow q$
- $p \rightarrow (q \vee \neg q)$
- $(p \rightarrow q) \vee (q \rightarrow p)$
- $(p \rightarrow q) \vee (q \rightarrow r)$
- $p \rightarrow (q \rightarrow q)$

All of these show that material implications are too easy to find – there are too many of them around. The problem with material implication, and classical logic more generally, is that it considers only the truth value of formulas in deciding whether to make an implication stand between them. It ignores everything else.

Relevant logics are subsystems of classical logic that reject the paradoxes of material implication. All relevant logics have the *variable sharing property*, that is, if a formula $A \rightarrow B$ is valid in a propositional relevant logic, then the formulas A and B share some non-logical content – they have at least one propositional variable in common. Note that the variable sharing property is only a necessary condition for being a relevant logic. The logic must also reject all the paradoxes of material implication.

In this section we will discuss only the relevant logic R of relevant implication. It is easiest to understand R through its natural deduction system. Consider the following classical proof of $p \rightarrow (q \rightarrow q)$:

1.	$\quad p$	*hyp.*
2.	$\quad\quad q$	*hyp.*
3.	$\quad\quad q$	2, *reit.*
4.	$\quad q \rightarrow q$	$2-3, \rightarrow I$
5.	$p \rightarrow (q \rightarrow q)$	$1-4, \rightarrow I$

The problem, from a relevant point of view, is that in the final step the first hypothesis, p, is discharged without ever having been used. The core concept of a relevant theory of deduction is that of the real use of hypotheses.[21] In the following subsections we will describe the natural deduction system for R and the behaviour of negation in it, and connect it with situated models.

7.2 Natural Deduction for Relevant Logic

In order to make sure that a hypothesis is really used in an inference, we label each hypothesis with a number and then we put a subscript on each line of the proof that indicates which hypotheses were used to infer that line. For example:

1.	$A \to B_{\{1\}}$	*hyp.*
2.	$A_{\{2\}}$	*hyp.*
3.	$A \to B_{\{1\}}$	1, *reit*
4.	$B_{\{1,2\}}$	$3, 4, \to E$

Here the rule for $\to E$ is: *From* $A \to B_\alpha$ and A_β we can infer $B_{\alpha \cup \beta}$. This proof shows that we can validly and relevantly infer B from $A \to B$ and A. The hypotheses that $A \to B$ and A are really used to infer B. We can see this because the hypotheses numbers for these premises show up in the subscript for the conclusion B.

The rule for implication introduction is: From a proof that B_α from the hypothesis $A_{\{k\}}$ (where k is a number), we can infer $A \to B_{\alpha-\{k\}}$, *where* k *really is in* α ($\alpha - \{k\}$ is just the set α with k removed from it). Here is a proof of $(A \to B) \to ((B \to C) \to (A \to C))$:

1.	$A \to B_{\{1\}}$	*hyp*
2.	$B \to C_{\{2\}}$	*hyp*
3.	$A_{\{3\}}$	*hyp*
4.	$A \to B_{\{1\}}$	1, *reit*
5.	$B_{\{1,3\}}$	$3, 4, \to E$
6.	$B \to C_{\{2\}}$	2, *reit*
7.	$C_{\{1,2,3\}}$	$5, 6, \to E$
8.	$A \to C_{\{1,2\}}$	$3 - 7, \to I$
9.	$(B \to C) \to (A \to C)_{\{1\}}$	$2 - 8, \to I$
10.	$(A \to B) \to ((B \to C) \to (A \to C))_\emptyset$	$1 - 9, \to I$

A valid formula in this system is just one that can be proven with the subscipt $\emptyset$ (the empty set).

But what do the subscripts mean? Consider again the hypothesis $A_{\{1\}}$. If this is hypothesized in a proof, what it means is 'suppose that there is a situation (call it s_1) in a world which contains the information that A'. Now, suppose that we make further hypotheses in the same proof, for example, $B_{\{2\}}$. We are now saying 'suppose that there is also a situation (call it s_2) in the same world which contains the information that B'. Consider the following proof:

1.	$A_{\{1\}}$	*hyp*
2.	$A \to B_{\{2\}}$	*hyp*
3.	$A_{\{1\}}$	1, *reit*
4.	$B_{\{1,2\}}$	$2, 3, \to E$
5.	$(A \to B) \to B_{\{1\}}$	$2 - 4, \to I$
6.	$A \to ((A \to B) \to B)_\emptyset$	$1 - 5, \to I$

Let's forget about the last line for a moment. The first line says 'suppose that there is a situation s_1 in a world in which A'. The second line says 'suppose there is a situation s_2 in the same world in which $A \rightarrow B$'. The third line just reiterates the first line, but the fourth line is interesting. It says that there is a situation s' in the same world in which B, and we know that there is this situation because we have derived that it is so by *really using* the information in s_1 and s_2.

The fifth line tells of course that we know (from the discharged subproof in steps 2–4) that in s_1 there is the information that $(A \rightarrow B) \rightarrow B$. The situational interpretation of the natural deduction system and the implication introduction rule together tell us that a s_1 situation contains the information that an implication $A \rightarrow B$ obtains if and only if it contains information that allows us to infer from the hypothesis that there is a situation s_2 in the same world in which A that there is also a situation s_2 in that world in which B. The basis for the inferential connections between situations are constraints like the ones discussed in Section 6.1 above. As we saw, not only do some constraints occur globally in a model, some also occur locally. This means that the information that a constraint holds may be information contained within some situations. Other constraints, such as that which links two propositions to their conjunction, also occurs globally, as a rule that dictates the behaviour of conjunction in the model itself. The constraints contained as information in a situation are employed as bases for inferences about what other situations exist in that world. A law of nature is such a constraint – it can be used as a licence for a situated inference – but so is the information that a particular telephone connection is reliable and free of noise. Situated inferences also use the structural rules of the logic R, such as the rule that it is permissible to use hypotheses as many times as we wish, the rule that we may reorder hypotheses as needed, and so on ([Mares, 2004, Chapter 3]).[22]

Now we turn to the final line of the proof. What does '$A \rightarrow ((A \rightarrow B) \rightarrow B)_{\emptyset}$' mean? As we know, it means that this formula is valid. But what does 'valid' mean here? It means that $A \rightarrow ((A \rightarrow B) \rightarrow B)$ is true in every *normal* situation. In the context of a particular model a law of logic is an implicational formula that describes a condition under which every situation in that model is closed. For example, if $A \rightarrow B$ is a law of logic in a model, then every situation in that model which satisfies A also satisfies B. If $A \rightarrow B$ is a law of logic for a particular model, then every normal situation contains the information that $A \rightarrow B$.

Certain actual concrete situations are normal. How do they contain information about every other situation? There may be different ways in which this is possible. One which seems reasonable is that a situation can contain a community of people whose use of language we are trying to model. Their use of language determines which situations are in the model and the semantic relationships between those situations. Thus, a situation which contains those people and the facts about the way they use language contains information about the laws of logic (see [Mares, tab]).

Now we add conjunction. Here's a proof using the conjunction rules:

1.	$(A \to B) \wedge A_{\{1\}}$	*hyp.*
2.	$A \to B_{\{1\}}$	$1, \wedge E$
3.	$A_{\{1\}}$	$1, \wedge E$
4.	$B_{\{1\}}$	$2, 3, \to E$
5.	$B \wedge A_{\{1\}}$	$3, 4, \wedge I$
6.	$((A \to B) \wedge A) \to (B \wedge A)_{\emptyset}$	$1-5, \to I$

The conjunction elimination rule ($\wedge E$) is: From $A \wedge B_\alpha$ we can infer A_α and B_α, which is what one would expect. The conjunction introduction rule is just the reverse. It says that from A_α and B_α we can infer $A \wedge B_\alpha$. Note that in order to do a conjunction introduction, the two formulas that you want to conjoin have to have the same subscript. If we do not require that they have the same subscript and change the rule to from A_α and B_β we can infer $A \wedge B_{\alpha \cup \beta}$, then we will have a natural deduction system for classical logic.[23] Here is a proof in that system of $p \to (q \to q)$:

1.	$p_{\{1\}}$	*hyp.*
2.	$q_{\{2\}}$	*hyp.*
3.	$p_{\{1\}}$	1, *reit.*
4.	$p \wedge q_{\{1,2\}}$	$2, 3, \wedge I$
5.	$q_{\{1,2\}}$	$4, \wedge E$
6.	$q \to q_{\{1\}}$	$2-5, \to I$
7.	$p \to (q \to q)_{\emptyset}$	$1-6, \to I$

So, to block proofs like this we restrict conjunction introduction to connecting formulas with the same subscript. Another reason for these rules for conjunction are that they correspond to the information conditions for conjunction given in Section 6.2. For more on conjunction in relevant logic see [Read, 1988] and [Mares, taa].

7.3 Negation in Relevant Logic

In our natural deduction system, we use a falsum to treat negation. Here f means 'a contradiction occurs'. Unlike intuitionist logic, relevant logic does not treat every contradiction as equivalent. Rather, the falsum can be understood as the (infinite) disjunction of all of the contradictions. In algebraic terms, it is the least upper bound of all the contradictions. Thus, the formula '$A \to f$' means 'A implies that there is a contradiction'. Like intuitionist logic, in relevant logic we take $A \to f$ to be equivalent to $\neg A$. Thus, in effect, to say that it is not the case that A is to say the same thing as A implies that there is a contradiction.

Thus, we start with the following rule of negation introduction:

> ($\neg I$) From a proof of f_α from the hypothesis that $A_{\{k\}}$,
> you may discharge the hypothesis and infer $\neg A_{\alpha-\{k\}}$

where k really is in α. Or, in more graphically:

$$\begin{array}{l} \left|\begin{array}{l} A_{\{k\}} \\ \quad\vdots \\ f_\alpha \end{array}\right. \\ \neg A_{\alpha-\{k\}} \end{array}$$

We also have the following version of negation elimination:

> ($\neg E_1$) From A_α and $\neg A_\alpha$ you may infer $f_{\alpha\cup\beta}$.

Our treatment of the falsum is more like that of minimal logic rather than intuitionist or classical logic. That is, we do not include the falsum elimination rule. So in relevant logic we cannot infer just anything from a contradiction. Thus, it is a paraconsistent logic.

To see how these rules are used, here is a relevant proof of $(A \rightarrow B) \rightarrow (\neg B \rightarrow \neg A)$:

$$\begin{array}{lll}
1. & \quad| \; A \rightarrow B_{\{1\}} & hyp \\
2. & \quad| \;\; | \; \neg B_{\{2\}} & hyp \\
3. & \quad| \;\; | \;\; | \; A_{\{3\}} & hyp \\
4. & \quad| \;\; | \;\; | \; A \rightarrow B_{\{1\}} & 1,\ reit \\
5. & \quad| \;\; | \;\; | \; B_{\{1,3\}} & 3, 4, \rightarrow E \\
6. & \quad| \;\; | \;\; | \; \neg B_{\{2\}} & 2,\ reit \\
7. & \quad| \;\; | \;\; | \; f_{\{1,2,3\}} & 5, 6, \neg E \\
8. & \quad| \;\; | \; \neg A_{\{1,2\}} & 3-7, \neg I \\
9. & \quad| \; \neg B \rightarrow \neg A_{\{1\}} & 2-8, \rightarrow I \\
10. & (A \rightarrow B) \rightarrow (\neg B \rightarrow \neg A)_\emptyset & 1-9, \rightarrow I
\end{array}$$

We can interpret the incompatibility semantics using the falsum. To do so we say that two situations s_1 and s_2 are *in*compatible if and only if we can infer (in the relevant manner) from the information in s_1 and s_2 that there is a situation in the same world as those which contains the information that f. The incompatibilities that we cited in Section 6.1 are then taken to be informational constraints.[24]

So far we have added a form of minimal negation to relevant logic. I prefer this sort of negation to formalize relevance, because I find its model theory and proof theory rather natural. But the usual sort of negation that is found in relevant logics is a 'De Morgan negation'. De Morgan negation obeys all of the De Morgan laws (of course) and the law of double negation elimination. In order to make $\neg$ into a De Morgan negation, we need to add one more rule to our natural deduction system. This a relevant version of the classical rule *Rd* that we met in Section 5.5.

> (*Rd*) From a proof of f_α on the hypothesis that $\neg A_{\{k\}}$,
>
> you may discharge the hypothesis and infer $A_{\alpha-\{k\}}$

where k really is in α. The most straightforward way of modifying our situated models to validate R is to replace the compatibility relation with the 'Routley star operator'. The Routley star operator was discovered by Richard and Val Routley in the early 1970s ([Routley and Routley, 1972]). We add the star, $*$, which is an operator on situations (that is, s^* is a situation, for any situation s). We now have the following information condition for negation:

$$s \models \neg A \text{ if and only if } s^* \not\models A.$$

We understand the star in terms of compatibility. For a situation s, s^*, is the maximal situation that is compatible with s. This means that any other situation that is compatible with s contains less information that s^*.[25]

8. Summing Up

We can see from this survey that negation really is a key connective in thinking about logic and especially in the way in which different logical systems are related to one another. It is natural to think that the central difference between classical logic and intuitionist logic, for example, lies in their treatments of negation. Classical logic, but not intuitionist logic, makes valid the law of excluded middle and double negation elimination. From the perspective of natural deduction, one way of viewing the difference between the two systems is that classical logic makes the reductio rule valid. Moreover, paraconsistent logics are understood most naturally in terms of their treatments of negation, since it is the central aim of paraconsistent logic to reject EFQ.

Relevant logic is a bit different from these other systems in this regard, since it was invented to provide a more natural treatment of implication. Its treatment of negation, however, could not be purely classical, since it rejects EFQ, but also the theses that say that all classical tautologies, such as instances of excluded middle, are implied by every formula. Thus relevant logic is forced to accept some weaker form of negation, such as De Morgan negation or a relevant version of minimal negation.

If we had more space, we could discuss even more issues related to the concept of negation. There are interesting connections between negation and the speech act of denial. The treatment of negation in sequent-style proof theories is also important and interesting. The role of negation in the history of logic, especially its role in the Aristotelean square of opposition is important as well. But to discuss all of these topics would take an entire book, and this is a book about philosophical logic, not just about negation![26]

Acknowledgements

I would like to thank Rob Goldblatt, Leon Horsten, Tim Irwin, Richard Pettigrew, Greg Restall, and Jeremy Seligman for discussions relating to the topic of this paper. Research for this paper was funded by grant 05-VUW-079 of the Marsden Fund of the Royal Society of New Zealand.

Notes

1. But, if not, here are some good textbooks that one can consult in order to learn the basic ideas: [Bergman et al., 1990], [Halbach, 2010].
2. There is a third perspective, that of algebraic logic, but this is not usually studied by philosophical logicians. We will discuss it briefly in Sections 2.1 and 4.2.
3. They do have *tableau*-style proof theories, but these I do not count as a form of proof theory that is independent of model theory. What a tableau system does is provide a means for generating counter-models for non-theorems of the logic and so can be looked at as part of the model theory for the system rather than a 'proof theory' properly so-called.
4. They do have natural deduction systems, but they are significantly flawed. Athough there is a sense in which they are natural, in my opinion they significantly distort our normal inferential practices. For example, they distinguish between a hypothesis that is assumed to be true and one that is assumed to be 'not false'. I doubt very much that people normally reason in this way. See [Woodruff, 1970] and [Roy, 2006]. These proof systems are reasonable.
5. The two-valued matrices make up only one of a great many possible classes of models for classical logic. Every boolean algebra is a model for classical logic and for each natural number n, there is a boolean algebra of size 2^n.
6. I have also assumed that the following rules are valid:

$$\frac{\begin{array}{c}\vdash A \leftrightarrow B \\ \vdash C \vee A\end{array}}{\therefore \vdash C \vee B}$$

and modus ponens for provable formulas. None of the logics that I discuss reject either of these rules, so it is not important that we discuss them here.

7. Some philosophers, such as Kripke ([Kripke, 1975a]), think of the 'third truth value', not as a real truth value, but as the absence of a truth value. Thus, a sentence that has the value .5 on this reading really is a sentence without any truth value.
8. The logic $Ł_\omega$ is sometimes called $Ł_{\aleph}$ (see [Rescher, 1969]).
9. Although Post's negation may seem odd to philosophical eyes, it has had applications in electronic engineering. Cyclic switches are useful in the design of electronic circuits.
10. The strengthened liar paradox is known as a 'revenge problem' against this K_3-based view of truth. It uses the resources of the K_3-view against the K_3-view itself. [Beall, 2007] is a good collection of papers largely about such revenge problems.
11. The formula $((p \to q) \wedge p) \to q$, however, is a tautology in LP!
12. For more on implication an other forms of conditionals, see Chapter 14.
13. Dunn developed this model for his logic D4 in the late 1960s but published it in the mid-1970s in [Dunn, 1976].
14. The logic D4 is usually called 'first-degree entailments' (or 'FDE'). But this is really a bad name for the system, since a first-degree entailment is a theorem of the relevant logic E the main connective of which is an entailment. The semantics for D4, on the other hand, captures the valid inferences of E in which no entailments occur.
15. I am re-using my negation symbols to formalize rather different forms of negation, since there are not that many symbols that look adequately like negation. I hope this does not cause any confusion.
16. This does not mean that what is constructively proven need correspond to what can be done by a *deterministic* program. As the father of intuitionism, L. E. J. Brouwer, stressed, there may be 'free choices' (non-deterministic steps) required in a mathematical construction.
17. In intuitionist maths, a set is sometimes called a 'species' to distinguish it from the classical notion of a set.
18. For good more recent accounts of the relation theory of meaning see [Bremer and Cohnitz, 2004, Chapter 4] and [Peréz-Montoro, 2007, Chapter 3].
19. For a different view of constraints, see [Barwise and Seligman, 1997], and for a comparison between that view and the view given here, see [Mares et al., ta].
20. I have recently begun to question the correctness of this information condition for disjunction. For an alternative treatment of disjunction see [Mares, tab].
21. The natural deduction system for R is due to Alan Anderson and Nuel Belnap (see [Anderson and Belnap, 1975] and [Anderson et al., 1992]).
22. This clearly is not a presentation of the mathematical model theory of relevant logic. In the early 1970s, Richard Routley and Robert Meyer constructed a model theory for relevant logic ([Routley and Meyer, 1973], [Routley and Meyer, 1972a], [Routley and Meyer, 1972b]). In the Routley Meyer semantics, there is a ternary relation, *R*, on situations. In [Mares, 2004, Chapters 2 and 3] this relation is interpreted in terms of my theory of situated inference. *R* is used to state their condition for implication, viz., $s \models A \to B$ iff for all *t* and *u* if *Rstu* and $t \models A$ then $u \models B$.
23. The resulting system is, in effect, the same as the system of [Lemmon, 1965].
24. In the context of the Routley-Meyer semantics we can either start with the falsum as primitive and then define the compatibility relation (as we have just done), or begin with the compatibility relation as primitive and define a falsum. To do so, we set $\mathbb{F} = \{u : \exists s \exists t(Rstu \wedge \neg Cst)\}$ and we make $s \models f$ iff $s \in \mathbb{F}$.
25. This is Dunn's interpretation of the star operator [Dunn, 1993]. There is, as far as I know, no existing argument that there is a unique maximal situation s^* for every situation *s*. Thus, at the moment, at best, we can only assume that there are such situations.
26. For a very nice book-length study on negation and its history, see [Horn, 1989].

9 Game-Theoretical Semantics

Gabriel Sandu

Chapter Overview

1. Introduction	217
2. Extensive Games of Perfect Information	219
2.1 Strategies	220
3. Game-Theoretical Semantics for First-Order Languages	221
3.1 Semantical Games	221
3.2 Negation	223
3.3 Truth and Falsity in a Structure	224
3.4 Logical Equivalence	226
3.5 Tarski Type Semantics	228
3.6 Satisfiability and Skolem Semantics	229
3.7 Falsifiability and Kreisel Counterexamples	232
4. IF Languages	234
4.1 Extensive Games of Imperfect Information	235
4.1.1 Indeterminacy	236
4.1.2 Dummy Quantifiers and Signalling	237
4.2 Generalizing Skolemization and Kreisel Counterexamples	238
4.2.1 Lewis' Signalling Games	241
4.3 Compositional Interpretation	242
4.4 Negation	247
4.5 Burgess' Separation Theorem	248
4.5.1 Game-Theoretical Negation versus Classical Negation	250
5. Strategic Games	251
5.1 Pure Strategies	251
5.1.1 Maximin Strategies	253
5.1.2 Pure Strategy Equilibria	255
5.2 Mixed Strategies	258

5.2.1 Mixed Strategy Equilibrium 262
5.2.2 A Criterion for Identifying Equilibria 264
6. Equilibrium Semantics 266
6.1 Equilibrium Semantics 267
Note 270

1. Introduction

One of the revolutionary aspects of modern logic consists in considering statements that involve multiple quantification like the following example from the mathematical vernacular. A function f is said to be *continuous* if, for all x in the domain of f and all $\varepsilon > 0$, there exists a $\delta > 0$ such that, for all y in the domain, we have

$$|x - y| < \delta \rightarrow |f(x) - f(y)| < \varepsilon.$$

In the symbolism of first-order logic, the definition is expressed by

$$\forall x \forall \varepsilon \exists \delta \forall y (|x - y| < \delta \rightarrow |f(x) - f(y)| < \varepsilon)$$

(we have ignored the restriction on the domain of quantification).

This chapter will be a systematic introduction to a tradition which emerged from the work of Leon Henkin and Jaakko Hintikka according to which the interpretation of a sequence of standard quantifiers is given in terms of the strategic interaction of two players in a semantical game. The players, *Eloise* and *Abelard* correspond to the existential and the universal quantifier, respectively. Each occurrence of a quantifier in a formula prompts a move by the respective player who chooses an individual from the relevant universe of discourse. This mode of thinking extends naturally to the logical connectives. Disjunction prompts a move by *Eloise* who will have to choose a disjunct, and conjunction will prompt a similar move by *Abelard*; negation prompts a switch of the players, etc. A play of the game ends up after a finite number of steps with an atomic formula. In the game associated with the sentence above (and a underlying structure which interprets its non-logical vocabulary), the choices of the players give rise to a sequence (play) (a, b, c, d) whose members are individuals in the universe of the structure, the first two and the fourth being chosen by *Abelard*, and the third one by *Eloise* (we disregard for the moment the choice associated with implication). If the sequence (a, b, c, d) verifies the matrix $(|x - y| < \delta \rightarrow |f(x) - f(y)| < \varepsilon)$, then *Eloise* wins the play; otherwise *Abelard* wins it. Our main interest will be in winning strategies rather than plays, as understood in the classical theory of games. Roughly, a strategy for a particular player is a function that is defined at all the possible positions reached in the game at which it is that player's turn to move.

The game-theoretical setting brings in a correlation between:

- material truth (falsity) of first-order formulas,
- winning strategies for *Eloise* (*Abelard*) in a certain subclass of games in classical game theory (i.e., strictly competitive two-person games of perfect information),
- Skolem functions (Kreisel's counterexamples).

These correlations allow for other reconceptualizations of notions and principles in logic in terms of game-theoretical principles:

- the notion of a quantifier *being in the scope of* other quantifiers corresponds to a move being *informationally dependent* on other moves;
- the counterpart of the law of excluded middle is the *principle of the determinacy* of games (Gale-Stewart theorem);
- the dependence of the semantic value of a formula on the current assignment has its counterpart in a strategy being *memoryless*; etc.

These questions will be treated in the first part of the chapter.

The correlations above trigger new ones. For instance, the notion of a move being infomationally dependent of other moves is akin to the notion of a move being informationally *independent* of others. They are two sides of the same coin. In classical game theory, informationally independent moves lead to games of imperfect information. The question that will occupy us in the second part of the chapter is how to represent informational independence in the logical language. This will lead us to *Independence-Friendly logic* (IF logic) introduced by Hintikka and Sandu. IF logic is an extension of first-order logic which allows for more patterns of dependence and independence of quantifiers and connectives than first-order languages. The main new ingredient are quantifiers of the form $(\exists x/W)$ and $(\forall y/V)$, where W and V are sets of variables. The interpretation of $\exists x/W$ is: there exists an x independent of the quantifiers which binds the variables in W. Similarly for $\forall y/V$.

To get an idea let us revisit our earlier definition of a continuous function. In this definition δ depends on (is in the scope of) both ε and the point x. Now we may want to consider a variant of continuity in which δ depends only on ε (and not on x). This will be represented in IF logic by

$$\forall x \forall \varepsilon (\exists \delta / \{x\}) \forall y (|x - y| < \delta \rightarrow |f(x) - f(y)| < \varepsilon). \tag{9.1}$$

The informational independence of $\exists\delta$ from $\forall x$ is implemented by the requirement of uniformity on *Eloise*'s strategies in the game of imperfect information which is the interpretation of (9.1). That is, whenever $a = a'$, then, for any c, any of *Eloise*'s strategies will have to assign the same value for the arguments (a, c) and (a', c). The resulting notion of continuity which corresponds to (9.1) is known as *uniform continuity*. Thus IF logic leads to a correlation between

- material truth (falsity) of IF formulas,
- *uniform* winning strategies for *Eloise* (*Abelard*) in a certain subclass of games in classical game theory (i.e., strictly competitive two-person games of imperfect information),
- generalized Skolem functions (Kreisel's counterexamples).

Apart from being a specification language for certain class of games of imperfect information, IF logic has certain interesting properties as compared to ordinary first-order languages:

- It leads to an increase in expressive power (for instance, IF logic defines its own truth predicate);
- It allows for a phenomenon known in classical game theory as *signalling* (the non-trivial role of dummy variables);
- It introduces indeterminacy into logic.

Obviously, we do not regard indeterminacy as pathological. From the perspective of our approach, the fact that certain sentences are neither true nor false (on certain structures) will be seen as the limit of a certain game-theoretical paradigm: the limitation to pure strategies in extensive games. To overcome it, in the third part of this chapter we switch from pure to mixed or randomized strategies and apply von Neumann's minimax theorem to IF logic. The result is a multi-valued semantics that we call *equilibrium semantics*. Hintikka's game-theoretical semantics is based on the notion of winning strategy; equilibrium semantics is based on the notion of equilibrium of (randomized) strategies.

2. Extensive Games of Perfect Information

It is customary to present games in classical game theory in *extensive form* (cf. [Osborne and Rubinstein, 1994]).

Definition 9.2.1 An *extensive game* G of perfect information is a tuple

$$G = (N, H, Z, P, (u_i)_{i \in N})$$

where

(i) N is the set of players.
(ii) H is a set of finite sequences $(a_1, \ldots, a_m)$ called *histories*, or *plays*.
(iii) Z is the set of terminal or maximal histories called *plays* of the game.
(iv) $P : H \backslash Z \to N$ is the *player function*, which assigns to every non-terminal history the player whose turn it is to move.
(v) For each $p \in N$, u_p is the *payoff function* for player p – that is, a function that specifies the payoffs of player p for each play of the game.

If h is a history then any non-empty initial segment of h is also a history. A member of a history is called an *action*. If $h = (a_1, \ldots, a_n)$ and $h' = (a_1, \ldots, a_n, a_{n+1})$ we say h' is a *successor* of h and we write $h' = h \frown a_{n+1}$. For a non-terminal history $h = (a_1, \ldots, a_m)$ the player $P(h)$ chooses an action to continue the play. The action is chosen from the set

$$A(h) = \{a : h \frown a = (a_1, \ldots, a_m, a) \in H\}$$

and the play continues from $h \frown a = (a_1, \ldots, a_m, a)$.

From the class of extensive games of perfect information, we single out a particular subclass: the class of finite, two person, strictly competitive *one-sum* (or win-loss) games. These are games played by two players (i.e., $N = \{1, 2\}$) for which there are only two payoffs 1 and 0. In addition, for all $h \in Z$, $u_1(h) + u_2(h) = 1$. Whenever $u_1(h) = 1$ and $u_2(h) = 0$ we say that player 1 wins the play h and player 2 loses it. These games are finite: every play in Z is finite. In addition, we are interested in one-sum games which have a tree structure with a unique root. The extensive form of a game may be thought of as a *tree structure*, having the initial position as its root, and the maximal histories as its maximal branches. Given that the payoffs of player 2 are completely determined by those of player 1, we can replace the the two payoff functions with one, $u = u_1 : Z \to N$.

2.1 Strategies

Let us write $P^{-1}(\{p\}) = H_p$ for the set of those histories in H at which it is player p's turn to move, as specified by the player function P. A *strategy* for a player p is standardly defined as a choice function

$$\sigma_p \in \prod_{h \in H_i} \to A(h)$$

that tells the player how to move whenever it is his or her turn.

A player *follows a strategy* σ during a history $h' = (a_1, \ldots, a_n)$ if for every $h = (a_1, \ldots, a_m) \in H_p$ which is a (proper) initial segment of h', $(a_1, \ldots, a_m, \sigma(h))$ is also an initial segment of h'.

We are interested in the following sets:

- H_σ, the plays in which a given strategy σ is followed;
- $Z_\sigma = H_\rho \cap Z$, the set of maximal plays in which σ is followed;
- $Z_p = u^{-1}(p)$, the maximal plays that player p wins.

We say that a strategy σ for player p is *winning* if $Z_\sigma \subseteq Z_p$, i.e., p wins every maximal play in which he or she follows σ.

Example 9.2.1 Consider the strictly competitive, one-sum game of perfect information in which player 1 can choose either a or b, after which player 2 can choose either c or d. The payoffs for the two players are given by

$$u_1(a,c) = 1 = u_1(b,d), \text{ and } u_1(a,d) = u_1(b,c) = 0$$
$$u_2(a,d) = 1 = u_2(b,c), \text{ and } u_2(a,c) = u_2(b,d) = 0$$

In this game player 1 has two strategies at his disposal, a and b, and player 2 has four strategies:

$$\tau_1(a) = c,\ \tau_1(b) = c$$
$$\tau_2(a) = c,\ \tau_2(b) = d$$
$$\tau_3(a) = d,\ \tau_3(b) = c$$
$$\tau_4(a) = d,\ \tau_4(b) = d$$

Player 2 has one winning strategy, namely, τ_3.

The following result is well known in game theory:

Theorem 9.2.1 (Gale, Stewart) *Every strictly competitive one-sum finite game of perfect information with a unique initial history is determined: exactly one of the players has a winning strategy in the game.*

For those two-player zero-sum games of perfect information where each player has only finitely many possible strategies, the result is proven in [von Neumann and Morgenstern, 1944, see esp. Section 15.6].

3. Game-Theoretical Semantics for First-Order Languages

3.1 Semantical Games

We fix a first-order language in a vocabulary L. An L-structure $\mathcal{M}$ is defined in the usual way: In addition to its universe M, it contains an individual $c^A \in M$ for each constant symbol c, a function $f^A : M^n \to M$ for each function symbol f of arity n, and a relation $R^M \subseteq M^n$ for each relation symbol R of arity n.

We take an assignment in M to be a function whose domain is a finite set of variables, and values in M. If s is an assignment in M, and $a \in M$, $s(x_i/a)$ denotes the assignment with domain $dom(s) \cup \{x_i\}$ defined by:

$$s(x_i/a)(x_j) = \begin{cases} s(x_j) & \text{if } i \neq j \\ a & \text{if } i = j \end{cases}$$

We use $s, s', \ldots$ to stand for assignments.

With each formula φ (in negation normal form), structure $\mathcal{M}$, and assignment s in Mm we associate a *semantical game* $G(\mathcal{M}, s, \varphi)$, which is played by *Eloise* ($\exists$) and *Abelard* ($\forall$). The rules of the game can be described informally as:

- The game has reached the position (s, φ), with φ an atomic formula or its negation (i.e., a literal): No move takes place. If $\mathcal{M}, s \models \varphi$, then *Eloise* wins right away; otherwise *Abelard* wins.
- The game has reached the position $(s, \psi \vee \theta)$: *Eloise* chooses $\chi \in \{\psi, \theta\}$, and the game continues from the position (s, χ).
- The game has reached the position $(s, \psi \wedge \theta)$: *Abelard* chooses $\chi \in \{\psi, \theta\}$ and the game continues from the position (s, χ).
- The game has reached the position $(s, \exists x \psi)$: *Eloise* chooses $a \in M$, and the game continues from the position $(s(x/a), \psi)$.
- The game has reached the position $(s, \forall x \psi)$: *Abelard* chooses $a \in M$, and the game continues from the position $(s(x/a), \psi)$.

It is obvious that every semantical game $G(\mathcal{M}, s, \varphi)$ can be reformulated as a one-sum extensive game of perfect information

$$G = (N, H, Z, P, (u_i)_{i \in N}).$$

where

- $N = \{\exists, \forall\}$,
- $H = \bigcup\{H_\psi : \psi \text{ is a subformula of } \varphi\}$, where H_ψ is defined recursively:

  (a) $H_\varphi = \{(s, \varphi)\}$
  (b) If ψ is $(\theta_1 \circ \theta_2)$, then $H_{\theta_i} = \{h \frown \theta_i : h \in H_{(\theta_1 \circ \theta_2)}\}$
  (c) If ψ is $Qx\chi$, then $H_\chi = \{h \frown (x, a) : h \in H_{Qx\chi} \text{ and } a \in M\}$.

  Observe that $\{(s, \varphi)\}$ is the unique initial history. The assignment s is called the *initial assignment*. Each history h induces an assignment s_h :

$$s_h = \begin{cases} s & \text{if } h = (s, \varphi) \\ s_{h'}(x/a) & \text{if } h = h' \frown (x, a) \\ s_{h'} & \text{if } h = h' \frown \chi \end{cases}$$

- Each play ends when an atomic formula is reached:

$$Z = \bigcup\{H_\chi : \chi \text{ is an atomic subformula of } \varphi\}$$

- P, the player function, is defined on every non-terminal history $h \in H$:

$$P(h) = \begin{cases} \exists & \text{if } h \in H_{\exists x\chi} \text{ or } h \in H_{\psi\vee\theta} \\ \forall & \text{if } h \in H_{\forall x\chi} \text{ or } h \in H_{\psi\wedge\theta} \end{cases}$$

- The payoff function u_p for player p is defined by:
  (a) $u_\exists(h) = 1$ and $u_\forall(h) = 0$, if $(M, s_h) \models \chi$
  (b) $u_\exists(h) = 0$ and $u_\forall(h) = 1$, if $(M, s_h) \not\models \chi$.

The extensive form of a game $G(\mathcal{M}, \varphi, s)$ has obviously a tree structure, having the initial position (s, φ) as its root, and the maximal histories as its maximal branches.

Example 9.3.1
(i) We consider the semantical game $G(\mathbb{N}, \varnothing, \varphi)$, where φ is $\exists x\forall y(x \leq y)$, $\varnothing$ is the empty initial assignment, and $\mathbb{N}$ is the standard structure of arithmetic with domain ω. Let ψ denote $\forall y(x \leq y)$. Then $H_\varphi = \{(\varnothing, \varphi)\}$. Eloise first chooses a value for x. Thus $H_\psi = \{(\varnothing, \varphi, (x, a)) : a \in \omega\}$. Then Abelard chooses a value for y, and the game ends:

$$Z = \{(\varnothing, \varphi, (x, a), (y, b)) : a, b \in \omega\}$$

Eloise wins if $a \leq^{\mathbb{N}} b$; otherwise Abelard wins. Eloise has a winning strategy: $\sigma(\varnothing, \varphi) = 0$.
(ii) Consider the semantical game $G(\mathbb{N}, \varnothing, \exists x\forall y(y \leq x))$. The collection of histories is the same as before, but now Eloise wins if $b \leq^{\mathbb{N}} a$. However, it is Abelard who has a winning strategy now:

$$\tau(\varnothing, \varphi, (x, a)) = (y, a + 1).$$

3.2 Negation

To deal with the case in which negation does not occur only in front of an atomic formula, but can occur in any position, we have to take into consideration the roles of the two players. At the beginning of each game, *Eloise* assumes the role of *verifier* and *Abelard* that of *falsifier*. The player function needs to be modified in order to account for possible role reversals. The semantical game in its extensive form is defined exactly as before except for the following changes.

- If ψ is $\neg\theta$ then $H_\theta = \{h \frown \theta : h \in H_{\neg\theta}\}$. We can tell which player is the verifier in the history by counting the number of changes from $\neg\theta$ to θ.
- Disjunctions and existential quantifiers prompt moves by the player who is the verifier; conjunctions and universal quantifiers are decision points for the player who is the falsifier.
- The rules of winning and losing are restated: if the atomic formula reached at the end of the play is satisfied by the current assignment, the player who is the verifier wins; otherwise the falsifier wins.

Example 9.3.2 Consider the semantical game $G(\mathbb{N}, \varnothing, \neg\varphi)$, where $\varphi = \exists x \forall y (y \leq x)$. Eloise has a winning strategy given by

$$\sigma(\varnothing, \neg\varphi, \varphi, (x, a)) = (y, a+1)$$

which is Abelard's strategy in the game $G(\mathbb{N}, \varnothing, \exists x \forall y (y \leq x))$ described in the previous example.

The example should make clear that for any first-order formula φ, structure $\mathcal{M}$ and assignment s, *Eloise* has a winning strategy in $G(\mathcal{M}, s, \neg\varphi)$ if and only if *Abelard* has a winning strategy in $G(\mathcal{M}, s, \varphi)$ and *vice versa*.

3.3 Truth and Falsity in a Structure

Definition 9.3.1 Let φ be a first-order formula, $\mathcal{M}$ a structure and s an assignment in M whose domain includes the set of free variables of φ. Then

$\mathcal{M}, s \models^+_{GTS} \varphi$ iff there is a winning strategy for Eloise in $G(\mathcal{M}, s, \varphi)$
$\mathcal{M}, s \models^-_{GTS} \varphi$ iff there is a winning strategy for Abelard in $G(\mathcal{M}, s, \varphi)$.

When φ is a sentence, and s is the empty assignment $\varnothing$, we write $\mathcal{M} \models^+_{GTS} \varphi$ whenever $\mathcal{M}, \varnothing \models^+_{GTS} \varphi$, and say that φ is true in $\mathcal{M}$. Symmetrically we write $\mathcal{M} \models^-_{GTS} \varphi$ whenever $\mathcal{M}, \varnothing \models^-_{GTS} \varphi$, and say that φ is false in $\mathcal{M}$.

It is straightforward to show that

$$\mathcal{M}, s \models^+_{GTS} \neg\varphi \quad \text{iff} \quad \mathcal{M}, s \models^-_{GTS} \varphi.$$

The game-theoretical negation is well behaved given that for any first-order formula φ, structure $\mathcal{M}$, and assigment s, we have

$$\mathcal{M}, s \models^+_{GTS} \neg\varphi \quad \text{iff} \quad \mathcal{M}, s \not\models^+_{GTS} \varphi$$

Indeed, if *Abelard* has a winning strategy for $G(\mathcal{M}, s, \varphi)$, *Eloise* cannot have one, because the game is strictly competitive. Conversely, if *Eloise* does not have a winning strategy for $G(\mathcal{M}, s, \varphi)$, then by the Gale-Stewart theorem, *Abelard* must have one.

Proposition 9.3.1 Let φ be a first-order formula, $\mathcal{M}$ a suitable structure, and s and s' assignments in M which agree on the free variables of φ. Then

$$\mathcal{M}, s \models^{+}_{GTS} \varphi \text{ iff } \mathcal{M}, s' \models^{+}_{GTS} \varphi$$

Proof. Suppose *Eloise* has a winning strategy σ in $G(\mathcal{M}, s, \varphi)$. Every history $h = (s, \varphi, \ldots)$ in G corresponds to a history $h' = (s', \varphi, \ldots)$ in $G(\mathcal{M}, s'\varphi)$ obtained by substituting s' for s and leaving the rest of the history unchanged. Define a strategy σ' for *Eloise* in $G(\mathcal{M}, s'\varphi)$ by $\sigma'(h') = \sigma(h)$. Now suppose $h' = (s', \varphi, \ldots, \chi)$ is a terminal history for $G(\mathcal{M}, s'\varphi)$ in which *Eloise* follows σ'. Then $h = (s, \varphi, \ldots, \chi)$ is a terminal history for $G(\mathcal{M}, s, \varphi)$ in which she follows σ. It is straightforward to show by induction that the assignments s_h and $s_{h'}$ agree on the free variables of χ. Therefore *Eloise* wins h' iff she wins h. But the she wins h because σ is a winning strategy. Thus σ' is a winning strategy in $G(\mathcal{M}, s'\varphi)$. The converse is similar. ■

A consequence of the preceding proposition is that the players can play semantical games without remembering every single move they make. For instance in the case of double quantification $\forall x \forall x \exists y(x = y)$, *Abelard* chooses a value for x twice but only his second choice matters. *Eloise* need only consider this second value of x when picking the value of y. The informal considerations are captured by the property of a strategy being *memoryless*.

A strategy σ in a semantical game $G(\mathcal{M}, s, \varphi)$ is said to be *memoryless* if for every history h, the action $\sigma(h)$ only depends on the current assignment and the current subformula, that is, for every non-atomic subformula ψ of φ, if $h, h' \in H_{\psi}$ and $s_h = s_{h'}$, then $\sigma(h) = \sigma(h')$.

Proposition 9.3.2 For every φ, s, and $\mathcal{M}$, if a player has a winning strategy in $G(\mathcal{M}, s, \varphi)$, then he or she has a memoryless winning strategy.

Proof. Suppose σ is a winning strategy for player p in the game $G(\mathcal{M}, s, \varphi)$. If φ is atomic then σ is the empty strategy which is memoryless. If φ is $\neg\psi$ the opponent $\overline{p}$ has a winning strategy τ in $G(\mathcal{M}, s, \psi)$, given by

$$\tau(s, \psi, \ldots) = \sigma(s, \neg\psi, \psi, \ldots).$$

That is, $\tau(h) = \sigma(h')$ where h' is the history of $G(\mathcal{M}, s, \neg\psi)$ that is identical to h except for the insertion of $\neg\psi$ after the initial assignment. By the inductive hypothesis, $\overline{p}$ has a memoryless winning strategy τ' in $G(\mathcal{M}, s, \psi)$. Hence p has a memoryless winning strategy in $G(\mathcal{M}, s, \neg\psi)$ given by

$$\sigma'(s, \neg\psi, \psi, \ldots) = \tau'(s, \psi, \ldots).$$

We consider one more case, where φ is $\exists x\psi$. Suppose $\sigma(s, \exists x\psi) = (x, a)$, where σ is a winning strategy for *Eloise*. We define

$$\sigma'(s(x/a), \psi) = \sigma(s, \exists x\psi, (x, a)).$$

Then σ' is a winning strategy for *Eloise* in $G(\mathcal{M}, s(x/a), \psi)$ so by the inductive hypothesis, *Eloise* has a memoryless winning strategy σ'' in $G(\mathcal{M}, s(x/a), \psi)$. Hence the strategy σ''' defined by

$$\begin{aligned} &\sigma'''(s, \exists x\psi) = (x, a), \\ &\sigma'''(s, \exists x\psi, (x, a) \ldots) = \sigma''(s(x/a), \psi, \ldots), \end{aligned}$$

is a memoryless winning strategy for *Eloise* in $G(\mathcal{M}, s, \exists x\psi)$. All the other cases are similar. ■

3.4 Logical Equivalence

Let φ and ψ be first-order formulas. We say that φ *entails* ψ, $\varphi \models \psi$, if for every structure $\mathcal{M}$ and assignment s we have

$$\mathcal{M}, s \models^{+}_{GTS} \varphi \text{ implies } \mathcal{M}, s \models^{+}_{GTS} \psi.$$

We say that φ and ψ are *logically equivalent* (written $\varphi \equiv \psi$) if $\varphi \models \psi$ and $\psi \models \varphi$. It is straightforward to check that the usual equivalences of propositional logic hold. To take one example, let us show that

$$\neg(\varphi \wedge \psi) \equiv \neg\varphi \vee \neg\psi.$$

Suppose *Eloise* has a winning strategy σ in $G(\mathcal{M}, s, \neg(\varphi \wedge \psi))$. Define a winning strategy σ' for *Eloise* in $G(\mathcal{M}, s, \neg\varphi \vee \neg\psi))$ as follows:

$$\sigma'(s, \neg\varphi \vee \neg\psi)) = \begin{cases} \neg\varphi & \text{if } \sigma(s, \neg(\varphi \wedge \psi), (\varphi \wedge \psi)) = \varphi \\ \neg\psi & \text{if } \sigma(s, \neg(\varphi \wedge \psi), (\varphi \wedge \psi)) = \psi \end{cases}$$

and then let σ' agree with σ on the rest of the game. For the converse, suppose *Eloise* has a winning strategy in $G(\mathcal{M}, s, \neg\varphi \vee \neg\psi))$. Define a winning strategy

σ' for *Eloise* in $G(\mathcal{M}, s, \neg(\varphi \wedge \psi))$ by

$$\sigma'(s, \neg(\varphi \wedge \psi)) = \begin{cases} \neg\varphi & \text{if } \sigma(s, \neg\varphi \vee \neg\psi)) = \neg\varphi \\ \neg\psi & \text{if } \sigma(s, \neg\varphi \vee \neg\psi)) = \neg\psi \end{cases}$$

and then, if *Eloise* chooses $\neg\varphi$, let σ' agree with σ on $\neg\varphi$; if *Eloise* chooses $\neg\psi$, let σ' agree with σ on $\neg\psi$.

Also the usual distribution laws for quantifiers hold. To take an example, consider

$$\exists x(\varphi \vee \psi) \equiv \exists x\varphi \vee \exists x\psi.$$

Suppose that *Eloise* has a winning strategy σ for $G(\mathcal{M}, s, \exists x(\varphi \vee \psi))$. Let $\sigma(s, \exists x(\varphi \vee \psi)) = (x, a)$ and $\sigma(s, \exists x(\varphi \vee \psi), (x, a)) = \chi$, where χ is φ or ψ. Define a strategy σ' in the game $G(\mathcal{M}, s, \exists x\varphi \vee \exists x\psi)$ as follows:

$$\begin{aligned} &\sigma'(s, \exists x\varphi \vee \exists x\psi) = \exists x\chi \\ &\sigma'(s, \exists x\varphi \vee \exists x\psi, \exists x\chi) = (x, a) \\ &\sigma'(s, \exists x\varphi \vee \exists x\psi, \exists x\chi, (x, a), \ldots) = \sigma(s, \exists x(\varphi \vee \psi), (x, a), \chi, \ldots). \end{aligned}$$

That is, σ' tells *Eloise* to choose $\exists x\varphi$ if she picks φ in $G(\mathcal{M}, s, \exists x(\varphi \vee \psi))$, to choose $\exists x\psi$ if she picks ψ, and to assign x the same value as she did in $G(\mathcal{M}, s, \exists x(\varphi \vee \psi))$.

Observe that in both games, after *Eloise*'s first two moves the current assignment is $s(x/a)$ and the current subformula is χ. The play proceeds as in the game $G(\mathcal{M}, s(x/a), \chi)$. Every terminal history

$$h' = (s, \exists x\varphi \vee \exists x\psi, \exists x\chi, (x, a), \ldots)$$

in $G(\mathcal{M}, s, \exists x\varphi \vee \exists x\psi)$ in which *Eloise* follows σ' corresponds to a terminal history

$$h = (s, \exists x(\varphi \vee \psi), (x, a), \chi, \ldots)$$

of $G(\mathcal{M}, s, \exists x(\varphi \vee \psi))$ in which *Eloise* follows the strategy σ that induces the same assignment and terminates with the same atomic formula. Thus *Eloise* wins h' if and only if she wins h. But she does win h given that σ is a winning strategy. Hence σ' is a winning strategy in $G(\mathcal{M}, s, \exists x\varphi \vee \exists x\psi)$. The converse is similar.

We can see that the existential quantifier distributes over disjunctions because they are both moves for the same player, whereas existential quantifiers fail to distribute over conjunctions because they are moves for different players. In the first case, *Eloise* can plan ahead and choose the value of x that will verify the appropriate disjunct, or choose the disjunct first and then choose the value of x. In the second case, she is forced to commit to a value of x before she knows which conjunct *Abelard* chooses.

3.5 Tarski Type Semantics

In the previous sections, we have construed first-order logic in a game-theoretical setting. We can now ask whether there is a method which determines the semantic value of a complex formula compositionally in terms of the semantic values of its subformulas and their mode of composition. The answer is well known: it is Tarski's notion of satisfaction. The next theorem recovers Tarski's compositional interpretation.

Theorem 9.3.1 *(Assuming the Axiom of Choice) Let φ and ψ be first-order formulas, $\mathcal{M}$ a suitable structure, and s an assignment in M whose domain contains the free variables of φ and ψ. Then*

$$\begin{array}{rcl} \mathcal{M},s \models^+_{GTS} \neg\varphi & \textit{iff} & \mathcal{M},s \not\models^+_{GTS} \varphi \\ \mathcal{M},s \models^+_{GTS} \varphi \vee \psi & \textit{iff} & \mathcal{M},s \models^+_{GTS} \varphi \text{ or } \mathcal{M},s \models^+_{GTS} \psi \\ \mathcal{M},s \models^+_{GTS} \varphi \wedge \psi & \textit{iff} & \mathcal{M},s \models^+_{GTS} \varphi \text{ and } \mathcal{M},s \models^+_{GTS} \psi \\ \mathcal{M},s \models^+_{GTS} \exists x\varphi & \textit{iff} & \mathcal{M},s(x/a) \models^+_{GTS} \varphi, \text{ for some } a \in M \\ \mathcal{M},s \models^+_{GTS} \forall x\varphi & \textit{iff} & \mathcal{M},s(x/a) \models^+_{GTS} \varphi, \text{ for every } a \in M. \end{array}$$

Proof. We have already established the case for negation. All the other cases are straightforward. For instance, suppose that *Eloise* has a winning strategy σ for the disjunction. Then $\sigma(s, \varphi \vee \psi) = \theta$, where θ is either φ or ψ. But then the strategy σ'

$$\sigma'(s,\theta,\ldots) = \sigma(s,\varphi \vee \psi,\theta,\ldots)$$

which mimics σ after the choice of θ is a winning strategy for *Eloise* in $G(M,s,\theta)$.

For the converse, suppose that $\theta \in \{\varphi,\psi\}$ and that *Eloise* has a winning strategy σ' in $G(M,s,\theta)$. Define a winning strategy σ for *Eloise* in $G(\mathfrak{M},s,\varphi\vee\psi)$ by

$$\sigma(s,\varphi \vee \psi) = \theta$$
$$\sigma(s,\varphi \vee \psi,\theta,\ldots) = \sigma'(s,\theta,\ldots).$$

Suppose now that *Eloise* has a winning strategy σ for $G(\mathcal{M},s,\forall x\varphi)$. For every $a \in M$, define

$$\sigma_a(s(x/a),\varphi,\ldots) = \sigma(s,\forall x\varphi,(x,a),\ldots)$$

That is, σ_a mimics σ after *Abelard* chooses a. But then σ_a is winning for $G(M,s(x/a),\varphi)$. Conversely, suppose that for every $a \in M$, *Eloise* has a winning strategy in $G(\mathcal{M},s(x/a),\varphi)$. Choose one, say σ_a (here we need the Axiom of Choice).[1] Define now a winning strategy for $G(\mathcal{M},s,\forall x\varphi)$ by

$$\sigma(s,\forall x\varphi,(x,a),\ldots) = \sigma_a(s(x/a),\varphi,\ldots)$$

That is, after the choice of a by *Abelard*, *Eloise* will mimic her winning strategy σ_a. ■

3.6 Satisfiability and Skolem Semantics

We often consider a first-order formula without having a particular structure in mind. A formula φ is *satisfiable* if there exists a structure $\mathcal{M}$ and an assignment s in M such that $\mathcal{M}, s \models \varphi$. When checking the satisfiability of a formula, we often look at a process called *Skolemization* to eliminate existential quantifiers.

Let φ be a first-order formula in negation normal form, in the vocabulary L, and let

$$L^* = L \cup \{f_\psi : \psi \text{ is an existential subformula of } \varphi\}$$

be the expansion of L by adding a new function symbol for each existentially quantified subformula of φ. The *Skolem form* or *Skolemization* of a subformula ψ of φ with variables in U is defined recursively:

$$\begin{aligned}
Sk_U(\psi) &:= \psi \quad \text{if } \psi \text{ is a literal} \\
Sk_U(\psi \vee \psi') &:= Sk_U(\psi) \vee Sk_U(\psi') \\
Sk_U(\psi \wedge \psi') &:= Sk_U(\psi) \wedge Sk_U(\psi') \\
Sk_U(\exists x \psi) &:= Subst(Sk_{U\cup\{x\}}(\psi), x, f_{\exists x\psi}(y_1, \ldots, y_n)) \\
Sk_U(\forall x \psi) &:= \forall x Sk_{U\cup\{x\}}(\psi)
\end{aligned}$$

where $y_1, \ldots, y_n$ enumerate the variables in U and where the substitution operation *Subst* is defined as follows: If φ is a first-order formula, x is a variable, and t is a term, $Subst(\varphi, x, t)$ denotes the first-order formula obtained from φ by replacing all free occurrences of x by the term t. If x does not occur free in φ, then $Subst(\varphi, x, t)$ is simply φ. Usually when substituting a term t for a free variable x, we must be careful that none of the variables in t become bound in the resulting formula. A term t which satisfies such a requirement is called *substitutible for the variable x in the formula φ*. The formal definition may be found in [Enderton, 1972, p. 105].

The term $f_{\exists x\psi}(y_1, \ldots, y_n)$ is called a *Skolem term*. For sentences φ, we abbreviate $Sk_\varnothing(\varphi)$ by $Sk(\varphi)$. The necessity to consider the Skolemization relativized to a set of variables U will become apparent later on.

Example 9.3.3 Let φ be the sentence

$$\forall x \exists y [x < y \vee \exists z (y < z)]$$

Then

$Sk_{\{x,y,z\}}(y < z)$	is	$y < z$
$Sk_{\{x,y\}}(\exists z(y < z))$	is	$y < g(x,y)$
$Sk_{\{x,y\}}(x < y)$	is	$x < y$
$Sk_{\{x,y\}}(x < y \vee \exists z(y < z))$	is	$x < y \vee y < g(x,y)$
$Sk_{\{x\}}[\exists y(x < y \vee \exists z(y < z))]$	is	$x < f(x) \vee f(x) < g(x,f(x))$
$Sk(\varphi)$	is	$\forall x[x < f(x) \vee f(x) < g(x,f(x))]$

Skolemizing a first-order sentence makes explicit the dependencies between quantified variables. Notice the difference in the Skolem form of $\forall x\exists yR(x,y)$ and $\exists y\forall xR(x,y)$. The Skolemization of the first is $\forall xR(x,f(x))$, whereas that of the second is $\forall xR(x,c)$, with c is a fresh constant symbol (nullary function symbol).

The next theorem establishes an equivalence between game-theoretical semantics as defined in the previous section and *Skolem semantics* defined below. The proof uses a well-known result in the meta-theory of first-order logic, the *Substitution Lemma* that will be given without proof.

Lemma 9.3.1 *(Substitution Lemma) Let φ be a first-order formula in the vocabulary L, $\mathcal{M}$ be a structure in the same vocabulary, s an assignment in M, and t an L-term substitutible for x in φ. Then*

$$\mathcal{M}, s \models Subst(\varphi, x, t) \quad iff \quad \mathcal{M}, s(x/s(t)) \models \varphi.$$

Before asking the question whether an L-structure $\mathcal{M}$ satisfies the Skolem form $Sk(\varphi)$ of a formula φ we must expand $\mathcal{M}$ to an L^*-structure that specifies how to interpret the new symbols of $Sk(\varphi)$.

Definition 9.3.2 Let φ be a first-order formula, $\mathcal{M}$ a structure in the same vocabulary, and s an assignment in $\mathcal{M}$ whose domain contains the free variables of φ. Define

$$\mathcal{M}, s \models^{+}_{Sk} \varphi \text{ iff } \mathcal{M}^*, s \models Sk_{dom(s)}(\varphi)$$

for some expansion $\mathcal{M}^*$ of $\mathcal{M}$ to the vocabulary $L^* = L \cup \{f_\psi : \psi \text{ is an existential subformula of } \varphi\}$.

The first thing we need to check is that the Skolem Semantics agrees with the game-theoretical semantics defined earlier.

Theorem 9.3.2 *Let φ be a first-order formula, $\mathcal{M}$ a structure and s an assignment in M whose domain dom(s) includes the free variables of φ. Then*

$$\mathcal{M}, s \models^{+}_{GTS} \varphi \text{ iff } \mathcal{M}, s \models^{+}_{Sk} \varphi.$$

Proof. Suppose *Eloise* has a winning strategy σ for $G(M, s, \varphi)$. Let $\mathcal{M}^*$ be an expansion of $\mathcal{M}$ to the vocabulary $L^* = L \cup \{f_\psi : \psi$ is an exist. subf. of $\varphi\}$ such that for every existential subformula $\exists x\psi'$ of φ and every history $h \in H_{\exists x\psi'}$

$$f^{\mathcal{M}^*}_{\exists x\psi'}(s_h(y_1), \ldots, s_h(y_n)) = a$$

where $y_1, \ldots, y_n$ enumerates the domain of s_h and $\sigma(h) = (x, a)$. It is easy to show that the function is well defined. We now show by induction on the subformulas ψ of φ that if *Eloise* follows σ in $h \in H_\psi$, then $\mathcal{M}^*, s_h \models Sk_{dom(s_h)}(\psi)$. If ψ is an atomic formula or its negation, then $Sk_{dom(s)}(\psi) = \psi$. If *Eloise* follows σ in $h \in H_\psi$, then given that σ is winning we have $\mathcal{M}, s_h \models \psi$. Hence $\mathcal{M}^*, s_h \models Sk_{dom(s_h)}(\psi)$.

Suppose φ is $\psi_1 \vee \psi_2$. If *Eloise* follows σ in $h \in H_{\psi_1 \vee \psi_2}$ and $\sigma(h) = \psi_i$, then *Eloise* follows σ in $h' = h \frown \psi_i \in H_{\psi_i}$. By the inductive hypothesis

$$\mathcal{M}^*, s_{h'} \models Sk_{dom(s_{h'})}(\psi_i)$$

whence

$$\mathcal{M}^*, s_{h'} \models Sk_{dom(s_{h'})}(\psi_1) \vee Sk_{dom(s_{h'})}(\psi_2).$$

Since $s_h = s_{h'}$, it follows that $\mathcal{M}^*, s_h \models Sk_{dom(s_h)}(\psi_1 \vee \psi_2)$.

The case for conjunction is similar.

Suppose that φ is $\exists x\psi'$. If *Eloise* follows σ in $h \in H_{\exists x\psi'}$ and $\sigma(h) = (x, a)$, then *Eloise* follows σ in $h' = h \frown (x, a) \in H_{\psi'}$. By the inductive hypothesis

$$\mathcal{M}^*, s_{h'} \models Sk_{dom(s_{h'})}(\psi')$$

which is the same as

$$\mathcal{M}^*, s_{h(x/a)} \models Sk_{dom(s_h(x/a))}(\psi').$$

By construction $f^{\mathcal{M}^*}_{\exists x\psi'}(s_h(y_1), \ldots, s_h(y_n)) = a$, where $y_1, \ldots, y_n$ enumerates the domain of s_h.Then by the *Substitution Lemma*

$$\mathcal{M}^*, s_h \models Subst(Sk_{dom(s_h(x/a)}(\psi'), x, f_{\exists x\psi'}(y_1, \ldots, y_n)).$$

Therefore $\mathcal{M}^*, s_h \models Sk_{dom(s_h)}(\exists x\psi')$.

Suppose that φ is $\forall x\psi'$. If *Eloise* follows σ in $h \in H_{\forall x\psi'}$, then she follows σ in every $h_a = h \frown (x, a) \in H_{\psi'}$. By the inductive hypothesis

$$\mathcal{M}^*, s_{h_a} \models Sk_{dom(s_{h_a})}(\psi').$$

Given that $s_{h_a} = s_h(x/a)$, it follows that

$$\mathcal{M}^*, s_h \models \forall x Sk_{dom(s_h)\cup\{x\}}(\psi')$$

which implies $\mathcal{M}^*, s_h \models \forall x Sk_{dom(s_h)}(\forall x \psi')$.

Finally observe that *Eloise* follows σ in the initial history $(s, \varphi) \in H_\varphi$. Therefore $\mathcal{M}^*, s \models Sk_{dom(s)}(\varphi)$.

Conversely, suppose there is an expansion $\mathcal{M}^*$ of $\mathcal{M}$ such that

$$\mathcal{M}^*, s \models Sk_{dom(s)}(\varphi).$$

Let σ be the strategy for *Eloise* defined as follows.

If $h \in H_{\psi_1 \vee \psi_2}$, then

$$\sigma(h) = \begin{cases} \psi_1 & \text{if } \mathcal{M}^*, s_h \models Sk_{dom(s_h)}(\psi_1) \\ \psi_2 & \text{otherwise.} \end{cases}$$

If $h \in H_{\exists x \psi'}$, then

$$\sigma(h) = (x, f^{\mathcal{M}^*}_{\exists x \psi'}(s_h(y_1), \ldots, s_h(y_n)))$$

where $y_1, \ldots, y_n$ enumerates the domain of s_h. It is straightforward to show by induction on the length of h that if *Eloise* follows σ in $h \in H_\psi$, then $\mathcal{M}^*, s_h \models Sk_{dom(s_h)}(\psi)$. The proof is left to the reader.

Finally observe that if *Eloise* follows σ in a terminal history $h \in H_\chi$, then $\mathcal{M}^*, s_h \models Sk_{dom(s_h)}(\chi)$. It follows that $\mathcal{M}, s_h \models \chi$, so *Eloise* wins in h. Therefore, σ is a winning strategy for *Eloise*. ■

3.7 Falsifiability and Kreisel Counterexamples

By analogy with the previous case where one can say that Skolem functions point out witnesses to existential formulas one can introduce *Kreisel counterexamples*, which point out falsifying instances to universal formulas. Let φ be a first-order sentence in the vocabulary L in negation normal form, and let

$$L^* = L \cup \{f_\psi : \psi \text{ is an universal subformula of } \varphi\}$$

be the expansion of L by adding a new function symbol for each universally quantified subformula of φ. The *Kreisel form* (or *Kreiselization*) $Kr_U(\varphi)$ of φ is defined recursively:

$$
\begin{aligned}
Kr_U(\psi) &:= \neg\psi \quad \text{if } \psi \text{ is a literal} \\
Kr_U(\psi \vee \psi') &:= Kr_U(\psi) \wedge Kr_U(\psi') \\
Kr_U(\psi \wedge \psi') &:= Kr_U(\psi) \vee Kr_U(\psi') \\
Kr_U(\exists x\psi) &:= \forall x Kr_{U\cup\{x\}}(\psi) \\
Kr_U(\forall x\psi) &:= Subst(Kr_{U\cup\{x\}}(\psi), x, f_{\forall x\psi}(y_1, \ldots, y_n))
\end{aligned}
$$

where $y_1, \ldots, y_n$ is the list of variables in U. An interpretation of $f_{\forall x\psi}(y_1, \ldots, y_n)$ is called a *Kreisel counterexample*. For sentences φ, we abbreviate $Kr_{\varnothing}(\varphi)$ by $Kr(\varphi)$.

Example 9.3.4 The Kreisel form of the sentence

$$\forall x(\exists y R(x,y) \vee \exists z R(x,z))$$

is obtained in the following stages:

$$
\begin{array}{lll}
Kr_{\{x,y\}}(R(x,y)) & \text{is} & \neg R(x,y) \\
Kr_{\{x,z\}}(R(x,z)) & \text{is} & \neg R(x,z) \\
Kr_{\{x\}}(\exists y R(x,y)) & \text{is} & \forall y \neg R(x,y) \\
Kr_{\{x\}}(\exists z R(x,z)) & \text{is} & \forall z \neg R(x,z) \\
Kr_{\{x\}}(\exists y R(x,y) \vee \exists z R(x,z)) & \text{is} & \forall y \neg R(x,y) \wedge \forall z \neg R(x,z) \\
Kr_{\varnothing}(\varphi) & \text{is} & \forall y \neg R(c,y) \wedge \forall z \neg R(c,z).
\end{array}
$$

Definition 9.3.3 Let φ be a first-order formula, $\mathcal{M}$ a structure in the same vocabulary, and s an assignment in $\mathcal{M}$ whose domain contains the free variables of φ. Define

$$\mathcal{M}, s \models^{-}_{Sk} \varphi \text{ iff } \mathcal{M}^*, s \models Kr_{dom(s)}(\varphi)$$

for some expansion $\mathcal{M}^*$ of $\mathcal{M}$ to the vocabulary $L^* = L \cup \{f_\psi : \psi$ is a universal subformula of $\varphi\}$.

It can be shown that falsity as the existence of Kreisel counterexamples coincides with game-theoretical falsity.

Theorem 9.3.3 *Let φ be a first-order formula, $\mathcal{M}$ a structure, and s an assignment in M whose domain $dom(s)$ includes the free variables of φ. Then*

$$\mathcal{M}, s \models^{-}_{GTS} \varphi \textit{ iff } \mathcal{M}, s \models^{-}_{Sk} \varphi.$$

Proof. Completely analogue to the previous theorem. ∎

Notes. Games in extensive form are discussed in [Osborne and Rubinstein, 1994]. The game-theoretical interpretation of connectives outside the framework of extensive forms of games is to be found in [Hintikka, 1974] and [Hintikka, 1983]. The representation of semantical games as two-person one-sum games appeared for the first time in [Sandu and Pietarinen, 2003]. The definition of Skolemization and Kreiselization together with the equivalence between the three semantical interpretations borrows from [Mann et al., ta], where one can also find the game-theoretical reformulations of other meta-theoretical properties of first-order logic.

4. IF Languages

In this section we give a short introduction to the syntax and semantics of *Independence-friendly logic* (IF logic).

Definition 9.4.1 The independence-friendly (IF-) formulas are generated by the following rules:

- If t_1 and t_2 are terms, then $t_1 = t_2$ and $\neg(t_1 = t_2)$ are IF-formulas
- If $t_1, \ldots, t_n$ are terms and R is an n-ary relation symbol, then $R(t_1, \ldots, t_n)$ and $\neg R(t_1, \ldots, t_n)$ are IF-formulas
- If φ and ψ are IF-formulas, then $(\varphi \vee \psi)$ and $(\varphi \wedge \psi)$ are IF-formulas
- If φ is an IF-formula, x is a variable and W is a finite set of variables, then $(\exists x/W)\varphi$ and $(\forall x/W)\varphi$ are IF-formulas.

To simplify things we let the negation symbol occur only in front of atomic formulas. The set W in $(\exists x/W)\varphi$ and $(\forall x/W)\varphi$ is called a *slashed* set. The intended interpretation of $(\exists x/W)$ is: there exists an x independent of the quantifiers that bind the variables in W. The intended meaning of $(\forall x/W)$ is similar. When $W = \varnothing$, we recover the classical quantifiers.

The notion of subformula is defined in the standard way. The set of free variables of an IF formula is defined as for ordinary first-order logic, except for quantified formulas:

$$Free((Qx/W)\varphi) = (Free(\varphi) - \{x\}) \cup W$$

As with ordinary first-order formulas, an occurrence of a variable x is bound by the innermost quantifier in the scope of which it occurs. For instance, in the formula

$$\forall x(\exists y/\{x\})R(x, y) \wedge \forall y(\exists z/\{x, y\})R(y, z),$$

the variables y and z are bound, while x is both free and bound. An IF formula with no free variables is called an IF *sentence*.

4.1 Extensive Games of Imperfect Information

Ordinary first-order languages have been interpreted by two-person, one-sum games of *perfect* information. In a similar spirit, IF first-order languages will be interpreted by two-person, one-sum games of *imperfect* information. An *extensive game of imperfect information* $G = (N, H, Z, P, (I_p)_{p \in N}, (u_i)_{i \in N})$ has the same components as an extensive game with perfect information plus the two *information sets* $I_p = \{I_h : h \in H_p\}$. If $h' \in I_h$, then we write $h \sim_p h'$ and say that h and h' are *indistinguishable* for player p. Indistinguishability (or $\sim_p$) is an equivalence relation in H_p, I_p is the set of equivalence classes of $\sim_p$, and I_h is the equivalence class that contains h.

Information sets specify how much information a player has at his or her disposal in a given position and a player may only use that information when deciding which action to take. This is reflected in the strategies of the players: a player must choose the same action in response to histories that are indistinguishable for him or her. That is, a strategy σ_p for the player p is defined exactly as before, except for the *requirement of uniformity*:

- If $h, h' \in H_p$ and $h \sim_p h'$, then $\sigma_p(h) = \sigma_p(h')$.

Example 9.4.1 Consider the imperfect information variant of the game in the first section in which player 1 can choose either a or b, after which player 2 can choose either c or d without knowing the choices of player 1. In other words, the histories a and b are indistinguishable for the second player. The payoffs for the two players are the same. Player 1 has two strategies at his disposal, a and b, but now player 2 has only two strategies (instead of four): $\tau(a) = c$ and $\tau(b) = c$, and $\tau'(a) = d$ and $\tau'(b) = d$.

It is easy to see that the Gale-Stewart theorem fails in this case: neither player has a winning strategy.

Semantical games of imperfect information are introduced in the definition below. We say that two assignments s and s' with common domain are W-equivalent (W is included in the common domain), denoted by $s \approx_W s'$ if s and s' agree on the variables not in W.

Definition 9.4.2 Let φ be an IF formula, $\mathcal{M}$ a structure, and s an assignment in M whose domain includes the free variables of φ. The semantical game $G(\mathcal{M}, s, \varphi)$ is a one-sum extensive game of imperfect information

$$G = (\{\exists, \forall\}, H, Z, P, I_\exists, I_\forall, (u_i)_{i \in \{\exists, \forall\}})$$

where H, Z, P, and u_i are exactly as before (the rules for $(\exists x/W)\varphi$ and $(\forall x/W)\varphi$ are exactly like the rules for $\exists x\varphi$ and $\forall x\varphi$, respectively). The equivalence classes

$$I_\exists = \{I_h : h \in H_\exists\} \text{ and } I_\forall = \{I_h : h \in H_\forall\}$$

are defined as follows:

- If $h \in H_{\psi \vee \psi'}$, then $I_h = \{h\}$.
- If $h \in H_{\psi \wedge \psi'}$, then $I_h = \{h\}$.
- If $h \in H_{(\exists y/W)\varphi}$, then

$$I_h = \{h' \in H_{(\exists y/W)\varphi} : s_h \approx_W s_{h'}\}$$

- If $h \in H_{(\forall y/W)\varphi}$, then

$$I_h = \{h' \in H_{(\forall y/W)\varphi} : s_h \approx_W s_{h'}\}$$

As pointed out, the information sets I_p specify how much information player p has at his or her disposal in a given position. For instance in the position corresponding to $(\exists x/W)\varphi$, *Eloise* must choose a value for x without having access to the values for the variables in W. And likewise for *Abelard*. This is reflected in the strategies of the players satisfying the above-mentioned *requirement of uniformity*.

The definitions of truth and falsity extend naturally to the new case:

Definition 9.4.3 Let $\mathcal{M}$ be a structure, φ an IF formula, and s an assignment in M whose domain includes Free(φ). Then

$\mathcal{M}, s \models^{+}_{GTS} \varphi$ iff There is a winning strategy for Eloise in $G(\mathcal{M}, s, \varphi)$
$\mathcal{M}, s \models^{-}_{GTS} \varphi$ iff There is a winning strategy for Abelard in $G(\mathcal{M}, s, \varphi)$.

4.1.1 Indeterminacy

We have given an example of a game of imperfect information that is not determined. So it is to be expected that there are formulas of IF logic that are indeterminate too.

Example 9.4.2 (Matching Pennies) In this game, two players choose simultaneously whether to show the Heads or the Tails of a coin. If they show the same side, player 1 wins; if they show different sides, player 2 wins.

We can express the Matching Pennies by using the IF sentence $\varphi := \forall x(\exists y/\{x\})(x = y)$ interpreted in a two element structure $\mathcal{M} = M = \{a, b\}$. We show that neither of the players has a winning strategy in the game $G(\mathcal{M}, \varnothing, \varphi)$. Let $\psi := (\exists y/\{x\})(x = y)$. $H_\varphi = \{(\varnothing, \varphi)\}$ and $H_\psi = \{(\varnothing, \varphi, (x, a)), (\varnothing, \varphi, (x, b))\}$. Let $h_a = (\varnothing, \varphi, (x, a))$ and $h_b = (\varnothing, \varphi, (x, b))$. In each position, Eloise chooses a

value for y, i.e., a strategy for her is any function $\sigma : H_\psi \to \{(y,a),(y,b)\}$. Given that $(\varnothing,\varphi,(x,a)) \sim_\exists (\varnothing,\varphi,(x,b))$, we must have $\sigma(h_a) = \sigma(h_b)$. And in order for σ to be a winning strategy, Eloise must win both terminal plays where she uses it, i.e., she must win both $(\varnothing,\varphi,(x,a),\sigma(h_a))$ and $(\varnothing,\varphi,(x,b),\sigma(h_b))$. But these conditions cannot be jointly satisfied. For instance, if $\sigma(h_a) = \sigma(h_b) = (y,a)$, then we must also have $a = b$, which is impossible.
A strategy for Abelard is any function $\tau : H_\varphi \to \{(x,a),(x,b)\}$. Let $\tau((\varnothing,\varphi)) = c \in \{(x,a),(x,b)\}$. In order for τ to be a winning strategy, Abelard must win both $(\varnothing,\varphi,(x,c),(y,a))$ and $(\varnothing,\varphi,(x,c),(y,b))$, which is again impossible.
We conclude that $\forall x(\exists y/\{x\})(x = y)$ is neither true nor false in any structure with at least two elements. A symmetrical argument shows that the same holds of $\forall x(\exists y/\{x\})(x \neq y)$.

4.1.2 Dummy Quantifiers and Signalling

We show that adding a dummy quantifier to the IF sentence $\varphi := \forall x(\exists y/\{x\})$ $(x = y)$ changes its semantical value from indeterminate to true.

Example 9.4.3 Let θ be the IF sentence

$$\forall x \exists z(\exists y/\{x\})(x = y).$$

Let χ abbreviate $(\exists y/\{x\})(x = y)$. Then H_χ contains the histories

$$\begin{aligned} h_{aa} &= (\varnothing,\theta,(x,a),(y,a)) \\ h_{ba} &= (\varnothing,\theta,(x,b),(z,a)) \\ h_{ab} &= (\varnothing,\theta,(x,a),(z,b)) \\ h_{bb} &= (\varnothing,\theta,(x,b),(z,b)). \end{aligned}$$

Observe that $h_{aa} \sim_\exists h_{ba}$ and $h_{ab} \sim_\exists h_{bb}$ because Eloise is not allowed to see the value of x. Therefore, by the requirement of uniformity, all her strategies σ must satisfy $\sigma(h_{aa}) = \sigma(h_{ba})$ and $\sigma(h_{ab}) = \sigma(h_{bb})$. Here is a winning strategy

$$\begin{aligned} &\sigma(h_a) = (z,a) \text{ and } \sigma(h_b) = (z,b) \\ &\sigma(h_{aa}) = \sigma(h_{ba}) = (y,a) \text{ and } \sigma(h_{ab}) = \sigma(h_{bb}) = (y,b). \end{aligned}$$

There are two terminal histories in which Eloise follows σ:

$$(\varnothing,\theta,(x,a),(z,a),(y,a)) \text{ and } (\varnothing,\theta,(x,b),(z,b),(y,b))$$

In both of these, Eloise wins.

The phenomena illustrated by this example are common in games of imperfect information. In bridge, partners can communicate with each other about their hands using only the cards they play. Playing according to a predetermined convention in order to circumvent informational restrictions is called *signalling*. We return to this topic later on.

4.2 Generalizing Skolemization and Kreisel Counterexamples

In this section, we give an alternative semantics for IF formulas by generalizing the Skolemization and Kreiselization procedure for first-order formulas to IF formulas. Let φ be a formula of IF logic in the vocabulary L. Then, the *skolemized form* or *skolemization* of φ, denoted $Sk_U(\varphi)$, is defined exactly as in the first-order case, with the exception of the clause for $(\exists x/W)\psi$. That is, the clauses for $Sk_U(\psi) = \psi$ (if ψ is a literal) and $Sk_U(\psi \circ \theta)$ are exactly like before; the clause for $Sk_U(\forall x/W)\psi$ is the same as the clause for ordinary universal quantifiers. The only new clause is

$$Sk_U((\exists x/W)\psi) = Subst(Sk_{U\cup\{x\}}(\psi), x, f_{(\exists x/W)\psi}(y_1, \ldots, y_n)),$$

where $y_1, \ldots, y_n$ is the list of all variables in $U - W$. Observe that at each stage $Sk_U(\psi)$ is an ordinary first-order formula.

Definition 9.4.4 Let φ be a formula of IF logic in the vocabulary L, $\mathcal{M}$ an L-structure, and s an assignment in M whose domain includes the free variables of φ. We define

$$\mathcal{M}, s \models^+_{Sk} \varphi \text{ iff } \mathcal{M}^*, s \models Sk_{dom(s)}(\varphi)$$

for some expansion $\mathcal{M}^*$ of $\mathcal{M}$ to the vocabulary $L^* = L \cup \{f_\psi : \psi \text{ is an existential subformula of } \varphi\}$.

When evaluating an IF formula under Skolem semantics, we implicitly assume that every variable that has been assigned a value is 'present' in the formula. Thus the Skolemization of an IF formula depends on the assignment used to evaluate it. For example, suppose s and s' are assignments in M such that $dom(s) = \{u, v\}$ and $dom(s') = \{u, v, w\}$. Then

$$\mathcal{M}, s \models^+_{Sk} (\exists x/\{u\})P(x) \text{ iff } \mathcal{M}^*, s \models P(f(v))$$

for some expansion $\mathcal{M}^*$ of $\mathcal{M}$, while

$$\mathcal{M}, s' \models^+_{Sk} (\exists x/\{u\})P(x) \text{ iff } \mathcal{M}^{**}, s' \models P(g(v, w))$$

for some expansion $\mathcal{M}^{**}$ of $\mathcal{M}$.

The next theorem states the equivalence between the Skolem semantics and the game-theoretical semantics.

Theorem 9.4.1 *Let φ be a formula of IF logic in the vocabulary L, $\mathcal{M}$ an L-structure, and s an assignment in M whose domain contains the free variables of φ. Then*

$$\mathcal{M}, s \models^{+}_{GTS} \varphi \textit{ iff } \mathcal{M}, s \models^{+}_{Sk} \varphi.$$

Proof. Analoguous to the first-order case. ∎

We use Skolem semantics to give an example of an IF sentence that expresses a concept – namely, Dedekind infinity – that is undefinable in ordinary first-order logic.

Example 9.4.4 Let φ be the IF formula

$$\exists w \forall x (\exists y/\{w\})(\exists z/\{w, x\})(x = z \wedge w \neq y)$$

and let ψ be the subformula $(x = z \wedge w \neq y)$. The skolemization of φ is obtained in the following stages:

$Sk_{\{w,x,y,z\}}(\psi)$	is	$x = z \wedge y \neq w$
$Sk_{\{w,x,y\}}[(\exists z/\{w, x\})\psi]$	is	$x = g(y) \wedge w \neq y$
$Sk_{\{w,x\}}[(\exists y/\{w\})(\exists z/\{w, x\})\psi]$	is	$(x = g(f(x)) \wedge w \neq f(x))$
$Sk_{\{w\}}[\forall x(\exists y/\{w\})(\exists z/\{w, x\})\psi]$	is	$\forall x(x = g(f(x)) \wedge w \neq f(x))$
$Sk(\varphi)$	is	$\forall x(x = g(f(x)) \wedge c \neq f(x))$

where f and g are unary function symbols and c is a nullary function symbol. $Sk(\varphi)$ asserts that f is a bijection from the universe to a proper subset of itself. Thus $Sk(\varphi)$ is true in an expansion of $\mathcal{M}$ iff the universe of $\mathcal{M}$ is Dedekind infinite.

The correspondence between *Abelard*'s strategies in semantical games of imperfect information and the generalized notion of Kreisel counterexamples extends also to the present case. As the reader might have guessed, the clauses for the Kreisel form for an IF formula are identical to their first-order counterparts, except for

$$Kr_U((\forall x/W)\psi) = Subst(Kr_{U \cup \{x\}}(\psi), x, f_{(\forall x/W)\psi}(y_1, \ldots, y_n))$$

where $y_1, \ldots, y_n$ is the list of all variables in $U - W$.

Definition 9.4.5 Let φ be a formula of IF logic in the vocabulary L, $\mathcal{M}$ an L-structure, and s an assignment in M whose domain includes the free variables in φ. We define

$$M, s \models^{-}_{Sk} \varphi \text{ iff } \mathcal{M}^{*}, s \models Kr_{dom(s)}(\varphi)$$

for some expansion $\mathcal{M}^*$ of $\mathcal{M}$ to the vocabulary $L^* = L \cup \{f_\psi : \psi$ is a universal subformula of $\varphi\}$.

The next theorem shows that we can find Kreisel counterexamples for a formula if, and only if, *Abelard* has a winning strategy for the semantic game. In fact the counterexamples may be thought of as 'local' strategies for *Abelard*.

Theorem 9.4.2 *Let φ be an IF-formula in the vocabulary L, $\mathcal{M}$ an L-structure, and s an assignment whose domain contains the free variables of φ. Then*

$$\mathcal{M}, s \models^-_{GTS} \varphi \text{ iff } \mathcal{M}, s \models^-_{Sk} \varphi.$$

Example 9.4.5 We return to the Matching Pennies example, i.e., the IF sentence $\forall x(\exists y/\{x\})(x = y)$. The Skolem form helps us to see why Eloise does not have a winning strategy on structures with at least two elements. $Sk(\forall x(\exists y/\{x\})(x = y))$ is obtained in the following stages:

$Sk_{\{x,y\}}(x = y)$	is	$x = y$
$Sk_{\{x\}}(\exists y/\{x\})(x = y))$	is	$x = c$
$Sk(\forall x(\exists y/\{x\})(x = y))$	is	$\forall x(x = c)$

where c is a fresh constant symbol. Let $\mathcal{M}$ be a structure that contains at least two elements. Now it should be obvious that no expansion of $\mathcal{M}$ to a model that interprets c will render $\forall x(x = c)$ true.

The Kreisel form helps us to see why Abelard does not have a winning strategy. We have:

$Kr_{\{x,y\}}(x = y)$	is	$x \neq y$
$Kr_{\{x\}}(\exists y/\{x\})(x = y))$	is	$\forall y(x \neq y)$
$Kr(\forall x(\exists y/\{x\})(x = y))$	is	$\forall x(c \neq y)$

where c is a fresh function symbol. Now it should be obvious that no expansion of M to a model which interprets c will render $\forall x(c \neq y)$ true.

We have seen that adding a dummy quantifier to $\forall x(\exists y/\{x\})x = y$ helps Eloise to win the game. The Skolem form of $\forall x \exists z(\exists y/\{x\})(x = y)$ helps us to see why. We have

$Sk_{\{x,y,z\}}(x = y)$	is	$x = y$
$Sk_{\{x,z\}}(\exists y/\{x\})(x = y))$	is	$x = g(z)$
$Sk_{\{x\}}(\exists z(\exists y/\{x\})(x = y))$	is	$x = g(f(x))$
$Sk(\forall x \exists z(\exists y/\{x\})(x = y))$	is	$\forall x(x = g(f(x)))$

where f and g are fresh unary function symbols. Now we can find an expansion M^* of M that satisfies $\forall x(x = g(f(x)))$: let the interpretation of f and g in M be the identity function on the universe.

The reader should now understand why Skolemization and Kreiselization are relativized to a set of variables U. In the example $\forall x(\exists y/\{x\})x = y$, the Skolemization of the atomic formula $x = y$ is done in the context $U = \{x,y\}$ while in the example $\forall x\exists z(\exists y/\{x\})x = y$, the Skolemization of $x = y$ is done in the context $U = \{x,y,z\}$. For ordinary first-order logic the change in context does not matter, but this is no longer so when we turn to IF logic.

4.2.1 Lewis' Signalling Games

In this section we revisit the example of signalling and give an application to David Lewis' signalling games.

Lewis considered signalling games to be useful in communication. A communication situation involves a *communicator* (C) and an *audience* (A). C observes one of several situations m, which he tries to communicate or 'signal' to A, who does not see m. After receiving the signal, A performs one of several alternative actions, called *responses*. Every situation m has a corresponding response $b(m)$ that the communicator and the audience agree is the best response to take when m holds. Lewis argues that a word acquires its meaning in virtue of its role in the solution of various signalling problems.

Let S be a set of situations or states of affairs, Σ a set of signals, and R a set of responses. Let $b : S \to R$ the function that maps each situation to its best response. C employs an encoding $f : S \to \Sigma$ to choose a signal for every situation. A employs a function $g : \Sigma \to R$ to decide which action to perform in response to the signal it receives. A *signalling system* is a pair (f,g) of encoding and decoding functions such that their composition $g \bullet f = b$.

For example, imagine a driver who is trying to back into a parking space. She has an assistant who gets out of the car and stands in a location where she can simultaneously see how much space there is behind the car and be seen by the driver. There are two states of affairs the assistant wishes to communicate, i.e., whether or not there is enough space behind the car for the driver to continue to back up. The assistant has two signals at her disposal: she can stand palms facing in or palms facing out. The driver has two possible responses: she can back up or she can stop.

There are two solutions to this signalling problem. The assistant can stand palms facing in when there is space, and palms facing out when there is no space, and *vice versa*. In the first case, the driving should continue backing up when she sees the assistant stand palms facing in, and stop when the assistant stands palms facing out. In the second case, the driver should stop when he sees the assistant stands palms facing in, and back up when the assistant stands palms facing out. Both systems work equally well in the sense that the composition of the two communicating and responding strategies realize the best response: the driver backs up when there is space, and he stops when there is not. The IF sentence $\forall x \exists z(\exists y/\{x\})(x = y)$ in our earlier example can be modified to express a Lewisian signalling situation. In the following sentence φ think of x as a situation, z as the signal sent by the communicator, and y as the audience's interpretation of the signal:

$$\forall x \exists z(\exists y/\{x\})[S(x) \rightarrow (\Sigma(x) \wedge R(y) \wedge y = b(x))].$$

The Skolemization of φ is

$$\forall x[S(x) \rightarrow (\Sigma(f(x)) \wedge R(g(f(x))) \wedge g(f(x)) = b(x))].$$

When $\mathcal{M}$ is structure for the language of φ, the signalling problem expressed by φ has a solution if, and only if, there is an expansion M^* of M such that $M^* \vDash Sk(\varphi)$. Thus a signalling system is just a pair of Skolem functions that encode a winning strategy for the semantical game of a certain IF sentence.

4.3 Compositional Interpretation

Neither of the two interpretations given so far is compositional, i.e., defines the meaning of a formula in terms of the meanings of its parts. We saw that for ordinary first-order logic the two interpretations are equivalent with the Tarski-type interpretation. In this section we shall give a compositional interpretation of IF logic. Compositionality does not come for free, however. The price we pay is that we must switch from thinking in terms of assignments to thinking in terms of sets of assignments.

A *team* X in $\mathcal{M}$ is a set of assignments in M that share the same domain, which we denote $dom(X)$.

Definition 9.4.6 Let X be a team in a structure $\mathcal{M}$. Let $a \in M$, $A \subseteq M$, and $f : X \rightarrow A$. Define

$$X[x,a] = \{s(x/a) : s \in X\}$$
$$X[x,A] = \{s(x/a) : s \in X, a \in A\}$$
$$X[x,f] = \{s(x/f(s)) : s \in X\}.$$

Given two assignments s and s', we say that s' extends s if $s \subseteq s'$. Given two teams X and Y, we say that Y extends X if every $s \in X$ has an extension $t \in Y$, and every $t \in Y$ is an extension of some $s \in X$. When $x \notin dom(X)$, $X[x,A]$ is the maximal extension of X to $dom(X) \cup \{x\}$, while $X[x,f]$ is a minimal extension of X to $dom(X) \cup \{x\}$.

Recall that two assignments s and s' over the same domain U are W-equivalent, if whenever $W \subseteq U$, then s and s' agree on the variables in $U - W$. In this case we write $s \approx_W s'$.

Definition 9.4.7 Let X be a team in a structure $\mathcal{M}$ and let $W \subseteq dom(X)$. A function $f : X \to A$ is W-uniform if for all $s, s' \in X$

$$s \approx_W s' \text{ implies } f(s) = f(s').$$

We are now ready for the compositional interpretation or *trump semantics*.

Definition 9.4.8 Let φ and φ' be IF-formulas, $\mathcal{M}$ a structure, and X a team whose domain contains the free variables of φ. We define $M^*, X \models^+_{Tr} \varphi$ by induction:

- If φ is a literal, $\mathcal{M}^*, X \models^+_{Tr} \varphi$ iff $\mathcal{M}, s \models \varphi$, for all $s \in X$.
- $\mathcal{M}, X \models^+_{Tr} (\varphi \vee \varphi')$ iff $\mathcal{M}, Y \models^+_{Tr} \varphi$ and $\mathcal{M}, Y' \models^+_{Tr} \varphi'$ for some $Y \cup Y' = X$.
- $\mathcal{M}, X \models^+_{Tr} (\varphi \wedge \varphi')$ iff $\mathcal{M}, X \models^+_{Tr} \varphi$ and $\mathcal{M}, X \models^+_{Tr} \varphi'$
- $\mathcal{M}, X \models^+_{Tr} (\exists x/W)\varphi$ iff there is a function $f : X \to A$ that is W-uniform and $\mathcal{M}, X[x,f] \models^+_{Tr} \varphi$,
- $\mathcal{M}, X \models^+_{Tr} (\forall x/W)\varphi$ iff $\mathcal{M}, X[x,M] \models^+_{Tr} \varphi$.

When $X = \{s\}$ and $\mathcal{M}, X \models^+_{Tr} \varphi$, we simply write $\mathcal{M}, s \models^+_{Tr} \varphi$. To avoid confusion we use $\mathcal{M} \models^+_{Tr} \varphi$ to abbreviate $\mathcal{M}, \{\varnothing\} \models^+_{Tr} \varphi$, and write $\mathcal{M}, \varnothing \models^+_{Tr} \varphi$ to indicate that the empty team $\varnothing$ of assignments satisfies φ.

Example 9.4.6 In a previous example we saw that the dummy quantifier $\exists z$ in $\forall x \exists z (\exists y/\{x\})(x = y)$ serves to signal the value of x to $\exists y$. Here we consider an example of signalling using disjunctions. Let ψ be the formula $(\exists y/\{x\})(x = y)$, $\mathcal{M} = \{a, b\}$, $s_a = \{(x,a)\}$, and $s_b = \{(x,b)\}$. We will show that $\mathcal{M} \not\models^+_{Tr} \forall x \psi$ but $\mathcal{M} \models^+_{Tr} \forall x(\psi \vee \psi)$. Suppose, for a contradiction, that $\mathcal{M}, \{\varnothing\} \models^+_{Tr} \forall x \psi$. Then

$$\mathcal{M}, \{s_a, s_b\} \models^+_{Tr} (\exists y/\{x\})(x = y)$$

which implies there is an $\{x\}$-uniform function $f : \{s_a, s_b\} \to M$ such that

$$\mathcal{M}, \{s_a(y/f(s_a)), s_b(y/f(s_b))\} \models^+_{Tr} x = y$$

Since f is $\{x\}$-uniform, we must have $a = f(s_a) = f(s_b) = b$, which is impossible. But we do have $\mathcal{M}, \{\varnothing\} \models^+_{Tr} \forall x(\psi \vee \psi)$ because Eloise can signal the value of x

to herself by choosing the left disjunct when Abelard chooses a and the right disjunct when Abelard chooses b. Let $s_{aa} = \{(x,a),(y,a)\}$, and $s_{bb} = \{(x,b),(y,b)\}$. Working from inside out,

$$\mathcal{M}, \{s_{aa}\} \models^{+}_{Tr} x = y \text{ and } \mathcal{M}, \{s_{bb}\} \models^{+}_{Tr} x = y.$$

Therefore

$$\mathcal{M}, \{s_a\} \models^{+}_{Tr} (\exists y/\{x\})(x = y) \text{ and } \mathcal{M}, \{s_b\} \models^{+}_{Tr} (\exists y/\{x\})(x = y)$$

because the functions $f : \{s_a\} \to M$ and $f : \{s_b\} \to M$ defined by $f(s_a) = a$ and $g(s_b) = b$ are both $\{x\}$–uniform. Since $\{s_a\} \cup \{s_b\} = \{s_a, s_b\}$ it follows that

$$\mathcal{M}, \{s_a, s_b\} \models^{+}_{Tr} \psi \vee \psi.$$

Finally $\{s_a, s_b\} = \varnothing[x, M]$, therefore $\mathcal{M}, \{\varnothing\} \models^{+}_{Tr} \forall x(\psi \vee \psi)$.

It remains to show that the team semantics is equivalent to one of the two semantics given earlier. We will show it is equivalent to the Skolem semantics. Before doing that we need to extend the latter to teams.

Definition 9.4.9 Let φ be an IF-formula, $\mathcal{M}$ a suitable structure, and X a team in $\mathcal{M}$ whose domain contains the free variables of φ. We define $\mathcal{M}, X \models^{+}_{Sk} \varphi$ to mean that there exists an expansion $\mathcal{M}^*$ of $\mathcal{M}$ to the vocabulary $L^* = L \cup \{f_\psi : \psi$ is an exist. subf. of $\varphi\}$ such that for all $s \in X$ we have $\mathcal{M}^*, s \models Sk_{dom(X)}(\varphi)$.

Theorem 9.4.3 *Let φ be an IF-formula, $\mathcal{M}$ be a suitable structure, and X be a team in $\mathcal{M}$ whose domain contains the free variables of φ. Then*

$$\mathcal{M}, X \models^{+}_{Tr} \varphi \textit{ iff } \mathcal{M}, X \models^{+}_{Sk} \varphi$$

Proof. We prove by induction on subformula ψ of φ, that for every team Y whose domain contains the free variables of ψ we have

$$\mathcal{M}, Y \models^{+}_{Tr} \psi \text{ iff } \mathcal{M}, Y \models^{+}_{Sk} \psi.$$

The basic step follows easily from the definitions.

Suppose ψ is $(\psi_1 \vee \psi_2)$. If $\mathcal{M}, Y \models^{+}_{Tr} (\psi_1 \vee \psi_2)$, then for some $Y_1 \cup Y_2 = Y$ we have $\mathcal{M}, Y_1 \models^{+}_{Tr} \psi_1$ and $\mathcal{M}, Y_2 \models^{+}_{Tr} \psi_2$. By the inductive hypothesis there exists an expansion $\mathcal{M}_1$ of $\mathcal{M}$ to the vocabullary $L_1^* = L \cup \{f_\psi : \psi$ is an exist. subf. of $\psi_1\}$ such that for all $s \in Y_1$ we have $\mathcal{M}_1, s \models Sk_{dom(Y_1)}(\psi_1)$ and an expansion $\mathcal{M}_2$ of $\mathcal{M}$ to the vocabullary $L_2^* = L \cup \{f_\chi : \chi$ is an exist subf. of $\psi_2\}$ such that for all $s \in Y_2$ we have $\mathcal{M}_2, s \models Sk_{dom(Y_2)}(\psi_2)$. Since $L_1 \cap L_2 = L$, there is a common expansion $\mathcal{M}^*$ of $\mathcal{M}$ to the vocabulary $L^* = L \cup \{f_\theta : \theta \text{ is an exist. subf. of } \psi\}$ such that for all $s \in Y$ we have

$$\mathcal{M}^*, s \models Sk_{dom(Y_1)}(\psi_1) \text{ or } \mathcal{M}^*, s \models Sk_{dom(Y_2)}(\psi_2)$$

which implies

$$\mathcal{M}^*, s \models Sk_{dom(Y)}(\psi_1 \vee \psi_2).$$

Hence $\mathcal{M}^*, s \models^+_{Sk} \psi_1 \vee \psi_2$.

Conversely, suppose there is an expansion $\mathcal{M}^*$ of $\mathcal{M}$ such that for all $s \in Y$

$$\mathcal{M}^*, s \models Sk_{dom(Y)}(\psi_1 \vee \psi_2).$$

Let

$$Y_i = \{s \in Y : \mathcal{M}^*, s \models Sk_{dom(Y_i)}(\psi_i)\}.$$

Then $Y_1 \cup Y_2 = Y$. In addition we have $\mathcal{M}, Y_1 \models^+_{Sk} \psi_1$ and $\mathcal{M}, Y_2 \models^+_{Sk} \psi_2$ so by the inductive hypothesis

$$\mathcal{M}, Y_1 \models^+_{Tr} \psi_1 \text{ and } \mathcal{M}, Y_2 \models^+_{Tr} \psi_2$$

Thus $\mathcal{M}, Y \models^+_{Tr} (\psi_1 \vee \psi_2)$.

Suppose ψ is $(\exists x/W)\psi'$. If $\mathcal{M}, Y \models^+_{Tr} (\exists x/W)\psi'$ then there exists a function $f : Y \to A$ such that f is W-uniform and $\mathcal{M}, Y[x,f] \models^+_{Tr} \psi'$. By the inductive hypothesis $\mathcal{M}, Y[x,f] \models^+_{Sk} \psi'$ hence there exists an expansion $\mathcal{M}'$ of $\mathcal{M}$ to the vocabulary $L^* = L \cup \{f_\chi : \chi \text{ is an exist. subf. of } \psi'\}$ such that for all $s \in Y$

$$\mathcal{M}', s(x/f(s)) \models Sk_{dom(Y)\cup\{x\}}(\psi').$$

Let $\mathcal{M}^*$ be an expansion of $\mathcal{M}'$ to the vocabulary L^* such that for all $s \in Y$

$$f^{\mathcal{M}^*}_{(\exists x/W)\psi'}(s(y_1), \ldots, s(y_n)) = f(s)$$

where $y_1, \ldots, y_n$ is the list of the variables in $dom(Y) - W$. Observe that $\mathcal{M}^*$ is well defined, because f is W-uniform. By the Substitution Lemma, we have for all $s \in Y$

$$\mathcal{M}^*, s \models Subst(Sk_{dom(Y)\cup\{x\}}(\psi'), x, f_{(\exists x/W)\psi'}(y_1, \ldots, y_n))$$

which implies $\mathcal{M}^*, s \models Sk_{dom(Y)}((\exists x/W)\psi')$. Thus $\mathcal{M}^*, Y \models^+_{Sk} (\exists x/W)\psi'$.

Conversely, suppose that there is an expansion $\mathcal{M}^*$ of $\mathcal{M}$ such that for all $s \in Y$

$$\mathcal{M}^*, s \models Subst(Sk_\varphi(\psi'), x, f_{(\exists x/W)\psi'}(y_1, \ldots, y_n))$$

Define a W-uniform function $f : Y \to M$ by

$$f(s) = f^{\mathcal{M}^*}_{(\exists x/W)\psi'}(s(y_1), \ldots, s(y_n))$$

where $y_1, \ldots, y_n$ is the list of the variables in $dom(Y) - W$ and let $\mathcal{M}'$ be the reduct of $\mathcal{M}^*$ to the vocabulary L'. Then for all $s \in Y$

$$\mathcal{M}', s(x/f(s)) \models Sk_{dom(Y)\cup\{x\}}(\psi')$$

which implies $\mathcal{M}, Y[x,f] \models^+_{Sk} \psi'$. By the inductive hypothesis $\mathcal{M}, Y[x,f] \models^+_{Tr} \psi'$. Thus $\mathcal{M}, X \models^+_{Tr} (\exists x/W)\psi'$.

The other cases are left to the reader. ■

The compositional semantics can be extended to cover falsity $\models^-_{Tr}$. The changes should be straightforward. In the compositional definition the clause for literals is changed to

$$\mathcal{M}, X \models^-_{Tr} \varphi \text{ iff } \mathcal{M}, s \models \neg\varphi, \text{ for all } s \in X.$$

The other clauses are as in the definition of the compositional interpretation except that we exchange everywhere $\vee$ with $\wedge$ and $(\exists x/W)$ with $(\forall x/W)$. In the same spirit we shall adopt the convention

$$\mathcal{M}, s \models^-_{Tr} \varphi \text{ iff } \mathcal{M}, \{s\} \models^-_{Tr} \varphi$$

Then the following analogue of the previous theorem can be proved:

Theorem 9.4.4 *Let φ be an IF-formula, $\mathcal{M}$ a suitable structure and X a team in $\mathcal{M}'$ whose domain contains the free variables of φ. Then*

$$\mathcal{M}, X \models^-_{Tr} \varphi \textit{ iff } \mathcal{M}, X \models^-_{Sk} \varphi.$$

We have defined three interpretations for IF logic. The first two were relative to assignments, whereas the third was relative to teams. By identifying assignments with singleton teams we were able to prove that

$$\begin{array}{lllll} \mathcal{M}, s \models^+_{GTS} \varphi & \text{iff} & \mathcal{M}, s \models^+_{Sk} \varphi & \text{iff} & \mathcal{M}, s \models^+_{Tr} \varphi \\ \mathcal{M}, s \models^-_{GTS} \varphi & \text{iff} & \mathcal{M}, s \models^-_{Sk} \varphi & \text{iff} & \mathcal{M}, s \models^-_{Tr} \varphi. \end{array}$$

From now on we shall often drop the subscript '*Tr*'.

Remark 9.4.1 It is easy to see that the empty team $\varnothing$ (to be distinguished from the empty assignment $\varnothing$) is winning for both players, that is, we have both

$\mathcal{M}, \varnothing \models^+ \varphi$ and $\mathcal{M}, \varnothing \models^- \varphi$ for every structure $\mathcal{M}$ and IF-formula φ. This may seem odd but it is necessary in order to properly interpret disjunctions and conjunctions. For any two formulas φ and ψ and any team X we want to have

$$\mathcal{M}, X \models^+ \varphi \quad \text{implies} \quad \mathcal{M}, X \models^+ \varphi \vee \psi$$

even when ψ is tautologically false, say $x \neq x$. The implication is guaranteed given that $X = X \cup \varnothing$.

4.4 Negation

Until now, in order to keep things simple, we have allowed the negation symbol $\neg$ to occur only infront of atomic formulas. We now relax this assumption and define

$$\begin{array}{lll} \neg(\neg\varphi) & \text{is} & \varphi, \text{for atomic } \varphi \\ \neg(\varphi \vee \varphi') & \text{is} & \neg\varphi \wedge \neg\varphi' \\ \neg(\varphi \wedge \varphi') & \text{is} & \neg\varphi \vee \neg\varphi' \\ \neg(\exists x/W)\varphi & \text{is} & (\forall x/W)\neg\varphi \\ \neg(\forall x/W)\varphi & \text{is} & (\exists x/W)\neg\varphi \end{array}$$

Lemma 9.4.1 *Let φ be an IF formula, $\mathbb{M}$ a suitable structure, and X a team of assignments whose domain contains the free variables of φ. Then*

$$\mathbb{M}, X \models^{\pm} \neg\varphi \textit{ iff } \mathbb{M}, X \models^{\mp} \varphi.$$

Proof. If φ is a literal, then

$$\begin{array}{lll} \mathbb{M}, X \models^+ \neg\varphi & \text{iff} & \mathbb{M}, s \models \neg\varphi \text{ (for all } s \in X) \\ & \text{iff} & \mathbb{M}, s \not\models \varphi \text{ (for all } s \in X) \\ & \text{iff} & \mathbb{M}, X \models^- \varphi \end{array}$$

$$\begin{array}{lll} \mathbb{M}, X \models^- \neg\varphi & \text{iff} & \mathbb{M}, s \not\models \neg\varphi \text{ (for all } s \in X) \\ & \text{iff} & \mathbb{M}, s \models \varphi \text{ (for all } s \in X) \\ & \text{iff} & \mathbb{M}, X \models^+ \varphi \end{array}$$

Suppose φ is $\psi \vee \psi'$. Then by inductive hypothesis,

$$\begin{array}{lll} \mathbb{M}, X \models^+ \neg(\psi \vee \psi') & \text{iff} & \mathbb{M}, X \models \neg\psi \wedge \neg\psi' \\ & \text{iff} & \mathbb{M}, X \models^+ \neg\psi \text{ and } \mathbb{M}, X \models^+ \neg\psi' \\ & \text{iff} & \mathbb{M}, X \models^- \psi \text{ and } \mathbb{M}, X \models^- \psi' \\ & \text{iff} & \mathbb{M}, X \models^- \psi \vee \psi'. \end{array}$$

Suppose φ is $\psi \wedge \psi'$. Then by the inductive hypothesis, $\mathbb{M}, X \models^{+} \neg(\psi \wedge \psi')$ if and only if there is a cover $Y \cup Y' = X$ such that $\mathbb{M}, Y \models^{+} \neg\psi$ and $\mathbb{M}, Y' \models^{+} \neg\psi'$ if and only if there is a cover $Y \cup Y' = X$ such that $\mathbb{M}, Y \models^{-} \psi$ and $\mathbb{M}, Y' \models^{-} \psi'$ if and only if $\mathbb{M}, X \models^{-} \psi \wedge \psi'$.

We leave the other clauses to the reader. ■

4.5 Burgess' Separation Theorem

We give an application of some of the notions introduced so far. We start with a few definitions. A set Γ of IF sentences is *satisfiable* if there is a suitable structure $\mathcal{M}$ such that $\mathcal{M} \models^{+} \varphi$ for every $\varphi \in \Gamma$. We leave the notation $\mathcal{M} \models^{+} \varphi$ unspecified on purpose. In the game-theoretical semantics, it means $\mathcal{M}, \varnothing \models^{+} \varphi$, with $\varnothing$ the empty assignment. This is, by the last theorem, equivalent with $\mathcal{M}, \varnothing \models^{+}_{Sk} \varphi$ in the Skolem semantics, and with $\mathcal{M}, \{\varnothing\} \models^{+}_{Tr} \varphi$ in the trump semantics.

When φ and ψ are IF sentences, we say that φ *truth entails* ψ (written $\varphi \models^{+} \psi$) if for every structure $\mathcal{M}$: $\mathcal{M}, \{\varnothing\} \models^{+} \varphi$ implies $\mathcal{M}, \{\varnothing\} \models^{+} \psi$. Again, by the previous theorem, the last clause is equivalent to ($\mathcal{M}, \varnothing \models^{+}_{Sk} \varphi$ implies $\mathcal{M}, \varnothing \models^{+}_{Sk} \psi$) in the Skolem semantics, and to ($\mathcal{M}, \varnothing \models^{+}_{GTS} \varphi$ implies $\mathcal{M}, \varnothing \models^{+}_{GTS} \psi$) in the game-theoretical semantics. Two IF sentences φ and ψ are said to be *truth equivalent* (written $\varphi \equiv^{+} \psi$) if $\varphi \models^{+} \psi$, and $\varphi \models^{+} \psi$. A set Γ of IF sentences *truth entails* the IF sentence ψ (written $\Gamma \models^{+} \varphi$) if, for every suitable structure $\mathcal{M}$, if $\mathcal{M} \models^{+} \psi$ for every $\psi \in \Gamma$, then $\mathcal{M} \models^{+} \psi$. We say that a class K of L-structures is *definable* in IF logic, if there is an IF L-sentence φ such that $K = \{\mathcal{M} : \mathcal{M} \models^{+} \varphi\}$.

Theorem 9.4.5 *(Separation Theorem). Let K_1 and K_2 be two classes of L-structures definable in IF logic by the IF sentences φ_1 and φ_2, respectively. If K_1 and K_2 are disjoint (i.e., φ_1 and φ_2 are incompatible), then there is a class K of L-structures definable by a first-order sentence θ such that $K_1 \subseteq K$ and $K \subseteq \overline{K_2}$, where $\overline{K_2}$ is the complement of K_2.*

Proof. Recall that $Sk(\varphi_1)$ is a first-order sentence in the language $L \cup \{f_\psi : \psi$ is an exist. subf. of $\varphi_1\}$ and $Sk(\varphi_2)$ is a first-order sentence in the language $L \cup \{f_\chi : \chi$ is an exist. subf. of $\varphi_2\}$. Let $L_1 = \{f_\psi : \psi$ is an exist. subf. of $\varphi_1\}$ and $L_2 = \{f_\chi : \chi$ is an exist. subf. of $\varphi_2\}$. We may assume that L_1 and L_2 are disjoint. Thus $K_1 = \{\mathcal{M} : \mathcal{M} \models^{+}_{Sk} \varphi_1\} = \{\mathcal{M} : \mathcal{M}^* \models Sk(\varphi_1)$, for some expansion $\mathcal{M}^*$ of $\mathcal{M}$ to $L_1\}$ and $K_2 = \{\mathcal{M} : \mathcal{M}, \models^{+}_{Sk} \varphi_2\} = \{\mathcal{M} : \mathcal{M}^* \models Sk(\varphi_2)$, for some expansion $\mathcal{M}^*$ of $\mathcal{M}$ to $L_2\}$. We must have $Sk(\varphi_1) \models \neg Sk(\varphi_2)$, for otherwise there is an $L \cup L_1 \cup L_2$-structure $\mathcal{M}'$ such that $\mathcal{M}' \models Sk(\varphi_1)$ and $\mathcal{M}' \models Sk(\varphi_2)$, which implies that $\mathcal{M}' \upharpoonright L \cup L_1 \in K_1$ and $\mathcal{M}' \upharpoonright L \cup L_2 \in K_2$. That is, $\mathcal{M}' \upharpoonright L \in K_1$ and $\mathcal{M}' \upharpoonright L \in K_2$ but this contradicts the assumption of the theorem. We now apply the Craig Interpolation Theorem for first-order logic to $Sk(\varphi_1) \models \neg Sk(\varphi_2)$ in order to get an L-sentence θ such that $Sk(\varphi_1) \models \theta$ and $\theta \models \neg Sk(\varphi_2)$. Let $K = \{\mathcal{M} : \mathcal{M} \models \theta\}$. From $Sk(\varphi_1) \models \theta$ we get $K_1 \subseteq K$ and from $\theta \models \neg Sk(\varphi_2)$ we get $K \subseteq \overline{K_2}$. ■

To avoid trivialities, we exclude structures with empty universes in first-order logic. For the same reason, we exclude structures with universes that contain less than two elements in IF logic. Recall our earlier example $\theta_0 := \forall x(\exists y/\{x\})(x = y)$ and its negation $\neg\theta_0$. Both are indeterminate on all structures that contain at least two elements, so if we adopt our convention, the classes of structures in which the sentences θ_0 and $\neg\theta_0$ are true are empty. Here is a strengthening of the Separation Theorem.

Theorem 9.4.6 ([Burgess, 2003]) *Let φ_0 and φ_1 be two incompatible IF sentences. Then we can find an IF sentence θ such that $\theta \equiv^+ \varphi_0$ and $\neg\theta \equiv^+ \varphi_1$.*

Proof. Let ψ_0 be $\varphi_0 \vee \theta_0$ and ψ_1 be $\varphi_1 \vee \theta_0$. We observe first that (i) $\psi_0 \equiv^+ \varphi_0$, for we have for any structure $\mathcal{M}$: $\mathcal{M}, \{\varnothing\} \models^+ \varphi_0 \vee \theta_0$ iff $\mathcal{M}, \{\varnothing\} \models^+ \varphi_0$ and $\mathcal{M}, \varnothing \models^+ \theta_0$, where the second $\varnothing$ is the empty team. But $\mathcal{M}, \varnothing \models^+ \theta_0$ always holds (see our last remark in the previous section), so

$$\mathcal{M}, \{\varnothing\} \models^+ \varphi_0 \vee \theta_0 \text{ iff } \mathcal{M}, \{\varnothing\} \models^+ \varphi_0.$$

By similar reasoning, we have (ii) $\psi_1 \equiv^+ \varphi_1$, (iii) $\neg\psi_0 \equiv^+ (\neg\varphi_0 \wedge \neg\theta_0)$, and (iv) $\neg\psi_1 \equiv^+ (\neg\varphi_1 \wedge \neg\theta_0)$. This shows that the class of structures in which $\neg\psi_0$ is true is empty and so is the class of structures in which $\neg\psi_1$ is true. Given (i) and (ii) and the fact that φ_0 and φ_1 are incompatible, we have that ψ_0 and ψ_1 are incompatible too. Whence, by the Separation Theorem, there is a first-order sentence ψ such that the class of structures in which ψ_0 is true is included in the class of structures in which ψ is true, and the class of structures in which ψ_1 is true is included in the class of structures in which $\neg\psi$ is true.

The sentence θ we are looking for is $\theta := \psi_0 \wedge (\neg\psi_1 \vee \psi) = (\varphi_0 \vee \theta_0) \wedge (\psi \vee \neg(\varphi_1 \vee \theta_0))$. It may be checked that (a) φ_0 and θ are truth equivalent; and (b) $\neg\theta$ and φ_1 are truth equivalent.

For (a), notice that for any structure $\mathcal{M}$:

$$\mathcal{M}, \{\varnothing\} \models^+ (\varphi_0 \vee \theta_0) \wedge (\psi \vee \neg(\varphi_1 \vee \theta_0))$$
$$\text{iff}$$
$$\mathcal{M}, \{\varnothing\} \models^+ (\varphi_0 \vee \theta_0) \text{ and } \mathcal{M}, \{\varnothing\} \models^+ (\psi \vee \neg(\varphi_1 \vee \theta_0))$$
$$\text{iff}$$
$$\mathcal{M}, \{\varnothing\} \models^+ \varphi_0 \text{ and } \mathcal{M}, \varnothing \models^+ \theta_0 \text{ and } \mathcal{M}, \{\varnothing\} \models^+ \psi \text{ and } \mathcal{M}, \varnothing \models^+ \neg(\varphi_1 \vee \theta_0)$$
$$\text{iff}$$
$$\mathcal{M}, \{\varnothing\} \models^+ \varphi_0$$

given that the empty team trivially satisfies an IF formula and that the class of structures in which ψ_0 is true is included in the class of structures in which ψ is true.

The claim (b) is established in a similar way. ■

Next we consider an application of the Separation Theorem.

4.5.1 Game-Theoretical Negation versus Classical Negation

We expand IF languages with a new clause that allows for contradictory negation $\sim$ to appear prefixed to IF sentences:

- If φ is an IF sentence, so is $\sim\varphi$.

This meta-theoretical negation is interpreted by the clause:

$$\mathcal{M}, \varnothing \models^{+}_{GTS} \sim\varphi \text{ iff } \mathcal{M}, \varnothing \not\models^{+}_{GTS} \varphi$$

That is, $\sim\varphi$ expresses the contradictory of φ: there is no winning strategy for *Eloise* in the game $G(M, \varnothing, \varphi)$, where $\varnothing$ is the empty assignment. We chose the game-theoretical interpretation here because it best highlights the distinction between the contrary negation $\neg$ which is interpreted semantically by a game rule (role swapping), and the contradictory negation $\sim$ which is expresses a fact *about* semantical games. Obviously the contradictory negation can be equivalently defined by

$$\mathcal{M}, \{\varnothing\} \models^{+}_{Tr} \sim\varphi \text{ iff } \mathcal{M}, \{\varnothing\} \not\models^{+}_{Tr} \varphi$$

etc. We shall simply write

$$\mathcal{M} \models^{+} \sim\varphi \text{ iff } \mathcal{M} \not\models^{+} \varphi.$$

Now it is a straightforward consequence of the Separation Theorem that if the contradictory negation of an IF sentence is truth equivalent to an ordinary IF sentence, then the sentence itself is truth equivalent to an ordinary first-order sentence.

Proposition 9.4.1 Let φ be a (contradictory negation free) IF sentence. Suppose there exists a contradictory negation free IF sentence ψ such that $\sim\varphi$ is truth equivalent to ψ. Then φ is truth equivalent to an ordinary IF sentence θ and ψ is truth equivalent to $\neg\theta$.

Proof. Obviously φ and ψ are incompatible. Let $K_1 = \{\mathcal{M} : \mathcal{M} \models^{+} \varphi\}$ and $K_2 = \{\mathcal{M} : \mathcal{M} \models^{+} \psi\}$. By the Separation Theorem, there is a class K of structures definable by a first-order sentence θ such that $K_1 \subseteq K$ and $K \subseteq \overline{K_2}$. This gives us, for every structure $\mathcal{M}$:

$$\mathcal{M} \models^+ \varphi \text{ iff } \mathcal{M} \models^+ \theta$$
$$\mathcal{M} \models^+ \psi \text{ iff } \mathcal{M} \models^+ \neg\theta.$$

■

Notes. IF languages were introduced in [Hintikka and Sandu, 1989]. Their source of inspiration lie in Henkin quantifiers, which were introduced in [Henkin, 1961]. Hintikka [Hintikka, 1996] discusses the relevance of IF first-order logic for the philosophy of logic and mathematics. For a critical assessment of IF logic the reader is referred to [Feferman, 2006]. Hodges ([Hodges, 1997]) provides a compositional interpretation for IF languages, that has inspired Caicedo and M. Krynicki ([Caicedo and Krynicki, 1999]) and Caicedo, Dechesne, and Janssen ([Caicedo et al., 2009]). The possibility of signalling in IF logic was first observed by Hodges ([Hodges, 1997]). The reader is referred to [Janssen and Dechesne, 2006] for the discussion of signalling in IF logic. Sandu ([Sandu, 1998]) shows that IF languages define their own truth predicate. A general evaluation of this result and of its philosophical consequences may be found in [de Rouilhan and Bozon, 2006]. For the equivalence of the three semantical interpretations for IF logic, and a systematical investigation of the meta-theoretical properties of IF logic, the reader is referred to [Mann et al., ta]. An alternative formulation of the syntax of IF logic may be found in [Abramsky and Väänänen, 2008]. Van Benthem ([van Benthem, 2006]) treats IF semantical games in the frame of epistemic logic.

5. Strategic Games

The interpretation of IF languages by winning strategies in extensive games of imperfect information – or, equivalently, by generalized Skolem and Kreisel counterexamples – introduced indeterminacy into the logic: recall the sentences $\forall x(\exists y/x)(x = y)$ and $\forall x(\exists y/x)(x \neq y)$, which are neither true nor false in any structures with at least two elements. There have been proposals to overcome the indeterminacy of such sentences by borrowing solutions from classical game theory. This is what we are going to do in this part of the chapter. The central notion in this approach is that of *equilibrium* of strategies in strategic games. In what follows we will give a systematic presentation of the results in the theory of strategic games that are relevant for the semantical interpretation of IF formulas.

5.1 Pure Strategies

In a strategic game between, say, two players, each can choose an element from S_1 and S_2, respectively. An element s_i from S_i is called a strategy or an action. It is harmless to think of s_i as the type of strategy from extensive games. However, in strategic games, strategies are not defined relative to histories reached in the game. Once the two players have chosen simultaneously their respective strategies $s_1 \in S_1$ and $s_2 \in S_2$, the game terminates and each player i receives an outcome $u_i(s_1, s_2)$ that is determined by s_1 and s_2.

One of the most discussed examples of such games in the literature is *Prisoners' dilemma*:

> Two suspects in a major crime are held in separate cells. There is enough evidence to convict each of them of a minor offense, but not enough evidence to convict either of them of the major crime unless one them acts as an informer against the other (finks). If they both stay quiet, each will be convicted of the minor offense and spend one year in prison. If one and only one of them finks, she will be freed and used as a witness against the other, who will spend four years in prison. If they both fink, each will spend three years in prison. [Osborne, 2004, p. 14]

This is a typical decision situation, described in Table 9.1.

TABLE 9.1 The payoff matrix of the *Prisoner's Dilemma* game

	t_1	t_2
s_1	$(-1, -1)$	$(-4, 0)$
s_2	$(0, -4)$	$(-3, -3)$

where $s_1 = t_1 = Fink$, and $s_2 = t_2 = Quiet$. The matrix makes explicit that one of the two prisoners chooses a strategy from $\{s_1, s_2\}$, and the other chooses a strategy from $\{t_1, t_2\}$. Each choice of the row player, and each choice of the column player results in an outcome whose payoff is marked in the game matrix. For instance, the pair $(-4, 0)$ represents the pair of payoffs when the 'row player' chooses s_1 and the 'column player' chooses t_2. An equivalent way to express payoffs is through *utility functions* $u_i : \{s_1, s_2\} \times \{t_1, t_2\} \to \{-4, -3, -1, 0\}$. For each pair (s, t), the utility function u_i gives the payoff $u_i(s, t)$ for player i. In our example, $u_1(s_1, t_2) = -4$, $u_2(s_1, t_2) = 0$, etc.

We are now ready for the general definition.

Definition 9.5.1 A *strategic game* is a triple $\Gamma = (N, (S_i)_{i \in N}, (u_i)_{i \in N})$ such that

- N is the set of players of the game.
- S_i is the set of choices or *pure strategies* of player i.
- $u_i : (S_j)_{j \in N} \to \mathbb{R}$, is the *utility function* of player i.

A strategic game Γ is finite, if S_i is finite, for all $i \in N$. If N contains n elements, then we say that Γ is an n-player game. Since our overall goal is to import results from strategic games to bear on semantic games, and the latter are played by two players, we shall be mostly interested in two-player strategic games. Furthermore, semantic games have the special property that if one player loses, the other player wins, that is, the payoffs of the players are diametrically opposed. This imposes a further restriction on the class of strategic games we shall consider. The following definition is standard in game theory.

Definition 9.5.2 Let $\Gamma = (N, (S_i)_{i \in N}, (u_i)_{i \in N})$ be a two-player game and let c be a real number ($c \in \mathbb{R}$.) Γ is *strictly competitive*, if for all $s, s' \in S_1$ and $t, t' \in S_2$ we have $u_1(s,t) \geq u_1(s',t')$ iff $u_2(s',t') \geq u_2(s,t)$. Γ is *c-sum*, if for all $s \in S_1$ and $t \in S_2$, $u_1(s,t) + u_2(s,t) = c$. Γ is *constant sum*, if it is c-sum, for some $c \in \mathbb{R}$.

In the special case where Γ is a zero-sum game, $u_1(s,t) = -u_2(s,t)$, for all $s \in S_1$ and $t \in S_2$. We will see below that strategic IF games are one-sum games. Observe that if a strategic game is constant sum, then it is strictly competitive. The converse is not true. Consider for instance a game in which $u_1(s,t) = -2u_2(s,t)$, for all $s \in S_1$ and $t \in S_2$.

The *Prisoners' dilemma* is not a constant game, unlike the *Matching Pennies* that we have encountered earlier in which the two players choose simultaneously whether to show the heads or the tails of a coin. Here we shall detail the game a bit more. If the players show the same side, player 1 wins one dollar; if they show different sides, player 1 pays player 2 one dollar. The utility function is depicted in matrix form in Table 9.2. In this table, the first player controls $S_1 = \{s_1, s_2\}$, the second controls $S_2 = \{t_1, t_2\}$. For each pair of strategies $s \in S_1$ and $t \in S_2$, the corresponding cell in the matrix denotes $(u_1(s,t), u_2(s,t))$.

TABLE 9.2 The payoff matrix of *Matching Pennies*

	t_1	t_2
s_1	$(1, -1)$	$(-1, 1)$
s_2	$(-1, 1)$	$(1, -1)$

5.1.1 Maximin Strategies

In this section and the next, we focus on solution concepts in strategic games: what is the strategy a 'rational' player should play in a strategic game? The overall conclusion will be that it is rational for *Eloise* and *Abelard* to seek for strategies that are in equilibrium. As we shall see, strategies that are in equilibrium have the property that they maximize the 'security' of the players.

Consider a two-player, strictly competitive game $\Gamma = (N, (S_i)_{i \in N}, (u_i)_{i \in N})$, where $S = \{s_1, s_2, \ldots, s_m\}$ and $T = \{t_1, t_2, \ldots, t_n\}$. For $s \in S$, the *security level* of player 1 for his strategy s, denoted by $v_1(s)$ is the least payoff he can receive when player 2 chooses to play any of his strategies:

$$v_1(s) = \min\{u_1(s, t_1), \ldots, u_1(s, t_n)\} = \min_{t \in T} u_1(s, t).$$

We shall write $\min_t u_1(s, t)$ instead of $\min_{t \in T} u_1(s, t)$. Thus playing s *guarantees* a payoff to player 1 of at least $v_1(s)$.

The corresponding notion for player 2 is

$$v_2(t) = \min\{u_2(s_1, t), \ldots, u_2(s_m, t)\} = \min_{s \in S} u_2(s, t).$$

We shall adopt a similar notation and write $\min_s u_2(s, t)$ instead of $\min_{s \in S} u_2(s, t)$.

Definition 9.5.3 Let $\Gamma = (N, (S_i)_{i \in N}, (u_i)_{i \in N})$ be a finite two-person strictly competitive strategic game.

(i) For $s^* \in S$ we say that s^* is a *maximin strategy* for player 1 if it maximizes player 1's security level,

$$v_1(s^*) = \max_s v_1(s) = \max_s \min_t u_1(s, t).$$

(ii) For $t^* \in T$ we say that t^* is a *maximin strategy* for player 2 if it maximizes player 2's security level,

$$v_2(t^*) = \max_t v_2(t) = \max_t \min_s u_2(s, t).$$

Notice that if s^* is a maximin strategy for player 1, then

$$\text{for every } t \in T,\ u_1(s^*, t) \geq \max_s \min_t u_1(s, t). \tag{9.2}$$

To see this, first note that for every $t \in T$, $u_1(s^*, t) \geq \min_t u_1(s^*, t)$. By the definition of security level, $v_1(s^*) = \min_t u_1(s^*, t)$. Therefore, for every $t \in T$, $u_1(s^*, t) \geq v_1(s^*)$, which together with the definition of a maximin strategy implies the desired result. A symmetrical reasoning shows that if t^* is a maximin strategy for player 2, then

$$\text{for every } s \in S,\ u_2(s, t^*) \geq \max_t \min_s u_2(s, t). \tag{9.3}$$

Maximin strategies always exist but they need not be unique. The following lemma shows that, in a zero-sum game, the maximinimization of player 2's payoff is equivalent to the minimaximization of player 1's payoff.

Lemma 9.5.1 *Let* $\Gamma = (N, (S_i)_{i \in N}, (u_i)_{i \in N})$ *be a two-player, zero-sum game. Then* $\max_t \min_s u_2(s,t) = -\min_t \max_s u_1(s,t)$.

Proof. Since Γ is a zero-sum game, we have $u_2 = -u_1$. Then $\min_s u_2(s,t) = \min_s -u_1(s,t) = -\max_s u_1(s,t)$. It follows that $\max_t \min_s u_2(s,t) = \max_t - \max_s u_1(s,t) = -\min_t \max_s u_1(s,t)$. ∎

5.1.2 Pure Strategy Equilibria

Consider the following two-player, zero-sum game:

	t_1	t_2	t_3	t_4
s_1	7	2	0	1
s_2	0	2	5	8
s_3	4	3	4	4
s_4	6	3	1	9
s_5	5	2	0	8

(We marked only the payoffs of player 1.)

We notice that s_3 is a maximin strategy for player 1, and t_2 is a maximin strategy for player 2. Thus it appears that there is a good reason for player 1 to choose the maximin strategy s_3 and for player 2 to choose the maximin strategy t_2: Each of them maximizes that player's security level. In addition we notice another property of the pair (s_3, t_2): when t_2 is fixed, player 1 is not better off choosing any other of his strategies in S; and when s_3 is fixed, player 2 is not better off choosing any other of his strategies. We say that the pair (s_3, t_2) is an *equilibrium*. The definition for the general case is given below.

Definition 9.5.4 Let $\Gamma = (N, (S_i)_{i \in N}, (u_i)_{i \in N})$ be a two-player strategic game, where $N = \{1, 2\}$. The pair (s', t') is an *equilibrium* if it satisfies the following two conditions:

- for every strategy s in S_1: $u_1(s', t') \geq u_1(s, t')$
- for every strategy t in S_2: $u_2(s', t') \geq u_2(s', t)$.

The two conditions say that if (s', t') is an equilibrium pair, then $u_1(s', t')$ is the maximum of player 1's payoffs in the column determined by t', and $u_2(s', t')$ is the maximum of player 2's payoffs in the row determined by s'. Equivalently:

$$u_1(s', t') = \max_s u_1(s, t') \quad \text{and} \quad u_2(s', t') = \max_t u_2(s', t).$$

If the game Γ is strictly competitive, the second condition in the definition can be rewritten as:

- for every t in S_2, $u_1(s',t') \leq u_1(s',t)$.

Then we have another equivalent way to specify an equilibrium. The pair (s',t') is an equilibrium in a strictly competitive game Γ, if

$$u_1(s,t') \leq u_1(s',t') \leq u_1(s',t) \tag{9.4}$$

for every $s \in S$ and every $t \in T$. Equivalently,

$$u_1(s',t') = \max_s u_1(s,t') = \min_t u_1(s',t). \tag{9.5}$$

The next theorem establishes a connection between between maximin strategies and equilibria.

Theorem 9.5.1 *Let* $\Gamma = (N, (S_i)_{i \in N}, (u_i)_{i \in N})$ *be a two-player, zero-sum game. Then the following hold:*

(1) If (s',t') is an equilibrium, then
- s' is a maximin strategy for player 1,
- t' is a maximin strategy for player 2, and
- $\max_s \min_t u_1(s,t) = \min_t \max_s u_1(s,t) = u_1(s',t')$.

(2) If
- $\max_s \min_t u_1(s,t) = \min_t \max_s u_1(s,t)$,
- s' is a maximin strategy for player 1, and
- t' is a maximin strategy for player 2,

then (s',t') is an equilibrium.

Proof. (1) Suppose (s',t') is an equilibrium. Then, by (9.5), $u_1(s',t') = \min_t u_1(s',t)$. Since s' is among the strategies in S, $\min_t u_1(s',t) \leq \max_s \min_t u_1(s,t)$, which establishes that

$$u_1(s',t') \leq \max_s \min_t u_1(s,t). \tag{9.6}$$

On the other side, by (9.4), $u_1(s',t') \geq u_1(s,t')$, for all $s \in S$. Since t' is among the strategies in T, we have that for each $s \in S$, $u_1(s,t') \geq \min_t u_1(s,t)$. But then $u_1(s,t') \geq \min_t u_1(s,t)$ also for the strategy $s \in S$ that maximizes $\min_t u_1(s,t)$. Hence,

$$u_1(s',t') \geq \max_s \min_t u_1(s,t). \tag{9.7}$$

(9.6) and (9.7) imply that

$$u_1(s', t') = \max_s \min_t u_1(s, t). \tag{9.8}$$

By (9.5), $u_1(s', t') = \min_t u_1(s', t)$. The definition of security level says that $v_1(s') = \min_t u_1(s', t)$. We just concluded that $u_1(s', t') = \max_s \min_t u_1(s, t)$. Hence, $u_1(s', t') = v_1(s') = \max_s \min_t u_1(s, t)$, that is, according to Definition 9.5.3, s' is a maximin strategy for player 1.

A symmetrical argument shows that

$$u_2(s', t') = \max_t \min_s u_2(s, t) \tag{9.9}$$

and that t' is a maximin strategy for player 2.

Since Γ is zero-sum, $u_1(s', t') = -u_2(s', t')$. It follows from (9.8) and (9.9) that $\max_s \min_t u_1(s, t) = -\max_t \min_s u_2(s, t)$. From this and Lemma 9.5.1 we derive $\max_s \min_t u_1(s, t) = \min_t \max_s u_1(s, t)$.

(2) Let v^* denote $\max_s \min_t u_1(s, t) = \min_t \max_s u_1(s, t)$. From the latter equivalence and Lemma 9.5.1 we get $\max_t \min_s u_2(s, t) = -v^*$. Given that s' is a maximin strategy for player 1, it follows from (9.2) that $u_1(s', t) \geq v^*$ for every $t \in T$. And given that t' is a maximin strategy for player 2, it follows by the same reasoning, using (9.3), that $u_2(s, t') \geq -v^*$, for every $s \in S$. Putting $s = s'$ and $t = t'$, we get $u_1(s', t') \geq v^*$ and $u_2(s', t') \geq -v^*$, which together with $u_2(s', t') = -u_1(s', t')$ yield $u_1(s', t') = v^*$.

The fact that $u_1(s', t) \geq v^*$ for every $t \in T$, and that $u_2(s, t') \geq -v^*$, for every $s \in S$ together with the fact that $u_1 = -u_2$ imply that

$$u_1(s, t') \leq u_1(s', t') \leq u_1(s', t), \tag{9.10}$$

for every $s \in S$ and $t \in T$. By (9.4), (s', t') is an equilibrium. ■

Corollary 9.5.1 *Let $\Gamma = (N, (S_i)_{i \in N}, (u_i)_{i \in N})$ be a zero-sum game. If (s, t) and (s', t') are equilibria in Γ, then*

- *(s, t') and (s', t) are also equilibria, and*
- *$u_1(s, t) = u_1(s', t') = u_1(s, t') = u_1(s', t)$.*

Proof. Let (s, t) and (s', t') be equilibria in Γ. Then by Theorem 9.5.1(1), the strategies s and s' are maximin strategies for player 1 and t and t' are maximin strategies for player 2. Further, it follows that $\max_s \min_t u_1(s, t) = \min_t \max_s u_1(s, t)$. But then, by Theorem 9.5.1.2, (s, t') and (s', t) are equilibria as well.

For any of the four equilibrium pairs (s^*, t^*), it follows from Theorem 9.5.1(1) that $u_1(s^*, t^*) = \max_s \min_t u_1(s, t)$. Since the latter expression is independent of s^* and t^* it follows that all equilibria have the same payoff for player 1. ∎

In virtue of the corollary, two equilibria in a strictly competitive strategic game return the same payoffs to the players. Accordingly, *when* there is an equilibrium (s', t') in the game, we can talk about *the value of the game*:

$$u_1(s', t') = \max_s \min_t u_1(s, t) = \min_t \max_s u_1(s, t).$$

Theorem 9.5.1(1) shows that $\max_s \min_t u_1(s, t) = \min_t \max_s u_1(s, t)$ for any strictly competitive game that has an equilibrium. It should be noted that $\max_s \min_t u_1(s, t) \leq \min_t \max_s u_1(s, t)$ holds for any game, independently of whether the game has an equilibrium and independently of whether the game is strictly competitive or not. For any $s' \in S$ and any $t \in T$ we have $u_1(s', t) \leq \max_s u_1(s, t)$. So for any $s' \in S$, $v_1(s') = \min_t u_1(s', t) \leq \min_t \max_s u_1(s, t)$. Thus we see that in any game the security level of player 1 for any of his strategies s' is at most the amount that player 2 can hold her down to. The hypothesis that the game has an equilibrium is needed in order to prove the other direction. It is instructive in this connection to look at games without an equilibrium such as the *Matching Pennies*: In this case $\max_s \min_t u_1(s, t) = -1 < \min_t \max_s u_1(s, t) = 1$.

In the next section we shall see that when *mixed* strategies are allowed, an equilibrium always exists.

5.2 Mixed Strategies

Mixed strategies may be used for finite strictly competitive games without equilibria to increase the security level of the players and to obtain equilibria. We return to the *Matching Pennies*:

Suppose that player 2 chooses each of her strategies with probability $\frac{1}{2}$. Then if player 1 chooses s_1 with probability p and s_2 with probability $1 - p$, the outcomes (s_1, t_1) and (s_1, t_2) occur each with probability $\frac{1}{2}p$, and the outcomes (s_2, t_1) and (s_2, t_2) occur with probability $\frac{1}{2}(1 - p)$. Thus the probability that the outcome is either (s_1, t_1) or (s_2, t_2) so that player 1 gains 1 is $\frac{1}{2}p + \frac{1}{2}(1 - p) = \frac{1}{2}$. And the probability that the outcome is either (s_1, t_2) or (s_2, t_1) in which case player 1 loses 1 is also $\frac{1}{2}$. Notice that the probability distribution over outcomes is independent of p. The strategy of choosing s_1 with probability x_1 and s_2 with probability x_2 is called a *mixed* or *randomized* strategy. A symmetrical argument shows that if we assume that player 1 chooses each of her strategies with probability $\frac{1}{2}$, then the probability that player 2 gains 1 equals the probability that she loses 1, which is $\frac{1}{2}$. We now show that this is a mixed strategy equilibrium, a generalization of the notion of equilibrium introduced earlier.

Definition 9.5.5 Let $\Gamma = (N, (S_i)_{i\in N}, (u_i)_{i\in N})$ be a strategic game. A *mixed strategy* σ_p for player $p \in N$ is a probability distribution over S_p. That is, σ_p is a function of type $S_p \to [0,1]$ such that $\sum_{s\in S_i} \sigma_p(s) = 1$.

To distinguish the strategies in S_p from mixed strategies, we shall sometimes call them *pure strategies*. If σ is a mixed strategy, it may still behave like a pure strategy in the sense that it assigns probability 1 to some pure strategy. Conversely, we can identify a pure strategy s with the mixed strategy σ such that $\sigma(s) = 1$, and thus $\sigma(s') = 0$, for each strategy $s' \neq s$ belonging to the owner of s.

The *uniform mixed strategy* of player p is the mixed strategy that assigns equal probability to all pure strategies in S_p. Note that the uniform mixed strategy does not exist if S_p contains infinitely many strategies.

A pair of mixed strategies (σ_1, σ_2) defines a probability distribution, or lottery, over $S_1 \times S_2$. The outcome of (σ_1, σ_2) for player p will be quantified in terms of p's expected payoff of the lottery it defines. Let Γ be a two-player strategic game and let σ_1 and σ_2 be mixed strategies of player 1 and 2 respectively. The *expected utility* for player p is given by

$$U_p(\sigma_1, \sigma_2) = \sum_{s\in S}\sum_{t\in T} \sigma_1(s)\sigma_2(t)u_p(s,t).$$

If Γ is a zero-sum game, then it can be checked that $U_2(\sigma, \tau) = -U_1(\sigma, \tau)$; if it is a c-sum game, then $U_2(\sigma, \tau) = c - U_1(\sigma, \tau)$. From now on, we shall denote by $\Delta(S_p)$ the set of mixed strategies of player p over S_p.

Example 9.5.1 Return to the Matching Pennies and consider the mixed strategy σ for player 1 such that $\sigma(s_1) = \frac{1}{2} = \sigma(s_2)$ and the mixed strategy τ for player 2 such that $\tau(t_1) = \frac{1}{2} = \tau(t_2)$. We compute

$$\begin{aligned}
U_1(\sigma, \tau) &= \sum_{s\in S}\sum_{t\in T} \sigma(s)\tau(t)u_1(s,t) \\
&= \sum_{t\in T} \sigma(s_1)\tau(t)u_1(s_1,t) + \sum_{t\in T} \sigma(s_2)\tau(t)u_1(s_2,t) \\
&= \sigma(s_1)\tau(t_1)u_1(s_1,t_1) + \sigma(s_1)\tau(t_2)u_1(s_1,t_2) + \\
&\quad \sigma(s_2)\tau(t_1)u_1(s_2,t_1) + \sigma(s_2)\tau(t_2)u_1(s_2,t_2) \\
&= \frac{1}{4}(1 - 1 + 1 - 1) = 0
\end{aligned}$$

That is, the expected utility for player 1 for the strategy pair (σ, τ) is 0 and that for player 2 is 0.

We shall introduce a couple of auxiliary notions. Let $s \in S$ and let $\tau \in \Delta(S_2)$ be a mixed strategy for player 2. Then we let $U_1(s, \tau)$ be the expected utility for player 1 when he uses the pure strategy s and player 2 uses the mixed strategy τ. More exactly

$$U_1(s, \tau) = \textstyle\sum_{t \in T} \tau(t) u_1(s, t)$$

By analogy for $t \in T$ and $\sigma \in \Delta(S_1)$, we let $U_1(\sigma, t)$ be the expected utility when player 1 uses σ and player 2 uses the pure strategy t.

$$U_1(\sigma, t) \;=\; \sum_{s \in S} \sigma(s) u_1(s, t)$$

Symmetrical notions can be defined for player 2.

It follows directly from the definitions that

$$U_p(\sigma, \tau) \;=\; \sum_{s \in S} \sigma(s) U_p(s, \tau) \;=\; \sum_{t \in T} \tau(t) U_p(\sigma, t). \tag{9.11}$$

The following simple facts will be useful later on.

Proposition 9.5.1 Let $\Gamma = (N, (S_i)_{i \in N}, (u_i)_{i \in N})$ be a strictly competitive strategic game, $\sigma \in \Delta(S_1)$, and $\tau \in \Delta(S_2)$. Then we have for $p \in N$

$$U_p(\sigma, \tau) \;=\; \sum_{s \in S} \sigma(s) U_p(\sigma_s, \tau) \;=\; \sum_{t \in T} \tau(t) U_p(\sigma, \tau_t).$$

Here σ_s denotes the mixed strategy that assigns 1 to s and 0 to all other strategies; and analogously for τ_t.

Proof. By (9.11),

$$U_p(\sigma, \tau) \;=\; \sum_{s \in S} \sigma(s) U_p(s, \tau).$$

Since σ_s and s are effectively the same strategy,

$$\sum_{s \in S} \sigma(s) U_p(s, \tau) \;=\; \sum_{s \in S} \sigma(s) U_p(\sigma_s, \tau).$$

■

The notions of security level and equilibrium for mixed strategies are the obvious analogues of the same notions for the pure strategy case.

The security level of player 1 when he uses the strategy $\sigma \in \Delta(S_1)$ is defined by

$$v_1(\sigma) = \min\{U_1(\sigma, \tau) : \tau \in \Delta(S_2)\} = \min_{\tau \in \Delta(S_2)} U_1(\sigma, \tau)$$

We shall write $\min_\tau U_1(\sigma, \tau)$ instead of $\min\limits_{\tau \in \Delta(S_2)} U_1(\sigma, \tau)$.

The security level of player 2 when he uses the strategy $\tau \in \Delta(S_2)$ is defined by:

$$v_2(\tau) = \min\{U_2(\sigma,\tau) : \sigma \in \Delta(S_1)\} = \min_{\sigma\in\Delta(S_1)} U_2(\sigma,\tau)$$

We shall write $\min_\sigma U_2(\sigma,\tau)$ instead of $\min_{\sigma\in\Delta(S_1)} U_2(\sigma,\tau)$.

By analogy with the pure strategy case, we define a maximin strategy σ^* for player 1 to be such that $v_1(\sigma^*) = \max_\sigma v_1(\sigma) = \max_\sigma \min_\tau U_1(\sigma,\tau)$, and a maximin strategy τ^* for player 2 to be such that $v_2(\tau^*) = \max_\tau v_2(\tau) = \max_\tau \min_\sigma U_2(\sigma,\tau)$.

Then a maximin strategy σ^* for player 1 ensures that $U_1(\sigma^*,\tau) \geq \max_\sigma \min_\tau U_1(\sigma,\tau)$, for all $\tau \in \Delta(S_2)$. And a maximin strategy τ^* for player 2 ensures that $U_2(\sigma,\tau^*) \geq \max_\tau \min_\sigma U_2(\sigma,\tau)$, for all $\sigma \in \Delta(S_1)$.

When $\Gamma = (N,(S_i)_{i\in N},(u_i)_{i\in N})$ is a zero-sum game, the equation

$$\max_\tau \min_\sigma U_2(\sigma,\tau) = -\min_\tau \max_\sigma U_1(\sigma,\tau) \tag{9.12}$$

holds for the mixed strategy case as well.

Example 9.5.2 We return to the Matching Pennies, which does not have an equilibrium in pure strategies. Let σ be the uniform probability distribution over $S_1 = \{s_1,s_2\}$ and τ the uniform probability distribution over $S_2 = \{t_1,t_2\}$.

It is easy to see that the security level of σ is $v_1(\sigma) = 0$.

To show that σ is a maximin strategy for player 1, it suffices to show that for every strategy $\sigma' \in \Delta(S_1)$, $v_1(\sigma) \geq v_1(\sigma')$. Without loss of generality, assume that $\sigma'(s_1) = p > 1/2$, that is, player 1 is more likely to play s_1 than s_2. The security level of σ' is defined as $\min_\tau U_1(\sigma',\tau)$. Let τ_{t_2} be the mixed strategy of player 2 that assigns probability 1 to t_2. We get

$$U_1(\sigma',\tau_{t_2}) = -p + (1-p) = 1-2p.$$

Since $p > 1/2$, it follows that $v_1(\sigma') < 0 = v_1(\sigma)$. So we see that if player plays s_1 more frequently than s_2, then player 2 can exploit this by always playing s_2.

An identical computation shows that $v_2(\tau) = 0$ and that τ is a maximin strategy for player 2.

This situation should be compared to the pure strategy case where

$$\max_s \min_t u_1(s,t) = -1 < \min_t \max_s u_1(s,t) = 1.$$

Proposition 9.5.2

(a) For each σ in $\Delta(S_1)$:

$$v_1(\sigma) = \min\{U_1(\sigma,t_1),\ldots,U_1(\sigma,t_n)\} = \min_t U_1(\sigma,t).$$

(a) For each τ in $\Delta(S_2)$:

$$v_2(\tau) = \min\{U_2(s_1, \tau), \ldots, U_2(s_m, \tau)\} = \min_s U_2(s, \tau)$$

Proof. (a) Let $\min_t U_1(\sigma, t)$ be $U_1(\sigma, t_j)$, and consider the strategy τ_j, that is, the mixed strategy that assigns 1 to t_j and 0 to all other strategies in T. Obviously $\tau_j \in \Delta(S_2)$ and thus $U_1(\sigma, \tau_j) \in \{U_1(\sigma, \tau) : \tau \in \Delta(S_2)\}$. We know already that $U_1(\sigma, \tau_j) = U_1(\sigma, t_j)$. It is straightforward to show that $U_1(\sigma, t_j) \leq U_1(\sigma, \tau)$, for every $\tau \in \Delta(S_2)$, which establishes (a) when we recall that $U_1(\sigma, t_j) = \min_t U(\sigma, t)$. The proof of (b) is entirely analogous. ■

5.2.1 Mixed Strategy Equilibrium

The notion of equilibrium for pure strategies extends quite naturally to the mixed strategies.

Definition 9.5.6 Let $\Gamma = (N, (S_i)_{i \in N}, (u_i)_{i \in N})$ be a two-player strategic game. Let $\sigma \in \Delta(S_1)$ and let $\tau \in \Delta(S_2)$. The pair (σ', τ') is an equilibrium if

(i) for every mixed strategy σ in $\Delta(S_1)$: $U_1(\sigma', \tau') \geq U_1(\sigma, \tau')$
(ii) for every mixed strategy τ in $\Delta(S_2)$: $U_2(\sigma', \tau') \geq U_2(\sigma', \tau)$.

If Γ is strictly competitive, we have that (σ', τ') is an equilibrium if, and only if,

$$U_1(\sigma, \tau') \leq U_1(\sigma', \tau') \leq U_1(\sigma', \tau)$$

for all $\sigma \in \Delta(S_1)$ and $\tau \in \Delta(S_2)$. Equivalently

$$U_1(\sigma', \tau') = \max_\sigma U_1(\sigma, \tau') = \min_\tau U_1(\sigma', \tau).$$

Now when we look at Theorem 9.5.1 – which establishes the equivalence between a pair (s', t') being an equilibrium in pure strategies, on one side, and s' being a maximin strategy for player 1, t' being a minimax strategy for player 2, and $v_1 = u(s', t') = v_2$, on the other – we observe that its proof depends entirely on the definitions of security levels and the definitions of minimax and maximin. The proof carries on unmodified to the present case. We shall add to it a third clause, which reflects more the present context.

Theorem 9.5.2 *Let $\Gamma = (N, (S_i)_{i \in N}, (u_i)_{i \in N})$ be a two-player, zero-sum strategic game. Let $\Delta(S_1)$ the set of mixed strategies of player 1 and let $\Delta(S_2)$ the set of mixed strategies of player 2. Then the following hold:*

1. If (σ', τ') is an equilibrium, then
   i. σ' is a maximin strategy for player 1,

    ii. τ' is a maximin strategy for player 2, and
    iii. $\max_\sigma \min_\tau U_1(\sigma,\tau) = \min_\tau \max_\sigma U_1(\sigma,\tau) = U_1(\sigma',\tau')$.
2. If
    i. $\max_\sigma \min_\tau U_1(\sigma,\tau) = \min_\tau \max_\sigma U_1(\sigma,\tau)$,
    ii. σ' is a maximin strategy for player 1, and
    iii. τ' is a maximin strategy for player 2,
    then (σ',τ') is an equilibrium.
3. (σ',τ') is an equilibrium iff both
    (a) $U_1(\sigma',t) \geq v^*$, for each $t \in S_2$, and
    (b) $U_2(s,\tau') \geq -v^*$, for each $s \in S_1$.
    where $v^* = \max_\sigma \min_\tau U_1(\sigma,\tau)$

Proof. (1) and (2) are exactly like in the pure strategy case. For (3), suppose that (σ',τ') is an equilibrium. Applying (1) to (σ',τ') yields $U_1(\sigma',\tau') = \max_\sigma \min_\tau U_1(\sigma,\tau)$. Hence, $U_1(\sigma',\tau') = \min_\tau U_1(\sigma',\tau)$. From Proposition 9.5.2, we know that $\min_\tau U_1(\sigma',\tau) = \min_t U_1(\sigma',t)$. Fix an arbitrary $t \in S_2$. By the property of the minimum, $U_1(\sigma',t) \geq \min_t U_1(\sigma',t)$. From (1) it also follows that $U_1(\sigma',\tau') = v^*$. We conclude that $U_1(\sigma',t) \geq v^*$.

A symmetrical argument shows that for an arbitrary $s \in S_1$, $U_2(s,\tau') \geq \max_\tau \min_\sigma U_2(\sigma,\tau)$. By (9.12), $\max_\tau \min_\sigma U_2(\sigma,\tau) = -\min_\tau \max_\sigma U_1(\sigma,\tau)$. Since (σ',τ') is an equilibrium, it follows from (1) that $\min_\tau \max_\sigma U_1(\sigma,\tau) = v^*$. Hence, $U_2(s,\tau') \geq -v^*$.

For the converse, assume that (a) and (b) hold. Let τ be an arbitrary strategy in $\Delta(S_2)$. By definition, for each $t \in S_2$, $U_1(\sigma',t) = \sum_{s\in S} \sigma'(s)u_1(s,t)$. By (a)

$$U_1(\sigma',t_1) \geq v^*, \ldots, U_1(\sigma',t_n) \geq v^*$$

and given that $\tau(t) \geq 0$ for each $t \in S_2$, we also have

$$\tau(t_1)U_1(\sigma',t_1) \geq \tau(t_1)v^*, \ldots, \tau(t_n)U_1(\sigma',t_n) \geq \tau(t_n)v^*.$$

Therefore

$$\tau(t_1)U_1(\sigma',t_1) + \ldots + \tau(t_n)U_1(\sigma',t_n) \geq v^*(\tau(t_1) + \ldots + \tau(t_n)) = v^*.$$

But

$$\tau(t_1)U_1(\sigma',t_1) + \ldots + \tau(t_n)U_1(\sigma',t_n) = \sum_{t\in T} \tau(t)U_1(\sigma',t) = U_1(\sigma',\tau).$$

So $U_1(\sigma',\tau) \geq v^*$, for every τ. A similar argument shows that for any σ in $\Delta(S_1)$ we have $U_2(\sigma,\tau') \geq -v^*$. But $U_2(\sigma,\tau') = -U_1(\sigma,\tau')$ so $v^* \geq U_1(\sigma,\tau')$ for every σ. We conclude that $U_1(\sigma',\tau) \geq v^* \geq U_1(\sigma,\tau')$, for all σ and τ. Putting $\sigma = \sigma'$ and $\tau = \tau'$ we get $U_1(\sigma',\tau') \geq v^* \geq U_1(\sigma',\tau')$. Then $v^* = U_1(\sigma',\tau')$ and thus (σ',τ') is an equilibrium pair. ∎

The next corollary should be compared to Corollary 9.5.1.

Corollary 9.5.2 *Let* $\Gamma = (N, (S_i)_{i\in N}, (u_i)_{i\in N})$ *be a two-player, zero-sum game. If* (σ, τ) *and* (σ', τ') *are equilibria in* Γ*, then*

- (σ, τ') *and* (σ', τ) *are also equilibria, and*
- $U_1(\sigma, \tau) = U_1(\sigma', \tau') = U_1(\sigma, \tau') = U_1(\sigma', \tau)$.

We have now seen a number of characterizations of equilibria in zero-sum games. We do not know yet under what conditions they exist. This is the content of the following result that is considered by many as the first important result of game theory.

Theorem 9.5.3 ([von Neumann, 1928]) *Let* Γ *be a finite, two-person, zero-sum strategic game. Then* Γ *has an equilibrium.*

So far the results we presented on strategic games mostly focused on zero-sum games. Since strategic games for IF logic will be one-sum games, we need to prove a simple result that helps us reduce constant-sum games to zero-sum games: equilibria are preserved under taking linear transformations of utility functions.

Proposition 9.5.3 Let $\Gamma = (N, (S_i)_{i\in N}, (u_i)_{i\in N})$ be a two-player strategic game, where $N = \{1, 2\}$. Let $f(x) = a \cdot x + b$, for some reals $a > 0$ and b. Let $\Gamma' = (N, (S_i)_{i\in N}, (u'_i)_{i\in N})$ be the two-player strategic game in which $u'_p(s, t) = f(u_p(s, t))$, for all $s \in S_1$ and $t \in S_2$. Then, every equilibrium in Γ is an equilibrium in Γ'.

Proof. We write U'_p for the expected utility of player p in Γ'. It is easy to see that $U'_p(\sigma, \tau) = f(U_p(\sigma, \tau)) = aU_p(\sigma, \tau) + b$, for every $\sigma \in \Delta(S_1)$ and $\tau \in \Delta(S_2)$. Let (σ^*, τ^*) be an equilibrium in Γ. This implies that for every $\sigma \in \Delta(S_1)$, $U_1(\sigma^*, \tau^*) \geq U_1(\sigma, \tau^*)$. Since $a > 0$, it follows that for every $\sigma \in \Delta(S_1)$, $aU_1(\sigma^*, \tau^*) + b \geq aU_1(\sigma, \tau^*) + b$. Hence, for every $\sigma \in \Delta(S_1)$, $U'_1(\sigma^*, \tau^*) \geq U'_1(\sigma, \tau^*)$.

Similarly, we can show that for every $\tau \in \Delta(S_2)$, $U'_2(\sigma^*, \tau^*) \geq U'_2(\sigma^*, \tau)$. Hence, (σ^*, τ^*) is also an equilibrium in Γ'. ■

5.2.2 A Criterion for Identifying Equilibria

Let $\Gamma = (N, (S_i)_{i\in N}, (u_i)_{i\in N})$ be a finite strategic, zero-sum game, $S_1 = \{s_1, s_2, \ldots, s_m\}$ and $S_2 = \{t_1, t_2, \ldots, t_n\}$.

Given a mixed strategy σ of player 1, the *support* of σ is the set of strategies $s \in S_1$ of player 1 such that $\sigma(s) > 0$, and the support of τ is the set of strategies $t \in S_2$ of player 2 such that $\tau(t) > 0$.

We review a result that will help us to identify equilibriums, see also [Osborne, 2004, p. 116].

Proposition 9.5.4 Let $\Gamma = (N, (S_i)_{i \in N}, (u_i)_{i \in N})$ be a two-player strategic game, where $N = \{1, 2\}$. Then (σ_1^*, σ_2^*) is an equilibrium in Γ iff all of the following conditions are met:

1. for every $s \in S_1$ such that $\sigma_1^*(s) > 0$, $U_1(s, \tau^*) = U_1(\sigma_1^*, \sigma_2^*)$;
2. for every $t \in S_2$ such that $\sigma_2^*(s) > 0$, $U_2(\sigma_1^*, t) = U_2(\sigma_1^*, \sigma_2^*)$;
3. for every $s \in S_1$ such that $\sigma_1^*(s) = 0$, $U_1(s, \sigma_2^*) \leq U_1(\sigma_1^*, \sigma_2^*)$;
4. for every $s \in S_2$ such that $\sigma_2^*(s) = 0$, $U_2(\sigma_1^*, t) \leq U_2(\sigma_1^*, \sigma_2^*)$.

Proof. (1) Write S for S_1. Suppose that (σ_1^*, σ_2^*) is an equilibrium. Let us consider only the strategies in the support of σ_1^*, i.e., $S^* = \{s \in S : \sigma_1^*(s) > 0\}$. It follows from Theorem 9.5.2(1) that $U_1(\sigma_1^*, \sigma_2^*) = \max_\sigma \min_\tau U_1(\sigma, \tau)$ and from Theorem 9.5.2(3)(b) that $U_1(s, \sigma_2^*) \leq \max_\sigma \min_\tau U_1(\sigma, \tau)$, for each $s \in S^*$. Hence, for each $s \in S^*$, $U_1(s, \sigma_2^*) \leq U_1(\sigma_1^*, \sigma_2^*)$. (9.11) implies that $\sum_{s \in S} \sigma_1^*(s) U(s, \sigma_2^*) = U_1(\sigma_1^*, \sigma_2^*)$. From this we get $U_1(s, \sigma_2^*) = U(\sigma_1^*, \sigma_2^*)$, for each $s \in S^*$.

(2) is completely analogous.

(3) and (4) are straightforward from the fact that (σ^*, τ^*) is an equilibrium.

For the converse, suppose that (σ^*, τ^*) satisfies conditions (1)–(4). Consider a strategy $\sigma \in \Delta(S_1)$. It suffices to show that $U_1(\sigma, \tau^*) \leq U_1(\sigma^*, \tau^*)$.

We divide S into $S_1 = \{s \in S : \sigma^*(s) > 0\}$ and $S_2 = \{s \in S : \sigma^*(s) = 0\}$. Obviously $S_1 \cup S_2 = S$ and $S_1 \cap S_2 = \varnothing$. Then, by (9.11),

$$U_1(\sigma, \tau^*) = \sum_{s \in S_1} \sigma(s) U_1(s, \tau^*) + \sum_{s \in S_2} \sigma(s) U_1(s, \tau^*).$$

By (1), $U_1(s, \tau^*) = U_1(\sigma^*, \tau^*)$, for each $s \in S_1$. By (3), $U_1(s, \tau^*) \leq U_1(\sigma^*, \tau^*)$, for each $s \in S_2$. Whence $U_1(\sigma, \tau^*) \leq U_1(\sigma^*, \tau^*)$.

A similar argument establishes that $U_1(\sigma^*, \tau^*) \leq U_1(\sigma^*, \tau)$ for every $\tau \in \Delta(S_2)$. ■

The above proposition is quite significant for it gives conditions for a pair of mixed strategies to be an equilibrium in pure strategies.

Example 9.5.3 Consider the two-player, one-sum game Γ of which player 1's payoff function is given as a matrix in Table 9.3. Player 1 controls strategies $S = \{s_1, \ldots, s_4\}$. Consider the pair (σ^*, τ^*), where σ^* is the mixed strategy

$$\sigma^*(s_i) = \begin{cases} \frac{1}{5} & \text{if } s_i \in \{s_1, s_2, s_3\} \\ \frac{2}{5} & \text{if } s_i \in \{s_4\} \end{cases}$$

and τ^* is the mixed strategy

$$\tau^*(t_j) = \begin{cases} \frac{1}{5} & \text{if } t_j \in \{t_1, t_2, t_3\} \\ \frac{2}{5} & \text{if } t_j \in \{t_4\}. \end{cases}$$

We leave it to the reader to compute the value of Γ which is 2/5. To see that (σ^*, τ^*) is an equilibrium, consider a strategy s_i from the support of σ^*. Suppose that s_i is s_1. Then, $U_1(s_1, \tau^*) = \sum_{t_j} \tau^*(t_j)u_1(s_1, t_j) = \tau^*(t_1)+\tau^*(t_3)$. Since $\tau^*(t_1) = \tau^*(t_3) = 1/5$, we get $U_1(s_1, \tau^*) = 2/5$. Suppose that s_i is s_4. Then, $U_1(s_4, \tau^*) = \tau^*(t_4) = 2/5$ and we are done.

A similar reasoning shows that for every t_j, $U_2(\sigma^*, t_j) = 3/5$. Hence, by Proposition 9.5.4, (σ^*, τ^*) is an equilibrium.

6. Equilibrium Semantics

Recall the *Matching Pennies* sentence $\varphi_{\mathrm{MP}} = \forall x(\exists y/x)(x = y)$, and its relative $\varphi_{\mathrm{IMP}} = \forall x(\exists y/x)(x \neq y)$. Both are undetermined on every structure $\mathcal{M}$ whose universe M contains at least two elements. Yet there is a difference between the two. When the universe increases it becomes easier for *Eloise* to verify φ_{IMP} and more difficult to verify φ_{MP}. The interpretation in terms of pure strategies does not do justice to these intuitions. Below the left column registers the increasing size of the universe and the two other columns indicate the probability that *Eloise* picks up an element y identical to (distinct from) the element x chosen by *Abelard*.

TABLE 9.3 The payoff matrix of Γ in Example 9.5.3

	t_1	t_2	t_3	t_4
s_1	1	0	1	0
s_2	1	1	0	0
s_3	0	1	1	0
s_4	0	0	0	1

Cardinality of M	φ_{MP}	φ_{IMP}
1	1	0
2	$\frac{1}{2}$	$\frac{1}{2}$
3	$\frac{1}{3}$	$\frac{2}{3}$
$\vdots$	$\vdots$	$\vdots$
n	$\frac{1}{n}$	$\frac{n-1}{n}$

To account for these facts, we will switch from pure to mixed strategies and take the value of φ_{MP} and φ_{IMP} to be the expected utility returned to player 1 by the equilibrium strategy pair guaranteed to exist by von Neumann's minimax theorem.

In view of our exposition in the earlier section, this move should appear as no surprise: it is the standard practice in game theory described above to obtain an equilibrium in games like *Matching Pennies*, which do not have one in pure strategies. We shall revisit the *Matching Pennies* sentences φ_{MP} and φ_{IMP} after defining the notion of strategic IF game.

6.1 Equilibrium Semantics

The two previous examples should be enough to motivate the following more general definition.

Definition 9.6.1 Let $\mathcal{M}$ be a structure, let s be an assignment in $\mathcal{M}$ and let φ be an IF formula. Let $G(\mathcal{M}, s, \varphi) = (N, H, Z, P, (I_i)_{i\in N}, (v_p)_{p\in N})$ (i.e., $G(\mathcal{M}, s, \varphi)$ is the extensive game of imperfect information introduced in our earlier section.) Then $\Gamma(\mathcal{M}, s, \varphi) = (N, (S_i)_{i\in N}, (u_i)_{i\in N})$ is the *strategic IF game* associated with $\mathcal{M}$, s, and φ, where

- $N = \{\exists, \forall\}$ is the set of players;
- S_p is the set of strategies of player p in $G(\mathcal{M}, s, \varphi)$;
- u_p is the utility function of player p, so that $u_p(s,t) = v_p(h)$, where h is the terminal history resulting from *Eloise* playing s and *Abelard* playing t, that is, h is the single element in $H_s \cap H_t$.

If φ is a sentence and s is the empty assignment, we write $\Gamma(\mathcal{M}, \varphi)$ instead of $\Gamma(\mathcal{M}, s, \varphi)$.

We shall often write $S_\exists = S = \{s_1, \ldots, s_m\}$ and $S_\forall = T = \{t_1, \ldots, t_n\}$

We recall that every strategic IF game is a *one-sum game*: for every $s \in S_\exists$ and $t \in S_\forall$, $u_\exists(s,t) + u_\forall(s,t) = 1$. On account of Proposition 9.5.3, we can reduce every strategic IF game Γ to a zero-sum game Γ' whose utility function is defined on the basis of Γ's utility function u_i by $u_i'(s,t) = 2(u_i(s,t)) - 1$, for every $s \in S_\exists$ and $t \in S_\forall$. Thus, by Proposition 9.5.3, if (σ^*, τ^*) is an equilibrium in Γ', then it is an equilibrium in our strategic IF game.

Let $(\sigma_1, \tau_1), \ldots, (\sigma_i, \tau_i), \ldots$ be the equilibria in the semantic IF game Γ. By Theorem 9.5.3 (and Proposition 9.5.3), Γ has at least one equilibrium: $i \geq 1$. By Corollary 9.5.2, $U_\exists(\sigma_1, \tau_1) = U_\exists(\sigma_i, \tau_i)$, for all i. Hence, it makes sense to refer to $U_\exists(\sigma_i, \tau_i)$ as *the value* of the game Γ, for any equilibrium (σ_i, τ_i). We write $V(\Gamma)$ for the value of Γ. It is obvious that $V(\Gamma)$ takes values in the closed unit interval $[0, 1]$. If $\Gamma = \Gamma(\mathcal{M}, s, \varphi)$, then we refer to $V(\Gamma)$ as the *truth value* of φ on $\mathcal{M}$ and s.

Note that if $\mathcal{M}$ is finite, then every strategic IF game $\Gamma(\mathcal{M}, s, \varphi)$ that is based on $\mathcal{M}$ is also finite. If $\mathcal{M}$ is infinite, however, its semantic IF games are infinite and are not covered by the Minimax theorem (Theorem 9.5.3). That is, if $\mathcal{M}$ in $\Gamma(\mathcal{M}, s, \varphi)$ is infinite, then it is not guaranteed that $\Gamma(\mathcal{M}, s, \varphi)$ has an equilibrium.

Example 9.6.1 Let $\mathcal{M}$ be a finite structure, consisting of n objects $\mathcal{M} = \{a_1, \ldots, a_n\}$. Let φ_{MP} be the IF sentence $\forall x(\exists y/x)x = y$, and let $G(\mathcal{M}, \varphi_{\mathrm{MP}})$ be the extensive game determined by $\mathcal{M}$ and φ_{MP}. The Skolemization of φ_{MP} is $\forall x(x = c)$, where c is a nullary function symbol; the Kreisel form of φ_{MP} is $\forall y(d \neq y)$, where d is a nullary function symbol. Thus, in $G(\mathcal{M}, \varphi)$, each player has one strategy that picks up the object a_i, for every $a_i \in M$. Let us write $S = T = \{a_1, \ldots, a_n\}$ for the strategies of *Eloise* and *Abelard*, respectively. The payoff functions in $G(\mathcal{M}, \varphi_{\mathrm{MP}})$ are given by

$$u_{\exists}(a_i, a_j) = \begin{cases} 1 & \text{if } i = j \\ 0 & \text{otherwise} \end{cases}$$

$$u_{\forall}(a_i, a_j) = 1 - u_{\forall}(a_i, a_j).$$

Eloise's payoff function is shown in Table 9.2. Let σ^* be the uniform strategy over S and let τ^* be the uniform strategy over T. We claim that (σ^*, τ^*) is an equilibrium in $\Gamma(\mathcal{M}, \varphi_{\mathrm{MP}})$. First observe that $U_{\exists}(\sigma^*, \tau^*) = 1/n$ and that $U_{\forall}(\sigma^*, \tau^*) = (n-1)/n$. Then, for any strategy $a_i \in S$, consider $U_1(a_i, \tau^*) = \sum_{a_j} \tau^*(a_j) u_{\exists}(a_i, a_j)$. *Eloise*'s payoff function $u_{\exists}$ returns 1 for $a_j = a_i$; otherwise it returns 0. Hence, $U_1(a_i, \tau^*) = \tau^*(a_i) = 1/n$. A similar reasoning shows that for each $a_j \in T$, $U_{\forall}(\sigma^*, a_j) = (n-1)/n$. Hence, by Proposition 9.5.4, (σ^*, τ^*) is an equilibrium.

Example 9.6.2 Let $\mathcal{M}$ be the structure in the previous example and φ_{IMP} the inverted *Matching Pennies* sentence $\forall x(\exists y/x)(x \neq y)$. In the extensive game $G(\mathcal{M}, \varphi_{\mathrm{IMP}})$, the set of strategies of *Eloise* and *Abelard* are the same as in the game $G(\mathcal{M}, \varphi_{\mathrm{MP}})$. The payoff function of *Eloise* in $G(\mathcal{M}, \varphi_{\mathrm{IMP}})$ is the inverse of the payoff function of $G(\mathcal{M}, \varphi_{\mathrm{IMP}})$, see Table 9.4.

TABLE 9.4 The payoff matrix of *Eloise* in the inverted *Matching Pennies* game

	a_1	a_2	a_3	$\cdots$
a_1	0	1	1	1
a_2	1	0	1	1
a_3	1	1	0	1
$\vdots$	1	1	1	$\ddots$

The two uniform strategies σ^* and τ^* are also in equilibrium in this case. However, in this game they yield an expected payoff for *Eloise* of $(n-1)/n$. That is, the value of $\Gamma(\mathcal{M}, \varphi_{\mathrm{IMP}})$ is $(n-1)/n$.

Comparing the two examples we notice that as the size of $\mathcal{M}$ increases, the truth value of $\forall x(\exists y/x)(x = y)$ on $\mathcal{M}$ asymptotically approaches 0 and that of $\forall x(\exists y/x)(x \neq y)$ asymptotically approaches 1.

The following result compares the truth value of strategic IF games to the three-valued semantic values of extensive IF games.

Proposition 9.6.1 Let $\mathcal{M}$ be a finite structure, let s be an assignment in $\mathcal{M}$, and let φ be an IF formula. Let G be the semantic game $G(\mathcal{M}, s, \varphi)$ and let Γ be the strategic IF game $\Gamma(\mathcal{M}, s, \varphi)$. Then

1. *Eloise* has a winning strategy in G iff the value of Γ is 1;
2. *Abelard* has a winning strategy in G iff the value of Γ is 0;

Proof. Let $S = S_\exists$ be *Eloise*'s strategies in G and let $T = S_\forall$ be *Abelard*'s.

We prove the first claim. Let s be a winning strategy in G. Since s is winning, it follows that for every strategy $t \in T$ in G of *Abelard*, $u(s,t) = 1$. Consequently, for each mixed strategy τ over T, $U(s,\tau) = 1$. Let σ be the mixed strategy in Γ that assigns probability 1 to s. We have that $U(\sigma,\tau) = 1$. Hence, condition 1 of Proposition 9.5.4 is met. To see that also condition 2 is satisfied, we observe that for each $t \in T$, $U(\sigma, t) = 1$. This is again a direct consequence of the fact that s is winning. Conditions 3 and 4 are immediate since $U(\sigma,\tau) = 1$ is the maximal value that can be secured in Γ.

For the converse direction, suppose that (σ,τ) is an equilibrium in Γ with value 1. Let $s \in S$ be a strategy of *Eloise* so that $\sigma(s) > 0$. By condition 1 of Proposition 9.5.4, $U(s,\tau) = U(\sigma,\tau) = 1$. That is, s is winning against every strategy t in the support in τ. For the strategies that are not in the support of τ, we derive from condition 4 of Proposition 9.5.4 that $U(\sigma,t) \geq 1$. Since the maximal value in Γ is 1, this reduces to $U(\sigma,t) = 1$. Hence, for every $t \in T$, $u(s,t) = 1$, and we conclude that s is a winning strategy in G. ■

The previous result shows that the truth of an IF formula corresponds to the value 1, and its falsity corresponds to the value 0. We will now introduce a new satisfaction relation $\models_\varepsilon$ that is based on the values of strategic IF games.

Definition 9.6.2 Let $0 \leq \varepsilon \leq 1$. Let $\mathcal{M}$ be a finite structure, s be an assignment and φ be an IF formula. Let Γ be the strategic IF game $\Gamma(\mathcal{M}, s, \varphi)$. We define the satisfaction relation $\models_\varepsilon$ by:

$$\mathcal{M} \models_\varepsilon \varphi \text{ iff } V(\Gamma) \geq \varepsilon.$$

We call the semantics defined by $\models_\varepsilon$ the *equilibrium semantics* for IF logic.

Example 9.6.3 The *Matching Pennies* sentence φ from Example 9.6.1, $\varphi_{\mathrm{MP}} := \forall x(\exists y/\{x\})(x = y)$, has truth value $1/n$ on every finite structure with n elements. Hence, $\mathcal{M} \models_\varepsilon \varphi_{\mathrm{MP}}$ iff $\varepsilon \le 1/n$. The inverted *Matching Pennies* sentence φ_{IMP} from Example 9.6.2 has truth value $(n-1)/n$. Hence, $\mathcal{M} \models_\varepsilon \varphi_{\mathrm{IMP}}$ iff $\varepsilon \le (n-1)/n$.

Note that the definition of equilibrium semantics is not symmetric. We have that $\mathcal{M}, s \models_\varepsilon \varphi$, if the value of the semantic IF game of $\mathcal{M}$, s, and φ is greater than *or equal to* ε. As a consequence, we have that $\mathcal{M}, s \models_\varepsilon \varphi$, for every IF formula φ, if $\varepsilon = 0$. A convenient property of the 'inclusive formulation' of equilibrium semantics is that it is a 'conservative extension' of GTS as introduced in the first part of this study. It may be proved that the following holds.

Corollary 9.6.1 *Let $\mathcal{M}$ be a finite structure, let s be an assignment in $\mathcal{M}$ and let φ be an IF formula. Then*

$$\mathcal{M}, s \models^{+}_{GTS} \varphi \quad \textit{iff} \quad \mathcal{M}, s \models_1 \varphi$$

Proof. Immediate from Proposition 9.6.1. ■

Corollary 9.6.1 shows that in the special case in which $\varepsilon = 1$, finding an equilibrium coincides with finding a winning strategy. Note that, by contrast with previous semantics, this semantics is not symmetric. That is, we do not have

$$\mathcal{M}, s \models^{-}_{GTS} \varphi \text{ iff } \quad \mathcal{M}, s \models_0 \varphi.$$

This follows from the observation above that $\mathcal{M}, s \models_\varepsilon \varphi$, for every IF formula φ, if $\varepsilon = 0$.

Notes. The idea of applying von Neumann's Minimax theorem to undetermined games (of Henkin quantifiers) goes back to Ajtai who suggested that the truth value of the undetermined IF sentence $\forall x(\exists y/x)(x = y)$ is $1/n$ in structures of cardinality n. Ajtai's suggestion, discussed in [Blass and Gurevich, 1986] has been developed in [Sevenster, 2006], and in [Galliani, 2009], and generalized in [Sevenster and Sandu, 2010]. We have drawn extensively from [Mann et al., ta], where the reader may found other applications of the strategic paradigm to IF logic. Theorem 9.5.3 is known in the literature as von Neumann's *Minimax theorem*. Later John Nash proved the same theorem for arbitrary finite strategic games. The notion of equilibrium has been associated henceforth with Nash's name. However, for the theory developed in this chapter we only need von Neumann's theorem as stated in Theorem 9.5.3.

Note

1. If we give up the requirement that strategies be deterministic, then only a weaker form of AC is needed, namely, Axiom of Dependent Choices. As the number of strategies may be infinite, these principles cannot be proved in ZF.

10 Mereology

Karl-Georg Niebergall

Chapter Overview

1. Introduction 271
2. Mereological Theories 274
   2.1 The Language L[∘] and the Mereological Core Axiom System Ax(CI) 274
   2.2 Optional Mereological Axioms and Further Sentences in L[∘] 276
   2.3 A Synopsis of Mereological Theories in L[∘] 279
   2.4 What is a Mereological Theory? History and Systematics 280
3. Models for L[∘] 284
   3.1 Boolean Algebras and Mereological Algebras 284
   3.2 Applications 285
4. The Main Meta-Theoretical Results 286
5. On the 'Strength' of Mereological Theories 288
   5.1 Natural Numbers 290
   5.2 Sets 291
6. Extensions of the Mereological Framework 291
Notes 295

1. Introduction

The expression 'mereology' has its roots in the Greek word '$\mu\epsilon\rho o\sigma$', meaning *part*. Thus, mereology is, roughly put, about the *part-whole relation*. While playing a role comparable in relevance to that of 'is an element of' in set theory, the predicate 'is a part of' has to be emphatically distinguished from the former. This manifests itself already in their different formal characteristics: by contrast with the relation *is an element of*, the relation *is a part of* is guaranteed to be transitive and reflexive, while neither density nor ill-foundedness is excluded for it.[1] Another distinguishing feature: informally, the common understanding is that whereas elements of a set x are of lower type than x itself, the part-whole relation always obtains between objects of the same type.[2] In particular, parts and *sums* (or *fusions*) of concrete objects are naively conceived of as concrete objects too (see Section 2.1).

Both of these rather general features can readily be illustrated by examples from real life. Thus, consider the United States (cf. [Quine, 1940]). It may be construed as a set or as a concrete object (certainly a scattered one; but it makes sense to say that, e.g., you can travel through it). Construed as a set, the states and counties of the United States will not be (spatial) parts of it. Instead, they will be, e.g., elements or subclasses of it. And the United States will not be identical with both the set of its states and the set of its counties (since these sets are different from each other). Construed as a concrete object, it is natural to regard the states and counties of the United States as parts of it. Then, the fusion of the states and the fusion of the counties of the United States turn out to be the same object, which, moreover, is just the United States.

Although the part-whole relation had relevance already in ancient Greek philosophy, its systematic development belongs to the twentieth century. It is commonly agreed that its treatment by means of formal theories originates with Stanisław Leśniewski: see [Leśniewski, 1916].[3] Being integrated into his idiosyncratic logical system, however, Leśniewski's version of mereology was investigated primarily by his followers.[4] A reformulation of it by Tarski [Tarski, 1929] was, as far as I am aware, the first version of a mereological theory in the (now common) framework of quantificational languages. A similar theory was put forward in [Leonard and Goodman, 1940] (who acknowledged the priority of Leśniewski and Tarski), but there it was called the 'calculus of individuals'. Both of these theories are higher-order theories or include set-theoretical notions. The first-order theories of the part-whole relation, which are currently preferred, were introduced in [Goodman, 1951], which is thus the defining text of the field (and this article).[5]

The term 'mereology' is not free from ambiguity.[6] It is used as a term referring to a discipline, as a term referring to a specific theory or as a predicate applying to this theory and similar ones, and as a predicate applying to structures that can be models of such theories.[7] In this chapter, terminology is (eventually) straightened as follows: 'mereological theory' is ascribed to certain theories stated in a specific first-order language $L[\circ]$ (see Section 2.4); models appropriate to $L[\circ]$ are called 'mereological algebras'.

Mereological theories are closely related to (the already mentioned) *calculi of individuals*. Intuitively, the former should fix the use of 'part of' and the latter should provide for an explication of 'individual'.[8] But this seems to leave us with two classes of theories that need not be very closely related. Now, it has to be granted that, although each of the predicates 'mereology' (in the sense of 'mereological theory') and 'calculus of individuals' is in common use, neither of them has found a definition that is both rigorous and widely accepted. This is not to say that they are not understood at all; it is only that it is difficult to pin down their precise meaning. Furthermore, there are only a few texts where both of these predicates are in use:[9] in addition to Leśniewski, philosophers like Simons, Smith, and Varzi favour expressions such as 'mereology'; others, notably Goodman, but also Eberle and Hendry prefer 'calculus of individuals' (cf. Section 2.4 for more details). We seem to have two communities of research here, working in frameworks which terminologically (and in part even philosophically) are only loosely connected.[10]

In spite of that, in their respective communities, mereological theories and calculi of individuals play a similar role. From a *logical point of view*, it is simply that the formal theories presented as mereological ones are the same – or almost the same – as the calculi of individuals: all of them are theories of the part-whole relation. From a *methodological point of view*, the development and investigation of mereological theories and calculi of individuals are motivated by the same considerations. In particular, from the beginning – that is, in the work of Leśniewski and Goodman – these theories were conceived of as alternatives to set theory. Only lately, especially in the context of *mereotopology*, have other goals become more prominent (see [Bochman, 1990] for comments). A common reason for avoiding the adoption of set theories is that, by contrast with mereological theories, the former quite naturally may lead (and have lead) to paradox (or so it may be claimed from the sceptics; see, e.g., [Goodman and Quine, 1947]).

There are further considerations for doing mereology that have found adherents in both communities. To start with, 'x is a part of y' may simply be regarded as a philosophically important predicate and its explication to be interesting in its own right. In particular, similarly to what many will claim of '$x \in y$', 'x is a part of y' is often viewed as widely applicable and as intuitively basic.[11] Accordingly, on the level of theories, mereological theories (maybe in conjunction with 'geometrical' and 'topological' ones) are and should be considered as the core of many empirical theories.[12] Finally, in some cases mereological theories may be just of the appropriate strength and richness (whereas set theories such as ZF tend to be unnecessarily strong).

Only when it comes to the *ontological point of view*, deeper differences seem to emerge. In general, the search for alternatives to set theories often rests on *nominalistic* grounds. Now calculi of individuals are regarded as the prototypical nominalistic theories in their community, as can be seen particularly clearly in [Goodman, 1951], [Eberle, 1970], and [Lewis, 1991]. Indeed, the research done on them is largely motivated by nominalism. In the mereological community, however, explicit commitments to nominalism can only seldom be found (apart from Leśniewski's program).[13] It seems that here, the classical ontological dispute on the 'problem of universals' simply is not that important.

Now this consideration may have consequences as to what is admitted as a mereological theory; cf. Section 2.4. Yet, since nominalistic concerns are only a side issue of this text, I am content in using 'calculus of individuals' as a synonym for 'mereological theory'.

This article is primarily about *pure mereological theories* (in formal languages).[14] Theories that are not exclusively about the part-whole relation are mentioned, but only sketchily dealt with (Section 6); and mereological algebras are considered mainly as a means by which to get a better grasp of the formal theories (Section 3.1). In its informal sections, the Chapter focuses on discussions of how 'mereological theory' and 'calculus of individuals' could be explicated (Section 2.4). Its style, however, is often more technical: it is to a large extent a report (and hence contains no proofs) of the not-so-many *meta-logical* (or: *meta-theoretical*) results that have been obtained for mereological theories (Sections 2.1–2.3 and, most importantly, Section 4).[15] Many of the more important meta-theorems of this article are eventually motivated by the question: are mereological theories a reasonable alternative to set theories (relative to the tasks accepted for the latter)? It will turn out that the answer is *No*. It is not that mereological theories fail to be ontologically and conceptually preferable to set theories; but what they lack is proof-theoretic strength (Sections 3.2 and 5).

2. Mereological Theories

2.1 The Language L[$\circ$] and the Mereological Core Axiom System Ax(CI)

Mereological theories and calculi of individuals T are most naturally formulated in a language that contains a 2-place predicate standing for 'is part of'. However, often a 2-place predicate '$\circ$', which is read 'overlaps', is preferred as a primitive. In this Chapter, I join this latter approach and deal with the first-order language L[$\circ$] with '$\circ$' as its sole non-logical primitive. Thus, although pre-theoretically, 'x overlaps y' is best understood as 'there is a z which is a part of x and of y', here the formulas '$x \sqsubseteq y$' and '$x \sqsubset y$', which are intended to express 'x is a part of y' and 'x is a proper part of y', are introduced by definition.

Definition 10.2.1

- $x \sqsubseteq y :\leftrightarrow \forall z(z \circ x \rightarrow z \circ y)$
- $x \sqsubset y :\leftrightarrow x \sqsubseteq y \wedge y \not\sqsubseteq x.$

L[$\circ$] is supplied with classical first-order logic. In addition, '=' is either treated as a logical sign (for identity) the use of which is fixed by usual axioms – namely, reflexivity and substitutivity in L[$\circ$];[16] or it is defined. In this article, I choose the latter alternative: see below.

Lemma 10.2.1 *The following are provable in first-order logic:*

1. $\forall x(x \sqsubseteq x)$
2. $\forall x, y, z(x \sqsubseteq y \wedge y \sqsubseteq z \rightarrow x \sqsubseteq z)$
3. $\forall x(\exists y \forall v(v \sqsubseteq y \leftrightarrow \neg v \circ x) \rightarrow \neg \forall v\, v \circ x)$
4. $\forall x(\forall w\, w \circ x \rightarrow \forall w\, w \sqsubseteq x)$.

Given the intended reading of '$\circ$' (and also of '$\sqsubseteq$'), some sentences from L[$\circ$] should be sound and are, moreover, so simple that they suggest themselves as axioms for 'overlaps'. One of them is (O):

- (O): $\forall x, y(x \circ y \leftrightarrow \exists z(z \sqsubseteq x \wedge z \sqsubseteq y))$

(O) alone yields interesting theorems: see especially (5) – (7) below, where (7) says that there is no null object (in non-trivial circumstances).

Lemma 10.2.2 *The following are derivable from (O):*

1. $\forall x(x \circ x)$
2. $\forall x(x \circ y \rightarrow y \circ x)$
3. $\forall x, y(x \sqsubseteq y \rightarrow y \circ x)$
4. $\forall x, y(\exists z \forall u(u \sqsubseteq z \leftrightarrow u \sqsubseteq x \wedge u \sqsubseteq y) \rightarrow x \circ y)$
5. $\forall y, z(\forall x(x \sqsubseteq y \rightarrow x \circ z) \rightarrow y \sqsubseteq z)$
6. $\forall x, y(x \sqsubset y \rightarrow \exists z(z \sqsubseteq y \wedge \neg x \circ z))$
7. $\exists x \neg \forall v\, v \circ x \rightarrow \neg \exists x \forall y(x \sqsubseteq y)$.[17]

With (O) in the background, it is reasonable to define '=' by

- (D=): $x = y :\leftrightarrow \forall z(z \circ x \leftrightarrow z \circ y)$

The usual principles of identity – reflexivity and substitutivity in L[$\circ$] for '=' – are consequences of (O) and (D=).

(O) seems to be universally accepted as a mereological principle (also when '$\sqsubseteq$' is assumed as a primitive; see Section 2.4). Two other L[$\circ$]-sentences which are often adopted as mereological axioms are:

- SUM: $\forall x, y \exists z \forall u(u \circ z \leftrightarrow u \circ x \vee u \circ y)$
- NEG: $\forall x(\neg \forall v\, v \circ x \rightarrow \exists z \forall v(v \sqsubseteq z \leftrightarrow \neg v \circ x))$.[18]

In SUM, the existence of the *sum* (or: the *fusion*) of x and y is postulated: this is the object z that consists exactly of x and y.[19] In NEG, given an object x that is not the universal object, the existence of the complement of x is postulated.

As far as I know, theories implying SUM and NEG have been accepted in all texts belonging to the calculus-of-individuals approach. SUM, for example, is for some philosophers simply intuitively sound, given their understanding of 'part of' and 'overlaps'. Others attempt to give additional reasons for it: thus Goodman ([Goodman, 1951]) appeals to an analogy with the comprehension or separation schema of set theory. There, sets are assumed to exist even if their elements have, intuitively speaking, nothing in common and are not contiguous; sums should be understand as being similar in this respect. Moreover, the mere fact that, if an object were to exist, its parts would be scattered and disconnected does not speak against its existence:[20] consider the United States, but also every non-atomic concrete object you might consider.

But SUM and NEG are not beyond dispute. Lewis' ([Lewis, 1991]) claim of the *ontological innocence* of SUM, in particular, has been severely criticized; see especially [van Inwagen, 1994]. 'Ontological innocence' may be understood in three ways: as (i) for all x and y, the sum of x and y exists; (ii) for all concrete x and y (alternatively: individuals), given that the sum of x and y exists, it is a concrete object (alternatively: an individual); (iii) for all x and y, the sum of x and y exists and is nothing over and above x and y. (iii) is the position held by Lewis ([Lewis, 1991]); and I fully agree with van Inwagen ([van Inwagen, 1994]) that the formulations Lewis employs to express it are hard to understand (if meaningful at all). But this does not constitute a refutation of (i) and (ii).

We let CI be the first-order theory axiomatized by Ax(CI) := {O, SUM, NEG}.[21] CI is the core of the theories investigated here.

2.2 Optional Mereological Axioms and Further Sentences in L[$\circ$]

Several L[$\circ$]-sentences not in Ax(CI) have been considered as possible further mereological axioms. Thus, there is a variant of NEG guaranteeing relative complements instead of absolute ones.

- NEG′: $\forall x, y[\exists w(w \sqsubseteq x \wedge \neg w \circ y) \rightarrow \exists z \forall w(w \sqsubseteq z \leftrightarrow w \sqsubseteq x \wedge \neg w \circ y)]$

Then, there is the product-principle PROD, which expresses that 'meets' of overlapping objects exist (keeping in mind that, in general, there is no null object):

- PROD: $\forall x, y(x \circ y \rightarrow \exists z \forall u(u \sqsubseteq z \leftrightarrow u \sqsubseteq x \wedge u \sqsubseteq y))$.

SUM and PROD have also infinitary extensions: the so-called *fusion-schema* FUS and *nucleus-schema* NUC (see [Goodman, 1951], [Breitkopf, 1978], [Simons,

1987]). Let's start with FUS, which is more often taken into consideration than NUC. It roughly states that for any non-empty set, there exists the sum or fusion of its elements. This statement, which contains set-theoretic or second-order terminology, is approximated in L[$\circ$] in the usual style – i.e., by a first-order schema.[22] Utilizing the common procedure of identifying a schema with the set of 'its instances', FUS can be precisely formulated as follows. Let ψ be a formula in L[$\circ$]; then let

- FUS_ψ: $\exists x\, \psi \rightarrow \exists z \forall y (z \circ y \leftrightarrow \exists x (x \circ y \wedge \psi))$.[23]
- FUS := $\{\text{FUS}_\psi \mid \psi$ is a L[$\circ$]-formula$\}$.

FUS is a highly important mereological schema: it seems to provide most of the power of mereological theories. Clearly, any criticism of SUM extends to FUS; but the type of reasoning advanced against SUM may lead to even graver doubts about FUS. Nonetheless, like SUM, FUS is more often than not accepted as a mereological axiom schema.[24]

NUC is explained similarly to FUS. Let ψ be a formula in L[$\circ$]; then let

- NUC_ψ: $\exists y \forall x (\psi(x) \rightarrow y \sqsubseteq x) \rightarrow \exists z \forall y (y \sqsubseteq z \leftrightarrow \forall x (\psi(x) \rightarrow y \sqsubseteq x))$
- NUC := $\{\text{NUC}_\psi \mid \psi$ is a L[$\circ$]-formula$\}$.

The sentences taken into account as axioms so far are multiply related.

Lemma 10.2.3

1. *FUS* $\vdash$ *SUM*.
2. *FUS* $\vdash$ *NUC*.
3. $\{O, FUS\} \vdash NEG$.
4. $\{O, NUC\} \vdash PROD$.

Corollary 10.2.1 $O + FUS = CI + FUS$.

Lemma 10.2.4

1. $O, PROD, NEG \vdash NEG'$.
2. $O, SUM, PROD, NEG' \nvdash NEG$.[25]

Lemma 10.2.5 $CI \vdash PROD$.

Another consequence of CI is the existence of a universal object.

Lemma 10.2.6

1. $CI \vdash \exists x \forall y\, y \circ x$.
2. $CI \vdash \exists x \forall y\, y \sqsubseteq x$.

The L[$\circ$]-sentences dealt with in this subsection *may* be considered as mereological axioms. Other sentences in L[$\circ$] seem to be indeterminate in this respect: neither they nor their negation seem to be a good choice as possible axioms. Informal examples are 'Each object has an atomic part' (the statement of atomicity) and 'Each object has a proper part' (the statement that there are no atoms). In L[$\circ$], they become

- AT: $\forall x \exists y(y \sqsubseteq x \wedge At(y))$.
- AF: $\forall x \exists y(y \sqsubset x)$,

where we use the abbreviation

Definition 10.2.2 $At(x) :\leftrightarrow \forall y(y \sqsubseteq x \rightarrow x \sqsubseteq y)$ (read 'x is an atom').

Lemma 10.2.7 *The following are provable in first-order logic:*

1. $\neg AT \leftrightarrow \exists x \forall y(y \sqsubseteq x \rightarrow \exists z\, z \sqsubset y)$
2. $\neg AF \leftrightarrow \exists x\, At(x)$
3. $AT \rightarrow \neg AF$.

Then we have several ways to express that objects are determined by their atomic parts:

- HYPEXT: $\forall x, y(\forall z(At(z) \wedge z \sqsubseteq x \leftrightarrow At(z) \wedge z \sqsubseteq y) \rightarrow x = y)$.[26]
- HYPEXT': $\forall x, y(\forall z(At(z) \rightarrow (z \sqsubseteq x \leftrightarrow z \sqsubseteq y)) \rightarrow x = y)$.
- HYPEXT'': $\forall x, y(\forall z(At(z) \wedge z \sqsubseteq x \rightarrow z \circ y) \rightarrow x \sqsubseteq y)$.

Lemma 10.2.8 *Relative to O, the sentences AT, HYPEXT, HYPEXT' and HYPEXT'' are equivalent.*

There seems to be general agreement that neither AT nor AF should be viewed as a necessary component of a mereological theory (see, e.g., [Goodman, 1951] and [Varzi, 1996]). As a matter of fact, each of them is consistent with CI, but CI + ¬ AT + ¬ AF is consistent, too. However, perhaps for reasons of technical simplicity, AT has probably more often been included in mereological theories. For example, relative to AT, it makes sense to state a version AT-FUS of the fusion schema which may look simpler then FUS (cf. [Eberle, 1970]): Let ψ be a formula in L[$\circ$]; then let

- AT-FUS$_\psi$: $\exists x(At(x) \wedge \psi) \rightarrow \exists y \forall x(At(x) \rightarrow (x \sqsubseteq y \leftrightarrow \psi))$
- AT-FUS := {AT-FUS$_\psi$ | ψ is a L[$\circ$]-formula}.

Then it can be shown that (cf. [Eberle, 1970]):

Corollary 10.2.2 $O + AT + FUS = O + AT + AT\text{-}FUS$.

AT guarantees the existence of atoms, but it remains silent about their number: it could be some natural number, but there could also be infinitely many of them. This can of course be expressed by using counting formulas. That is, by

- $\exists_{\geq n+1} At :\leftrightarrow$
  $\exists x_1 \ldots x_{n+1}(At(x_1) \wedge \ldots \wedge At(x_{n+1}) \wedge x_1 \neq x_2 \wedge \ldots \wedge x_n \neq x_{n+1})$
  ('there are more than n atoms')
- $\exists_{n+1} At :\leftrightarrow \exists_{\geq n+1} At \wedge \neg\exists_{\geq n+2} At$ ('there are $n + 1$ atoms').

2.3 A Synopsis of Mereological Theories in L[∘]

If from the domain of L[∘]-formulas considered in Sections 2.1 and 2.2 the superfluous ones are deleted, the list of the theories which extend CI and contain combinations of the remaining sentences is as follows.

First there are the extensions of CI by AT, AF, and their negations; here the axiom sets are

- Ax(ACI) := Ax(CI) ∪ {AT} ('atomic calculus of individuals')
- Ax(FCI) := Ax(CI) ∪ {AF} ('atom-free calculus of individuals')
- Ax(MCI) := Ax(CI) ∪ {¬AT, ¬AF} ('mixed calculus of individuals').

Second there are extensions of ACI in which the number of the atoms is addressed; here the axiom sets are

- $\text{Ax}(\text{ACI}_{\geq n+1}) := \text{Ax}(\text{ACI}) \cup \{\exists_{\geq n+1} At\}$ $(n \in \mathbb{N})$
- $\text{Ax}(\text{ACI}_{n+1}) := \text{Ax}(\text{ACI}) \cup \{\exists_{n+1} At\}$ $(n \in \mathbb{N})$
- $\text{Ax}(\text{ACI}_{\infty}) := \text{Ax}(\text{ACI}) \cup \{\exists_{\geq n+1} At \mid n \in \mathbb{N}\}$.

Third there are extensions of MCI in which the number of the atoms is addressed; here the axiom sets are

- $\text{Ax}(\text{MCI}_{\geq n+1}) := \text{Ax}(\text{MCI}) \cup \{\exists_{\geq n+1} At\}$ $(n \in \mathbb{N})$
- $\text{Ax}(\text{MCI}_{n+1}) := \text{Ax}(\text{MCI}) \cup \{\exists_{n+1} At\}$ $(n \in \mathbb{N})$
- $\text{Ax}(\text{MCI}_{\infty}) := \text{Ax}(\text{MCI}) \cup \{\exists_{\geq n} At \mid n \in \mathbb{N}\}$.

Moreover, arbitrary instances of FUS may be added to each of these sets as further axioms.

It is not so easy to envisage L[∘]-sentences that are independent from each of these theories. Here is a suggestion:

- DE: $\forall x, y(y \sqsubset x \rightarrow \exists z(y \sqsubset z \wedge z \sqsubset x))$

In non-trivial circumstances, DE expresses density. Yet, it does not deliver anything new:

Lemma 10.2.9 *FCI* $\vdash$ *DE*.

For a general meta-theorem that is relevant here, see Section 4.

2.4 What is a Mereological Theory? History and Systematics

Both 'mereology' and 'calculus of individuals' build on an 'ordinary' understanding of the expressions that are parts of them; but they are nonetheless technical terms, invented by philosophers for specific purposes. Thus, in order to understand these predicates, one should first try to pin down how their inventors and promotors actually used them. In doing this, I present a short history of the work on mereological theories and calculi of individuals.

In [Goodman, 1951], we encounter talk of *the* calculus of individuals. Goodman presents this theory only tentatively, mentioning O, FUS, and NUC as its possible axioms; this amounts to CI + FUS. In some presentations of his work, this axiomatization is adopted: see, e.g., [Shepard, 1973], [Breitkopf, 1978], perhaps [Hottinger, 1988] (which is less explicit than [Goodman, 1951]). In other texts, the definite article is applied, too, but to theories different from CI + FUS, such as CI in [Hodges and Lewis, 1968]. There we also find the constant 'the atomic calculus of individuals' (for ACI); in this respect, [Hellman, 1969] and [Hendry, 1980] concur.

[Eberle, 1967] may be the first text where the indefinite article is used; he talks of 'a calculus of individuals'. Among his calculi of individuals are CI + FUS, ACI + FUS, and a few subtheories of ACI. A collection of the same axiom sets is also put forward in [Eberle, 1970], though this time resting on a version of free logic.[27]

From the 1960s to the 1970s, the majority of the publications on the part-whole relation belonged to the calculus-of-individuals framework. During this time, contributions from the mereology community were primarily comments on or advancements of Leśniewski's theories and thus tended to share their pecularities. It seems that the use of 'mereology' resurfaced, now freed from its commitment to Leśniewski, in the 1980s with two approaches extending the calculus-of-individuals framework. Indeed, in the last 20 years or so, 'mereology' has been much more often used than 'calculus of individuals'.

First, in addition to the theories collected in Section 2.3, proper subtheories of CI – and their extensions – have been systematically investigated and classified as mereologies or mereological theories. It seems to be easier to find interesting examples of such theories in L[$\sqsubseteq$], the first-order language with the two-place predicate '$\sqsubseteq$' as its sole non-logical primitive. Here, (O) is transformed into a definition

- D$\circ$: $x \circ y :\leftrightarrow \exists z(z \sqsubseteq x \wedge z \sqsubseteq y)$.

The identity symbol '=' is treated as a primitive, axiomatized by reflexivity and substitutivity (in L[$\sqsubseteq$]). As mereology-specific base axioms, those for partial orderings[28] are adopted, resulting in a theory usually called 'M' here (see, e.g., [Varzi, 1996], [Hovda, 2009]).

Further examples of theories put forward as mereological ones are (cf. also [Simons, 1987]):

- the theory obtained from M by adding a principle called 'WSP' (called 'MM' in [Hovda, 2009]);
- the theory obtained from M by adding a principle called 'SSP' (called 'EM' in [Varzi, 1996]);
- the theory obtained from EM by adding SUM, PROD, and (NEG′) (called 'CEM' in [Pontow and Schubert, 2006]);
- the theory obtained from EM by adding FUS (called 'GEM' in [Varzi, 1996], [Pontow and Schubert, 2006]).[29]

Lemma 10.2.10

- *EM + (D$\circ$) is equivalent to O + (D=)*
- *CEM is a proper subtheory of CI*
- *GEM + (D$\circ$) is equivalent to CI + FUS + (D=).*

Second, theories that are stated in languages including L[$\sqsubseteq$] (or L[$\circ$]) and extensions of these languages – e.g., CI – were developed and studied. Most of the early[30] examples were conceived of as nominalistic theories (see [Lewis, 1970b], [Shepard, 1973]) or as calculi of individuals (see [Clarke, 1981], [Clarke, 1985]). But from the 1990s onwards, a wealth of papers connecting mereological notions with topological ones has been produced in the mereology framework, resulting in the flourishing area of so called *mereotopology*; see Section 6 for more on this.

In sum, the expressions 'mereological theory' and 'calculus of individuals' are now established as *predicates*. Many theories have been accepted as falling under them. I am not aware of any attempts to lay down general explications for either of these predicates, however. It is merely by examples that their extensions are (partly) determined.

Let me suggest this explication:

Definition 10.2.3 T is a mereological theory (or calculus of individuals) : $\Longleftrightarrow$ T is formulated in L[$\circ$] (or L[$\sqsubseteq$]) and CI $\subseteq T$.

Definition 10.2.3 may be employed only as a convenient abbreviation. As an explication, however, it should not only be faithful to the actual use of 'mereological theory' and 'calculus of individuals', but it should moreover be fruitful, supporting non-trivial meta-theorems. As to the latter, see Sections 3.2, 4, and 5. In addition, some of the possible competitors to Definition 10.2.3 that are built along the same lines are inferior to it.[31] Thus, consider the following theories in L[$\circ$]:

(I) $\mathrm{PL}_1^\circ := \{\psi \mid \psi$ is a sentence from L[$\circ$] and ψ is logically true$\}$.
(II) ZF$^\circ$ is the theory obtained from ZF by replacing (everywhere in L[$\in$]) '$\in$' by '$\circ$'.

Both PL_1° and ZF$^\circ$ are stated in L[$\circ$]; but I regard neither as a mereological theory nor a calculus of individuals. In my view, in order for a theory T to be rightfully called a 'mereological theory' or 'calculus of individuals', two conditions have to be satisfied: (i) many sentences containing '$\circ$' must belong to T which are supposed to be true if '$a \circ b$' is read as 'a overlaps b'; (ii) not too many sentences involving '$\circ$' should belong to T which are not compatible with our reading '$a \circ b$' as 'a overlaps b'. Moreover, we should be disposed to accept and reject these sentences already because of our usual understanding of 'a overlaps b'.

'Many' and 'too many' are vague; nonetheless, (i) and (ii) should suffice to dispose of both PL_1° and ZF$^\circ$ as mereological theories and calculi of individuals. This is in harmony with Definition 10.2.3 . But a similar reasoning suggests that the following definition, for example, should be rejected:

(a) T is a mereological theory : $\Longleftrightarrow$ T is formulated in L[$\circ$] and $\overline{\{O\}} \subseteq T$,

For define '$x \circ y :\leftrightarrow x = y$'. Then $\overline{\{O\}}$ turns out to be a subtheory of a definitional extension of the set of logical truths of first-order logic with identity (in the language L[=] with '=' as its sole predicate).[32] That means that the reading of '$\circ$' as *overlaps* is not at all specified by O. In the light of (i) and (ii), $\overline{\{O\}}$ should not be regarded as a mereological theory.

Such considerations suggest that, although the choice of CI as a base theory in Definition 10.2.3 may seem arbitrary, alternatives to CI should at least not be much weaker than CI. Could they be stronger? If so, L[$\circ$]-sentences that are unprovable in CI should be regarded as evident under their intended reading.

Such sentences may exist, but I wouldn't know which ones they are. In addition, it might be wondered whether all extensions of CI (in L[$\circ$]) should really be classified as mereological theories. Perhaps not. But the only reason for excluding a consistent proper extension T of CI from this domain is that T contains a sentence φ such that $\neg\varphi$ is acceptable as a mereological axiom. But then CI $\cup \{\neg\phi\}$, a proper extension of CI, could replace CI as our core system – contrary to what I have assumed.[33]

Two other modifications of Definition 10.2.3 are obtained by dropping the restriction to L[$\circ$] occuring in it.

(b) T is a calculus of individuals : $\Longleftrightarrow$ CI $\subseteq T$.
(c) T is a mereological theory : $\Longleftrightarrow$ CI $\subseteq T$.

Another example:

(III) Let L[$\circ$] be extended by '$\in$' to L and consider CI + ZF, formulated in L.

According to (b), CI + ZF is a calculus of individuals. Now, one thing seems clear to me: CI + ZF is not a nominalistic theory.[34] In addition, calculi of individuals are accepted as being nominalistic: from Goodman's perspective, where nominalism is conceived of as the rejection of all non-individuals (see [Goodman, 1951]), this assessment is trivial; but it is also plausible if, as, e.g., from Quine's viewpoint, nominalism is taken to admit only what is concrete.[35] Thus, CI + ZF cannot be a calculus of individuals. Thus, (b) is unacceptable. However, (c) may be sustained. After all, CI + ZF *is* a theory incorporating 'part of'; and in the mereology framework, mereological theories are not bound to nominalism.

Nonetheless, I doubt that the mereology community would classify CI + ZF as a mereological theory. If this assessment is right, a *definiens* in between those from Definition 10.2.3 and the alternative explication (c) may still seem plausible. Take an appropriate L extending L[$\circ$]; then:

(d) T is a mereological theory : $\Longleftrightarrow$ T is formulated in L and CI $\subseteq T$.[36]

Now, there is an obvious problem: *For which extensions L of, say, L[$\circ$] and theories T stated in L which extend, say, CI, do such T deserve to be classified as mereological theories?*

Again, no general answer to this question has been formulated, let alone accepted. More seriously, there may be a lack of stable intuitions as to what a convincing answer could be.

3. Models for L[∘]

Mereological algebras are the structures in which the expressions, in particular the formulas, from L[∘] can be evaluated. They are of the form $\langle M, \circ^M \rangle$, with M a non-empty set and $\circ^M$ a two-place relation over M, which is the interpretation of '∘' in $\mathcal{M}$.[37] In this section, mereological algebras are employed to obtain information about mereological theories.

3.1 Boolean Algebras and Mereological Algebras

A *Boolean algebra* is a structure of the form

$$\mathcal{B} := \langle B, \sqcap^B, \sqcup^B, -^B, 0^B, 1^B \rangle$$

Let L[BA] be the first-order language that contains the two-place function symbols '⊓' and '⊔', the one-place function symbol '−', and the constants '0' and '1'. Sentences of L[BA] are evaluated in Boolean algebras. For an axiomatization of BA, the theory of Boolean algebras, in this language, see [Chang and Keisler, 1973]. Given this, a correspondence between Boolean algebras and mereological algebras can be set up as follows:[38]

Definition 10.3.1 Let $(\mathcal{M} =)\ \langle M, \circ^M \rangle \models \mathrm{CI}$, $n \notin M$. Then let

$$\mathcal{M}^{+n} := \langle M^{+n}, \sqcap^{M^{+n}}, \sqcup^{M^{+n}}, -^{M^{+n}}, 0^{M^{+n}}, 1^{M^{+n}} \rangle$$

where

- $M^{+n} := M \cup \{n\}$;
- $0^{M^{+n}} := n$;
- $1^{M^{+n}} :=$ the maximal element (relative to $\circ^M$) of M;
- $a \sqcap^{M^{+n}} b :=$ the product of a and b (in $\mathcal{M}$), if $a, b \in M$ and $a \circ^M b$; n, else;
- $a \sqcup^{M^{+n}} b :=$ the sum of a and b (in $\mathcal{M}$), if $a, b \in M$; a, if $b = n$; b, if $a = n$;
- $-^{M^{+n}} a :=$ the complement of a (in $\mathcal{M}$), if $a \in M$ and $a \neq 1^{M^{+n}}$; n, if $a \in M$ and $a = 1^{M^{+n}}$; $1^{M^{+n}}$, if $a^{M^{+n}} = n$.

Definition 10.3.2 Let $\mathcal{B}$ have the same signature as L[BA]. Then let

$$\mathcal{B}^- := \langle B^-, \circ^{B^-} \rangle$$

where

- $B^- := B \setminus \{0^B\}$,
- $a \circ^{B^-} b :\Longleftrightarrow a \sqcap^B b \neq 0^B$, for $a, b \in B^-$.

Lemma 10.3.1

1. *If $\mathcal{M} \models CI$, $n \notin M$, then $\mathcal{M}^{+n} \models BA$.*
2. *If $\mathcal{B} \models BA$, then $\mathcal{B}^{-} \models CI$.*

This correspondence of models induces a translation from L[$\circ$] to L[BA] which, eventually, leads to a faithful relative interpretation of CI in BA.[39] More explicitly, let the function $\mathcal{J}$ from Fml[L[$\circ$]], the set of formulas of L[$\circ$], to Fml[L[BA]], the set of formulas of L[BA] be inductively defined as follows:[40]

Definition 10.3.3

- $\mathcal{J}(\text{'}x \circ y\text{'}) := \text{'}x \sqcap y \neq 0\text{'}$,
- $\mathcal{J}$ commutes with the propositional operators,
- $\mathcal{J}(\forall x\varphi) := \forall x(x \neq 0 \rightarrow \mathcal{J}(\varphi))$.

Lemma 10.3.2 *If $\mathcal{B} \models BA$, and β an assignment over B^{-}, then for all L[$\circ$]-formulas ψ*

$$\mathcal{B}^{-}, \beta \models \psi \iff \mathcal{B}, \beta \models \mathcal{J}(\psi).$$

Lemma 10.3.3 *For all L[$\circ$]-formulas ψ: $CI \vdash \psi \rightarrow BA \vdash \mathcal{J}(\psi)$.*

The converse holds, too; this rests mainly on the following observation:

Lemma 10.3.4 *If $\mathcal{M} \models CI$ and $n \notin M$, then $(\mathcal{M}^{+n})^{-} = \mathcal{M}$.*

Lemma 10.3.5 *If $\mathcal{M} \models CI$, and β is an assignment over M and $n \notin M$, then for all L[$\circ$]-formulas ψ*

$$\mathcal{M}, \beta \models \psi \iff \mathcal{M}^{+n}, \beta \models \mathcal{J}(\psi).$$

Theorem 10.3.1 *For all L[$\circ$]-formulas ψ: $CI \vdash \psi \iff BA \vdash \mathcal{J}(\psi)$.*

3.2 Applications

By combining these results with pre-existing knowledge about Boolean algebras and the theory BA, several important meta-theoretical results can be established. One is that all the theories listed in Section 2.3 are consistent; another is that each finite extension of CI is decidable.

First application:

Lemma 10.3.6 *Each of the theories ACI_{n+1} and MCI_{n+1} ($n \in \mathbb{N}$), ACI_{∞}, FCI and $MCI_{\infty} + FUS$ is consistent.*

It is intuitively obvious that the theories ACI_{n+1}, ACI_∞, and FCI are consistent. Power-set algebras and the Boolean algebra of the regular open sets of $\mathbb{R}$, supplied with the usual Euclidean topology, establish this result on a formal level. With more complex constructions of the same type, the consistency of the MCI_{n+1} ($n \in \mathbb{N}$) and MCI_∞ + FUS can also be shown.

Second application:

Theorem 10.3.2 *CI is decidable.*

Tarski has shown the decidability of BA (see [Tarski, 1949]). This in conjunction with Theorem 10.3.1 and the recursiveness of $\mathcal{J}$ immediately yields Theorem 10.3.2.

Corollary 10.3.1 *Each finite extension of CI (in L[$\circ$]) is decidable.*

Third application:

Lemma 10.3.7 *FCI is $\aleph_0$-categorical.*

The reason is that the theory of atom-free Boolean algebras is $\aleph_0$-categorical.[41]

4. The Main Meta-Theoretical Results

In this section, some of the main meta-theoretical results on mereological theories are collected.[42] They concern variants of categoricity, maximal consistency, and decidability of these theories. Some of the meta-theorems seem to be known only for extensions of CI + FUS_{AT}, where FUS_{AT} is the following instance of FUS:

- FUS_{AT}: $\exists x\, At(x) \rightarrow \exists z \forall y (z \circ y \leftrightarrow \exists x (At(x) \wedge x \circ y))$.

For the atomistic mereological theories, it is not difficult to get a good grasp of the situation.

Lemma 10.4.1 [43]

1. *For each $n \in \mathbb{N}$, ACI_{n+1} is categorical.*
2. *For each $n \in \mathbb{N}$, ACI_{n+1} is maximally consistent and decidable.*
3. *ACI_∞ is maximally consistent and decidable.*
4. *ACI_∞ is not $\aleph_0$-categorical and not finitely axiomatizable.*

Lemma 10.4.2 *The theories ACI_{n+1} ($n \in \mathbb{N}$) and ACI_∞ are the only maximally consistent extensions of ACI (in* L[$\circ$]*).*

Corollary 10.4.1 *ACI proves each instance of FUS.*

Lemma 10.4.3

1. *For each* L[$\circ$]*-sentence ψ: if for each $n \in \mathbb{N}$, $ACI_{n+1} \vdash \psi$, then $ACI \vdash \psi$.*
2. *Let $\mathcal{E} := \{\mathcal{M} \mid M \text{ is finite } \wedge \mathcal{M} \models CI\}$. Then $Th(\mathcal{E}) = ACI$.*
3. *$Th(\mathcal{E})$ is decidable.*

The situation for FCI is not very different.

Lemma 10.4.4

1. *FCI is $\aleph_0$-categorical.*
2. *FCI is maximally consistent and decidable.*
3. *FCI proves each instance of FUS.*

When it comes to the theories MCI_{n+1}, the composition of models of ACI_{n+1} and FCI is helpful. By this technique, one obtains:

Lemma 10.4.5 *For each $n \in \mathbb{N}$, $MCI_{n+1} + FUS_{AT}$ is $\aleph_0$-categorical.*[44]

Since for each $n \in \mathbb{N}$, $\mathrm{CI} \vdash \exists_{\leq n+1} At \rightarrow \mathrm{FUS}_{AT}$, we even have

Lemma 10.4.6

1. *For each $n \in \mathbb{N}$, MCI_{n+1} is $\aleph_0$-categorical.*
2. *For each $n \in \mathbb{N}$, MCI_{n+1} is maximally consistent and decidable.*
3. *For each $n \in \mathbb{N}$, MCI_{n+1} proves each instance of FUS.*

The theories that are most recalcitrant are extensions of MCI_∞. What can be shown here is this:

Lemma 10.4.7

1. *$MCI_\infty + FUS_{AT}$ is maximally consistent and decidable.*
2. *$MCI_\infty + FUS_{AT}$ is not $\aleph_0$-categorical and not finitely axiomatizable.*

Lemma 10.4.8 *The theories MCI_{n+1} ($n \in \mathbb{N}$) and $MCI_\infty + FUS_{AT}$ are the only maximally consistent extensions of $MCI + FUS_{AT}$ (in* L[$\circ$]*).*

Lemma 10.4.9

1. *For each* L[$\circ$]*-sentence* ψ*: if for each* $n \in \mathbb{N}$, $MCI_{n+1} \vdash \psi$*, then* $MCI + FUS_{AT} \vdash \psi$.
2. $MCI + FUS_{AT}$ *proves each instance of FUS.*

Some of these lemmata can be conjoined to obtain a sort of classification result:

Theorem 10.4.1 *The maximally consistent extensions of* $CI + FUS_{AT}$ *in* L[$\circ$] *are exactly the* ACI_{n+1} *and the* MCI_{n+1} ($n \in \mathbb{N}$)*, plus* ACI_{∞}*, FCI and* $MCI_{\infty} + FUS_{AT}$.

Theorem 10.4.1 has various consequences, some of which are somewhat surprising:

Corollary 10.4.2

1. *For each model* $\mathcal{M}$ *of* $CI + FUS_{AT}$ *there is a complete Boolean algebra* $\mathcal{B}$ *such that* $\mathcal{B}^{-} \equiv \mathcal{M}$.[45]
2. *Each maximally consistent extension of* $CI + FUS_{AT}$ *is decidable.*
3. $CI + FUS_{AT}$ *proves each instance of FUS.*
4. $CI + FUS + DE = ACI_1 \cap FCI$.

5. On the 'Strength' of Mereological Theories

I do not know if talk of *measuring the strength of a theory* makes sense. But theories can certainly be *compared* with respect to their strength: in particular, some can be stronger than others. Now, there are several suggestions for an *explicans* of 'T is at least as strong as S' – or 'S is reducible to T' – which are well known: ordinals may be assigned to the theories and compared, and proof-theoretic reducibility and (provable) relative consistency are options; but relative interpretability with its many variants also comes to mind.[46]

Roughly stated, a relative interpretation of a theory S in a theory T is a function $\mathcal{I}$ from L[S] to L[T] that preserves the quantificational structure of the L[S]-formulas (while relativizing quantifiers) and that maps S-theorems to T-theorems. More precisely, a somewhat restricted version (which suffices here) can be defined as follows:[47]

Definition 10.5.1 Let S, T be theories in first-order languages L[S] and L[T] that contain finitely many relation signs. Assume that for each k-place relation sign 'R' in L[S] there is a k-place formula ψ_{R} in L[T], such that for all relation signs R, R$'$, if $\psi_{\mathrm{R}} = \psi_{\mathrm{R}'}$, then R = R$'$. Let δ be a fixed one-place formula in L[T]. Then

I is a relative interpretation of S in T with respect to δ if I :Fml[L[S]] $\rightarrow$ Fml[L[T]] and I is primitive recursive and

1. for all n, m, $I(v_n = v_m) = (v_n = v_m)$ (if '=' belongs to L[S] and L[T]),
2. for each k-place relation sign R in L[S], $I(\mathrm{R}(v_{i_1}, .., v_{i_k})) = \psi_{\mathrm{R}}(v_{i_1}, .., v_{i_k})$,
3. for all formulas φ, ψ in L[S], $I(\neg\varphi) = \neg I(\varphi)$ and $I(\varphi \rightarrow \psi) = I(\varphi) \rightarrow I(\psi)$,
4. for all formulas φ in L[S] and all variables u, $I(\forall u \varphi) = \forall u\, (\delta(u) \rightarrow I(\varphi))$,
5. for all sentences φ in L[S], if $S \vdash \varphi$, then $T \vdash I(\varphi)$,
6. $T \vdash \exists x \delta(x)$.

   In addition, *I is a faithful relative interpretation of S in T with respect to δ* if I is a relative interpretation of S in T with respect to δ and, for all sentences φ in L[S], if $T \vdash I(\varphi)$, then $S \vdash \varphi$.

Definition 10.5.2

- $S \preceq^{\delta} T :\Longleftrightarrow \exists \mathcal{I}$ ($\mathcal{I}$ is a relative interpretation of S in T w.r. δ).
- $S \preceq T :\Longleftrightarrow$ S is relatively interpretable in $T :\Longleftrightarrow$ there is a formula δ with $S \preceq^{\delta} T$.

The mapping $\mathcal{J}$ treated in Section 3.1 is a relative interpretation of CI in BA. Of the inter-theoretic relations considered above, it is only relative interpretability and its variants that are of any use when comparing mereological theories with each other and with other theories. Moreover, I think that in general, relative interpretability (in particular) is preferable to its alternatives as a relation of reducibility: see [Niebergall, 2000].

It has already been mentioned in the introduction that for the research on mereological theories, the question whether a mereological treatment (or foundation) of mathematics is possible is of particular importance. To give a positive answer, it must be possible to develop at least sets and natural numbers in a mereologically admissible way.[48] Given the above remarks, I propose the following claims as precise renderings of this aim:[49]

- (MRset) For each consistent set theory S there is a consistent mereological theory T such that S is relatively interpretable in T,
- (MRnumber) For each consistent number theory S there is a consistent mereological theory T such that S is relatively interpretable in T.

In order to argue *for* (MRset) or (MRnumber), an explication of 'S is a set theory' or 'S is a number theory' has to be provided. But it may be conjectured that, for example, (MRnumber) is false. In this case, one may attempt to show that it is *very false*. Now this can be done by exhibiting a particularly weak theory which, intuitively, is classified as a number theory, even if one does not have a general explication of 'number theory' at one's disposal, and by showing that no consistent mereological theory exists that interprets this weak theory.

The next subsection contains examples of such number theories S for which, indeed, no consistent mereological theory T exists such that S is relatively interpretable in T. And in the subsequent subsection, set theories are presented for which analoguous meta-theorems hold. I regard these results as sort of a 'proof' that a mereological foundation of mathematics is impossible. Let me emphasize that this by no means implies the impossibility of *nominalistic* foundation of mathematics.[50]

5.1 Natural Numbers

The paradigmatic number theory is PA; for its axioms, see [Hájek and Pudlák, 1993]. An important subtheory of PA is Q (i.e., *Robinson arithmetic*; see [Tarski et al., 1953], [Monk, 1976]), which is axiomatized by

- $\forall x(Sx \neq 0)$
- $\forall x, y(Sx = Sy \rightarrow x = y)$
- $\forall x(x + 0 = x)$
- $\forall x(x + Sy = S(x + y))$
- $\forall x(x \times 0 = 0)$
- $\forall x(x \times Sy = x \times y + x)$
- $\forall x(x \neq 0 \rightarrow \exists y\, Sy = x)$.

Experience teaches that Q is pretty much the greatest lower bound for those theories that are not only taken as the object of investigation, but also as the means to do number theory.[51] Therefore, the following result is of relevance for (MRnumber) and its variants:

Theorem 10.5.1 There is no consistent mereological theory in which Q is relatively interpretable.

This result can be extended in several ways. *First*, theories weaker than Q can be taken into account. Thus, consider the theory of discrete linear orderings with minimum and no maximum, which is sometimes called 'DIL'. DIL is $\mathrm{Th}(\langle \mathbb{N}, \leq \rangle)$ in the appropriate language, whence maximally consistent and decidable. We therefore have DIL $\preceq$ Q, yet not Q $\preceq$ DIL.

Theorem 10.5.2 There is no consistent extension T of CI + FUS_{AT} (in $\mathrm{L}[\circ]$) in which DIL is relatively interpretable.

Second, relative interpretability may be replaced by wider intertheoretic relations. Thus, consider the liberalization of relative interpretability obtained by deleting its quantifier-clause. That is, let the preconditions of Definition 10.5.1 be given (with the exception of the assumption of δ); then I *is a* $\neg$-$\wedge$-*translation from* S *to* T is defined by clauses (1)–(3) and (5) from Definition 10.5.1. And S *is* $\neg$-$\wedge$-*translatable in* T if, and only if, $\exists I(I$ is a $\neg$-$\wedge$-translation from S to $T)$.

$\neg$-$\wedge$-translatability is very liberal: ZF, for example, is $\neg$-$\wedge$-translatable in Q (see [Pour-El and Kripke, 1967]), but, of course, it is far from being relatively interpretable into Q. Yet even $\neg$-$\wedge$-translatability does not identify Q with extensions of CI + FUS_{AT}.

Theorem 10.5.3 There is no consistent extension T of CI + FUS_{AT} (in L[$\circ$]) in which Q is $\neg$-$\wedge$-translatable.

5.2 Sets

The paradigmatic set theories are Z, ZF, and ZFC; for their axioms, see [Kunen, 1980]. A weak subtheory of ZF, called 'S' here (following [Monk, 1976]), is axiomatized by:

- $\exists x \forall y \neg(y \in x)$
- $\forall x, y(\forall z(z \in x \leftrightarrow z \in y) \rightarrow x = y)$
- $\forall x \exists z \forall u(u \in z \leftrightarrow u \in x \vee u = x)$.

Like Q, S has no finite models; but intuitively, it does not prove the existence of infinitely large sets. It seems to be among the weakest theories which still deserve to be called 'set theory'.

Lemma 10.5.1 [52] Q is relatively interpretable in S.

We therefore also have:

Theorem 10.5.4

1. There is no consistent mereological theory in which S is relatively interpretable.
2. There is no consistent extension T of CI + FUS_{AT} (in L[$\circ$]) in which S is $\neg$-$\wedge$-translatable.

6. Extensions of the Mereological Framework

The domain of the theories treated above as mereological ones can be and has been extended. There are essentially two ways of carrying out this idea:

(I) allow T to be stated in a language L obtained from, say, L[$\circ$] through the addition of new vocabulary: in particular, (i) add new propositional operators, (ii) add new quantifiers, (iii) add new (non-logical, descriptive) vocabulary or, (iv) extend L[∘] to a higher-order language;

(II) allow *T* to be an extension of a theory 'weaker' than CI.

Indeed, for each of these options, theories have been put forward that their authors have classified as calculi of individuals, as mereological, or as nominalistic theories. Whether this is appropriate has been partly discussed in Sections 1 and 2.4. This concluding section consists primarily of pointers to the relevant literature and contains some sketchy comments on other themes.

As to (I)(i), one may look at modal and temporal operators: see, e.g., [Simons, 1987]. Contributions to (I)(ii) are [Martin, 1943] and [Field, 1980], though the latter is more about nominalistic theories. Coming to (I)(iv), relevant examples for theories stated in higher-order languages can be found in [Leonard and Goodman, 1940], [Field, 1980], and [Lewis, 1991], but also in, e.g., [Clarke, 1981], [Clarke, 1985], [Biacino and Gerla, 1991], and [Pontow and Schubert, 2006]. [Niebergall, 2009b] contains an approach to a general treatment of extensions of CI formulated in monadic second-order languages containing '∘'.

Among all the suggestions mentioned under (I), (I)(iii) has most often been dealt with. Among others, L[∘] has been extended by:[53]

- topological vocabulary (resulting in mereotopological theories):[54] '*x* is a sphere' ([Tarski, 1929]), '*x* is next to y' ([Lewis, 1970b]), '*x* is connected to/with y' ([Clarke, 1981], [Clarke, 1985], [Biacino and Gerla, 1991], [Roeper, 1997]), '*x* is a connection' ([Bochman, 1990]), '*x* is connected' ([Pratt and Lemon, 1997], [Pratt and Schoop, 1998], [Pratt and Schoop, 2000]), '*x* and *y* are in contact' ([Pratt and Schoop, 2000], [Pratt-Hartmann and Schoop, 2002]), '*x* is an interior part of y' ([Kleinknecht, 1992], [Smith, 1996], [Forrest, 2010]), '*x* is a region' ([Eschenbach, 1994], [Varzi, 1996], [Ridder, 2002]), '*x* is a boundary for y' ([Smith and Varzi, 2000]), '*x* coincides with y' ([Smith and Varzi, 2000]), '*x* is limited' ([Roeper, 1997]);
- geometrical predicates: '*x* is a sector (i.e., segment) of y' ([Glibowski, 1969]),[55] '*x* precedes y' ([Mortensen and Nerlich, 1978], [van Benthem, 1983]);[56]
- predicates dealing with size: equivalences in general ([Janicki, 2005]), '*x* is the size of y' ([Shepard, 1973]), '*x* is of equal (aggregate) size as y' ([Goodman, 1951], [Breitkopf, 1978]), '*x* is bigger than y' ([Goodman and Quine, 1947]), '*x* is longer than y' ([Martin, 1958]), '*x* contains fewer points than y' ([Field, 1980]);
- means of *composition* different from fusion: token-concatenation belongs here (see [Goodman and Quine, 1947], [Martin, 1958], [Niebergall, 2005]),

but there may be other forms as well (see [Fine, 1994], [Fine, 1999], and [Janicki, 2005]);

- specific predicates: for example, 'is with' ([Goodman, 1951], [Breitkopf, 1978]), 'x is the singleton of y' ([Lewis, 1991]) and 'x is the unicle of y' ([Bunt, 1979]), syntactical predicates like 'x is a variable', 'x is a left parenthesis', 'x is a stroke' (see [Goodman and Quine, 1947] and [Martin, 1958]), '(ontological) dependence' and 'foundation' (in various forms; see [Simons, 1987], [Fine, 1995], and [Ridder, 2002]), temporally relativized part-of relations ([Simons, 1987], [Fine, 1999]).

The contributions to (II) fall into two groups: we find weakenings of the background logic and, more importantly, weakenings of the specifically mereological axioms when compared with Ax(CI). The latter approach is addressed in Section 2.4. Concerning the former, both free logic[57] (see [Eberle, 1970] and [Simons, 1991]) and intuitionistic logic (see [Tennant, forthcoming]) have been suggested.

Of course, all of these ways of extending the austere framework of pure mereological theories can be combined, and some have been. In fact, already in [Tarski, 1929] we encounter a higher-order theory containing mereotopological vocabulary, built on mereogeometrical axioms. Since mereotopology seems to provide the most successful extended framework, let me close this article with a few remarks on mereotopological theories.

First, againt the background of the set-theoretical definition of 'x is a topological space' discovered by Kuratowski, it would be most natural to take a closure operator (or predicate) as the new primitive for mereotopological theories. Thus, let L[$\circ, c$] be the extension of L[$\circ$] by the one-place function sign 'c' (read 'the closure of') and consider these axioms:

- (AxTop): $\forall x(x \sqsubseteq cx); \forall x(ccx = cx); \forall x, y(c(x \sqcup y) = cx \sqcup cy)$.[58]

Intuitively, (AxTop) should be convincing; and it could well serve as the core of the topological component of mereotopological theories. Yet, (AxTop) has found only few adherents (perhaps [Grzegorczyk, 1951], [Smith, 1996], [Smith and Varzi, 2000]). As can be seen from the above list of predicates, other topological primitives are usually adopted; but there still seems to be no agreement as to which ones should be chosen.

Second, being formulated in different languages, mereotopological theories can often be compared only *via* the subtheory-of-a-definitional-extension relation or *via* relative interpretability. Fortunately, the above-mentioned 'x is an interior part of y', 'the closure of x', and 'x is connected to y', for example, seem informally to be interdefinable. In fact, formalized versions of such definitions yield that many of the theories addressed in the above list are subtheories of each other (modulo the *intended* definitions). But more can be shown.

Third, a repeatedly presented motivation for the development of mereotopological languages and theories is an alleged lack of expressive power of mereological theories: the added topological vocabulary belonging to L should allow for distinctions that seem to be unattainable in L[$\circ$] (see, for example, [Varzi, 1996]). But then, topological predicates should not be definable purely mereologically. More explicitly:

Let L be any of the mereotopological extensions of L[$\circ$] considered above, α being the newly introduced predicate, and let Δ be a set of axioms containing α; then there should exist a mereological theory S (in L[$\circ$]) such that $S + \Delta$ (in L) is no subtheory of a definitional extension of S.

In what follows, let S be a consistent mereological theory. Now set α := 'c' and Δ := (AxTop), and define $cx := x$. Or set α := 'C' (for 'is connected with') and Δ := the axioms C1–C5, C7 from [Varzi, 1996] (this is also relevant for [Clarke, 1981]), and define '$Cxy :\leftrightarrow x \circ y$'. Alternatively, set α := 'IP' (for 'is an interior part') and Δ := the axioms AIP1–AIP6 from [Smith, 1996], and define '$IPxy :\leftrightarrow x \sqsubseteq y$'. Finally, one may set α := '$<<$' (for 'is an interior part') and Δ := the axioms A1–A6 from [Forrest, 2010], and define '$x << y :\leftrightarrow x = y$'.

In each of these cases, $S + \Delta$ turns out to be a subtheory of a definitional extension of S.

It is granted that the definitions employed here are not the intended ones. But they are compatible with the mereotopological axiom systems. Surely, their authors wanted them to add something new to the mereological theories; but the mereotopological calculi do not deliver.

Fourth, additional mereotopological axioms can hardly be as evident as the ones belonging to the various core systems. In this situation, one may aim at mereotopological theories of specific topological spaces (spaces, for example, which are like the space-time we live in): this is done, in particular, in [Pratt and Lemon, 1997], [Pratt and Schoop, 1998], [Pratt and Schoop, 2000], and [Pratt-Hartmann and Schoop, 2002].[59] As I understand most of the philosophers and logicians working in the area of mereotopology, however, they are interested in theories of an intermediate level of generality: theories that on the one hand hold in many structures that might intuitively be regarded as spaces (and, thus, should not be maximally consistent) and theories that on the other hand go distinctly beyond the core systems considered above. Yet, it may not be easy to find good examples for such theories. Actually, only a few have been developed: see [Kleinknecht, 1992], [Smith, 1996], and [Smith and Varzi, 2000]. Now even among these, not all can be sound (relative to the above-mentioned preferred readings of their vocabularies); for (apart from merely notational differences) some of them are inconsistent with each other: take the theories from [Kleinknecht, 1992] and [Smith, 1996].

If linguistic intuitions differ so markedly at this point, it could be that we eventually have to stay with the core systems. But in this case, what is the sense of the project of mereotopology?

Notes

1. As with 'x is a subset of y', it is common to understand 'x is a part of y' as 'x is a proper part of y or $x = y$'.
2. This reminds us of the subset-relation. Indeed, there is a close formal connection between 'x is a part of y' and 'x is a non-empty subset of y' – so close that one may view non-empty subclasses of classes as parts of classes; see [Lewis, 1991] for an elaboration of the idea. It has, however, no effect on this article.
3. Occasionally, Husserl ([Husserl, 1913]) and Whitehead (e.g., [Whitehead, 1929]) are also mentioned in this connection; but they did not carry out a formalized treatment.
4. [Simons, 1987] provides a convenient route into Leśniewski's mereology.
5. This is the potted history that is usually told. It is, however, clouded by an early paper by Martin ([Martin, 1943]), which already contains a theory stated in a first-order theory with additional quantifiers, which may be interpreted as a mereological theory.
6. This reminds us of the term 'logic', which unfortunately is also usually not defined and, furthermore, used in various non-equivalent ways: for instance, as referring to a discipline, as synonymous with 'formal language', and as synonymous with 'set of logical truths' – or perhaps even as something else.
7. Sometimes, this ambiguity can be found in a single text: see, e.g., [Janicki, 2005].
8. With respect to the latter, this is indeed Goodman's conception from [Goodman, 1951] onwards.
9. In these writings, both predicates are mentioned in the same breath, however: cf. [Bochman, 1990], [Janicki, 2005] and, in particular, [Leonard and Goodman, 1940]. For an explicit discussion of their relation, see [Simons, 1991].
10. For example, in [Pontow and Schubert, 2006], which belongs to the mereological community, Goodman is not even mentioned.
11. This implies neither that 'part of' is a logical or a mathematical expression, nor that all the conceptual truths containing it are logical truths or mathematical truths. All the same, mereological theories may be regarded as in some sense non-empirical theories: they are not tested by observation (in the long run), but are rather reports of conceptual reflections.
12. Even if one is not a nominalist, one may regard it as contrived that so often empirical theories should include mathematical theories as components. For an elaboration of what science without numbers could be, see especially [Field, 1980].
13. This is also noticed in [Varzi, 1996].
14. From the many contributions to the subject of mereology, some are not even sketchily treated in this article. Among them are texts on its history (see [Burkhardt and Dufour, 1991], [Henry, 1991], [Libardi, 1994]), on its application to mass terms (see [Pelletier, 1979], [Bunt, 1979], [Simons, 1982]), and writings of a purely mathematical and writings of a non-formal character. For the 'modern classics' (i.e., de Laguna, Husserl, Leśniewski, Tarski, and Whitehead), [Ridder, 2002] is a good starting point.
15. Actually, the few existing proof- and model-theoretical investigations on mereological theories are scattered throughout the literature. Thus, even in the quite comprehensive monographs from [Simons, 1987] and [Ridder, 2002], several of them are not even mentioned.

16. In this case,
$$x = y \leftrightarrow \forall z(z \circ x \leftrightarrow z \circ y)$$
must be added.
17. This should be satisfying, at least in the case of concrete objects. Occasionally, a null object is nonetheless postulated. This seems to be for technical simplification, however: see, e.g., [Martin, 1965], [Roeper, 1997].
18. The objects whose existence is claimed in these sentences are, moreover, uniquely determined.
19. Note that a seemingly plausible variant of SUM, i.e.,
$$\forall x, y \exists z \forall u(u \sqsubseteq z \leftrightarrow u \sqsubseteq x \vee u \sqsubseteq y),$$
is inadequate.
20. Thus, we have no principled caveat against the assumption of such sums of concrete objects. For a sum to exist, it seems to be sufficient that we have a name which we regard as referring to it. But we allow ourselves to introduce such a name whenever we want it.
21. If T is a first-order theory, Ax(T) is a set of sentences whose deductive closure is T; i.e., $\overline{\mathrm{Ax}(T)} = T$. Furthermore, $\Sigma + \Sigma' := \overline{\Sigma \cup \Sigma'}$, if Σ and Σ' are sets of first-order formulas.
22. This is similar to the relation of a general, language-transcendent induction principle to the induction schema in the language of PA (first-order Peano arithmetic).
23. If sentences are preferred as axioms, the universal closures of such formulas may be taken.
24. '$\exists x\, \psi$' may look like an awkward premise in FUS$_\psi$, but it cannot be deleted: for
$$\mathrm{O} + \{\exists z \forall y(z \circ y \leftrightarrow \exists x(x \circ y \wedge \psi)) \mid \psi \text{ is a L}[\circ]\text{-formula}\}$$
is inconsistent.
25. See, in principle, [Hellman, 1969].
26. See [Goodman, 1958]. HYPEXT is for 'hyperextensionality'.
27. Hendry ([Hendry, 1982]) also notices that several theories are called calculi of individuals. Unfortunately, the theory eventually adopted by him as *the* calculus of individuals is not precisely presented.
28. That is, '$\forall x(x \sqsubseteq x)$', '$\forall x, y, z(x \sqsubseteq y \wedge y \sqsubseteq z \rightarrow x \sqsubseteq z)$', and '$\forall x, y(x \sqsubseteq y \wedge y \sqsubseteq x \rightarrow x = y)$'.
29. In [Pontow, 2004] and [Hovda, 2009], variants of FUS are discussed which are provably equivalent to FUS relative to CI, but fail to be so in, e.g., MM.
30. 'Classical' predecessors are [De Laguna, 1922] and [Whitehead, 1929]; see [Ridder, 2002].
31. See [Niebergall, 2005] for further remarks, in particular on the futility of model-theoretic explications.
32. (D=) is also correct under this definition.
33. Actually, I take it that in the discipline of mereology, one does not aim at the investigation of just one structure (up to isomorphism), but rather tries to ascertain what is common to various structures that share features that are realized in a variety of cases.
34. It is granted, though, that a precise explication of 'T is a nominalistic theory' is unfortunately missing. *Explicantia* have been suggested by Goodman and his followers (see [Goodman, 1956], [Yoes Jr., 1967], [Eberle, 1968], [Schuldenfrei, 1969], [Eberle, 1969]). I find them unconvincing: some are *ad hoc* (see [Rosenberg, 1970]), some are not clear, and some may have the consequence that CI + ZF is a nominalistic theory.
35. Actually, I think that calculi of individuals are perfectly plausible when interpreted as theories about the part–whole relation as a relation between concrete objects; but I doubt that they provide a similarly convincing explication of 'individual'.

36. Let me remark that, in [Pratt and Schoop, 2000] and [Pratt-Hartmann and Schoop, 2002], theories stated in languages L obtained from, e.g., L[$\circ$] through the addition of 'topological' vocabulary are expressly classified as *mereotopological* (and not mereological) ones.
37. Given $\mathcal{M}$, '=' is always assumed to be evaluated as the identity relation (over M).
38. As far as I know, Tarski was already aware of this; an early presentation can be found in [Grzegorczyk, 1955].
39. See Section 5 for the general definition.
40. It can be assumed that L[$\circ$] and L[BA] have the same variables.
41. I think that this reasoning was first noticed by Pühringer.
42. When it comes to the intertheoretic relations involving such theories, see Section 5.
43. For (1) to (3), see [Hodges and Lewis, 1968] and [Hendry, 1982]. (4) is contained in [Hellman, 1969].
44. As I understand [Hendry, 1982], it contains a sketch of this result for $\mathrm{MCI}_{n+1} + \mathrm{FUS}$.
45. A Boolean algebra $\langle B, \sqcap^B, \sqcup^B, -^B, 0^B, 1^B \rangle$ is *complete* if each non-empty subset A of B has a least upper bound in $\langle B, \sqcap^B, \sqcup^B, -^B, 0^B, 1^B \rangle$. In some texts, the following claim can be found (see, e.g., [Varzi, 1996]):

    - For each model $\mathcal{M}$ of $\mathrm{CI} + \mathrm{FUS}_{AT}$ there is a complete Boolean algebra $\mathcal{B}$ such that $\mathcal{B}^- \cong \mathcal{M}$.

    Since complete Boolean algebras that are infinite cannot be countable, the Löwenheim–Skolem Theorem entails that this claim is false.
46. See [Niebergall, 2000] for a more detailed presentation and comparison of these proposals. Let me remark that I certainly do not claim that 'T is at least as strong as S' and 'S is reducible to T' are in any sense equivalent with each other.
47. This rests on [Feferman, 1960]; for the original treatment, see [Tarski et al., 1953]; for a recent overview, see [Joosten and Visser, 2000], and see [Hájek and Pudlák, 1993]. Some simplifying assumptions are made here: for example, expressions of L[S] or L[T] are at times taken to be (Gödel-) numbers; and only '$\neg$', '$\rightarrow$', and '$\forall$' are adopted as primitive logical operators.
48. The list could be extended by functions and ordered pairs; but a mereological treatment of inductive definitions and infinity is also desirable. In the approach suggested here, that means that convincing axioms for functions, ordered pairs, theories of inductive definitions and the (in-)finite should be found that are relatively interpretable in mereological theories. For ordered pairs, the axioms are more or less clear; and for finiteness, axioms stated in an extension of L[$\circ$] are proposed in [Niebergall, 2009a]. In the other cases, it is not that evident what plausible choices of axiom-systems could be; see [Martin, 1943] and [Goodman and Quine, 1947] for further remarks.
49. See [Niebergall, 2007] for more on this.
50. Token-concatenation theories (see [Goodman and Quine, 1947]) and perhaps also geometrical theories (e.g., theories of space-time; see [Schwabhaüser et al., 1983]) should be regarded as nominalistic theories which are neither mereological theories nor calculi of individuals.
51. This can be supported by meta-mathematical results; see [Hájek and Pudlák, 1993].
52. See [Monk, 1976], [Montagna and Mancini, 1994].
53. Several of the following predicates may be classified under different headings, and the classification itself is vague.
54. To be precise, in some cases 'part' has been introduced by definition; see, e.g., the theories put forward in [Clarke, 1981], [Clarke, 1985] and [Forrest, 2010], plus some of the theories considered in [Varzi, 1996].
55. Initiated by Tarski, there is a large amount of important work on geometrical theories which are stated in a first-order language L with the primitives 'x lies between y and

z′ and 'the distance between x and y is the same as the distance between u and v' (see [Schwabhaüser et al., 1983]). These theories can be regarded as nominalistic (cf. [Field, 1980]); yet L does not extend L[∘].

56. Whereas for most of the mereotopological predicates a spatial interpretation is natural, 'precedes' may well receive a temporal reading.
57. When it comes to the explication of 'logical truth', I am certainly a supporter of one of the main ingredients of the *free logic* project: that existence claims cannot be logical truths (see Chapter 4). But here, where the mereological theories investigated contain specific existence assumptions anyway, I regard a restriction to free logic as superfluous. Moreover, it leads to theoretical complications (e.g., which one to choose?) and does not even simplify the notation.
58. I use '⊔' for the mereological sum.
59. Here, [Grzegorczyk, 1951] is also of interest: in this paper, a theory T in L[∘, c] extending CI + (AxTop) is presented (which is not supposed to be intuitively sound) for which Grzegorczyk claims that it relatively interprets Q. Thus, the conceptual framework given by CI + (AxTop) in L[∘, c] is much richer, has much more potential with respect to proof-theoretic strength, than that of the mereological theories.

11 The Logic of Necessity

John Burgess

Chapter Overview

1. Senses of Necessity	299
2. Propositional Modal Logic: Proofs	303
3. Propositional Modal Logic: Models	307
4. Propositional Modal Logic: Interpretation	315
5. Quantified Modal Logic	317
6. Books and Papers of Particular Note	322
Notes	323

1. Senses of Necessity

Modal logic is concerned with the distinction between what merely is and what in one or another sense necessarily must be. It was pursued throughout the ancient and medieval periods, but modality is ignored by classical logic (modern textbook logic), which was developed for the analysis of mathematical arguments, where modality plays no role. The creation of modern modal logic was nonetheless a response to the development of classical logic rather than a revival of ancient and medieval tradition, which was not well understood until the advance of modern modal logic inspired historical scholars to reexamine it. This chapter will treat only modern developments.

Modal logic adds a symbol $\Box$ for necessity to classical logic's list of symbols $\neg$ and $\wedge$ and $\vee$ and $\rightarrow$ for negation and conjunction and disjunction and conditional. Other modal notions, some with symbols of their own, may then be defined, as in Table 11.1. (In the table, A and B appear in places grammatically appropriate for sentences except in the last two lines, where they appear in places appropriate for nominalizations of sentences. In principle, $A \circ B$ ought to be 'its being the case that A is compossible with its being the case that B' and similarly for $A \Rightarrow B$. In practice, the nominalizing phrase required by grammar is left tacit.)

TABLE 11.1 Definitions of some modal notions

Modal Notion	Definition	Symbol
it is necessary that A	$\Box A$	
it is impossible that A	$\Box\neg A$	
it is contingent whether A	$\neg\Box\neg A \wedge \neg\Box A$	
it is possible that A	$\neg\Box\neg A$	$\Diamond A$
A is compossible with B	$\neg\Box\neg(A \wedge B)$	$A \circ B$
A necessitates B	$\Box(A \to B)$	$A \Rightarrow B$

In natural languages, modal distinctions are expressed through verb-inflections making grammatical moods such as indicative and subjunctive, and/or through modal auxiliaries such as the English 'must' and 'may'. Different senses of necessity are often expressed similarly. We need to distinguish a few.

Epistemic Necessity: With 'She must have gone, but he may have stayed', meaning 'Given what we know, she must have gone, but for all we know, he may have stayed', we have knowledge-related or *epistemic* possibility and necessity. Their logic, though in principle part of modal logic in the broad sense, has its own flavour, and is in practice treated separately as 'epistemic logic'. It has its own chapter in this book.[1]

Deontic Necessity or Obligation: With 'He must stay, but she may go', meaning 'He is obligated to stay, but she is permitted to go', we have duty-related or *deontic* modality. Deontic logic, too, is generally treated as a separate subject. Modal logic in the narrower sense, concerned with 'vanilla' modality as opposed to the epistemic and deontic flavours, is called truth-related or *alethic* modal logic. The terminology epistemic/deontic/alethic was popularized by [von Wright, 1951].

Necessity *Tout Court* or Metaphysical Necessity: The label 'alethic' conceals distinctions. Since [Kripke, 1972b], many use 'necessity' *sans phrase* for what both is and inevitably would have been (could not have failed to be) even if the world had been otherwise, and 'possibility' *sans phrase* for what either is or isn't but potentially might have been (need not have failed to be) if only the world had been otherwise. When a distinguishing epithet is wanted, these are called *metaphysical* modalities. Within the 'alethic' category, they contrast with logical modalities, which concern not the question 'What if the world had been otherwise?' but the question 'What can without contradiction be supposed about how the world actually is?' But the distinction has often been overlooked. (Stock example: It is logically possible that water is not H_2O, since there is no internal contradiction in the traditional view that water is an element, but it is metaphysically necessary that water is H_2O, since imagining a world in which some liquid with a composition other than H_2O fills lakes and is called 'water' is imagining a world in which some liquid other than water fills lakes and is called 'water', not a world in which water has some composition other than H_2O.)

Linguistic Necessity or Analyticity: Logical necessitation is distinctively called *implication*. But the label 'logical' itself conceals distinctions. Logicians tend to use 'logical necessity' more narrowly than philosophers, to cover 'No unmarried man is married', where all that matters is logical form, but not 'No bachelor is married', where meaning as well as form is pertinent. But there is an established expression for the broader notion, *analyticity* or *linguistic* necessity. The narrower can be called *formal* necessity.

Validity *versus* Demonstrability: This last label, too, conceals a distinction, between the nonepistemological notion of what is *true* by virtue of form alone and the epistemological notion of what is *verifiable* by virtue of form alone. Logical theory analyses the former as being true in all models, the latter as having a proof. No other labels being on offer for the contrasting pair of intuitive notions, we may use the ones for the contrasting pair of technical notions, as in Table 11.2. (In the table, consequence and deducibility are properly speaking the *converses* of implication in the model-related and proof-related senses. For classical first-order logic, the completeness theorem guarantees that validity and demonstrability coincide; in other cases, notably second-order logic, there is no completeness theorem.)

TABLE 11.2 Model-related and proof-related counterparts to a few modal notions

Modal Notion	Model-Related	Proof-Related
necessity	validity	demonstrability
impossibility	unsatisfiability	inconsistency
possibility	satisfiability	consistency
compossibility	(joint) satisfiability	(joint) consistency
implication	consequence	deducibility

Some terminological pedantry is justifiable with modern modal logic, since it historically has been plagued by terminological misunderstandings, and indeed originated with one. Today $\neg$ and $\wedge$ and $\vee$ and $\rightarrow$ are generally pronounced 'not' and 'and' and 'or' and 'if', but about a century ago some of the founders of classical logic were erroneously pronouncing $\rightarrow$ as 'implies'. Modern modal logic began with C. I. Lewis, correctly noting that $\rightarrow$ does not express logical necessitation, and fatefully proposed adding a symbol $\Rightarrow$ that would. (The first major work of modern modal logic was [Lewis, 1918]. Here the single and double arrow, $\rightarrow$ and $\Rightarrow$, are being used as substitutes for older symbols, the horseshoe and fishhook, $\supset$ and $\prec$, used there.)

In classical logic, logical modalities are not notions for which there are symbols used in formulas, but jargon used in technical English in speaking *about* formulas. They belong not to the 'object language' but to the 'metalanguage'. Lewis differed from classical logicians by wanting to move into the object language something classical logicians wanted to leave in the metalanguage. That was the only real difference, but Lewis read $A \rightarrow B$ as '*A materially implies B*' and $A \Rightarrow B$ as '*A strictly implies B*', and thus created an appearance of a conflict between two theories about a single topic.

There are genuine dissenters who hold that $A \rightarrow B$ as understood by classical logicians does not adequately represent the conditional in ordinary language outside mathematics, and who have developed non-classical *conditional* logics important enough to have their own chapter in this book.[2] There are also genuine dissenters who object to the classical identification of A implying B with its being logically necessary that $A \rightarrow B$, since this counts in the degenerate cases where premise A is logically impossible and/or conclusion B logically necessary. They have developed so-called *relevance* or *relevant* logics.[3] But Lewis was neither kind of dissenter. He seems never to have evaluated or even considered the reading of $\rightarrow$ as 'if' rather than 'implies', and he strongly defended counting in the degenerate cases of implication. Each principle that classical logic would formulate metalinguistically has an object-language counterpart in modal logic, and Lewis accepted all of them, including the counterparts of the principles admitting the degenerate cases, $\Box\neg A \rightarrow (A \Rightarrow B)$ and $\Box B \rightarrow (A \Rightarrow B)$.

Since formulas involving $\Box$ or $\Diamond$ or $\Rightarrow$ do not occur in the classical object language, claims about logical necessity or impossibility or implication for such formulas do not occur in the classical metalanguage. Hence the object-language counterparts of classical principles do not include formulas with *nested* modalities, boxes or diamonds or double arrows inside boxes or diamonds or double arrows. Lewis differed from classical logicians in being willing to consider such formulas, and in claiming to have intuitions about, for instance, $(A \Rightarrow B) \Rightarrow ((B \Rightarrow C) \Rightarrow (A \Rightarrow C))$. Lewis sought to codify his intuitions in an axiomatic system. But his intuitions (about this example among others) wavered, and the intuitions of his followers did not always agree. Soon there were five axiomatic systems, numbered **S1** through **S5**, and later works in the Lewis tradition such as [Zeman, 1973] list dozens.

An extensive mathematical theory developed, of which there will be space here to present only the rudiments. The proliferation of axiomatic systems ran ahead of their intuitive interpretation – not always a bad thing, since, just as some of the many geometries that proliferated after the discovery of hyperbolic geometry turned out to have applications having nothing to do with the original conception of geometry as a theory of the space around us, so some of the technical side of modal logic has turned out to be useful in unexpected ways, beginning with the modal interpretation of intuitionistic logic in [Gödel, 1933a].

2. Propositional Modal Logic: Proofs

The formulas of the language of propositional modal logic will officially comprise *atoms* $p_0, p_1, p_2, \ldots$ or $p, q, r, \ldots$ and formulas build up from these using $\neg$ and $\wedge$ and $\Box$ and parentheses for punctuation. We think of $\vee$ and $\rightarrow$ and $\Diamond$ and $\Rightarrow$ as unofficial abbreviations. This language is adequate to formalize such an example as

> If I could have been in Tibet, but could not have been in Tibet without having a special visa, then I could have had a special visa. (11.1)

in the sense of representing its logical form by a formula, thus:

$$\Diamond p \wedge \neg\Diamond(p \wedge \neg q) \rightarrow \Diamond q \tag{11.2}$$

Here p and q stand for 'I am in Tibet' and 'I have a special visa'. The diamond, pronounced 'possibly', really represents the transformation of the verb from the indicative to a nonindicative mood.

In an axiomatic system, certain formulas are adopted as *axioms*, and certain forms of transition from premises to conclusion as primitive *rules* of inference. A *demonstration* or *proof* is a sequence of formulas, called *steps*, each either one of the axioms or following from earlier steps by one of the rules. A formula is *demonstrable* or a *theorem* if it is the last step of some demonstration.

Other proof-related notions are defined in terms of demonstrability or theoremhood just as with classical logic. A set Γ of formulas is *inconsistent* if, for some $A_1, \ldots, A_n$ in Γ, the following is a theorem:

$$\neg(A_1 \wedge \ldots \wedge A_n) \tag{11.3}$$

Formula B is *deducible* from Γ if, for some $A_1, \ldots, A_n$ in Γ, the following is a theorem:

$$A_1 \wedge \ldots \wedge A_n \rightarrow B \tag{11.4a}$$

$$\neg(A_1 \wedge \ldots \wedge A_n \wedge \neg B) \tag{11.4b}$$

(Here (11.4a) abbreviates (11.4b).) Inconsistency of and deducibility from a formula A are inconsistency of and deducibility from the set $\{A\}$.

In any axiom system, the result of making any substitution of formulas for atoms in a theorem is a theorem. In one style of system, a rule to this effect is among the primitive rules of inference. In another style, instead of the axioms being specific formulas, such as perhaps (11.2), what are informally called 'axioms' are properly speaking axiom *schemes*, rules to the effect that all formulas of a specified form, such as perhaps

$$\Diamond A \wedge \neg\Diamond(A \wedge \neg B) \rightarrow \Diamond B \tag{11.2'}$$

count as theorems. With this style, a substitution in an axiom is still an axiom, so a substitution in all the steps of a demonstration is still a demonstration, and a substitution in a theorem is still a theorem – without our having to adopt a special rule to this effect. When this style is adopted, as here, few specific formulas are seen, since what are informally called 'theorems' are properly speaking theorem *schemes*, or results to the effect that every formula of a specified form is a theorem.

By a *tautology* we mean any substitution in a theorem of classical logic. We say B follows *tautologically* from $A_1, \ldots, A_n$ if (11.4) above is a tautology. Every modal system we consider will have every tautology as an axiom, and the rule allowing inference from any premises to any conclusion that follows tautologically. Every such system will have as one of its axioms the following:

$$\Box(A \rightarrow B) \rightarrow (\Box A \rightarrow \Box B) \tag{11.5}$$

Every such system will have exactly one modal rule, *necessitation*, permitting inference from A to $\Box A$. (Necessitation expresses, not the absurd assumption that every truth is necessary, but the reasonable one that every logically demonstrable truth is.) Both axiom and rule seem intuitively correct for any of the notions of necessity considered in the preceding section, though the reader may wish to stop to think this through.

The style of axiomatization used here, replacing clumsy axiomatizations of Lewis, originated with [Gödel, 1933a] and was developed and popularized in notes belatedly published as [Lemmon et al., 1977]. Alternate proof procedures (sequent calculi or tableaux rather than axiomatic systems) are also available, as for classical logic; see [Zeman, 1973].

The axioms and rules we have so far are those of the system called **K** or *minimal* modal logic. A basic result is the following (wherein a *modality* is a sequence of boxes and diamonds):

Theorem 11.2.1 (Becker's rule) *If $A \rightarrow B$ is a theorem, then $\triangle A \rightarrow \triangle B$ is a theorem, for any modality $\triangle$.*

Proof. First, consider the case $\triangle = \Box$. Suppose we have a demonstration of $A \rightarrow B$. We will still have a demonstration if we add $\Box(A \rightarrow B)$ at the end, since it follows by necessitation. Likewise if we then add (11.5), since it is an axiom.

Likewise if we finally add $\Box A \to \Box B$, since it follows tautologically. We can abbreviate the foregoing argument thus:

(i)	$A \to B$	given
(ii)	$\Box(A \to B)$	Nec, (i)
(iii)	$\Box(A \to B) \to (\Box A \to \Box B)$	Ax
(iv)	$\Box A \to \Box B$	Taut, (ii), (iii)

Second, consider the case $\triangle = \Diamond = \neg\Box\neg$. We have:

(i)	$A \to B$	given
(ii)	$\neg B \to \neg A$	Taut, (i)
(iii)	$\Box\neg B \to \Box\neg A$	Case $\triangle = \Box$, (ii)
(iv)	$\neg\Box\neg A \to \neg\Box\neg B$	Taut, (iii)
(v)	$\Diamond A \to \Diamond B$	Abbrev, (iv)

Finally, the general case is obtained by repeated application of the box and diamond cases. ■

Another basic result:

Theorem 11.2.2 (Replacement rule) *If $A \to B$ and $B \to A$ are theorems, $C(p)$ any formula, and $C(A)$ and $C(B)$ the results of substituting A and B for p in it, then $C(A) \to C(B)$ and $C(B) \to C(A)$ are theorems.*

Proof. Since every formula is built up from atoms using $\neg$ and $\wedge$ and $\Box$, it will be enough to prove that:

1. replacement holds for C an atom;
2. if it holds for C, it holds for $\neg C$;
3. if it holds for C and C', it holds for $C \wedge C'$;
4. if it holds for C, it holds for $\Box C$.

This method of proof is called *induction on complexity*. As to (i), if the atom is $p, C(A)$ is $A, C(B)$ is B, and $C(A) \to C(B)$ is $A \to B$, a theorem by hypothesis; if the atom is $q \neq p, C(A) \to C(B)$ is $q \to q$, a tautology, hence a theorem; similarly for the converse. As to (ii), if $C(A) \to C(B)$ and $C(B) \to C(A)$ are theorems, so are $\neg C(B) \to \neg C(A)$ and $\neg C(A) \to \neg C(B)$, which follow tautologically. As to (iii), it resembles (ii) and is left to the reader. As to (iv), if $C(A) \to C(B)$ is a theorem, $\Box C(A) \to \Box C(B)$ is a theorem by Becker; similarly for the converse. ■

It follows that, under the hypotheses of the theorem, if $C(A)$ is a theorem, so is $C(B)$ (which follows tautologically given the theorem $C(A) \to C(B)$). Since $A \to \neg\neg A$ and $\neg\neg A \to A$ are tautologies, hence theorems, replacement implies that we can in any theorem switch A for $\neg\neg A$ or vice versa – in other words, put in or take out a double negation. In particular, we can switch $\neg\Box$ and $\Diamond\neg$ (= $\neg\Box\neg\neg$) or $\Box\neg$ and $\neg\Diamond$ (= $\neg\neg\Box\neg$).

Now five theorems:

Theorem 11.2.3 $(\Box A \wedge \Box B) \to \Box(A \wedge B)$

Proof.

(i)	$A \to (B \to A \wedge B)$	Taut
(ii)	$\Box A \to \Box(B \to A \wedge B)$	Beck, (i)
(iii)	$\Box(B \to A \wedge B) \to (\Box B \to \Box(A \wedge B))$	Ax
(iv)	$(\Box A \wedge \Box B) \to \Box(A \wedge B)$	Taut, (ii), (iii)

■

Theorem 11.2.4 $\Box(A \wedge B) \to (\Box A \wedge \Box B)$

Proof.

(i)	$(A \wedge B) \to A$	Taut
(ii)	$(A \wedge B) \to B$	Taut
(iii)	$\Box(A \wedge B) \to \Box A$	Beck, (i)
(iv)	$\Box(A \wedge B) \to \Box B$	Beck, (ii)
(v)	$\Box(A \wedge B) \to (\Box A \wedge \Box B)$	Taut, (iii), (iv)

■

Theorem 11.2.5 $\Box(A \to B) \to (\Diamond A \to \Diamond B)$

Proof.

(i)	$(A \to B) \to (\neg B \to \neg A)$	Taut
(ii)	$\Box(A \to B) \to \Box(\neg B \to \neg A)$	Beck, (i)
(iii)	$\Box(\neg B \to \neg A) \to (\Box\neg B \to \Box\neg A)$	Ax
(iv)	$\Box(A \to B) \to (\neg\Box\neg A \to \neg\Box\neg B)$	Taut, (ii), (iii)
(v)	$\Box(A \to B) \to (\Diamond A \to \Diamond B)$	Abbrev, (iv)

■

Theorem 11.2.6 $(\Box A \wedge \Diamond B) \to \Diamond(A \wedge B)$

Proof.

(i)	$A \to (B \to A \wedge B)$	Taut
(ii)	$\Box A \to \Box(B \to A \wedge B)$	Beck, (i)
(iii)	$\Box(B \to A \wedge B) \to (\Diamond B \to \Diamond(A \wedge B))$	(11.2.5)
(iv)	$(\Box A \wedge \Diamond B) \to \Diamond(A \wedge B)$	Taut, (ii), (iii)

■

Theorem 11.2.7 $(\Box A_1 \wedge \Box A_2 \wedge \Diamond B) \to \Diamond(A_1 \wedge A_2 \wedge B)$

Proof.

(i)	$(\Box A_1 \wedge \Box A_2) \to \Box(A_1 \wedge A_2)$	(11.2.3)
(ii)	$\Box(A_1 \wedge A_2) \wedge \Diamond B \to \Diamond(A_1 \wedge A_2 \wedge B)$	(11.2.6)
(iii)	$(\Box A_1 \wedge \Box A_2 \wedge \Diamond B) \to \Diamond(A_1 \wedge A_2 \wedge B)$	Taut, (i), (ii)

■

This last generalizes:

Theorem 11.2.8 $(\Box A_1 \wedge \ldots \wedge \Box A_n \wedge \Diamond B) \rightarrow \Diamond(A_1 \wedge \ldots \wedge A_n \wedge B)$

3. Propositional Modal Logic: Models

In an attempt to develop a notion of model to go with our notion of proof, we start from the idea that the technical notion of validity of a formula, or truth in all models, is intended to analyse the intuitive idea of truth by virtue of form alone, or truth in all instances, truth no matter what specific sentences are put in for the atoms in a formula.

In classical logic, for the truth of an instance of a formula what matters about the sentences put in for atoms is not their meaning, but only their truth value. So for a model we may simply take a *valuation* or assignment of truth values to atoms. The valuation is then extended to other formulas by the usual rules, which (symbolizing 'A is true in model V' as '$V \models A$', and abbreviating 'if and only if' to 'iff'), read as follows:

$$V \models \neg A \qquad \text{iff} \qquad \text{not} \quad V \models A \tag{11.6a}$$

$$V \models A \wedge B \qquad \text{iff} \qquad V \models A \text{ and } V \models B \tag{11.6b}$$

With just two atoms p and q, though there are infinitely many pairs of sentences that might be put in for them, there are only four combinations of truth values, and so only four models. In each, one of the four combinations

$$A_1 = p \wedge q \tag{11.7a}$$

$$A_2 = p \wedge \neg q \tag{11.7b}$$

$$A_3 = \neg p \wedge q \tag{11.7c}$$

$$A_4 = \neg p \wedge \neg q \tag{11.7d}$$

is true, the rest false. For modal logic we cannot do anything so simple-minded as add the clause

$$V \models \Box A \qquad \text{iff} \qquad \text{necessarily,} \quad V \models A \tag{11.6c}$$

since our models are mathematical, and what is true in them is presumably necessary, so (11.6c) would make the truth of $\Box A$ equivalent to that of A. We will need a more complicated notion of model.

Consider the following examples:

$$\neg A_1 \wedge \Diamond A_1 \tag{11.8a}$$

$$\neg A_1 \wedge \neg\Diamond A_1 \wedge \Diamond\Diamond A_1 \tag{11.8b}$$

$$\Diamond(A_1 \wedge \Diamond A_2) \wedge \Diamond A_3 \wedge \neg\Diamond(A_3 \wedge \Diamond A_2) \tag{11.8c}$$

$$\Diamond(A_1 \wedge \Diamond A_2) \wedge \Diamond(A_1 \wedge \neg\Diamond A_2) \tag{11.8d}$$

(11.8a) suggests that if truth in all instances is to agree with truth in all models, a model must represent not only which combination of atoms is actually true in a given instance, but also which combinations are possible. (11.8b) suggests that a model indeed must represent not only actual possibilities but possible possibilities. (11.8c) suggests that the model must represent not only possibilities of various orders, but also which possible possibilities are possible relative to which actual possibilities (as there is a possible possibility that A_2 possible relative to the actual possibility that A_1, but not relative to the actual possibility that A_3). (11.8d) suggests that the model must allow distinct possibilities at which the same combination of atoms is true (as there is one actual possibility that A_1 with, and another without, a possible possibility that A_2 possible relative to it), so possibilities must be not just valuations but objects having valuations associated with them.

All this suggests a model consisting of a set U of elements representing possibilities of all orders, a relation $\prec$ representing relative possibility, plus a function V associating with each element u in U a valuation of atoms, telling us for each p_i whether it is true under possibility u, or as is said, true 'at' u. A little thought shows that we may define what it is for a formula A other than an atom to be true at an element u in a model M, symbolized $M \models A[u]$, as follows:

$M \models \neg A[u]$	iff	not $M \models A[u]$	(11.9a)
$M \models (A \wedge B)[u]$	iff	$M \models A[u]$ and $M \models B[u]$	(11.9b)
$M \models \Box A[u]$	iff	for all v with $u \prec v$, $M \models A[v]$	(11.9c)

The definitions of $\vee$ and $\rightarrow$ and $\Diamond$ and $\Rightarrow$ in terms of $\neg$ and $\wedge$ and $\Box$ then give

$M \models (A \vee B)[u]$	iff	$M \models A[u]$ or $M \models B[u]$	(11.9d)
$M \models (A \rightarrow B)[u]$	iff	if $M \models A[u]$ then $M \models B[u]$	(11.9e)
$M \models \Diamond A[u]$	iff	for some v with $u \prec v$, $M \models A[v]$	(11.9f)
$M \models (A \Rightarrow B)[u]$	iff	for all v with $u \prec v$, if $M \models A[v]$ then $M \models B[v]$	(11.9g)

We could require a model to have one element distinguished as representing actuality, define truth in the model as a whole to be truth at that element, and validity as truth in all models. Instead we just define validity as truth at all elements in all models.

Other proof-related notions are defined in terms of models much as with classical logic. A set Γ of formulas is *satisfiable* if all its members are true at some element in some model. A formula B is a *consequence* of Γ if B is true at any element in any model where all members of Γ are true. Satisfiability and consequence for a formula A are satisfiability and consequence for the set $\{A\}$.

Does this notion of model fit our earlier notion of proof? It is not hard to establish the following:

Theorem 11.3.1 (Soundness theorem) *Every theorem is valid.*

Proof. Let $M = (U, \prec, V)$ be any model. It is enough to show that every axiom has the property of being true in M at every u in U, and that each rule preserves this property. For then it follows that in every demonstration each step has the property, including the last. The tautology axiom and tautological-following rule will be left to the reader. For axiom (11.5), unpacking the definitions (11.9) we see that

$$M \models (\Box(A \to B) \to (\Box A \to \Box B))[u] \tag{11.10}$$

amounts to

$$\begin{aligned}&\text{if for every } v \text{ with } u \prec v, \text{ if } M \models A[v], \text{ then } M \models B[v],\\ &\text{then if for every } v \text{ with } u \prec v \text{ we have } M \models A[v],\\ &\text{then for every } v \text{ with } u \prec v, \text{ we have } M \models B[v]\end{aligned} \tag{11.11}$$

which is clear. For the necessitation rule, unpacking the definitions we see that what we need to show is

$$\begin{aligned}&\text{If for every } u \text{ we have } M \models A[u],\\ &\text{then for every } u \text{ and every } v \text{ with } u \prec v \text{ we have } M \models A[v]\end{aligned} \tag{11.12}$$

which is equally clear (since v is as much an element of M as u is, so to speak). ■

Soundness can be used to show that formulas are *not* theorems. For $\Box p \to \Diamond p$, consider a model with a single element u, not having $u \prec u$. Since there is no v with $u \prec v$, the condition that p is true at all such v holds vacuously, and $\Box p$ is true at u, while the condition that p is true at some such v fails trivially, and $\Diamond p$ is not true at u. Hence $\Box p \to \Diamond p$ is not true at u, and so by soundness not a theorem.

It is harder to establish the following:

Theorem 11.3.2 (Completeness theorem) *Every valid formula is a theorem.*

This is equivalent to saying that every consistent formula is satisfiable. Given a consistent formula, rather than going off immediately to search for a model, let us consider what a model would look like if we had one. Our formula would be an element of the set Γ of sentences true at some element u of the model. What does such a set look like?

For one thing, such a set Γ must be consistent. For if (11.3) is a theorem, it will be true at u by soundness, and so by (11.9a) and (11.9b) not all the A_i will be true at u, or elements of Γ. For another thing, Γ must be a *maximal* consistent set in the sense that adding any formula not in Γ will produce inconsistency. For if A is not in Γ, not true at u, then $\neg A$ is true at u, and in Γ, and adding A to Γ would produce a set that contains both A and $\neg A$, hence is inconsistent (since $\neg(A \wedge \neg A)$ is a tautology, hence a theorem).

Further, Γ will be *deductively closed*, in the sense that any formula deducible from Γ will be in Γ. But this is not really a *further* property: it follows from maximal consistency. For if A is deducible from Γ, there are B_i in Γ such that

$$\neg(B_1 \wedge \ldots \wedge B_m \wedge \neg A) \tag{11.13}$$

is a theorem. And if Γ is a maximal consistent set and A not in Γ, then adding A to Γ produces inconsistency, and there are C_j in Γ such that

$$\neg(C_1 \wedge \ldots \wedge C_n \wedge A) \tag{11.14}$$

is a theorem. But then

$$\neg(B_1 \wedge \ldots \wedge B_m \wedge C_1 \wedge \ldots \wedge C_n) \tag{11.15}$$

is a theorem since it follows tautologically from (11.13) and (11.14), and since (11.15) is the negation of a conjunction of formulas in Γ, Γ is inconsistent, contrary to hypothesis.

Further, we will have the following:

$$\neg A \text{ is in } \Gamma \quad \text{iff} \quad A \text{ is not in } \Gamma \tag{11.16a}$$

$$A \wedge B \text{ is in } \Gamma \quad \text{iff} \quad A \text{ is in } \Gamma \text{ and } B \text{ is in } \Gamma \tag{11.16b}$$

These also follow from maximal consistency. The 'only if' direction of (11.16a) we have already seen to follow from consistency. For the 'if' direction, if neither A nor $\neg A$ is in a maximal consistent set Γ, adding either will produce inconsistency, and hence there will be B_i and C_j in Γ such that (11.13) and (11.14) are theorems, which we have already seen to be contrary to hypothesis. (11.16b) follows easily from deductive closure, and is left to the reader.

If follows from all this that if a formula A is to be satisfiable, it must belong to a set Γ that has the property of maximal consistency and the other properties we have just seen to follow from that. But it turns out that any consistent A *does* belong to a maximal consistent set, applying to $\{A\}$ the following lemma (wherein 'can be extended to' just means 'is a subset of'):

Lemma 11.3.1 (Lindenbaum's lemma) *Any consistent set can be extended to a maximal consistent set.*

Proof. We first show

$$\text{The formulas of our language can be enumerated } A_1, A_2, A_3, \ldots \qquad (11.17)$$

For each formula is a finite sequence of symbols, each having an ASCII number of no more than four digits. Add zeros at the front to make it exactly four if it is less. Any n-symbol formula can be given a code number of $4n + 1$ digits, with numeral consisting of a one followed by four-digit blocks representing its symbols. Formulas can then be listed in order of increasing code number.

Having (11.17), starting with a given consistent set Γ_0, go through the A_i in order, adding each when one comes to it iff this can be done without producing inconsistency, producing in the end a set Γ. Any finitely many formulas in Γ will have gotten in by some stage along the way, and since we maintained consistency at each stage, the negation of their conjunction will not be a theorem, and Γ will be consistent. But if A_i is not in Γ, it is because to add it to what we had at that stage when its turn came would have produced inconsistency. Hence adding it to Δ would certainly produce inconsistency, showing that Δ is a maximal consistent set. ∎

We have seen some properties a set Γ must have if it is to be the set of formulas true at some u in some model. What properties must a pair of sets Γ and Δ have if they are to be the sets of formulas true at some u and v in some model, where $u \prec v$? From (11.9c) it must be that for any formula A, if $\Box A$ is in Γ, then A is in Δ. When this condition holds, let us say Δ is *potential* relative to Γ, and write $\Gamma \lhd \Delta$. With this notation, (11.9c) gives us a further property a set Γ must have if it is to be the set of formulas true at some element in some model. The following lemma says that this property, too, follows from maximal consistency:

Lemma 11.3.2 (Main lemma) *If Γ is a maximal consistent set, then for any formula B, if $\Box B$ is not in Γ, then there is a maximal consistent Δ such that $\Gamma \lhd \Delta$ and B is not in Δ.*

Proof. If $\Box B$ is not in Γ, then $\neg\Box B$ is (by (11.16a)) and hence $\Diamond\neg B$ is (by replacement and deductive closure). It will be enough to show that the set Δ_0 consisting of (i) all A such that $\Box A$ is in Γ, plus (ii) $\neg B$ is consistent. For then Lindenbaum's lemma will imply that it can be extended to a maximal consistent Δ, and (i) will guarantee that $\Gamma \lhd \Delta$ while (ii) will guarantee that $\neg B$ is in Δ, hence B is not (by (11.16a)).

Well, if Δ_0 were inconsistent, there would be $\Box A_i$ in Γ such that (11.4) is a theorem. But then the following would be theorems as well:

$$\Box\neg(A_1 \wedge \ldots \wedge A_n \wedge \neg B) \tag{11.18a}$$

$$\neg\Diamond(A_1 \wedge \ldots \wedge A_n \wedge \neg B) \tag{11.18b}$$

using necessitation to get (11.18a) and then replacement to get (11.18b). But since the $\Box A_i$ and $\Diamond\neg B$ are in Γ, and (11.2.8) is a theorem,

$$\Diamond(A_1 \wedge \ldots \wedge A_n \wedge \neg B) \tag{11.18b'}$$

is deducible from Γ, hence by deductive closure in Γ. So by consistency (11.18b) *cannot* be in Γ, and Δ_0 must be consistent. ■

We now have everything we need to put together the *canonical model* $M = (U, \lhd, V)$, which consists of the set U of all maximal consistent sets, the relation of relative potentiality $\lhd$, and the valuation V that makes p_i true at u (as an element of the model) iff p_i belongs to u (as a set of formulas). For this model we have, for all elements u and formulas A, the following:

Theorem 11.3.3 *$M \models A[u]$ iff A is in u.*

Proof. (11.3.3) holds by definition for atoms p_i. To prove by induction on complexity that it holds for all formulas, we must prove that:

1. if (11.3.3) holds for A, it holds for $\neg A$;
2. if (11.3.3) holds for A and for B, it holds for $A \wedge B$;
3. if (11.3.3) holds for A, it holds for $\Box A$.

As for (i), by (11.9a), $\neg A$ is true at u iff A is not true at u, which supposing (11.3.3) holds for A means iff A is not in u, which by (11.16a) means iff $\neg A$ is in u. As for (ii), it is similar, using (11.9b) and (11.16b). As for (iii), in one direction, if $\Box A$ is in u, then for any v with $u \lhd v$, A is in v by definition of $\lhd$, which supposing (11.3.3) holds for A means A is true at v; hence $\Box A$ is true at u. In the other direction, if $\Box A$ is not in u, by the main lemma there is a v with $u \lhd v$ such that A is not in v, which supposing (11.3.3) holds for A means that A is not true at v; hence $\Box A$ is not true at u. ■

Thus any consistent formula is true in the canonical model at any maximal consistent set to which it belongs, completing the proof of completeness. This theorem and the notion of model it involves are due to Saul Kripke, though the method of proof used here is that popularized in [Lemmon et al., 1977]. By soundness and completeness, our notions of proof and model for **K** agree with each other. The methods used are flexible, and can be applied to other systems. There will be space here to consider just three systems, with successively larger sets of theorems, and successively smaller classes of models.

Consider these axioms:

$$\Box A \to A \tag{11.19}$$

$$\Box A \to \Box\Box A \tag{11.20}$$

$$A \to \Box\Diamond A \tag{11.21}$$

The system **T** is obtained by adding to **K** the axiom (11.19); the system **S4** is obtained by adding to **T** the axiom (11.20); the system **S5** is obtained by adding to **S4** the axiom (11.21).

Consider these conditions (to hold for all elements in a model):

Reflexivity: $u \prec u$
Transitivity: if $u \prec v$ and $v \prec w$, then $u \prec w$
Symmetry: $u \prec v$, then $v \prec u$

Axioms and conditions match, thus:

Theorem 11.3.4 (Correspondence theorem)

(a) *A formula is a theorem of* **T** *iff it is true at all elements of all reflexive models.*
(b) *A formula is a theorem of* **S4** *iff it is true at all elements of all reflexive, transitive models.*
(c) *A formula is a theorem of* **S5** *iff it is true at all elements of all reflexive, transitive, symmetric models.*

Proof. For soundness it suffices to show that if **Reflexivity** holds for a model, then (11.19) is true at every element of it, and similarly for **Transitivity** and (11.20) and for **Symmetry** and (11.21). Well, $M \models (11.19)[u]$ and $M \models (11.20)[u]$ and $M \models (11.21)[u]$ respectively amount to

$$\text{if, for every } v \text{ with } u \prec v,\ M \models A[v], \text{ then } M \models A[u] \tag{11.22}$$

$$\text{if, for every } v \text{ with } u \prec v, M \models A[v], \text{ then}$$
$$\text{for every } w \text{ with } u \prec w \text{ and } w' \text{ with } w \prec w', M \models A[w'] \qquad (11.23)$$

$$\text{if } M \models A[u], \text{ then for every } v \text{ with } u \prec v$$
$$\text{there is a } w \text{ with } v \prec w \text{ such that } M \models A[w] \qquad (11.24)$$

(11.22) follows from **Reflexivity** because reflexivity guarantees that u itself is among the v with $u \prec v$. (11.23) follows from **Transitivity** because transitivity guarantees that w', too, is among the v with $u \prec v$. (11.24) follows from **Symmetry** because symmetry guarantees that u itself is among the w with $v \prec w$.

For completeness, it suffices to show that

1. if (11.19) is a theorem, $\lhd$ is reflexive;
2. if (11.20) is a theorem, $\lhd$ is transitive;
3. if (11.21) is a theorem, $\lhd$ is symmetric.

So let Γ and Δ and Θ be maximally consistent. Given (11.19), for any $\Box A$ in Γ, A is in Γ by deductive closure, so $\Gamma \lhd \Gamma$. Given (11.20), if $\Gamma \lhd \Delta$ and $\Delta \lhd \Theta$, then for any $\Box A$ in Γ, $\Box\Box A$ is in Γ by deductive closure, so $\Box A$ is in Δ and A is in Θ, and $\Gamma \lhd \Theta$. Given (11.21), if $\Gamma \lhd \Delta$, then for any $\neg A$ in Γ, $\Box\Diamond\neg A$ is in Γ by deductive closure and $\Diamond\neg A$ and equivalently (by replacement and deductive closure) $\neg\Box A$ is in Δ, and (by (11.16a) above) $\Box A$ is *not* in Δ; so if $\Box A$ is in Δ, $\neg A$ is *not* in Γ and (by (11.16a) again) A is in Γ, so $\Delta \lhd \Gamma$. ■

Alternate axiomatizations are possible. For instance, for **S5**, it is an easy exercise to show that (11.21) could be replaced by

$$\Diamond\Box A \to A \qquad (11.21^*)$$

and a not-so-easy exercise to show that (11.20) and (11.21) or (11.21*) could be replaced by

$$\neg\Box A \to \Box\neg\Box A \qquad (11.25)$$

Many more examples of correspondence are treated in standard textbooks such as [Hughes and Cresswell, 1996], which take up questions of decidability as well as completeness, and besides this 'basic' modal logic there is now an 'advanced' modal logic, concerned not with specific instances but with a general theory of correspondence, as in [Blackburn et al., 2002].

For **S5**, the model theory simplifies. By (11.9) whether a formula is true or not at an element in a model depends only on what atoms are true at that element, or elements possible relative to it, or elements possible relative to such elements, and so on; and for reflexive and transitive models this just means what axioms are true at elements possible relative to the given element. Any other elements are irrelevant, and might as well be thrown out. If they are, then if the model is also symmetric, all elements left will be possible relative to each other, and there will be no need to mention relative possibility at all: (11.9c) can be replaced by

$$M \models \Box A[u] \quad \text{iff} \quad \text{for all } v, M \models A[v] \tag{11.9c$'$}$$

Further, if two elements agree about the truth values of all atoms – or if we are only interested in a single formula, all atoms in it – we do not need both. When duplicates are deleted – which if we are only interested in finitely many atoms will leave only finitely many elements – there is no longer any reason to distinguish an element u from the valuation of atoms associated with it. We can take a model to be simply a finite set $\mathbf{V}$ of valuations, and define truth as follows:

$$\mathbf{V} \models p_i[V] \quad \text{iff} \quad V(p_i) = \text{true} \tag{11.26a}$$

$$\mathbf{V} \models \neg A[V] \quad \text{iff} \quad \text{not } \mathbf{V} \models A[V] \tag{11.26b}$$

$$\mathbf{V} \models (A \wedge B)[V] \quad \text{iff} \quad \mathbf{V} \models A[V] \text{ and } \mathbf{V} \models B[V] \tag{11.26c}$$

$$\mathbf{V} \models \Box A[V] \quad \text{iff} \quad \text{for all } V' \text{ in } \mathbf{V}, \mathbf{V} \models A[V'] \tag{11.26d}$$

Validity of a formula will be truth at all elements in all such models. In fanciful language the valuations can be spoken of as representing *possible worlds*, and (11.26d) then is a kind of embodiment of the old notion of Leibniz that necessity is 'truth in all possible worlds'.

The fact that if there is no proof of A in **S5** there is such a finite model for $\neg A$ gives a *decision procedure* for **S5**, a method that, applied to any formula A, will in principle enable one to determine in a finite amount of time whether or not it is a theorem. The key point is that one can make a list of all proofs and all finite models, much as one can make a list of all formulas according to (11.17), and then go down the list until one finds either a proof of A, showing that A is a theorem, or a finite model of $\neg A$, showing that A is not a theorem. Such *decidability* (the existence of a decision procedure) can also be established for **K** and **T** and **S4**, though the proof of the *finite model property* (the existence of a *finite* model for the negation of any non-theorem) is more difficult than for **S5**.

4. Propositional Modal Logic: Interpretation

But which of the many modal logics is the right one? That depends on the sense of necessity in question. The label 'semantics' is used sometimes for model theory and sometimes for meaning theory, but the models we have been considering tell us little about meaning, even if we adopt the common fanciful usage that calls the elements of a model 'worlds' and the relation of the model 'accessibility'. For difference senses of 'necessary' do not correspond in any obvious way to different assumptions about accessibility of worlds. Technical soundness and completeness results have no obvious direct implications for the intuitive question which axioms are appropriate for which sense of necessity. The right modal logic for a more technical sense of necessity than any on our original list in Section 1 has been identified in the specialized branch of modal logic called *provability logic*, as in [Boolos, 1993]. For most other senses the problem is open.

With epistemic or deontic or metaphysical or linguistic necessity, all that can be legitimately asked of the logician is to lay out the options for the epistemologist or deontologist or metaphysician or linguist. For instance, with deontic necessity the logician may point out that the candidate axiom $\Box p \rightarrow \Diamond p$ or $\neg(\Box p \wedge \Box\neg p)$ stands or falls with the principle that there are no conflicts of all-things-considered obligations. For truth or verifiability by virtue of form, however, there is no other discipline but logic involved. If there is a prevailing view among logicians, it is that of [Halldén, 1963], which associates **S5** with truth by virtue of form, **S4** with verifiability by virtue of form. As these opinions are about intuitive notions, not notions with mathematically rigorous definitions, they cannot have mathematically rigorous proofs. The cases for the two opinions are examined in [Burgess, 1999], where the case for the one about **S5**, to be presented here, is found stronger than that for the one about **S4**. The basic idea goes back to [Carnap, 1946].

In one direction, we must argue one by one that each axiom or rule of **S5** is intuitively correct, reading the box as 'it is true by virtue of form that' (from which it will follow that all theorems are). We earlier invited the reader to think about the axioms and rules of **K**, and will now discuss only the further axioms of **S5**, which we may take to be (11.19) and (11.25). Imagine putting in specific sentences for the atoms in A, to obtain a specific sentence α, and writing out logical symbols in words, (11.19) and (11.25) become:

If it is true by virtue of form that α, then α. (11.19′)

If it is not true by virtue of form that α, then

it is true by virtue of form that it is not true by virtue of form that α. (11.25′)

The first is obvious. For the second, suppose the antecedent holds. Then there is some β of the same form as α such that β is not true. To show the consequent holds, we must show that anything of the same form as 'it is not true by virtue of form that α' is true. Well, any such thing will be 'it is not true by virtue of form that γ' for some γ that is of the same form as α, hence of the same form as β, and the thing will be true since β is not.

In the other direction, we must argue that if a given formula is not a theorem of **S5**, then some instance is intuitively incorrect when the box is read as indicated. Instances are produced by putting in specific sentences for the atoms, so we will need some assumption about what sentences are available, and we make the reasonable one that there are indefinitely many that are logically independent, with no relations holding among their truth values by virtue of logical form alone. Suppose now that A is not a theorem. Then it fails at some valuation V_0 in some finite model $\mathbf{V}$. To produce an instance α of A that is not true, we must find sentences π_i to put in for the atoms p_i that will have two properties: (i) the combinations of truth values that the π_i are not precluded by their logical form from having are precisely the combinations assigned to the corresponding p_i by valuations in $\mathbf{V}$; and (ii) the π_i actually have the truth values assigned to the corresponding p_i by V_0. Towards producing such π_i, let $\mathbf{V}$ contain n valuations, and let k be such that $n \leq 2^k$. Take k logically independent sentences $\tau_1, \ldots, \tau_k$. There are 2^k conjunctions of the form

$$(\neg)\tau_1 \wedge (\neg)\tau_2 \wedge \ldots \wedge (\neg)\tau_k \tag{11.27}$$

where each bracketed negation may be present or absent. Independence means the truth of none is precluded by logical form, but logical form does preclude the truth of the conjunction of any two, and guarantees the truth of the disjunction of all: they are *mutually exclusive* but *jointly exhaustive*. Enumerate these conjunctions as $\sigma_1, \sigma_2, \ldots, \sigma_K$ where $K = 2^k$. Let ρ_i for $i < n$ just be σ_i, and let ρ_n be the disjunction of the σ_j for $j \geq n$. Then the ρ_i are also mutually exclusive and jointly exhaustive, and there are exactly as many of them as there are valuations in $\mathbf{V}$. Associate to each such valuation V a distinct ρ_V. Now for each p_i occurring in A, let π_i be the disjunction of the ρ_V for those V that make p_i true. Then (i) holds. As for (ii), consider the ρ corresponding to V_0. There is some assignment of truth values to the τ_i that would make ρ true. We may suppose that the τ_i actually have those truth values, since the only fact about them we have used is their logical independence, and that would not be affected if we replaced each that does not have the truth value we want by its negation, which does. The actual truth of ρ guarantees (ii).

5. Quantified Modal Logic

Quantifiers were introduced into modal logic in [Marcus, 1946], [Marcus, 1947] and [Carnap, 1946]. One usual axiomatic system classical quantification theory with identity adds to classical propositional logic four axioms, thus:

$$\forall x(A \to B) \to (A \to \forall xB) \qquad x \text{ not free in } A \qquad (11.28)$$

$$\forall xA \to A(y/x) \qquad \text{if } y \text{ free for } x \text{ in } A \qquad (11.29)$$

$$x = x \qquad (11.30)$$

$$x = y \to (A(x/z) \to A(y/z)) \qquad x, y \text{ free for } z \text{ in } A \qquad (11.31)$$

Also added is one rule, *universal generalization* (UG), permitting inference from A to $\forall xA$. Adding the axioms and rules of **K** delivers two interesting theorems:

Theorem 11.5.1 (Converse Barcan formula) $\Box\forall xA \to \forall x\Box A$

Proof.

(i)	$\forall xA \to A$	Ax
(ii)	$\Box\forall xA \to \Box A$	Beck, i
(iii)	$\forall x(\Box\forall xA \to \Box A)$	UG, ii
(iv)	$\forall x(\Box\forall xA \to \Box A) \to (\Box\forall xA \to \forall x\Box A)$	Ax
(v)	$\Box\forall xA \to \forall x\Box A$	Taut, iii, iv

■

Theorem 11.5.2 (Necessity of identity) $x = y \to \Box x = y$

Proof.

(i)	$x = x$	Ax
(ii)	$\Box x = x$	Nec, i
(iii)	$x = y \to (\Box x = x \to \Box x = y)$	Ax
(iv)	$x = y \to \Box x = y$	Taut, ii, iii

■

Adding the extra axioms of **S5** gives two more:

Theorem 11.5.3 (Direct Barcan formula) $\forall x\Box A \to \Box\forall xA$

Proof.

(i)	$\forall x\Box A \to \Box A$	Ax
(ii)	$\Diamond\forall x\Box A \to \Diamond\Box A$	Beck, i
(iii)	$\Diamond\Box A \to A$	(11.21*)
(iv)	$\Diamond\forall x\Box A \to A$	Taut, ii, iii
(v)	$\forall x(\Diamond\forall x\Box A \to A)$	UG, iv
(vi)	$\forall x(\Diamond\forall x\Box A \to A) \to (\Diamond\forall x\Box A \to \forall xA)$	Ax
(vii)	$\Diamond\forall x\Box A \to \forall xA$	Taut, v, vi
(viii)	$\Box\Diamond\forall x\Box A \to \Box\forall xA$	Beck, vii
(ix)	$\forall x\Box A \to \Box\Diamond\forall x\Box A$	(11.21)
(x)	$\forall x\Box A \to \Box\forall xA$	Taut, viii, ix

■

Theorem 11.5.4 (Necessity of distinctness) $\neg x = y \rightarrow \Box\neg x = y$

Proof.

(i)	$\Diamond\neg x = y \rightarrow \neg x = y$	Taut, (11.5.2)
(ii)	$\Box\Diamond\neg x = y \rightarrow \Box\neg x = y$	Beck, i
(iii)	$\neg x = y \rightarrow \Box\Diamond\neg x = y$	(11.21)
(iv)	$\neg x = y \rightarrow \Box\neg x = y$	Taut, ii, iii

■

Here (11.5.1) seem dubious: Necessarily, everything that exists exists, but it does not follow that everything that exists necessarily exists. (Don't worry here about Kant's doctrine that existence is not a predicate; for present purposes 'x exists' can be expressed by the formula $\exists y(y = x)$.) And if (11.5.2) seems plausible, still the value of deriving it as we have is questionable so long as the same method of derivation produces dubious results as well as plausible ones. It appears that the logic either (i) has abandoned the leading idea of modal logic, to take seriously the distinction between 'is' and 'could have been', and is reading 'there is' as short for 'there is or could have been', or (ii) has failed to maintain metaphysical neutrality and is building in some extravagant metaphysical hypothesis implying that all existence is necessary (perhaps the fantastic doctrine known as 'contingent concreteness', according to which if anything that does exist hadn't existed as a concrete reality, it would still have existed as an abstract idea).

Model theory tends to confirm such suspicions. As with modal propositional logic, a model for modal quantification theory consists in a set of 'possibilities' or 'worlds' and a relation of 'relative possibility' or 'accessibility', with a classical model attached to each element. At the propositional level the classical model attached is just a valuation of atoms, but at the quantificational level it consists of a universe together with an assignment to each predicate of a relation thereon. When this idea is implemented appropriately, the converse (respectively, direct) Barcan formula is found to be valid only if it is assumed that everything in the universe attached to u is also in the universe attached to v whenever $u \prec v$ (respectively, $v \prec u$). A neutral notion of proof, sound and complete for models without such special assumptions, in which the Barcan formulas become optional extras, can be developed by replacing the above version of classical quantification theory by another in which only formulas without free variables appear in proofs; but the resulting system is not perspicuous. Generally speaking, results on quantified modal logic found in standard textbooks such as [Hughes and Cresswell, 1996] are less systematic than those for propositional modal logic.

All this aside, the very meaningfulness of quantified modal logic was famously challenged by W. V. O. Quine. Relevant papers by and about Quine are collected in [Linsky, 1971], and the issue reviewed in [Burgess, 1998]. To make sense of $\Box\exists xPx$ one need only make sense of a sentence $\exists xPx$ being necessarily true. But to make sense of $\exists x\Box Px$ being true, one must make sense of an *open* sentence $\Box Px$ being satisfied by or true of an *object*. In jargon, the former involves only *de dicto* ('of a saying') modality, while the latter involves *de re* ('of a thing') modality. The obvious way to attempt to reduce de re to de dicto would be to define $\Box Px$ to be true of an object a iff $\Box Pc$ is true where c is a term denoting a. The trouble is that we may have two terms c_1 and c_2 denoting the same object but with $\Box Pc_1$ true and $\Box Pc_2$ false, as in these much discussed putative examples from Quine:

Necessarily, eight is a perfect cube. (11.32a)

Necessarily, the number of planets is a perfect cube. (11.32b)

Necessarily, Hesperus is identical with Hesperus. (11.33a)

Necessarily, Hesperus is identical with Phosphorus. (11.33b)

In order for the reduction to work, we would need a class of *privileged* terms such that, whatever may happen with other terms, for any privileged c_1 and c_2, $\Box Pc_1$ and $\Box Pc_2$ would have the same truth value. Whether such a class can be identified depends on the nature of the objects and sense of necessity involved.

For linguistic necessity, numbers are among the few objects with a natural choice of canonical terms: numerals. With this choice, it is the truth of (11.32a), where the number is denoted by a numeral, and not the falsehood of (11.32b), where the number is denoted in another way, that matters when trying to decide whether 'Necessarily, x is a perfect cube' is true of the number in question. There are different systems of numerals, but intuitively it seems that

Necessarily, VIII is a perfect cube. (11.32a′)

is just as true as (11.32a).

For metaphysical necessity, since [Kripke, 1972b] it has been widely accepted proper names may be chosen as canonical terms for any sorts of object that have them. Or at least, it has been widely accepted that though (11.33b) may not be *a priori* or analytic like (11.33a), it is metaphysically necessary. No matter what, the planet Venus, alias Hesperus, a.k.a. Phosphorus, would have been identical to itself, the planet Venus, alias Phosphorus, a.k.a. Hesperus. The leading idea behind the metaphysical necessity of identities connecting proper names is that when using a name to discuss what might have been, it denotes the same thing it denotes when discussing what is; therefore, if two names denote the same thing when discussing what is (as 'Hesperus' and 'Phosphorus' both denote the planet Venus), they continue to do so when discussing what might have been. This property of names is called 'rigidity'.

Most descriptions do not share it, but rather are 'flexible', and become ambiguous in modal contexts. Thus

If Bill Gates had given all his wealth to Ivana Trump, the richest person in the world would have been female. (11.34)

is ambiguous between a true and a false reading:

If Bill Gates had given all his wealth to Ivana Trump, the person who then would then have been the richest in the world (namely, Ms Trump) would have been female. (11.34a)

If Bill Gates had given all his wealth to Ivana Trump, the person who now is the richest in the world (namely, Mr Gates) would have been female. (11.34b)

This example brings out a deficiency in the formalism of quantified modal logic. Its operators apply to predicates, whereas the natural language operation of changing from indicative to nonindicative applies to verbs, and a single predicate may have more than one verb. The formalism does not have the resources to distinguish the following:

If Bill Gates had given half his wealth to Ivana Trump, she would have had more money than he had. (11.35a)

If Bill Gates had given half his wealth to Ivana Trump, she would have had more money than he has. (11.35b)

For attempts to deal with the limitations of expressive power of box-diamond modal logic by adding further operators (for 'actually' and the like), see [Cresswell, 1990].

The fact that in mathematics nothing could have been other than as it is, which accounts for the neglect of 'could have' by classical logic, does not prevent the application of mathematics to possibility in mathematical statistics. It is just that, to apply mathematics, we have to depart from our usual ways of speaking and thinking. We have to conceive of a 'space' whose 'points' are 'possibilities' or 'states'. Instead of saying that 'I am not in Tibet' or 'I could have been in Tibet' we can say 'In the actual state, I am not in Tibet' and 'In some possible state, I am Tibet'. In so speaking we are in effect shifting from an indicative-mood 'am' as contrasted with the non-indicative mood 'could have been' to a moodless 'am' as short for 'am or could have been'. Once this shift is made, (11.1) can be formalized in classical quantification theory, thus:

$$\exists tPt \wedge \neg\exists t(Pt \wedge \neg Qt) \rightarrow \exists tQt \quad (11.36)$$

The variable t ranges over states, one of which we may suppose to be the actual state, and the rest merely possible states. The predicates Pt and Qt stand for 'in state t, I am in Tibet' and 'in state t, I have a special visa'.

Thus formalizing in classical logic what does not naturally invite such formalization is called *regimentation*. On this approach there are modal applications of classical logic, but no *autonomous* modal logic. Mathematical statistics provides no obvious precedent for distinguishing possible possibilities from actual possibilities, but the model theory of modal logic suggests adding to the classical language, besides the variables $t_0, t_1, t_2, \ldots$ for 'possibilities' or 'states', and predicates $P_0, P_1, P_2, \ldots$ or $P, Q, R, \ldots$ corresponding to atoms $p, q, r, \ldots$, a symbol $<$ for relative possibility. We can then translate every formula A of the autonomous modal propositional language into a formula $A^*(t_0)$ of the regimented classical language thus:

$$p_i^* = P_i t_0 \tag{11.37a}$$

$$(\neg A)^* = \neg A^* \tag{11.37b}$$

$$(A \wedge B)^* = A^* \wedge B^* \tag{11.37c}$$

$$(\Box A)^* = \forall t_1 (t_0 < t_1 \rightarrow A^{*+}) \tag{11.37d}$$

where + indicates increasing the subscript on each variable by one. For instance, $(\Diamond\Box\Diamond p)^*$ will amount to the following:

$$\exists t_1 (t_0 < t_1 \wedge \forall t_2 (t_1 < t_2 \rightarrow \exists t_3 (t_2 < t_3 \wedge P t_3))) \tag{11.38}$$

A will be true at every element in every modal model iff $\forall t_0 A^*$ is true in every classical model.

Everything that can be said in the autonomous language can be said in the regimented language, and more also, since there are many classical formulas that are not $*$-translations of modal formulas. Even the kind of distinction seen in (11.35) can be expressed, by introducing, besides or instead of the predicate 'in state u, x has more money than y has', a predicate 'x has more money in state u than y has in state v'. Greater *expressiveness*, in the sense of the ability to say more, is not always an unmixed blessing, however, since features of *tractability*, meaning nice properties such as decidability, are sometimes lost when moving from a less expressive to a more expressive framework.

6. Books and Papers of Particular Note

In addition to other references cited above, mention should be made of: [Lewis and Langford, 1932], which gives the mature views of the founder of modern modal logic; [McKinsey, 1941], whose decidability results show the highest level reached prior to [Kripke, 1963a] and [Kripke, 1963b], which revolutionized the subject; and [Goldblatt, 2006], which gives an authoritative history of the field on its mathematical side, something lacking for the philosophical side.

Notes

1. See Chapter 18.
2. See Chapter 14.
3. See Chapter 8.

12 Tense or Temporal Logic

Thomas Müller

Chapter Overview

1. Introduction	324
1.1 Motivation and Terminology	325
1.2 Tense Logic and the History of Formal Modal Logic	326
2. Basic Formal Tense Logic	327
2.1 Priorean (Past, Future) Tense Logic	328
2.1.1 The Minimal Tense Logic $\mathbf{K}_t$	330
2.1.2 Frame Conditions	332
2.1.3 The Semantics of the Future Operator	337
2.1.4 The Indexical 'Now'	342
2.1.5 Metric Operators and At_t	343
2.2 Temporal Logic with 'Since' and 'Until'	344
3. Some Further Topics	345
3.1 Temporal Logic and Natural Languages	345
3.2 Tense Logic and Relativity Theory	346
3.3 Temporal Predicate Logic	347
3.4 Branching Time and the Logic of Agency	348
4. Suggested Further Reading	349
Notes	349

1. Introduction

The logic of time is a subfield of logic of special philosophical interest. Its development in the twentieth century is connected with issues in metaphysics, the philosophy of language, philosophy of science, the philosophy of logic, and action theory. But tense logic, also called temporal logic, is interesting from a purely formal point of view as well. It played a decisive role in the development of modal logic, and it continues to play an important role in theoretical computer science and in many applications.

This introduction provides some motivational and historical background. Basic formal issues of propositional tense logic are treated in Section 2. Section 3 gives a glimpse of a number of related further topics.

1.1 Motivation and Terminology

Arthur Prior, who initiated the formal study of the logic of time in the 1950s, called his logic *tense logic* to reflect his underlying motivation: to advance the philosophy of time by establishing formal systems that treat the tenses of natural languages like English as sentence-modifying operators. While many other formal approaches to temporal issues have been proposed in the meantime, Priorean tense logic is still the natural starting point. Prior introduced four *tense operators* that are traditionally written as follows:

P it was the case that (**P**ast)
H it **h**as always been the case that
F it will be the case that (**F**uture)
G it is always **g**oing to be the case that

Thus, a sentence in the past tense like 'John was happy' is analysed as a temporal modification of the sentence 'John is happy' via the operator 'it was the case that', and can be formally expressed as 'P Happy(John)'.

While the syntax of the tense operators is analogous to that of a simple one-place operator like negation, their semantical interpretation transcends the territory of standard extensional languages: tense operators are not truth functional (obviously one cannot compute the truth value of a temporally modified sentence from the *truth value* of the unmodified sentence alone), and tense logic is accordingly an intensional logic, or a modal logic in the wider sense.

Typical foundational questions of intensional logic acquire a specific twist when applied to tense logic. Once one considers tense to be part of the *content* of a proposition modified by a tense operator, one has to become clear on the meaning of an unmodified sentence. Basically there are two options: the tense operators could form something that is tensed out of material that is not yet tensed, or they could modify something that is already tensed. The plausibility of iterated tense operators ('it was the case that it was the case that it is raining') points towards option two: unmodified propositions should be considered to be in the present tense, not tenseless. Such a view was held by a number of ancient and medieval logicians starting with Aristotle (in fact, an additional motivation for the development of tense-logical systems was the hope that they could contribute to the exegesis and understanding of historical texts, especially on the much-discussed issue of future contingents). Furthermore, the view that unmodified propositions are already tensed accords well with the grammatical form of the sentences expressing unmodified propositions like 'John is happy' in English, which are in the present tense. However, this seems to make trouble for the traditional idea that a proposition has a truth value *simpliciter*. Frege for one took this as so basic that he argued that all common unmodified sentences are elliptical, expressing a proposition that in fact contains a time index ('John is happy on January 8, 2010, 1:07 pm GMT', with the 'is' interpreted as a *tenseless* copula). The idea that an ideally rational language should not make a difference between past, present, and future is indeed a motif found in a variety of philosophers ranging from Spinoza to the logical empiricists of the twentieth century. Tense logic, on the other hand, invites one to view *one and the same proposition* as true at one time, yet false at another. This was certainly Prior's view. Writers who wish to make a metaphysical point about this are perhaps more likely to use Prior's original term 'tense logic' than others, who may prefer the somewhat more neutral terms 'temporal logic' or 'the logic of time'.

1.2 Tense Logic and the History of Formal Modal Logic

As just mentioned, modern logicians did not commonly take modalities and tenses to be elements of the logical form of a proposition. Frege, continuing the Kantian tradition, completely banished modalities from that part of a sentence that was considered amenable to logical analysis. The late nineteenth century saw some work on the formalization of modalities; formal modal logic proper is mostly considered to begin with C. I. Lewis' early twentieth century investigations into the logic of conditionals leading, e.g., to the well-known systems **S4** and **S5**.[1] These systems of modalities were characterized *syntactically*.

A *semantic* interpretation for these systems was not established until much later. The key step towards an adequate semantics, which occurred to a number of philosophers and mathematicians more or less simultaneously in the 1950s, was to allow for different states of affairs (often called 'possible worlds') in one single semantic model, and to impose a relational structure on them. Metaphorically this relation is often called an 'accessibility relation', but it just expresses the relation of relative possibility: given a set W of worlds and the accessibility relation R (set-theoretically, a subset of $W \times W$), if wRw', then w' is possible relative to w. Necessity is then taken to be truth in all accessible worlds, and accordingly, a sentence $\Box\phi$ ('it is necessary that ϕ') is said to be true at w in a model iff for all w' for which wRw', ϕ is true at w'. $\Box$ is called a *strong* modal operator since it corresponds to universal quantification over accessible worlds. Dually, the *weak* operator $\diamond$ ('it is possible that') is defined via existential quantification: $\diamond\phi$ is true at w iff there is some w' with wRw' s.t. ϕ is true at w'.

In general, the accessibility relation R is hard to make concrete sense of; this conceptual difficulty may account for the fact that formal semantics for modal languages were developed rather late. The relation R is however easy to interpret if the modalities involved are in fact tenses and the 'possible worlds' contain information about the state of things at different moments. In that case, the 'accessibility' relation is simply the relation > of temporal succession: Pϕ is true at m iff there is some moment m' for which $m > m'$ and at which ϕ is true. It is no surprise, then, that some of the first ideas about a relational semantics for a modal language came from tense logic. Many interesting correspondences between modal formulae and conditions on relational structures – like the condition of transitivity and the validity of FF$\phi \rightarrow$ Fϕ, to be discussed in Section 2.1.2 below – can be intuitively well motivated by considering the structure of time, cf. e.g., [Prior, 1967]. In this way, tense logic facilitated the development of formal modal logic by providing a plausible reading of the relational semantics. On the other hand, tense logic is more complicated than the logic of necessity and possibility (so-called alethic modal logic) because it contains *two* interacting sets of modal operators, one for the future and one for the past, and because the interpretation of the future operator poses special problems (see Section 2.1.3 below). For a detailed overview of the development of the semantics of temporal and modal logic, cf. [Copeland, 2002] and [Goldblatt, 2005]. This history certainly contains a wealth of material for a case study of the phenomenon of 'multiple discovery'.

2. Basic Formal Tense Logic

Above we have already described informally the basic ingredients of the syntax and semantics of tense logic. In fact there are many variants. Here we give some basic information about *propositional* tense logic based on point structures. (There are also good motivations for building systems based on interval structures; cf., e.g., [van Benthem, 1991] for a balanced overview. Reichenbach introduced a somewhat different analysis of tenses; for references see Section 3.1. The predicate logic will be discussed very briefly in Section 3.3 below.) A more complete overview of propositional tense logic can be found, e.g., in [Burgess, 2002] and [Finger et al., 2002].

Syntax: Our common background for the following considerations is to assume for the syntax

- a countable set *Atoms* of atomic propositions $p, q, r, \ldots$;
- the usual propositional operators: negation ($\neg$) and conjunction ($\wedge$), and defined from these, disjunction ($\vee$), the conditional ($\rightarrow$) and the biconditional ($\leftrightarrow$); we will also use the (defined) propositional constants TRUE ($\top$) and FALSE ($\bot$);
- a choice of temporal operators: Prior's original one-place strong operators for past (H) and future (G) and, usually defined from these, their weak duals P and F (see Section 2.1), and/or the two-place

connectives for 'since' (S) and 'until' (U) introduced by Kamp (see Section 2.2).[2]

Semantics: On the semantic side, we will be working with

- a non-empty relational structure $\langle W, R\rangle$, also called a *frame* and often written as $\langle T, <\rangle$ to suggest 'time' and the earlier–later relation.

Note that the latter notation suggests, e.g., transitivity of $<$, but in the minimal setting with which we will start this is not yet required. In most cases, however, we will assume a structure which could intuitively be a *flow of time*, i.e., we will be working with

- a non-empty *strict partial ordering* of moments $\langle T, <\rangle$ as the underlying frame; being a partial ordering, $<$ is transitive (for all $m, m', m'' \in T$: if $m < m'$ and $m' < m''$, then $m < m''$) and irreflexive (for all $m \in T$, $m \not< m$), hence also asymmetric (for all m and m': if $m < m'$, then $m' \not< m$); the non-strict ordering $\leq$, defined in the obvious way ($m \leq m'$ iff $m < m'$ or $m = m'$), is antisymmetric (if $m \leq m'$ and $m' \leq m$, then $m = m'$);
- a *model* $\mathcal{M} = \langle T, <, V\rangle$ based on the frame $\langle T, <\rangle$, which is that frame together with a *valuation* V, which assigns to each $p \in Atoms$ the set $V(p) \subseteq T$ of moments at which that atom is true in the model.

In many cases the frame is assumed to have additional properties: most commonly, one only considers such partial orderings that are

- *backwards-linear* (also called 'left-linear' or 'future-branching', also simply 'branching'), i.e., if $m < n$ and $m' < n$, then ($m < m'$ or $m = m'$ or $m' < m$)

and very often one narrows this down further to orderings that are

- *linear* (or: trichotomous), i.e., for all $m, m' \in T$, we have one of the three following cases: $m < m'$ or $m = m'$ or $m' < m$.

The following discussion is split into a longer part on Priorean (P / F) tense logic (Section 2.1) and a much shorter part on tense logic with 'since' and 'until' (Section 2.2).

2.1 Priorean (Past, Future) Tense Logic

As mentioned, Prior's motivation was to capture the notion of tense as a sentence-modifying operator. He introduced the weak one-place modal operators P (past) and F (future) and their strong duals H (always in the past;

$H\phi \Leftrightarrow \neg P \neg \phi$) and G (always in the future; $G\phi \Leftrightarrow \neg F \neg \phi$). Prior was of the opinion that the semantics of these operators could not be *given* via model-theoretic conditions, because the only way we could understand these conditions and the underlying models was via our natural language and its informal counterparts to these operators. He did however acknowledge the usefulness of a model-theoretic approach for getting clear about frame conditions (see Section 2.1.2 below), and it has become customary to lay out the semantics of tense operators in that fashion too. Prior's own 'internal' approach, which sticks to modal languages and increases their expressive power via special propositions (so-called nominals) that characterize individual moments, ultimately led to the now flourishing field of so-called hybrid logic; cf. [Prior, 1967, Chapter V], [Prior and Fine, 1977] and, for an overview, [Blackburn, 2000] and [Areces and ten Cate, 2007].

Given a model $\mathcal{M} = \langle T, <, V \rangle$ based on some frame $\langle T, < \rangle$ and a moment of evaluation $m \in T$, the truth conditions for a sentence ϕ are as follows (where $\mathcal{M}, m \models \phi$ is to be read 'ϕ is true in model $\mathcal{M}$ at moment m'):[3]

- If $\phi = p \in Atoms$: $\mathcal{M}, m \models \phi$ iff $m \in V(p)$.
- If $\phi = \neg\psi$: $\mathcal{M}, m \models \phi$ iff $\mathcal{M}, m \not\models \psi$.
- If $\phi = \psi_1 \wedge \psi_2$: $\mathcal{M}, m \models \phi$ iff $\mathcal{M}, m \models \psi_1$ and $\mathcal{M}, m \models \psi_2$.
- If $\phi = H\psi$: $\mathcal{M}, m \models \phi$ iff for all $m' \in T$ s.t. $m' < m$, we have $\mathcal{M}, m' \models \psi$.
- If $\phi = G\psi$: $\mathcal{M}, m \models \phi$ iff for all $m' \in T$ s.t. $m < m'$, we have $\mathcal{M}, m' \models \psi$.

The more general notions of global truth, validity, and logical consequence are then defined in the usual way:

- ϕ is globally true in a model $\mathcal{M} = \langle T, <, V \rangle$ iff for all $m \in T$, $\mathcal{M}, m \models \phi$. We write $\mathcal{M} \models \phi$.
- ϕ is valid in a frame $F = \langle T, < \rangle$ iff ϕ is globally true in every model based on that frame. We write $F \models \phi$. Generalizing, ϕ is valid in a class of frames $\mathcal{F}$ iff it is valid in any frame from that class. We write $\mathcal{F} \models \phi$
- ϕ is a logical consequence of a set Ψ of formulae in a model $\mathcal{M} = \langle T, <, V \rangle$ iff for all $m \in T$, if $\mathcal{M}, m \models \psi$ for all $\psi \in \Psi$, then $\mathcal{M}, m \models \phi$ as well ($\Psi \models_{\mathcal{M}} \phi$). Again, this is generalized to frames ($\Psi \models_F \phi$) and classes of frames ($\Psi \models_{\mathcal{F}} \phi$).

Once a tense-logical language with the above semantics is available, one can rephrase certain questions in the philosophy of time in a formal manner. For example, does time have a beginning? Certainly some frames have a first moment, while others don't. Can one express this fact tense-logically, as it were 'from within' the temporal structure? (It turns out that one can.) More generally, it is interesting to see how much the tense-logical language can reflect about the structure of the frames on which it is defined. One of the early questions in the development of tense logic, mirrored by similar questions in alethic modal logic, was about useful axiomatic systems and the correspondence of the validity of tense-logical formulae and general properties of frames: such correspondences can lead the way to *complete* axiomatizations, in which the formulae derivable syntactically from the axioms are exactly the formulae valid in a suitably characterized class of models. Most interestingly, this will be a class of models singled out by simple conditions on the underlying frame, e.g., first-order conditions like transitivity, density, the existence of a first moment, or the like. This issue, which pertains to modal logic generally, will be discussed in Section 2.1.2. In tense logic, a further point of interest is the question of the adequacy of the definition of the F operator; for discussion see Section 2.1.3 below. We will also discuss the treatment of the indexical 'now' (Section 2.1.4) and metric aspects of tense logic (Section 2.1.5). But first, let us introduce the simplest tense logic, $\mathbf{K}_t$.

2.1.1 The Minimal Tense Logic $\mathbf{K}_t$

Above we have laid out the semantics of tense logic. What would be a proper axiomatic basis? Note that this style of approach turns the general historical development upside down: historically, syntactic systems of axioms and rules came much earlier than the semantics. Tense logic, however, has from the beginning been a rather semantically-driven field, in which intuitions about sensible temporal structures take precedence over the choice of particular axioms. Still one wants to know what that syntactic side looks like.

By the definition of the semantics given above, it is clear that all propositional tautologies are true in any model at any moment – thus, any useful axiomatic basis for a tense logic should contain a set of axioms for the propositional calculus. As in basic propositional logic, so in propositional tense logic, an axiomatization can be given either by specifying axiom schemata (which are not themselves formulae of the language, but characterize an infinite set of such formulae), or concrete axioms (which are formulae) together with a substitution rule (allowing for the uniform substitution of atomic propositions by arbitrary formulae). For simplicity's sake we choose to work with axiom schemata here, even though we will be lax in sometimes calling them 'axioms'.[4]

Over and above the propositional calculus, the semantics laid out above guarantees, even in the most general case in which there are no conditions on the relation $<$ (not even transitivity), the validity of the *K*-axiom (the modal distribution axiom) for each of the strong operators, like in alethic modal logic (cf. Chapter 11):

$$\mathsf{H}(\phi \to \psi) \to (\mathsf{H}\phi \to \mathsf{H}\psi);$$

$$\mathsf{G}(\phi \to \psi) \to (\mathsf{G}\phi \to \mathsf{G}\psi).$$

Furthermore, the semantics supports rules of temporal generalization (corresponding to the necessitation rule in alethic modal logic) for both strong operators, i.e.,:

- if ϕ is a theorem, then so is $\mathsf{H}\phi$;
- if ϕ is a theorem, then so is $\mathsf{G}\phi$.

These are, twice, the standard axioms and rules for a normal modal logic: tense logic as laid out above is a *normal bimodal logic*. The fact that two sets of modal operators are involved immediately raises the question whether these modalities interact in any specific way. After all, their semantics involves not two independent relations, but just one relation and its converse. As one can check easily, there are indeed two characteristic interaction axioms that follow from the semantics given above (again these are valid in any model based on any frame):[5]

$$\phi \rightarrow \mathsf{HF}\phi$$

and

$$\phi \rightarrow \mathsf{GP}\phi.$$

The minimal tense logic comprising just the mentioned axioms (a set of axioms for propositional calculus, the two *K* axioms for H and G, and the two interaction axioms) and rules (modus ponens and the two rules of temporal generalization) is called $\mathbf{K}_t$, mimicking the name for the minimal normal monomodal logic, **K** (so-called after its single characteristic axiom, which in turn is named after one of the pioneers of the field, Saul Kripke). We write '$\vdash_{K_t} \phi$' for 'ϕ is a theorem of $\mathbf{K}_t$', and '$\Psi \vdash_{K_t} \phi$' for 'ϕ can be proved from Ψ in $\mathbf{K}_t$'.

One can show that the logic $\mathbf{K}_t$ is sound and complete with respect to the class of all models based on all frames, i.e., that all theorems (formulae derivable from the axioms) of the logic are validities (true at every moment in every model based on any frame) – the soundness direction – and that every validity is also in fact a theorem (weak completeness). Weak completeness corresponds to the property that every single $\mathbf{K}_t$-consistent formula (i.e., every formula from which one cannot derive a contradiction in $\mathbf{K}_t$) is satisfiable. $\mathbf{K}_t$ is also *strongly complete* w.r.t. the class of all frames: if ϕ is a logical consequence of Ψ, then ϕ is also $\mathbf{K}_t$-provable from Ψ. Strong completeness corresponds to the property that every $\mathbf{K}_t$-consistent set of formulae is satisfiable.

Soundness can be proved formally, as usually, by induction on the length of a proof, i.e., by showing that the mentioned axioms are validities and that the mentioned inference rules preserve validity; we have motivated such a proof informally above. Completeness can be proved in a number of ways, e.g., via the canonical model technique (cf., e.g., [Blackburn et al., 2002, pp. 205f.]). The first completeness proof for $\mathbf{K}_t$ is due to Lemmon; it is laid out, e.g., in

[Burgess, 2002].[6] Lemmon's proof in fact establishes that every $\mathbf{K}_t$-consistent set of formulae is satisfiable on an *irreflexive* frame; we will comment on this below.

Like many other useful modal logics, $\mathbf{K}_t$ also has the *finite model property*, from which it follows that $\mathbf{K}_t$ is *decidable*. The investigation of the computational complexity of the satisfiability problem for various tense logics is an important topic for theoretical computer science, with obvious repercussions for applications. To mention just one relevant result, the satisfiability problem for linear transitive flows of time is NP-complete. For further results, cf. [Gabbay et al., 1994, Chapter 15.8].

2.1.2 Frame Conditions

Having studied the basic semantics, one can ask which modal formulae can be used to express certain metaphysically interesting assumptions about a flow of time – a task that provided much motivation for the early tense logicians in the 1950s and 1960s. As Lemmon's completeness proof for $\mathbf{K}_t$ shows that any consistent set of $\mathbf{K}_t$-formulae is in fact satisfiable in a model based on an *irreflexive* frame, the class of all frames gives rise to the same tense logic as the class of irreflexive frames. Thus, tense-logical formulae cannot single out the irreflexive frames. This is one sign of the expressive limitations of tense logic that was noticed early on. Insofar as irreflexivity is metaphysically interesting (it certainly is in connection with avoiding circularity: circular time is reflexive), this is also a philosophically important negative result.[7] Many other metaphysically interesting conditions on frames can however be expressed tense-logically; we will discuss the most prominent cases. More generally, one can ask what the fundamental connections are between validity of formulae and conditions on a class of frames; we will comment briefly on this at the end of this section.

Transitivity: As one of the simplest cases, consider transitivity. It is easy to check that the formula

$$\mathsf{FF}\phi \rightarrow \mathsf{F}\phi \tag{4}$$

(or in the contrapositive, $\mathsf{G}\psi \rightarrow \mathsf{GG}\psi$) is valid in all transitive frames. In fact it defines the class of transitive frames: (4) is valid on a frame iff that frame is transitive. To prove the direction from right to left, let $F = \langle T, < \rangle$ be a transitive frame and $\mathcal{M} = \langle T, <, V \rangle$ a model based on F, and let $m \in T$. If $\mathcal{M}, m \models \mathsf{FF}\phi$, then by the semantic clause for F, there must be $m', m'' \in T$ s.t. $m < m'$, $m' < m''$, and $\mathcal{M}, m'' \models \phi$. By transitivity, we also have $m < m''$. (Note that while the notation suggests this anyway, this does not hold for all relations $<$; transitivity *is* needed at this step.) Applying the semantic clause for F once more, we get $\mathcal{M}, m \models \mathsf{F}\phi$, and thus (4). The direction from left to right can be proved contrapositively.

Thus, let $F = \langle T, < \rangle$ be a frame that is not transitive, i.e., in which there are $m, m', m'' \in T$ for which $m < m'$, $m' < m''$, but $m \not< m''$. Now, let p be an atomic proposition, and let V be a valuation for which $V(p) = \{m''\}$, i.e., according to V, p is true only at m'' and false at all other moments. By the semantic clause for F, we have $\mathcal{M}, m \models \mathsf{FF}p$, but $\mathcal{M}, m \not\models \mathsf{F}p$; thus, for $\phi = p$, we have a counterexample to (4).

The fact that a tense-logical formula defines a certain class of frames does not yet amount to a completeness proof for the respective extension of $\mathbf{K}_t$. In the present case, however, a completeness proof for $\mathbf{K}_t$ can be extended fairly easily to show that the logic $\mathbf{K}_t\mathbf{4}$, i.e., the system of $\mathbf{K}_t$ enriched by the axiom (4),[8] is complete for the class of all transitive frames (and, by the remark about irreflexivity above, also for all partial orders). As the mirror image of (4),

$$\mathsf{PP}\phi \rightarrow \mathsf{P}\phi \tag{4'}$$

is also valid in all transitive frames, it must (by completeness) be provable from (4) given the other axioms. The proof amounts to about half a page (cf., e.g., [McArthur, 1976, Chapter 2] for this and other elementary results).

In a similar vein, there are characteristic axioms and corresponding completeness results for many other theoretically interesting and/or metaphysically significant frame conditions.

Left Linearity (Past Non-Branching) / Right Linearity: Frames that are left linear (for all n, m, m', if $m < n$ and $m' < n$, then either $m < m'$ or $m = m'$ or $m' < m$) capture the widespread idea that the past is fixed while the future is open. Left linearity, or past non-branching, is defined by the axiom

$$\mathsf{FP}\phi \rightarrow (\mathsf{F}\phi \vee \phi \vee \mathsf{P}\phi). \tag{LLIN}$$

Many applications of tense logic in computer science as well as in philosophy presuppose left linearity, but allow for future branching (often just called 'branching'); e.g., left-linear frames form the basis for the Ockhamist tempo-modal language to be discussed below. We will give a rather philosophically tinted overview of branching time in Section 2.1.3.

Right linearity (future non-branching) is defined by the mirror image of (LLIN),

$$\mathsf{PF}\phi \rightarrow (\mathsf{P}\phi \vee \phi \vee \mathsf{F}\phi). \tag{RLIN}$$

Linearity (Trichotomy): Linearity amounts to backward and forward linearity.[9] Thus, linear frames are defined, e.g., by the above axioms taken together, i.e.,

$$LLIN \wedge RLIN. \tag{LIN}$$

It seems fair to say that linear tense logics are the most well-studied ones, and many specific applications have been developed on the basis of linear time. [Kamp, 1968] established a number of significant results early on.

First/Last Moment: The formula

$$G\bot \vee FG\bot \qquad \text{(LASTMO)}$$

defines frames with a last moment (some $m \in T$ for which there is no $m' \in T$ s.t. $m < m'$): at a moment that has no successors, in fact *any* formula starting with 'G' is vacuously true. Similarly, the mirror image

$$H\bot \vee PH\bot \qquad \text{(FIRSTMO)}$$

defines frames in which there is a first moment. These two conditions are obviously independent, i.e., one can hold in a frame without the other also holding.

No First/Last Moment: Frames in which there is no last moment (for all $m \in T$ there is some $m' \in T$ such that $m < m'$), also called *serial* (or, more precisely, right serial), are defined by the formula

$$G\phi \rightarrow F\phi; \qquad \text{(NOLAST)}$$

the mirror image

$$H\phi \rightarrow P\phi \qquad \text{(NOFIRST)}$$

defines frames without a first moment, also called left-serial. (The shorter formulae $F\top$ and $P\top$, respectively, also do the job.) In alethic modal logic, the axiom of this form, $\Box\phi \rightarrow \Diamond\phi$, is sometimes called D (for 'deontic': given a deontic reading of the modalities, it expresses the desirable condition that what is obligatory is also permitted). Again, the two conditions for past and future are independent.

Density: A *dense* frame is one in which there is a moment between any two given ones: for all $m, m' \in T$ there is some $n \in T$ s.t. $m < n < m'$. The rational numbers are a well-known example of a dense ordering (in fact, any countable dense linear ordering without minima or maxima is isomorphic to the rationals). The axiom

$$F\phi \rightarrow FF\phi \qquad \text{(DENSE)}$$

defines dense frames. In fact $\mathbf{K}_t\mathbf{4}$ together with (LIN) and (DENSE) is sound and complete w.r.t. the class of dense linear orders. Again, like in the case of transitivity, this means that there must be a proof for the temporal mirror image of the axiom,

$$P\phi \rightarrow PP\phi, \qquad \text{(DENSE')}$$

which is also valid in all dense linear orders; as in the case of transitivity, the proof involves a little work.

Discreteness: In contradistinction to density, discreteness again splits into two cases, one claiming the existence of immediate successors (for any moment $m \in T$ that is not maximal, there is some $m' \in T$ for which $m < m'$ and there is no $n \in T$ s.t. $m < n < m'$) and the existence of immediate predecessors (the mirror image). The formula

$$(\phi \wedge \mathsf{H}\phi) \rightarrow \mathsf{FH}\phi \qquad \text{(RDISCR)}$$

defines right-discrete frames that have immediate successors, as is easy to check: if a moment doesn't have an immediate successor, like the number 0 in the reals, then it is possible to verify the antecedent (setting ϕ to be true up until and including the moment 0) and falsify the consequent (setting ϕ to be false at every moment after 0). The mirror image

$$(\phi \wedge \mathsf{G}\phi) \rightarrow \mathsf{PG}\phi \qquad \text{(LDISCR)}$$

defines left-discrete frames that have immediate predecessors.

Characterizing Number Systems: All of the common number systems are linear orders. Can they be captured tense-logically? From what was said above it is clear that we can characterize dense linear orders without endpoints, like the rationals ($\mathbb{Q}$). In fact the extension of the axiom system $\mathbf{K}_t$ by (LIN), (NOFIRST), (NOLAST), and (DENSE) is sound and strongly complete with respect to the class of dense linear orders without endpoints.[10] In that tense-logical system, a nice result by Hamblin shows that there are only 15 different combinations of the four tense operators, i.e., any sequence of tense operators, no matter its length, reduces to one of these 15 basic tenses [Prior, 1967, Chapter 3].[11]

Similarly to the case of the rationals, tense-logical characterizations exist for the natural numbers, the integers, and the reals. The relevant property of the reals, over and above linearity, density, and no endpoints, is Dedekind continuity, i.e., the absence of 'gaps' in the ordering. (Dedekind continuity is a second-order notion.) If a formula ϕ is first true and then false, then there has to be either a last moment at which it is true, or a first moment at which it is false. That property is characterized by the axiom

$$(\mathsf{F}\phi \wedge \mathsf{FG}\neg\phi) \rightarrow \mathsf{F}(\mathsf{HF}\phi \wedge \mathsf{G}\neg\phi) \qquad \text{(DEDEKIND)}$$

Indeed, this formula is refutable in the rationals, which has gaps at irrational numbers like $\sqrt{2}$. Let ϕ be true for all $m \in \mathbb{Q}$ for which $m < \sqrt{2}$ and false for all m with $m > \sqrt{2}$ (so a truth value has been defined for all $m \in \mathbb{Q}$). Then at 0, the antecedent is true (ϕ will be true, e.g., at 1, and forever false from, e.g., 2 on), but the consequent is false, as one can check easily. In a Dedekind continuous ordering, on the other hand, this cannot happen; if the antecedent is true, then so is the consequent.

In order to characterize the natural numbers, one needs, over and above the presence of a first and the absence of a last moment, an axiom guaranteeing that between any two moments there is only a finite number of moments. This splits in two parts, the condition of future finite intervals

$$\mathrm{G}(\mathrm{G}\phi \rightarrow \phi) \rightarrow (\mathrm{FG}\phi \rightarrow \mathrm{G}\phi) \qquad \text{(FFIN)}$$

and its temporal mirror image,

$$\mathrm{H}(\mathrm{H}\phi \rightarrow \phi) \rightarrow (\mathrm{PH}\phi \rightarrow \mathrm{H}\phi). \qquad \text{(PFIN)}$$

In order to refute (FFIN), one needs a moment at which $\mathrm{G}(\mathrm{G}\phi \rightarrow \phi)$ and $\mathrm{FG}\phi$ are true but $\mathrm{G}\phi$ is false. The latter means that there has to be at least one future moment at which ϕ is false. Given the truth of $\mathrm{FG}\phi$, on a frame with finite future intervals there has to be a *last* such moment; at that moment, $\mathrm{G}\phi$ holds. But then by $\mathrm{G}(\mathrm{G}\phi \rightarrow \phi)$, ϕ has to be *true* at that moment, a contradiction. Thus (FFIN) holds on all frames with finite future intervals; and similarly for (PFIN).

Summing up, the following extensions of $\mathbf{K}_t$ characterize the common number systems in the sense that they provide sound and strongly complete axiomatizations of the respective frames (see, e.g., [Segerberg, 1971] and [Hodkinson and Reynolds, 2007]):

- The natural numbers ($\mathbb{N}$): (LIN), (FIRSTMO), (NOLAST), (PFIN), (FFIN).
- The integers ($\mathbb{Z}$): (LIN), (NOFIRST), (NOLAST), (PFIN), (FFIN).
- The rationals ($\mathbb{Q}$): (LIN), (NOFIRST), (NOLAST), (DENSE).
- The reals ($\mathbb{R}$): (LIN), (NOFIRST), (NOLAST), (DENSE), (DEDEKIND).

Some Difficulties: The discussion of the correspondence between temporal formulae and simple frame conditions so far may suggest that this is all smooth sailing. This would however be too optimistic. First, there is no guarantee of completeness – in fact, the first example of an incomplete modal logic, given by [Thomason, 1972], was tense-logical. Second, we have already discussed the fact that irreflexivity cannot be captured by a temporal formula – and in fact, Priorean tense logic is severely limited in its expressiveness. A systematic reason for this limitation can be seen from the interrelation between tense logic and first-order logic.

General Remarks on Correspondence: The metalogical properties of tense logic can be best understood by relating tense logic explicitly to first-order logic. Indeed the semantic clauses given above can be used to establish a direct correspondence via the so-called *standard translation*, which associates a monadic first-order formula $ST_x(\phi)$ with one free variable x with each tense-logical formula ϕ. Corresponding to every atomic proposition $p, q, r, \ldots$, the first-order language has a one-place predicate $P, Q, R, \ldots$:

- If $\phi = p$, p atomic: $ST_x(\phi) = P(x)$.
- If $\phi = \neg\psi$: $ST_x(\phi) = \neg ST_x(\psi)$.
- If $\phi = \psi_1 \wedge \psi_2$: $ST_x(\phi) = ST_x(\psi_1) \wedge ST_x(\psi_2)$.
- If $\phi = \mathsf{H}\psi$: $ST_x(\phi) = \forall y\, (y < x \rightarrow ST_y(\psi))$, where y is a new variable.
- If $\phi = \mathsf{G}\psi$: $ST_x(\phi) = \forall y\, (x < y \rightarrow ST_y(\psi))$, where y is a new variable.

This translation establishes a link between the 'local' tense-logical perspective of evaluating formulae at moments, with formulae that are true or false *at* moments, and the 'global' first-order perspective on a relational structure (the model), with open formulae that are true or false *of* the elements of the structure. This shift of perspective allows one to bring to bear many tools of standard model theory and algebraic semantics.

The two most important observations resulting from this shift of perspective are:

- In the standard translation, one only really needs *two* different variables. Priorean tense logic is therefore contained in the two-variable fragment of first-order logic. This explains its expressive limitations, on the one hand, and its decidability, on the other.
- Completeness for a class of frames is a *second-order* notion, as it forces one to quantify over relational structures. This explains the incompleteness of many temporal logics.

This perspective is developed nicely in [Blackburn et al., 2002].

The technical considerations mentioned in this section all presuppose the adequacy of the model-theoretic semantics for tense logic laid out above. As mentioned, Prior had qualms about the model-theoretic presentation for metaphysical reasons. But he also raised important doubts about the semantics of the F operator, which led to a number of alternative semantics.

2.1.3 The Semantics of the Future Operator

If the future operator is to mimic the use of the future tense in natural language, or to provide a basis for discussions in the philosophy of time, the symmetrical nature of the interaction axioms of $\mathbf{K}_t$ may strike one as inappropriate. Consider first the axiom

$$\phi \rightarrow \mathsf{GP}\phi.$$

The validity of this formula follows directly from the semantics of $\mathbf{K}_t$: if ϕ is true at m, then either there is no $m' > m$, in which case the consequent is vacuously true, or one picks some $m' > m$, in which case m itself provides the sought-for moment in the past of m' at which ϕ is true. And metaphysically speaking, there seems to be nothing wrong with the following paraphrase according to the suggestions of Section 1.1: if ϕ is true now, it is always going to be the case that it was the case that ϕ is true. Intuitively, we think of the present as well as the past as fixed and unalterable, and the formula seems to express just such a sentiment. But what about the temporal mirror image,

$$\phi \rightarrow \mathsf{HF}\phi?$$

The validity of this formula can be established in exactly the same way as for its mirror image: given that ϕ is true at m, either there is no $m' < m$ (so that the consequent $\mathsf{HF}\phi$ is vacuously true), or one chooses some such $m' < m$, for which $m > m'$ then witnesses the truth of the consequent. But what about the corresponding paraphrase: if ϕ is true now, it has always been the case that it will be the case that ϕ is true? This seems to suggest the dubious view that whatever is so now, was going to be so anyway. And that is not in accord with our view of the world, in which we acknowledge indeterminism of various forms.[12] The point here is not that we have to assume indeterminism and therefore the formula is inadequate. Indeterminism is a strong metaphysical thesis, and it should be open to discussion. The point is just that the formula seems to make such a discussion impossible.

But *is* the formula really metaphysically inappropriate? Is the paraphrase appropriate at all? *Does* F express something close to the future tense – and not merely something like 'it may be the case that'? The point is important and subtle. Since the formula is simply a validity, there are two ways to go: either to adhere to the interpretation of F as 'it will be the case that' and to show that the formula is not in fact philosophically inappropriate (or at least, not inappropriate in certain important cases), or to interpret F differently and to give a different semantics for a more adequate future operator.

As to the first option, once one restricts attention to linear frames, the problems vanish: there is no branching, just a single linear chain of future moments. Both interaction axioms have exactly the same status, there is no remaining philosophical issue. Linear frames are important in many applications, and they are especially well studied in computer science. Are non-linear, branching frames ever needed? There is certainly room for debate. However, one fairly obvious philosophical argument will not do: the fact that we experience the flow of time as linear, and that we give names (viz., dates and times) to moments in a linear fashion, proves nothing. It is simply a conceptual truth that we experience one moment at a time, not multiple incompatible moments at once.

In the absence of a conclusive argument for restricting attention to linear frames describing only a single possible future (thus, a deterministic flow of time), tense logic should be open towards other flows of time – and especially towards ones that incorporate the idea of an open future. The struggle for a suitable semantics for the future operator occupied a central place in the early development of tense logic. Prior, who was also interested in questions of freedom and predestination, rejected the basic semantics for $\mathbf{K}_t$ given above. Indeed his first approach to the topic ([Prior, 1957]) resulted in a fairly complicated predicate logical system in which reference to nonexistent entities was a major issue. (The problem is still there to haunt temporal predicate logic.)

In the propositional case that we are dealing with here, there are two main options for alternative semantics, which are known as 'Peircean' and 'Ockhamist', after the nineteenth-century American philosopher C. S. Peirce and the medieval logician William of Ockham whom Prior ([Prior, 1967]) cited as sources of inspiration. Both alternative semantics for F are based on frames that are backwards linear (left linear), so that the past operator P, based on a linear past, poses no additional difficulties.

In a backwards linear frame $\langle T, < \rangle$, also called *branching time*, one can single out the set *Hist* of maximal linear subsets (maximal chains) of T. These maximal linear subsets are called *histories*, the idea being that a maximal chain pictures a complete possible course of events: a possible history of the world. The set of all histories containing a moment m, which is a subset of *Hist*, is written H_m; this set may be viewed as the set of all possible futures of m complemented by the unique past of m and m itself. Single histories are denoted h, h' etc. Histories h, h' are called *undivided* at a moment m ($h \equiv_m h'$) iff they share a moment m' that is properly later than that moment ($m' > m$). On plausible topological conditions on $\langle T, < \rangle$, $\equiv_m$ is an equivalence relation inducing a partition of H_m that may be thought of as the set of immediate future possibilities at m.

From backwards linearity it follows that the set of histories gets smaller as time progresses: if $m < m'$, then $H_m \supseteq H_{m'}$. This accords well with our intuitions about the flow of time excluding ever more of the once open possibilities; witness, e.g., the notion of a missed opportunity, or the wisdom in 'no use crying over spilt milk'. The Peircean and the Ockhamist semantics for F both make use of the notion of histories in branching time models, but they differ as to the role that histories play.[13]

Peircean F: The discussion of the controversial interaction axiom $\phi \rightarrow \mathsf{HF}\phi$ also reveals another counterintuitive feature of the $\mathbf{K}_t$ semantics for F: too many future-tensed sentences turn out to be true. If ϕ will be true in one history from H_m and $\neg\phi$ in another, then both $\mathsf{F}\phi$ and $\mathsf{F}\neg\phi$ count as true at m – even if ϕ explicitly pertains to a specific date. This clashes with our idea that at a certain date, things will be one way or other, but not both ways. How then could both $\mathsf{F}\phi$ and $\mathsf{F}\neg\phi$ be true? The Peircean idea is to make many fewer future-tensed sentences true by considering a future-tensed sentence $\mathsf{F}_{\text{Peirce}}\phi$ to be true exactly if ϕ will be true *no matter what* – i.e., if it will be true in all (currently still) possible histories. Formally:[14]

- If $\phi = \mathsf{F}_{\text{Peirce}}\psi$: $\mathcal{M}, m \models \phi$ iff for all $h \in H_m$ there is some $m' \in h$ s.t. $m < m'$ and $\mathcal{M}, m' \models \psi$.

While this move resolves the problem of inconsistent truths about the future, it makes nonsense of many of our other ideas about the future. E.g., when I bet that it will rain tomorrow, I am not saying that there is no chance that it won't rain – I am just claiming that it is *actually* going to rain. Indeed, in betting, one of the main points seems to be that both parties agree that the outcome is contingent, not fixed one way or other – that's part of the fun! But if I say that it will rain tomorrow, and I acknowledge that it could fail to rain, then on the Peircean semantics I am contradicting myself: for a future-tensed sentence to be true, there cannot be a possible history on which it is false. It seems that too few future-tensed sentences are true on the Peircean reading.

Ockhamist F: The Ockhamist semantics was sketched by Prior ([Prior, 1967]) and developed fully by Thomason ([Thomason, 1970], [Thomason, 1984]). It amounts to a substantial revision of the standard semantics. The basic idea is to evaluate a formula not just at a moment, but at a moment/history pair m/h (where $m \in h$).[15] As histories are linear by definition, the past and future tense operators are again symmetrical, like in $\mathbf{K}_t$:

- If $\phi = \mathsf{H}\psi$: $\mathcal{M}, m/h \models \phi$ iff for all $m' \in h$ s.t. $m' < m$, we have $\mathcal{M}, m' \models \psi$.
- If $\phi = \mathsf{G}\psi$: $\mathcal{M}, m/h \models \phi$ iff for all $m' \in h$ s.t. $m < m'$, we have $\mathcal{M}, m' \models \psi$.

In the clause for H, the requirement that $m' \in h$ is in fact superfluous since T is backwards linear anyway – but in the clause for G, the requirement does some useful work. The derived semantic clause for F (remember that $\mathsf{F}\phi \Leftrightarrow \neg\mathsf{G}\neg\phi$),

- if $\phi = \mathsf{F}\psi$: $\mathcal{M}, m/h \models \phi$ iff there is some $m' \in h$ s.t. $m < m'$ and $\mathcal{M}, m' \models \psi$,

shows that the problem of inconsistent futures for the same date vanishes: histories, being linear, are consistent. Furthermore, it becomes possible to combine tense and modality by defining the so-called *historical modalities* 'Poss' and 'Sett' (for 'possibly' and 'it is settled that'):

- If $\phi = \mathsf{Sett}\,\psi$: $\mathcal{M}, m/h \models \phi$ iff for all $h' \in H_m$, $\mathcal{M}, m/h' \models \psi$.
- If $\phi = \mathsf{Poss}\,\psi$: $\mathcal{M}, m/h \models \phi$ iff there is some $h' \in H_m$ s.t. $\mathcal{M}, m/h' \models \psi$.

As can be seen from these semantic clauses, Poss is the dual of Sett, and could as well be introduced as an abbreviation ($\mathsf{Poss}\,\phi :\Leftrightarrow \neg\mathsf{Sett}\,\neg\phi$).

On the Ockhamist semantics, $\phi \rightarrow \mathsf{HF}\phi$ is valid, but that doesn't cause any interpretational difficulties anymore, since a (linear) history has to be specified. The deterministic sentiment connected with that formula in basic tense logic is now expressed more clearly as $\phi \rightarrow \mathsf{HSett}\,\mathsf{F}\phi$, and that formula is not a validity.

Given the historical modalities, it is possible to define the Peircean F in the Ockhamist framework:

$$\mathsf{F}_{\mathrm{Peirce}}\phi \quad :\Leftrightarrow \quad \mathsf{Sett}\,\mathsf{F}\phi.$$

So, if we do want to talk about inevitable future happenings in the Ockhamist framework, we can do that – but the framework is vastly more expressive. Does the Ockhamist solution therefore deliver *the* adequate semantics for the future tense? Opinions are divided. The main technical *and* philosophical issue is the question of the initialization of the h parameter in a stand-alone sentence. In standard tense logic, there is one *mobile parameter*, m. This parameter is shifted by the tense operators. In a stand-alone sentence, uttered at a moment of context m_C, m gets its initial value from that context: one starts evaluating with $m = m_C$. Now for a stand-alone sentence containing an Ockhamist future operator, we also need a history h to evaluate its truth value – but where do we get that history from? One idea is that the context also furnishes a 'history of the context': *the* real future. But arguably that will lead to a collapse of branching time to linear time: if there is a unique real future, isn't the rest of the model just sugaring? (See [Belnap et al., 2001] for discussion, and [Belnap, 2007] for a detailed and systematic, general overview of parameters of truth.) Thomason ([Thomason, 1970]) suggests to use the technique of supervaluations to resolve the issue. In supervaluational semantics, uninitialized parameters are quantified over, and a truth value is assigned only if the verdict is uniform; otherwise, no truth value is assigned. Accordingly, if a stand-alone sentence is declared to be true, it is also settled true. In this respect, supervaluational semantics agrees with Peirceanism to a certain extent, but some of the unwelcome consequences of the semantics of $\mathsf{F}_{\mathrm{Peirce}}$ are avoided – e.g., unlike in Peirceanism, $\mathsf{F}\phi \vee \mathsf{F}\neg\phi$ does turn out to be a validity (the disjunction is true on every history). Still there seems to be a difference between truth and settled truth that the supervaluational framework is unable to capture: the supervaluational framework is arguably unable to account for our assessment of bets. A better option may therefore be to say that a future-tensed sentence can simply fail to have a truth value at its time of utterance, but later on (when more things have become settled) it will be true to say that at the time of utterance, the sentence was true (or false). That seems to accord better with the way we handle bets: once the outcome is settled, we assess the bet – not earlier. (See [Belnap, 2001] for technical details of such so-called double time references.) For some more comments on the formal analysis of tense in natural languages, see Section 3.1 below.

Time and Modality: We have seen that the Ockhamist framework affords a well-integrated notion of (historical) possibility and necessity based on branching time. For the record, there are also *purely temporal* interpretations of possibility and necessity, and these played an important role in the development of tense logic and in its application to historical sources. One idea is to call that necessary which is now and will always be:

$$\Box_D\phi :\Leftrightarrow (\phi \wedge \mathrm{G}\phi).$$

This interpretation has been claimed to capture the notion of necessity in the writings of the Stoic philosopher Diodorus Cronus, and is accordingly called the Diodorean modality. Note that this modality also makes sense in linear time.

A similar notion of necessity that has been related to the works of Aristotle, the so-called Aristotelian modality, identifies necessity with truth at all times; in linear time:

$$\Box_A\phi :\Leftrightarrow (\mathrm{H}\phi \wedge \phi \wedge \mathrm{G}\phi).$$

Cf., e.g., [Rescher and Urquhart, 1971, Chapter 1] for some pointers to source texts.

2.1.4 The Indexical 'Now'

In English, words like 'I', 'you', 'here' and 'now' are special because their reference depends on the context of utterance. The semantical analysis of such indexical expressions went hand in hand with the development of tense logic, because the tenses themselves are indexical: past and future localize events relative to the present moment that 'now' refers to. Whether the sentence 'It was raining yesterday' is true or false depends on the weather on the day before the utterance is made, which may differ for different utterances of the same sentence. Can one get rid of this indexical context dependence? As mentioned, Frege suggested that an explicit date could always be substituted; Russell considered a token-reflexive analysis in which the content would refer back to the utterance act producing the sentence token. These approaches were however later seen to be inadequate vis-à-vis the phenomenon of so-called essential indexicality. Prior ([Prior, 1959]) gave the following example: when I say 'thank goodness that's over', what I am thankful for is not that something is finished before a certain date (I knew *that* all along), nor that something is finished before a certain utterance of mine (what a weird thing to be thankful for). Rather, it seems to be the thing's being past, an indexical feature, that I am thankful for. (Cf., e.g., [Perry, 1977] for more on essential inexicality, and [Mellor, 1998] and [Oaklander and Smith, 1994] for a discussion of consequences for the philosophy of time.)

Given that indexicality won't go away, semantic analysis has to make room for it. The simple rule for 'now' is the following:

- If $\phi = \mathsf{Now}\ \psi$: $\mathcal{M}, m_C, m \models \phi$ iff $\mathcal{M}, m_C, m_C \models \psi$.

Thus, the 'Now' operator (sometimes also written 'J', from the German 'jetzt') simply shifts back the moment of evaluation m to the moment of the context, m_C (which has here been mentioned explicitly as a parameter of evaluation as well).

Does the presence of this operator enhance the expressive power of tense logic? Kamp ([Kamp, 1971]) gave a proof that 'Now' can in fact be eliminated from propositional tense logic – the relevant observation is that the operator is redundant at the beginning of a stand-alone sentence, and that there are transformations of propositional tense-logical formulae into a normal form that always puts 'Now' up front. The operator is however not eliminable in quantified tense logic, nor in the context of propositional attitudes (cf. [Burgess, 2002, pp. 32f.]).

The insights gained from the formal study of 'Now' have been generalized in various treatments of indexical semantics, sometimes known as 'two-dimensional semantics'; note that that term is used in a number of different ways.

2.1.5 Metric Operators and At_t

The P and F operators, Ockhamist or not, do not distinguish between the near and the far past, or future. We do, for obvious reasons – we want to be able say that it will rain tomorrow, or on New Year's Eve 2025; not just that it will rain sometime or other.

If time is discrete, as is presupposed in many computer science applications, then there is a natural way of defining a 'previous time' and a 'next time' operator (sometimes written 'Y' and 'X', for 'yesterday' and 'next'), and one can build up further operators from these. In general, however, there is no natural 'next time'. Still, metric tense operators have some popularity; they were extensively used by Prior ([Prior, 1967]). Such operators are often written P_x and F_x, and there are suggestive combination axioms such as, e.g.,

$$\mathsf{F}_x \mathsf{F}_y \phi \quad \leftrightarrow \quad \mathsf{F}_{x+y} \phi,$$

sometimes with real numbers, sometimes with integers as index. In order to have a well-defined semantics for these operators, the underlying models need to have an appropriate metric structure – e.g., all of the model's histories can be required to be isomorphic to the reals. Such structure can also be used to define 'At_t' operators for indicating dates and times. In an Ockhamist framework, these operators will shift the moment of evaluation along the current history to the moment on that history that has the appropriate clock time. As one can see, it pays to distinguish the elements of the partial ordering $\langle T, < \rangle$ from clock times: there can be many incompatible elements of T at one and the same clock time, in different histories. Calling the elements of T 'times' tends to blur this distinction, which is why we call the elements of T 'moments'.

Formal details about 'At_t' are given, e.g., in [Belnap et al., 2001, Chapter 8].

2.2 Temporal Logic with 'Since' and 'Until'

Hans Kamp not only contributed to the formal study of indexicals starting with 'now', but also introduced two more general operators into temporal logic: 'Since' (S) and 'Until' (U). Due to their greater expressive power these operators have found widespread application and are often taken as basic in advanced approaches to temporal logic. (Details of the semantics differ somewhat between the computer science and philosophy communities; we'll stick to the philosophy convention of [Kamp, 1968]. Cf. [Pnueli, 1977] for the origin of the other tradition.)

S and U are two-place operators; unlike Prior's connectives, they do not modify a single sentence, but combine two sentences in a temporally specific way, in analogy with conjunction. While one can also find them written in infix notation like conjunction, they are mostly written in prefix notation with explicit parentheses, and we adopt that convention. $\mathsf{S}(\psi_1, \psi_2)$ means that there was a past moment at which ψ_1 was true, and that ψ_2 has been true since then; U is the temporal mirror image. Formally, the clauses are the following:

- If $\phi = \mathsf{S}(\psi_1, \psi_2)$: $\mathcal{M}, m \models \phi$ iff there is some $m' \in T$ s.t. $m' < m$ for which $\mathcal{M}, m' \models \psi_1$, and for all n for which $m' < n < m$, we have $\mathcal{M}, n \models \psi_2$.
- If $\phi = \mathsf{U}(\psi_1, \psi_2)$: $\mathcal{M}, m \models \phi$ iff there is some $m' \in T$ s.t. $m < m'$ for which $\mathcal{M}, m' \models \psi_1$, and for all n for which $m < n < m'$, we have $\mathcal{M}, n \models \psi_2$.

If the S and U connectives are present, the Priorean connectives can be defined easily:

- $\mathsf{P}\phi :\Leftrightarrow \mathsf{S}(\phi, \top)$;
- $\mathsf{F}\phi :\Leftrightarrow \mathsf{U}(\phi, \top)$.

Also the 'previous time' and 'next time' operators mentioned in Section 2.1.5 above are definable (obviously these definitions make good sense only if the underlying flow of time is discrete):

- $\mathsf{Y}\phi :\Leftrightarrow \mathsf{S}(\phi, \bot)$;
- $\mathsf{X}\phi :\Leftrightarrow \mathsf{U}(\phi, \bot)$.

Kamp introduced these connectives in his study of linear orders [Kamp, 1968]. It is customary to presuppose linear time (i.e., a model based on a linear frame) when discussing S and U. For the tense logic of Dedekind complete linear orders, Kamp showed that in fact any temporal operator that has a first-order truth definition can be defined on the basis of S and U. This important, early expressiveness result can be extended to linear orders generally, using an alternative set of connectives named after Jonathan Stavi (cf. [Gabbay et al., 1994, Chapter 6.3.9]).

The rich logic of S and U has been studied closely, especially with respect to computer science applications. For a detailed overview cf., e.g., [Finger et al., 2002].

In terms of viewing temporal logic as a fragment of first-order logic via the standard translation, it turns out that the tense logic with 'since' and 'until' is contained in the three-variable fragment. (A look at the semantic clauses already suggests as much.) As Priorean tense logic is already contained in the two-variable fragment, the greater expressive power of 'since' and 'until' (as well as of the Stavi connectives) can readily be accounted for. Since that greater expressive power is bought at the price of a rather modest increase in complexity, since-until-logics are used in many applications.

3. Some Further Topics

The overview of temporal logic given here is necessarily limited. We briefly look at a number of selected further topics.

3.1 Temporal Logic and Natural Languages

Is the tense-logical approach laid out here a useful approach to capturing the temporal phenomena of natural language? Certainly consideration of the English or Latin system of tenses has provided important motivation for the development of tense logic, but those logical systems also have an independent systematic interest – adequacy with respect to the linguistic facts is not necessarily crucial. Still it is interesting to ask how close the match really is between the formal systems and natural languages. (In what follows we stick to a discussion of English, which is sufficiently rich to point out the main issues.)

There are at least two ways in which one can understand the mentioned question. Is tense logic an adequate representation of the English verbal tenses? And if that is so, is tense logic also adequate vis-à-vis all the other temporal features of English?

The answer to the first question in fact depends on what one understands by 'tense'. Traditionally, the English simple past, past continuous, perfect, present tense, the 'will' future, etc., are all called tenses. Among linguists it has however become customary to limit the notion of tense to the specification of temporal location relative to the present. In this narrow sense, tense logic is indeed adequate as a formal means of expressing the English tenses – however, one could almost say that this holds by definition. Still even here there is room for debate. Reichenbach ([Reichenbach, 1947]), in an influential chapter on 'The tenses of verbs', proposed a scheme in which a fixed number of three moments – point of speech, point of reference, and point of event – are used to analyse all verbal tenses. In comparison, tense logic appears to be both too weak (it only allows for two moments to play a role in the analysis) and too strong (its recursive machinery allows tense operators to be iterated). Prior ([Prior, 1967, Ch. I.6]), on the other hand, while mentioning Reichenbach as one of the 'precursors of tense logic', criticizes the non-recursiveness of the Reichenbachian scheme.

Considering the wider sense of 'tense', it is clear that tense logic by itself is unable to express the difference between, e.g., the simple past ('I made tea') and the past continuous ('I was making tea'). Linguistically, this difference is considered to be not one of tense proper, but one of *aspect*: the simple past pictures a happening from the point of view of its completion (perfective aspect), while the past continuous pictures a happening as incomplete (imperfective aspect). Thus, an action that was never completed may be reported in the imperfective aspect ('I was making tea when the phone rang, I completely forgot about it and in fact never made tea'), but not in the perfective aspect.

Tense logic is not, and was not meant to be, a logic of aspect. For two rather different approaches to such a logic, see [Galton, 1984] and [van Lambalgen and Hamm, 2005]. At any rate, it is clear that there is more to the temporal structure of English – and most, if not all other languages – than what is captured by the formal framework of tense logic.

3.2 Tense Logic and Relativity Theory

Early on, tense logic was confronted with a scientific objection along the following lines: Tense logic, with its notion of past and future, is wedded to a Newtonian picture of time and space that has become obsolete through the advent of relativity theory. We have come to see that there isn't space and time separately, but just a space-time manifold that can be foliated (divided into space and time) in different ways, none of which is physically distinguished. Your rest frame is as good as mine. So, why bother developing a tense logic at all [Massey, 1969]?

One reaction to this worry is to develop a tense logic that is based on the Minkowskian partial ordering of causal connectability in special relativistic space-time. In that ordering, which is branching to the past as well as to the future, incomparability is not a sign of incompatibility like in branching time, but of space-like relatedness, or causal isolatedness. A number of interesting results in this area are surveyed by [Uckelman and Uckelman, 2007].

While this body of work manages to reconcile temporal logic and relativity in some fashion, one may wonder whether it isn't possible to establish a *purely temporal* tense logic before a relativistic background. This is still an area of debate, especially among those who want logic to have some metaphysical import. One option is to argue for a physically distinguished rest frame after all (e.g., based on some notion of cosmic time), which would single out a preferred foliation and thus a preferred time coordinate; another option is to work on the assumption of a *metaphysically* distinguished time coordinate. Rakić ([Rakić, 1997]) has shown that this move is logically unobjectionable; both options would allow one to stay within the standard Priorean paradigm. A further option is to extend tense logic to cover relativistic space-time in another way: somewhat parallel to the Ockhamist move of introducing a history as an additional parameter in the semantic clauses, one can include a reference frame as a further mobile parameter of truth. This move is unproblematic in so far as a given utterance context supplies a reference frame of the context, viz., the speaker's rest frame.

Even though Minkowski space-time is only partially ordered, the discussion so far parallels the case of linear time. A semantic parallel to branching time, *branching space-times*, has been developed by Belnap [Belnap, 1992]. The logical properties of branching space-times are the subject of ongoing research; cf., e.g., [Müller, tab] for pointers to the recent literature.

3.3 Temporal Predicate Logic

The extension of a propositional modal logic to a system of quantified modal logic is always an intricate matter. In fact, many choices need to be made – cf., e.g., [Garson, 2006] for a broad, philosophically oriented overview, and Chapter 11 for a succinct presentation of axiomatic and interpretational issues. If anything, the construction of a system of quantified *tense* logic is conceptually even more difficult than in the case of alethic modal logic. This may to a large part be due to the fact that we have fairly firm but not completely coherent intuitions about things and their persistence through time.

Many questions need to be answered. What are we talking about? What should the universe of discourse for a temporal predicate logic look like? Should it contain ordinary individuals like you and me, which, as we all know, come to be and pass away, leading to different domains at different moments and the difficult question of what to do with names of mere have-beens and mere will-bes? Or should we try to capture (maybe as a first step) a temporal predicate logic of some eternal and perhaps ultimately simple objects? And if so, what would these be? Anyone who has tried to come to grips with atomistic philosophies like Wittgenstein's *Tractatus* knows that these are deep and troubling questions. The issue may look 'merely metaphysical', but it has obvious and far-reaching consequences for the logical formalism. Logical modelling and metaphysical argument tend to go hand in hand here, again substantiating the claim made at the beginning, viz., that tense logic is a philosophically especially rich subject.

A key question revolves around the so-called temporal Barcan formula,[16]

$$\forall x\, \mathsf{G}\phi(x) \rightarrow \mathsf{G}\forall x\, \phi(x),$$

or equivalently,

$$\mathsf{F}\exists x\, \psi(x) \rightarrow \exists x\, \mathsf{F}\psi(x).$$

While this formula helps one to simplify the semantics, it seems intuitively questionable: If it will be the case that there is someone who is flying to Mars, does this imply that there is someone for whom it will be the case that she is flying to Mars?

It seems safe to say that no consensus has been reached with respect to predicate tense logic, and a pragmatic attitude towards the construction of such systems may often be the most satisfactory choice.

3.4 Branching Time and the Logic of Agency

The tempo-modal framework of Ockhamist branching time has been used as a basis for logics of agency, again bringing formal and philosophical issues very closely together. The idea behind such logics is to view truth due to agency as a modus of a proposition, an idea that can be traced back to Anselm of Canterbury. In *stit* logics, developed by Belnap and collaborators, a branching time model is enriched with agent's choices at moments: Given a set A of agents, for each agent $\alpha \in A$ and each moment $m \in T$ there is a partition $Choice^{\alpha}_{m}$ of H_m that respects undividedness at m ('no choice between undivided histories') and that forms the basis for the semantics of the *stit* ('seeing to it that') connective. $Choice^{\alpha}_{m}(h)$ indicates that element of the partition that contains h (presupposing $h \in H_m$). The semantic clause then is as follows:

- If $\phi = [\alpha\ \text{Stit}]\psi$: $\mathcal{M}, m/h \models \phi$ iff
  (1) for all $h' \in Choice^{\alpha}_{m}(h)$, $\mathcal{M}, m/h' \models \psi$; and
  (2) there is some $h^* \in H_m$ for which $\mathcal{M}, m/h^* \not\models \psi$.

Here (1) is a positive condition requiring *effectiveness* of the choice: the embedded sentence ψ has to be true on all histories belonging to the choice under consideration (the choice has to secure the outcome, so to speak). The negative condition (2) ensures that, e.g., nobody sees to it that $2 + 2 = 4$: agency, on that view, presupposes contingency of the outcome. Horty ([Horty, 2001]) bases his development of a *stit* logic on a connective without condition (2); this move is popular in computer science applications. Cf., e.g., [Broersen, 2009]; cf. [Belnap et al., 2001] for a philosophy-centered overview.

4. Suggested Further Reading

The classic of the field is [Prior, 1967], which contains a wealth of background material and philosophical discussions. A more modern and more mathematically oriented treatment is given by [van Benthem, 1991]. [Belnap et al., 2001] has a strong focus on philosophical issues of branching time. The computer science oriented development of temporal logic is discussed in depth in [Gabbay et al., 1994, Gabbay et al., 2000].

Notes

1. For the terminology of modal systems as well as for more on Lewis' motivation cf. Chapter 11.
2. In fact, Prior and other early tense logicians used Polish prefix notation, in which one-letter connectives are always put *before* their arguments, making parentheses and special symbols unnecessary. The connectives are written *N* ('negation'), *C* ('conditional'), *K* (think 'konjunction'), *A* ('alternation', or disjunction), and *E* ('equivalence'). Tense operators are as stated; alethic modal operators are often symbolized as *M* (for the weak modality, possibility, 'möglich') and *L* (necessity). Handling that notation requires a little training. (As an exercise, you may wish to translate the formula *CACCKppqpNKpqq* into standard infix notation, or transcribe some of the proofs in Prior's original writings.)
3. We give the official semantic rules for the *strong* operators here, treating the weak operators as defined. This way we avoid having to discuss a fine point about axiomatization later; cf. [Blackburn et al., 2002, p. 34].
4. As in the propositional calculus, the use of axiom schemata is thus just a small convenience. In a similar vein, many formal developments of propositional modal or tense logic do not hesitate to simply add the whole set of propositional tautologies to the axiom set. The important point to keep in mind is that such moves are just a convenience – in contradistinction to, e.g., the axiomatization of Peano arithmetic in first-order logic, where an axiom schema is needed essentially. First-order Peano arithmetic is *not* finitely axiomatizable; the logics we are dealing with here are finitely axiomatizable. (Thus in fact, a single axiom, the conjunction of the separate axioms, would be sufficient.)
5. In fact, if one views temporal logic as a bimodal logic based on two *independent* accessibility relations, these two axioms are strong enough to force the accessibility relations to be each other's converse.

6. A tableaux-based proof is given, e.g., in [Rescher and Urquhart, 1971, Chapter VI].
7. It is possible to tackle irreflexivity syntactically at the level of *rules*: The so-called Gabbay rule, from an unconditional proof of $\neg q \wedge \mathsf{H}q \rightarrow \phi$ to infer ϕ (where q is an atomic proposition not occurring in ϕ), is valid for $\mathbf{K}_t$. While this adds no new theorems in $\mathbf{K}_t$, it forces irreflexivity for extensions of $\mathbf{K}_t$. Cf., e.g., [Gabbay et al., 1994, Chapter 3.2].
8. As in modal logic (cf. Chapter 11) it is customary to specify names of tense-logical systems by concatenating the names of the characteristic axioms.
9. This assertion has to be taken with a grain of salt. Certainly every linear frame is both backward and forward linear. But full trichotomy also requires that there be only *one* line – any two points have to be comparable – while both backward and forward linearity holds, e.g., if there are two unconnected lines. In general, modal formulae cannot 'see' disjoint unions. Cf. [Blackburn et al., 2002, Ch. 3] for a survey of this and similar results leading to the celebrated Goldblatt–Thomason theorem.
10. Due to the 'local' nature of the tense-logical modalities, global characteristics like the cardinality of such a flow are beyond the power of tense-logical characterization. To give a very simple example, no tense-logical characterization can distinguish between a single copy of the rationals and any number of disjoint copies (where that number could be any infinite cardinal). See also note 9.
11. From here it is a short step to speculations that the limited amount of explicit tenses in English may be due to some such collapsing property.
12. To cite Prior on the matter: 'Before God, or whoever it is that is responsible for these things, had decided to make parrots, "There are going to be parrots" wasn't true at all' [Prior, 1976, p. 128] .
13. The philosophy-driven discussion given in what follows is parallelled in computer science by a discussion of the relative merits of (Peircean) 'computational tree logic' CTL and its Ockhamist alternative, CTL*. We will keep the discussion at the philosophical level, referring the computer-science interested reader instead to, e.g., [Gabbay et al., 2000, Chapter 3]. Cf. also [Øhrstrøm and Hasle, 1995] for more historical background information.
14. A simplification is possible in discrete time: One can simply do away with the F operator, leave the semantics for G as in $\mathbf{K}_t$ above, and give a specific semantics to the 'next time' modality X (see Section 2.1.5): instead of the F-analogue existential quantification over immediate successors, one can read X as 'inevitably, at the next time step', employing universal quantification over immediate successors. This move avoids the second-order aspects of quantification over branches, and thus leads to a much more tractable logic. Cf. [Gabbay et al., 2000, p. 65]. At any rate, the interpretational problems mentioned below remain.
15. The question arises whether the valuation V should remain, like in standard tense logic, a function from the atomic propositions to the powerset of T. Prior ([Prior, 1967, p. 123]) discusses the issue; Thomason ([Thomason, 2002, p. 214]) gives an argument to the effect that V should map the atomic propositions to subsets of $T \times Hist$ instead: this way, all instances of the future necessitation rule could be preserved. Lacking a full axiomatic framework for Ockhamist tense logic, we leave this matter open.
16. Interesting questions can also be asked with respect to the formula's converse; again, see Chapter 11 for more information.

13 Truth and Paradox

Leon Horsten and Volker Halbach

Chapter Overview

1. The Problem of the Liar Paradox	352
2. Axiomatic versus Semantic Truth Theories	353
3. Typed Disquotational Theories	355
3.1 Tarski's Undefinability Theorem and the Naive Theory of Truth	355
3.2 The Disquotational Theory	357
3.3 The Soundness of the Disquotational Theory	358
3.4 The Tarskian Hierarchy	359
3.5 Contextualist Theories	361
4. Typed Compositional Theories	363
4.1 The Compositional Theory of Truth	363
4.2 Truth and Satisfaction	365
4.3 The Power of Truth	366
5. Type-Free Disquotation	368
5.1 Type-Free Truth	368
5.2 The Strength of Type-Free Disquotation	368
5.3 Positive Disquotation	369
6. Kripke's Theory of Truth	370
6.1 Partial Models for Reflexive Truth	370
6.2 Properties and Variations	373
6.3 Axiomatising Kripke's Theory	376
7. The Revision Theory of Truth	378
7.1 Two Revision-Theoretic Notions of Truth	378
7.2 The Friedman-Sheard Theory	380
8. Other Approaches and Further Reading	382
Notes	382

1. The Problem of the Liar Paradox

The immediate and most natural suggestion for at least a minimal theory of truth consists in the following axiom schema:

A sentence 'ϕ' is true if and only if ϕ.

Sentences of this form are known as *Tarski-biconditionals*, named after Alfred Tarski, who put principles of this form centre stage in the discipline of truth theories. The Tarski-biconditionals seem eminently natural and plausible. They seem to latch onto what we may call the disquotational intuition, which can be canvassed as follows. If you are prepared to sincerely assert in a suppositional or non-suppositional mode a sentence ϕ, then you had better also be prepared to assert (in the same mode) that ϕ is true. And if you are prepared to sincerely assert in that mode that ϕ is true, then you had better also be prepared to assert ϕ in that mode. What can be more obvious than that?

But now consider the liar sentence L which says of itself that it is not true. The 'naive' axiom scheme which we have just proposed tells us that L is true if and only if L. But L if and only if L is not true – for this is what L says of itself. So L is true if and only if L is not true: a short truth table calculation convinces us that we have lapsed into inconsistency. This line of reasoning is known as the argument of the *liar paradox*.[1]

The liar sentence refers to itself, and one might suspect that self-reference is ultimately incoherent. So it is not immediately clear how convincing this argument against the naive theory of truth really is. But Gödel has *shown* that self-reference is coherent. He articulated a mathematically precise way in which sentences in a sufficiently expressive language can talk about themselves (via coding). And, as we will see, Tarski showed how in such a self-referential language, the argumentation of the liar paradox can be carried out. So we must do better. And it turns out that it is hard to do well. Formulating a satisfactory list of axioms for the notion of truth is a very difficult task. But it is an important one, for the concept of truth has been a key concept in philosophy since Plato.

Before embarking on our mission, let us discuss some preliminary matters. In the formal approach that we shall be pursuing, we shall mostly work in the language of first-order Peano arithmetic, augmented with one or more truth predicates. This language is intended to be a toy version of a much more complicated and interesting language: English. The reason why aside from logical notions and one or more truth predicates, arithmetical notions are present in the language, is the following. The truth of sentences depends to some extent on the syntactical structure of sentences. It is reasonable to assume, for instance, that if two sentences ϕ and ψ are true, that then their conjunction $\phi \wedge \psi$ is true. Also, in order to investigate the argument of the liar paradox we have to at least simulate self-reference. Both tasks can be carried out admirably in an arithmetical setting. The syntactical structure of sentences is ultimately a finite combinatorial structure which can be described in the language of arithmetic. And Gödel discovered how, given a coding scheme, the language of arithmetic contains sentences which talk about themselves.

The most natural axiomatic theory of arithmetic is first-order Peano arithmetic (*PA*), formulated in the language $\mathcal{L}_{PA}$. In this chapter we shall assume familiarity with this system. Thus, given a coding scheme, Peano arithmetic can reason about the syntactical structure of sentences, and even about self-referential sentences. It is important to keep firmly in mind that for our purposes Peano arithmetic serves first and foremost as a way of talking and reasoning about syntax. We are not interested in the natural numbers *per se*. As our background theory, we *could* also have used a theory which *directly* describes the structure of expressions. This idea was worked out in detail in [Quine, 1946] and in [Smullyan, 1957]. Our reasons for opting for Peano arithmetic instead are twofold. First, in the literature on theories of truth and the semantic paradoxes, it is standard practice to take Peano arithmetic as a background theory. And one of the aims of this chapter is to function as an inroad to the logico-philosophical literature on truth. Second, theories about expressions are not as simple and elegant as Peano arithmetic.

Since we work in the language of arithmetic, truth predicates really apply to codes of sentences, i.e., numbers, rather than to sentences themselves. And if we want to be scrupulously precise, this will show up in the notation. Nevertheless, for the benefit of readability, in this chapter we will forsake the details of coding. Thus we will write, for example, $T(\phi \wedge \psi)$ ('it is true that $\phi \wedge \psi$'), even though this is strictly speaking not syntactically well-formed: the truth predicate takes terms and not sentences as arguments. In the sequel, we shall assume that our sole primitive logical connectives are $\neg$, $\wedge$, $\forall$ (and the identity symbol). The other logical connectives (such as $\rightarrow$, $\exists$, $\leftrightarrow$) are taken to be defined in terms of the primitive connectives in the customary, classical way.

In the sequel we will not spell out the proofs of all the theorems that we discuss. But we will do our best to direct the reader to places in the literature where good expositions of the relevant proofs can be found.

2. Axiomatic versus Semantic Truth Theories

In the aftermath of logical empiricism, where truth was regarded as a relic of a metaphysical past, Tarski rehabilitated truth as a respectable notion in his seminal article 'The concept of truth in formalized languages' [Tarski, 1935]. In this article he gave a *definition* of truth for a formal language in purely logical and mathematical terms. Tarski never attempted to define truth for mathematical English as a whole. Instead, he defined truth for fragments of (mathematical) English that do not themselves contain the truth predicate. And, most importantly, he imposed an adequacy condition on any definition of truth: a definition of truth for a fragment of English should imply all the sentences of the form 'It is true that ϕ if and only if ϕ' for ϕ being any sentence of the fragment in question. This is his condition of *material adequacy* for truth definitions.

It is well known that Tarski also invented the logical notion of a *model* for a formal language, and explained what it means for a sentence of a formal language to be *true in a model* (cf. Chapter 3). Giving a model for a formal language can be seen as giving a *semantic* theory of truth for that formal language.

Some of the most popular contemporary semantic theories of truth and the semantic paradoxes are Kripke's theory of truth and the revision theory of truth ([Kripke, 1975b], [Gupta and Belnap, 1993]). These theories describe or define a class of models for languages with a truth predicate. There is an important difference between contemporary semantic theories of truth and Tarski's semantic theory of truth. Contemporary semantic theories attempt to describe interesting models or classes of models for formal languages that contain a (unique) truth predicate, whereas Tarski was in the first place interested in constructing models for formal languages that do not themselves contain the truth predicate. The former are called *type-free truth theories*; the latter are called *typed truth theories*. We shall in the next sections describe Tarski's work on typed truth theories, and only afterwards turn our attention to type-free theories.

Aside from defining truth and from describing models for languages containing truth predicates, one can also write down axioms for the notion of truth. This is the proof-theoretic approach. For the following reasons, this latter approach can be seen as in some sense more fundamental than the other two.

First, we shall see how Tarski demonstrated that in general a sufficiently expressive formal language cannot contain its own definition of truth. Yet, ideally, we want a definition of truth for *our* language: English. But it appears that English is the most encompassing language that we have. So it is unclear, to say the least, in which language a definition of truth should be stated. If we go the definitional way, then we enter a regress.

Second, there is a close connection between attempting to define truth for a language with a truth predicate and giving a semantical theory of truth for a language with a truth predicate. Semantic theories of truth describe a class of models for a language with a truth predicate, but they single out one or more individual models as somehow preferred. These models are presented as candidates for being the intended interpretation of a simplified version of English. These intended models can only be given in words. One can 'give' a model only by describing it. And this description can be seen as a definition of truth. We will see how Tarski has taught us that this description will on pain of contradiction have to be given in a more encompassing framework than the language for which the models are intended. So again the question arises how the semantics for this more encompassing language is to be expressed. In sum, if we go the model-theoretic way, then we enter a regress as well.

The axiomatic approach does not suffer from this annoying regress problem: an axiomatic truth theory at least partially gives the meaning of the truth predicate of the language in which it is stated. This is why the axiomatic approach is preferred. Yet semantic theories of truth and definitions of truth have turned out to contain great heuristic value. We will see that some of the most interesting axiomatic theories of truth have been obtained as attempts to axiomatize (in the sense of 'describe') semantical theories of truth.

Another problem for the model-theoretic approach is the following. On the model-theoretic approach, we seek to construct a nice model for a formal language which includes the truth predicate. Such a nice model ought to emulate (in a simplified fashion) our informal interpretation of English. But the domain of discourse of English does not form a set, for the simple reason that every set is included in it. But by the definition of the notion of a model, its domain must form a set (cf. Chapters 3 and 6). And many of the mathematical results concerning models crucially depend on their domain forming a set. So we are confronted with a sense in which models are radically unlike an interpreted informal language such as English.

3. Typed Disquotational Theories

In this section, we shall look at the first family of successful typed theories of truth, which trace back to [Tarski, 1935].

3.1 Tarski's Undefinability Theorem and the Naive Theory of Truth

Let us denote the language of first-order arithmetic augmented with a truth predicate T as $\mathcal{L}_T$. In $\mathcal{L}_T$ infinitely many Tarski-biconditionals, i.e., biconditional statements of the form

$$T(\phi) \leftrightarrow \phi,$$

can be formulated. Let NT (the naive theory of truth) be the system consisting of PA^T, which is Peano arithmetic formulated in the extended language $\mathcal{L}_T$ with the truth predicate allowed in instances of the induction scheme, plus all the Tarski-biconditionals.

We have remarked that these Tarski-biconditionals enjoy a prima facie plausibility. At the same time, the argument of the liar paradox casts doubt on some of them. Indeed, it can be formally demonstrated that the Tarski-biconditionals cannot all be correct, as we will presently see.

In order to show this, we must appeal to what is called the *diagonal lemma*[2]:

Theorem 13.3.1 (Gödel) *For each formula $\phi(x) \in \mathcal{L}_T$, there is a sentence $\lambda \in \mathcal{L}_T$ such that PA^T proves:*

$$\lambda \leftrightarrow \phi(\lambda).$$

This lemma says that (modulo coding) for every property $\phi(x)$ there is a sentence that 'says of itself' that it has that property, and that this is provably so. This is how self-reference is effected (or, to be more precise: simulated) in the language of arithmetic.

Tarski's celebrated *undefinability theorem* builds on Gödel's diagonal lemma [Tarski, 1935]:

Theorem 13.3.2 (Undefinability theorem) *No consistent extension S of PA^T proves*

$$T(\phi) \leftrightarrow \phi$$

for all $\phi \in \mathcal{L}_T$.

Proof. The proof has the form of a reductio. We use the diagonal lemma to produce a (liar) sentence λ such that:

$$PA^T \vdash \lambda \leftrightarrow \neg T(\lambda).$$

If the theory S in question indeed proves $T(\phi) \leftrightarrow \phi$ for all $\phi \in \mathcal{L}_T$, then in particular S proves $T(\lambda) \leftrightarrow \lambda$. Putting these two equivalences together, we obtain a contradiction in S:

$$S \vdash T(\lambda) \leftrightarrow \neg T(\lambda).$$

■

In other words, no consistent theory extending Peano arithmetic can prove all the Tarski-biconditionals. So the system *NT* must be inconsistent. In particular, the Tarski-biconditional for the liar sentence is inconsistent. This is a strange result, for we have seen how we are intuitively inclined to regard the Tarski-biconditionals as utterly unproblematic.

At this point it is sometimes argued that the naive theory of truth might nevertheless be *our* theory of truth. Perhaps the notion of truth that is used in ordinary discourse is ultimately simply incoherent. This may or may not be the case. In the final analysis, this is an empirical question. But even if it is incoherent, then it is incumbent upon us as philosophers to excise this inconsistency from our ordinary use of the notion of truth and to replace our ordinary concept with a natural concept of truth that is as close as possible to our pre-theoretic notion without being inconsistent. Indeed, we will require more. Our theory of truth will have to be sound. After all, we may want to make use of the concept of truth in our philosophical and perhaps even scientific argumentation. So in any case we must and will do better than the naive theory of truth.

3.2 The Disquotational Theory

We have seen that the naive theory of truth is inconsistent. It is instructive to inspect closely the derivation of the contradiction from the Tarski-biconditionals that form the core of the naive theory. We see that the crucial axiom in the derivation of the contradiction is the Tarski-biconditional: $T(L) \leftrightarrow L$, where L is the liar sentence. If we reflect on the construction of L, we see that it contains the truth predicate. So in the fateful Tarski-biconditional $T(L) \leftrightarrow L$, there is a subformula (namely, $T(L)$) in which an occurrence of the truth predicate occurs in the scope of the truth predicate.

This inspired Tarski to offer an immensely insightful *diagnosis* of what went wrong. He conjectured that the root of the disease lies in allowing the Tarski-biconditionals to regulate the truth conditions of sentences that themselves contain the truth predicate.

If Tarski's diagnosis is correct, then one possible cure for the disease of the liar paradox is to excise and discard, for every formula ϕ which contains occurrences of the truth predicate, the corresponding Tarski-biconditional.

What we are left with is a new axiomatic theory of truth. This theory is called the *disquotational theory*, and it will be abbreviated as *DT*. It contains the following axioms:

DT1 PA^T, which is Peano arithmetic with the truth predicate T allowed in the induction scheme
DT2 $T(\phi) \leftrightarrow \phi$ for all $\phi \in \mathcal{L}_{PA}$

The sentences of the form $T(\phi) \leftrightarrow \phi$ with $\phi \in \mathcal{L}_{PA}$ are called the *restricted Tarski-biconditionals*. Thus *DT* has as its sole truth axioms the restricted Tarski-biconditionals.

Note that *DT* is not a *definition* of truth; it is not even a definition of truth for $\mathcal{L}_{PA}$. A definition allows a defined term to be eliminated in every context in which it appears. But suppose that we have a sentence of the form

$$\exists x \in \mathcal{L}_{PA} : T(x) \wedge \phi(x),$$

with $\phi(x)$ an arithmetical formula. This sentence says that there is at least one (code of an) arithmetical sentence which is true and which has the property ϕ.

Then DT does not provide, in general, a way of finding a truth-free sentence that is provably equivalent with this sentence. So DT can only be understood as a *theory* of truth. And since we are looking for a theory of truth for our whole language, DT should be evaluated as a theory of truth not just for $\mathcal{L}_{PA}$ but for the entire language $\mathcal{L}_T$.

3.3 The Soundness of the Disquotational Theory

The reader will appreciate that in the light of unpleasant past experiences with the naive truth theory we are at this point worried about the soundness of DT. After all, all we have done so far is block the argument that we used to prove NT inconsistent. Who knows, maybe another argument can be found to show that DT suffers the same fate as NT?

Fortunately, this is not the case. We know that by the completeness theorem, finding a model for DT is enough to prove it consistent. In fact, we shall do more than that: we shall find a *nice* model for DT.

A *nice* model is a model that is based on the natural numbers, i.e., a model $\mathcal{M}$ which is of the form $\langle \mathbb{N}, \mathcal{E} \rangle$ for some collection $\mathcal{E}$ of (codes of) formulae. Here $\mathbb{N}$, the standard natural number structure, of course serves as the interpretation of the mathematical vocabulary, and $\mathcal{E}$ is the set of (codes of) sentences that serves as an interpretation of the truth predicate.

Proposition 13.3.1 DT has a nice model.

Proof. Consider the model

$$\mathcal{M} =: \langle \mathbb{N}, \{\phi \mid \phi \in \mathcal{L}_{PA} \wedge \mathbb{N} \models \phi\} \rangle,$$

i.e., the model in which as the extension of the truth predicate we take all arithmetical truths. An induction on the length of proofs in DT verifies that

$$\mathcal{M} \models DT.$$

■

$\mathcal{M}$ is the *minimal* nice model for DT: it contains the minimal extension of the truth predicate that is needed to make all the restricted Tarski-biconditionals true.

Corollary 13.3.1 *DT is consistent.*

Proof. This follows from Proposition 13.3.1 via the soundness theorem for first-order logic. ■

But why are models based on the standard natural numbers structure nice? They are nice because they prove more than just consistency of the theories that they model. A moment's reflection shows that our proposition gives us a stronger corollary:

Corollary 13.3.2 *DT is arithmetically sound.*

Proof. Suppose $DT \vdash \phi$ for an arithmetical ϕ. Then by Proposition 13.3.1,

$$\langle \mathbb{N}, \{\phi \mid \phi \in \mathcal{L}_{PA} \wedge \mathbb{N} \models \phi\}\rangle \models \phi.$$

But because ϕ does not contain the truth predicate, this entails that $\mathbb{N} \models \phi$. ■

Note that this second corollary is strictly stronger than the first one. For it is a lesson of Gödel's first incompleteness theorem that consistency does not entail soundness.

We can go further and state with some confidence that DT is a truth-theoretically *sound* theory of truth. Not only does DT prove only true arithmetical sentences; also all the sentences containing the truth predicate that it proves seem correct. After all, the restricted Tarski-biconditionals seem completely unproblematic: they make no claims about paradoxical sentences like the liar sentence.

3.4 The Tarskian Hierarchy

Thus we have obtained our first successful axiomatic theory of truth. Let us now take the first steps in investigating how close DT comes to truth-theoretic completeness.

Suppose that you overhear the following exciting conversation:

A: It is true that $0 = 0$.
B: What you have just said is true.

The assertions of A and B seem equally correct. A and B appear to assert trivial truths which are not in any way paradoxical. A's assertion can be expressed in $\mathcal{L}_T$ as $T(0 = 0)$; B's assertion can be expressed as $T(T(0 = 0))$. So it seems that in English we can truthfully predicate truth of sentences which themselves contain the concept of truth.

In this scenario B might just as well have repeated what A said. So in a sense, B does not really have to iterate truth to get their message across. But avoiding truth iteration is not always possible. B might know that A has asserted the truth of some mathematical sentence without knowing what this mathematical sentence is. If B believes that A is very reliable in matters mathematical, then B might still be prepared to assert that what A said is true. But now B is not in a position to avoid at least implicit truth iteration by repeating what A has said.

Unfortunately DT does not prove $T(T(0 = 0))$. Indeed, the nice model that we have constructed for DT shows that DT does not prove *any* sentence of the form $T(\phi)$ where ϕ contains an occurrence of the truth predicate. It does not even prove $\exists x \in \mathcal{L}_{PA} : T(T(x))$.

So the question arises whether the Tarskian treatment has been too severe. Can the Tarskian approach validate the feeling that some truth iterations are genuinely unproblematic? Well, yes, to some extent.

Nothing prevents us from taking DT, instead of PA, as the theory against the background of which we formulate our truth theory. Let $\mathcal{L}_{T,T_1}$ be defined as $\mathcal{L}_T \cup \{T_1\}$. T_1 will serve as our new, more encompassing truth predicate. And let PA^{T,T_1} be Peano arithmetic formulated in the extended language $\mathcal{L}_{T,T_1}$. Now we define DT_1 to consist of:

1. PA^{T,T_1};
2. $T(\phi) \leftrightarrow \phi$ for all $\phi \in \mathcal{L}_{PA}$;
3. $T_1(\phi) \leftrightarrow \phi$ for all $\phi \in \mathcal{L}_T$.

Just like DT, the theory DT_1 will prove truth-ascribing sentences such as $T(0 = 0)$. But then the restricted Tarski-biconditionals for T_1 can be used to derive from $T(0 = 0)$ the sentence $T_1(T(0 = 0))$. This looks like just the sort of truth iteration which we wanted to assert.

As in the case of DT we want to satisfy ourselves that DT_1 has nice models. Since DT_1 is formulated in the language $\mathcal{L}_{T,T_1}$, nice models of it will be of the form $\langle \mathbb{N}, \mathcal{E}, \mathcal{E}_1 \rangle$, where $\mathcal{E}$ as before serves as an extension of T, and $\mathcal{E}_1$ serves as an extension of T_1.

Proposition 13.3.2 DT_1 has a nice model.

Proof. Consider the model

$$\mathcal{M}_1 =: \langle \mathbb{N}, \{\phi \mid \phi \in \mathcal{L}_{PA} \wedge \mathbb{N} \models \phi\}, \{\phi \mid \phi \in \mathcal{L}_T \wedge \mathcal{M} \models \phi\} \rangle,$$

where $\mathcal{M}$ is the model of Proposition 13.3.1. An induction on the length of proofs in DT verifies that $\mathcal{M}_1 \models DT_1$. This proof is similar to that of Proposition 13.3.1. ∎

Now suppose the conversation above were to be extended thus:

A: It is true that $0 = 0$.
B: What you have just said is true.
C: Yes, B, that is very true.

It seems in effect that C asserts $T(T(T(0 = 0)))$. And this appears just as acceptable as A and B's assertions.

If we want a truth theory which even proves a threefold truth iteration, then we must repeat the trick. We have to add a third truth predicate to our language, and construct in the appropriate way a theory DT_2 which proves $T_2(T_1(T(0 = 0)))$.

And so we can go on, up the *Tarskian hierarchy*. Let

$$\mathcal{L}_{T,T_1,T_2,\ldots} = \mathcal{L}_{PA} \cup \{T, T_1, T_2, \ldots\}.$$

And let DT_ω be DT plus for every $n > 0$

$$T_n\,(\phi) \leftrightarrow \phi \text{ for all } \phi \in \mathcal{L}_{T,T_1 \ldots T_{n-1}}.$$

Proposition 13.3.3 DT_ω has a nice model.

Proof. The obvious generalization of the model which we have constructed for DT_1 will make DT_ω true. ■

Must we stop here? No, we can go on into the transfinite, and formulate:

$$DT_{\omega+1}, DT_{\omega+2}, \ldots$$

If we want to do this in $\mathcal{L}_T$, then we somehow have to code transfinite ordinal numbers such as $\omega + 2$ as natural numbers. Up to some transfinite ordinals, this can indeed be done.

3.5 Contextualist Theories

Let us recapitulate. We have seen that $DT_1 \vdash T_1(T(0 = 0))$. But the fact that $\mathcal{M}_1 \models DT_1$ entails that $DT_1 \nvdash T_1(T_1(0 = 0))$. This follows via the completeness theorem from the construction of the model $\mathcal{M}_1$, for $\mathcal{M}_1 \not\models T_1(T_1(0 = 0))$. For similar reasons, $DT_1 \nvdash T(T(0 = 0))$.

So in some sense, DT_1 gives rise to truthful iteration of truth predicates. But in a strict sense, Tarski's diagnosis is upheld: no truth predicate can be truthfully predicated of a sentence containing that *same* truth predicate.

In other words, according to the Tarskian hierarchical conception, truth is not a *uniform* notion. There is in reality not one property of truth, but there are many properties of truth, ordered in levels in a linear way.

There are objections to the Tarskian hierarchy account. For one thing, *our* notion of truth seems at first sight to be a uniform notion. If there were more than one property of truth, then one would expect this to be reflected in natural language. A second objection is due to Kripke. He has emphasized that the level of a token use of the truth predicate can depend on contingent factors: it can depend on what things have been (or will be) said by the speaker or by others [Kripke, 1975b, Section I]. And because of that, it is in certain situations practically impossible for a speaker to determine which level his truth predicate should take in order for his utterance to express the intended proposition.

Some philosophers have developed a theory of truth which admits that truth is a uniform notion but still makes use of a Tarski-like hierarchy to hold the semantic paradoxes at bay. (See [Burge, 1979]. Structurally similar theories of truth have been proposed by Barwise and Etchemendy [Barwise and Etchemendy, 1987] and Gaifman [Gaifman, 1992].) On these theories, the truth predicate indeed has a uniform meaning. But its extension varies over contexts. In this respect, the truth predicate semantically behaves like indexical expressions such as 'I' and 'here': these words have a uniform meaning, but their reference differs from context to context.

From the Kripkean considerations above it is clear that the speaker's intentions cannot in general determine the extension of the truth predicate in a given context. If anything fixes the level of the property of truth which my use of the truth predicate expresses, it must be the conversational context (in the widest sense of the word) plus the world as we find it.

Contextualist truth theories deal with the liar paradox in roughly the following manner. Consider the liar sentence once again:

$$\text{Sentence } L \text{ is not true.} \qquad (L)$$

Let us evaluate sentence L. The truth predicate occurring in L must be 'indexed' to a particular context, which we shall call context 0. So, semi-formally, we can say that sentence L expresses that L is not true_0. For the familiar liar argument reasons, L cannot be true_0. If L were true_0, then what it says of itself, namely that it is *not* true_0, would have to be the case, and this would yield a contradiction. But if sentence L is not true_0, then it ought to be in some sense *true* that it is not true_0. This is where, in the original argument of the liar paradox, we were led into trouble. Indexical theories of truth have it that when we assert that it is true that L is not true_0, we are shifting to a new context. And it is not that we intentionally make the shift: it happens automatically. The occurrence of 'true' in 'it is true that L is not true_0' must be given an index different from that of 'true_0'; let its index be 1. Then we have both:

Sentence L is not true_0.
Sentence L is true_1.

Because of the indexical shift in extension of the truth predicate between context 0 and context 1, this is not a contradiction.

Thus we have a way of maintaining the uniformity of the notion of truth while at the same time helping ourselves to Tarski's hierarchy. Of course, strictly speaking, we do not have a hierarchy of *languages* any more – our language contains only one truth predicate. We have so to speak pushed Tarski's hierarchy into semantics. Or you may say that we have pushed the hierarchy into pragmatics: it depends on where the line between semantics and pragmatics is drawn.

Unfortunately, it appears that the liar paradox can strike back. Consider the sentence:

$$\text{Sentence } L' \text{ is not true in any context.} \qquad (L')$$

This is called a *strengthened liar sentence*. It is not hard to figure out that there is no context in which L' can be coherently evaluated.

Burge himself has always been keenly aware of this temptation to try to produce a 'super-liar' sentence. In his view, it is simply impossible to successfully quantify into the index of the truth predicate. He says that an attempt to quantify out the indexical character of 'true' 'has some of the incongruity of "here at some place"' ([Burge, 1979, p. 108]). But it is not clear how convincing this reply really is. For the indexical 'here', the phrase 'at some place' does the job: a sentence such as 'it rains at some place' is, in contrast to 'it rains (here)', not sensitive to spatial context shifts. But apparently, for the indexical notion of truth, no qualifier can successfully carry out the corresponding task. The only reason why this is so appears to be that if there were one, the liar paradox would rear its head. A more principled reason would surely be more satisfying.

4. Typed Compositional Theories

In this section we shall discuss typed truth theories that are stronger than the typed disquotational theories, but which nevertheless appear to be equally sound. These theories also find their origins in the work of Tarski.

4.1 The Compositional Theory of Truth

The following two propositions are typical illustrations of the proof-theoretic strength of the disquotational theory DT.

Proposition 13.4.1 For all $\phi \in \mathcal{L}_{PA}$: $DT \vdash T(\phi) \vee T(\neg\phi)$

Proof. Already propositional logic alone proves $\phi \vee \neg\phi$. Two restricted Tarski-biconditionals are $T(\phi) \leftrightarrow \phi$ and $T(\neg\phi) \leftrightarrow \neg\phi$. Combining these facts yields the desired result. ■

Proposition 13.4.2 **(Tarski)** $DT \nvdash \forall\phi \in \mathcal{L}_{PA} : T(\phi) \vee T(\neg\phi)$

So somehow *DT* proves all the instances of an intuitively plausible logical principle concerning the notion of truth, but is unable to collect all these instances together into a general theorem. In sum, *DT* fails to fully validate our intuitions concerning the *compositional nature* of the notion of truth, i.e., the fact that the property of truth 'distributes' over the logical connectives. In fact, our intuition that truth is compositional is (perhaps independent from but) just as basic as our intuition that truth is a disquotational device. Our truth theory has an obligation either to do justice to it or to explain what is wrong with it.

The inability of *DT* to fully explicate the compositional nature of truth is a motivation for taking the principles that do explicate it as *axioms* of a theory of truth. We already know beforehand that if this is done, then there is a sense in which the resulting theory is stronger than *DT*. These principles of composition are of course contained in Tarski's clauses for recursively explicating the notion of truth of a formula in a model.

Tarski's axiomatic compositional theory of truth is denoted as *TC*. In the literature, the theory *TC* is often referred to as *T(PA)*. The axiomatic theory of truth *TC* is formulated in $\mathcal{L}_T$. It makes use of the fact that even though the collection of arithmetical truths cannot be defined in the language of arithmetic (by Tarski's undefinability theorem), the collection of atomic sentences that are true can be defined in the language of first-order arithmetic by a complicated first-order arithmetical formula val^+. Similarly, there will be an arithmetical formula val^- defining the atomic arithmetical falsehoods. Given this fact, the compositional theory of truth can be expressed as follows:

TC1 PA^T
TC2 $\forall$ atomic $\phi \in \mathcal{L}_{PA} : T(\phi) \leftrightarrow val^+(\phi)$
TC3 $\forall \phi \in \mathcal{L}_{PA} : T(\neg\phi) \leftrightarrow \neg T(\phi)$
TC4 $\forall \phi, \psi \in \mathcal{L}_{PA} : T(\phi \wedge \psi) \leftrightarrow (T(\phi) \wedge T(\psi))$
TC5 $\forall \phi(x) \in \mathcal{L}_{PA} : T(\forall x \phi(x)) \leftrightarrow \forall x T(\phi(x))$

These formulae are to be read again in the 'natural' way. TC5, for instance, says that for all arithmetical properties $\phi(x)$, it is true that all numbers have this property if and only if the result of plugging in any standard numeral for x in $\phi(x)$ results in a true arithmetical sentence. (Thus 'quantifying in' the context of the truth predicate is to be understood in a substitutional way.)

So the idea behind this axiom system is straightforward. An explicit definition of the class of true atomic arithmetical sentences can be given by means of an arithmetical formula val^+. Truth for complex arithmetical sentences can be reduced to truth of atomic arithmetical formulae through the compositional truth axioms.

It is important that each compositional truth axiom is expressed as a universally quantified sentence rather than as an axiom scheme. For from the axiom scheme the corresponding universally quantified sentence cannot be derived whereas each instance can be derived from DT, whereby the schematic version of TC is a consequence of DT. The whole motivation of TC lies in our desire for the universally quantified truth axioms to be provable by our truth theory.

As in the case of the disquotational theory, we can make a case for the soundness of TC. First of all, TC has models which are as nice as those of DT:

Proposition 13.4.3 TC has nice models.

Proof. It can be shown that the model $\mathcal{M}$ that was constructed in the proof of Proposition 13.3.1 is also a model of TC. ∎

Tarski's diagnosis of the paradoxes is upheld in TC. As before, no sentences of the form $T(\ldots T \ldots)$ are provable in TC. Also as before, we can construct a whole hierarchy $TC, TC_1, \ldots, TC_n, \ldots$ of axiomatic compositional theories of truths.

As in the case of DT, it seems plausible that TC is *truth-theoretically* sound. It appears to accurately explicate the common-sense view that the truth value of a complex sentence is determined by the truth values of its component parts. We might call this the *compositionality intuition*. In sum, TC is a very attractive theory of truth.

4.2 Truth and Satisfaction

In the formulation of the compositional theory of truth, a simplifying assumption has been made which is not completely innocuous. The assumption is related to the compositional axiom $TC5$ which says that truth commutes with the universal quantifier. But to see this, we have to look 'through' our non-well-formed notation.

We have alluded to the fact that when axiom $TC5$ is made grammatical, we see that a *substitution function* appears. It *really* says that a universally quantified sentence is true if and only if all the instantiations by standard numerals are true. Now this works fine for arithmetic: every number is named by a standard numeral. But of course the compositional theory of truth is intended to be quite general. It should also work for an interpreted ground language such that not every element of its domain of discourse has a name, let alone a standard name. Indeed, the number of English expressions is denumerably infinite. But a theorem of Cantor shows that there are non-denumerably many real numbers. So not every real number has a (simple or complex) name in English. Perhaps there are even real numbers which are somehow inherently unnameable even in extensions of English that humans can master.

This problem is well known, and there is a standard solution. The solution is due, not surprisingly, to Tarski. It says that the truth *predicate* should be defined in terms of the more primitive *satisfaction relation*: the relation of being *true of*.

The fundamental notion is that of a formula (containing free variables) being true of a sequence of objects (that serve as values of these variables).

A *rough* sketch of the solution goes as follows. Let $Sat(x, y)$ be the satisfaction predicate, where now x stands for a number rather than for a sequence of numbers. Then a satisfaction axiom for (one-place) atomic formulae looks roughly like this:

$$\forall \text{ atomic } \phi(x), \forall y : Sat(y, \phi(x)) \leftrightarrow \phi(y)$$

The compositional axioms for the propositional logical connectives are the analogues of $TC3$ and $TC4$. The compositional axiom governing the interaction between satisfaction and universal quantification is (roughly):

$$\forall x, \forall \phi(y) : Sat(x, \forall y \phi(y)) \leftrightarrow \forall z Sat(z, \phi(y))$$

Truth is then defined as satisfaction by all sequences of objects of the domain.

In sum, the situation is not *very* different from the case where all objects have standard names. Nevertheless, one should be mindful. In the general case, satisfaction is a more basic notion than truth. It is also even more susceptible to semantical paradox than the truth predicate: even stronger no-go theorems can be proved for axiomatic theories of satisfaction than can be proved for axiomatic theories of truth [Horsten, 2004]. This means that one has to be even more cautious in the formulation of a theory of satisfaction than in the formulation of a formal theory of truth.

We shall leave this complication behind us from now on. This means that we shall effectively be working under the assumption that every element of the domain has a standard name. But as a matter of fact, the truth theories that we shall consider can be fairly straightforwardly reformulated as theories of satisfaction.

4.3 The Power of Truth

The concept of *conservativeness* has played an important role in recent philosophical discussions of axiomatic theories of truth.

Intuitively, we say that a theory of truth is conservative over a background theory S if and only if no sentences in the language of S can be proved using the axioms of the truth theory and the axioms of the background theory that cannot already be proved in S alone. We have taken a theory of syntax as our background theory. And we have for convenience identified this theory of syntax with the arithmetical theory PA. So we focus on the notion of arithmetical conservativeness of a truth theory:

Definition 13.4.1 A theory of truth S is *arithmetically conservative* over PA if for every sentence $\phi \in \mathcal{L}_{PA}$, if $S \vdash \phi$, then already $PA \vdash \phi$.

The question then arises whether truth theories are arithmetically conservative over *PA*. It can be shown that:

Proposition 13.4.4 *DT* is arithmetically conservative over *PA*.

This is not really a surprising result. *DT* differs from *PA* only in having in addition logical axioms concerning the notion of truth. One feels inclined to believe that truth is a *philosophical* notion, and not a mathematical notion. And logical axioms concerning philosophical notions will not help with the solution of *mathematical* problems that *PA* cannot solve.

We have seen that *TC* proves intuitively valid principles concerning the notion of truth that *DT* fails to prove. In fact, it is not hard to see that *DT* is a subtheory of *TC*:

Proposition 13.4.5 $TC \vdash T(\phi) \leftrightarrow \phi$ for all $\phi \in \mathcal{L}_{PA}$.

This is as it should be. Proving the restricted Tarski-biconditionals is an adequacy condition for truth theories.

In fact, *TC* is even *arithmetically* stronger than *DT*: the collection of arithmetical theorems of *DT* is a proper subset of the collection of theorems of *TC*. This implies, of course, that *TC* is *not* arithmetically conservative over *PA*.

First, it is shown how *TC* proves the *global reflection principle* for *PA* [Halbach, ta, Chapter 2, Section 8.6]:

Theorem 13.4.1 $TC \vdash \forall \phi \in \mathcal{L}_{PA} : Bew_{PA}(\phi) \rightarrow T(\phi)$, *where* $Bew_{PA}(\ldots)$ *is an arithmetical predicate that expresses provability in Peano arithmetic in a natural way.*

Corollary 13.4.1 $TC \vdash \neg Bew_{PA}(0 = 1)$.

Proof. This follows from Theorem 13.4.1 and Proposition 13.4.5 by instantiating $0 = 1$ for ϕ. ■

This result concerning the power of *TC* is really surprising. *PA* fails to prove certain arithmetical sentences, such as $\neg Bew_{PA}(0 = 1)$, which expresses the consistency of *PA*. It is very surprising that by just adding to *PA* principles concerning the notion of *truth* we increase the *mathematical* strength of *PA*. This means that contrary to our expectations, the 'philosophical' notion of truth has real mathematical content.

Gödel's second incompleteness phenomenon shows that *PA* cannot prove its own consistency. So *TC* is mathematically stronger than *PA*, for it can prove the consistency of *PA*. All this does not imply that *TC* escapes the incompleteness phenomena. The moral of the second incompleteness theorem is that a (sufficiently strong) consistent theory cannot prove its own consistency. Since we know TC to be consistent, this moral holds also for TC: it cannot prove *its own* consistency.

The phenomenon that we have just described is robust. If we move one step up the Tarskian hierarchy and consider TC_1, for instance, we will see that we have again gained arithmetical strength. TC_1 proves the consistency of TC for pretty much the same reasons as TC proves the consistency of PA. And so on, up the hierarchy.

5. Type-Free Disquotation

5.1 Type-Free Truth

We have seen that Tarski proposed languages that contain a hierarchy of truth predicates. Tarski's proposal has inspired axiomatic theories which do not prove sentences such as $T(T(0 = 0))$, but which do prove sentences such as $T_1(T_0(0 = 0))$. Here T_1 is a 'higher-level' truth predicate for the language $\mathcal{L}_T$.

We have also seen that truth theories which are formulated in languages that contain truth predicates of different levels, and which prove iterated truth ascriptions only if the hierarchy constraints are satisfied, are called *typed theories of truth*. There also exist truth systems which contain a single truth predicate but which do validate sentences of the form $T(T(0 = 0))$. These systems are called *untyped* or *type-free* theories of truth. Sometimes they are also called *reflexive* or *semantically closed* truth theories.

Untyped theories of truth abandon Tarski's strictures on truth iteration. We have seen that how the self-referential paradoxes are related to *particular* iterations of the truth predicate. So the idea is to distinguish carefully between sentences that contain problematic iterations of the truth predicate, such as $T(L) \leftrightarrow L$ where L is the liar sentence, and sentences that contain innocuous truth iterations, such as $T(T(0 = 0))$. The latter may be validated by an untyped truth theory, but not the former.

5.2 The Strength of Type-Free Disquotation

Generally, disquotational theories of truth are thought to be weak. This is correct for the typed disquotational theory DT, which is conservative over its base theory and properly contained in the much stronger typed compositional theory TC.

Once the restriction to T-free instances of the Tarski schema $T(\phi) \leftrightarrow \phi$ is relaxed, disquotational theories can become very strong. Of course, *some* restriction on the instances is needed to avoid an inconsistency, but there are many – in fact, uncountably many – sets of Tarski-biconditionals that are consistent with PA (see [McGee, 1992]). Finding a sensible restriction remains a challenge.

To substantiate the claim that untyped disquotational theories can be very strong, we adapt an observation due to [McGee, 1992] in the proof of the following result:

Theorem 13.5.1 Any theory extending *PA* can be re-axiomatized by the axioms of *PA* and a set of Tarski-biconditionals. That is, for any theory $S \supseteq PA$, there is a set D of sentences of the form $T(\phi) \leftrightarrow \phi$ such that S and $PA \cup D$ prove the same theorems.

Proof. Consider an axiom ψ of S. Using the diagonal lemma (Theorem 13.3.1) one can find a sentence λ such that

$$\lambda \leftrightarrow (T(\lambda) \leftrightarrow \psi)$$

is provable in *PA*. This equivalence is logically equivalent to the following sentence:

$$\psi \leftrightarrow (T(\lambda) \leftrightarrow \lambda)$$

So ψ is *PA*-provably equivalent to the Tarski-biconditional $T(\lambda) \leftrightarrow \lambda$. Using this method one can re-express every axiom of S that is not an axiom of *PA* as a Tarski-biconditional. If S is given by finitely many axioms beyond those of *PA*, S can be re-axiomatized as *PA* plus a single Tarski-biconditional. ■

In particular, any theory of truth can be re-axiomatized as a disquotational theory of truth. This observation does not really help the disquotationalist: it rather shows that the restriction to sentences $T(\phi) \leftrightarrow \phi$ as axioms for truth is not a real restriction at all.

5.3 Positive Disquotation

Theorem 13.5.1 makes disquotationalism look like an idle doctrine. The claim that the notion of truth is given by a consistent set of Tarski-biconditionals is completely vacuous as any theory containing *PA* can be axiomatized as such a disquotational truth theory.

To defend his doctrine against the charge of vacuity, the disquotationalist may reject certain Tarski-biconditionals as axioms for truth even though they are consistent with Peano arithmetic and propose a specific set of Tarski-biconditionals as axioms for truth. Without such a specification, disquotationalism is as interesting as a position where one insists that all axioms for truth should contain a closing bracket.

Tarski did provide such a specification; it leads to the theory *DT*. But if typing is rejected, a more liberal restriction is needed. There are other ways than

Tarski's method of distinguishing between an object- and a meta-language for resolving the paradoxes.

Essential to the derivation of the liar paradox is not only that truth is applied to a sentence containing the truth predicate but also that truth is applied to a sentence that contains a *negated* occurrence of the truth predicate. One must be careful here: If the language contained the connective $\rightarrow$, the sentence $\neg T(\lambda)$ could be re-expressed as $T(\lambda) \rightarrow 0 = 1$. Such occurrences would count as 'negative' occurrences of the truth predicate. In the present setting, with only $\forall$ and $\neg$ as connectives, we can take those Tarski-biconditionals $T(\phi) \leftrightarrow \phi$ where T does not occur in the scope of an odd number of negation symbols in ϕ. [Halbach, 2009] considered a slightly strengthened version of this: *PUTB* is the theory given by PA and the set of all sentences

$$\forall x\,(T(\phi(x)) \leftrightarrow \phi(x))$$

In the formula $\phi(x)$, T must not occur in the scope of an odd number of negation symbols in the formula ϕ, but ϕ may contain free variables. *PUTB* stands for *positive uniform Tarski-biconditionals.*

This restriction rules out the Tarski-biconditionals used in the proof of Theorem 13.5.1 as legitimate instances of the Tarski schema. The theory *PUTB* has some pleasing properties. It is as strong as (but not equivalent to) the Kripke-Feferman theory, which is one of the strongest untyped theories of truth found in the literature (see Section 6.3 below).

The example of *PUTB* shows that there are *natural* restrictions to the general Tarski schema $T(\phi) \leftrightarrow \phi$ that admit sentences ϕ containing the truth predicate as instances and that yield interesting theories of truth.

6. Kripke's Theory of Truth

Now we will leave typed truth theories behind: we will henceforth be concerned with untyped theories. In this arena, Kripke's theory of truth, in one or other of its incarnations, must surely be counted as one of the best truth theories available to date.

6.1 Partial Models for Reflexive Truth

Kripke has constructed a semantical theory of self-referential truth. He is not occupied with formulating an axiomatic theory of truth. For Kripke, models come first. His aim is to construct particularly nice models of the language $\mathcal{L}_T$, which do justice to the self-applicative nature of the concept of truth.

The argument of the liar paradox puts us in the awkward position of recognizing that both the supposition that the liar sentence L holds and the supposition that its negation holds, lead to a contradiction. Kripke takes the moral of the argument to show that neither L nor its negation holds.

In ordinary Tarskian models for $\mathcal{L}_T$, a given sentence either holds or its negation holds. So if we are to respect Kripke's diagnosis of the liar paradox, then we must modify our notion of model so as to leave room for the situation that for some sentences, neither they nor their negation hold. This leads us to the notion of a *partial model*, to which we now turn.

We want to build a model for the language $\mathcal{L}_T$. The arithmetical vocabulary is interpreted throughout as in the standard model $\mathbb{N}$. The truth predicate T will be the only partially interpreted symbol: it will receive, at each ordinal stage, an *extension* $\mathcal{E}$ and an *anti-extension* $\mathcal{A}$. $\mathcal{E} \cup \mathcal{A}$ does not exhaust the domain, for otherwise T would be a total predicate. The extension $\mathcal{E}$ of T is the collection of (codes of) sentences which are (at the given stage) determinately true; the anti-extension $\mathcal{A}$ of T is the collection of (codes of) sentences which are (at the given stage) determinately false. Since we do not want to allow for the possibility that a sentence is both true and false, we insist that $\mathcal{E} \cap \mathcal{A} = \emptyset$.

A partial model $\mathcal{M}$ for $\mathcal{L}_T$ can then be identified with an ordered pair $(\mathcal{E}, \mathcal{A})$. (We denote partial models as $(\mathcal{E}, \mathcal{A})$ and not as $\langle \mathcal{E}, \mathcal{A} \rangle$ in order to clearly distinguish them from classical models of the form $\langle \mathbb{N}, \mathcal{E} \rangle$.) In general, the union of the extension and the anti-extension will not exhaust the collection of all sentences. Some sentences will at each ordinal stage retain their indeterminate status. An example of an eternally indeterminate sentence is the liar sentence L. The intuition that the liar argument shows that the liar sentence L cannot have a determinate truth value is the basic motivation for constructing a theory of truth in which T is treated as a partial predicate.

At stage 0, the extension and the anti-extension are empty. This yields a partial model,

$$\mathcal{M}_0 = (\mathcal{E}_0, \mathcal{A}_0) := (\emptyset, \emptyset) .$$

Next, a popular evaluation scheme for partial logic is used: the so-called *strong Kleene scheme*. The strong Kleene evaluation scheme $\models_{sk}$ is defined as follows:

- For any atomic formula $Fx_1 \ldots x_n$:
  1. $\mathcal{M} \models_{sk} Fk_1 \ldots k_n$ if the n-tuple $(k_1, \ldots, k_n)$ belongs to the extension of F;
  2. $\mathcal{M} \models_{sk} \neg Fk_1 \ldots k_n$ if the n-tuple $(k_1, \ldots, k_n)$ belongs to the anti-extension of F.
- For any formulae ϕ, ψ :
  1. $\mathcal{M} \models_{sk} \phi \wedge \psi$ if and only if $\mathcal{M} \models_{sk} \phi$ and $\mathcal{M} \models_{sk} \psi$;
  2. $\mathcal{M} \models_{sk} \neg (\phi \wedge \psi)$ if and only if either $\mathcal{M} \models_{sk} \neg\phi$ or $\mathcal{M} \models_{sk} \neg\psi$ (or both);
  3. $\mathcal{M} \models_{sk} \forall x\phi$ if and only if for all n, $\mathcal{M} \models_{sk} \phi(n/x)$;

4. $\mathcal{M} \models_{sk} \neg\forall x\phi$ if and only if for at least one n, $\mathcal{M} \models_{sk} \neg\phi(n/x)$;
5. $\mathcal{M} \models_{sk} \neg\neg\phi$ if and only if $\mathcal{M} \models_{sk} \phi$.

The strong Kleene scheme is used to determine the collection $\mathcal{E}_1$ of sentences of $\mathcal{L}_\mathrm{T}$ that are made true by $\mathcal{M}_0$, and the collection $\mathcal{A}_1$ of sentences of $\mathcal{L}_\mathrm{T}$ the negation of which are made true by $\mathcal{M}_0$. Thus a new partial model $\mathcal{M}_1 = (\mathcal{E}_1, \mathcal{A}_1)$ is obtained. Using $\mathcal{M}_1$, a new extension $\mathcal{E}_2$ and a new anti-extension $\mathcal{A}_2$ are then determined, and so on. In general, for any ordinal α,

$$\mathcal{E}_{\alpha+1} =: \{\phi \in \mathcal{L}_\mathrm{T} \mid \mathcal{M}_\alpha \models_{sk} \phi\} \text{ and}$$

$$\mathcal{A}_{\alpha+1} =: \{\phi \in \mathcal{L}_\mathrm{T} \mid \mathcal{M}_\alpha \models_{sk} \neg\phi\}.$$

For limit stages λ, we set

$$\mathcal{E}_\lambda =: \bigcup_{\kappa<\lambda} \mathcal{E}_\kappa,$$

$$\mathcal{A}_\lambda =: \bigcup_{\kappa<\lambda} \mathcal{A}_\kappa.$$

Now we are going to consider the resulting transfinite sequence of models:

$$(\mathcal{E}_0, \mathcal{A}_0), (\mathcal{E}_1, \mathcal{A}_1), \ldots, (\mathcal{E}_\omega, \mathcal{A}_\omega), \ldots$$

It is not hard to see that the strong Kleene valuation scheme has the following (important) *monotonicity* property:

Theorem 13.6.1 *For any two partial models $(\mathcal{E}_a, \mathcal{A}_a), (\mathcal{E}_b, \mathcal{A}_b)$, if $\mathcal{E}_a \subseteq \mathcal{E}_b$ and $\mathcal{A}_a \subseteq \mathcal{A}_b$, then*

$$\{\phi \mid (\mathcal{E}_a, \mathcal{A}_a) \models_{sk} \phi\} \subseteq \{\phi \mid (\mathcal{E}_b, \mathcal{A}_b) \models_{sk} \phi\}.$$

Proof. This follows by an induction on the complexity of formulae of $\mathcal{L}_T$. ■

A consequence of this is that:

Corollary 13.6.1 For all α, β with $\alpha < \beta$, we have:

$$\{\phi \mid (\mathcal{E}_\alpha, \mathcal{A}_\alpha) \models_{sk} \phi\} \subseteq \{\phi \mid (\mathcal{E}_\beta, \mathcal{A}_\beta) \models_{sk} \phi\}.$$

In other words, as we proceed in the sequence of models, more and more sentences of $\mathcal{L}_T$ end up in the extension or in the anti-extension of T, and once a sentence is in the extension (anti-extension) of T, it stays in forever onward. But by elementary cardinality considerations, this process must eventually come to an end. Intuitively, what happens is that at some ordinal stage the basket of sentences which can be put in the extension or in the anti-extension of T is exhausted.

Proposition 13.6.1 For some ordinal ρ, $\mathcal{E}_\rho = \mathcal{E}_{\rho+1}$ and $\mathcal{A}_\rho = \mathcal{A}_{\rho+1}$.

Proof. This is an elementary argument in transfinite set theory.

Remember that for every ordinal α, $\mathcal{E}_\alpha$ is just a set of codes of sentences, and thus countable. There are uncountably many ordinal numbers. Suppose that for uncountably many ordinals α at least one new sentence ϕ_α enters in $\mathcal{E}_\alpha$. Then there would be a one-one correspondence between an uncountable set O of ordinals and the union of all the $\mathcal{E}_\alpha$ for $\alpha \in O$. But because the language $\mathcal{L}_T$ is countable, the union of all the $\mathcal{E}_\alpha$ is countable. So by Cantor's Theorem, no such one-to-one correspondence can exist.

The argument for the anti-extension of the truth predicate is symmetrical. ■

The partial model $\mathcal{M}_\rho = (\mathcal{E}_\rho, \mathcal{A}_\rho)$ is called the *least fixed point* model, or the least fixed point, for short. It is the particularly nice model for the language $\mathcal{L}_T$ we have been looking for.

6.2 Properties and Variations

This model is nice because it is based on the natural numbers. But why is this model *particularly* nice?

First, very long truth iterations hold in the least fixed point. For one thing, the sentence $\forall x T^x(0 = 0)$ for the first time enters the extension of T at stage ω. And the sentence $T\forall xT(T^x(0 = 0))$ (here T^x stands for an x-fold iteration of T) enters the extension only at the next stage $\omega + 1$.

Second, the least fixed point is 'consistent' in the sense that:

Proposition 13.6.2 There is no sentence ϕ of $\mathcal{L}_T$ such that $\mathcal{M}_\rho \models_{sk} T(\phi)$ and $\mathcal{M}_\rho \models_{sk} T(\neg\phi)$.

Proof. Suppose there were such a sentence ϕ. Then there would be an ordinal $\alpha < \rho$ such that $\mathcal{M}_\alpha \models_{sk} \phi$ and $\mathcal{M}_\alpha \not\models_{sk} \neg\phi$. Or conversely, but that case is symmetric. But then by monotonicity and the clauses of truth in a partial model there can be no $\beta > \alpha$ such that $\mathcal{M}_\beta \models_{sk} \neg\phi$. Contradiction. ■

Third, and most importantly, *in a certain (partial) sense* the unrestricted Tarski-biconditionals hold in $\mathcal{M}_\rho$:

Theorem 13.6.2 *For all sentences ϕ in $\mathcal{L}_T$:*

$$\mathcal{M}_\rho \models_{sk} \phi \Leftrightarrow \mathcal{M}_\rho \models_{sk} T(\phi).$$

Proof. First, suppose $\mathcal{M}_\rho \models_{sk} \phi$. Then by the definition of the sequence of partial models, $\mathcal{M}_{\rho+1} \models_{sk} T(\phi)$. But since $\mathcal{M}_\rho$ is a fixed point, we have $\mathcal{M}_{\rho+1} = \mathcal{M}_\rho$. So $\mathcal{M}_\rho \models_{sk} T(\phi)$.

Second, suppose $\mathcal{M}_\rho \models_{sk} T(\phi)$. Then there must be an ordinal $\alpha < \rho$ such that $\mathcal{M}_\alpha \models_{sk} \phi$. And therefore, by monotonicity, $\mathcal{M}_\rho \models_{sk} \phi$. ■

This is the clever insight that Kripke had: Tarski's undefinability theorem does *not* hold in an unqualified way for partial logic. Recall that Tarski's undefinability theorem says that no *classical*, Tarskian model $\mathcal{M}$ is such that for all sentences ϕ of $\mathcal{L}_T$, $\mathcal{M} \models \phi$ exactly if $\mathcal{M} \models T(\phi)$, or, equivalently, is such that for all sentences ϕ of $\mathcal{L}_T$, $\mathcal{M} \models \phi \leftrightarrow T(\phi)$. But the previous theorem tells us that we *have* constructed a *partial* model $\mathcal{M}_\rho$ such that for all sentences ϕ in $\mathcal{L}_T$: $\mathcal{M}_\rho \models_{sk} \phi$ exactly if $\mathcal{M}_\rho \models_{sk} T(\phi)$. However, this is *not* equivalent to saying that for all sentences ϕ in $\mathcal{L}_T$:

$$\mathcal{M}_\rho \models_{sk} \phi \leftrightarrow T(\phi).$$

In fact – and this is a *fourth* reason why the least fixed point is a particularly nice model – this cannot be the case, for the liar sentence L is left indeterminate by the least fixed point. To prove this, we first describe a way of obtaining a classical model for $\mathcal{L}_T$ from a partial model by *closing off* the partial model:

Definition 13.6.1 For any partial model $(\mathcal{E}, \mathcal{A})$ with $\mathcal{E} \cap \mathcal{A} = \emptyset$, the closed off model corresponding to $(\mathcal{E}, \mathcal{A})$ is the classical model $(\mathcal{E}, \mathbb{N} \setminus \mathcal{E})$.

Intuitively, a partial model is closed off by absorbing the gap into the anti-extension of the truth predicate.

Theorem 13.6.3 $\mathcal{M}_\rho \not\models_{sk} T(L)$ and $\mathcal{M}_\rho \not\models_{sk} \neg T(L)$.

Proof. The second part is proved in a similar way, so we concentrate on the first part of the theorem.

Since $\mathcal{M}_\rho$ has the fixed point property (Theorem 13.6.2), we have $\mathcal{M}_\rho \models_{sk} L$ exactly if $\mathcal{M}_\rho \models_{sk} T(L)$. Now by closing off the model $\mathcal{M}_\rho$ we obtain a classical model $\mathcal{M}^c_\rho$. Suppose that $\mathcal{M}_\rho \models_{sk} T(L)$, and thereby $\mathcal{M}_\rho \models_{sk} L$. Then by monotonicity, also $\mathcal{M}^c_\rho \models_{sk} T(L)$ and $\mathcal{M}^c_\rho \models_{sk} L$. (Note that because $\mathcal{M}^c_\rho$ is a classical model, $\mathcal{M}^c_\rho \models_{sk} \ldots$ amounts to the same as $\mathcal{M}^c_\rho \models \ldots$.) But since $\mathcal{M}^c_\rho$ is just a classical model, the diagonal lemma holds in it. So we have

$$\mathcal{M}^c_\rho \models_{sk} L \leftrightarrow \neg T(L).$$

But putting these three facts together gives us a contradiction. So we deny our supposition and conclude that $\mathcal{M}_\rho \not\models_{sk} T(L)$. ■

Since L and $T(L)$ are left undecided by $\mathcal{M}_\rho$, so is the sentence $L \leftrightarrow T(L)$, by the clauses of the strong Kleene valuation scheme. So in *this* sense, the unrestricted

Tarski-biconditionals are not satisfied by $\mathcal{M}_\rho$. But we have come very close to getting what we want.

Some philosophers are nevertheless not content. They insist that there is a sense in which the Tarki-biconditionals should all be outright assertible. One way in which one can try to implement this is by adding a new primitive conditional operator to the language $\mathcal{L}_T$ which differs in its logical properties from the material implication that we have hitherto used. Then the naive unrestricted Tarski-biconditionals can be expressed in terms of this new operator, and coherently upheld. This line of research is explored in [Field, 2008]. It should by now not come as a surprise that such a new conditional operator will have some unexpected logical properties.

So far we have discussed the least fixed point, the smallest fixed point model that exists. But there is a plethora of partial models that have the fixed point property, and these fixed point models have been intensively studied. It is no easy matter to decide which of these models should be preferred as the 'intended' model(s) of the language $\mathcal{L}_T$.

The liar sentence is of course not the only paradoxical sentence of $\mathcal{L}_T$. And not all paradoxical sentences behave in the same way in all fixed points. Consider the sentence J such that

$$PA^T \vdash J \leftrightarrow T(J).$$

The diagonal lemma guarantees that J exists. This sentence says of itself (modulo coding) that it is true. It is called the *truth-teller* sentence. It can be shown that like the liar sentence, the truth-teller is gappy in the least fixed point model. But whereas the liar sentence is gappy in *all* fixed point models, there are fixed point models in which J is true and fixed point models in which J is false. If we would have put J in the extension (anti-extension) of the truth predicate at the outset, it would have remained there throughout the process.

Nevertheless, the least fixed point model is special. In the first stage, the extension and anti-extension of the truth predicate were taken to be empty. Then we looked at the world (the world of the natural numbers, in this case) to include sentences in the extension and in the anti-extension. And then we built on that. As a result, the sentences in the extension and anti-extension of the least fixed point are all *grounded* in the world. (For more about the notion of groundedness, see [Yablo, 1982] and [Leitgeb, 2005].) Their truth value is ultimately determined by facts in the (mathematical) world. For other fixed points, this is not always the case.

It was mentioned above that beside the strong Kleene scheme, there are other valuation schemes for partial logic. One of these schemes deserves special attention. This is the *supervaluation scheme*, which is due to van Fraassen. In the supervaluation approach, a formula $\phi \in \mathcal{L}_T$ is regarded as true in a partial model $\mathcal{M} = (\mathcal{E}, \mathcal{A})$ if and only if ϕ is true in all total (or classical) models $\mathcal{M}_c = \langle \mathbb{N}, \mathcal{C} \rangle$

for which the interpretation $\mathcal{C}$ of the truth predicate is such that $\mathcal{E} \subseteq \mathcal{C}$ and $\mathcal{A} \subseteq \mathbb{N}\backslash\mathcal{C}$. Intuitively, this means that ϕ is regarded as supervaluation-true in a partial model if ϕ comes out true in every way of extending this partial model to a total, classical model. Similarly, we say that a formula $\phi \in \mathcal{L}_T$ is regarded as false in a partial model $\mathcal{M} = (\mathcal{E}, \mathcal{A})$ if and only if ϕ is false in all total (or classical) models $\mathcal{M}_c = \langle \mathbb{N}, \mathcal{C} \rangle$ for which the interpretation $\mathcal{C}$ of the truth predicate is such that $\mathcal{E} \subseteq \mathcal{C}$ and $\mathcal{A} \subseteq \mathbb{N}\backslash\mathcal{C}$.

Of course the notion of supervaluation-truth leaves room for a formula to be neither supervaluation-true, nor supervaluation-false. But it is clear that every classical logical truth will come out supervaluation-true in all partial models. So in this sense, supervaluation logic is closer to classical logic than strong Kleene logic.

The considerations that we have gone through in this section establish that a minimal supervaluation fixed point model can be built for $\mathcal{L}_T$ in the same way as for the strong Kleene scheme. This minimal supervaluation fixed point model will also judge neither the liar sentence nor its negation to be true. But unlike the strong Kleene fixed point, it will judge $L \vee \neg L$ to be determinately true, for this is an instance of the law of excluded third. So the supervaluation-incarnation of Kripke's theory of truth is non-compositional.

There is one pressing philosophical objection to Kripke's theory of truth, both in its strong Kleene incarnation and in its supervaluation guise. In the least fixed point model, the liar sentence L ends up in the gap: it is neither in the extension nor in the anti-extension of T. This may be expressed as: L is neither made true nor made false in the least fixed point. But if it is made neither true nor false, then it is not made true. But that is exactly what L says of itself. So L seems true after all! Thus the liar paradox strikes again – or so it seems. In Kripke's own words: 'The ghost of the Tarski hierarchy is still with us' [Kripke, 1975b, p. 80]. In this way, we have arrived at another variant of the strengthened liar problem. The reader will recall that contextualist theories of truth were plagued by it. It is one of the most recalcitrant problems for theories of truth. Even a cursory look at the essays in [Beall, 2007] will convince the reader of the seriousness of this problem for many contemporary formal theories of truth. Almost every truth theorist alleges that *his* theory escapes the strengthened liar problem, but all the other truth theories succumb to it.

6.3 Axiomatising Kripke's Theory

It was argued earlier that semantic theories of truth in general are not wholly satisfactory. Thus Kripke's theory of truth cannot be accepted as it stands. But Kripke's theory of truth has inspired strong and natural axiomatic theories of truth. To these we now turn.

There exists an axiomatic theory of self-referential truth which is due to Feferman [Feferman, 1991]. This theory is called *KF* (for 'Kripke-Feferman'). It can be seen as an attempt to axiomatically describe the construction of Kripke's fixed point models. Aside from PA^T, the theory *KF* consists of the following axioms:

KF1 $\forall$ atomic $\phi \in \mathcal{L}_{PA} : T(\phi) \leftrightarrow val^+(\phi)$
KF2 $\forall$ atomic $\phi \in \mathcal{L}_{PA} : T(\neg\phi) \leftrightarrow val^-(\phi)$
KF3 $\forall\phi \in \mathcal{L}_T : T(\neg\neg\phi) \leftrightarrow T(\phi)$
KF4 $\forall\phi, \psi \in \mathcal{L}_T : T(\phi \wedge \psi) \leftrightarrow (T(\phi) \wedge T(\psi))$
KF5 $\forall\phi, \psi \in \mathcal{L}_T : T(\neg(\phi \wedge \psi)) \leftrightarrow (T(\neg\phi) \vee T(\neg\psi))$
KF6 $\forall\phi(x) \in \mathcal{L}_T : T(\forall x\phi(x)) \leftrightarrow \forall y T(\phi(y))$
KF7 $\forall\phi(x) \in \mathcal{L}_T : T(\neg\forall x\phi(x)) \leftrightarrow \exists y T(\neg\phi(y))$
KF8 $\forall\phi \in \mathcal{L}_T : T(T(\phi)) \leftrightarrow T(\phi)$
KF9 $\forall\phi \in \mathcal{L}_T : T(\neg T(\phi)) \leftrightarrow T(\neg\phi)$
KF10 $\forall\phi \in \mathcal{L}_T : \neg(T\phi \wedge T\neg\phi)$

Thus *KF* is a strongly compositional type-free theory of truth (KF1–KF7) that includes truth iteration axioms (KF8–KF9).

KF10 expresses the consistency of the extension of the truth predicate. Unlike the other truth axioms, it does not reduce the truth of statements to the truth of other statements or to elementary facts. But it allows *KF* to prove that truth is closed under modus ponens:

Proposition 13.6.3 $KF \vdash \forall\phi, \psi \in \mathcal{L}_T : [T(\phi) \wedge T(\phi \rightarrow \psi)] \rightarrow T(\psi)$

This proposition cannot be proved without making use of axiom KF10.

Let us have a look at the formal properties of *KF*.

Theorem 13.6.4 *KF has nice models.*

Proof. We start with a partial model: any fixed point model as constructed in Kripke's semantical theory of truth will do. Then we close this model off so as to obtain a classical model. It is then routine to verify by an induction on the length of proofs that this model verifies all the axioms of *KF*. ■

So far, it seems that *KF* is an attractive theory of truth. However, we will now turn to properties of *KF* which disqualify it from ever becoming our favourite theory of truth.

Lemma 13.6.1 *For all sentences* $\phi \in \mathcal{L}_T$, $KF \vdash T(\phi) \rightarrow \phi$.

Proof. By induction on the complexity of ϕ. ■

This important lemma has a very unwelcome consequence:

Corollary 13.6.2 *KF* $\vdash L \wedge \neg T(L)$, *where L is the liar sentence.*

Proof. We reason in *KF*. We know by the extended diagonal lemma that L is such that *KF* $\vdash L \leftrightarrow \neg T(L)$. So assume $\neg L$. Then $T(L)$. So by Lemma 13.6.1, we obtain L, which gives us a contradiction. So we reject our assumption and conclude L. Then we appeal once again to $L \leftrightarrow \neg T(L)$ to obtain $\neg T(L)$. ■

In other words, *KF* proves sentences which *by its own lights* are untrue.

The problem is that *KF* is formulated in classical logic, whereas Kripke's semantical theory is partial. Instead of axiomatizing the closing off of fixed point models (which are classical models), perhaps we should try to capture, in a proof-theoretical manner, and to the maximal extent possible, the sentences that are made true (in the strong Kleene sense) by the minimal fixed point. This line of research was pursued in [Halbach and Horsten, 2006].

7. The Revision Theory of Truth

The revision theory of truth [Gupta and Belnap, 1993] is a semantical untyped truth theory which aims to classify many sentences that express truth-iterations as true.

7.1 Two Revision-Theoretic Notions of Truth

The general idea is this. We start with a classical model for $\mathcal{L}_T$. This model is transformed into a new model again and again, thus yielding a long sequence of classical models for $\mathcal{L}_T$, which are indexed by ordinal numbers. The official notion of truth for a formula of $\mathcal{L}_T$ is then distilled from this long sequence of models.

As before, we only consider models that are based on the standard natural number structure. For simplicity, let us start with the model

$$\mathcal{M}_0 := \langle \mathbb{N}, \emptyset \rangle$$

the model which regards no sentence whatsoever as true. Suppose we have a model $\mathcal{M}_\alpha$. Then the next model in the sequence is defined as follows:

$$\mathcal{M}_{\alpha+1} := \langle \mathbb{N}, \{\phi \in \mathcal{L}_T \mid \mathcal{M}_\alpha \models \phi\} \rangle.$$

In other words, the next model is always obtained by putting those sentences in the extension of the truth predicate which are made true by the last model that has already been obtained.

Now suppose that λ is a limit ordinal, and that all models $\mathcal{M}_\beta$ for $\beta < \lambda$ have already been defined. Then

$$\mathcal{M}_\lambda := \langle \mathbb{N}, \{\phi \in \mathcal{L}_T \mid \exists\beta\forall\gamma : (\gamma \geq \beta \wedge \gamma < \lambda) \Rightarrow \mathcal{M}_\gamma \models \phi\}\rangle.$$

In words: we put a sentence ϕ in the extension of the truth predicate of $\mathcal{M}_\lambda$ if there is a 'stage' β before λ such that from $\mathcal{M}_\beta$ onwards, ϕ is *always* in the extension of the truth predicate.

This yields a chain of models that is as long as the chain of the ordinal numbers. Elementary cardinality considerations (Cantor's theorem) tell us that there must be ordinals α and β such that $\mathcal{M}_\alpha$ and $\mathcal{M}_\beta$ are identical. In other words, the chain of models must be periodic.

On the basis of this long sequence of model, we can now define the notion of *stable truth* for the language $\mathcal{L}_T$. A sentence $\phi \in \mathcal{L}_T$ is said to be stably true if at some ordinal stage α, ϕ enters in the extension of the truth predicate of $\mathcal{M}_\alpha$ and stays in the extension of the truth predicate in all later models. A sentence $\phi \in \mathcal{L}_T$ is said to be stably false if at some ordinal stage α, ϕ is outside the extension of the truth predicate of $\mathcal{M}_\alpha$ and stays out forever thereafter. A sentence that is neither stably true nor stably false is said to be *paradoxical*.

The reader will easily verify that the sentence $T(T(0 = 0))$, for instance, enters the extension of the truth predicate in $\mathcal{M}_3$ and stays in for ever after. So this sentence is stably true. The reader will also verify that the liar sentence L, governed by the principle $L \leftrightarrow \neg T(L)$, vacillates. It enters the truth predicate in $\mathcal{M}_1$, but jumps out again in $\mathcal{M}_2$, then comes back, jumps out again, ... In brief, it never settles down.

Now revision theorists tentatively propose to identify truth simpliciter with stable truth, and falsehood simpliciter with stable falsehood. Sentences that never stabilise, such as the liar, are classified as paradoxical. But they hesitate to fully endorse this identification. Another strong contender for identification with truth simpliciter (falsehood simpliciter) is the slightly more complicated notion of *nearly stable truth* (nearly stable falsehood). A sentence $\phi \in \mathcal{L}_T$ is said to be nearly stably true if for every stage α, there is a natural number n such that for all natural numbers $m \geq n$, ϕ is in the extension of the truth predicate of $\mathcal{M}_{\alpha+m}$. And a sentence $\phi \in \mathcal{L}_T$ is said to be nearly stably false if for every stage α, there is a natural number n such that for all natural numbers $m \geq n$, ϕ is outside the extension of the truth predicate of $\mathcal{M}_{\alpha+m}$. In other words, for this notion of truth we do not care what happens before any fixed finite number of steps after any limit ordinal. (Below, we will give an example of a nearly stable truth that is not a stable truth.)

The revision theory of truth was developed as an attempt to explicate our naive patterns of reasoning with self-referential sentences containing the truth predicate. Indeed, when we naively go about trying to decide whether a sentence ϕ is *true*, we consider the situation we think we find ourselves in, and evaluate whether ϕ holds in this situation. Such a situation is like a model. So what we in fact do is to use the naive Tarski-biconditional $T(\phi) \leftrightarrow \phi$. When we find out that ϕ indeed holds in the 'model' we are considering, we add ϕ to the extension of T, thus generating a revised model. In other words, we use the naive Tarski-biconditionals to keep on adjusting our view of the situation we find ourselves in. And this results in *diachronic inconsistency*: we keep changing our minds about the truth status of the liar sentence, for example. It must be conceded that it is not so easy to explain how the limit rule for constructing models is connected to our naive reasoning practice. But at the same time the limit rule appears to be the only natural one that springs to mind.

The notions of stable truth and of nearly stable truth seem, from an extensional point of view, to match up nicely with the class of sentences that we are intuitively inclined to regard as true. One objection that is often raised to the revision theory of truth is that its notions of truth (stable truth, nearly stable truth) are too complicated: it is difficult to believe that *our* notion of truth is that complex. In particular, the revision theory of truth is not accompanied with a story about how we acquire the concept of truth in a way that Kripke's theory of truth is.

7.2 The Friedman-Sheard Theory

Friedman and Sheard have proposed an axiomatic theory of self-referential truth which is called *FS* [Friedman and Sheard, 1987]. Friedman and Sheard gave a slightly different list of axioms, but the following list is equivalent to their system:

FS1 PA^T
FS2 $\forall$ atomic $\phi \in \mathcal{L}_{PA} : T(\phi) \leftrightarrow val^+(\phi)$
FS3 $\forall \phi \in \mathcal{L}_T : T(\neg\phi) \leftrightarrow \neg T(\phi)$
FS4 $\forall \phi, \psi \in \mathcal{L}_T : T(\phi \wedge \psi) \leftrightarrow T(\phi) \wedge T(\psi)$
FS5 $\forall \phi(x) \in \mathcal{L}_T : T(\forall x \phi(x)) \leftrightarrow \forall x T(\phi(x))$

Moreover, *FS* contains two extra rules of inference, which are called *Necessitation* (NEC) and *Co-Necessitation* (CONEC), respectively:

NEC From a proof of ϕ, infer $T(\phi)$
CONEC From a proof of $T(\phi)$, infer ϕ

Let us consider the axioms and rules of *FS*. The compositional axioms of *FS* show that it seeks to reflect, like *TC*, the intuition of the *compositionality* of truth.

In this sense, *FS* can be seen as a natural extension of *TC*. In fact, the *axioms* are exactly like the axioms of *TC*, except that the compositional axioms quantify over the entire language of truth instead of only over $\mathcal{L}_{PA}$. But if we disregard the rules of inference NEC and CONEC, this does not help us in any way in proving *iterated* truth statements. The reason is that the truth axiom for atomic sentences, only quantifies over atomic *arithmetical* sentences.

FS is the result of maximizing the intuition of the compositionality of truth. Nevertheless, the truth of truth attribution statements is in *FS* only in a weaker sense compositionally determined than the truth of other statements. For *FS* only claims that if a truth attribution has been *proved*, then this truth attribution can be regarded as true (and conversely), whereas for a conjunctive statement, for instance, *FS* makes the stronger hypothetical claim that *if* it is true, then both its conjuncts are true also (and conversely). But it is necessarily that way. If we replace NEC and the CONEC by the corresponding *axiom schemes*, an inconsistent theory results.

It is comforting to know that *FS* meets a minimal coherence constraint:

Theorem 13.7.1 (Friedman and Sheard) *FS is consistent.*

Observe that this consistency theorem shows that the *inference rules* NEC and CONEC together are in the context of the other axioms of *FS* weaker than their corresponding *axioms* $\phi \rightarrow T(\phi)$ and $T(\phi) \rightarrow \phi$. Indeed, as mentioned above, including the latter axioms amounts to including the unrestricted Tarski-biconditionals, which results in inconsistency.

Not all of *FS* is stably true. Indeed, it is not hard to see that every limit model fails to make axiom FS3 true, for instance. Nevertheless, the following does hold:

Proposition 13.7.1 *FS* is nearly stably true.

Proof. This is shown by a straightforward induction on the length of proofs in *FS*. ■

So *FS* is indeed closely connected to the 'nearly stable truth'-variant of the revision theory of truth. And this also shows that the notions of stable truth and nearly stable truth do not coincide.

Nevertheless, not all is well with *FS*. Even though *FS* is consistent and indeed even arithmetically sound, it is in some sense 'almost inconsistent':

Definition 13.7.1 An arithmetical theory T is ω-inconsistent if for some formula $\phi(x)$, the theory T proves $\exists x\phi(x)$ while at the same time for every $n \in \mathbb{N}$, T proves $\neg\phi(\overline{n})$.

McGee has proved that *FS* is indeed ω-inconsistent [McGee, 1985b]:

Theorem 13.7.2 (McGee) *For some formula* $\phi(x) \in \mathcal{L}_T : FS \vdash \exists x \phi(x)$ *and* $FS \vdash \neg\phi(n)$ *for all* $n \in \mathbb{N}$.

In other words, *FS* asserts that there is a property that some number has, but at the same time it asserts that 0 does not have it, 1 does not have it, ... So *FS* cannot be the final word on axiomatic theories of truth. One obvious strategy at this point is to try to weaken *FS* in such a way that an ω-consistent theory is obtained which still preserves the main virtues of *FS*. This strategy is pursued in [Horsten, ta, Chapter 8].

8. Other Approaches and Further Reading

For further reading on the topics that have been covered in this chapter, we refer the reader to the following works. [Visser, 1989] gives a good exposition of semantical approaches to type-free truth. For this purpose, also [McGee, 1991] is still a very good source. The authoritative reference work for axiomatic theories of truth is [Cantini, 1996]. [Horsten, ta] is an introductory textbook on proof-theoretic approaches to truth and their applications to philosophy. [Halbach, ta] is a slightly more advanced and more comprehensive work on the same subject.

We should also mention an approach that has been gaining in popularity in recent years, but which was not discussed in the present chapter. Graham Priest has developed a theory of truth according to which paradoxical sentences such as the liar sentence are true and false at the same time. The seminal work for this line of research is [Priest, 1987].

Notes

1. There are similar paradoxes for intensional notions such as knowledge, necessity, past, and future. See [Kaplan and Montague, 1960], [Montague, 1963], [Horsten and Leitgeb, 2001], [Halbach et al., 2003]. These paradoxes will not be discussed in this chapter.
2. A proof of this lemma can be found in any good exposition of Gödel's incompleteness theorems, such as [Boolos and Jeffrey, 1989].

14 Indicative Conditionals

Igor Douven

Chapter Overview

1. Truth Conditions	384
1.1 Two Truth-Conditional Accounts of Conditionals	385
1.2 Why Believe that Conditionals Have Truth Conditions?	386
1.3 Arguments against Truth-Conditionality	389
1.3.1 Arguments against the Material Conditional Account	389
1.3.2 Arguments against the Possible Worlds Account	392
1.3.3 Two General Arguments against Truth-Conditionality	393
2. Assertability and Acceptability	397
Acknowledgements	401
Notes	401

Conditionals are sentences of the forms 'If *A*, [then] *B*' and '*B* if *A*', such as

(1) a. If the village is flooded, then the dam must have broken.
    b. If Henry had come to the party, Sue would have come, too.
    c. Paul would have bought the house if it hadn't been so expensive.

Some authors also classify as conditionals sentences that can be naturally put in the above forms, such as

(2) a. They will leave in an hour, unless John changes his mind.
    b. No guts, no glory.

which can be rephrased as, respectively, (3a) and (3b):

(3) a. If John does not change his mind, they will leave in an hour.
    b. If a person lacks courage, there will be no glory for him or her.

In 'If *A*, *B*', *A* is called the 'antecedent' and *B*, the 'consequent'.

It is common practice to group conditionals into two major types, to wit, *indicative* conditionals and *subjunctive* conditionals. Typically, the antecedents of subjunctive conditionals, but not those of indicative conditionals, are known to be false or at least strongly suspected to be false.[1] The difference tends to be reflected grammatically by the use of the indicative and subjunctive mood, respectively, in the main clause of the conditional.[2] Of the above examples, (1b) and (1c) are subjunctive conditionals; the others are indicative conditionals. The present chapter will be exclusively concerned with conditionals of the latter type. The noun 'conditional' refers throughout to indicative conditionals, unless specified otherwise.

The importance of the role or roles conditionals play in both everyday and scientific discourse and reasoning is hard to miss. Perhaps, then, it is no surprise that, for some decades now, conditionals have been a central area of investigation not only in philosophy but also in both linguistics and psychology. What *is* surprising, however, is that despite the considerable expenditure of time and effort of many researchers from those fields, there is still little one can say about conditionals that is *not* controversial. Even with respect to the most fundamental questions about conditionals – do conditionals have truth conditions and, if so, what are they?, what are the acceptability and assertability conditions of conditionals? – there is no unanimity or even something that one could rightly designate as a majority position on the issue.

This chapter focusses on the aforementioned fundamental questions, in the order in which they were mentioned, and it reviews the main answers that have been put forth in the literature. It thereby leaves entirely out of consideration worthwhile work on conditionals that has been done by linguists, in particular work centring on the classification of conditionals.[3] And although we will briefly touch upon some relevant research carried out by experimental psychologists, most of the psychological research on conditionals will also remain undiscussed here.[4] Finally, while below we will come across some of the better-known inferential principles for conditionals, the interested reader may want to consult [Cross and Nute, 2001] and [Arló Costa, 2007] for systematic treatments of conditional logics.

1. Truth Conditions

This section starts by presenting the two main truth-conditional accounts of conditionals. It then considers the question of why it is prima facie reasonable to think that conditionals have truth conditions to begin with. Finally, it discusses the most powerful objections that have been brought against the truth-conditionality of conditionals.

1.1 Two Truth-Conditional Accounts of Conditionals

Among those who hold that conditionals have truth conditions, some think these truth conditions are truth-functional, while others think they are not. Those who hold the former view all agree that the truth conditions of a conditional are those of the corresponding material conditional; that is to say, according to them, 'If *A*, *B*' is false if *A* is true and *B* false, and is true in the other cases. Equivalently, on this proposal a conditional expresses the proposition that either the conditional's antecedent is false or its consequent is true (or both).[5]

The other position takes for granted the notion of a possible world, roughly understood as a way the world might be or might have been. On this position, a conditional is true iff its consequent is true in the 'closest' possible world in which the conditional's antecedent is true, provided there *is* a world in which the antecedent is true, and where 'closest' means 'most similar to the actual world'.[6] Accordingly, the proposition expressed by a conditional 'If *A*, *B*' corresponds to the set of worlds whose closest *A*-worlds (the worlds where *A* holds) are *B*-worlds. Given that if *A* is true at a world, it is its own closest *A*-world, we can put the foregoing differently by saying that the proposition expressed by 'If *A*, *B*' consists of the union of the set of *A*-and-*B*-worlds with the set of not-*A*-worlds whose closest *A*-world is a *B*-world.

It might appear that, because of its reference to the actual world in the truth condition, this proposal allows one to evaluate a conditional only if one knows which world is actual (which we typically do not know). But that is not so. For instance, to be certain that 'If *A*, *B*' is true it is enough to be certain that each world we take to be a candidate for being the actual world has a closest *A*-world that is also a *B*-world.

According to Robert Stalnaker – its main proponent – the possible worlds account applies to both indicative and subjunctive conditionals.[7,8] However, in the case of indicative conditionals there is a pragmatic constraint on which worlds can count as 'closest', to wit, that they must be among the ones that are not ruled out by anything that has been accepted or is presupposed in the context in which the conditional is asserted or being evaluated.[9] As a result, on the possible worlds account, indicative conditionals have truth conditions only relative to contexts.

Restricted to indicative conditionals, the material conditional account and the possible worlds account partially agree: if a conditional's antecedent is true and its consequent false, then both assign the value 'false' to the conditional; and if both the conditional's antecedent and its consequent are true, then both accounts assign it the value 'true'. In the remaining cases, however, the accounts may diverge in their truth-value assignments. On the possible worlds account, it is not necessarily the case that a conditional is true if its antecedent is false; perhaps its consequent fails to hold in the closest world in which its antecedent is true. Differently put, it is not necessarily the case that the set of possible worlds in which either A is false or B is true coincides with the set of worlds whose closest A-worlds are B-worlds.

It is fair to say that, while the possible worlds account is currently regarded as the best semantics around for subjunctive conditionals, the material conditional account is more widely accepted as a semantics for indicative conditionals (although it is no longer as popular as it once was).

1.2 Why Believe that Conditionals Have Truth Conditions?

The uninitiated may have been surprised to read that there is a controversy about the truth conditions of conditionals; it is not as though there were any disagreement over the truth conditions of conjunctions or disjunctions, after all. They will then find it even more surprising that there is a genuine dispute about the question of whether conditionals have truth conditions to begin with. Of course, not all sentences have truth conditions; commands and questions do not, for instance. But conditionals are not commands or questions; at least pretheoretically, they would seem to be grouped most naturally with the declarative sentences. So, it might seem already a point in favour of truth-conditional accounts of conditionals that they do not make conditionals appear oddballs among the declarative sentences. However, those who hold that conditionals lack truth conditions may retort that conditionals must appear to be special from any perspective. Indeed, it might be said, even the newcomer to the field should have expected as much, finding on the shelves books entitled *If* [Evans and Over, 2004], *Ifs* [Harper et al., 1981], *A Philosophical Guide to Conditionals* [Bennett, 2003], and several books with the title *Conditionals* (e.g., [Jackson, 1987], [Jackson, 1991], [Woods, 1997]), whereas there is no book called *And*, or *Ands*, or *A Philosophical Guide to Conjunctions*, or *Conjunctions* (or *Or*, etc.).[10]

The advocates of the truth-conditional accounts of conditionals have advanced more compelling considerations in favour of their commitment to truth conditions than simply the contention that non-truth-conditional views make conditionals appear too special. Here is one well-known objection: On the known conceptions of propositions, sentences count as expressing propositions only if they have truth conditions. But if conditional sentences do not express propositions, then it should be puzzling how we can make sense of Boolean embeddings of such sentences, such as

(4) a. Joan won't come, but if Henry comes, he will bring Sue.
b. It is not the case that if she fails the exam, she will be allowed to resit.
c. Either he will change his mind if he hears my arguments or he will leave the department.

After all, the Boolean connectives are *propositional* connectives. Clearly, though, these sentences pose not the least interpretational difficulties.[11]

Those who deny that conditionals have truth conditions may argue that in ordinary language, grammar does not always transparently reveal logical form, and that for instance (4a) is not really a conjunction but is to be thought of as consisting of two sentences, as follows:

(5) Joan won't come. If Henry comes, he will bring Sue.

For pragmatic or stylistic reasons we may concatenate these sentence by dint of 'but', which however does not function as a propositional operator in that case.

Similarly, it may be suggested that the conditional in (4b) is only apparently in the scope of the negation operator, and that the sentence can plausibly be rephrased as

(6) If she fails, then she won't be allowed to resit.

As Dorothy Edgington [Edgington, 1995b, p. 283] argues more generally, '*A* is to ¬*A* as "If *A*, *B*" is to "If *A*, ¬*B*" [and] "It's not the case that if *A*, *B*" has no clear established sense distinguishable from ["If *A*, ¬*B*"].'[12,13]

The third example, (4c), requires more explaining. It may, to begin with, be paraphrased as

(7) If he hears my arguments, then if he does not change his mind, he will leave the department.

which is a right-nested conditional, that is, a conditional of the form 'If *A*, then if *B*, then *C*.' It has been argued – convincingly, most think – that such conditionals can be reduced to simple conditionals – conditionals whose antecedent and consequent are not themselves conditional in form – via the following principle:

Import–Export (IE) 'If *A*, then if *B*, then *C*' and 'If *A* and *B*, then *C*' are logically equivalent.

In the present case, this principle indisputably yields the right result, for (4c) can be naturally understood as meaning

(8) If he hears my arguments and does not change his mind, he will leave the department.

However, those favouring a non-truth-conditional view on conditionals will have to show that *all* embedded conditionals that we can intuitively make sense of are somehow reducible to simple conditionals or that they can be otherwise reformulated to avoid the embedding. And that this can be accomplished is far from obvious. Consider the following sentence:

(9) If the match is cancelled if it starts raining, then the match is cancelled if it starts snowing.

It is hopeless to try and apply (IE) here. One might first make a right-nested conditional out of (9),

(10) If it starts raining, then if the match is cancelled and it starts snowing, the match is cancelled.

and then apply (IE) to obtain

(11) If it starts raining and the match is cancelled and it starts snowing, then the match is cancelled.

But this sounds trivial and nonsensical at the same time, and in any event (11) is not an adequate rephrasing of (9). It cannot be excluded that, with enough ingenuity, sentences like (9) *can* be rephrased as simple conditionals, or even as non-conditional sentences. Still, the proponents of the non-truth-conditional view on conditionals have not shown us how to do it.

Note that if, by contrast, conditionals express propositions, (9) should pose no special interpretational difficulties – which is in accordance with how we intuitively assess the sentence, to wit, as being readily interpretable.

A second reason for thinking that conditionals have truth conditions is that conditionals appear to be bona fide candidates for belief. For example, many of us believe the following:

(12) If no measures are taken to reduce the emission of greenhouse gases, then sea levels will keep rising.

But how else can we make sense of believing something than as believing the thing to be *true*? And how can we believe a conditional to be true if it does not have truth conditions?

In fairness, it must be noted that some philosophers who hold that conditionals do not have truth conditions in the ordinary sense think there are conditions under which conditionals are true – namely, when the antecedent and consequent are both true – and also conditions under which they are false, namely, when the antecedent is true and the consequent false. But this does not help with the second problem: we may believe that (12) is true regardless of whether measures will be taken to reduce the emission of greenhouse gases, so in particular also if its antecedent is false. Nor, in effect, does it help with the first problem, because to say that, in some special cases, conditionals are true and, in other special cases, false, does not suffice to turn conditionals into propositions.

A seemingly more effective reply is to claim that we do not really believe conditionals to be true but still have the intuition that sometimes we do because we are prone to mistake

(13) [It is true that sea levels will keep rising] if no measures are taken to reduce the emission of greenhouse gases.

for

(14) It is true that [sea levels will keep rising if no measures are taken to reduce the emission of greenhouse gases].

Because we are insufficiently sensitive to this scope ambiguity of the truth predicate, we may take ourselves to have a belief expressed by (14), while what we actually have is the *conditional* belief that we could express by (13): we believe something to be true, or better perhaps, we are prepared to believe something to be true (to wit, that sea levels will keep rising), on the condition that something else is true (that no measures are taken to reduce the emission of greenhouse gases).[14] Still, one may not find this reply very satisfactory, given that it assumes people to be massively mistaken about the correct interpretation of a sizeable portion of their beliefs.

1.3 Arguments against Truth-Conditionality

It is not as though proponents of the view that conditionals lack truth conditions are unaware of the above problems.[15] Nevertheless, they see serious problems for the opposite position as well and believe that, on balance, their own position does best. Some of their arguments are specifically directed against one of the two truth-conditional accounts that were summarized above. But they have also levelled some general and arguably more forceful arguments against truth-conditional accounts. The remainder of this section reviews the best known of these arguments.

1.3.1 Arguments against the Material Conditional Account

While the material conditional account validates many inferential principles that we deem pretheoretically valid, it is a common complaint about the account that it overgenerates 'validities', that is, some inferential principles that are intuitively *rejectable* also come out as valid on this account. These principles are generally known as 'the paradoxes of material implication', and as such they are often discussed in introductory logic courses. To give a famous example, on the material conditional account a conditional is true whenever its antecedent is false. Yet we do not think that the falsity of

(15) Jim has retired.

entails

(16) If Jim has retired, then he will teach next year's epistemology course.

Equally, the account rules true 'If *A* and *B*, then *C*' whenever it rules true 'If *A*, then *C*'. But it would strike us as a mistake were one to infer

(17) If John works hard and is shot on his way to the exam, he will pass the exam.

from

(18) If John works hard, he will pass the exam.

In defence of the material conditional account, several attempts have been made to explain away along pragmatic lines the unintuitiveness of the above and related seemingly problematic inferences. It has been argued, for instance, that while the falsity of (15) does entail (16), we still have the intuition that the inference from the falsity of (15) to (16) is invalid because, typically, when the negation of (15) is assertable, (16) is *not*: by asserting the latter when one is in a position to deny the former one would say less than one could with more words, thus violating a fundamental principle of conversational practice.[16] The details of these attempts vary somewhat, for not all authors taking this line of defence agree on the question of which conditions make a conditional assertable.[17]

Above, we saw that embedded conditionals have been adduced to argue in favour of at least a truth-conditional semantics for conditionals. However, embedded conditionals have also been adduced *against* the material conditional account.[18] Allan Gibbard [Gibbard, 1981, p. 235] asks us to consider this conditional:

(19) If the cup broke if dropped, then it was fragile.

As he points out, (19) is assertable even if we deem it unlikely that the cup was dropped or that it is fragile. This is a problem for the material conditional account, Gibbard claims, for if the cup was not dropped and is not fragile, then, on the said account, (19) is false. As a result, (19) would be assertable even if one thinks it unlikely to be true. According to Gibbard (p. 236), invoking pragmatic principles of conversation will not help one out here, for (he contends) such principles explain only why a believed truth may be unassertable, not why a sentence one believes to be false may be nonetheless assertable.

First off, the claim about pragmatic principles is contestable. The following sentences may be perfectly assertable:

(20) a. It's two hours by train from Brussels to London.
b. There is no more beer.
c. There ain't no sunshine when she's gone.

even though they are false: it takes (currently) *slightly* over two hours to get from Brussels to London by train; there will be beer somewhere in the universe even if there is no more beer in the fridge (which is what, we may suppose, (20b) is meant to convey); and for sure there will be sunshine (somewhere, at least) no matter who is or is not gone. Authors have suggested various pragmatic explanations of the phenomenon at issue here. In fact, Grice [Grice, 1989b, p. 34] was very outspoken about the role of pragmatic principles in understanding irony and metaphor, which often involve the assertion of manifest falsehoods.

But grant that, at least in the case of (19), pragmatic principles will not help to explain why it is assertable if false. Then it still holds that (19) poses a threat to the material conditional account only if it must further be granted that, on this account, (19) is false if the cup was not dropped and is not fragile. And why should that be so? Gibbard's argument supposes that (19) is to be analysed as a left-nested conditional, that is, as

(21) If {the cup broke if it was dropped}, then [it was fragile].

where the part between curly brackets is (21)'s antecedent and the part between square brackets, its consequent. On this analysis, (21) is false indeed if the cup was not dropped – so that the antecedent is true – and the cup is not fragile, so that the consequent is false. But (19) does not carry its logical form on its sleeve, and there is room for disputing that the sentence is to be analysed as (21) and for claiming that (19) is properly analysed as follows:

(22) If {the cup was dropped}, then [if it broke, it was fragile].

At a minimum, it is far from evident that (22) misconstrues (19). For Gibbard's argument to go through, however, (22) must be ruled out as a reasonable interpretation of (19), for read as a right-nested conditional, (19) is *true* if, as Gibbard supposes, the cup was not dropped. In fact, even if Gibbard could argue that (19) cannot properly be read as (22), those in favour of the material conditional account might still point to the fact that, pretheoretically, it seems all right to read (19) as (22), and that this (then mistaken) reading may fuel our intuition that (19) is assertable under the circumstances Gibbard assumes, even if, under those circumstances, it comes out false (which, from the perspective of the material conditional account, it does if (21) is the only legitimate analysis of (19)).

1.3.2 Arguments against the Possible Worlds Account

Whereas, as we saw, the material conditional account has been said to make too many inferences involving conditionals valid, critics of the possible worlds account have pointed at inferential principles that we deem intuitively valid but that are not validated by the possible worlds accounts. For instance, this is true of what Robert Stalnaker [Stalnaker, 1975] calls the 'direct argument', that is, the argument from '*A* or *B*' to 'If not *A*, then *B*'. To see why it does not hold on this account, note that if *A* is true in the actual world, then so is the premise of the direct argument. However, nothing follows from this about any not-*A*-world, so in particular it does not follow that *B* holds in the not-*A*-world that is closest to the actual world. In reply to this objection, Stalnaker argues that although the direct argument is not valid on his preferred semantics for conditionals, it is nevertheless (what he calls) a reasonable inference, where an inference is reasonable if one cannot reasonably assert or accept the premise without being committed to the conclusion. This, he claims, is what creates the illusion that the direct argument is valid.

More troubling for the possible worlds account is that it does not validate (IE), a principle that is prima facie plausible and that, for indicative conditionals at least, has no known counterexamples in natural language. This principle, recall, postulates the equivalence of 'If *A*, then if *B*, then *C*' and 'If *A* and *B*, then *C*'. However, given the possible worlds account, there is nothing to guarantee that the equivalence holds: for all the defenders of the account have said, the *B*-world that is closest to the *A*-world that is closest to the actual world need not be identical to the *A*-and-*B*-world that is closest to the actual world. And if the two are non-identical, then the one might be a *C*-world while the other is a not-*C*-world. On the present account, this would yield a counterexample to (IE).[19]

What is generally considered to be the most serious shortcoming of the possible worlds account is that it appears to make the propositional content of a conditional much more sensitive to context than one would reasonably expect it to be. The point was most forcefully made in [Gibbard, 1981]. Gibbard starts by observing that the possible worlds account entails the principle of Conditional Non-Contradiction (CNC), according to which 'If *A*, then *B*' is inconsistent with 'If *A*, then not *B*': the *A*-world closest to the actual world cannot be both a B-world and a not-B-world. He then invites us to consider the following story:

> Sly Pete and Mr. Stone are playing poker on a Mississippi riverboat. It is now up to Pete to call or fold. My henchman Zack sees Stone's hand, which is quite good, and signals its content to Pete. My henchman Jack sees both hands, and sees that Pete's hand is rather low, so that Stone's is the winning hand. At this point, the room is cleared. A few minutes later, Zack slips me a note which says 'If Pete called, he won,' and Jack slips me a note which says 'If Pete called, he lost.' I know that these notes both come from my trusted henchmen, but do not know which of them sent which note. I conclude that Pete folded. [Gibbard, 1981, p. 231]

Gibbard then plausibly argues that if

(23) If Pete called, he won.

and

(24) If Pete called, he lost.

express propositions, they must both express *true* propositions: Zack and Jack are both warranted in their assertions, and their warrants do not rest on any false beliefs about relevant matters of fact. However, this creates a problem for the possible worlds account, for it would seem that, by (CNC), these conditionals are jointly inconsistent. The only escape route for the proponent of the possible worlds account is to claim that (23) does not express the same proposition when it is uttered by Zack as when it is uttered by Jack, so that we would equivocate by taking (CNC) to apply to (23) and (24). But that route leads to a rather extreme kind of context-sensitivity that has little intuitive support. For consider that, in the setting of Gibbard's story, we would have no difficulty interpreting Zack's and Jack's notes, even if – like the first-person narrator of the story – we had no idea which of them slipped us which note.[20]

1.3.3 Two General Arguments against Truth-Conditionality

Stalnaker [Stalnaker, 1970] presented the following thesis as an adequacy condition for any semantics of conditionals:

Stalnaker's Hypothesis (SH) $\Pr(\text{If } A, B) = \Pr(B \mid A)$, for all probability functions Pr such that $\Pr(A) > 0$.

That is to say, if conditionals express propositions, the probability of the proposition expressed by a given conditional must equal the conditional probability of the conditional's consequent given its antecedent. However, Lewis [Lewis,

1976] famously showed that (SH) cannot hold generally, and in effect can hold only for very special, 'trivial', probability functions, that is, probability functions that make any two propositions probabilistically independent of each other (provided these propositions have positive probability) and that have various other features that make them unrealistic as representations of people's states of graded belief.[21]

Instead of going into the details of Lewis' arguments, we consider a simpler and more intuitive argument showing that (SH) is incompatible with the principle (IE) that we already encountered and that, as was said, virtually all who have thought about the issue endorse.

First, by (IE) and probability theory the following holds:

$$\Pr\big(\text{If } A\text{, then if } B\text{, then } C\big) \quad = \quad \Pr(\text{If } A \text{ and } B\text{, then } C).$$

Thus, in particular,

$$\Pr\big(\text{If } \neg B\text{, then if } A\text{, then } B\big) \quad = \quad \Pr(\text{If } A \text{ and } \neg B\text{, then } B).$$

Applying (SH) to both sides of this equation yields

$$\Pr(\text{If } A\text{, then } B \mid \neg B) \quad = \quad \Pr(B \mid A \text{ and } \neg B).$$

So, given that $\Pr(B \mid A \text{ and } \neg B) = 0$, it holds that also $\Pr(\text{If } A\text{, then } B \mid \neg B) = 0$. And $\Pr(\text{If } A\text{, then } B \mid \neg B) = 0$ iff $\Pr(\neg B \mid \text{If } A\text{, then } B) = 0$ iff $\Pr(B \mid \text{If } A\text{, then } B) = 1$.[22] Surely this is absurd. For it means that one cannot be certain of a conditional without being certain of the conditional's consequent, whereas it is pretheoretically clear that one *can* be certain that if John comes to the party, then Mary will come too, even if one doubts that Mary will come to the party.[23]

Insofar as one wants to stick to (IE) and to save at least the gist of (SH), the foregoing result gives one grounds for denying that conditionals express propositions. For if conditionals do not express propositions, then the conditional probability figuring in the left-hand side of the previous equation is not well defined. After all, by the ratio definition of conditional probability,[24]

$$\Pr(\text{If } A\text{, then } B \mid \neg B) \quad = \quad \frac{\Pr\big((\text{If } A\text{, then } B) \wedge \neg B\big)}{\Pr(\neg B)},$$

and if 'If A, then B' does not express a proposition, then, as was remarked in Section 1.2, it cannot occur in Boolean combinations and thus cannot occur as a conjunct in a conjunction. That is enough to block the above argument.

In the face of (IE), to deny that conditionals express propositions amounts effectively to limiting the scope of (SH) to simple conditionals.[25] But, as was said in Section 1.2, those who hold that conditionals do not express propositions are committed, anyway, to showing that all conditionals we can make sense of are reducible to simple conditionals. There it was further noted that if conditionals do not express propositions, they cannot be objects of belief in any straightforward sense. For the same reason, denying that conditionals express propositions also affects the *interpretation* of (SH) in that probabilities of conditionals cannot be probabilities in the ordinary sense of the word: normally, to say that something is probable to a certain extent is to say that the thing has a probability of *truth* to that extent. But to say this of a conditional presupposes that it *can* be true.[26] Well aware of this fact, those who deny that conditionals have truth conditions have claimed that probabilities of conditionals are not probabilities in the usual sense, but are instead to be conceived as degrees of acceptability or assertability. Restricted to simple conditionals, and interpreted as a thesis about the acceptability or assertability of conditionals, (SH) is now commonly referred to as 'Adams' Thesis'.[27]

While the above argument against truth-conditionality depends not only on (SH) but also on (IE), Lewis' and other triviality results do not rest on the latter principle.[28] How strong one thinks the argument is will thus depend on how compelling one finds (SH), at least in the restricted version originally proffered by Ernest Adams. It is noteworthy that, contrary to what one might expect, no normative argument has been given for the thesis. For instance, there is no known Dutch book argument showing that a cunning bookmaker can consistently win money from a person whose degrees of belief fail to obey (SH). Some have taken (SH) to be pretheoretically obvious. Bas van Fraassen [van Fraassen, 1976, pp. 272f.] suggests that it is when he says that 'the English statement of a conditional probability sounds exactly like that of the probability of a conditional. What is the probability that I throw a six if I throw an even number, if not the probability that: if I throw an even number, it will be a six?' But although authors have assumed that (SH) also enjoys massive support from the linguistic data when it is interpreted in terms of acceptability or assertability (rather than in terms of probability of truth), this may just be wishful thinking. Indeed, Jonathan Lowe [Lowe, 1996, pp. 611ff.] points to the following problem for Adams' Thesis: Suppose we know that a lottery is to be held, and know that the lottery is fair and consists of a hundred thousand tickets, only one of which will win. Then, while the conditional probability of 'We won't win' given 'We buy ticket no. 1' is close to 1, (25) appears to be neither highly acceptable nor highly assertable:

(25) If we buy ticket no. 1, then we won't win.

Lowe took this fact to refute Adams' Thesis. That was rash, though, for (25) might be an exception, a linguistic curiosity, and therefore of little significance, given that few (if any) theses about natural language (such as Adams' Thesis claims to be) can be expected to hold unexceptionably.[29] However, empirical results presented in Douven and Verbrugge [Douven and Verbrugge, 2010] show that (25) is no exception, and that, for large classes of conditionals, acceptability fails to match conditional probability. In view of these results, the advocates of a truth-conditional view on conditionals may have an easy escape route from the above argument, namely, to reject Adams' Thesis on the grounds of material inadequacy.

We now turn to another general argument against truth-conditional semantics for conditionals, this one being due to Edgington. The argument is pointed and seemingly very effective. According to Edgington [Edgington, 1995b, p. 279], no truth-conditional semantics for conditionals can satisfy both (26) and (27):

(26) Minimal certainty that $\neg A \vee B$ (ruling out just $A \wedge \neg B$) is enough for certainty that if *A*, *B*.

(27) It is not necessarily irrational to disbelieve *A* yet disbelieve that if *A*, *B*.

For – Edgington claims – the material conditional account satisfies (26) but not (27); and while accounts that attribute stronger truth conditions may satisfy (27), they cannot satisfy (26). So, granting that (26) and (27) are 'desirable properties of indicative conditional judgements' (ibid.), it follows that conditionals do not have truth conditions.[30]

Why is it that, according to Edgington, the material conditional account fails to satisfy (27)? This is her argument (the symbol $\supset$ designates the material conditional operator):

> Someone who believes $\neg A$ but disbelieves 'If *A*, *B*' is [on the material conditional account] making an Incredibly Gross Logical Error. For to disbelieve $A \supset B$, i.e. $\neg(A \wedge \neg B)$, is to believe its negation, $A \wedge \neg B$. How can anyone be so stupid as to believe $A \wedge \neg B$ yet disbelieve *A*, i.e. believe $\neg A$? (p. 244)

There may seem little to contest here. The underlying assumption that for the material conditional loyalist to disbelieve that if *A*, *B* is to disbelieve $A \supset B$ may seem particularly unassailable. On closer inspection, however, this assumption appears doubtful.

To see why, first note that to disbelieve something, at least in the presently relevant sense, is to believe the thing to be false, or equivalently, to believe that it is not the case.[31] If we further grant Edgington's claim, cited in Section 1.2, that 'It is not the case that if *A*, *B*' has no clear established sense distinguishable from 'If *A*, $\neg B$', it would seem that the natural interpretation of 'disbelieving that if *A*, *B*' is this: 'believing that if *A*, $\neg B$'. And believing that if *A*, $\neg B$ is quite evidently not the same as believing $A \supset B$ not to be the case, that is, as believing $A \wedge \neg B$.

If the advocates of the material conditional account can go along with this, they have a perfectly good reply to the charge that their account fails to satisfy (27). For they can then claim that, since to disbelieve that if A, B is to believe that if A, $\neg B$, the former does not amount to believing $A \wedge \neg B$ but rather to believing $A \supset \neg B$; and surely it can be rational to believe both $\neg A$ and $A \supset \neg B$.

To undermine Edgington's argument, it thus suffices to notice that there is no reason why the advocates of the material conditional account could not go along with the suggested natural interpretation of 'disbelieving that if A, B'. It is not the exclusive privilege of those who, like Edgington, think that conditionals do not have truth conditions to hold that surface grammar does not always reflect logical form. In particular, those advocating the material conditional account can consistently maintain that, to reveal the logical form of sentences containing the expression 'disbelieves that if A, B', this expression is to be replaced by 'believes $A \supset \neg B$' (and similarly for similar expressions).[32]

In sum, the material conditional account satisfies not only (26); on the natural interpretation of the phrase 'to disbelieve that if A, B', the material conditional account satisfies (27) as well. That gives the lie to Edgington's claim that no truth-conditional semantics for conditionals satisfies (26) and (27).[33]

2. Assertability and Acceptability

Philosophers of language and epistemologists have devoted a fair amount of work to the general question of under which conditions a sentence is assertable. And the epistemological literature is rife with theories purporting to state, also quite generally, the conditions under which it is rational to accept a given proposition. Why should we pay *special* attention to the assertability or acceptability conditions of conditionals? Do the more general rules of assertion, or theories of rational acceptability, not equally apply in the case of conditionals?

To see why conditionals might indeed require special treatment in these respects, consider first the issue of assertion. Many believe that the practice of assertion is governed by the so-called knowledge rule, according to which one must assert only what one knows.[34] But *can* we know a conditional? Not, one would think, if conditionals fail to express propositions.

According to a rival rule of assertion, one must assert only what is rationally credible, or acceptable, to one.[35] While there is no consensus on what is required for acceptability, there is at least widespread agreement that high probability is *close* to being sufficient for acceptability (most think it cannot be *quite* correct as a sufficient condition for reasons related to Henry Kyburg's [Kyburg Jr., 1961] lottery paradox, which we shall not go into here). But, first, 'high probability' is here taken to mean 'high probability of truth', which, as already remarked, is not something we can attribute to conditionals if they do not express propositions. Second, even if conditionals do express propositions, there may be a problem. Suppose conditionals express the proposition expressed by the corresponding material conditional. Then a conditional is highly probable if its antecedent is highly improbable. But consider this:

(28) If Manchester United ends last in this year's Premier League, they will shoot their coach.

Although it is exceedingly unlikely that Manchester United will end last in this year's Premier League, we do not find (28) acceptable; to the contrary, (28) seems highly *un*acceptable. This suggests that, assuming the material conditional account, high probability is not even nearly correct as a sufficient condition for the acceptability of conditionals.[36]

In this section, we will look at a number of proposals concerning the assertability and/or acceptability of conditionals. We begin with what many regard as the ur-proposal of acceptability conditions, and which in any case inspired much recent thinking on the issue.

In a footnote to his paper 'General propositions and causality', Frank Ramsey says:

> If two people are arguing 'If p will q?' and are both in doubt as to p, they are adding p hypothetically to their stock of knowledge and arguing on that basis about q .... We can say they are fixing their degrees of belief in q given p. [Ramsey, 1990, p. 155n.]

To be sure, if this is to state the acceptability conditions of conditionals, then the proposal is not as clear as one might wish. Is the idea that a conditional is acceptable if (and only if?) the conditional probability of the consequent given the antecedent is high? Or is it that a conditional is acceptable to a degree corresponding to the corresponding conditional probability? Also, one wonders why, as part of the procedure, the conditional's antecedent should be hypothetically added to one's *knowledge* and not just to one's outright beliefs, or to the things of which one is certain. What difference would it make to the resulting hypothetical probability function, relative to which the probability of the consequent is to be assessed, whether I suppose that I know the antecedent or believe it or am certain of it? Or is Ramsey using the word 'knowledge' loosely here, as meaning 'knowledge or outright belief', as we sometimes do in quotidian speech?

Most later proposals inspired by Ramsey's footnote do not suffer from these unclarities. This is especially true of what is generally considered to be the most direct descendant of Ramsey's proposal, to wit, Adams' Thesis, according to which, we saw in Section 1.3.3, the degree of assertability/acceptability of a conditional equals the probability of its consequent given its antecedent. Nevertheless, though this claim itself is as unambiguous as can be, the usage in the literature of the label 'Adams' Thesis' may still give rise to some confusion. In particular, one should be aware that not all authors who proclaim to avow Adams' Thesis may subscribe to the foregoing strict claim. For instance, Lewis [Lewis, 1976, p. 133], purportedly stating Adams' Thesis, says that, according to it, assertability of a conditional 'goes by' the relevant conditional probability, which admits of less strict readings. The same holds true of Jonathan Bennett's [Bennett, 2003, p. 46] formulation of the thesis, according to which 'the assertability or acceptability of $A \rightarrow C$ for a person at a time is governed by the probability the person then assigns to C on the supposition of A' ($\rightarrow$ is Bennett's symbol for the indicative conditional); the phrase 'is governed by' can also be interpreted in a number of ways. The thesis that Vann McGee [McGee, 1989] refers to as 'Adams' Thesis', and which he aims to defend, is quite explicitly intended to be less strict than Adams' canonical statement: McGee's version amounts to the claim that the assertability/acceptability of a conditional is high/middling/low iff the corresponding conditional probability is high/middling/low.[37]

It was already mentioned that, in Adams' view, conditionals lack truth conditions. However, endorsing Adams' Thesis, in whichever of the versions just suggested, does not commit one to that view. Most notably, Frank Jackson and David Lewis defend the material conditional account when it comes to specifying the truth conditions of conditionals. This does not mean that according to them a conditional is highly assertable or acceptable if the corresponding material conditional is true. Rather, they hold that 'If A, B' is highly assertable/acceptable by one iff $A \supset B$ is highly probable on one's degrees of belief function *and* is 'robust' with respect to A, meaning that the conditional probability of $A \supset B$ given A is (i) high and (ii) close to the unconditional probability of $A \supset B$. As Jackson [Jackson, 1987, p. 31] notes, the latter conditions boil down to requiring that B be highly probable conditional on A. He further argues that the natural way to generalize these assertability/acceptability conditions to degrees of assertability/acceptability other than 'high' yields Adams' Thesis [Jackson, 1987, p. 32].[38]

One might wonder how conditionals possess these assertability/acceptability conditions if not by virtue of their truth conditions. According to Jackson and Lewis, the answer lies in the conventional meaning of the word 'if': just as it is due to the conventional meaning of 'but' that 'A but B' is unacceptable/unassertable unless there is some sort of contrast between A and B, it is due to the conventional meaning of 'if' that 'If A, B' is unacceptable/unassertable unless B is highly probable conditional on A.

Douven [Douven, 2008] argues that both Adams' account and Jackson's and Lewis' account of the assertability/acceptability conditions of conditionals are materially inadequate. Consider the sentences (29α) and (29β) about a fair coin that is to be tossed at least 1,000,000 times:

(29) There will be at least one head in the first 1,000,000 tosses of this fair coin

$$\text{if} \begin{cases} \text{there is a head in the first ten tosses.} & (\alpha) \\ \text{Chelsea wins the Champions League.} & (\beta) \end{cases}$$

If no special assumptions about the context of utterance are made, it appears that (29α) is perfectly assertable/acceptable and that (29β) is assertable/acceptable, if at all, to a vastly lesser extent. According to Douven, there is little hope that these differences between the two sentences can be explained along Gricean lines. Nor, however, can they be explained in terms of the aforementioned proposals.

To see this, first note that it is a priori already highly probable that there will be at least one head in the first 1,000,000 tosses. Naturally, conditional on (α), the truth of the consequent of (29) is certain. But Chelsea's winning the Champions League is, we may assume, probabilistically independent of there being at least one head in the first 1,000,000 tosses. So conditional on either (α) or (β), the probability that the consequent of (29) is true is at most marginally different from 1. Moreover, we can bring the probability of (variants of) the consequent of (29) conditional on (β) as close as we like to that of the consequent of (29) conditional on (α) by supplanting the number 1,000,000 in (29) by a larger one.

Thus, Adams' account predicts that (i) both (29α) and (29β) are highly assertable/acceptable for any person who is able to take account of the foregoing facts; (ii) the differences in assertability/acceptability between the sentences are only minute; and (iii) by making some suitable substitution (e.g., substituting 10^{20} for 1,000,000 in (29)) the assertability/acceptability of (variants of) (29β) can be brought arbitrarily close to the perfect assertability/acceptability of (29α). All three predictions are manifestly wrong. The example creates no less a problem for Jackson and Lewis, given that (29)'s probability is high conditional on (β), so that (29β) comes out as being highly assertable/acceptable on their account.

To accommodate this datum, Douven proposes what he calls the 'evidential support theory of conditionals'. According to this, a conditional is assertable/acceptable iff (roughly) the antecedent would constitute sufficient evidence to warrant the acceptance of the consequent (if it is not warrantedly acceptable already). Modulo some considerations having to do with the lottery paradox, this is spelt out as the requirement that the conditional probability of the consequent given the antecedent not only be *high*, but that it also be *higher* than the unconditional probability of the consequent. It is easy to verify that this theory yields the correct verdict about both (29α) and (29β).

In closing, it should be emphasized that not all proposals for explicating the acceptability conditions of conditionals that take their cue from Ramsey's footnote refer to probabilities. Peter Gärdenfors [Gärdenfors, 1986] cashes out Ramsey's suggestion in purely qualitative terms by stipulating that a conditional is acceptable by one iff the consequent is accepted in a revision of one's current belief state by the antecedent. Here, 'revision' is a technical term. To revise one's current belief state by the antecedent, one first adds the antecedent to one's beliefs and then makes the minimally required changes (if any) to secure consistency of the new belief state (while keeping on board the hypothetical belief in the antecedent). Belief states are supposed to be deductively closed. So, if the consequent follows from the belief state revised by the antecedent, the conditional is acceptable. However plausible this explication may sound, Gärdenfors only presents it to prove that it is inconsistent with what he takes to be an intuitively compelling epistemic principle, to wit, that revising a belief state B by something that is consistent with it will result in a belief state $B' \supseteq B$. But Edgington [Edgington, 1995a, pp. 73f.] seems right to note that this so-called preservation principle only looks plausible as long as we think about beliefs in qualitative terms. From a quantitative perspective, it is natural to think that adding a proposition to our belief state may make some of our current beliefs less probable, even if the added proposition is consistent with everything we now believe. If probability above a given threshold value (perhaps .5) is at least necessary for categorical belief, then some propositions that qualify as categorical beliefs in the current belief state may no longer do so in the revised belief state.[39,40]

Acknowledgements

I am greatly indebted to Filip Buekens, Richard Dietz, David Etlin, Leon Horsten, and Richard Pettigrew for very helpful comments and discussions.

Notes

1. Typically, though not universally. For instance, one may reasonably suspect to be false the antecedent of the indicative conditional 'If I win the lottery, I will be rich', yet reasonably suspect to be true the antecedent of the subjunctive conditional 'If I were to lose the lottery, I would need to get back to work on my job search.' (Thanks to David Etlin here.)
2. Bennett [Bennett, 2003, p. 10] offers this grammatical difference as a demarcation criterion, but that seems to overstate matters, as some examples presented and discussed in [DeRose, ta] show.
3. For useful guides to the linguistics literature on conditionals, the reader is referred to [Dancygier, 1998] and [Declerck and Reed, 2001].

4. See [Evans and Over, 2004] for an excellent overview of the main results in this area.
5. Proponents of this view include Lewis ([Lewis, 1976]), Jackson ([Jackson, 1979], [Jackson, 1987]), and Grice ([Grice, 1989a]).
6. See [Stalnaker, 1968]. For a recent defense of a modified version of the possible worlds account, see [Nolan, 2003]. See [Lewis, 1973, pp. 91–95] for a thorough discussion of the notion of similarity between worlds.
7. On Lewis' ([Lewis, 1973]) view, the possible worlds account only applies to subjunctive conditionals. Davis ([Davis, 1979]) disagrees with both Lewis and Stalnaker in holding that the possible worlds account gives the right semantics for indicative conditionals and *only* for these conditionals.
8. It will be noted that the material conditional account is not even a prima facie plausible candidate for giving the semantics of subjunctive conditionals. For that would have us evaluate as true all so-called counterfactual conditionals, that is, subjunctive conditionals with antecedents we take to be false; and we have no difficulty thinking of counterfactuals we deem to be false.
9. See [Stalnaker, 1968, pp. 109ff.] for details. Subjunctive conditionals are obviously exempt from this constraint.
10. Though, admittedly, there is *A Natural History of Negation* ([Horn, 1989]), *The Syntax of Negation* ([Haegeman, 2005]), and *The Genealogy of Disjunction* ([Jennings, 1994]).
11. See [Lewis, 1976, pp. 141f.] for an objection along these lines.
12. See in the same vein [Ramsey, 1990, pp. 147f.].
13. This general claim may appear too strong, for it seems that sometimes we deny a conditional 'If *A*, *B*' to indicate that *B* might be false even if *A* holds true (cf. [Grice, 1989a, p. 81]). However, Edgington [Edgington, 2001, p. 24] seems right that in such a case the denial would not properly be expressed by asserting 'It's not the case that if *A*, *B*' but rather by something like 'It might well not be' or 'I wouldn't be so sure.'
14. This parallels von Wright's ([von Wright, 1957, p. 131]) view that in asserting a conditional one conditionally asserts a proposition rather than asserts a conditional proposition; see also [Quine, 1982, p. 21].
15. For instance, Adams ([Adams, 1998, p. 273]), one of the major proponents of the non-truth-conditional view, admits that ' "the problem of iterated conditionals" [i.e., the problem of how to account for conditionals such as (9)] is still very much an open one.'
16. See [Grice, 1989a], [Grice, 1989b].
17. A very general objection that has been raised against pragmatic defences of the material conditional account is that we do not only think that (18) is an insufficient basis for *asserting* (17) but also for *accepting* that sentence, and pragmatics concerns only assertion and not acceptance (see, e.g., [Edgington, 1995b, p. 245]). The latter claim is dubitable, however. See [Douven, 2010], where it is argued that much the same pragmatic principles that apply to assertion also apply to acceptance.
18. In fact, they have been adduced against truth-conditional semantics for conditionals generally, the claim being that there are some embedded conditionals of which we cannot make sense, which is then alleged to be puzzling if conditionals have truth conditions. First, however, it takes little effort to construct syntactically very complex non-conditional declarative sentences that we cannot make sense of. Second, even if they express propositions, conditionals may be harder to process mentally than non-conditional sentences, if only (perhaps) because their assertability/acceptability conditions deviate somewhat from those of non-conditional sentences; see Section 2. So, there may be a natural explanation – namely, in terms of limitations on people's processing capacities – of why we cannot make sense of all embedded conditionals.

19. McGee ([McGee, 1985a, pp. 469f.], [McGee, 1989, Sect. 7]) proposes a modified version of Stalnaker's semantics for conditionals that does validate (IE). It is probably fair to say, however, that the modified semantics lacks the intuitive plausibility of Stalnaker's original proposal.
20. It is worth remarking that the Gibbard example does not jeopardize the material conditional account, given that (CNC) is no consequence of this account. In fact, on that account, (CNC) is false, given that, on the material reading of the conditional, 'If *A*, *B*' and 'If *A*, not *B*' are both true if *A* is false. Some may see the failure of (CNC) as being itself a reason for doubting the material conditional account. For instance, Bennett ([Bennett, 2003, p. 84]) thinks that denying (CNC) commits one to holding that 'sometimes [If *A*, *B*] and [If *A*, not *B*] are both true, in which case one person could coherently accept both' (where by 'coherently accept' he means 'rationally accept'). And surely these conditionals are not both rationally acceptable by one and the same person. Note, however, that it is unclear why the fact that both conditionals can be true at the same time should entail that they can both be rationally accepted by a person. It may well happen that 'Ticket no. 467 will lose' and 'Ticket no. 7298 will lose' are both true, yet, according to most who have thought about the lottery paradox, neither is rationally acceptable by anyone at any time prior to the drawing of the relevant lottery. This is not to deny that there exist conceptual connections between truth and acceptability, but these are not so strong that truth entails acceptability. In fact, in the final section we will encounter a number of views on the assertability/acceptability conditions of conditionals that are compatible with the material conditional account yet on which at most one of 'If *A*, *B*' and 'If *A*, not *B*' can be acceptable to a person at a given time. While these views differ in their details, they all require that a conditional's consequent be highly probable conditional on its antecedent for the conditional to be acceptable. Clearly, *B* and not-*B* cannot both be highly probable conditional on *A*.
21. For some later, stronger triviality results, see, e.g., [Hájek, 1989], [Hájek, 1994], [Döring, 1994], [Hall, 1994], and [Etlin, 2009]. Van Fraassen ([van Fraassen, 1976]) famously argued that Lewis' triviality results implicitly assume that conditionals have their interpretation independent of people's belief states and that, absent that assumption, (SH) is tenable. In response, Lewis ([Lewis, 1976, p. 138]) dismissed the possibility that the semantics of conditionals is relativized to belief states because – he claimed – that would make it hard to explain how people can genuinely disagree about conditionals. Douven and Dietz [Douven and Dietz, ta] argue that Lewis' response to van Fraassen may have been rash, but they also give another argument against van Fraassen's so-called tenability result.
22. The 'iff's hold provided the conditional probabilities are defined – which here can be assumed without loss of generality, at least on the assumption that conditionals express propositions.
23. For very similar triviality arguments also relying on (IE), see [Blackburn, 1986, pp. 218ff.] and [Jeffrey, 2004, pp. 15f.].
24. And given that, as we are assuming, $\Pr(\neg B) > 0$; see the footnote above.
25. At least there is no obvious way to relax this restriction; see Dietz and Douven ([Dietz and Douven, 2010]).
26. As noted earlier, some of those who deny that conditionals express propositions still hold that conditionals can be true, namely, if both their antecedent and consequent are true. So, by 'the probability of a conditional' one might mean in the ordinary sense of the word 'probability' the probability of the conjunction of the given conditional's antecedent and its consequent. However, when conjoined with (SH) this has the absurd consequence that $\Pr(B \mid A) = \Pr(\text{If } A, B) = \Pr(A \wedge B) = \Pr(A)\Pr(B \mid A)$ and thus that the probability of the antecedent of every conditional must be 1.
27. The thesis was already defended – although not under this name – in [Adams, 1965].

28. Significantly, though, no triviality results exist that assume *only* (SH) next to probability theory.
29. If Lackey ([Lackey, 2007, p. 618]) is right that sentences such as 'Your ticket won't win' conversationally implicate that the speaker has insider knowledge about the outcome of the relevant lottery, and are therefore unassertable on Gricean grounds, the same would seem to hold for sentences such as (25). That such sentences are not highly acceptable either could then again be explained by appeal to the idea, mentioned in note 17, that Gricean considerations have some bearing on what we can accept, too.
30. She supposes, quite rightly, that accounts which attribute truth conditions weaker than the material conditional account are not worth considering; everyone agrees that the material conditional account is right at least insofar as it rules a conditional false if its antecedent is true and its consequent false.
31. According to Webster's Dictionary, 'disbelieve' can mean both 'deem false' and 'refuse to believe'. Given the latter, more uncommon and in the present context obviously unintended, interpretation, it is straightforward that the material conditional account does satisfy (27): surely it can be rational to be agnostic both about a conditional and about the conditional's antecedent.
32. More generally, they can maintain that the truth conditions of 'It's not the case that if A, B' are those of $A \supset \neg B$. Edgington also seems to miss this when in her [Edgington, 2001, p. 393] she presents what she thinks of as another problem for the material conditional account: '$\neg(A \supset B)$ is equivalent to $A \wedge \neg B$. Intuitively, one may safely say, of an unseen figure, "It's not the case that if it's a pentagon, it has six sides." But by [the lights of the proponents of the material conditional account], one may well be wrong; for it may not be a pentagon'. Quite patently, this has as a hidden premise that the proponents of the material conditional account ought to accept that 'It's not the case that if it is a pentagon, it has six sides' is of the logical form $\neg(A \supset B)$, a premise which appears to be false; they can insist that the sentence is naturally interpreted as saying that if the figure is a pentagon, then it does not have six sides, and thus is of the logical form $A \supset \neg B$.
33. For another general argument against truth-conditional semantics for conditionals, see [Bradley, 2000]. For a critique, see [Douven, 2007]. Bennett ([Bennett, 2003, p. 102]) seems to regard Gibbard's poker example to provide still another argument against truth-conditionality, the idea being that, given that (23) and (24) would appear to be both true if conditionals have truth conditions, and given that they cannot both be true, conditionals do not have truth conditions. As stated in note 20, however, there is no reason why we should go along with Bennett's premise that the conditionals at issue cannot both be true.
34. See, e.g., [Williamson, 1996a], [Adler, 2002], and [DeRose, 2002].
35. See [Douven, 2006], [Douven, 2009]; also [Lackey, 2007].
36. As noted, high probability is not generally thought to be quite sufficient for acceptability. But high probability plus some condition that keeps the lottery paradox at bay *is* thought by many to be sufficient. And lottery paradox considerations seem to play no role in the case of (28).
37. Incidentally, the results of Douven and Verbrugge ([Douven and Verbrugge, 2010]) mentioned earlier show that even McGee's version and other prima facie plausible weaker versions of Adams' Thesis are not generally correct as descriptive claims about people's assessments of the assertability or acceptability of conditionals.
38. Some may regard it as a problem for Jackson's and Lewis' account, and also for Douven's account (to be presented shortly) if that is wedded to the material conditional account of the truth conditions of conditionals (as is possible), that, on these accounts, acceptability is not closed under logical consequence. For take any proposition A such that $\neg A$ is acceptable. Given that, on the said accounts, at most one of 'If A, B'

and 'If A, $\neg B$' can be acceptable, and given that, assuming the material conditional account, $\neg A$ entails *both* conditionals, we will have a violation of closure. However, to regard this as a problem is to assume that acceptability must be closed under logical consequence, and this assumption is contentious. See, for instance, [Christensen, 2004] for a recent extended argument against this assumption. See [Douven, 2010] for a quite different argument against the same assumption. (As an aside, I note that this undercuts Edgington's ([Edgington, 1995b, p. 245]) objection that the material conditional account does not allow one 'to discriminate believable from unbelievable conditionals whose antecedents we think false'.)

39. For more on this, see Chapter 17.
40. Adam Rieger has recently presented a theory of conditionals that is in the spirit of Grice's work on conditionals rather than in that of Ramsey's footnote; see [Rieger, 2006]. Rieger agrees with Jackson and Lewis insofar as truth conditions are concerned, but on his proposal 'If A, B' is assertable by one only if (i) one knows the corresponding material conditional, and (ii) one knows none of the following: A, $\neg A$, B, and $\neg B$. Rieger argues that this theory can handle problem cases that Grice's ([Grice, 1989a]) theory cannot. Be that as it may, Rieger's proposal is hard to assess, given that it only offers necessary conditions for assertability. Who knows which conditionals come out as being unassertable once a full account is on the table? The necessary conditions Rieger propounds can, in principle, be supplemented by an indefinite number of further conditions, each of which might render the theory materially inadequate by ruling unassertable conditionals that are incontrovertibly assertable.

15 Probability

Richard Pettigrew

Chapter Overview

1. Mathematical Probability	408
2. Physical Probability	409
2.1 Actualist Frequentism	409
2.1.1 The Interpretation	410
2.1.2 Assessing the Interpretation	410
2.2 Hypothetical Frequentism	412
2.2.1 The Interpretation	412
2.2.2 Assessing the Interpretation	413
2.3 Single-Case Propensity Theories	415
2.3.1 The Interpretation	415
2.3.2 Assessing the Interpretation	416
2.4 The Best-System Analysis of Probability	417
2.4.1 The Interpretation	417
2.4.2 Assessing the Interpretation	418
3. Epistemic Probability	419
3.1 Norms for Epistemic Probabilities	420
3.2 Epistemic Probabilities as Fair (or Favourable) Betting Odds	420
3.2.1 The Interpretation	421
3.2.2 Assessing the Interpretation	422
3.3 Epistemic Probabilities and Expected Utility	422
3.3.1 The Interpretation	423
3.3.2 Assessing the Interpretation	424
3.4 *Sui Generis* Epistemic Probabilities	424
3.4.1 The Interpretation	424
3.4.2 Assessing the Interpretation	426
4. Prospects	426
Notes	427

Our task in this article is to discover what we do when we assign a probability to something. That is, we wish to give a semantics for our talk of probabilities. Thus, I begin with a representative sample of such talk:

(1) The probability of this die coming up three on the next roll is $\frac{1}{6}$.
(2) The probability of this sample of plutonium-236 emitting an alpha particle in the next ten minutes is 0.0001.
(3) For me, the probability that there is an external world that is approximately as I believe it to be is extremely high.
(4) When it was observed that the perihelion of Mercury's orbit precesses at an anomalous rate, this observation rendered Einstein's theory of relativity more probable than it had been before.

In each case, we assign a probability to something. But it seems that very different features of reality are responsible for the truth or falsity of the claims. More precisely, the claims fall into three categories:

Physical Probability: In examples (1) and (2), the truth or falsity of our probability assignment seems to be determined by a feature of the physical world. If the die or the sample of radioactive material were different physically, a different probability assignment might hold. In these cases, we say that we have assigned a *physical probability*. And, in the standard cases, we assign physical probabilities to types of events. For instance, in (1), we assign a probability to one of the following event-types: *Next roll of this die comes up three* or *Next roll of a die comes up three*.

Epistemic Probability: In example (3), the probability assignment seems to depend not on the physical world, but on the epistemic state of the speaker. The existence of the external world is not relevant to the claim that I assign its existence a high probability. In these cases, we say that we have assigned an *epistemic probability*. And, in the standard cases, we assign epistemic probabilities to those propositions towards which the speaker in question has an epistemic attitude. For instance, in (3), we assign an epistemic probability to the following proposition: *The external world is as I represent it to be.*

Logical Probability: In example (4), the probability assignment seems to depend neither on any actual fact about the physical world, nor on the epistemic state of any agent. Rather, it seems to depend on a purely logical relation between one proposition – the proposition that the perihelion of Mercury's orbit precesses at an anomalous rate – and another – the proposition that Einstein's theory of relativity is true. Irrespective of whether either of these propositions is true, and irrespective of whether either is believed to any degree by any epistemic agent, the statement seems to hold true. In these cases, we say that we have assigned a *logical probability*. And, in the standard cases, we assign logical probabilities to all propositions, or sentences in a given language.

In what follows I consider attempts to say what we mean when we assign *physical* or *epistemic probabilities*. For an extended discussion of *logical probability*, I refer the reader to Chapter 16 in this volume.

However, there is a fourth way in which we talk about probabilities that is not represented in the examples above. It is the way that mathematicians talk about probability. I turn to this in the next section, mainly in order to fix terminology. There are no peculiarly philosophical questions about mathematical talk of probabilities since the terms are introduced by stipulative definition, as they are in all parts of modern mathematics.

1. Mathematical Probability

In mathematics, probability theory is the branch of real analysis that studies the behaviour of what mathematicians call *probability spaces*. A probability space has three components: a set Ω, a set $\mathcal{F}$ of subsets of Ω, and a function $p : \mathcal{F} \to \mathbb{R}$. Intuitively, we think of Ω as the set of basic possibilities, and $\mathcal{F}$ as the collection of sets of those possibilities to which p assigns probabilities.

Definition 15.1.1 We say that $\mathcal{F}$ *is an algebra on* Ω if

(1) $\varnothing \in \mathcal{F}$
(2) If $E \in \mathcal{F}$, then $\Omega - E \in \mathcal{F}$
(3) If $E_1, E_2 \in \mathcal{F}$, then $E_1 \cup E_2 \in \mathcal{F}$.

Definition 15.1.2 We say further that $\mathcal{F}$ *is a* σ*-algebra on* Ω if

(3′) If $E_1, E_2, \ldots \in \mathcal{F}$, then $\bigcup_{i=1}^{\infty} E_i \in \mathcal{F}$.

Given an algebra $\mathcal{F}$ (whether a σ-algebra or not), and given $A \in \mathcal{F}$, we say that *A is an atom of* $\mathcal{F}$ if (i) $A \neq \varnothing$ and (ii) there is no $E \in \mathcal{F}$ such that $\varnothing \subsetneq E \subsetneq A$. We say that $\mathcal{F}$ *is atomic* if, for all $E \in \mathcal{F}$, there is an atom A of $\mathcal{F}$ such that $A \subseteq E$. If $\mathcal{F}$ is atomic, then the set Atoms($\mathcal{F}$) of atoms of $\mathcal{F}$ forms a partition of Ω.

Definition 15.1.3 We say that $(\Omega, \mathcal{F}, p)$ *is a probability space* if

(1) $\mathcal{F}$ is an algebra on Ω.
(2) p is a probability measure on $\mathcal{F}$: that is,
    (a) $0 \leq p(E) \leq 1$ for all $E \in \mathcal{F}$.

(b) $p(\varnothing) = 0$ and $p(\Omega) = 1$.
(c) $p(E_1 \cup E_2) = p(E_1) + p(E_2)$ whenever $E_1 \cap E_2 = \varnothing$.

Definition 15.1.4 We say that $(\Omega, \mathcal{F}, p)$ *is a countably additive probability space* if, furthermore, $\mathcal{F}$ is a σ-algebra and p is *countably additive*: that is,

(2c′) If $E_1, E_2, \ldots \in \mathcal{F}$ are pairwise disjoint, then

$$p\left(\bigcup_{i=1}^{\infty} E_i\right) = \sum_{i=1}^{\infty} p(E_i).$$

We also introduce the notion of a *restricted conditional probability space*. This is not standard terminology, but it will prove useful below.

Definition 15.1.5 Suppose Ω is a set; suppose $\mathcal{F} \subseteq \mathcal{P}(\Omega) \times \mathcal{P}(\Omega)$; and suppose $p : \mathcal{F} \to \mathbb{R}$. Then we say that $(\Omega, \mathcal{F}, p)$ *is a restricted conditional probability space* if, wherever the following identities and inequalities are defined, they hold:

(1) $0 \leq p(E|A) \leq 1$
(2) $p(\varnothing|A) = 0$ and $p(\Omega|A) = 1$
(3) $p(E \cup F|A) = p(E|A) + p(F|A)$ for disjoint E and F
(4) $p(E|E) = 1$
(5) $p(E \cap F|A) = p(E|A)p(F|E \cap A)$

If $\mathcal{F} = \mathcal{F}' \times \mathcal{F}'$ and $\mathcal{F}'$ is an algebra, then p is called a *Popper function*. This is standard terminology.

2. Physical Probability

We begin with physical probabilities, which we assign to types of event or event-types. When we assign a physical probability to an event-type, it seems that we associate a quantity with that event-type – to wit, its 'probability' – and give a measure of that quantity. That is, assigning a physical probability to an event-type is analogous to assigning a mass to a body. In both cases, we introduce a theoretical term into our language. In this section, we consider the most prominent accounts of this theoretical term.

2.1 Actualist Frequentism

The first interpretation we consider is *actualist frequentism*.[1] This interpretation attempts to provide an explicit definition of probability.

2.1.1 The Interpretation

Let's begin by asking what sort of quantity the actualist frequentist takes probability to be. In the examples of our talk about probability given at the beginning of this article, it seems that physical probability is a quantity that is associated with a single event-type E. That is, it seems that there is a set Ω of basic possible events, and physical probability can be represented by a one-place function $p(\_)$ that assigns a real number to each member of a certain collection $\mathcal{F}$ of event-types represented by subsets of Ω. However, according to the actualist frequentist, this is a mistake. Rather, on her account, a physical probability is always assigned to an event-type E *relative to an event-type A*. We call A *the reference type* of the probability assignment. Thus, physical probability is represented by a two-place function $p(\_|\_)$ that assigns a real number to certain types of events and reference types (both represented by subsets of Ω). For instance, when I ask for the probability that the next roll of the die will come up three, I am not asking for the probability of the event-type *Roll of the die that comes up three*. According to the actualist frequentist, that would make no sense. Rather, I am asking for the probability of that event-type relative to a reference type. That reference type might be, for instance, *Roll of the die* or *Roll of the die under such-and-such atmospheric conditions* or *Roll of the die on such-and-such a surface*. And relative to these different reference types, the probability of *Roll of the die that comes up three* might be radically different.

Having seen what sort of quantity the actualist frequentist takes probability to be, we can give the explicit definition of probability that she proposes. She says that the physical probability of E relative to A is the relative frequency of the *actual* events of type E among the *actual* events of type A. That is,

$$p(E|A) = \frac{\text{Number of } \textit{actual} \text{ events that are of type } E \text{ and type } A}{\text{Number of } \textit{actual} \text{ events that are of type } A}$$

Thus, the actualist frequentist probability of *Roll of the die that comes up three* relative to *Roll of the die* is the ratio of the number of actual rolls of the die that come up three to the number of actual rolls of the die.

It follows that, on this interpretation, physical probability $p(E|A)$ is defined just in case the set of actual events of type A is finite and non-empty. Let $\mathcal{F}_2$ be the set of subsets of Ω for which this is the case. Then, for any algebra $\mathcal{F}_1$ on Ω, $(\Omega, \mathcal{F}_1 \times \mathcal{F}_2, p)$ is a restricted conditional probability space. (Exercise.) It also follows that $p(E|A)$ is always a rational number.

2.1.2 Assessing the Interpretation

Strongly in favour of actualist frequentism are its parsimonious ontology and tractable epistemology. It provides a semantics for our talk of probabilities without positing entities beyond the actual world. And the propositions by which it interprets that talk admit of straightforward empirical investigation. These are extremely attractive features of an interpretation, and they are worth paying for.

However, the disadvantages are so numerous that most philosophers consider the price too high. I catalogue only the most serious problems here. These not only expose the difficulties with the actualist frequentist interpretation; they also provide constraints on any attempt to give a better interpretation. The objections may be categorized by the aspect of the actualist frequentist interpretation that they attack:

(I) *The probability of E is defined only relative to a reference type A.*

The problem is that we often seem to assign probabilities to event-types without relativizing them in this way. For instance, when I ask for the probability that a particular individual will contract a given disease in the coming five years, I do not mean to assign that individual to a particular class of individuals – such as the class of all joggers, or the class of all smokers – and then to ask for the incidence of that disease amongst the members of that class. I simply ask for the probability that she will contract the disease within that time. Call this the *unrelativized probability objection.*

(II) *The probability of E relative to A is the frequency of events of type E among events of type A.*

Many event-types are so specific that they can only be sensibly assigned a reference type with at most one actual instance. For instance, I might ask for the probability that a particular coin tossed at a particular time will land heads. But the natural reference type for this event is the event-type whose only instance is the particular coin toss at the particular time. Thus, according to the actualist frequentist, the probability will then be 0 if the coin lands heads and 1 if it lands tails. And this conflicts with our intuitions. Call this is the *one-off event objection.*

(III) *The probability of E relative to A is the frequency of actual events of type E among actual events of type A.*

The most obvious problem with the restriction to actual frequencies is that it seems possible that probabilities and actual frequencies could come apart. For instance, it seems possible that the probability of a particular coin landing heads is $\frac{1}{2}$ while the coin land heads every time it is thrown. For the actualist frequentist, however, such a situation is ruled out as impossible by the definition of probability. That is, for her, it is an analytic truth that probabilities and relative frequencies never come apart. Call this the *coming apart objection.*

If no events of type *A* actually occur, the probability of *E* relative to *A* is undefined. But consider a coin that is never tossed. We wish to

> say that it has a probability of landing heads, and yet for the actualist frequentist the probability of it landing heads relative to the reference type of its being tossed is undefined. Call this the *probabilities of non-actual events objection*.
>
> Also, by restricting our attention only to actual events, the actualist frequentist creates a further problem. Intuitively, it seems that the probability of a particular event depends only on facts that are, in some sense, local to the event in question. However, on the actualist frequentist account, the probability of a coin landing heads when it is next thrown depends on other throws of that coin at other times. And these facts are not local to the next coin toss. It seems counterintuitive that the probability of the next coin toss coming up heads should depend upon all tosses of the coin, including future ones; including, indeed, the next coin toss itself. Call this the *locality of probabilities objection*.

Of course, each objection just enumerated relies on an intuition about the nature of physical probabilities. Thus, the actualist frequentist can reply that it is simply not possible to provide an interpretation of such probabilities that respects these intuitions about them while also making them metaphysically acceptable and epistemically accessible. The only response to this reply is to provide an acceptable interpretation that does respect the intuitions. Let us see whether there is such an interpretation.

2.2 Hypothetical Frequentism

2.2.1 The Interpretation

The *hypothetical frequentist* adapts actualist frequentism by allowing her interpretation of probability to appeal not only to the actual world but to other possible worlds as well.[2] The objections listed under item (III) in the previous section all turn on the fact that the actual world may happen not to have sufficiently many events of particular types to produce relative frequencies that match our intuitive physical probabilities. The hypothetical frequentist seeks to solve these problems as follows: when we ask for the probability of event-type E relative to event-type A, we do not ask what the relative frequency of the Es among the As *actually is*; rather we ask what it *would be* if there *were* denumerably many events of each type. Now, this immediately raises a problem for a frequentist definition of probability, since relative frequencies are defined only for event-types with finitely many instances. The solution is to replace the notion of *relative frequency* with the notion of *limiting relative frequency*. Suppose we are at a world w at which there are infinitely many events of type A. Then we order the events of type A in an infinite sequence $a_1, \ldots, a_n, \ldots$, and we let A_n be the event-type to which only $a_1, \ldots, a_n$ belong. Then we define the limiting relative frequency of the Es among the As at w to be

$$\lim_{n \to \infty} \frac{\text{Number of events of type } E \cap A_n \text{ at world } w}{\text{Number of events of type } A_n \text{ at world } w}$$

if this limit exists. Thus, given event-types E and A, we define the probability of E relative to A as follows: $p(E|A) = r$ iff the limiting relative frequency of the Es among the As would be r if there were denumerably many As. In possible worlds language, $p(E|A)$ is the limiting relative frequency of the As among the Es at the nearest possible world at which there are infinitely many As.

It follows that, on this interpretation, $p(E|A)$ is defined only if the limit of the relative frequencies of the $(E \cup A_n)$s among the A_ns exists. A serious consequence of this is that there is no longer any guarantee that the set of event-types E for which $p(E|A)$ is defined is even an algebra – see ([de Finetti, 1972, p. 75]). Nonetheless, it is straightforward to use elementary theorems concerning limits to show that $(\Omega, \mathcal{F}, p)$ is a restricted conditional probability space.

2.2.2 Assessing the Interpretation

Hypothetical frequentism retains the actualist frequentist's identity of probabilities and frequencies, but it seems at first to avoid many of the problems of that position. However, as we shall see, in each case, either it does not avoid these problems, or it does so only by making itself vulnerable to a new, equally devastating objection. Let's consider each in turn.

First, the unrelativized probability objection. The hypothetical frequentist not only fails to solve this problem; in fact, she faces a more serious version of it. After all, according to hypothetical frequentism, the probability of an event is not only relativized to a reference type A; it is also relativized to an ordering a_1, $a_2, \ldots$ of the instances of A that occur at the nearest possible world at which infinitely many As occur. We might call this the *reference sequence* of the probability assignment. But it seems to conflict even more strongly with our intuitions about physical probabilities that they are assigned relative to a reference type *and* a reference sequence than just that they are made relative to a reference type. After all, it is very implausible that the As at a particular world have a unique natural ordering that can act as the required reference sequence: one might order them temporally, or by some spatial ordering, and neither would be more natural, yet each might give rise to a different limiting relative frequency.[3]

We might think initially that hypothetical frequentism can solve the *one-off event objection* for the following reason: however specific the event-type in which we are interested, we can always find a possible world at which that event-type occurs infinitely often. Indeed, this is the solution offered by what are sometimes called the *long-run propensity interpretations of probability*. Such interpretations, which were first proposed by Karl Popper ([Popper, 1957], [Popper, 1959]), take probabilities to be only those limiting relative frequencies that are defined relative to a reference sequence that results from repeating a particular experiment, such as the tossing of a coin, or the alpha particle emission of similarly-sized samples of radioactive isotopes.

However, while this might work for coin tosses and radioactive decay, which by their nature are repeatable, there are one-off event-types for which it will not work: for instance, the universe taking particular values for its fundamental constants; or the obtaining of any universal or existential quantification, where the quantifier ranges over everything in the universe. In each case, these events are unrepeatable by their very nature. Thus, again, the frequentist position faces a one-off event objection. The proponent of the long-run propensity theory must simply limit the ambitions of her interpretation to probabilities of event-types of repeatable events.

It might seem that hypothetical frequentism is best placed to solve the coming apart objection. If the problem raised by this objection is that the actual world might happen not to match the true probability of an event-type in the relative frequencies it displays, this discrepancy should be removed by turning to a possible world in which the reference type has infinitely many tokens. However, upon reflection, it is not clear why this should be the case. After all, there are possible worlds at which an unbiased coin is tossed infinitely often and comes up heads every time, just as there are possible worlds at which it comes up tails every time, and possible worlds at which it comes up heads half of the time, and so on through all possible limiting relative frequencies. That's just the nature of probability. Thus, the hypothetical frequentist must claim that, among this array of possible worlds, one is the closest and the probability is given by the limiting relativity frequency at that world. But there seems to be no reason to assume this. Indeed, it seems an ad hoc assumption that is made to save the hypothetical frequentist's account. What's more, it renders the account less attractive by appealing to a modal fact without giving any account of how we might hope to discover this fact. This is a serious problem for the hypothetical frequentist: since her definition of the probability of E relative to A appeals to the closest world at which A is infinite, the probability is not even well defined unless this problem can be solved.

However, if these problems can be solved, it is clear that the hypothetical frequentist does avoid the remaining two objections to actualist frequentism, namely, the probabilities of non-actual events objection and the locality of probabilities objection. Thus, I close this section by considering a serious attempt to solve the problem. This is Popper's long-run propensity theory mentioned above.

Popper's theory is a slight, but important variant on hypothetical frequentism.[4] For Popper, the probability of E relative to A is not the value that the limiting relative frequency would take if there were infinitely many *A*s. Rather, if events of type *A* are those that come about when a particular set of conditions obtain, then the probability of *E* relative to *A* is the limiting relative frequency of the *E*s among the *A*s that those conditions have the strongest propensity to produce in a situation in which the conditions are repeated infinitely often. In this account, a propensity is something like a disposition, but it is a sort of disposition that can come in degrees of strength. This account has the advantage of avoiding the implausible modal claims of hypothetical frequentism. We no longer have to say that an unbiased coin *would* land heads with limiting relative frequency $\frac{1}{2}$ if it *were* tossed infinitely often. Rather, we say that it is *most strongly disposed* to produce that limiting relative frequency if it is tossed infinitely often. However, it pays for this advantage by introducing a new metaphysical component to the theory, namely, the notion of a propensity, or a disposition that comes in degrees. Moreover, it is often complained that, just as Molière's notorious 'dormative virtue' is introduced to explain why opium puts a person to sleep, but does so only by ascribing to it the property of putting people to sleep, so propensities are introduced to explain the relative frequencies that we observe in the world, but do so only by positing a propensity to produce them! And clearly this is no explanation. What's more, on our best attempt to explicate the notion of a propensity, it seems to be a sort of physical probability, namely, the probability with which a particular limiting relative frequency will be displayed. But if this is the case, surely just such a notion of probability could be employed from the outset. In fact, just such a notion of probability is employed from the outset in the so-called *single-case propensity theory*, which we consider in the next section.

I conclude this section with a final objection, which applies only to the hypothetical version of frequentism. The problem is that any non-actual world that contains infinitely many events of the reference type *A* will have to be very different from the actual world. For instance, in a world that contains infinitely many coin tosses, the laws of physics concerning entropy might have to be very different. Thus, it is not clear that the modal facts about what the limiting relative frequency would be if there were infinitely many tosses, or what limiting relative frequency the experimental set-up has the strongest propensity to produce, will accord with what we intuitively think of as the probability that the coin lands tails.

2.3 Single-Case Propensity Theories

2.3.1 The Interpretation

In the previous section, we introduced the notion of a propensity, which is a sort of disposition that can come in degrees of strength. And I observed that its most natural place in an account of probability is not as a means by which to pick out the limiting relative frequency that gives the probability, as the long-run propensity theorists would have it, but as the quantity of which probabilities are the measure. This is the starting point for the single-case propensity interpretation of probability. On this account, the probability of E relative to A is the propensity of events of type A to produce events of type E. Thus, the probability of a particular coin toss landing heads measures the strength of the disposition of that particular coin to land heads when tossed.

Single-case propensity theorists disagree on which event-types A are to be permitted as reference types. According to David Miller ([Miller, 1996]) and the later Popper ([Popper, 1990]), only event-types that determine the complete state of the world at a particular time are permitted. Thus, the probability of a coin toss landing heads is not a quantity associated only with the coin and the hand that tosses it. Rather, it is associated with the complete state of the world at the time of the toss. On the other hand, according to James Fetzer ([Fetzer, 1982]), the only legitimate reference types are those whose instances are generated by a complete set of causally or nomologically relevant conditions. Thus, for Fetzer, the probability of a coin toss landing heads is a quantity associated with a complete set of relevant conditions – the weight of the coin, the atmospheric conditions, etc.

2.3.2 Assessing the Interpretation

A number of major problems face the single-case propensity theory. Again, it is objected that the notion of propensity is mysterious. When the long-run propensity theorist posits propensities to account for frequencies, she is accused of positing something akin to a dormative virtue. But when the single-case propensity theorist posit propensities she does not even do so in order to account for anything else. Thus, this central metaphysical component of her theory is doubly mysterious.

One popular response to this charge is to say that the notion of propensity is a generalization of the notion of cause. That is, a propensity is a degree of causal power. Thus, assuming the Miller-late Popper account for the moment, the entire state of the world at the current moment does not stand in the full causal relation to the emission of an alpha particle from a particular sample of plutonium-236 in the next five minutes; however, it does stand in a weakened version of that relation. That is what we mean when we say that the entire state of the world has a propensity of a certain strength to bring about an emission of an alpha particle from that sample.

However, a famous objection has been raised against this attempt to give substance to the single-case propensity interpretation. It is known as *Humphrey's paradox* and runs as follows. Many mathematical theories of conditional probability have the following consequence: if $p(E|A)$ is defined, then $p(A|E)$ is defined. But suppose that the single-case propensity account is correct and $p(E|A)$ is defined. Then $p(E|A)$ measures the strength of the causal relation running from events of type A to events of type E. But it is generally accepted that a causal relation (of whatever strength) is asymmetric. Thus, if $p(E|A)$ is defined, and if it measures the strength of the causal relations that run from As to Es, then $p(A|E)$ is not defined, since it would measure the strength of the causal relations that run from Es to As.

Of course, this only shows that the formal features of probability on a single-case propensity interpretation cannot entail that $p(A|E)$ and $p(E|A)$ are both defined for events E and A. But, having seen how we must limit the domain of our probability function in the actual and hypothetical frequency case, it should come as no surprise that we might have to limit it also in this case.[5] However, while this clarification of propensities as degrees of causation serves to introduce them by appealing to a more familiar metaphysical notion, it does not provide an account of how we can detect the existence and strength of propensities in the world, and this is a strong desiderata for an interpretation of probabilities.

The conclusion of the previous three sections is that it is difficult to interpret probabilities in a way that respects our pre-philosophical intuitions and yet makes them amenable to empirical investigation. The closer the connection we posit between probabilities and frequencies, the easier it becomes to investigate the former. But we pay for this by sacrificing many of our intuitive beliefs about probabilities. On the other hand, when we do not include frequencies in our account – as in the single-case propensity interpretation – we sacrifice our ability to investigate probabilities. In the next section, we turn to an account that attempts to avoid this dilemma.

2.4 The Best-System Analysis of Probability

According to David Lewis, there is just one desiderata that our interpretation of probability must satisfy: whatever it is that is the probability of an event E, it must be such that, were an agent to know the value $p(E)$ of that probability, she would be rationally compelled to adjust her confidence in the proposition *E occurs* to match it. Lewis calls this the *Principal Principle* ([Lewis, 1980]). Thus, suppose we model an agent's epistemic state by a function $b(\_)$ that assigns to a proposition X the agent's confidence in X. And suppose we define $b(X|Y) = \frac{b(X \,\&\, Y)}{b(Y)}$. Then the Principal Principle says that, if X is a proposition that entails $p(E) = r$ (and X does not contain inadmissible information, such as information from the future about the actual occurrence of E), then rationality requires that $b(E\ occurs|X) = r$.

2.4.1 The Interpretation

Having stated his guiding principle, Lewis then claims that only one interpretation of probability satisfies it, namely, the *best systems analysis* ([Lewis, 1980],

[Lewis, 1994]). On this account, the probability of an event E is the chance assigned to E by our best theory of the world, where *chance* is one of the theoretical terms in terms of which the theory is stated. Let me explain. Consider all the theories that concern the world. These may include generalizations, descriptions of particular local facts, and claims about the chances of events. There are four desiderata for such a theory: it must be true, simple, and strong, and it must assign to the actual world a high chance. There can be no compromises on truth, so we say that our best theory of the world must have only true consequences. But the other desiderata come in degrees. And the world might be such that these other requirements pull in different directions. Nonetheless, suppose there is a single theory that is true and which optimally combines the other three desiderata. Then, on Lewis' best systems analysis, the probability of E is the chance that this theory assigns to E.

On Lewis' interpretation, we derive the formal features of probability from the Principal Principle. As we will see in the second half of this paper, there are various arguments that purport to show that, if we model an agent's epistemic state at a particular time t by a function b that assigns to a proposition X the agent's confidence in X at t, then rationality requires that b should be a probability function. It follows from this together with the Principal Principle that, on the best systems analysis, p must also be a probability function. (Exercise.)

2.4.2 Assessing the Interpretation

Lewis' theory has a number of advantages. First, it avoids the unrelativized probability objection. On Lewis' theory, probabilities are assigned to events without relativizing them to reference types. Thus, a probability function is a one-place function $p(\_)$. It also avoids the one-off event objection since probabilities of particular one-off events will follow from many theories of the world and thus hopefully from the best such theory. And, for similar reasons, it is not vulnerable to the non-actual events objections. It might seem that the coming apart objection may be raised against this account since it is surely possible that a type of event displays a frequency so different from its true probability that we are led to crown as the best theory a theory that does not entail its true probability. But Lewis might well contend that no sense can be made of a notion of true probability that is undetectable by all scientific investigation. However, Lewis' theory is certainly vulnerable to the locality of probabilities objection as well as two objections that are unique to his account.

The first is the *Big Bad Bug objection*. It is closely related to the locality of probabilities objection. The problem is that, on Lewis' account, the probability of a particular event E at the actual world is determined in part by the entire history of the actual world. Thus, for any particular time, there are ways that the future might be such that, if it were that way, the probability of E would be different. Let F be one of those futures. And let X be a proposition that specifies both the probability of E and the probability of F at the actual world. Then both of the following facts must hold. First, by hypothesis, F is possible; and many philosophers take this to entail that $p(F) > 0$; thus, by the Principal Principle, since X specifies the probability of F, we have $b(F\ occurs|X) = p(F) > 0$. Second, since the occurrence of F would entail that the actual probability of E is different, we have that X and F *occurs* are contradictory, so $b(F\ occurs|X) = 0$. For Lewis' attempt to solve this problem, see ([Lewis, 1994, 485–9]).

The second objection is the *zero fit objection*. Suppose there is a denumerably infinite set of event-types $\{E_i : i = 1, 2, \ldots\}$ such that our evidence imposes the following three conditions on any theory that purports to be a good theory of the world. First, E_i may or may not occur at the actual world, so if p is the physical probability function of our theory, then again according to many philosophers $0 < p(E_i) < 1$. Second, the physical probability of each E_i is the same: that is, for all $i, j = 1, 2, \ldots, p(E_i) = p(E_j)$. Third, perhaps due to evidence about the causal structure of the world, we require that the occurrence of E_i is independent of the occurrence of E_j (when $i \neq j$): that is, $p(E_i|E_j) =_{df.} \frac{p(E_i \cap E_j)}{p(E_j)} = p(E_i)$; or, equivalently, $p(E_i \cap E_j) = p(E_i)p(E_j)$. For instance, perhaps a fair coin is tossed infinitely many times and E_i is the event of it landing heads on the ith toss. Then the probability assigned to the actual world by any theory that satisfies these conditions is zero. (Exercise.) Thus, for any theory that is a candidate for being the best theory of such a world, there are infinitely many other theories that are equally good according to Lewis' criteria, but which assign different probabilities to events. After all, any two theories that both satisfy the three conditions imposed by our evidence are equally good with respect to the probability they assign to the actual world – they all assign it probability zero. Thus, if the world has these features, then, on Lewis' account, the probabilities of the E_is are not well-defined. For, on that account, the probability of an event is the probability assigned to it by the *single* best theory. But there is no single best theory, and the infinitely many equally good theories assign different probabilities to the E_is.

3. Epistemic Probability

We turn now to epistemic probability. It is a philosophical commonplace to note that beliefs comes in degrees. For instance, my degree of belief in the proposition *Chelsea Clinton is the President in 2040* is greater than my degree of belief in the proposition *Chelsea Clinton is the President in 2080*. What's more, there are norms that govern my degrees of belief. For instance, intuitively, it is irrational to have a higher degree of belief in a conjunction than in either of its conjuncts. The epistemic probability of a proposition for a particular agent is her degree of belief in that proposition. In this section, I survey the most prominent accounts of epistemic probabilities, I describe some of the norms that are thought to govern them, and I consider the justifications we might give for those norms.

3.1 Norms for Epistemic Probabilities

I begin by describing two of the central norms that have been thought to govern epistemic probabilities. Given an agent and given a time t during her epistemic life, let b_t be the function that assigns to each proposition in an algebra of propositions the agent's degree of belief in that proposition at time t.[6] We call b_t the agent's *belief function at t*.

The first norm is *Probabilism*. It is often thought to be the analogue for degrees of belief of the requirement that an agent have a consistent set of full beliefs.

Probabilism: For each time t during her epistemic life, an agent's belief function b_t at t ought to be a probability measure.

Some philosophers require further that b_t is countably additive.

The second norm is *Bayesian Conditionalization*. It governs how an agent ought to respond to a certain sort of evidence.

Bayesian Conditionalization: Suppose that, between times t and t', an agent learns the proposition E with certainty. And suppose that $b_t(E) > 0$. Then, her belief function $b_{t'}$ at time t' ought to be

$$b_{t'}(\cdot) = b_t(\cdot|E) =_{df.} \frac{b_t(\cdot \mathbin{\&} E)}{b_t(E)}$$

Many other norms have been proposed, such as *regularity* [Kemeny, 1955], *reflection* [van Fraassen, 1984], *Jeffrey conditionalization* [Jeffrey, 1983], and *MaxEnt* [Jaynes, 1957a], [Jaynes, 1957b]. I will not consider these here.

3.2 Epistemic Probabilities as Fair (or Favourable) Betting Odds

Our first attempt to give an interpretation of epistemic probability is motivated by the same considerations that motivate actualist frequentism, namely, ontological parsimony and epistemological tractability. *Prima facie*, degrees of belief are mental phenomena. Thus, philosophers who are sceptical of the reality of these phenomena, such as behaviourists, seek a reduction of facts about these mental entities to facts about an epistemically more accessible realm; in particular, the realm of observable behaviour. Just as the actualist frequentist reduces physical probability to actual frequencies, so the *fair betting odds* interpretation of degrees of belief reduces epistemic probabilities to observable behaviour.

3.2.1 The Interpretation

Suppose X is a proposition. Then consider the following bet on that proposition: it pays \$1 if X is true, and \$0 if X is false – we call this the *dollar bet on* X. (In what follows, we assume that utility is linear in dollars; if it is not, replace dollars by units of utility throughout. See Chapter 19 for a discussion of utility generally, and what it means to say that it is linear in a given quantity.) According to the fair betting odds interpretation, an agent's degree of belief in proposition X is p if that agent is prepared to buy a dollar bet on X for \$$p$ or less, and sell a dollar bet on X for \$$p$ or more. Thus, if her degree of belief in *Chelsea Clinton is the President in 2040* is 0.1, she would be prepared to pay 10¢ (or less) for a dollar bet on that proposition, and she would be prepared to accept 10¢ (or more) for a dollar bet on that proposition.

One of the main advantages of this account of degrees of belief is that it provides a simple form of argument that purports to justify the sort of norms governing degrees of belief described in Section 3.1: this form of argument is called the *Dutch Book argument*. A *Dutch Book* is a series of bets that will result in a loss for the bettor however the world turns out. For instance, suppose I pay 70¢ for a dollar bet on X and suppose I pay 50¢ for a dollar bet on $\neg X$. Then, however the world turns out, I will lose 20¢. Thus, this series of bets is a Dutch Book. A Dutch Book argument for a particular norm Q runs as follows:

1. (*Fair betting odds interpretation*) If an agent has degree of belief r in proposition X, she is prepared to pay \$$r$ (or more) for a dollar bet on X and she is prepared to accept \$$r$ (or less) for a dollar bet on X.
2. (*Dutch Book Theorem for* Q) If an agent violates Q, there is a series of bets that she will be prepared to accept individually, but which together form a Dutch Book.
3. (*Converse Dutch Book Theorem for* Q) If an agent obeys Q, there is *no* series of bets that she will be prepared to accept individually, but which together form a Dutch Book.
4. An agent who is prepared to accept individually a series of bets that together must result in loss is irrational.
5. Therefore, an agent ought to obey norm Q.

Dutch Book and Converse Dutch Book theorems, and thus Dutch Book arguments, are available for the norms of Probabilism (both the version that requires countable additivity [Adams, 1962] and the version that does not ([Ramsey, 1931b], [Kemeny, 1955])) and Bayesian Conditionalization [Lewis, 1999], as well as many other putative norms for degrees of belief, such as regularity, reflection, and Jeffrey conditionalization mentioned above.[7] Indeed, our example of a Dutch Book from above generalizes to show that an agent who violates the norm $p(A) + p(\neg A) = 1$ is vulnerable to a Dutch Book. (Exercise.)

3.2.2 Assessing the Interpretation

Since the fair betting odds interpretation of epistemic probabilities is reductive, it faces an objection that is analogous to the coming apart objection that was raised against actualist frequentism above. It seems possible that our degrees of belief and the betting behaviour we exhibit come apart. For instance, it seems possible that an agent who is extremely risk averse might have degree of belief 0.6 in proposition X and yet be willing to pay only a very small price for a dollar bet on X. On the fair betting odds interpretation, this is not possible. Indeed, on this interpretation, risk aversion is synonymous with scepticism: agents prepared to pay only a very small price for dollar bet on X must, by definition, have a correspondingly low degree of belief in X. However, just as the actualist frequentist is inclined to bite the bullet in response to the coming apart objection, so is the proponent of the fair betting odds interpretation.

However, further problems arise when we try to use a Dutch Book argument to establish a particular norm. We have just seen the problems that arise with the first premise of these arguments. The second and third premises are mathematical theorems. However, some philosophers have raised objections against the final premise, which is a strong normative claim. The problem is that it is not clear what is irrational about an agent who is willing to accept individually a series of bets that together must result in a loss for her. Initially, one might think that such an agent will end up losing a lot of money. But this is only true if she is actually offered all of the bets in the Dutch Book that can be made against her. Thus, one might say instead that it is irrational for an agent to be *vulnerable* to such exploitation, even if she is never actually *subject* to it. But again, it isn't clear that she is vulnerable in this way. To conclude that she is, we require the following principle, which is sometimes called the *package principle*: an agent willing to accept each of a set of bets individually is prepared to accept all of those bets when offered together as a package. But this principle is controversial. A rational agent whose degrees of belief make her willing to accept individually each bet in a Dutch Book might nonetheless refuse to accept all of the bets if they are offered, since she will be able to calculate that they will result in a sure loss. In order to save the package principle, we must alter the fair betting odds interpretation to say that an agent has degree of belief p in X if she would be prepared to buy a dollar bet for \$$p$ or more and sell for \$$p$ of less X *regardless of the other bets she has accepted*. But this merely serves to strengthen the version of the coming apart objection raised in the previous paragraph.

3.3 Epistemic Probabilities and Expected Utility

For those sceptical of the metaphysical reality of degrees of belief, but who nonetheless recognize their usefulness in theorizing, there is an alternative to the fair betting odds interpretation. I will call it the *expected utility interpretation*. It is the interpretation favoured by most economists.

3.3.1 The Interpretation

On the fair betting interpretation, we begin with the bets that an agent would be prepared to accept, and we define her degrees of belief explicitly in terms of these. In this sense, it is analogous to the frequentist interpretations of physical probability, which begin with frequencies (either actual or hypothetical) and define physical probabilities explicitly in terms of them. On the expected utility interpretation, on the other hand, we begin with all possible bets ordered by the agent's preferences for them, and we define her degrees of belief implicitly in terms of these. In this sense, it is analogous to Lewis' best system analysis interpretation of physical probability, which begins with the Humean matters of fact about the world, and defines physical probabilities implicitly in terms of the best theory to account for these facts.

Thus, suppose $\mathcal{B} = \{B_i : i \in I\}$ is the set of all possible bets on an atomic algebra of propositions $\mathcal{F}$. And suppose that $\preceq$ is the agent's ordering of the bets in $\mathcal{B}$ by preference: that is, $B_i \preceq B_j$ if the agent does not prefer B_i to B_j. Then we say that her degrees of belief are given by the unique belief function b (if such exists) for which there is a utility function U such that the ordering of the bets B_i given by their expected utilities relative to b and U is $\preceq$. That is,

$$B_i \preceq B_j \text{ iff } \sum_{A \in \text{Atoms}(\mathcal{F})} b(A)U(A, B_i) \leq \sum_{A \in \text{Atoms}(\mathcal{F})} b(A)U(A, B_j)$$

where $U(A, B_i)$ is the utility that would result from bet B_i if A were true. When this holds, we say that *b and U generate $\preceq$*.

On the expected utility interpretation, then, degrees of belief are theoretical terms that are introduced, along with utilities, to explain or analyse or interpret an agent's betting preferences. The appeal of this interpretation lies in the possibility of justifying norms that govern epistemic probabilities by appealing to a sort of mathematical result called a *representation theorem*. A *representation theorem argument* begins by listing constraints on an agent's preference ordering $\preceq$ that are held to be necessary conditions on rationality. For instance, it is proposed that $\preceq$ ought to be transitive. It then proceeds to show that, for any agent whose preference ordering satisfies these constraints, there is a unique belief function b for which there is a utility function U such that b and U generate $\preceq$, and that unique b is a probability function. Moreover, conversely, any probability function b and utility function U will generate an ordering $\preceq$ that satisfies the rational constraints. Thus, just as there is a Dutch Book argument that purports to establish Probabilism on the fair betting odds interpretation, there is a representation theorem argument that purports to establish that norm on the expected utility interpretation ([Savage, 1954], [Maher, 1993]).

3.3.2 Assessing the Interpretation

Again, this interpretation faces a coming apart objection. Degrees of belief certainly guide betting preferences. And perhaps in ideally rational agents those preferences are generated by an agent's degrees of belief and her utility function in the way described by expected utility theory. But it is surely not *constitutive* of degrees of belief that they generate preferences in this way. But again, the proponent of the expected utility interpretation might simply bite the bullet.

Another problem with this interpretation is that it fails to assign degrees of belief to any agent for whom there is not a unique b and U that generate $\preceq$. For instance, an agent whose preferences are intransitive is not assigned degrees of belief on this interpretation. However, intuitively, we would like to say that such an agent has degrees of belief, even if they are irrational in some way.

A further problem lies in the definition of expected utility. Unless there is an argument that shows that the only rational way to combine degrees of belief and utilities to give preferences is via expected utility defined as above, the argument will not work. After all, there may be equally rational alternatives to the definition of expected utility that result in a representation theorem for non-probability functions. That is, it might be that, for any agent whose preference ordering satisfies the rational constraints, there is a unique *non*-probability function b for which there is a utility function U such that b and U generate $\preceq$ when combined using this alternative version of expected utility. (See [Zynda, 2000] for an example of such an alternative.)

3.4 *Sui Generis* Epistemic Probabilities

3.4.1 The Interpretation

If the fair betting interpretation of epistemic probabilities is the natural cousin of the frequentist interpretation of physical probabilities, and the expected utility interpretation is the natural cousin of the best system analysis, then the *sui generis interpretation* of epistemic probabilities is the natural cousin of the single case propensity interpretation. For, on this interpretation, degrees of belief are introduced as real irreducible features of mental states.

It might seem that such an interpretation will suffer from the same problems as the single case propensity interpretation. If we don't reduce facts about degrees of belief to some further facts, there will be no way for us to discover anything about them. But this isn't so. Instead, it turns out that we know enough about degrees of belief to establish some norms that govern them.

To do this, we need to introduce the notion of an *epistemic utility function* (sometimes called a *scoring rule*). Suppose our agent assigns a degree of belief to each proposition in an atomic algebra $\mathcal{F}$. Then an *epistemic utility function on* $\mathcal{F}$ is a function EU that takes a belief function b defined on the algebra $\mathcal{F}$, and an atom $A \in \text{Atoms}(\mathcal{F})$ of $\mathcal{F}$, and returns a real number $EU(b, A)$ that purports to measure the purely epistemic, non-pragmatic utility that the belief function b would have if A were true. For instance, the epistemic utility of a belief function were A true might be a function of its accuracy or verisimilitude were A true, or its simplicity or explanatory power, and so on.

Given an epistemic utility function EU, we introduce the following definitions. If b and b' are belief functions we say that

Definition 15.3.1 *b weakly EU-dominates* b' if, for all $A \in \text{Atoms}(\mathcal{F})$,

$$EU(b', A) \leq EU(b, A)$$

with inequality for at least one atom A.

Definition 15.3.2 *b strongly EU-dominates* b' if, for all $A \in \text{Atoms}(\mathcal{F})$,

$$EU(b', A) < EU(b, A).$$

It is claimed that it is epistemically irrational to have a belief function b that is strongly dominated by another belief function b', since there is a belief function that would be better however the world turns out, namely, b'.

Using this norm, it is thought that we can give a justification of Probabilism ([Joyce, 2009], [Leitgeb and Pettigrew, 2010a], [Leitgeb and Pettigrew, 2010b]). The strategy is not to specify a particular epistemic utility function. Rather, we enumerate plausible features that such a function must have. And we show that, for any epistemic utility function EU with these features, the following holds:

1. For every non-probability function b, there is a probability function p that strongly EU-dominates b.
2. There is no probability function p for which there is another belief function b (whether a probability function or not) that even weakly EU-dominates p.

It is claimed that Probabilism follows.

To justify Bayesian Conditionalization, a slightly different style of argument is required. Suppose that an agent has belief function b_t at time t. (I assume in this paragraph that all belief functions are probability functions.) And suppose that, between time t and a later time t', she learns with certainty the truth of the proposition E and no more. Which belief function ought such an agent to adopt at t'? The idea is that we should treat the choice of which belief function to adopt just as we treat the choice of which action to perform in decision theory. That is, given the epistemic utility function EU, she ought to adopt a belief function $b_{t'}$ that maximizes her expected value of EU when that expected value is calculated relative to her belief function b_t at time t, and over the atoms $A \in \text{Atoms}(\mathcal{F})$ such that $A \subseteq E$. That is, she ought to choose b that maximizes

$$\sum_{A \in \text{Atoms}(\mathcal{F}):A \subseteq E} b_t(A)EU(b, A)$$

As before, we must make some assumptions about EU. Having done that, we show that the belief function $b_{t'}$ that is demanded by Bayesian Conditionalization is the unique belief function that maximizes this expected value.

Arguments from considerations of epistemic utility have been given for other norms as well [Leitgeb and Pettigrew, 2010b].

3.4.2 Assessing the Interpretation

When we attempt to justify a particular norm by appealing to epistemic utilities, the plausibility of that justification depends on the plausibility of the features that epistemic utilities must be assumed to have in order to mobilize the mathematical theorems. In Joyce's most recent justification of Probabilism [Joyce, 2009] and Greaves and Wallace's justification of Bayesian Conditionalization [Greaves and Wallace, 2006], it is assumed that certain belief functions ought not to be ruled out *a priori* by considerations of epistemic utility: Joyce holds that no probability function ought to be ruled out *a priori*, while Greaves and Wallace assume that the particular probability function $b_t(\cdot|E)$ advocated by Bayesian Conditionalization ought not to be ruled out *a priori*. On the other hand, Leitgeb and Pettigrew try to avoid assumptions that identify particular sorts of belief functions that ought not to be ruled out as irrational *a priori*. Instead, they provide a characterization of the legitimate epistemic utility functions by appealing to the relationship between the epistemic utility enjoyed by a belief function at a particular world, and the epistemic utility enjoyed by a particular degree of belief in a proposition at that world ([Leitgeb and Pettigrew, 2010a], [Leitgeb and Pettigrew, 2010b]).

4. Prospects

In the case of physical probabilities and in the case of epistemic probabilities, we have accounts in which probabilities are introduced reductively by explicit or implicit definitions, and accounts in which probabilities are *sui generis*.

In the case of physical probabilities, it seems that it is the implicit definition account given by Lewis' best system analysis interpretation that offers the greatest hope. However, we require solutions to the Big Bad Bug objection and the zero fit objection. Lewis has suggested a solution to the former. Does it work? Regarding the latter: Is there a notion of fit that can discriminate between the infinitely many competing theories that cannot be distinguished by the current notion?

In the case of epistemic probabilities, both explicit and implicit definition interpretations are vulnerable to coming apart objections. A fruitful current line of research seeks to exploit the Dutch Book arguments and representation theorem arguments on which these interpretations depend, but in such a way that they do not rely on interpretations that are vulnerable to these objections [Christensen, 1996]. Another ongoing project belongs to the *sui generis* interpretation of epistemic probabilities. Here, we seek versions of the arguments from epistemic utility functions that impose weaker and more evident conditions on those functions.

All of this is work for the future.

Notes

1. For an endorsement of this position, see [Venn, 1876]. For criticisms, see [Hájek, 1997].
2. This interpretation is proposed by [Howson and Urbach, 1993]. Closely related positions that are vulnerable to the same objections are found in [Reichenbach, 1949], [von Mises, 1957], and [van Fraassen, 1980]. For criticisms, see [Hájek, 2009] and [Jeffrey, 1977].
3. Note that von Mises' well-known attempt to solve this problem by appealing to his theory of *collectives* does not completely solve this problem. See ([Hájek, 2009, p. 225]).
4. For a more detailed consideration of this position and the single-case propensity interpretation considered in the next section, see [Gillies, 2000] and [Eagle, 2004].
5. Nonetheless, Fetzer and Donald Nute have developed a formal account that avoids this consequence ([Fetzer, 1981]).
6. Let Ω be the set of possible worlds, and represent propositions as sets of possible worlds. Then an algebra of propositions is an algebra on Ω.
7. See [Hájek, 2008] for a survey.

16 Pure Inductive Logic

J. B. Paris

Chapter Overview

1. Introduction	428
2. Context	430
3. Probability Functions	433
4. Rational Principles	435
5. Consequences of the Principles for Unary L	438
6. de Finetti's Theorem	441
7. Polyadic Inductive Logic	442
8. Symmetry	444
9. Analogical Reasoning	446
10. Universal Certainty	447
11. Conclusion	447
Acknowledgements	448
Notes	448

1. Introduction

To what extent does my evidence[1] determine my beliefs? Putting it another way if one could somehow extract all my available evidence is there some logic or calculus which could be applied to this to yield my beliefs?

Most of us I imagine would say that the total impracticality of ever carrying out such an experiment, even if we could formalize what was meant by 'evidence' and 'belief', makes the question so hypothetical as to be meaningless. However there are some extreme situations where the question does seem to make some sense. One is where 'I' am an artificial agent which has been programmed with a particular knowledge base. In this case we can have access to the agent's total knowledge or evidence. Another is when we agree to put aside all evidence outside of some fixed set of assumptions (as in a thought experiment) and then argue on the basis of these assumptions alone. In these cases it seems that there might be an argument for some such logic, at least when we add suitable simplifying assumptions about what we mean formally by 'evidence' and 'belief'.

The version of 'Inductive Logic' described in this chapter is based on one such formalization within first-order probability logic and is a natural continuation of Carnap's Inductive Logic (see [Carnap, 1950], [Carnap, 1952], [Carnap and Jeffrey, 1971], [Carnap, 1980], [Fitelson, 2006]) and an earlier approach along similar lines by Johnson [Johnson, 1932]. It is important however to emphasize the limited scope of Inductive Logic as presented here compared with the original aspirations of Carnap and Johnson that it might provide a *practical* guide to everyday inductive reasoning. Carnap's vision of Inductive Logic as applicable in the real world is now judged by the majority of Philosophers to have received a death blow with the publication in 1946 of what subsequently became known as Goodman's 'grue' Paradox (see [Goodman, 1946], [Goodman, 1947], and more recently [Stalker, 1994]). Here we are presented with 'isomorphic' premises with different (contradictory even) conclusions so the conclusion cannot simply be a logical function of the premises. Consequently Carnap's hope of determining such beliefs by purely logical/rational considerations cannot succeed.

In his initial response to this paradox, Carnap argued that it in no way derailed his programme because it transgressed the standing requirement that *all* the available evidence is to be taken into account, indeed it is exactly this additional evidence which we need to employ in order to conclude that there is a paradox there in the first place (see [Carnap, 1947a], [Carnap, 1947b], [Carnap, 1980]). Nevertheless it would appear that Carnap eventually capitulated because of the general impracticality of fulfilling this requirement. Since then, some effort has been made to temper the requirement of total evidence by proposing some sort of ring fencing on the 'relevant evidence'. The notion of a *projectible predicate* is one such proposal. Nonetheless, the general opinion is that the *Applied* Inductive Logic programme as Carnap envisaged it is dead. As a result the further development of 'Carnapian Inductive Logic' was essentially halted for a long period towards the end of the twentieth century. In its place a number of off-shoot approaches to the practical problem of how our evidence influences our beliefs have been investigated (see for example [Earman, 1992], [Earman, 1985], [Fitelson, 2004], [Hájek and Hall, 2002], [Maher, 2006]).

However, as argued in [Nix and Paris, 2007], for the interpretation of 'Pure'[2] Inductive Logic as presented here Goodman's Paradox is simply no obstacle whatsoever. For instead of aiming at a practically applicable logic to guide our everyday actions we aim to present Inductive Logic as a formal study of 'rational uncertain reasoning', an investigation into putatively rational principles of belief formation and their mathematical consequences. Within this framework we can blithely make the assumption that all the available evidence is up front and any criticism based on the impractically of this assumption is irrelevant as far as the pure theory goes, just as in studying the Classical Propositional Calculus we lay no restrictions on the number of premises which we may consider.

This approach to Inductive Logic then harkens back to Carnap in the sense that it is to be seen as an extension of First-Order Predicate Reasoning. However its intended scope is far reduced from that of its source. For rather than providing guidance and laws for practical human reasoning it should be seen as applicable to certain tightly controlled 'toy' situations such as are encountered by agents in Artificial Intelligence (where there is much interest in such matters under the heading of uncertain reasoning) or where there is an agreed tight ring fence on what evidence is to be allowed. Nevertheless the underlying requirement, that this reasoning should be 'logical', or putting it another way that the belief forming agent should be 'rational', remains in accord with the original aims of Johnson and Carnap.

One might reasonably question the value within Philosophy (as opposed to Mathematics and Artificial Intelligence) of such an enquiry. The reward, we would claim, is that this relatively simple context in which we shall work allows us to formulate and study various aspects and principles of 'rationality', as it applies to belief formation, analytically, with mathematical precision. Given the simplicity of the framework we might at least hope by this device to gain some understanding of the local notion of 'rationality' – where else if not in this simplest of contexts?[3]

Up to this point we have been inserting quotes around the word 'rational' to indicate the contentious status of this notion. Henceforth we will drop the quotes though without wishing at all to imply that the status has in any way altered. However, as indicated above, we might hope that the endeavour of Inductive Logic may ultimately lead to some semblance of clarification and understanding. For now, let it suffice that the notions we dub rational may at least be entertained to have some claim to that title.

2. Context

We shall assume that the 'evidence' applies to a world populated by a countable set of individuals $a_1, a_2, a_3, \ldots$ and in which there are a finite number of relations $R_1, R_2, \ldots, R_q$ which may or may not hold of these individuals. All the information we have about these relations and constants is to be included in the evidence, we should have no preferred or intended interpretations except in as far as these are fully captured in the evidence. So if we have zero evidence (the main situation which we shall consider) then the R_i are just relations and the a_j just constants about which we make no assumptions whatsoever about meaning, projectibility, etc.[4] In this context we are interested in what belief on a scale between 0 and 1 a 'rational agent' should assign to the assertion that some sentence θ is true in this world when all the evidence we have about this world is, say, that some other sentences $\phi_1, \ldots, \phi_m$ are true in this world[5] – here 1 denotes absolute surety and we assume beliefs can be specified by a single figure.

For example suppose these a_i were runs of an experiment and for the unary relation (i.e. predicate) R_1 the sentence $R_1(a_i)$ is true if the outcome of the experiment is a 1 and is false if the outcome is the only other alternative 0. We run the experiment four times and on each occasion the outcome is a 1. So in this case our evidence $\phi_1, \ldots, \phi_m$ is just $R_1(a_1), R_1(a_2), R_1(a_3), R_1(a_4)$ (assuming we know nothing else about the experiments). Then if we are to act rationally what belief should we give to $R_1(a_5)$, that the next run of the experiment will also yield outcome 1? (So in this case θ above would be $R_1(a_5)$). Similarly what belief should we rationally give on the basis of this evidence to all future runs of the experiment yielding outcome 1, i.e., to $\forall x\, R_1(x)$ being true in the world?

The methodology of Pure Inductive Logic for addressing such questions is to propose ostensibly rational, or logical, principles that we, being rational, should observe and to investigate their consequences for such questions. Observance of these rational principles constrains the possible answers we can proffer, and the ideal situation is that there is just one precisely determined answer. Before we can take this path however we need to make the context a little more formal.

Let L be a first-order predicate language with relation symbols $R_1, \ldots, R_q$, of arities $r_1, r_2, \ldots, r_q$ respectively,[6] constant symbols $a_1, a_2, a_3, \ldots$ but no function symbols nor (as far as this introductory account is concerned) equality. The intention is that these a_i exhaust the universe. Let SL/FL denote the set of first-order sentences/formulae of L formed in the usual way and let $QFSL$ denote the set of quantifier-free sentences of L.

Definition 16.2.1 A *probability function on L* is a function w from SL into $[0,1]$ such that for $\theta, \phi, \exists x\, \psi(x) \in SL$:

(P1) If $\vdash \theta$ then $w(\theta) = 1$.
(P2) If $\vdash \neg(\theta \wedge \phi)$ then $w(\theta \vee \phi) = w(\theta) + w(\phi)$.
(P3) $w(\exists x\, \psi(x)) = \lim_{n\to\infty} w(\bigvee_{i=1}^{n} \psi(a_i))$.

Condition (P3), which is due to Gaifman [Gaifman, 1964], reflects the intention that the a_i exhaust the universe, and is peculiar to the definition of a probability function in this context (the standard definition consisting of just (P1) and (P2)).

As is the common practice in Inductive Logic we shall assume throughout that our degree of belief in a sentence θ of L is to be equated with the subjective probability $w(\theta)$ that we would assign to θ.[7]

Having set up this framework the basic question we are interested in is:

> Given evidence $\phi_1, \phi_2, \ldots, \phi_m \in SL$ what probability $w(\theta)$ should rationally be assigned to $\theta \in SL$? More generally, given evidence $\phi_1, \phi_2, \ldots, \phi_m \in SL$ what is the rational choice of probability function w on L?

Notice that this question also subsumes what is often referred to as the 'problem of induction': *Why* should the evidence that $\psi(a_1), \psi(a_2), \ldots, \psi(a_m) \in SL$ influence my belief in $\psi(a_{m+1})$, or in $\forall x \psi(x)$?

For evidence sets like these $\phi_1, \phi_2, \ldots, \phi_m \in SL$ this question can be further simplified once one is willing to accept the 'received wisdom'[8] that the probability one should give to $\theta \in SL$ given $\phi_1, \phi_2, \ldots, \phi_m$ should be the *conditional probability*

$$w(\theta \,|\, \phi_1, \phi_2, \ldots, \phi_m) = \frac{w\left(\theta \wedge \bigwedge_{i=1}^{m} \phi_i\right)}{w\left(\bigwedge_{i=1}^{m} \phi_i\right)} \tag{16.1}$$

where w is the rational choice of probability function on L *in the absence of any evidence at all*, at least provided that the denominator here is non-zero.[9]

In consequence the key question[10] above now reduces to:

> What is the rational choice of probability function w on L in the absence of any evidence?

Inductive Logic, as far as this account is concerned, is the formulation and investigation of various arguably rational principles which bear on this question by reducing the choice of w from just *any* probability function on L, ideally reducing it to a single 'perfectly rational' choice. For the most part this goal can rarely be attained, and moreover even some apparently reasonable principles turn out to point in different directions as we shall later demonstrate.

In a way the situation here resembles that current in Set Theory where various axioms are proposed for their intuitive appeal and their relationships, and the nature of the universes they allow, are investigated. In our case here various principles are proposed or mooted, now on the grounds of their intuitive rationality, and the relationships between them and the probability functions that they allow are investigated. As with Set Theory it is not necessary to believe these proposed principles unconditionally; we are still at the 'long list' stage in this selection process with the 'short list' currently just a project for future research.

We shall shortly introduce some of the main rational principles which have been considered to date. Before that however we need to say something about the structure and properties of probability functions on L.

3. Probability Functions

On the face of it, it may seem that the conditions (P1-3) are rather tolerant and that apart from the obvious properties, given for example in [Paris, 1994, p. 10], probability functions on L might be a rather disparate bunch. However it turns out that structurally they are in fact relatively easy to describe. To do so requires us to introduce a little notation.

Definition 16.3.1 Let $b_1, b_2, \ldots, b_n$ be some distinct constants from L, i.e., distinct a_i, a notation we shall use throughout. A *state description*, $\Theta(b_1, b_2, \ldots, b_n)$, for $b_1, b_2, \ldots, b_m$ is a sentence of the form

$$\bigwedge_{s=1}^{q} \bigwedge_{i_1, i_2, \ldots, i_{r_s} \in \{1, \ldots, n\}} \pm R_s(b_{i_1}, b_{i_2}, \ldots, b_{i_{r_s}}), \tag{16.2}$$

where $\pm R$ stands for one of R or $\neg R$. In other words this state description $\Theta(b_1, b_2, \ldots, b_n)$ tells us precisely which of $R_s(b_{i_1}, b_{i_2}, \ldots, b_{i_{r_s}})$ or $\neg R_s(b_{i_1}, b_{i_2}, \ldots, b_{i_{r_s}})$ holds for each relation symbol R_s and each choice (possibly with repeats) $b_{i_1}, b_{i_2}, \ldots, b_{i_{r_s}}$ of r_s constants from $\{b_1, b_2, \ldots, b_n\}$.

A particularly important special case of this is when the language L consists only of predicates, that is when the $R_1, \ldots, R_q$ are all unary. In that case the state description can be written in the special form

$$\bigwedge_{i=1}^{n} \alpha_{h_i}(b_i)$$

where

$$\alpha_1(x), \alpha_2(x), \ldots, \alpha_{2^q}(x)$$

are the *atoms* of L,[11] that is 2^q formulae of the form

$$\pm R_1(x) \wedge \pm R_2(x) \wedge \ldots \wedge \pm R_q(x).$$

Notice that by the Disjunctive Normal Form Theorem any $\theta(b_1, b_2, \ldots, b_n) \in QFSL$ is logically equivalent to a disjunction of state descriptions for $b_1, b_2, \ldots, b_n$, and so since distinct state descriptions for $b_1, b_2, \ldots, b_n$ are disjoint the probability of $\theta(b_1, b_2, \ldots, b_n)$ will be the sum of the probabilities of these state descriptions. Indeed this determinacy extends also to all of SL as the following result explains (see [Gaifman, 1964]).

Theorem 16.3.1 *Let w be a probability function on L. Then w is uniquely determined by its values on the state descriptions $\Theta(a_1, a_2, \ldots, a_n)$ for $n = 1, 2, 3, \ldots$. Furthermore the only constraint on these values $w(\Theta(a_1, a_2, \ldots, a_n))$ is that they satisfy $w(\top) = 1$,* *where* $\top$ *is the state description for an empty sequence* $b_1, b_2, \ldots, b_n$*, i.e. a tautology, and for any state description*

$$w(\Theta(a_1, a_2, \ldots, a_n)) = \sum_{\Phi(a_1, \ldots, a_{n+1})} w(\Phi(a_1, a_2, \ldots, a_n, a_{n+1}))$$

where the $\Phi(a_1, \ldots, a_{n+1})$ *range over all state descriptions extending* $\Theta(a_1, a_2, \ldots, a_n)$*, equivalently such that*

$$\Phi(a_1, a_2, \ldots, a_n, a_{n+1}) \models \Theta(a_1, a_2, \ldots, a_n).$$

From this theorem it follows that the battle over which is the most rational probability function w to adopt in the absence of any evidence is essentially being fought on the quantifier free sentences, even just the state descriptions, of L. This explains the immediate importance of such sentences in Inductive Logic.

Theorem 16.3.1 also shows that probability functions on L are rather easy to construct and there are very many of them. What we need now are some criteria to weed out those which are illogical or irrational. The method of achieving that, which defines what Inductive Logic as presented here is all about, is to require that w satisfy some (arguably) rational principles. That will be the subject of the next section. Before that however it will be useful to mention two particular 'extreme' probability functions on L.

The first,[12] which we shall call w_∞ (or w_∞^L if we wish to exhibit its dependence on L) just gives each state description for $a_1, a_2, \ldots, a_n$ the same probability, which must of course be $1/K_n$ where K_n is the number of possible state descriptions for $a_1, a_2, \ldots, a_n$. Alternatively w_∞ is the probability function such that

$$w_\infty(R_i(c_1, c_2, \ldots, c_{r_i})) = w_\infty(\neg R_i(c_1, c_2, \ldots, c_{r_i})) = 1/2$$

for any of the relation symbols R_i of L and constants $c_1, c_2, \ldots, c_{r_i}$ of L, in this case not necessarily distinct, and treats all such $R_i(c_1, c_2, \ldots, c_{r_i})$ as stochastically independent.

In a way, w_∞ looks a rather natural choice of probability function on L in the absence of any evidence, after all why should one treat $\neg R_i$ differently from R_i or different $R_i(c_1, c_2, \ldots, c_{r_i})$, $R_j(d_1, d_2, \ldots, d_{r_j})$ as stochastically dependent in the total absence of any evidence at all? Certainly that is a position one could adopt, though w_∞ is commonly criticized for not positively supporting induction (or learning). For example in the case of the experiment described in the second section making w_∞ one's rational choice would lead to giving $R_1(a_5)$ probability 1/2 on the evidence of $R_1(a_1), R_1(a_2), R_1(a_3), R_1(a_4)$, i.e.,

$$w_\infty(R_1(a_5) | R_1(a_1) \wedge R_1(a_2) \wedge R_1(a_3) \wedge R_1(a_4)) = 1/2,$$

which is no different from the unconditional probability w_∞ would give to $R_1(a_5)$ prior to any experimenting having taken place.

The probability function w_0 is in a sense the exact opposite of w_∞ though they start off looking the same in that for each state description $\Theta(a_1)$,

$$w_0(\Theta(a_1)) = 1/K_1.$$

However for a state description $\Phi(a_1,a_2)$, w_0 will only give this a non-zero value – value $1/K_1$, in fact – if, according to the information in $\Phi(a_1,a_2)$, a_1 and a_2 are indistinguishable. This means that, if $R_i(c_1,c_2,\ldots,c_{r_i})$ (respectively $\neg R_i(c_1,c_2,\ldots,c_{r_i})$) is a conjunct in $\Phi(a_1,a_2)$, so $c_1,c_2,\ldots,c_{r_i} \in \{a_1,a_2\}$, and $R_i(d_1,d_2,\ldots,d_{r_i})$ is the result of replacing some of these occurrences of a_1 by a_2 and vice versa, then $R_i(d_1,d_2,\ldots,d_{r_i})$ (respectively $\neg R_i(d_1,d_2,\ldots,d_{r_i})$) is also a conjunct in $\Phi(a_1,a_2)$. Equivalently, this means that $\Phi(a_1,a_1)$ is consistent, and so logically equivalent to a state description $\Theta(a_1)$ for a_1.

More generally, for a state description $\Psi(a_1,a_2,\ldots,a_n)$,

$$w_0(\Psi(a_1,a_2,\ldots,a_n)) = \begin{cases} 1/K_1 & \text{if } \Psi(a_1,a_1,\ldots,a_1) \text{ is consistent} \\ 0 & \text{otherwise} \end{cases}$$

and either way this equals $w_0(\Psi(a_1,a_1,\ldots,a_1))$ since $\Psi(a_1,a_1,\ldots,a_1)$ is either inconsistent, so has probability zero, or is logically equivalent to a state description for a_1.

The probability function w_0 is in a sense the exact opposite of w_∞ in that it will unequivocally give $R_1(a_5)$, and all other $R_1(a_i)$, the highest possible probability 1 on the evidence of just $R_1(a_1)$. The unfortunate aspect of this is that evidence such as $R_1(a_1) \wedge \neg R_1(a_2)$ confronts us with the problem of how to condition on a sentence of probability zero.

4. Rational Principles

To date it seems that almost all of the rational principles proposed in Inductive Logic are based on three somewhat overlapping considerations: Symmetry, Relevance, and Irrelevance. We now briefly consider each of these in turn.

Principles based on symmetry are justified by the idea that if the context possesses a symmetry then it would be irrational for one's assigned probabilities to break that symmetry. An example of this in the case of no evidence is when we take a permutation σ of the set $\mathbb{N} = \{1,2,3,\ldots\}$ of positive natural numbers and extend this to SL by setting:

$$\sigma(\theta(a_{i_1},a_{i_2},\ldots,a_{i_n})) = \theta(a_{\sigma(i_1)},a_{\sigma(i_2)},\ldots,a_{\sigma(i_n)}).$$

Then σ provides an evident symmetry of SL[13] which a rational choice of probability function w on L should respect. That is w should satisfy:

Principle 16.4.1 (Constant Exchangeability Principle, Ex). For σ a permutation of $\mathbb{N}$ and $\theta(a_{i_1}, a_{i_2}, \ldots, a_{i_n}) \in SL$,

$$w(\theta(a_{i_1}, a_{i_2}, \ldots, a_{i_n})) = w(\theta(a_{\sigma(i_1)}, a_{\sigma(i_2)}, \ldots, a_{\sigma(i_n)})).$$

In an exactly similar fashion we can justify the *Predicate Exchangeability Principle*, where we permute predicates of the same arity, the *Variable Exchangeability Principle*, where for a relation symbol R_i and τ a permutation of $\{1, 2, \ldots, r_i\}$ we replace $R_i(t_1, t_2, \ldots, t_{r_i})$ (here the t_j are constants or variables) everywhere by $R_i(t_{\tau(1)}, t_{\tau(2)}, \ldots, t_{\tau(r_i)})$ etc.

For L a purely unary language (so $r_1 = r_2 = \ldots = r_q = 1$) a somewhat strong symmetry principle can similarly be obtained by permuting the atoms $\alpha_1(x), \alpha_2(x), \ldots, \alpha_{2^q}(x)$:

Principle 16.4.2 (Atom Exchangeability Principle, Ax). For σ a permutation of $\{1, 2, \ldots, 2^q\}$

$$w\left(\bigwedge_{i=1}^{n} \alpha_{h_i}(b_i)\right) = w\left(\bigwedge_{i=1}^{n} \alpha_{\sigma(h_i)}(b_i)\right).$$

We shall say more later about why this should still be considered a 'symmetry' but for the moment we remark that in the original formulation by Carnap the classifying role we have for atoms could be taken instead by simply a finite set of exclusive and exhaustive *attributes*, $Q_1(x), Q_2(x), \ldots, Q_k(x)$, commonly illustrated as colours or shapes. In that case just permuting the names given to colours appears, in the absence of any other information, to be a symmetry entirely on a par with permuting constants and predicates.

Irrelevance Principles are of the form that we should have[14]

$$w(\theta|\phi \wedge \psi) = w(\theta|\phi)$$

because, in the presence of ϕ, ψ is thought to be irrelevant to θ.

One example of such a principle is (see [Hill et al., 2002] for this and others):

Principle 16.4.3 (Weak Irrelevance Principle, WIP). If $\theta, \psi \in QFSL$ and θ, ψ have no relation or constant symbols in common then

$$w(\theta|\psi) = w(\theta).$$

In this case ϕ is just a tautology and the perception that ψ is irrelevant to θ is based on the fact that they share no common language whatsoever.

A second irrelevance principle *for purely unary languages* which was central to the endeavors of both Johnson and Carnap was:

Principle 16.4.4 (Johnson's Sufficientness Principle, JSP). The value

$$w(\alpha_k(b_{n+1}) \mid \bigwedge_{i=1}^{n} \alpha_{h_i}(b_i))$$

depends only on n and the number of times that $\alpha_k(x)$ appears among $\alpha_{h_1}(x), \alpha_{h_2}(x), \ldots, \alpha_{h_n}(x)$.

Notice that this does have the above general form since it is equivalent to the the assertion that

$$w(\alpha_k(b_{n+1}) \mid \phi \wedge \bigwedge_{i=1}^{n} \alpha_{h_i}(b_i)) = w(\alpha_k(b_{n+1}) \mid \phi)$$

where

$$\phi = \bigvee_{\vec{g}} \bigwedge_{i=1}^{n} \alpha_{g_i}(b_i)$$

and the $\vec{g}$ range over all sequences $g_1, g_2, \ldots, g_n$ from $\{1, 2, \ldots, 2^q\}$ in which k appears as many times as it does in $h_1, h_2, \ldots, h_n$.

Atom Exchangeability, Ax, and hence Constant Exchangeability, are both straightforward consequences of JSP. This is of interest because, for example, it shows there are two seemingly separate justifications for Ax, one directly from symmetry considerations and the other through irrelevance and the intermediary of JSP.

We shall return to these principles shortly but first we give two examples of principles based on *relevance*. In direct contrast to irrelevance such principles are of the form that under certain specified conditions on θ, ϕ, ψ we should have

$$w(\theta \mid \phi \wedge \psi) \geq w(\theta \mid \phi).$$

i.e., that, in the presence of ϕ, ψ should be positively, or more precisely not negatively, relevant to θ. The best-known version of this is:[15]

Principle 16.4.5 (Principle of Instantial Relevance, PIR). For $\theta(x) \in FL$ and $\phi \in SL$ not mentioning the constants a_i, a_j,

$$w(\theta(a_i) \mid \theta(a_j) \wedge \phi) \geq w(\theta(a_i) \mid \phi).$$

The intuition here is that given ϕ the additional evidence $\theta(a_j)$ that the constant a_j satisfies $\theta(x)$ should enhance (or at least not decrease) one's belief that a_i also satisfies $\theta(x)$ (i.e., that $\theta(x)$ is projectible). On the other hand stating this as a principle might appear rather heavy handed, after all is not the purpose here to investigate why this in particular is a rational principle? Fortunately as we shall see in the next section this is not a serious objection.

The second relevance principle we shall mention is a direct generalization of PIR. Whilst PIR says that $\theta(a_j)$ should enhance $\theta(a_i)$ the following generalization says the same thing should hold even if we only have evidence that $\psi(a_j)$ holds where $\psi(x)$ is a consequence of $\theta(x)$. That is, consequences as well as instances should be relevant.

Principle 16.4.6 (Generalized Principle of Instantial Relevance, GPIR). For $\theta(x), \psi(x) \in FL$ and $\phi \in SL$ not mentioning the constants a_i, a_j, if $\theta(x) \models \psi(x)$ then

$$w(\theta(a_i)|\psi(a_j) \wedge \phi) \geq w(\theta(a_i)|\phi).$$

There are a number of other rational principles which have been suggested in the literature, some of which we shall introduce in the following section where we consider the relationships between these principles when L is purely unary.

5. Consequences of the Principles for Unary L

For this section we shall assume that the language L is purely unary, in other words that $R_1, R_2, \ldots, R_q$ are actually just predicates. This was, up to the use of properties rather than predicates, the version of Inductive Logic studied by Johnson, Carnap et al. and remains among philosophers the main area of interest to this day.

To repeat ourselves, the goal in Inductive Logic as presented here is to formulate rational principles, which by their nature should be acceptable to any rational agent, and whose imposition reduces the available choice of a probability function on the basis of zero evidence, ideally to single possibility. While not quite achieving such complete unanimity, Johnson's Sufficientness Principle is remarkably successful in this regard since as shown by Johnson [Johnson, 1932], and independently later by Kemeny [Kemeny, 1963], provided the number q of predicates in the language is at least two, the only probability functions satisfying JSP are those comprising a one parameter family $\{c_\lambda^L : \lambda \in [0, \infty]\}$.

This family is referred to as *Carnap's Continuum of Inductive Methods* and its members are rather easy to describe. Firstly c_0^L is just the probability function w_0 on L given earlier. Since we are now restricting ourselves to this unary L this means that for a state description $\bigwedge_{i=1}^{n} \alpha_{h_i}(b_i)$,

$$c_0^L\left(\bigwedge_{i=1}^{n} \alpha_{h_i}(b_i)\right) = \begin{cases} 2^{-q} & \text{if } h_1 = h_2 = \ldots = h_n, \\ 0 & \text{otherwise.} \end{cases}$$

For $\lambda = \infty$, c_∞^L is just the w_∞ (for L) given earlier, so

$$c_\infty^L\left(\bigwedge_{i=1}^{n} \alpha_{h_i}(b_i)\right) = 2^{-nq}.$$

Finally for $0 < \lambda < \infty$

$$c_\infty^L\left(\alpha_k(b_{n+1}) | \bigwedge_{i=1}^{n} \alpha_{h_i}(b_i)\right) = \frac{s_k + \lambda/2^q}{n + \lambda},$$

where $s_k = |\{i | h_i = k\}|$, from which it follows that

$$c_\lambda^L\left(\bigwedge_{i=1}^{n} \alpha_{h_i}(b_i)\right) = \frac{\prod_{k=1}^{2^q} \prod_{m=0}^{s_k-1}(m + \lambda/2^q)}{\prod_{m=0}^{n-1}(m + \lambda)}.$$

Carnap's Continuum continues to the present to be highly influential in Inductive Logic with a number of attractive properties. For example, through JSP it has just the sort of rational justification we are seeking, the c_λ^L can be specified as above by simple algebraic identities which generally makes calculating values comparatively easy (for example in verifying that they satisfy PIR), and, again through satisfying JSP, the c_λ^L satisfy Ax and Ex.

For $0 < \lambda < \infty$ they also satisfy two other principles which were not obviously covered by the considerations of symmetry, relevance, and irrelevance discussed in the previous section. The first of these is Reichenbach's Axiom which asserts that as we successively accumulate more and more evidence, $\alpha_{h_1}(a_1), \alpha_{h_2}(a_2), \alpha_{h_3}(a_3), \ldots$ the conditional probability assigned to $\alpha_k(a_n)$ on the basis of the $\alpha_{h_1}(a_1), \alpha_{h_2}(a_2), \alpha_{h_3}(a_3), \ldots, \alpha(a_{n-1})$ should converge to the proportion of these earlier instances which were $k = h_i$. Precisely:

Principle 16.5.1 (Reichenbach's Axiom, RA). Let $\alpha_{h_i}(x)$ for $i = 1, 2, 3, \ldots$ be an infinite sequence of atoms of L. Then for $\alpha_k(x)$ an atom of L,

$$\lim_{n\to\infty}\left(w(\alpha_k(a_{n+1}) | \bigwedge_{i=1}^{n} \alpha_{h_i}(a_i)) - \frac{u(n)}{n}\right) = 0$$

where $u(n) = |\{i | 1 \le i \le n \text{ and } h_i = k\}|$.[16]

A second principle that the c_λ^L satisfy, now for the whole range $[0, \infty]$ of λ is that they are members of a *unary language invariant family for JSP*. In order to explain this desiderata we need to take a step back.

Suppose we have settled on a probability function w on a language L because it satisfies a certain principle(s), $\mathcal{P}$ say, and we now enlarge L to L^+, an apparently reasonable possibility since we would have no reason to suppose that L encompassed all the relations there could ever be. In that case it would seem to be a serious weakness on the part of w if it did not have an extension w^+ to L^+ (meaning that w^+ restricted to $SL \subseteq SL^+$ was w) which also satisfied $\mathcal{P}$ as it applies to this extended language L^+.

Call a class of probability functions w^L on L for each language L a *language invariant family*[17] if whenever languages L_1, L_2 are such that L_1 is a sublanguage of L_2 then w^{L_1} is w^{L_2} restricted to SL_1. We say this is a language invariant family for $\mathcal{P}$ if all the w^L satisfy $\mathcal{P}$.

For w a probability function on L satisfying $\mathcal{P}$, w satisfies *Language Invariance for $\mathcal{P}$* if there is a language invariant family satisfying $\mathcal{P}$ which includes w (i.e. w is the w^L in this family). Similarly *Unary Language Invariance for $\mathcal{P}$* is defined in the same way except that we restrict ourselves throughout to unary languages. Clearly then the argument for w satisfying Language Invariance for $\mathcal{P}$ is hardly less forceful than the argument that w in isolation should satisfy $\mathcal{P}$.

Following that diversion we can now clarify our earlier remarks, for each $\lambda \in [0, \infty]$ and unary language L c_λ^L satisfies Language Invariance for JSP, or Ax, namely a suitable language invariant family is just the class of all probability functions c_λ^L for this same λ and unary L.

The c_λ^L do not however satisfy Weak Irrelevance or GPIR in the case $0 < \lambda < \infty$, though in the context it is debatable whether this failure of Weak Irrelevance is not actually desirable (see [Hill et al., 2002], [Nix and Paris, 2006]).

Somewhat surprisingly the c_λ are not the only 'continuum of inductive methods' based on arguably rational principles: The requirement that the probability function w on the unary language L satisfies Unary Language Invariance for GPIR + Ax + Regularity, where *Regularity* means that w does not give probability zero to any consistent sentence of L, forces w to be a member of a different continuum of inductive methods, w_L^δ for $\delta \in [0, 1)$, and conversely (see [Nix and Paris, 2006]).

Again as with the c_λ^L these probability functions have a simple form:

$$w_L^\delta\left(\bigwedge_{i=1}^{n} \alpha_{h_i}(a_i)\right) = \frac{\gamma^n}{2^q} \sum_{k=1}^{2^q} \left(1 + \frac{\delta}{\gamma}\right)^{s_k}$$

where $2^q\gamma = 1 - \delta$. They furthermore satisfy the Weak Irrelevance Principle, WIP, but fail to satisfy Reichenbach Axiom (see [Nix and Paris, 2006]).

The purpose of highlighting this w_L^δ continuum here is not particularly to promote it but to note firstly that Carnap's Continuum is not alone in being derivable from seemingly rational principles and second that such principles, despite the apparent intuition behind them, may well turn out to contradict each other.

6. de Finetti's Theorem

In this section we shall continue to restrict attention to purely unary languages and discuss a theorem due to de Finetti, ([de Finetti, 1974]), which has proved to be of inestimable value in the context of Inductive Logic.

Before stating the theorem it will be useful to develop a little notation. As usual let L be a unary language with q predicates and let

$$\mathbb{D}_q = \{\langle x_1, x_2, \ldots, x_{2^q}\rangle \mid x_i \geq 0 \text{ for } i = 1,2,\ldots,2^q \text{ and } \sum_{i=1}^{2^q} x_i = 1\}.$$

For $\vec{e} = \langle e_1, e_2, \ldots, e_{2^q}\rangle \in \mathbb{D}_q$ let $y^{\vec{e}}$ be the probability function on L defined by

$$y^{\vec{e}}\left(\bigwedge_{i=1}^{n} \alpha_{h_i}(b_i)\right) = e_{h_1} e_{h_2} e_{h_3} \ldots e_{h_n} = \prod_{k=1}^{2^q} e_k^{s_k}$$

where for $1 \leq k \leq 2^q$, $s_k = |\{i \mid h_i = k\}|$. In other words $y^{\vec{e}}$ just corresponds to a Bernoulli process where each $\alpha_k(b_i)$ has probability e_k and for different i these are stochastically independent.

These $y^{\vec{e}}$ satisfy Ex and de Finetti's Theorem says that in fact *any* probability function on L satisfying Ex must be a mixture of these very simple $y^{\vec{e}}$:

Theorem 16.6.1 (de Finetti's Representation Theorem). *If the probability function w on the unary language L satisfies Ex then there is a (normalized) probability measure μ on $\mathbb{D}_q$, the de Finetti prior for w, such that*

$$w\left(\bigwedge_{i=1}^{n} \alpha_{h_i}(b_i)\right) = \int_{\mathbb{D}_q} y^{\vec{x}}\left(\bigwedge_{i=1}^{n} \alpha_{h_i}(b_i)\right) d\mu(\vec{x})$$

$$= \int_{\mathbb{D}_q} \prod_{k=1}^{2^q} x_k^{s_k} d\mu(\vec{x}),$$

where for $1 \leq k \leq 2^q$, $s_k = |\{i \mid h_i = k\}|$. Conversely any probability function w on L defined in this way satisfies Ex.[18]

The value of this result is firstly that it tells us precisely what probability functions on L satisfying Ex look like, and how to make them to suit particular needs, and second it can enable us to answer questions about Ex by translating them into questions about integrals where we already have the well-developed theory of the integral calculus to call on.

An example of this is Gaifman's result, see [Gaifman, 1971], that in fact Ex alone implies the Principle of Instantial Relevance (in other words that all predicates are projectible). This follows because once the inequality is expressed in terms of integrals it simply becomes a version of the well known Schwartz Inequality. Similarly in the same paper Gaifman uses this representation theorem to elucidate when a probability function satisfying Ex can give non-zero probability to a non-tautological universal sentence $\forall\vec{x}\,\theta(\vec{x})$ with $\theta(\vec{x})$ quantifier free and not mentioning any constants, an issue we shall return to in a later section.

In the case of the member c_λ^L of Carnap's Continuum, for $0 < \lambda < \infty$ the measure μ in the de Finetti Representation turns out to be given by

$$d\mu(\vec{x}) = \kappa \prod_{k=1}^{2^q} x_k^{\lambda 2^{-q}-1}\, d\vec{x}$$

where κ is a normalizing constant. This may be seen as shedding some light on a question which Carnap considered at length, 'given this continuum which value of λ should we settle on in order to make the final step to a unique rational probability function?'. For on some vague grounds of 'indifference' one might feel that the fairest or least informative μ here would be the uniform distribution, which corresponds to $\lambda = 2^q$. Unfortunately however if we want language invariance we have to keep λ fixed, so any such argument *for* $\lambda = 2^q$ for a unary L with q predicate symbols is itself an argument *against* the corresponding choice for a language with any other number of predicate symbols! To put it another way if our language L has q predicates and we take the choice of measure μ in de Finetti's Representation Theorem to be the uniform measure then we will obtain the Carnap's $c_{2^q}^L$, which would seem to give this choice some *special status*. However if we consider the restriction of this probability function $c_{2^q}^L$ to a sublanguage L^- of L with $q-1$ predicates then we obtain Carnap's $c_{2^q}^{L^-}$, which is not the same as the corresponding *special status* $c_{2^{q-1}}^{L^-}$ for L^-.

We shall briefly mention some further de Finetti style representation theorems in the sections to come but first we need to move out of purely unary languages.

7. Polyadic Inductive Logic

The development of Inductive Logic by Johnson, Carnap et al., (see for example [Carnap, 1950], [Carnap, 1952], [Carnap and Jeffrey, 1971], [Carnap, 1980],

[Johnson, 1932]) was almost entirely set in the context of purely unary languages, though it is clear from brief comments by Carnap [Carnap, 1950, pp. 123–4] and Kemeny [Kemeny, 1963] that extending these results to the polyadic, where there were binary, ternary etc. relations as well as unary predicates, was a future intention of the programme. However, apart from somewhat isolated papers by Hoover [Hoover, 1979] and Krauss [Krauss, 1969] around 1970 the 'challenge' of Polyadic Inductive Logic remained largely unaddressed until the work in [Nix and Paris, 2007] from the start of this millennium.

The primary reason for that hiatus was, as previously mentioned, the disheartening effect of Goodman's 'grue' Paradox on the programme as a whole. A second possible reason for the slow development of Polyadic Inductive Logic however is its relative isolation from mainstream Philosophy. For not only does it require much more technical mathematics but practical examples of inductive reasoning with higher arities are far less frequent, so that in turn our intuitions about what is rational are less finely developed.

Nevertheless on occasions we do appear to happily apply some such reasoning. For example if Adam the Gardener knows that apple trees of variety X are good pollinators and apples of variety Y are easily pollinated he might well conclude that planting them together is likely to be fruitful.

Again as with the unary case we seek to propose rational principles that a probability function w on a, now polyadic, language L should satisfy. Several such principles based on symmetry considerations, for example Constant[19], Predicate, and Variable Exchangeability, have already been mentioned, see for example [Nix and Paris, 2007].

To pursue this generalization of the unary case any further however we need to consider generalizations of Atom Exchangeability to the polyadic. A key difference between the unary and (properly) polyadic at this juncture is that in the former knowing the state description

$$\bigwedge_{i=1}^{n} \alpha_{h_i}(b_i) \tag{16.3}$$

satisfied by $b_1, b_2, \ldots, b_n$ tells us all there is to know about $b_1, b_2, \ldots, b_n$, at least as far as quantifier free sentences are concerned. However once L contains, say, a binary relation symbol R, knowing the state description $\Theta(b_1, b_2, \ldots, b_n)$ tells us nothing about whether or not $R(b_1, b_{n+1})$ etc. holds.

One such generalization can be motivated as follows. Given a state description $\Phi(b_1, b_2, \ldots, b_n)$ as in (16.3) define $\sim_\Phi$ to be the equivalence relation on $\{b_1, b_2, \ldots, b_n\}$ given by

$$b_i \sim_\Phi b_j \iff b_i, b_j \text{ are indistinguishable according to } \Phi(b_1, b_2, \ldots, b_n),$$

where indistinguishability was defined earlier when the probability function w_0 was introduced. Let the *Spectrum of* $\Phi(b_1, b_2, \ldots, b_n)$ be the multiset of sizes of the equivalence classes with respect to $\sim_\Phi$.

For example in the case of a language with a single binary relation R, the state description $\Psi(a_1, a_2, a_3, a_4)$ given by the conjunctions of

$$\begin{array}{rrrr} \neg R(b_1,b_1) & R(b_1,b_2) & R(b_1,b_3) & \neg R(b_1,b_4) \\ R(b_2,b_1) & \neg R(b_2,b_2) & \neg R(b_2,b_3) & R(b_2,b_4) \\ R(b_3,b_1) & \neg R(b_3,b_2) & \neg R(b_3,b_3) & R(b_3,b_4) \\ R(b_4,b_1) & R(b_4,b_2) & R(b_4,b_3) & R(b_4,b_4) \end{array}$$

has spectrum $\{2,1,1\}$, since b_2, b_3 are indistinguishable according to $\Psi(b_1, b_2, b_3, b_4)$ but all the rest are distinguishable.

Now for purely unary languages Ax + Ex is equivalent to the assertion that for any two state descriptions $\Theta(b_1, b_2, \ldots, b_n), \Phi(b'_1, b'_2, \ldots, b'_n)$ with the same spectra,

$$w(\Theta(b_1, b_2, \ldots, b_n)) = w(\Phi(b'_1, b'_2, \ldots, b'_n)).$$

Simply generalizing this to the polyadic language L gives:

Principle 16.7.1 (Spectrum Exchangeability Principle, Sx). For state descriptions $\Theta(b_1, b_2, \ldots, b_n), \Phi(b'_1, b'_2, \ldots, b'_n)$ with the same spectra,

$$w(\Theta(b_1, b_2, \ldots, b_n)) = w(\Phi(b'_1, b'_2, \ldots, b'_n)).$$

Unlike the earlier exchangeability principles it is not known if Spectrum Exchangeability can be justified in terms of symmetry (in a sense to be made clear shortly) and its current primary justification is that in the presence of Ex it generalizes Ax. A secondary 'justification' however is that it has a number of nice properties which considerably simplify[20] Polyadic Inductive Logic (see [Landes et al., 2008], [Landes et al., ta] for recent surveys). For example there are de Finetti style representation theorems (see [Landes et al., 2009b], [Paris and Vencovská, 2009]) and an Instantial Relevance Property (see [Landes et al., 2009a]). Furthermore for fixed λ and δ, both Carnap's c^L_λ and the w^δ_L extend to language invariant families for Sx for polyadic as well as unary languages L though the obvious generalization of Johnson's Sufficientness Principle to polyadic L now has but two solutions, w^L_0 and w^L_∞, and so no longer characterizes these probability functions (see [Landes, 2009], [Vencovská, 2006]).

8. Symmetry

In the early discussion we treated symmetry on a par with relevance and irrelevance. However whereas these latter appear to require digging into one's intuitions, symmetry seems to be an altogether more formal notion. A symmetry is an 'isomorphism of the language' and it begets a principle by imposing the condition that a rational probability function should be invariant under this isomorphism.

One version of what we might mean by 'an isomorphism of the language' can be explained as follows. Firstly we have tacitly assumed that the constant symbols $a_1, a_2, a_3, \ldots$ exhaust the universe so the overlying worlds we have in mind should be structures M for the language L with universe these $a_1, a_2, a_3, \ldots$ and each constant symbol a_i interpreted in M as itself. Let $\mathcal{T}$ be the set of such structures for L. Then one could argue that an 'isomorphism σ of L' should be a bijection on $\mathcal{T}$, mapping structures in $\mathcal{T}$ one to one onto the structures in $\mathcal{T}$ and should preserve the semantics in the sense that any subset of $\mathcal{T}$ of the form

$$\{M \in \mathcal{T} | M \models \theta\}$$

for some $\theta \in SL$ should be mapped by σ to a set of the same form, i.e. for some $\phi \in SL$ we should have

$$\{\sigma(M) \in \mathcal{T} | M \models \theta\} = \{M \in \mathcal{T} | M \models \phi\}, \tag{16.4}$$

and conversely for any $\phi \in SL$ there should be some $\theta \in SL$ such that (16.4) holds.

In this case we can, unambiguously up to logical equivalence, write $\sigma(\theta) = \phi$.

Having settled on this formulation of an isomorphism of L we can propose a very general Invariance Principle:

Principle 16.8.1 (The Invariance Principle, INV). For σ an isomorphism of L and $\theta \in SL$,

$$w(\theta) = w(\sigma(\theta))$$

its rationality being based, as with Ex, Ax etc, on the ground that it would be irrational for assigned probabilities to break such a symmetry.

A natural question to ask at this point is whether INV is actually even consistent, might it not be that the conditions imposed by INV are actually so strong that *no* probability function could satisfy them all? Fortunately the answer to that (in the case dealt with here of zero evidence) is that INV is consistent (see [Paris and Vencovská, ta]): the probability function w_0^L satisfies INV.

Each of the previously proposed symmetry principles are special cases of INV, which raises the question whether they exhaust the possibilities or whether there are other symmetry principles waiting to emerge. In the case of purely unary L the answer is yes, indeed they set such demands that in that case w_0^L is the *only* probability function satisfying INV (see [Paris and Vencovská, ta]).

Here then we again find that imposing rational principles has cut down the possibilities to a small family, in this case a singleton, though it is perhaps not the choice we would most like to have been left with.[21]

These results in the unary case suggest INV is too strong a principle, though it is hard to see what extra conditions one could reasonably impose on an 'automorphism of L' to address that complaint. It is currently not clear if this situation is replicated in the case of genuinely polyadic L. As in the unary case particular families of automorphisms of L have led to the formulation of new symmetry principles for the polyadic though these have not yet been seriously studied (see [Paris and Vencovská, shed]).

Overall then it would appear that we may have some way to go in understanding symmetry principles. In the next section we mention another notion, this time related to relevance, which despite some effort seems to be proving problematic to properly formalize.

9. Analogical Reasoning

Recall that in the case of a unary language L the Principle of Instantial Relevance gives us that for atoms $\alpha(x), \beta(x)$, and $\phi \in QFSL$ not mentioning the distinct constants a_i, a_j,

$$w(\alpha(a_i)|\beta(a_j) \wedge \phi) \geq w(\alpha(a_i)|\phi) \tag{16.5}$$

when $\alpha(x) = \beta(x)$. Indeed this is a consequence of Ex since PIR follows from that symmetry principle.

However one could argue that even if $\beta(x)$ is not actually equal to $\alpha(x)$, $\beta(a_j)$ should nevertheless 'by analogy' provide more support for $\alpha(a_i)$ if $\beta(x)$ is close to $\alpha(x)$ than if it is far away, where (say) the distance between them is the number of predicates $R_i(x)$ which $\alpha(x), \beta(x)$ decide differently: i.e., if say $q = 5$ and

$$\alpha(x) = R_1(x) \wedge \neg R_2(x) \wedge R_3(x) \wedge R_4(x) \wedge \neg R_5(x),$$

$$\beta(x) = R_1(x) \wedge R_2(x) \wedge \neg R_3(x) \wedge R_4(x) \wedge \neg R_5(x),$$

then this distance (commonly referred to as the *Hamming distance* and written $|\alpha(x) - \beta(x)|$) would be 2 since $\alpha(x), \beta(x)$ differ just on the two predicates $R_2(x), R_3(x)$. This suggests the following principle for unary L:

Principle 16.9.1 (The Analogy Principle, AP). For atoms $\alpha(x), \beta(x), \gamma(x)$, and $\phi \in QFSL$ not mentioning the distinct constants a_i, a_j, if $|\alpha(x) - \beta(x)| < |\alpha(x) - \gamma(x)|$ then

$$w(\alpha(a_i)|\beta(a_j) \wedge \phi) > w(\alpha(a_i)|\gamma(a_j) \wedge \phi).$$

Notice that this principle is inconsistent with Ax (and hence JSP) since Ax would give straight equality in the conclusion when $\beta(x), \gamma(x) \neq \alpha(x)$ and ϕ was a state description with no instances of $\beta(x)$ or $\gamma(x)$. In [Hill and Paris, shed] it is shown that AP *is* consistent when L has at most two predicates but conjectured that this fails for three or more predicates. Together with AP a number of other attempts have been made to elucidate the intuitively appealing idea of 'analogical support', for example [Festa, 1996], [Maher, 2001], [di Maio, 1995], [Romeijn, 2006], [Skyrms, 1993], mostly involving variations on the functions in Carnap's Continuum, but currently we still seem short of properly capturing this notion, if it is even possible at all.

10. Universal Certainty

Given a consistent formula $\theta(x)$ one's natural feeling might be that even in the absence of any evidence there should be a non-zero probability that $\forall x \theta(x)$ held. However for unary L and $\theta(x)$ not mentioning any constants this fails for Carnap's c_λ^L when $0 < \lambda \leq \infty$. Indeed this will be the case for *any* probability function w on L whose de Finetti prior μ gives measure zero to the sets

$$\{\langle x_1, x_2, \ldots, x_{2^q} \rangle \in \mathbb{D}_q | x_i = 0\} \tag{16.6}$$

for $i = 1, 2, \ldots, 2^q$ (see [Dimitracopoulos et al., 1999, p. 36] for a discussion in the present notation). But from this angle it is the condition (16.6) which looks rather natural since to flout it would require μ to give non-zero measure to a set of points with dimension less than that of $\mathbb{D}_q$. Indeed if w is to go all the way to addressing the problem of not giving zero probability to such sentences $\forall x\, \theta(x)$ then μ would have to put non-zero measure on the single points $\langle 0, 0, \ldots, 0, 1, 0, \ldots, 0, 0 \rangle$ in $\mathbb{D}_q$.

Proposals have been made concerning families of probability functions fulfilling this requirement but a stronger case for their justification, or for that of alternatives, on grounds of rationality would be welcome (see [Dimitracopoulos et al., 1999], [Earman, 1992, p. 87] [Hintikka, 1965], [Hintikka, 1966], [Paris, 2001]).

11. Conclusion

We have presented here a view of Inductive Logic that develops the original programme of Carnap, but with a somewhat different emphasis: Our aim is to investigate rational principles for assigning subjective probabilities and the relationships between them rather than seeking a practical method or formula for assigning or estimating possibly even objective probabilities. Most of the present-day developments in 'Inductive Logic' continue to follow that latter path, and perhaps to avoid future confusion it is time for a division into Pure and Applied Inductive Logic.

Despite the differences the current outstanding problems are shared. One is tidying up our understanding of analogical reasoning, a second is elucidating some widely acceptable insights into the problem of universal generalizations as explained in the previous section. At the same time Polyadic Inductive Logic is in its infancy. Currently only a handful of principles there have been studied in any depth; surely there are further insights and principles awaiting discovery and investigation. There is also the wider question of finding grand, overarching, principles which capture generic considerations such as symmetry, relevance, and irrelevance. For symmetry we have speculated the principle INV, but relevance and irrelevance remain at present elusive.

Acknowledgements

I would like to thank Alex Hill, Richard Pettigrew, and Alena Vencovská for reading and improving earlier drafts of this chapter.

Notes

1. Generally we have employed the term 'knowledge' rather than 'evidence' in this context. However to avoid the possibility of any distracting epistemological side issue we shall use the latter expression in this account.
2. As opposed to Applied Inductive Logic, in the same fashion that Pure Mathematics relates to Applied Mathematics.
3. This aspiration is well illustrated in Propositional Uncertain Reasoning where for an analogously simplified framework there are a number of arguments to the effect that if an agent is to be 'rational' then its inferences should necessarily be made according to Maximizing Entropy, see for example [Cox, 1979], [Grove et al., 1994], [Paris, 1999], [Paris and Vencovská, 1989], [Paris and Vencovská, 1990], [Paris and Vencovská, 2001], [Shore and Johnson, 1980], [Williamson, 2010]. However these results are as much advisory as prescriptive, namely advising that if the agent does not use Maximum Entropy then it must be flouting some 'rationality' requirement.
4. Thus it will be invalid to criticize subsequent conclusions by saying, 'Well what if R stands for ...?' when this bears on properties of R not already included in the initial evidence, and similarly for the constants a_j.
5. On this point see also Chapter 15.
6. So if $r_1 = 1$ then R_1 is a unary relation or predicate symbol, if $r_1 = 2$ then R_1 is a binary relation symbol etc.
7. Various justifications for this involving appeals to accuracy, scoring rules, and the Dutch Book Argument may be found in Chapter 15.
8. Again this may be justified by a diachronic Dutch Book argument, see for example [Lewis, 1980] or [Teller, 1976].

9. To avoid worrying about zero denominators we henceforth adopt the convention that an identity such as (16.1) actually stands for the well defined

$$w(\theta \mid \phi_1, \phi_2, \ldots, \phi_m) \cdot w\left(\bigwedge_{i=1}^{m} \phi_i\right) = w\left(\theta \wedge \bigwedge_{i=1}^{m} \phi_i\right).$$

10. Even if one was unwilling to accept the 'received wisdom' this would still figure as the obvious first issue to resolve.
11. Called *Q-predicates* by Carnap.
12. This is referred to as the *Completely Independent Probability Function* in [Paris, 1994] and as we shall see corresponds for unary languages to Carnap's c_∞.
13. We shall later say more precisely what might be meant by this assertion, for the moment we will simply take it as intuitively clear.
14. Or

$$w(\theta \wedge \phi \wedge \psi) \cdot w(\phi) = w(\theta \wedge \phi) \cdot w(\phi \wedge \psi)$$

if we wish to avoid any danger that these conditional probabilities are not well defined.
15. This is a slightly simplified version of the original Principle of Instantial Relevance given by Carnap in [Carnap and Jeffrey, 1971, Section 13].
16. Notice that unless $u(n) = n/2^q, 0, n$ none of the c_λ^L can satisfy that $c_\lambda^L(\alpha_k(a_{n+1}) \mid \bigwedge_{i=1}^{n} \alpha_{h_i}(a_i))$ is *exactly* $u(n)/n$ since that would require $\lambda = 0$ in which case the conditional probability would not even be defined.
17. These L are still assumed to be of the form considered in this chapter, namely having just constants $a_1, a_2, \ldots$ and finitely many relations.
18. The same result holds for Ax in place of Ex, provided we require the measure μ to be invariant under permutation of the 2^q coordinates.
19. In [Hoover, 1979] (or see the more easily available [Kallenberg, 2005, Section 7.6]) Hoover gives a representation theorem for probability functions satisfying Constant Exchangeability (or Array Exchangeability as it is more usually referred to within Probability Theory).
20. In many branches of mathematics axioms or principles are esteemed for their widespread applications and power to clarify and bring order to the area. For example the Riemann Hypothesis.
21. It is worth pointing out here that if instead we had started with the evidence $R_1(a_1) \wedge \neg R_1(a_2)$ and considered only automorphisms which fix the set of $M \in T$ such that $M \models R_1(a_1) \wedge \neg R_1(a_2)$ then the corresponding Invariance Principle would have been inconsistent, no probability function could satisfy it.

17 Belief Revision

Horacio Arló Costa and Arthur Paul Pedersen

Chapter Overview

1. Introduction	451
1.1 Historical Remarks	456
1.2 The AGM Model	456
1.3 Technical Preliminaries	457
2. Contraction	458
2.1 Partial Meet Contraction	458
2.2 Entrenchment-Based Models	461
3. Revision	462
3.1 Partial Meet Revision	463
3.2 Propositional Models	465
3.2.1 Sphere-Based Revision	466
3.2.2 The Grove Connection, and Geometric Depictions of Belief Change	467
3.2.3 Persistent Revision	471
3.3 Belief Change and Rational Choice	472
4. Doubts about Recovery, and Some Reactions	478
4.1 Levi Contractions	479
4.2 Mild Contractions and Severe Withdrawals	482
4.3 Belief Base Contraction	486
5. Doubts about Other Postulates	488
6. Probability, Belief; Belief Change and Supposition	492
6.1 Core Dynamics and Matter-Of-Fact Supposition	494
6.2 Update, Imaging and Subjunctive Supposition	495
7. Epistemic States vs. Belief Sets: The Problem of Iteration	496
7.1 Special Axioms for Iteration	497

7.2 Other Approaches to Iteration	498
7.3 Which Axioms are Correct?	500
Notes	501

1. Introduction

David is a professor at Carnegie Mellon University. This summer he has committed himself to his chilled, quiet office to prepare the final chapters of a draft of his book. He will get the damn thing done.

Still, just as much now as with the rest of year, David often seeks conversation with colleagues both for pleasure and information, to keep his mind relaxed but sharp, motivated but deliberate. Unfortunately, two of David's most valuable interlocutors, Kevin K. and Kevin Z., are out of town. David knows that Kevin K. spends his summers in San Francisco with his family, while Kevin Z., a year-round Pittsburgh resident, happens to be visiting Irvine. When David hears from the department chairman that Kevin Z. has just arrived on campus, he is delighted. He remembers that they were having a conversation that was interrupted and never finished. Some important issues about his plans for the last chapter of his book were at stake, so he looks forward to continuing where they left off.

It turns out that the chairman was wrong. David learns that Kevin K. is on campus, visiting for the weekend for a conference. David asks around about the whereabouts of Kevin Z. and is told that he will be in Irvine for at least another week. David thereby abandons his recently acquired belief that Kevin Z. is on campus, reverting to his initial belief that Kevin Z. is in Irvine. Last time he spoke with Kevin K. they spent most of their conversation talking about subtle and interesting connections between their work.

The story is mundane and simple. But actually there are representations of belief according to which this epistemic story is impossible. Suppose that we represent David's beliefs using a probability measure, a mapping from propositions in some field to $[0, 1]$ measuring David's degrees of belief. Thus, say that David assigns a high degree of belief to the proposition expressed by 'Kevin K. is currently residing in San Francisco.' According to the orthodox Bayesian story, there is a precise number that measures David's degree of belief in this proposition, say, 0.935. According to a less orthodox Bayesian account, there is at least a probability interval measuring David's degrees of belief, say, the interval $[.8, .95]$.

When David learns that Kevin Z. is in town he modifies his beliefs using an operation called *conditionalization* according to which the new probability of the proposition expressed by 'Kevin Z. is currently in Pittsburgh' shifts from a low value to exactly one. Unfortunately, one of the properties of conditionalization is that when a proposition acquires value one there is no proposition one can subsequently learn by conditionalization that can modify this value. After you acquire certainty, you will remain certain forever. Why is this so? We need some definitions to explain this peculiar feature of conditionalization.

Let's start with the basic idea that propositions are sets of possibilities selected from a primitive space W of possibilities. We shall remain silent about the nature of the points in W.[1] Propositions, denoted by the letters A, B, C, etc., are subsets of W.

What basic structural features should we require a collection of propositions to satisfy? A mild requirement is that the set of propositions in question is closed under logical operations – that it forms an algebra. We will use the notation $\overline{A}$ to denote the absolute complement of a proposition A; $\subseteq$ to denote subset inclusion; and $\subset$ to denote proper subset inclusion. We appeal to the usual symbols for intersection and union. We can now make some of the foregoing ideas more precise.

Definition 17.1.1 A collection $\mathscr{A}$ of subsets of a set W is called an *algebra* of sets (or *field* of sets) over W if it contains W itself and is closed under the formation of complements and finite unions:

(i) $W \in \mathscr{A}$;
(ii) If $A \in \mathscr{A}$, then $\overline{A} \in \mathscr{A}$;
(iii) If $A, B \in \mathscr{A}$, then $A \cup B \in \mathscr{A}$.

The collection $\mathscr{A}$ is called a σ*-algebra* of sets (or a σ*-field* of sets) over W if it is an algebra and it is also closed under countable unions:

(iv) For every collection $\{A_n\}_{n=1}^{\infty}$ with $A_n \subseteq \mathscr{A}$, $\bigcup_{n=1}^{\infty} A_n \in \mathscr{A}$.

We call an element A of $\mathscr{A}$ a *proposition* (or an *event*) from $\mathscr{A}$.

Of course, we may omit reference to the underlying set W or collection of sets $\mathscr{A}$ when there is no danger of confusion. The distinction between algebra and σ-algebra is relevant when W is infinite, collapsing otherwise.

Now that we have established how we can represent objects of belief, we can introduce the classical axioms of probability.

Definition 17.1.2 Let $\mathscr{A}$ be an algebra over W. A *probability measure* on $\mathscr{A}$ is a non-negative, normalized, and finitely-additive real-valued function P on $\mathscr{A}$:

Non-Negativity $P(A) \geq 0$ for every $A \in \mathscr{A}$;
Normalization $P(W) = 1$;
Finite Additivity For every $A, B \in \mathscr{A}$ such that $A \cap B = \emptyset$, $P(A \cup B) = P(A) + P(B)$.

If $\mathscr{A}$ is in addition a σ-algebra over W, then P is a *σ-additive probability measure* on $\mathscr{A}$ if it is a probability measure and for every sequence $\{A_n\}_{n=1}^{\infty}$ of pairwise disjoint propositions in $\mathscr{A}$,

(σ-additivity) $P(\bigcup_{n=1}^{\infty} A_n) = \sum_{n=1}^{\infty} P(A_n)$.

These axioms (first proposed by Kolmogorov) characterize a monadic notion of probability. Conditional probability can then be defined in terms of monadic probability:

Definition 17.1.3 Let P be a probability measure on an algebra $\mathscr{A}$, and let $A, B \in \mathscr{A}$ be such that. Then the *conditional probability of B given A*, $P(B|A)$, is defined as

$$P(B|A) := \frac{P(A \cap B)}{P(A)},$$

provided $P(A) > 0$ and is undefined otherwise.

Both notions of probability are purely synchronic. Why should one adopt these axioms and definitions? There are ingenious arguments offering justification for these axioms if one interprets probability as degrees of belief but we cannot enter into this issue here.

What about learning? Many Bayesians would propose that one learns by conditioning. So, the result of updating a probability function P with a proposition A, denoted P_A, can be defined as follows: $P_A(B) = P(B|A)$ and in general for conditional probability: $P_A(X|Y) = P(X|Y \cap A)$.

It is clear from this definition that $P_A(A) = 1$. So, after updating with a proposition A, the probability of A is raised to exactly the value one. Suppose now that you want to update P_A with an arbitrary proposition C. Then we will have that for any proposition B, its value will be $P_A(B|C)$, i.e., we will have $P(B|C \cap A)$. In particular when C is A we have: $P_A(A|C) = P(A|C \cap A) = 1$. So, after learning A its value is raised to 1 and after that the result of updating P_A with any other proposition will not change this fact. You will continue to be certain that A is the case. Moreover updating with A and then with its complement is tantamount to learning a contradiction. And this either leads to incoherence or is undefined.

In spite of that it seems that in many circumstances, for example as a result of an error, one can receive information saying that A is the case, and then learn that this is false. Unfortunately this is not representable by using probability functions. In general one limitation of the notion of probability we just presented is that one cannot learn a proposition of probability zero. Conditioning is just undefined in this case.

There are some remedies for this problem within the boundaries of a probabilistic framework. One of them (perhaps the most fruitful) is to assume conditional probability as a primitive rather than deriving it from monadic probability. This makes possible to condition with events of measure zero but still most accounts of this type will assume that updating a conditional probability function is defined as follows: $P_A(X|Y) = P(X|Y \cap A)$. And this puts constraints on possible iterated updates as we explained above.

Alternatively Richard Jeffrey proposed a modification of conditioning that is a generalization of conditioning. The main epistemological idea is that when we receive information from the environment the probabilities might increase or decrease but never increase to one or decrease to zero. So, when you learn that Kevin just arrived to campus the probability that Kevin Z. is on campus shifts to a high value strictly less than one. This is more flexible than conditioning but ultimately Jeffrey's proposal does extend conditioning. If your probabilities increase up to one, then this is irreversible. Jeffrey conditioning has other problems as well: for example, unlike conditioning, it is path dependent.

The limitations of the probabilistic model of learning and supposing motivated researchers to think about the problem of belief change in a non-probabilistic setting. Consider again the previous example. One can represent David's beliefs in a purely qualitative way. For example one can focus on a propositional language L and one can use sentences of L to represent beliefs. So, for example one can use the sentence A to represent the fact that David believes that Kevin Z. is not in Pittsburgh at the moment and we can use the sentence B to represent the fact that Kevin K. is not in Pittsburgh at the moment. More generally, David's *belief set* K will contain all sentences that David believes at a certain time t.

There are certain decisions one should make about the structure of K. The simplest assumption is that this set contains all the sentences explicitly believed by David at t. Presumably this is a finite set rather unstructured logically. If instead we use K to represent David's *doxastic commitments* then one can argue that this set should be logically closed. If I believe A and A entails B then I might not be aware of B but in certain sense I am committed to believing B.

Let's abstract for the moment from the problem of finding a relation between this type of qualitative model and the probabilistic model presented above. This is a complicated problem that we will consider below. To give the reader an idea of why this is a complicated problem, let's consider $\neg A$. Previously we said that David attributes a high probability to this sentence (or to the proposition expressed by this sentence). Should we include in K exactly the sentences that carry high probability? We could do so, but then K will not be closed under logical consequence. It is easy to see that even when A and B might carry high probability their conjunction might not carry high probability. Should we include in K only the sentences carrying probability one? It is unclear whether belief (even full belief or certainty) corresponds exactly with measure one sets. Many philosophers think that full beliefs carry probability one but that there are sentences carrying measure one that are not necessarily fully believed.

The relations between belief (full belief) and probability are not straightforward. So, many researchers in belief revision have proceeded independently of probability when they use belief sets. They assume some notion of belief as a primitive (full belief, plain belief) and use belief sets to represent the corresponding doxastic commitments. There are today models capable of providing bridges between the probabilistic model and this qualitative model. We will review them at the end of this note.

Let's go back to belief sets then and in particular to David's belief set K. A is in K representing the fact that David believes that Kevin Z. is not in Pittsburgh at certain time t. Then the chairman (an authoritative oracle we can suppose) tells David that $\neg A$. Obviously this sentence is inconsistent with K. Moreover this sentence might be entailed by a number of other sentences in K (for example, the sentence stating that Kevin is in Irvine attending a conference, that the conference will last for one week and that he departed yesterday). If David wants to introduce $\neg A$ in his belief set preserving consistency it seems that he needs to eliminate A from it. But simply deleting A would not do. K is logically closed and A is entailed by other sentences. So, the operation of *contracting* A from K is not straightforward. It seems that in order to perform it David has to make some choices that are not completely determined by logic.

Notice that once one manages to remove A from K the introduction of $\neg A$ to this contracted set (which we can denote by $K \dot{-} A$) is indeed straightforward. One just has to add $\neg A$ set-theoretically to $K \dot{-} A$ and take the corresponding logical closure. This addition operation is usually called *expansion* and the composition of the contraction of K with A and the expansion with $\neg A$ is usually called *revision*.

The theory of belief change is largely the corresponding theory of contraction and revision (taken as an epistemological primitive). Are there interesting axioms that are obeyed by these operations? Are there clear procedures to construct revisions and contractions? Is it possible to prove representation results for a given axiomatic base in terms of these constructive procedures (contractions)?

Obviously in order to construct a concrete theory of contraction (revision) one has to make crucial assumptions as to what is an epistemic state and what is its logical structure. If we decide to represent the dynamic of explicit belief presumably we will work with *belief bases*, i.e., mere sets of sentences. Commitment sets for various attitudes would be logically closed. Moreover, one might think that an epistemic state is something more complex than a belief set of a belief base. Perhaps one should add to the representation other elements like an entrenchment ordering or a plausibility ordering, for example. Theories of this sort would be richer and logically distinct from the simpler theories.

We will consider some of the most salient epistemological and logical options below.

1.1 Historical Remarks

Perhaps the earliest fully formalized version of a theory of belief change appears in the writings of William Harper in the mid-1970s. For example, [Harper, 1975] presents various crucial axioms of revision that later on were employed by logicians. Harper's ideas were influenced by Bayesian insights and the appeal to various forms of probability kinematics. He was also one of the first researchers to investigate the use of primitive conditional probability and its dynamics. Unfortunately his work remains unknown to many logicians working in belief change. But his contributions to belief change were very important and they antedated much of the logical and probabilistic work in the field.

Isaac Levi made important philosophical contributions to belief change in the early 1980s. In [Levi, 1980], Levi presents original work on belief change. Unlike Harper, Levi did not offer an axiomatic account of belief change. But he characterized various operations of belief change in a decision-theoretic manner. More recent work includes [Levi, 1991, Levi, 1996, Levi, 2004].

The logical work on belief change starts in 1985 with the publication of an influential paper by Alchourrón, Gärdenfors, and Makinson ([Alchourrón et al., 1985]). The AGM paper offers axiomatizations of the notions of contraction and revision and proves completeness results for these axiomatizations.

Three years later, Wolfgang Spohn published an article [Spohn, 1988] in which he presents a theory of belief change based on the use of *ordinal conditional functions*, which today tend to be known as *ranking functions*. The account has some advantages over AGM. For example, AGM is silent about iterated change, while the theory of ranking functions is able to deal with iteration. A representation result for ranking functions has been obtained only recently ([Hild and Spohn, 2008]).

During the 1990s there was a fair amount of work in computer science devoted to the topic of belief change. Spohn's ideas have been very influential among computer scientists especially taking into account the problem of how to characterize iterated change. A very influential paper articulating a theory of iterated change ([Darwiche and Pearl, 1997]) offers an account compatible with the use of ranking functions, although it is more general.

1.2 The AGM Model

After almost 25 years of research, the model of belief change proposed by Alchourrón, Gärdenfors, and Makinson ([Alchourrón et al., 1985]) in their classic paper remains influential. Even when the axiomatic base for contraction has been revised, expanded, and contracted, the basic formal techniques used in the paper have passed the test of time.

In the AGM framework, an agent's belief state is represented by a logically closed set of sentences K, called a *belief set*. The sentences of K are intended to represent the *beliefs* held by the agent. Belief change then comes in three flavours: *expansion*, *revision*, and *contraction*.

In expansion, a sentence ϕ is added to a belief set K to obtain an expanded belief set $K + \phi$. Since in the AGM framework $K + \phi$ is simply the logical closure of the set-theoretic sum of ϕ with K, the resulting expansion might be logically inconsistent. In revision, by contrast, a sentence ϕ is added to a belief set K to obtain a revised belief set $K * \phi$ in a way that preserves logical consistency. To ensure that $K * \phi$ is consistent, some sentences from K might be removed. In contraction, a sentence ϕ is removed from K to obtain a contracted belief set $K \dot{-} \phi$ that does not include ϕ. In the AGM framework, revision can be reduced to contraction via the so-called *Levi identity*, according to which the revision of a belief set K with a sentence ϕ is identical to the contraction $K \dot{-} \neg\phi$ expanded by ϕ. We will first focus on contraction, later discussing revision.

1.3 Technical Preliminaries

We presuppose a propositional language $\mathcal{L}$ with the connectives $\neg, \wedge, \vee, \rightarrow, \leftrightarrow$. We let For($\mathcal{L}$) denote the set of formulae of $\mathcal{L}$; $a, b, c, \ldots p, q, r, \ldots$ denote propositional variables of $\mathcal{L}$; $\alpha, \beta, \delta, \ldots, \phi, \psi, \chi, \ldots$ denote arbitrary formulae of $\mathcal{L}$; and $\Gamma, \Delta, \Theta, \ldots, \Lambda, \Sigma, \Psi, \ldots$ denote arbitrary sets of formulae. Sometimes we assume that the underlying language $\mathcal{L}$ is *finite*. By this we mean that $\mathcal{L}$ has only finitely many propositional variables.

As is customary, we assume that $\mathcal{L}$ is governed by a Tarskian consequence operation $\text{Cn} : \mathscr{P}(\text{For}(\mathcal{L})) \rightarrow \mathscr{P}(\text{For}(\mathcal{L}))$ such that ([Hansson, 1999, p. 26]):

(i) (*Inclusion*) $\Gamma \subseteq \text{Cn}(\Gamma)$.
(ii) (*Monotony*) If $\Gamma \subseteq \Delta$, then $\text{Cn}(\Gamma) \subseteq \text{Cn}(\Delta)$.
(iii) (*Idempotence*) $\text{Cn}(\text{Cn}(\Gamma)) \subseteq \text{Cn}(\Gamma)$.

In addition, the operator Cn is assumed to satisfy the following conditions:

(iv) (*Supraclassicality*) $\text{Cn}_0(\Gamma) \subseteq \text{Cn}(\Gamma)$, where Cn_0 is the classical consequence operation.
(v) (*Compactness*) If $\phi \in \text{Cn}(\Gamma)$, then there is some finite $\Gamma_0 \subseteq \Gamma$ such that $\phi \in \text{Cn}(\Gamma_0)$.
(vi) (*Deduction*) If $\phi \in \text{Cn}(\Gamma \cup \{\psi\})$, then $\psi \rightarrow \phi \in \text{Cn}(\Gamma)$.

As usual, Γ is called *logically closed* with respect to Cn if $\text{Cn}(\Gamma) = \Gamma$, and $\Gamma \vdash \phi$ is an abbreviation for $\phi \in \text{Cn}(\Gamma)$. While in logical parlance logically closed sets are called *theories*, the belief revision literature has adopted its own terminology, calling theories *belief sets*. The usual epistemological interpretation of theories is as *commitment sets*, representing the doxastic commitments of a rational agent ([Levi, 1991]). We let $\mathbb{K}$ denote the collection of logically closed sets in $\mathcal{L}$, an arbitrary element of which we usually denote by K.

2. Contraction

We first discuss an influential model of belief contraction due to [Alchourrón et al., 1985], called *partial meet contraction*. We will then turn to so-called *entrenchment*-based models of contraction due to [Gärdenfors, 1988, Gärdenfors and Makinson, 1988] and [Rott, 1991].

2.1 Partial Meet Contraction

A central notion used to construct an AGM contraction function of a set of formulae Θ is the concept of an α-remainder set of Θ, the collection of maximal subsets of Θ which do not imply α. Such a set guarantees minimal loss of information in the sense of subset inclusion.

Definition 17.2.1 Let Θ be a collection of formulae and α be a formula. The *α-remainder set* of Θ, $\Theta\bot\alpha$, is the collection of subsets Γ of $\mathrm{For}(\mathcal{L})$ such that:

(i) $\Gamma \subseteq \Theta$;
(ii) $\alpha \notin \mathrm{Cn}(\Gamma)$;
(iii) There is no set Δ such that $\Gamma \subset \Delta \subseteq \Theta$ and $\alpha \notin \mathrm{Cn}(\Delta)$.

A member of $\Theta\bot\alpha$ is called an *α-remainder* of Θ. We let $\Theta\bot\mathcal{L} := \{\Theta\bot\alpha : \alpha \in \mathrm{For}(\mathcal{L})\}$.

From this definition, we can immediately derive the following two properties of remainder sets:

(a) $\Theta\bot\alpha = \{\Theta\}$ if and only if $\alpha \notin \mathrm{Cn}(\Theta)$;
(b) $\Theta\bot\alpha = \emptyset$ if and only if $\alpha \in \mathrm{Cn}(\emptyset)$.

Established straightforwardly using Zorn's Lemma, the so-called *Upper Bound Property* specifies natural conditions which guarantee the existence of α-remainders:

(c) If $\Gamma \subseteq \Theta$ and $\alpha \notin \mathrm{Cn}(\Gamma)$, then there is some Δ such that $\Gamma \subseteq \Delta \in \Theta\bot\alpha$.

It is well known that remainder sets of belief sets behave quite well from several perspectives, enjoying many nice and useful properties, such as the following:

Proposition 17.2.1 Let K be a belief set. Then:

(i) If $\Gamma \in K\bot\alpha$, then for every $\beta \in K\backslash\Gamma$, $\Gamma \in K\bot\beta$;
(ii) If $\alpha, \beta \in K$, $K\bot(\alpha \wedge \beta) = K\bot\alpha \cup K\bot\beta$.
(iii) If $\alpha, \beta \in K$, $K\bot(\alpha \vee \beta) = K\bot\alpha \cap K\bot\beta$.

We now have enough elements to introduce the main operation of contraction proposed by AGM, called *partial meet contraction*. The idea is to select a subset of the collection of maximal consistent subsets of a belief set K that do not imply α, thereupon identifying the intersection of the selected α-remainders with the contraction of K by α. A selection function is introduced in order to make the selection. Here generalized for arbitrary sets of formulae, the notion of a selection function utilized by AGM can be defined as follows:

Definition 17.2.2 Let Θ be a set of formulae. A *selection function* for Θ is a function γ on $\Theta\bot\mathcal{L}$ such that for all formulae α:

(i) If $\Theta\bot\alpha \neq \emptyset$, then:
  (a) $\gamma(\Theta\bot\alpha) \subseteq \Theta\bot\alpha$, and
  (b) $\gamma(\Theta\bot\alpha) \neq \emptyset$;
(ii) If $\Theta\bot\alpha = \emptyset$, then $\gamma(\Theta\bot\alpha) = \{\Theta\}$.

Partial meet contraction for arbitrary sets of formulae Θ can then be defined as follows:

Definition 17.2.3 Let Θ be a set of formulae. A function $\dot{-}$ on For($\mathcal{L}$) is a *partial meet contraction* for Θ if there is a selection function γ for Θ such that for all formulae α,

$$\Theta \dot{-} \alpha = \bigcap \gamma(\Theta\bot\alpha).$$

A partial meet contraction for a *belief set* K is a contraction operation in the sense of AGM.

It follows from these three definitions that if α is a logical truth or $\alpha \notin \Theta$, then Θ remains unchanged after contraction by α; in symbols, $\Theta \dot{-} \alpha = \Theta$. Two limiting cases of partial meet contraction are of special interest: The case in which the selection function selects (i) exactly one element of $\Theta\bot\alpha$, and the case in which it selects (ii) the entire set $\Theta\bot\alpha$. These two special cases are now known as *maxichoice contraction* and *full meet contraction*, respectively ([Gärdenfors, 1988]).

Actually, the general approach behind AGM is concerned not only to provide semantic characterizations of belief change but also to supply postulates contraction operations must obey. Accordingly, the main logical goal of this approach is a representation result for a set of compelling postulates. AGM show that partial meet contraction for *belief sets* is characterized by the following postulates:

$(K \dot{-} 1)$ $K \dot{-} \alpha = \mathrm{Cn}(K \dot{-} \alpha)$. (*Closure*)

$(K \dot{-} 2)$ $K \dot{-} \alpha \subseteq K$. (*Inclusion*)

$(K \dot{-} 3)$ If $\alpha \notin K$ or $\alpha \in \mathrm{Cn}(\emptyset)$, then $K \dot{-} \alpha = K$. (*Vacuity*)

$(K \dot{-} 4)$ If $\alpha \notin \mathrm{Cn}(\emptyset)$, then $\alpha \notin K \dot{-} \alpha$. (*Success*)

$(K \dot{-} 5)$ If $\mathrm{Cn}(\{\alpha\}) = \mathrm{Cn}(\{\beta\})$, then $K \dot{-} \alpha = K \dot{-} \beta$. (*Extensionality*)

$(K \dot{-} 6)$ $K \subseteq \mathrm{Cn}((K \dot{-} \alpha) \cup \{\alpha\})$. (*Recovery*)

By *characterized* we mean that a function $\dot{-}$ on $\mathrm{For}(\mathcal{L})$ satisfies the above postulates just in case it is a partial meet contraction for K.

These postulates are commonly referred to as the *basic* AGM postulates. All the conditions except perhaps Recovery seem reasonable. There is a relatively large literature on the adequacy of Recovery (the following articles are perhaps salient: [Makinson, 1987], [Levi, 1991]). Several competing operations of contraction which do not obey the Recovery postulate have been proposed in the literature, such as *saturatable contractions* ([Levi, 1991]), *severe withdrawals* ([Rott and Pagnucco, 1999]), and *systematic withdrawals* ([Meyer et al., 2002]). We will discuss some of these operations later when we consider the work of Isaac Levi in this area.

It is possible to strengthen the notion of partial meet contraction by requiring that the selected members of the remainder set are the 'best' elements with respect to an underlying relation defined on the collection of remainders.

Definition 17.2.4 Let Θ be a set of formulae. A function $\dot{-}$ on $\mathrm{For}(\mathcal{L})$ is a *relational partial meet contraction* for Θ if there is a selection function γ for Θ and a binary relation $\preceq$ on $\Theta \bot \mathcal{L}$ such that for every formula α:

(i) $\Theta \dot{-} \alpha = \bigcap \gamma(\Theta \bot \alpha)$;
(ii) If $\Theta \bot \alpha \neq \emptyset$, then $\gamma(\Theta \bot \alpha) = \{\Gamma \in \Theta \bot \alpha : \Lambda \preceq \Gamma \text{ for all } \Lambda \in \Theta \bot \alpha\}$.

If such a relation $\preceq$ is in addition transitive, then we call such $\dot{-}$ *transitively relational*.[2]

This semantic requirement is reflected in two *supplementary postulates*:

$(K \dot{-} 7)$ $(K \dot{-} \alpha) \cap (K \dot{-} \beta) \subseteq K \dot{-} (\alpha \wedge \beta)$. (*Conjunctive Overlap*)

$(K \dot{-} 8)$ If $\alpha \notin K \dot{-} (\alpha \wedge \beta)$, then $K \dot{-} (\alpha \wedge \beta) \subseteq K \dot{-} \alpha$. (*Conjunctive Inclusion*)

The centrepiece of AGM's influential 1985 paper can now be stated as follows:

Theorem 17.2.1 ([Alchourrón et al., 1985]) *Let K be a belief set, and let* $\dot{-}$ *be a function on* For($\mathcal{L}$). *Then:*

(i) *The function* $\dot{-}$ *is a partial meet contraction for K if and only if it satisfies postulates* $(K \dot{-} 1)$ *to* $(K \dot{-} 6)$.
(ii) *The function* $\dot{-}$ *is a transitively relational partial meet contraction for K if and only if it satisfies postulates* $(K \dot{-} 1)$ *to* $(K \dot{-} 8)$.

2.2 Entrenchment-Based Models

Several other procedures for constructing contractions have been shown to coincide with transitively relational partial meet contraction. Perhaps one of the most important is based on a notion of *epistemic entrenchment*. The idea behind the notion of entrenchment is that when one says that 'one sentence β is more entrenched than a sentence α in the current belief set', this means that β is more useful in inquiry and deliberation, or has more 'epistemic value' than α. In symbols we may write $\alpha < \beta$.

Let us first introduce a relation of entrenchment formally. Let $\leq$ be a binary relation on the sentences of the underlying language. We call $\leq$ an *entrenchment* relation for a theory K if the following conditions are satisfied:

Transitivity If $\alpha \leq \beta$ and $\beta \leq \gamma$, then $\alpha \leq \gamma$.
Dominance If $\beta \in \mathrm{Cn}(\alpha)$, then $\alpha \leq \beta$.
Conjunctiveness $\alpha \leq \alpha \wedge \beta$ or $\beta \leq \alpha \wedge \beta$.
Minimality If the belief set K is consistent, then $\alpha \leq \beta$ for every formula β if and only if $\alpha \notin K$.
Maximality If $\beta \leq \alpha$ for every β, then $\alpha \in \mathrm{Cn}(\emptyset)$.

A natural and reasonable principle of entrenchment says that in giving up a non-tautological sentence α from the current view one should preserve the sentences better entrenched than α. [Gärdenfors, 1988] and [Gärdenfors and Makinson, 1988] pursued this principle, offering the following definition.

Definition 17.2.5 Let K be a belief set. We say that a function $\dot{-}$ on $\mathrm{For}(\mathcal{L})$ is a Gärdenfors' *entrenchment-based contraction* for K if there is an entrenchment relation $\leq$ such that for every formula α:

$$K \dot{-} \alpha = \begin{cases} K \cap \{\beta : \alpha < \alpha \vee \beta\} & \text{if } \alpha \notin \mathrm{Cn}(\emptyset); \\ K & \text{otherwise.} \end{cases}$$

As reported in the following theorem, Gärdenfors' entrenchment-based contraction is characterized by the AGM postulates for contraction.

Theorem 17.2.2 ([Gärdenfors, 1988; Gärdenfors and Makinson, 1988]) *Let K be a belief set, and let* $\dot{-}$ *be a function on* $\mathrm{For}(\mathcal{L})$*. Then* $\dot{-}$ *is a Gärdenfors' entrenchment-based contraction for K if and only if it satisfies postulates* $(K \dot{-} 1)$ *to* $(K \dot{-} 8)$.

To establish the 'if' direction, one defines an entrenchment relation $\leq$ on $\mathrm{For}(\mathcal{L})$ by setting for every formula α, β:

$$\alpha \leq \beta \quad \text{:iff} \quad \text{either } \alpha \notin K \dot{-} (\alpha \wedge \beta) \text{ or } \alpha \wedge \beta \in \mathrm{Cn}(\emptyset).$$

This definition is the 'right' definition in the sense that any Gärdenfors' entrenchment-based contraction must satisfy the above constraint when it is understood as a statement.

Hans ([Rott, 1991]) has suggested that Gärdenfors' entrenchment-based contraction has little motivation. He has proposed that contraction is more plausibly defined by setting for all formulae α:

$$K \dot{-} \alpha := \begin{cases} K \cap \{\beta : \alpha < \beta\} & \text{if } \alpha \notin \mathrm{Cn}(\emptyset); \\ K & \text{otherwise.} \end{cases}$$

However, a contraction function thus defined is not characterized by the AGM postulates of contraction. We will consider arguments concluding that this is a good thing later when we discuss doubts about the Recovery postulate.

3. Revision

As indicated above, the AGM framework admits a reduction of revision to contraction via the so-called *Levi identity*, in symbols expressed as:

$$K * \phi = (K \dot{-} \neg\phi) + \phi.$$

Here $(K \dot{-} \neg\phi) + \phi := \mathrm{Cn}((K \dot{-} \neg\phi) \cup \{\phi\})$. Thus, according to the Levi identity, the revision of a belief set K with a sentence ϕ can be divided into two steps: first, contract K by $\neg\phi$; second, expand the contracted belief set $K \dot{-} \neg\phi$ by ϕ. The composition of the contraction and expansion function ensures both that $K * \phi$ is consistent and that ϕ is a member of the revision $K * \phi$.

We first discuss *partial meet revision*, the dual of partial meet contraction. We will then discuss propositional models of belief revision, focusing on *sphere*-based revision and then on *persistent* revision. In between the latter two discussions we make a few remarks about the connection between propositional models and syntactical models of belief change. We illustrate how belief change within propositional models can be depicted geometrically. This sheds light on syntactical models of belief change.

3.1 Partial Meet Revision

As should be suspected, one can define *partial meet revision* by way of the Levi Identity. We define partial meet revision for arbitrary sets of formulae Θ:

Definition 17.3.1 Let Θ be a set of formulae. A function $*$ on $\mathrm{For}(\mathcal{L})$ is a *partial meet revision* for Θ if there is a selection function γ for Θ such that for all formulae α,

$$\Theta * \alpha = \mathrm{Cn}((\bigcap \gamma(\Theta \perp \neg\alpha)) \cup \{\alpha\}) :$$

A partial meet revision for a *belief set* K is a revision operation in the sense of AGM.

It is also possible to axiomatically characterize revision. The following *basic* revision postulates are analogues of the basic contraction postulates:

$(K * 1)$ $K * \phi = \mathrm{Cn}(K * \phi)$. (*Closure*)

$(K * 2)$ $\phi \in K * \phi$. (*Success*)

$(K * 3)$ $K * \phi \subseteq \mathrm{Cn}(K \cup \{\phi\})$. (*Inclusion*)

$(K * 4)$ If $\neg\phi \notin K$, then $\mathrm{Cn}(K \cup \{\phi\}) \subseteq K * \phi$. (*Vacuity*)

$(K * 5)$ If $\mathrm{Cn}(\{\phi\}) \neq \mathrm{For}(\mathcal{L})$, then $K * \phi \neq \mathrm{For}(\mathcal{L})$. (*Consistency*)

$(K * 6)$ If $\mathrm{Cn}(\{\phi\}) = \mathrm{Cn}(\{\psi\})$, then $K * \phi = K * \psi$. (*Extensionality*)

Partial meet revision for *belief sets* is *characterized* by these postulates, i.e., a function $*$ on $\mathrm{For}(\mathcal{L})$ satisfies the above postulates just in case it is a partial meet revision for K.

Attention can be turned from the larger class of partial meet revisions to the smaller class of functions derived from relational partial meet contractions.

Definition 17.3.2 Let Θ be a set of formulae. A function $*$ on For($\mathcal{L}$) is a *relational partial meet revision* for Θ if there is a selection function γ for Θ and a binary relation $\preceq$ on $\Theta\bot\mathcal{L}$ such that for every formula α:

(i) $\Theta * \alpha = \mathrm{Cn}((\bigcap \gamma(\Theta\bot\neg\alpha)) \cup \{\alpha\})$;
(ii) If $\Theta\bot\alpha \neq \emptyset$, then $\gamma(\Theta\bot\alpha) = \{\Gamma \in \Theta\bot\alpha : \Lambda \preceq \Gamma \text{ for all } \Lambda \in \Theta\bot\alpha\}$.

If such a relation $\preceq$ is in addition transitive, then we call such $*$ *transitively relational.*

As with contraction functions, the six basic postulates are elementary requirements of belief revision and taken by themselves are much too permissive, requiring additional postulates to rein in this permissiveness and to reflect the above semantic notion of relational belief revision.

($K * 7$) $K * (\phi \wedge \psi) \subseteq \mathrm{Cn}((K * \phi) \cup \{\psi\})$. (*Superexpansion*)

($K * 8$) $\neg\psi \notin K * \phi$, then $\mathrm{Cn}(K * \phi \cup \{\psi\}) \subseteq K * (\phi \wedge \psi)$. (*Subexpansion*)

As counterparts of the supplementary contraction postulates, such additional postulates are also called *supplementary postulates*. Together, the foregoing postulates are enough to characterize transitively relational partial meet revision. We state the aforementioned results in a theorem.

Theorem 17.3.1 *Let K be a belief set, and let $*$ be a function on* For($\mathcal{L}$). *Then:*

(i) The function $*$ is a partial meet revision for K if and only if it satisfies postulates ($K * 1$) *to* ($K * 6$).
(ii) The function $*$ is a transitively relational partial meet revision for K if and only if it satisfies postulates ($K * 1$) *to* ($K * 8$).

We wish to bring to the reader's attention another postulate – or some postulate at least as strong as it – often added to the mix:

($K * 8r$) $K * (\phi \vee \psi) \subseteq \mathrm{Cn}(K * \phi \cup K * \psi)$. (*Disjunction*)

We will see later on in the next section the significance of this postulate in belief change.

3.2 Propositional Models

The AGM framework for belief change uses the notion of a *remainder set* to define operators of belief change. As such, belief states and belief change have a syntactic character. An alternative and arguably more suitable and elegant framework for belief change uses *propositions*, or sets of possible worlds, instead. A belief state can then be represented in terms of a set of possible worlds rather than a collection of sentences. Accordingly, a set of sentences has a propositional representation as precisely those possible worlds in which all sentences in the set in question are true.

Propositional models of belief change can be connected to the syntactic models of belief change we have hereunto discussed, offering a useful visualization of the different operators of belief change. It is therefore somewhat unsurprising to find that several authors have utilized propositional models, including [Arló Costa and Pedersen, 2010], [Grove, 1988], [Harper, 1975, Harper, 1977], [Katsuno and Mendelzon, 1989, Katsuno and Mendelzon, 1991a, Katsuno and Mendelzon, 1991b], [Morreau, 1992], [Pedersen, 2008], [Rott, 1993, Rott, 2001], and [Spohn, 1988, Spohn, 1990, Spohn, 1998]. In his [Grove, 1988], Adam Grove famously connected a generalization of Lewis' semantics for conditional logic with the AGM model of belief change, and more recently Hans Rott ([Rott, 2001]) expanded upon this line of research with an eye towards the choice functional literature in rational choice, establishing a one-to-one correspondence between functional constraints on propositional models with postulates of belief change. In this section we discuss possible worlds approaches to modelling belief change, paying particular attention to the work of Grove and Rott.

Some notational remarks are in order. We let $\mathcal{W}_{\mathcal{L}}$ denote the collection of all maximal consistent sets of $\mathcal{L}$ with respect to Cn.[3] Members of $\mathcal{W}_{\mathcal{L}}$ are often called *states, possible worlds* or just *worlds*, and we denote an arbitrary member of $\mathcal{W}_{\mathcal{L}}$ by w. For a non-empty collection of worlds W of $\mathcal{W}_{\mathcal{L}}$, let $\mathrm{Th}(W)$ denote the set of formulae of $\mathcal{L}$ which are members of all worlds in W (briefly, $\mathrm{Th}(W) := \bigcap_{w \in W} w$); if W is empty, we define $\mathrm{Th}(W) := \mathrm{For}(\mathcal{L})$, by convention. If Γ is a set of formulae of $\mathcal{L}$, we let $[\![\Gamma]\!] := \{w \in \mathcal{W}_{\mathcal{L}} : \Gamma \subseteq w\}$. If ϕ is a formula of $\mathcal{L}$, we write $[\![\phi]\!]$ instead of $[\![\{\phi\}]\!]$. A member of $\mathscr{P}(\mathcal{W}_{\mathcal{L}})$ is often called a *proposition*, and $[\![\phi]\!]$ is often called the *proposition expressed by* ϕ. Intuitively, $[\![\Gamma]\!]$ consists of those worlds in which all formulae in Γ hold. Finally, let $\mathcal{E}_{\mathcal{L}}$ be the set of all elementary subsets of $\mathcal{W}_{\mathcal{L}}$, i.e., $\mathcal{E}_{\mathcal{L}} := \{W \in \mathscr{P}(\mathcal{W}_{\mathcal{L}}) : W = [\![\phi]\!] \text{ for some } \phi \in \mathrm{For}(\mathcal{L})\}$.

The major innovation in [Alchourrón et al., 1985] is the employment of selection functions to define operators of belief change. As we have seen, in the AGM framework selection functions take *remainder sets* as arguments. Analogously, many propositional models of belief change use selection functions which instead take *propositions* as arguments. We will call such selection functions *propositional selection functions*. Rott has shown in [Rott, 2001] that this approach is a fruitful generalization of the AGM approach. For our purposes, it will suffice to couch our discussion in terms of such functions.

Definition 17.3.3 A *propositional selection function* is a function f on $\mathscr{E}_{\mathcal{L}}$ such that $f(S) \subseteq S$ for every $S \in \mathscr{E}_{\mathcal{L}}$.

3.2.1 Sphere-Based Revision

Proposed by [Grove, 1988], so-called *sphere semantics* offers an elegant representation of belief change. We now introduce the notion of a *system of spheres* and *sphere-based revision*, the latter of which is completely characterized by the classical AGM postulates of belief revision.

Definition 17.3.4 Let $C \subseteq \mathcal{W}_{\mathcal{L}}$, and let $\mathscr{S} \subseteq \mathscr{P}(\mathcal{W}_{\mathcal{L}})$. We call $\mathscr{S}$ a *system of spheres centred on* C if it satisfies the following properties:

($\mathscr{S}$1) $\mathscr{S}$ is totally ordered by $\subseteq$;[4]
($\mathscr{S}$2) C is the $\subseteq$-minimum of $\mathscr{S}$;[5]
($\mathscr{S}$3) $\mathcal{W}_{\mathcal{L}} \in \mathscr{S}$;
($\mathscr{S}$4) For every formula ϕ and $S \in \mathscr{S}$, if $S \cap [\![\phi]\!] \neq \emptyset$, then there is a $\subseteq$-minimum $S_0 \in \mathscr{S}$ such that $S_0 \cap [\![\phi]\!] \neq \emptyset$.

Now for each formula ϕ, define the following set:

$$\mathscr{C}_\phi := \{S \in \mathscr{S} : S \cap [\![\phi]\!] \neq \emptyset\} \cup \{\mathcal{W}_{\mathcal{L}}\}.$$

Definition 17.3.5 Let $\mathscr{S}$ be a system of spheres centred on C. Define a propositional selection function $f_{\mathscr{S}} : \mathscr{E}_{\mathcal{L}} \to \mathscr{P}(\mathcal{W}_{\mathcal{L}})$ by setting for every formula ϕ:

$$f_{\mathscr{S}}([\![\phi]\!]) := \min_{\subseteq}(\mathscr{C}_\phi) \cap [\![\phi]\!]$$

where $\min_{\subseteq}(\mathscr{C}_\phi)$ is the minimum element of $\mathscr{C}_\phi$ when this set is ordered by $\subseteq$. We call $f_{\mathscr{S}}$ the *Grovean selection function* for $\mathscr{S}$.

We now introduce *sphere-based revision*.

Definition 17.3.6 Let K be a belief set. A function $*$ is a *sphere-based revision* for K if there is system of spheres $\mathscr{S}$ centred on $[\![K]\!]$ such that for all formulae ϕ:

$$K * \phi = \text{Th}(f_{\mathscr{S}}([\![\phi]\!]))$$

The idea behind sphere-based revision can be easily visualized geometrically as in Figure 17.1. The upper right region of Figure 17.1 consists of those worlds

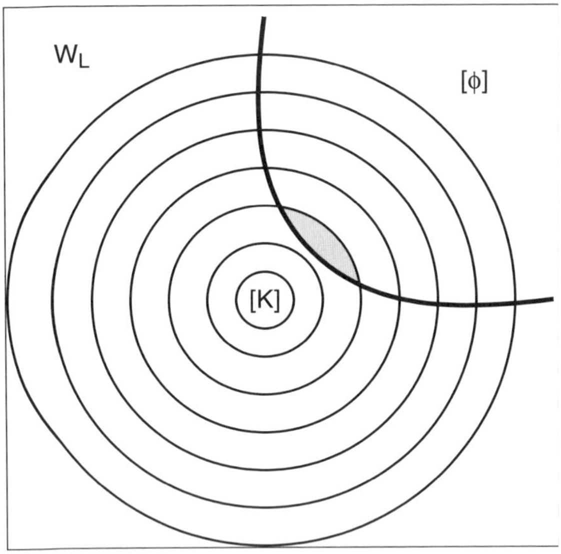

FIGURE 17.1 *Sphere-Based Revision* (the case in which $\phi \in K \backslash \mathrm{Cn}(\emptyset)$). The grey region represents $f_{\mathscr{S}}(\llbracket\phi\rrbracket)$, which generates the revision of K by ϕ, $K * \phi = \mathrm{Th}(f_{\mathscr{S}}(\llbracket\phi\rrbracket))$.

in which ϕ is true, while the centre disc, or sphere, consists of those worlds in which all sentences in K are true. The third sphere from the centre is the least sphere $\min_{\subseteq}(\mathscr{C}_\phi)$ intersecting $\llbracket\phi\rrbracket$, and the grey region is the area of the intersection of $\min_{\subseteq}(\mathscr{C}_\phi)$ and $\llbracket\phi\rrbracket$, representing the resulting belief state $f_{\mathscr{S}}(\phi)$. The corresponding syntactical representation of $f_{\mathscr{S}}(\phi)$ is given by $K * \phi = \mathrm{Th}(f_{\mathscr{S}}(\phi))$.

[Grove, 1988] establishes an important and useful connection between sphere-based revision and the AGM revision postulates.

Theorem 17.3.2 ([Grove, 1988]) *Let K be a belief set. Then:*

(i) Every sphere-based revision for K satisfies postulates $(K * 1)$ *to* $(K * 8)$.
(ii) Every function on For($\mathcal{L}$) *satisfying* $(K * 1)$ *to* $(K * 8)$ *is a sphere-based revision.*

Part (i) shows that the postulates are *sound* with respect to sphere-based revision, while part (ii) shows that the postulates are *complete* with respect to sphere-based revision.

3.2.2 The Grove Connection, and Geometric Depictions of Belief Change

In fact, [Grove, 1988] reveals a close connection between the AGM modelling and the sphere modelling of belief change. To see this, suppose that $\phi \in K \backslash \mathrm{Cn}(\emptyset)$. To define belief contraction and so belief revision, [Alchourrón et al., 1985] consider the ϕ-remainder set $K\perp\phi$ of maximal subsets Γ of K such that Γ does not imply ϕ. It is easily verified that on the one hand, for every $\Gamma \in K\perp\phi$ there is $w \in [\![\neg\phi]\!]$ such that $[\![\Gamma]\!] = [\![K]\!] \cup \{w\}$, and on the other hand, for every $w \in [\![\neg\phi]\!]$, $K \cap w \in K\perp\phi$. This establishes a one-to-one correspondence $g_\phi : [\![\neg\phi]\!] \to K\perp\phi$ given by $g_\phi(w) = K \cap w$. Putting $K\perp(K\backslash\mathrm{Cn}(\emptyset)) := \bigcup_{\phi\in K\backslash\mathrm{Cn}(\emptyset)} K\perp\phi$ and observing that $\mathcal{W}_\mathcal{L}\backslash[\![K]\!] = \bigcup_{\phi\in K\backslash\mathrm{Cn}(\emptyset)}[\![\neg\phi]\!]$, the family of bijections $(g_\phi)_{\phi\in K\backslash\mathrm{Cn}(\emptyset)}$ induces a one-to-one correspondence $G_K : (\mathcal{W}_\mathcal{L}\backslash[\![K]\!]) \to K\perp(K\backslash\mathrm{Cn}(\emptyset))$ given by $G_K(w) := K \cap w$. In light of its fundamental importance, we record the result in a proposition.

Proposition 17.3.1 *(The Grove Connection, [Grove, 1988])* Let K be a belief set. Then there is a bijection $G_K : (\mathcal{W}_\mathcal{L}\backslash[\![K]\!]) \to K\perp(K\backslash\mathrm{Cn}(\emptyset))$ such that for every $\phi \in K\backslash\mathrm{Cn}(\emptyset)$ and $w \in \mathcal{W}_\mathcal{L}\backslash[\![K]\!]$:

(1) $w \in [\![\neg\phi]\!]$ if and only if $G_K(w) = K \cap w$ and $G_K(w) \in K\perp\phi$;
(2) $[\![G_K(w)]\!] = [\![K]\!] \cup \{w\}$.

The Grove Connection facilitates the geometric visualization of contraction operators. Setting limit cases aside, the first modelling considered in [Alchourrón et al., 1985], *maxichoice* contraction, takes $K \dot{-} \phi$ to be some ϕ-remainder $K \cap w$ in $K\perp\phi$ furnished by a singleton-valued selection function γ. Thus, in terms of propositions, $[\![K \dot{-} \phi]\!] = [\![K]\!] \cup \{w\}$, where $w \in [\![\neg\phi]\!]$. If the values of γ are generated by a transitive relation $\preceq$ (as in Definition 17.2.4), the maxichoice operation $\dot{-}$ is of course also a transitively relational partial meet contraction (thereby satisfying postulates $(*7)$ and $(*8)$, among other, stronger postulates; see [Alchourrón et al., 1985]); yet more is true, as $\preceq$ must also be a total order because γ is singleton-valued. In light of the Grove Connection G_K, the ordering $\preceq$ induces a natural total ordering $\preceq'$ on $\mathcal{W}_\mathcal{L}$ and so a system of spheres centred on $[\![K]\!]$ as depicted in Figure 17.2, generating what we may call the *sphere-based maxichoice contraction* of K by ϕ.

The second modelling considered in [Alchourrón et al., 1985], *full meet* contraction, is the opposite extreme, taking $K \dot{-} \phi$ to be the intersection of all ϕ-remainders in $K\perp\phi$ furnished by the identity selection function $\gamma = \mathrm{id}$. This corresponds to amassing all worlds in $[\![\neg\phi]\!]$, resulting in $[\![K \dot{-} \phi]\!] = [\![K]\!] \cup [\![\neg\phi]\!]$. Since the selection function is the identity function, the Grove Connection G_K induces a 'flat' weak ordering on $\mathcal{W}_\mathcal{L}$ (for which all elements are equivalent) and so the 'coarsest' system of spheres consisting of $[\![K]\!]$ and $\mathcal{W}_\mathcal{L}$, as depicted in Figure 17.3. This results in what we may call the *sphere-based full meet contraction* of K by ϕ.

The final model considered in [Alchourrón et al., 1985], *partial meet* contraction, corresponds to the intermediate between the above two extremes. Instead of just a single ϕ-remainder of $K\perp\phi$, $K \dot{-} \phi$ takes the intersection of some subset

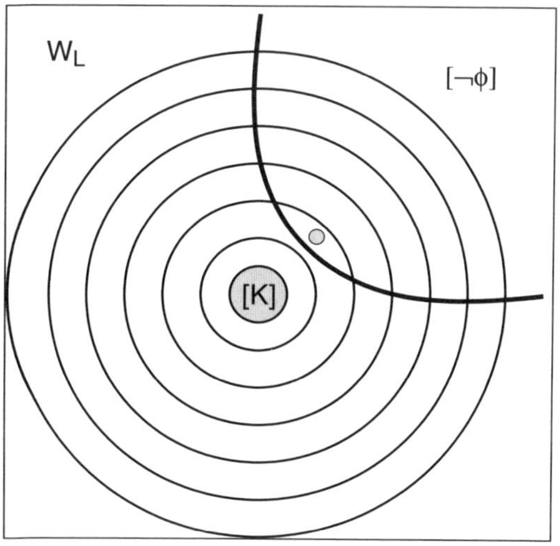

FIGURE 17.2 *Maxichoice Contraction* (the case in which $\phi \in K\backslash \mathrm{Cn}(\emptyset)$). The small grey disc represents the singleton proposition $\{w\}$ selected by $f_{\mathscr{S}}([\![\neg\phi]\!])$, generating the contraction of K by ϕ, $K \dot{-} \phi = K \cap \mathrm{Th}(f_{\mathscr{S}}([\![\neg\phi]\!])) = \mathrm{Th}([\![K]\!] \cup \{w\})$

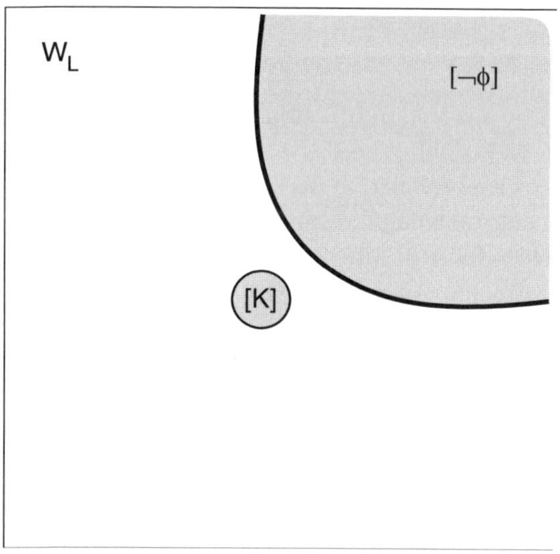

FIGURE 17.3 *Full Meet Contraction* (the case in which $\phi \in K\backslash \mathrm{Cn}(\emptyset)$). The large grey region in the upper right corner represents the proposition $[\![\neg\phi]\!]$ selected by $f_{\mathscr{S}}([\![\neg\phi]\!])$, generating the contraction of K by ϕ, $K \dot{-} \phi = K \cap \mathrm{Th}(f_{\mathscr{S}}([\![\neg\phi]\!])) = \mathrm{Th}([\![K]\!] \cup [\![\neg\phi]\!])$

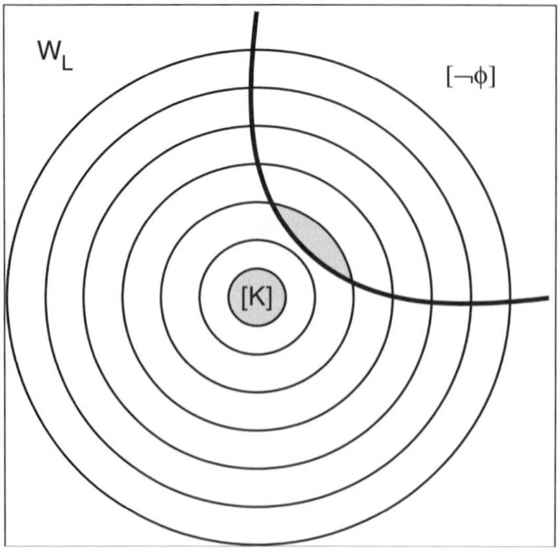

FIGURE 17.4 *Partial Meet Contraction* (the case in which $\phi \in K\backslash\mathrm{Cn}(\emptyset)$). The grey lens represents the proposition given by $f_{\mathscr{S}}([\![\neg\phi]\!])$, generating the contraction of K by ϕ, $K \dot{-} \phi = K \cap \mathrm{Th}(f_{\mathscr{S}}([\![\neg\phi]\!])) = \mathrm{Th}([\![K]\!] \cup f_{\mathscr{S}}([\![\neg\phi]\!]))$

of $K\bot\phi$ furnished by a selection function γ. So $[\![K \dot{-} \phi]\!]$ is the union of propositions of the form $[\![K]\!] \cup \{w\}$, where $w \in [\![\neg\phi]\!]$. As depicted in Figure 17.4, if γ is generated by a transitive relation (as in Definition 17.2.4), the Grove Connection G_K induces a natural weak ordering on $\mathcal{W}_{\mathcal{L}}$ and a system of spheres exactly intermediate between those of sphere-based maxichoice contraction and sphere-based full meet contraction, thereby generating what we may call the *sphere-based partial meet contraction* of K by ϕ.

The previous pictorial representation should make it clear that full meet contraction is a particular case of partial meet contraction. Full meet contraction is not mandatory but is *permissible*. Researchers have recently criticized the AGM approach for being too permissive because it admits the possibility of trivial updates of this sort. Perhaps the first to raise his voice against this feature of the AGM theory of belief change is Rohit Parikh in [Parikh, 1999]. Parikh offered in this article a model of revision that rules out trivial update by appealing to a syntactic model in which one can articulate the notion of relevance in belief change. The central idea proposed by Parikh, *language splitting*, has other applications in areas other than belief change. In particular, it is related to some of the literature related to the Beth interpolation theorem ([Parikh, 2008a]). George Kourousias and David Makinson also wrote a recent paper ([Kourousias and Makinson, 2007]) inspired by Parikh's work.

Another researcher who protested against the permissibility of the trivial update in AGM is Neil Tennant. In his [Tennant, 2006], Tennant tackles this issue, but his account is quite different than the one offered by Parikh. He offers a relational model of belief change (instead of the usual functional account), and one of the byproducts of his account is a principle of minimal mutilation in belief change that rules out the trivial update. The idea of a relational approach in belief change is not new (see, for example, [Rabinowicz and Lindström, 1994]).

3.2.3 Persistent Revision

The above discussion of contraction functions naturally led to our considering orderings over $\mathcal{W}_{\mathcal{L}}$ supplied by the Grove Connection. We will now briefly discuss propositional models of belief revision which take this as the starting point, focusing in particular on the material of [Katsuno and Mendelzon, 1989, Katsuno and Mendelzon, 1991a, Katsuno and Mendelzon, 1991b]. We now introduce the notion of a *persistent* binary relation, a measure of how 'compatible' alternative worlds are with the current beliefs of an agent, or how 'close' such worlds are to those beliefs.

Definition 17.3.7 Let $C \subseteq \mathcal{W}_{\mathcal{L}}$, and let $\leq$ be a binary relation $\mathcal{W}_{\mathcal{L}}$. We say that $\leq$ is *C-persistent* if it satisfies the following properties:

($\leq$ 1) $\leq$ is a weak order;[6]
($\leq$ 2) For every formula ϕ, if $[\![\phi]\!] \neq \emptyset$, then $\{w \in [\![\phi]\!] : v \leq w \text{ for all } v \in [\![\phi]\!]\} \neq \emptyset$;
($\leq$ 3) For every $w \in \mathcal{W}_{\mathcal{L}}$, w is a $\leq$-maxima if and only if $w \in C$.[7]

We define the notion of a selection function based on a persistent binary relation.

Definition 17.3.8 Let $\leq$ be a C-persistent binary relation. Define a propositional selection function $f_{\leq} : \mathscr{E}_{\mathcal{L}} \to \mathscr{P}(\mathcal{W}_{\mathcal{L}})$ by setting for every formula ϕ:

$$f_{\leq}([\![\phi]\!]) := \{w \in [\![\phi]\!] : v \leq w \text{ for all } v \in [\![\phi]\!]\}.$$

We call $f_{\leq}$ the *persistent selection function* based on $\leq$.

We now offer a definition of what we call *persistent revision*.

Definition 17.3.9 Let K be a belief set. A function $*$ is a K-*persistent revision* if there is a $[\![K]\!]$-persistent binary relation $\leq$ such that, for all formulae ϕ:

$$K * \phi = \mathrm{Th}(f_{\leq}([\![\phi]\!]))$$

Among other very useful results, [Katsuno and Mendelzon, 1989], [Katsuno and Mendelzon, 1991a], [Katsuno and Mendelzon, 1991b] show that the expected should be unsurprising.

Theorem 17.3.3 ([Katsuno and Mendelzon, 1991b]) *Let K be a belief set. Then:*

(i) Every K-persistent revision satisfies postulates $(K * 1)$ *to* $(K * 8)$.
(ii) Every function on For($\mathcal{L}$) *satisfying* $(K*1)$ *to* $(K*8)$ *is a K-persistent revision.*

Indeed, ignoring limit cases, we can easily fill in the lacuna concerning the relationship between systems of spheres and persistent relations.[8] On the one hand, given a system of spheres $\mathscr{S}$ centred on $[\![K]\!]$, we can define a $[\![K]\!]$-persistent relation by setting for all $w, v \in \mathcal{W}_{\mathcal{L}}$, $w \leq v$:iff for every $T \in \mathscr{S}$, if $w \in T$, then there is some sphere $S \subseteq T$ such that $v \in S$. The latter definition is a useful simplification of the intuition that $w \leq v$ should hold just in case either there are $S, T \in \mathscr{S}$ such that $S \subseteq T$ and $w \in T \backslash S$ and $v \in S$ or for every $S \in \mathscr{S}$, $w \in S$ iff $v \in S$. On the other hand, given a $[\![K]\!]$-persistent relation, we can define a system of spheres $\mathscr{S}$ centred on $[\![K]\!]$ by setting $\mathscr{S} := \{S_w : w \in \mathcal{W}_{\mathcal{L}}\} \cup \{\mathcal{W}_{\mathcal{L}}\}$, where $S_w := \{v \in \mathcal{W}_{\mathcal{L}} : w \leq v\}$.

3.3 Belief Change and Rational Choice

Grovean selection functions and persistent selection functions are but two equivalent ways to generate operators of belief change in line with the AGM paradigm. Such functions generate belief change operators characterized by the whole set of basic and supplementary AGM postulates. Exploiting results from the theory of choice, Sten ([Lindström, 1991]) and Hans ([Rott, 1993]) systematically studied the relationship between functional constraints placed on selection functions and postulates of belief change. Hans ([Rott, 2001]) continued these studies, generalizing and improving them in various ways. Among other things, Rott shows in [Rott, 2001] that certain functional constraints placed on propositional selection functions correspond in a one-to-one fashion to postulates of belief change. Rott's results forge a useful bridge between the mathematical theories of belief change and rational choice. We will discuss a small selection of the material from [Rott, 2001]. In rational choice theory, a selection function is a rule that associates with each menu S, or set of alternatives available for choice, a subset of S (see Chapter 19 of this volume). The subset of alternatives from S are those options which an agent regards as choosable when faced with the decision problem S. As such, a selection function is often called a *choice function* in the context of rational choice.

In the study of rational choice, so-called *coherence constraints* have been imposed on the form relationships may take among choices across varying menus. These requirements specify how choices must be made across different decision problems. Restricting our attention to propositional selection functions, some predominant coherence constraints are the following:

(α) For every $S, T \in \mathscr{E}_{\mathcal{L}}$, if $S \subseteq T$, then $S \cap f(T) \subseteq f(S)$.
(γ^*) For every $S, T \in \mathscr{E}_{\mathcal{L}}$ such that $S \cup T \in \mathscr{E}_{\mathcal{L}}$, $f(S) \cap f(T) \subseteq f(S \cup T)$.
(β^+) For every $S, T \in \mathscr{E}_{\mathcal{L}}$, if $S \subseteq T$ and $S \cap f(T) \neq \emptyset$, then $f(S) \subseteq f(T)$.

Condition α demands that whatever is rejected for choice from a menu must remain rejected if the menu is expanded. More formally, this means that for any menu S, if x is an alternative in S and x is not in $f(S)$ – that is, x is not chosen, i.e., is rejected, from S – then if S is expanded to a menu S' – that is, if S' is such that S is a subset of S' – then x is not in $f(S')$. Equivalently, this condition demands that whatever is admissible for choice from a menu must also be admissible from any smaller menu for which this choice is still available. This motivates calling condition α a 'contraction consistency' condition.[9]

While condition α is concerned with ensuring that an admissible alternative remains admissible as a menu is contracted, condition γ^* is concerned with ensuring that an admissible alternative remains admissible as a menu is expanded. As an 'expansion consistency' condition, condition γ^* requires that whatever is admissible for choice from each menu in a collection of menus must remain admissible from the union of the collection of menus.[10]

Condition β^+, another expansion consistency condition, demands that if *any* alternative from a menu is admissible for choice when the menu is expanded, then *every* admissible alternative from the menu must be admissible for choice in the expanded menu.[11]

Definition 17.3.10 Let f be a propositional selection function.

(i) We say that a binary relation R on $\mathcal{W}_{\mathcal{L}}$ *rationalizes* f if for every $S \in \mathscr{E}_{\mathcal{L}}$:

$$f(S) = \{x \in S : yRx \text{ for all } y \in S\}.$$

We call f *rational* (or *rationalizable*) if there is a binary relation R on $\mathcal{W}_{\mathcal{L}}$ that rationalizes f.

(ii) We say that f is (*transitive, complete, quasi-order, etc.*) *G-rational* (or *G-rationalizable*) if there is a reflexive (transitive, complete, quasi-order, etc.) binary relation $\preceq$ on $\mathcal{W}_{\mathcal{L}}$ that rationalizes f.[12]

A rational selection function captures the basic idea behind the principle of preference maximization: For each decision problem S, $f(S)$ represents those options from S which are optimal according to some underlying binary relation R. G-rational selection functions require more. Intuitively, a quasi-order G-rational selection function, for example, has the property that an agent's disposition f to choose reveals that he or she would maximize according to a reflexive and transitive relation $\preceq$ which represents his or her preferences.

It is well known from the theory of choice functions that under certain domain constraints conditions α and γ^* completely characterize rational selection functions (see, e.g., [Sen, 1971]). Stated in the context of belief change, we have the following theorem.

Theorem 17.3.4 *A propositional selection function f is rational if and only if it satisfies condition α and condition γ^*.*

In much of the literature on the theory of choice, selection functions are assumed to take the empty set as a value only if the menu under consideration is null:

($f_{>\emptyset}$) For every $S \in \mathscr{E}_{\mathcal{L}}$, if $S \neq \emptyset$, then $f(S) \neq \emptyset$. (*Regularity*)

Rott calls this condition *success* in [Rott, 2001, p. 150]. We will call a selection function that satisfies condition $f_{>\emptyset}$ *regular*. Added as a hypothesis, regularity guarantees that G-rational selection functions are characterized by α and γ^*.

Theorem 17.3.5 *A regular propositional selection function f is G-rational if and only if it satisfies condition α and condition γ^*.*

G-rationality alone is a weak rationality constraint on selection functions. Among other properties, often quasi-order G-rationality is an additional constraint imposed on selection functions, requiring the rationalizing relation $\preceq$ to be both reflexive and transitive.

Theorem 17.3.6 *A regular propositional selection function f is quasi-order G-rational if and only if it satisfies condition α and β^+.*

A straightforward application of Zorn's Lemma establishes a result due to Szpilrajn ([Szpilrajn, 1930]), which states that every quasi-order has a weak order extension.[13] With this result at hand, it is easily proved that a regular selection function is weak order G-rational just in case it is quasi-order G-rational, whereby the following result obtains.

Corollary 17.3.1 *A regular selection function f is weak order G-rational if and only if it satisfies condition α and β^+.*

Let us now turn to Rott's correspondence results. We first define the notion of a *complete* propositional selection function.

Definition 17.3.11 Let f be a propositional selection function on $\mathscr{E}$.

(i) We define a propositional selection function $\bar{f}$ on $\mathscr{E}$ by setting for all $S \in \mathscr{E}$:

$$\bar{f}(S) := [\![\mathrm{Th}(f(S))]\!].$$

We call $\bar{f}$ the *completion* of f.

(ii) We say that f is *complete* if $f = \bar{f}$.

Observe that for every $S \in \mathscr{E}_{\mathcal{L}}$, $\bar{f}(S) \subseteq S$, so $\bar{f}$ is a propositional selection function. Also observe that for all $S \in \mathscr{E}_{\mathcal{L}}$, $f(S) \subseteq \bar{f}(S)$. Finally, observe that if $\mathcal{L}$ is finite, then every propositional selection function is complete.[14]

We now define the notion of a choice-based revision function.

Definition 17.3.12 Let K be a belief set, and let f be a propositional selection function. The *propositional choice-based revision function* $*$ for K *generated by* f is defined by setting for every formula ϕ,

$$K * \phi := \mathrm{Th}(f([\![\phi]\!])).$$

We say that f *generates* $*$ or that $*$ is *generated by* f.

To bring the ideas concerning rationalizability to the foreground, we offer the following definition.

Definition 17.3.13 Let K be a belief set. We call a function $*$ a (*complete, regular, rational, G-rational, etc.*) *choice-based revision function* for K if there is a (complete, regular, rational, G-rational, etc.) propositional selection function f on $\mathscr{E}_{\mathcal{L}}$ that generates $*$.

Observe that every choice-based revision function for K satisfies postulates $(K*1)$, $(K*2)$, and $(K*6)$. It is an easy matter to check that the converse holds as well. If $*$ satisfies postulates $(K*1)$, $(K*2)$, and $(K*6)$, then $*$ is a choice-based revision function for K.

Also observe that $*$ is a choice-based revision function for K generated by f if and only if for every formula ψ,

$$\psi \in K * \phi \text{ if and only if } f([\![\phi]\!]) \subseteq [\![\psi]\!].$$

Intuitively, an agent believes a sentence ψ in the revision of K by ϕ just in case ψ is true in all the most 'plausible' worlds in which ϕ is true. Of course, the role of a propositional selection function – or any selection function – can be interpreted in various ways in different contexts.

In his [Rott, 2001], Rott discusses a handful of coherence constraints for selection functions, some of which are well known and others of which he introduces. We present two conditions of the latter sort without offering motivation (see [Rott, 2001, pp. 147–9] for such motivation):

($F1_B$) For every $S \in \mathscr{E}_{\mathcal{L}}$, if $S \cap B \neq \emptyset$, then $f(S) \subseteq B$. *(Faith 1 respect to B)*

($F2_B$) For every $S \in \mathscr{E}_{\mathcal{L}}$, $S \cap B \subseteq f(S)$. *(Faith 2 respect to B)*

We finally turn to Rott's recent correspondence results which establish a one-to-one correspondence between coherence constraints from rational choice and postulates of belief revision.[15] Presented in a form suitable for this article, the following theorem provides one part of the connection (cf. [Rott, 2001, p. 197]).

Theorem 17.3.7 *Let K be a belief set. For every propositional selection function f which satisfies a condition in Column I and the adjoining constraint in Column II, the propositional choice-based revision function $*$ for K generated by f satisfies ($K * 1$), ($K * 2$), and ($K * 6$) and the adjacent postulate in column III (see Table 17.1).*

TABLE 17.1 If f satisfies a condition in column I and the adjoining constraint in column II, then $*$ satisfies the adjacent postulate in column III

I	II	III
-	-	-
$F2_{[K]}$	-	$(K * 3)$
$F1_{[K]}$	-	$(K * 4)$
$f_{>\emptyset}$	-	$(K * 5)$
α	$f = \bar{f}$	$(K * 7)$
γ^*	$f = \bar{f}$	$(K * 8r)$
β^+	-	$(K * 8)$

Theorem 17.3.7 is a 'soundness' result, and it is accompanied by a 'completeness' result. Also presented in a form suitable for this article, the following completeness result is the other part of the connection between coherence constraints of rational choice and rationality postulates of belief revision (cf. [Rott, 2001, p. 198]).

Theorem 17.3.8 *Every function $*$ satisfying ($K * 1$), ($K * 2$), and ($K * 6$) is a propositional choice-based revision function for K generated by a propositional selection*

function f, such that if $*$ satisfies a postulate in column I, *then f satisfies the adjacent condition in column* II *(see Table 17.2).*

TABLE 17.2 If $*$ satisfies a postulate in column I, then f satisfies the adjacent condition in column II

I	II
-	-
$(K * 3)$	$F2_{[K]}$
$(K * 4)$	$F1_{[K]}$
$(K * 5)$	$f_{>\emptyset}$
$(K * 7)$	α
$(K * 8r)$	γ^*
$(K * 8)$	β^+

The reader should observe the modular character of Theorem 17.3.7 as well as Theorem 17.3.8 below. Theorem 17.3.7, for example, says that for every belief set K and propositional selection function f, *if* f satisfies condition $F1_{[K]}$, *then* the choice-based revision function $*$ for K generated by f satisfies postulate $(K * 4)$ (as well as postulates $(K * 1)$, $(K * 2)$, and $(K * 6)$). Theorem 17.3.7 *also* says that for every belief set K and propositional selection function f, *if* f is complete and satisfies condition α, *then* the propositional choice-based revision function $*$ for K generated by f satisfies postulate $(K * 7)$ (again, as well as $(K * 1)$, $(K * 2)$, and $(K * 6)$).

The preceding theorems do not presuppose any basic postulates other than $(K * 1)$, $(K * 2)$, and $(K * 6)$. We can apply the results from the theory of choice functions to obtain the following corollary.

Corollary 17.3.2 *Let $*$ be a function on* For($\mathcal{L}$) *satisfying $(K*1)$, $(K*2)$, and $(K*6)$. Then:*

- *(i) The function $*$ is a rational complete choice-based revision function for K if and only if it satisfies $(K * 7)$ and $(K * 8r)$.*
- *(ii) The function $*$ is a regular G-rational complete choice-based revision function for K if and only if it satisfies $(K * 5)$, $(K * 7)$, and $(K * 8r)$.*
- *(iii) The function $*$ is a regular weak order (quasi-order) G-rational complete choice-based revision function for K if and only if satisfies $(K * 5)$, $(K * 7)$, and $(K * 8)$.*

The preceding corollary reveals the close connection between rationalizability and postulates of belief change. One can add or subtract postulates of belief change to obtain corresponding coherence constraints which characterize various notions of rationalizability, thereby exploiting results from the theory of choice functions. Thus, the foregoing discussion of Rott's results should serve to indicate the depth and utility of the connection between rational choice and belief change. Indeed, Rott's work in [Rott, 2001] has initiated a new and exciting area of research in the study of belief change.[16]

4. Doubts about Recovery, and Some Reactions

We now return to belief contraction. We anticipated before that Recovery is one of the most controversial postulates proposed by AGM. Around 1991 researchers offered various counterexamples to Recovery. For example, Sven Ove Hansson offers the following alleged counterexamples:

Example 17.4.1 ([Hansson, 1991]) While reading a book about Cleopatra I learned that she had both a son and a daughter. I therefore believe both that Cleopatra had a son (s) and Cleopatra had a daughter (d). Later I learn from a well-informed friend that the book in question is just a historical novel, accordingly contracting my belief that Cleopatra had a child ($s \vee d$). However, shortly thereafter I learn from a reliable source that in fact Cleopatra had a child. I find it quite reasonable to thereby reintroduce $a \vee b$ to my collection of beliefs without also returning either s or d. This contradicts Recovery.

Example 17.4.2 ([Hansson, 1996]) I believed both that George is a criminal (c) and George is a mass murderer (m). Upon receiving certain information I am induced to retract my belief set K by my belief that George is a criminal (c). Of course, I therefore retract my belief set by my belief that George is a mass murderer (m). Later I learn that in fact George is a shoplifter (s), so I expand my contracted belief set $K \dot{-} c$ by s to obtain $(K \dot{-} c) + s$. As George's being a shoplifter (s) entails his being a criminal (c), $(K \dot{-} c) + c$ is a subset of $(K \dot{-} c) + s$. Yet by Recovery it follows that $K \subseteq (K \dot{-} c) + c$, so m is a member of the expanded belief set $(K \dot{-} c) + s$. But I do not believe that George is a mass murderer (m), contradicting the recommendation of Recovery.

While Peter Gärdenfors ([Gärdenfors, 1982]) has contended that Recovery is a reasonable principle, another member of the AGM trio, David Makinson, has expressed doubts about Recovery ([Makinson, 1987]) and at the same time has defended its use in certain contexts ([Makinson, 1997]). Indeed, [Makinson, 1997] argues that the examples presented above are persuasive only as a result of tacitly adding to the theory of contraction a justificatory structure that is not formally represented. For example, Makinson claims that in the second example above we are inclined to take for granted that $m \vee \neg s$ is in the belief set *only because* m is there. Makinson concludes:

> As soon as contraction makes use of the notion 'y is believed only because x,' we run into counterexamples to recovery [...] But when a theory is 'naked,' i.e. as a bare set $A = \mathrm{Cn}(A)$ of statements closed under consequence, then recovery appears to be free of intuitive counterexamples. [Makinson, 1997, p. 478]

Thus Makinson seemingly argues that Recovery can fail only in cases in which some justificatory structure is added to the belief set and used to determine the content of a contraction. More recently, however, Isaac ([Levi, 2003]) has argued that Recovery can fail even when belief sets are 'naked'. To appreciate Levi's point we need to introduce some salient aspects of his work in belief change. We will do this in the next subsection.

4.1 Levi Contractions

Levi's point of departure is based on the observation that remainder sets are too restrictive. He proposes instead to focus on supersets of remainder sets called *saturatable sets* ([Levi, 1991]).

Definition 17.4.1 Let K be a theory, and let α be a formula. The *α-saturatable set*, $S(K,\alpha)$, is the collection of subsets Γ of $\mathrm{For}(\mathcal{L})$ such that:

(i) $\Gamma \subseteq K$;
(ii) $\Gamma = \mathrm{Cn}(\Gamma)$;
(iii) $\mathrm{Cn}(\Gamma \cup \{\neg\alpha\})$ is maximal consistent with respect to Cn.[17]

We call a member of $S(K,\alpha)$ an *α-saturatable subset* of K. We let $S(K,\mathcal{L}) := \{S(K,\alpha) : \alpha \in \mathrm{For}(\mathcal{L})\}$.

In Levi's terminology, members of $S(K,\alpha)$ are *saturatable contractions* of K removing α. It follows from the above definition that a saturatable set indeed contains the corresponding remainder set:

Proposition 17.4.1 **(Hansson and Olsson [Hansson and Olsson, 1995])** Let K be a theory. Then for every formula $\alpha \in K$, $K \perp \alpha \subseteq S(K,\alpha)$.

In [Levi, 1991], Levi also reformulates the *Principle of Economy*, a maxim guiding the AGM theory according to which losses of information should be minimized in contraction. Levi instead adopts a principle according to which what is minimized in contraction are losses of informational *value* rather than information. To represent informational value, we can use a real-valued function $V : \mathbb{K} \rightarrow \mathbb{R}$, called a *value function*. Levi argues that an important requirement of informational value is that it is *weakly monotonic*:

Principle of Weak Monotony For every $\Gamma, \Delta \in \mathbb{K}$, if $\Gamma \subseteq \Delta$, then $V(\Gamma) \leq V(\Delta)$.

This principle does not exclude the possibility that a set contains strictly less *information* than another set, yet the *informational value* of both sets is the same. The extra information in the larger set might not be relevant or epistemically important.

Recall that partial meet contraction employs a selection function that selects among the elements of $K\bot\alpha$. In this setting, a selection function selects among elements of $S(K, \alpha)$.

Definition 17.4.2 Let K be a theory. A *selection function* for K is a function δ on $S(K, \mathcal{L})$ such that for all formulae α:

(i) If $S(K, \alpha) \neq \emptyset$, then:
   (a) $\delta(S(K, \alpha)) \subseteq S(K, \alpha)$, and
   (b) $\delta(S(K, \alpha)) \neq \emptyset$;
(ii) If $S(K, \alpha) = \emptyset$, then $\delta(S(K, \alpha)) = \{K\}$.

Now we have a feasible set $S(K, \alpha)$ that is larger than a remainder set and a notion of informational value that should at least obey the Principle of Weak Monotony. We can thereby define the notion of a *value-based Levi contraction*.[18]

Definition 17.4.3 Let K be a belief set. A function $\dot{-}$ is a *value-based Levi contraction* for K if there is a selection function δ for K and a weakly monotonic value function V such that for every formula α:

$$K \dot{-} \alpha = \begin{cases} \bigcap \delta(S(K, \alpha)) & \text{if } \alpha \in K; \\ K & \text{otherwise.} \end{cases} \tag{17.1}$$

If $\alpha \in K \backslash \mathrm{Cn}(\emptyset)$, then:

$$\delta(S(K, \alpha)) = \{\Gamma \in S(K, \alpha) : V(\Delta) \leq V(\Gamma) \text{ for all } \Delta \in S(K, \alpha)\}.^{19} \tag{17.2}$$

[Hansson and Olsson, 1995] have shown that every value-based Levi contraction satisfies postulates $(K \dot{-} 1)$ to $(K \dot{-} 5)$ as well as $(K \dot{-} 7)$ and $(K \dot{-} 8)$.

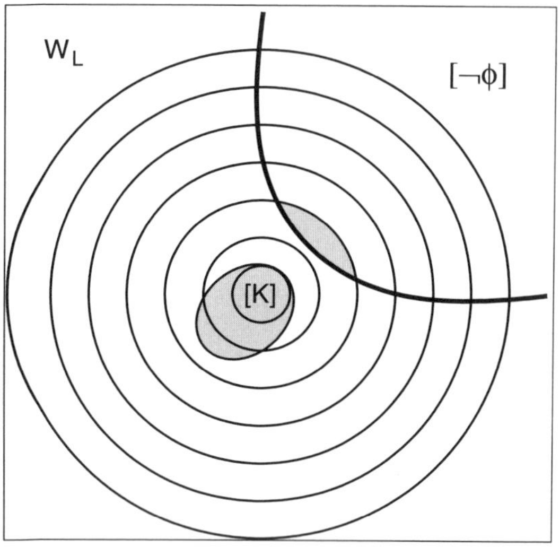

FIGURE 17.5 *Levi Contraction* (the case in which $\phi \in K\backslash\mathrm{Cn}(\emptyset)$). The grey region represents $[\![K \dot{-} \phi]\!]$

More recently, [Arló Costa and Liu, 2010] have proven that value-based Levi contraction is characterized by the above postulates and an additional postulate:

$(K \dot{-} 7c)$ If $\alpha \in K \dot{-} (\alpha \wedge \beta)$, then $K \dot{-} \beta \subseteq K \dot{-} (\alpha \wedge \beta)$. (*Conjunctive Reduction*)

We accordingly have the following theorem.

Theorem 17.4.1 ([Hansson and Olsson, 1995], [Arló Costa and Liu, 2010]) *Let K be a belief set, and let* $\dot{-}$ *be a function on* $\mathrm{For}(\mathcal{L})$. *Then* $\dot{-}$ *is a value-based Levi contraction for K if and only if it satisfies* $(K \dot{-} 1)$ *to* $(K \dot{-} 5)$, $(K \dot{-} 7)$, $(K \dot{-} 7c)$, *and* $(K \dot{-} 8)$.

Notice that Recovery does not appear among the list of axioms. It is not difficult to produce counterexamples to Recovery in this setting even when the theories used in this approach are 'naked' and no justificatory structure appears in the belief sets. Figure 17.5 is a geometrical depiction of a Levi contraction.

Makinson discusses saturatable contractions in [Makinson, 1987] (he calls these contractions *withdrawals*), arguing against recommending Levi contractions. He contends that any given saturatable but not maxichoice contraction removing α is always weaker than some maxichoice contraction removing α. As a consequence, he concludes, choosing the meet of saturatable but not maxichoice contractions always incurs a greater loss of information than choosing the meet of the associated maxichoice contractions. As [Levi, 2003] has argued, this argument is compelling if the sole aim of contraction is the minimization of *informational* loss. But we have seen above that such a principle is compromised in the AGM theory and cannot be taken as the sole aim of contraction. Levi plainly rejects the Principle of Economy, so the argument does not apply to his theory.

4.2 Mild Contractions and Severe Withdrawals

Levi's notion of contraction has a decision-theoretic flavour at least insofar as a relevant epistemic index is maximized (minimized) over a feasible set of potential contractions. As we have seen, the first approximation to the problem of maximization from the point of view of AGM is to appeal to the Principle of Economy. Yet if one were to apply this principle strictly, the only contractions that would be justified would apparently be maxichoice contractions. But this principle is compromised in partial meet contraction, which takes the *intersection* of a *subset* of maxichoice contractions. Clearly the intersection need not be optimal with respect to the Principle of Economy. Levi contractions face the same problem, since there is no guarantee that the intersection of of a subset of saturatable contractions is itself optimal. To solve this problem, Levi proposes a value index for which the intersection of optimal elements is itself optimal.

Accordingly, [Arló Costa and Levi, 2006] introduces a further constraint on the value function V by way of the principle of *Weak Min*:

Weak Min For every finite $\mathscr{F} \subseteq S(K,\alpha)$, $V(\bigcap_{\Gamma\in\mathscr{F}} \Gamma) = \min_{\Gamma\in\mathscr{F}} V(\Gamma)$.

More generally, for any two potential contractions K_0 and K_1 the value of their intersection is the minimum of the values of K_0 and K_1. [Arló Costa and Levi, 2006] derive these principles from more primitive axioms in an attempt to justify them in general (see the principles of Weak Monotony, Extended Weak Monotony, and Weak Intersection Equality presented in [Arló Costa and Levi, 2006]). An obvious justification of Weak Min must show that the intersection of optimal items is optimal. This is not present in the theory presented in [Levi, 1991]. So in this case one needs to assume a special Rule for Ties that is not directly derived from pure considerations of optimality. In his recent book [Levi, 2004], Levi offers another decision-theoretic justification of mild contractions.

[Arló Costa and Levi, 2006] present an argument showing that value-based Levi contractions obeying the aforementioned constraints on V are characterized by postulates $(K \dot{-} 1)$ to $(K \dot{-} 5)$, $(K \dot{-} 8)$, and the following postulate:

$(K \dot{-} 7a)$ If $\alpha \notin \mathrm{Cn}(\emptyset)$, then $K \dot{-} \alpha \subseteq K \dot{-} (\alpha \wedge \beta)$. *(Antitony)*

[Rott and Pagnucco, 1999] offer an independent representation result for the same set of postulates in terms of sphere semantics, calling an operation satisfying these postulates a *severe withdrawal* rather than a mild contraction

(Levi's opposing terminology reflects the idea that what might look severe from the point of view of pure informational loss might not look this way if one changes perspective and focuses on information value). Recall that for a system of spheres $\mathscr{S}$ and a formula ϕ, we have defined the following set:

$$\mathscr{C}_\phi := \{S \in \mathscr{S} : S \cap [\![\phi]\!] \neq \emptyset\} \cup \{\mathcal{W}_\mathcal{L}\}.$$

We now define Rott and Pagnucco's withdrawal operation in terms of sphere semantics.

Definition 17.4.4 Let K be a belief set. A function $\dot{-}$ is a *sphere-based severe withdrawal* for K if there is system of spheres $\mathscr{S}$ centred on $[\![K]\!]$ such that for all formulae ϕ:

$$K \dot{-} \phi = \begin{cases} \mathrm{Th}(\min_{\subseteq}(\mathscr{C}_{\neg\phi})) & \text{if } \phi \notin \mathrm{Cn}(\emptyset); \\ K & \text{otherwise.} \end{cases}$$

Figure 17.6 illustrates the situation with severe withdrawal. Observe that in contrast with partial meet contraction, a severe withdrawal is determined not only by worlds in $[\![\neg\phi]\!] \cap \min_{\subseteq}(\mathscr{C}_{\neg\phi})$ but also by worlds in $[\![\phi]\!] \cap \min_{\subseteq}(\mathscr{C}_{\neg\phi})$.

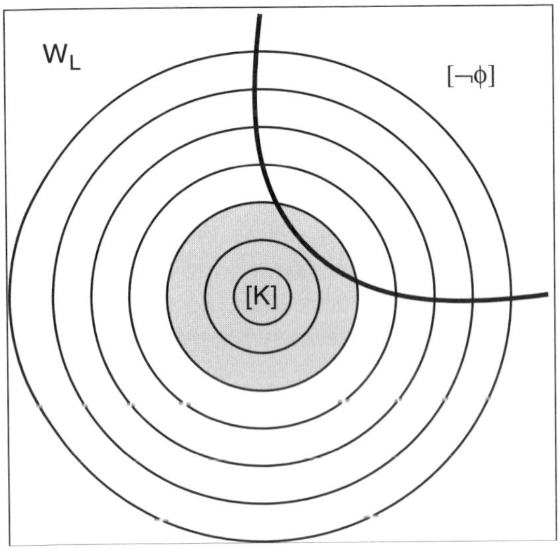

FIGURE 17.6 *Severe Withdrawal* (the case in which $\phi \in K\backslash\mathrm{Cn}(\emptyset)$). The grey disc represents $\min_{\subseteq}(\mathscr{C}_{\neg\phi})$, which generates the contraction of K by ϕ, $K \dot{-} \phi = \mathrm{Th}(\min_{\subseteq}(\mathscr{C}_{\neg\phi}))$

Rott and Pagnucco offer a general philosophical argument defending the coherence of severe withdrawal. With respect to sphere semantics, they contend that severe withdrawals obey the *Principle of Weak Preference*, according to which if a world w is considered at least as plausible as another w', then w should be admitted in the agents epistemic state if w' is admitted ([Rott and Pagnucco, 1999]). They write:

> The Principle of Informational Economy, in a weak form, can be viewed as limiting the extent of change to that sphere containing the closest $\neg\phi$-worlds and not beyond. The Principle of Weak Preference determines which worlds inside this limited region should be included in the new epistemic state. Without any further restrictions it suggests that all worlds inside this region should form part of the contracted epistemic state. In a way, even AGM appeal to this principle. There, however, the principle is only applied relative to $\neg\phi$-worlds, not all worlds in $\mathcal{W}$. However, no principle authorising a restricted imposition of this principle is established... The agent has determined a preference over worlds and does not prefer the (closest) $\neg\phi$-worlds over the (closer) ϕ-worlds just because it is giving up belief in ϕ. Its preferences are established prior to the change and we assume that there is no reason to alter them in light of the new information (epistemic input). ([Rott and Pagnucco, 1999, pp. 8–9])

For this reason, Rott and Pagnucco conclude that the Principle of Economy must give way.

Perhaps the simplest and most elegant way of introducing severe withdrawals is by way of epistemic entrenchment. Recall that in Section 2.2 we offered a definition of contraction in terms of entrenchment (Definition 17.2.5) due to [Gärdenfors, 1988] and [Gärdenfors and Makinson, 1988]. We then indicated that [Rott, 1991] has suggested that Gärdenfors' entrenchment-based contraction has little motivation. As we have seen, Rott has proposed an alternative definition of contraction in terms of entrenchment which seems better motivated and certainly more intuitive.

Definition 17.4.5 Let K be a belief set. We say that a function $\dot{-}$ on For($\mathcal{L}$) is an *entrenchment-based severe withdrawal* for K if there is an entrenchment relation $\leq$ such that for every formula α:

$$K \dot{-} \alpha = \begin{cases} K \cap \{\beta : \alpha < \beta\} & \text{if } \alpha \notin \mathrm{Cn}(\emptyset); \\ K & \text{otherwise.} \end{cases}$$

[Rott and Pagnucco, 1999] show that the postulates for severe withdrawal characterize this entrenchment-based operation.[20] In summary, we have the following theorem.

Theorem 17.4.2 ([Rott and Pagnucco, 1999]) *Let K be a belief set, and let* $\dot{-}$ *be a function on* For($\mathcal{L}$). *Then:*

(i) *The function* $\dot{-}$ *is a sphere-based severe withdrawal for K if and only if it satisfies postulates* $(K\dot{-}1)$ *to* $(K\dot{-}5)$, $(K\dot{-}7a)$, *and* $(K\dot{-}8)$.
(ii) *The function* $\dot{-}$ *is an entrenchment-based severe withdrawal for K if and only if it satisfies postulates* $(K\dot{-}1)$ *to* $(K\dot{-}5)$, $(K\dot{-}7a)$, *and* $(K\dot{-}8)$.

Despite the appeal of several withdrawals, some consequences of their characterizing postulates are puzzling. For example, one can derive that either $K\dot{-}\phi \subseteq K\dot{-}\psi$ or $K\dot{-}\psi \subseteq K\dot{-}\phi$. That is, severe withdrawals are nested. This suggests that severe withdrawals are too orderly: Any two contractions of a theory are such that either one of them entails the other, or vice versa. Perhaps this consequence is too strong, even while it is a trivial consequence of the sphere semantics used in [Rott and Pagnucco, 1999] and the semantics of shells of informational value used in [Arló Costa and Levi, 2006].

Other consequences of the postulates for severe withdrawals also seem rather unintuitive. For example, a property called *Expulsiveness* is a consequence of the postulates that has received criticism. Expulsiveness requires that for any two non-tautological sentences α and β that either $\alpha \notin K\dot{-}\beta$ or $\beta \notin K\dot{-}\alpha$. [Hansson, 2009] argues against this condition:

> This is a highly implausible property of belief contraction, since it does not allow unrelated beliefs to be undisturbed by each other's contraction. Consider a scholar who believes that her car is parked in front of the house. She also believes that Shakespeare wrote the Tempest. It should be possible for her to give up the first of these beliefs while retaining the second. She should also be able to give up the second without giving up the first. Expulsiveness does not allow this. The construction of a plausible operation of contraction for belief sets that does not satisfy Recovery is still an open issue. ([Hansson, 2009])

Expulsiveness seems implausible for *related* beliefs as well. Consider the same example but with two relevant beliefs, that her car is parked in front of the house and that the car contains a bomb. It seems that it should be plausible to give up the belief that the car is parked in front of her house with a bomb in it. It also seems perfectly possible to give up the belief that the car contains a bomb while preserving the belief that the car is parked in front of the scholar's house.

Antitony itself has also been criticized. For example, Hansson asserts that Antitony (without the proviso that the contracted sentence α is not a logical theorem) 'does not hold for any sensible operator of contraction' [Hansson, 1999,p. 117].[21]

None of the aforementioned problems arise for saturatable contraction. It seems that this notion of contraction is the best candidate currently available in the literature that can violate Recovery.

4.3 Belief Base Contraction

There is a separate and independently motivated way of avoiding Recovery. The idea is to appeal to *belief bases* rather than belief sets to represent explicit beliefs. A *belief base* is simply a set of formulae which is not required to be logically closed. The formulae comprising a belief base are intended to represent those beliefs that are held independently of any other belief or collection of beliefs. As such, logical consequences of a belief base that are not in the belief base are Òmerely derivedÓ, i.e., they have no independent standing ([Hansson, 2009]).

The central idea regarding belief dynamics is that changes are always performed on the belief base. While an agent might be committed to the logical consequences of a base, if a derived belief loses support it will be automatically discarded. The following example, due to Hansson, makes this explicit.

Example 17.4.3 ([Hansson, 2009]) I believe that Paris is the capital of France (p). I also believe that there is milk in the fridge (m). Therefore, I believe that Paris is the capital of France if and only if there is milk in the fridge ($p \leftrightarrow m$). I open the fridge and find it necessary to replace my belief in m with belief in $\neg m$. I cannot then, on pain of inconsistency, retain both my belief in p and my belief in $p \leftrightarrow m$.

If we were to represent the current epistemic state by a theory, then both p and $p \leftrightarrow m$ would be elements of the belief set. When one opens the fridge and finds no milk one has to choose between retaining p and retaining $p \leftrightarrow m$. The retraction of $p \leftrightarrow m$ is not automatic. But in the belief base approach, the option of retaining $p \leftrightarrow m$ does not even arise. Since m is a basic belief, while $p \leftrightarrow m$ is a derived belief, when m is removed, the biconditional is immediately removed.

Although Hansson's example is quite convincing, the situation can be reversed. Consider the following example:

Example 17.4.4 On March 12, 2008, I believe that governor Spitzer will resign effective on March 17, 2008 (s). I also believe that David Paterson will assume as governor of New York on March 17, 2008 (p), so I believe that governor Spitzer will resign effective on March 17, 2008 if and only if David Paterson will assume as governor on March 17, 2008 (s and $s \leftrightarrow p$). Now (say on March 13th) I learn that governor Spitzer has not resigned ($\neg s$). I cannot then, on pain of inconsistency, retain both my belief in p and my belief in s and $s \leftrightarrow p$.

Structurally the examples are similar, only that, in spite of the fact that p is a basic belief and $s \leftrightarrow p$ is a derived belief, it seems more reasonable to retain s and $s \leftrightarrow p$ and to reject p. At least this seems a permissible epistemic strategy. Notice, nevertheless, that if we were to use bases to represent this example, the strategy in question would not be available. The rejection of s and $s \leftrightarrow p$ would be automatic.

The previous example suggests that the representation of epistemic states using bases may be too rigid, limiting the epistemic options of an agent in an unreasonable manner. In spite of this and other problems, there is an important and interesting literature on bases. Many applications, for example in computer science, depend on representing epistemic states using belief bases.

The definitions of a remainder set and partial meet contraction from Section 2.1 apply to belief bases. One can thereby investigate the logical structure of partial meet contraction for belief bases rather than just belief sets. Most postulates for contraction hold in this new setting, with the exception of Recovery. The following example illustrates the failure of Recovery in this setting. The example was originally formulated by [Levi, 1991] and adapted with a different purpose by [Hansson, 2009].

Example 17.4.5 ([Hansson, 2009]) Let the belief set K include both a belief that the coin was tossed (c) and a belief that it landed heads (h). The epistemic agent wishes to consider whether on the supposition that the coin had been tossed, it would have landed heads. In order to do that, it would seem reasonable to remove c from the belief set and then reinsert it, i.e., to perform the series of operations $(K \dot{-} c) + c$.

(1) If partial meet contraction is performed directly on the belief set, then it follows from Recovery that $h \in (K \dot{-} c) + c$, i.e. h comes back with c. This is contrary to reasonable intuitions.

(2) If partial meet contraction is instead performed on a belief base for K, then Recovery can be avoided. Let the belief base be $\{p_1, \ldots, p_n, c, h\}$, where the background beliefs $p_1, \ldots, p_n$ are unrelated to c and h, whereas h logically implies c. Then $K = \text{Cn}(\{p_1, \ldots, p_n, c, h\})$. Since h implies c, it will have to go when c is removed, so that
$K \dot{-} c = \text{Cn}(\{p_1, \ldots, p_n\})$. When c is reinserted, the outcome is
$(K \dot{-} c) + c = \text{Cn}(\{p_1, \ldots, p_n, c\})$ that does not contain h, as desired.

An operator of partial meet contraction for an arbitrary set of formulae Θ is characterized by the following postulates ([Hansson, 1999]):

Success If $\alpha \notin \mathrm{Cn}(\emptyset)$, then $\alpha \notin \mathrm{Cn}(\Theta \dot{-} \alpha)$.
Inclusion $\Theta \dot{-} \alpha \subseteq \Theta$.
Relevance If $\beta \in \Theta$ and $\beta \notin \Theta \dot{-} \alpha$, then there is a set Θ' such that $\Theta \dot{-} \alpha \subseteq \Theta' \subseteq \Theta$ and that $\alpha \notin \mathrm{Cn}(\Theta')$ but $\alpha \in \mathrm{Cn}(\Theta' \cup \{\beta\})$.
Uniformity If it holds for all subsets Θ' of Θ that $\alpha \in \mathrm{Cn}(\Theta')$ if and only if $\beta \in \mathrm{Cn}(\Theta')$, then $\Theta \dot{-} \alpha = \Theta \dot{-} \beta$.

As the reader can see the postulate of Relevance has in this setting a role similar to that of Recovery in the theory of partial meet contraction for belief sets, without many of the undesirable consequences of adopting Recovery.

Hansson studied in a series of articles (see [Hansson, 1999] for a concise presentation) a different operation on belief bases called *kernel contraction*. For any sentence α, a α-kernel is a minimal α-implying set. A contraction operation $\dot{-}$ can be based on the simple principle that no α-kernel should be included in $K \dot{-} \alpha$. In order to implement this idea one can deploy an *incision function* selecting at least one element from each α-kernel. Hanson explains the relation between this operation with partial meet contraction in [Hansson, 2009]:

> An operation that removes exactly those elements that are selected for removal by an incision function is called an operation of kernel contraction. It turns out that all partial meet contractions on belief bases are kernel contractions, but the converse relationship does not hold, i.e. there are kernel contractions that are not partial meet contractions. In other words, kernel contraction is a generalization of partial meet contraction.

Another important application of kernel contraction is related to its use in the study of the form of contraction less understood in the literature so far: *safe contraction* [Alchourrón and Makinson, 1985]. Basically safe contractions can be seen as relational restrictions on certain type of kernel contractions. The problem of proving a characterization theorem for the class of safe contractions over theories remains open. Preliminary results towards finding such a characterization result can be found in the work of Alex Smith ([Smith, 2009]).

5. Doubts about Other Postulates

Thus far we have primarily focused on doubts about the Recovery postulate and several ways to accommodate these doubts within formal frameworks which still possess the spirit of that proposed by AGM. In this section we turn to doubts about other postulates, providing the reader with a glimpse of the formal and philosophical issues involved.

Let us begin with a simple purported counterexample to postulate ($K \dot{-} 7$).

Example 17.5.1 ([Hansson, 1999]) I believe that Accra is a national capital (a). I also believe that Bangui is a national capital (b). As a (logical) consequence of this, I also believe that either Accra or Bangui is a national capital ($a \vee b$).

Case 1. 'Give the name of an African capital' says my geography teacher.
'Accra' I say, confidently.
The teacher looks angrily at me without saying a word. I lose my belief in a. However, I still retain my belief in b, and consequently in $a \vee b$.

Case 2. I answer 'Bangui' to the same question. The teacher gives me the same wordless response. In this case, I lose my belief in b, but I retain my belief in a and consequently my belief in $a \vee b$.

Case 3. 'Give the names of two African capitals' say my geography teacher.
'Accra and Bangui' I say, confidently.
The teacher looks angrily at me without saying a word. I lose confidence in my answer, that is, I lose my belief in $a \wedge b$. Since my beliefs in a and b were equally strong, I cannot choose between them, so I lose both of them. After this, I no longer believe in $a \vee b$.

Since $a \vee b$ is an element of $K \dot{-} a \cap K \dot{-} b$ but not an element of $K \dot{-} (a \wedge b)$, clearly postulate ($K \dot{-} 7$) is violated. [Hansson, 1999, p. 79] argues that this postulate can be defended from the perspective of a belief base representation. Since $a \vee b$ is not a basic belief, it is not an element of $K \dot{-} a$, although it is an element of $\mathrm{Cn}(K \dot{-} a)$. Therefore, $a \vee b$ is not an element of $(K \dot{-} a) \cap (K \dot{-} b)$. Hansson concludes that the fact that $a \vee b$ is not a member of $K \dot{-} (a \wedge b)$ does not contradict ($K \dot{-} 7$).

Recently Hans ([Rott, 2004a]) has presented a single counterexample to several postulates of belief contraction and belief revision, most notably postulates ($K * 7$) and ($K * 8$). Rott takes his counterexample to suggest that many of the most cherished fundamental principles of belief change should not be regarded as valid for commonsense reasoning, explaining this in terms of a transformation of a familiar problem of rational choice to a problem of belief formation.

We will present Rott's counterexample here, focusing on its relevance to postulates of belief revision. The counterexample involves three hypothetical scenarios in which an agent accepts belief-contravening information. Each scenario describes a potential unfolding of events. The scenarios in the counterexamples are *not* consecutive stages of a single chain of events. Rather, each scenario describes one way things could turn out. Moreover, only one of these scenarios will be realized.

Example 17.5.2 ([Rott, 2004a]) A philosophy department has announced an open position in metaphysics. Tom, an interested bystander, happens to know a few of the applicants: Amanda Andrews, Bernice Becker, Carlos Cortez, and Don Doyle. Tom, just like everyone else, knows that Andrews is an outstanding specialist in metaphysics, whereas Becker, who is also a very good metaphysician, is not quite as excellent as Andrews. However, Becker has done some substantial work in logic. Cortez has a comparatively slim record in metaphysics, yet he is widely recognized as one of the most brilliant logicians of his generation. By contrast, Doyle is a star metaphysician, while Andrews has done close to no work in logic.

Now suppose Tom initially believes that neither Andrews, Becker, nor Cortez will be offered the position because he, like everyone else, believes that Doyle is the obvious candidate to be offered the position. Tom is well-aware that only one of the applicants will be offered the position. Let a, b, c, and d stand for the following sentences:

a: Andrews will be offered the position.
b: Becker will be offered the position.
c: Cortez will be offered the position.
d: Doyle will be offered the position.

Tom is having lunch with the dean. The dean is a very competent, serious, and honest man. He is also the chairman of the selection committee.

Scenario 1. The dean informs Tom that either Andrews or Becker will be offered the position. That is, the dean informs Tom that $a \vee b$. Because Tom presumes that expertise in metaphysics is the decisive criterion for the selection committee's decision, Tom concludes that Andrews will be offered the position (and of course that all other applicants will not be offered the position).

Scenario 2. The dean confides to Tom that either Andrews, Becker, or Cortez will be offered the position, thereby supplying him with $a \vee b \vee c$. Because Cortez is a brilliant logician, Tom realizes that he cannot sustain his presumption that metaphysics is *the* decisive criterion for the selection committee's decision.

From Tom's perspective, logic *also* appears to be regarded as a considerable asset by the selection committee. Nonetheless, because Cortez has such a slim record in metaphysics, Tom believes that Cortez will not be offered the position. But Tom sees that logic contributes to an applicant's chances of being offered a position. Tom thereby concludes that Becker will be offered the position (and so no other applicant will be offered the position).

Scenario 3. The dean tells Tom that Cortez will be offered the position, thereby supplying him with c. Tom is certainly surprised, yet he believes what the dean tells him.

Let us take stock of Tom's beliefs in these scenarios. Initially, Tom believes d, $\neg a$, $\neg b$, and $\neg c$. Thus, letting K denote Tom's initial belief set, $d, \neg a, \neg b$ and $\neg c$ are in K. In Scenario 1, Tom's revises his belief set K by $a \vee b$, and his revised belief set $K * (a \vee b)$ contains a and $\neg b$, as well as $\neg c$ and $\neg d$. In Scenario 2, Tom revises his belief set K by $a \vee b \vee c$. His revised belief set $K * (a \vee b \vee c)$ includes b, $\neg a$, $\neg c$, and $\neg d$. Finally, in Scenario 3, Tom revises his belief set K by c, whereby his revised belief set $K * c$ contains c, $\neg a$, $\neg b$, and $\neg d$.

We are now in a position to see that Example 17.5.2 constitutes a violation of postulates $(K * 7)$ and $(K * 8)$. Since $\neg b \in K * (a \vee b \vee c) \wedge (a \vee b) = K * (a \vee b)$ and $\neg b \notin \mathrm{Cn}((K * (a \vee b \vee c)) \cup \{a \vee b\}) = K * (a \vee b \vee c)$, postulate $(*7)$ is violated. Similarly, postulate $(K * 8)$ is violated.

In light of Theorem 17.3.7, we should be unsurprised to see that conditions α and β^+ are also violated. And they are. Rott argues that a well-known phenomenon from rational choice is responsible for these violations. This phenomenon turns on the *epistemic value* or *relevance* of the menu with which an agent is faced. We can explain Rott's idea as follows. When Tom faces the 'menu' represented by $a \vee b$, he does it under the presumption that metaphysics is the decisive criterion for the selection committee's decision. Therefore, when he has to judge the relative merits of Andrews and Becker as candidates, Tom concludes that Andrews will be offered the position. But the disclosure of certain facts about Cortez in Scenario 2 alters Tom's evaluation of the relative merits of Andrews and Becker as candidates and as a consequence Tom concludes that Becker will be offered the position instead. Since the information Tom receives includes certain facts about Cortez, and since this information has been acquired from a reliable source (viz., the dean), Tom *learns* something important about the selection criterion used by the selection committee (viz., that expertise in metaphysics is not the only decisive criterion used by the selection committee). Thus, Rott argues, Tom's revision when faced with $a \vee b \vee c$ has *epistemic relevance* for Tom's epistemic decision.

In his [Stalnaker, 2009], Robert Stalnaker scrutinizes Rott's example, contending that it does not threaten the principles of AGM and in particular the revision postulates in question. The principles, Stalnaker claims, should continue to apply. Nonetheless, Stalnaker agrees with Rott that the example in question shows that we need to take account of a richer body of information than done in the simple model supplied by AGM.

[Arló Costa and Pedersen, 2010] argue that the phenomenon pointed to above arises quite generally in the context of belief change, with particular attention given to the role norms play in belief formation. The authors propose a new theory of belief revision called *norm-inclusive belief revision*. As the name suggests, this theory is meant to accommodate the influence of norms in belief formation. The authors state and prove correspondence results in the style of Rott's results. This work is extended in various ways in [Pedersen, 2008].

6. Probability, Belief; Belief Change and Supposition

We return here to the topics considered at the beginning of this article. It has been pointed out rather frequently that the view of probability presented at the beginning of this essay is difficult to reconcile with the traditional notion of belief used in epistemology (both in its formal and informal variants). Some of the obvious options – such as adopting an acceptance rule that identifies highly probable propositions with believed propositions – either lead to paradox or require for their sound formulation abandoning basic logical principles. Nevertheless there are some recent attempts to derive both belief and monadic degree of belief from suppositions (i.e., from conditional probability assumed as a primitive). The idea that we will consider here is based on a slight reformulation of ideas presented by Bas van Fraassen in [van Fraassen, 1995].

Let's first introduce a notion of conditional probability that allows from conditioning on events of zero measure. We present a similar axiom system than the one proposed by [van Fraassen, 1995]. The idea is to introduce a function $P(\cdot|\cdot)$ defined on a σ-field $\mathscr{F}$ over some set W. The requirements are that

- (I) For every $A \in \mathscr{F}$, either:
  - (a) $P(\cdot|A)$ is a (countably additive) probability measure, or
  - (b) $P(\cdot|A)$ has constant value 1;
- (II) $P(A|A) = 1$;
- (III) $P(B \cap C|A) = P(B|A)P(C|B \cap A)$ for all $A, B, C \in \mathscr{F}$.

Axiom (I) allows for the representation of an inconsistent state, given by the constant function with value 1. The second axioms seems constitutive of the notion of conditional probability (any notion of probability that does not satisfy it cannot be properly called probability). The third axiom is very important. It has a long history going back at least to Jeffreys and to Keynes who used it in their books on probability.

For fixed A, if $P(\cdot|A)$ is a probability measure, then A is *normal*; otherwise it is *abnormal*, i.e., $P(\cdot|A)$ has constant value 1, so in particular, $P(\emptyset|A) = 1$.

Slightly modifying van Fraassen's definition we define a *core* as a set K which is normal and satisfies the *strong superiority condition* (SSC) i.e., if A is a nonempty subset of K and B is disjoint from K, then $P(B|A \cup B) = 0$ (and so $P(A|A \cup B) = 1$). Thus *any* non-empty subset of K is more 'believable' than any set disjoint from K. It can then be established that all non-empty subsets of a core are normal.

More importantly one can show that the family of cores that corresponds to a given probability function $P(\cdot|\cdot)$ is nested, i.e., that for any two cores for P, K_1, and K_1, either K_1 is included in K_2 or vice versa. In addition Arló-Costa showed in [Arló Costa, 1999] that the chain of belief cores induced by a 2-place function P cannot contain an infinitely descending chain of cores (countable additivity plays a central role in this proof).

Cores are well ordered under inclusion and closely resemble Grove spheres ([Grove, 1988]) and Spohn's *ranking functions* ([Spohn, 1998]). When the probabilistic space is countable one can show that there is a smallest as well as a largest core (the union of all cores).

The smallest core can be identified with (ordinary) beliefs or expectations and the largest core with full beliefs (i.e., a priori beliefs), so that in general probability 1 is not sufficient for full belief.

One can also see the smallest core (in the countable case) as the strongest proposition of measure one. One can establish that all points carrying non-zero measure constitute exactly the innermost core. So, the innermost core (and all cores) carry probability one, but any point outside of the smallest core carries measure zero. So, in a way the core system orders points of zero probability. A possible interpretation of this ordering is as a plausibility measure.

There is no consensus as to what is exactly the attitude that is revised or contracted in the standard theory of belief change. Many philosophers maintain that this attitude is full belief. Under that point of view the account of belief change emerging from this probabilistic framework does not fit with the received view in the field. But when one supposes a proposition that is compatible with the full beliefs for P, an operation of belief change occurs that can be seen as the revision of expectations rather than the revision of full beliefs. Seen from the point of view of the corpus of full beliefs for P these changes can be seen as inductive expansions of the body of full beliefs for P.

Hannes Leitgeb recently offered a very interesting model of belief in terms of degrees of belief [Leitgeb, 2010]. Starting from very different insights than the ones presented above he showed how to construct cores systems from standard monadic probability. Unlike the previous construction the innermost core might carry high probability that is less than one. So, his construction seems to derive a notion of plain belief rather than certainty or full belief. Arló-Costa and Pedersen have showed in an even more recent paper ([Arló Costa and Pedersen, 2010]) that Leitgeb's construction can be derived from an extension of the probabilistic theory of cores presented above. So, various different approaches seem to converge into an unified theory. This body of work seems to point in the direction of finally reconciling probabilistic and qualitative notions of belief.

6.1 Core Dynamics and Matter-Of-Fact Supposition

One natural question related to the previous proposal is the following: Given an initial two-place probability function $P(\cdot|\cdot)$ and its core system C, what is the shape of the core system that corresponds to $P[\![\alpha]\!](\cdot|\cdot)$, the update of the probability function $P(.|.)$ with the proposition expressed by α (denoted here as $[\![\alpha]\!]$)? We assume here the Bayesian characterization of update: $P[\![\alpha]\!](\cdot|\cdot) = P(\cdot| \cdot \cap [\![\alpha]\!])$.

The answer has a Bayesian flavour that nevertheless is difficult to reconcile with the dominant views about revision and contraction in the field of belief change: the core system $C_{[\alpha]}$ corresponding to $P[\![\alpha]\!](\cdot|\cdot)$ is obtained by the following operation $C_{[\alpha]} = \{X \cap [\![\alpha]\!] : X \in C\}$. So, basically one just takes the intersection of each core with the incoming proposition expressed by α and this is the new core system (see [Arló Costa, 2001b] for details).

The notion of belief change arising from this core dynamics can be axiomatized as follows ([Arló Costa, 2001a])

Entailment: $Ex(P) \subseteq F(P)$.
Full Belief Expansion: $F(P) \cap [\![\alpha]\!] = F(P[\![\alpha]\!])$.
Success: $Ex(P[\![\alpha]\!]) \subseteq [\![\alpha]\!]$.
Preservation: If $Ex(P[\![\alpha]\!]) \cap [\![\alpha]\!] \neq \emptyset$, then $Ex(P[\![\alpha]\!]) \cap [\![\alpha]\!] = Ex(P[\![\alpha]\!])$.
Restricted Consistency Preservation: If $F(P) \cap [\![\alpha]\!] \neq \emptyset$, then $Ex(P[\![\alpha]\!]) \neq \emptyset$.
Entertainability: If $F(P) \cap [\![\alpha]\!] = \emptyset$, then $P[\![\alpha]\!]$ is abnormal.
Fixity: If P is the abnormal function, then $Ex(P[\![\alpha]\!]) = F(P[\![\alpha]\!]) = \emptyset$ and $P[\![\alpha]\!] = P$.
Cumulativity: $Ex((P[\![\alpha]\!])[\![\beta]\!]) = Ex(P([\![\alpha]\!] \cap [\![\beta]\!]))$.

Here we use $Ex(P)$ and $F(P)$ to denote the expectations and full beliefs of P respectively (otherwise they can be seen as denoting the innermost and outermost core respectively).

Various axioms conflict directly with well-known AGM axioms. For example, fixity is incompatible with AGM which assumes that it is always possible to extricate oneself from inconsistency by updating with a consistent proposition. In this setting once one falls into inconsistency there is no possible repair and one will continue to be in an incoherent state no matter what. Cumulativity is not satisfied by any notion of revision we are aware of in the literature.[22]

In [Arló Costa, 2001a] an argument is presented indicating that this Bayesian notion of belief change can be used to model indicative or matter-of-fact supposition. In virtue of this interpretation the notion of change is called *hypothetical revision* in [Arló Costa, 2001a]. One of the conditional axioms that holds for this notion of supposing is the export–import axiom, which is validated by cumulativity.

6.2 Update, Imaging and Subjunctive Supposition

There is another notion of change that has both a suppositional and a probabilistic pedigree. In [Lewis, 1976] and [Lewis, 1986b] David Lewis proposed a notion of probabilistic update called imaging. In these articles Lewis proved that the probability of a conditional cannot be conditional probability. Nevertheless it is true that the probability of a conditional 'If A, then B' equals the value of $P([B]\backslash[A])$ where $P([B]\backslash[A])$ is the result of computing the probability of $[B]$ upon imaging on [A]. What is imaging? Suppose that there is a set of points F carrying positive probability in a space U. Then the result of imaging on $[A]$ should be computed as follows: (1) for every A-world in F its probability remains unchanged and (2) for every $\neg A$-world w in F one first identifies the most similar A-word to it and then transfers the probability rigidly to its most similar A-point (we assume here for simplicity that there is always a unique most similar A-point). This operation is rather different from conditioning.

In an important paper [Katsuno and Mendelzon, 1991a] the computer scientists Hirofumi Katsuno and Alberto Mendelzon axiomatized and proved a representation result for a qualitative counterpart of imaging. The properties of this notion of change are quite different from the ones that AGM has. For example, this notion of change has a property very similar to the notion of fixity proposed above. The update of an inconsistent state remains inconsistent. Moreover, unlike most notions of change, update is monotonic, in the sense that if $K \subseteq H$ then the update of K with an arbitrary sentence α is also included in the update of H with the same sentence. Both properties are incompatible with AGM and compatible with a form of the Ramsey test first proposed by Peter Gärdenfors. This test states that a conditional $\alpha > \beta$ belongs to a belief set K if and only if β belongs to the update of K with α. It is well known that this test is incompatible with AGM. It is not difficult to see that both monotony and the property that the update of an inconsistent belief set remains inconsistent are entailed by Gärdenfors' version of the Ramsey test. Moreover one can prove that when the notion of update obeys the axioms of Katsuno and Mendelzon the logic of conditionals validated by this version of Gärdenfors' test is exactly the system

VC of Lewis, which is Lewis' official axiomatization of the notion of counterfactual. So, many have proposed that the axioms of update encode the notion of supposition tacitly proposed by Lewis in his analysis of counterfactuals.

7. Epistemic States vs. Belief Sets: The Problem of Iteration

A belief set is a representation of the beliefs that a rational agent is committed to have. But perhaps an *epistemic state* is a more complex entity. Perhaps an epistemic state contains not only the beliefs of the agent but also a dynamic component useful to guide changes of these beliefs. We have seen above various possible dynamic components: plausibility orderings, entrenchment orderings, a probability measure. These examples do not exhaust the list of all possible dynamic components.

We can think abstractly about epistemic states as a complex entity that is associated with its belief set. But it is conceivable to have the same beliefs paired with different dynamic components. We can use here a minor variant of the notation employed by Adnan Darwiche and Judea Pearl in a classic paper on iterated belief change ([Darwiche and Pearl, 1997]). We denote epistemic states with upper case Greek letters (Φ, Ψ). Given an epistemic state Ψ its associated belief set is denoted by $Bel(\Psi)$. Of course $\Psi * \mu$ stands for an epistemic state, not a belief set. We can now introduce axioms that take into account the distinction between epistemic state and belief set:

$(R*0)$ $Bel(\Psi) = Cn(Bel(\Psi))$. *(Closure)*
$(R*1)$ $\mu \in Bel(\Psi * \mu)$. *(Success)*
$(R*2)$ If $\neg\mu \notin Bel(\Psi)$, then $Bel(\Psi * \mu) = Bel(\Psi) + \mu$. *(Inclusion + Vacuity)*
$(R*3)$ If $\nvdash \neg\mu$, then $\perp \notin Bel(\Psi * \mu)$. *(Consistency)*
$(R*4)$ If $\Psi_1 = \Psi_2$ and $\mu_1 \leftrightarrow \mu_2$, then $Bel(\Psi_1 * \mu_1) = Bel(\Psi_2 * \mu_2)$. *(Extensionality)*
$(R*5)$ $Bel(\Psi * \mu) + \phi \subseteq Bel(\Psi * (\mu \wedge \phi))$. *(Superexpansion)*
$(R*6)$ If $\neg\phi \notin Bel(\Psi * \mu)$, then $Bel(\Psi * (\mu \wedge \phi) \subseteq Bel(\Psi * \mu) + \phi$. *(Subexpansion)*

The axiom (R∗4) is a crucial axiom in this representation. The standard axiom of extensionality is quite different. In this notation it should be formulated as follows:

($R4$) If $Bel(\Psi_1) = Bel(\Psi_2)$ and $\mu_1 \leftrightarrow \mu_2$, then $Bel(\Psi_1 * \mu_1) = Bel(\Psi_2 * \mu_2)$. *(Extensionality)*

But it should be clear that ($R4$) can fail to be true in the case that the dynamic components of Ψ_1 and Ψ_2 are different.

7.1 Special Axioms for Iteration

Darwiche and Pearl propose in their paper special axioms for iteration. We will review these special axioms here.

(C1) If $\alpha \models \mu$, then $Bel((\Psi * \mu) * \alpha) = Bel(\Psi * \alpha)$.

Explanation : When two pieces of evidence arrive, the second being more specific than the first, the first is redundant; that is, the second evidence alone would yield the same belief set.

(C2) If $\alpha \models \neg\mu$, then $Bel((\Psi * \mu) * \alpha) = Bel(\Psi * \alpha)$.

Explanation : When two contradictory pieces of evidence arrive, the last one prevails; that is, the second evidence alone would yield the same belief set.

(C3) If $\mu \in Bel(\Psi * \alpha)$, then $\mu \in Bel((\Psi * \mu) * \alpha)$.

Explanation : Evidence μ should be retained after accommodating a more recent evidence α that implies μ given current beliefs.

(C4) If $\neg\mu \notin Bel(\Psi * \alpha)$, then $\neg\mu \notin Bel((\Psi * \mu) * \alpha)$.

Explanation : No evidence can contribute to its own demise. If μ is not contradicted after seeing α, then it should remain uncontradicted when α is preceded by μ itself.

Several useful examples are discussed in [Darwiche and Pearl, 1997]. For example epistemic states can be encoded as *rankings* (or *ordinal conditional functions*) first introduced by Wolfgang Spohn ([Spohn, 1988]). A ranking is a function κ from the set of all interpretations of the underlying language (worlds) into the natural numbers. A ranking is extended to propositions by requiring that the rank of a proposition be the smallest rank assigned to a world that satisfies:

$$\kappa(A) = \min_{w \models A} \kappa(w).$$

The set of models corresponding to the *belief set* $\rho(\kappa)$ associated with a ranking κ is the set $\{w : \kappa(w) = 0\}$. Darwiche and Pearl proved in [Darwiche and Pearl, 1997] that the following method for updating rankings satisfies their postulates:

$$(\kappa \bullet A)(w) = \begin{cases} \kappa(w) - \kappa(A) & \text{if w} \models \text{A} \\ \kappa(w) + 1 & \text{otherwise} \end{cases}$$

A representation result for ranking functions is offered in [Hild and Spohn, 2008]. The result requires the use of additional axioms for iterated contraction. In this notation the axioms entail at least:[23]

(C5) If $\models \mu \vee \phi$, then $Bel((\Psi \div \mu) \div \phi) = Bel((\Psi \div \phi) \div \mu)$. *(Restricted Commutativity)*

(C6) If $\mu \models \phi$ and $\phi \rightarrow \mu \notin Bel(\Psi \div \mu)$, then $Bel((\Psi \div (\phi \rightarrow \mu)) \div \phi) = Bel((\Psi \div \mu) \div \phi)$. *(Path Independence)*

7.2 Other Approaches to Iteration

The distinction between epistemic state and belief set can be applied in a slightly different way to make iteration possible. The epistemic state Ψ can be an entrenchment ordering. Then we have that:

$$Bel(\Psi) = \{q : r < q, \text{for some } r\}$$

where $<$ is the entrenchment ordering identical to the epistemic state Ψ. So, the challenge is to provide an algorithm for changing entrenchment orderings in the presence of new information (rather than belief sets). So, if one starts with an entrenchment ordering $\leq = \Psi$, when one learns α, the idea is to map $\leq$ to a new entrenchment ordering $\leq' = \Psi * \alpha$. The new belief set is calculated immediately as follows:

$$Bel(\Psi * \alpha) = \{q : r <' q, \text{for some } r\}$$

The crucial problem is therefore to indicate how epistemic states change when they are identified with an entrenchment ordering. There are various proposals in the literature suggesting how to do this (see, among others, [Nayak, 1994], [Rott, 2003], [Ferme and Rott, 2004]).

In spite of its apparent naturalness, this approach to iteration has not produced a breakthrough in this area. Hans Rott's proposal, for example, has some of the same fundamental problems as an approach proposed earlier by Craig Boutilier ([Boutilier, 1996]). Boutilier proposed to adopt the following iterated axiom:

(CB) If $\neg\alpha \in Bel(\Psi * \mu)$, then $Bel((\Psi * \mu) * \alpha) = Bel(\Psi * \alpha)$. *(Absolute Minimization)*

To visualize this axiom think that epistemic states are encoded as pre-orders. Then one has $\leq_\Psi$ as the pre-order corresponding to Ψ and $\leq_{\Psi*\mu}$ as the pre-order corresponding to $\Psi * \mu$. We need an additional notion to discuss the axiom. We can say that an epistemic state Ψ *supports* a conditional $\alpha > \beta$ if and only if the minimal α worlds according to $\leq_\Psi$ are β worlds.

Then the axiom *(CB)* recommends a minimizing changes in conditional beliefs due to a revision by making the pre-orders $\leq_\Psi$ and $\leq_{\Psi*\mu}$ as similar as possible. But this leads to unreasonable conclusions as the following example shows:

Example 17.7.1 We encounter a strange new animal and it appears to be a bird, so we believe the animal is a bird. As it comes closer to our hiding place, we see clearly that the animal is red, so we believe that it is a red bird. To remove further doubts about the animal birdness, we call in a bird expert who takes it for examination and concludes that it is not really a bird but some sort of mammal. The question now is whether we should still believe that the animal is red. Postulate (CB) tells us that we should no longer believe that the animal is red. [Darwiche and Pearl, 1997, p. 10]

The reason for this behaviour is that retaining the belief in the animal's colour means that we are implicitly acquiring a new conditional belief – that the animal is red given that it is not a bird – which we did not have before. So, the strategy of minimizing changes in conditional beliefs can lead to counterintuitive recommendations. As Darwiche and Pearl observe, once the animal is seen to be red, it should be presumed red no matter what ornithological classification results from further examination. And if this requires introducing new conditional beliefs, so be it.

The postulates offered by Darwiche and Pearl seem to avoid these problems and therefore they should be considered an improvement with respect to accounts of the sort defended by Rott and Boutilier.

The additional proposals that recommend to operate directly on entrenchment orderings have departed considerably from the AGM orthodoxy. Nayak has proposed to revise entrenchments by other entrenchments, changing therefore radically the way in which inputs tend to be understood in the traditional theories of belief change ([Nayak, 1994]). Fermé and Rott have proposed to investigate belief revision with inputs of the form 'accept q with a degree of plausibility that at least equals that of p' ([Ferme and Rott, 2004]). Again epistemic states are represented by entrenchment orderings, which are revised by this kind of input, yielding new entrenchment orderings.

When belief contraction and revision are constructed decision-theoretically (as in many proposals recently offered by Isaac Levi) the notion of iteration can be investigated as well. In this case the relevant contextual parameter is the value function used in the model. The type of iterated change that arises when the value function is kept fixed has been investigated in [Arló Costa, 2006]. The idea is analogous to the situation when iterated changes are modelled with respect to a fixed entrenchment ordering or a fixed ranking system. An open problem in this area is the determination of the dynamics of value functions.

7.3 Which Axioms are Correct?

Perhaps the axioms offered by Darwiche and Pearl (and extended by Hild and Spohn) are the least controversial set of axioms for iteration offered so far. But they do not enjoy the degree of consensus that the AGM axioms have in the one-shot case. At least this is so for the AGM axioms for revision (the situation is more nuanced in the case of contraction).

But the problem of iteration remains in a way unresolved. And we would like to argue that there is perhaps an unavoidable degree of indetermination associated with it. To appreciate the problem let's consider another article by Pearl, this time written in collaboration with Moises Goldszmidt ([Pearl and Goldszmidt, 1996]). In this article Pearl and Goldszmidt consider the often neglected problem of computational feasibility of belief revision. So, various algorithms designed to compute with rankings are offered and their computational complexity is investigated. Based on these considerations Pearl and Goldszmidt recommend the following algorithm to update ranking functions:

$$(\kappa \bullet \alpha)(w) = \begin{cases} \kappa(w) - \kappa(\alpha) & \text{if w} \models \alpha \\ \infty & \text{otherwise} \end{cases}$$

It is clear that this procedure violates the axiom (*C*2) proposed by Pearl himself in collaboration with Darwiche. So, the *C*-axioms for iteration are not a gold standard that has to be preserved in all forms of iterated belief change. In a way this should not be surprising. The meta-criterion used to propose the *C*-axioms is symmetry. The idea is that when revising with a sentence α the relative ordering of the α and $\neg\alpha$ worlds has to be preserved. Obviously the procedure for updating rankings proposed by Pearl and Goldszmidt violates this symmetry: when one updates with α the relative ordering of the $\neg\alpha$ worlds is destroyed and no memory is preserved of the previous ordering. But this procedure (which has a Bayesian flavor) might be very efficient. And if efficiency rather than symmetry is the dominant consideration one should not be constrained by the *C*-postulates.

Computational feasibility and symmetry need not be the only meta-criteria that matter. One can classify different methods for updating rankings in terms of their capacity to learn the truth in the long run, for example. Kevin Kelly did such a study in his [Kelly, 1998]. Or one can focus on the orthogonal goal of minimizing losses of informational value in the next step of inquiry, as Isaac Levi has proposed for years, and consequently deny the importance or interest of iterated change. Perhaps it only makes sense to elicit iterated axioms relative to a determinate understanding of inquiry. And one should not be surprised if two axioms systems corresponding to different views of inquiry conflict. Since the different philosophical positions about inquiry and rationality often conflict, one should expect that the axioms that reflect them syntactically conflict as well. In conclusion, perhaps it is foolish to expect the emergence of a consensus regarding the correct set of axioms that would apply across different views of inquiry and rationality. If such axioms exist, they will be very weak indeed.

Notes

1. A similar strategy is used by Wolfgang Spohn in his recent book [Spohn, 2010, Chapter 2].
2. A binary relation R on X is *transitive* if for every $x, y, z \in X$, if xRy and yRz, then xRz.
3. We say that $\Gamma \subseteq \text{For}(\mathcal{L})$ is *inconsistent* with respect to Cn if $\text{Cn}(\Gamma) = \text{For}(\mathcal{L})$ and *consistent* otherwise. We call $\Delta \subseteq \text{For}(\mathcal{L})$ *maximal consistent* with respect to Cn if Δ is consistent and for every $\Sigma \subseteq \text{For}(\mathcal{L})$, if $\Delta \subseteq \Sigma$ and Σ is consistent, then $\Delta = \Sigma$. A maximal consistent set Δ has the important property that for every formula ϕ, either $\phi \in \Delta$ or $\neg\phi \in \Delta$.
4. A binary relation R on a set X is a *total order* if it is antisymmetric (i.e., for every $x, y \in X$, if xRy and yRx, then $x = y$), transitive, and complete (i.e., total: for every $x, y \in X$, either xRy or yRx).
5. Given a total order R on X, we say that an element $x \in X$ is the *R-minimum* if for every $y \in X$, xRy. Note that the use of 'the' is justified because R is a total order.
6. A binary relation R on a set X is a *weak order* if it is transitive and complete.
7. Given a binary relation R on a set X, we say that an element $x \in X$ is an *R-maxima* if for every $y \in X$, yRx.
8. Also note that Part (ii) of Theorem 17.3.3 holds provided that K is consistent. One can of course modify the definition of a persistent relation to accommodate such limit cases.
9. Condition α, also known as Heritage or Chernoff's Axiom, was introduced in [Chernoff, 1954, p. 429]. Condition α should not be confused with another important condition, the so-called *Independence of Irrelevant Alternatives* [Arrow, 1951, p. 27]. See [Sen, 1977, pp. 78–80] for a vivid discussion of the difference between these two conditions. See also [Ray, 1973] for another clear discussion of this sort.
10. Condition γ^* was introduced in [Chernoff, 1954, p. 432]. A generalized constraint, condition γ, was introduced in [Sen, 1971, p. 314].
11. β is a close relative of condition β^+ [Sen, 1977, p. 66]. Introduced in [Sen, 1969], condition β demands that if $S \subseteq T$ and $f(S) \cap f(T) \neq \emptyset$, then $f(S) \subseteq f(T)$. Condition β^+ entails condition β, and in the presence of condition α, condition β and condition β^+ are logically equivalent.
12. A binary relation R on a set X is a *quasi-order* if it is a transitive and reflexive. Thus, a weak order is a complete quasi-order (see footnote 6).
13. If R_0 and R_1 are binary relations on a set X, we call R_1 an *extension* of R_0 (with respect to X) if $R_0 \subseteq R_1$ and $R_0 \cap ((X \times X)\backslash R_0^{-1}) \subseteq R_1 \cap ((X \times X)\backslash R_1^{-1})$, where $R^{-1} := \{(x, y) \in X \times X : (y, x) \in R\}$.
14. If $\mathcal{L}$ is infinite, there are propositional selection functions which are are not complete. For example, let $\mathcal{L}$ consist of countably infinite propositional variables $(p_i : i < \omega)$, and suppose that $f([\![p_0]\!]) = [\![p_0]\!]\backslash\{w_0\}$, where $w_0 := \text{Cn}(\{p_i : i < \omega\})$. Then $\bar{f}([\![p_0]\!]) = [\![p_0]\!]$, so $f \neq \bar{f}$.
It is an easy matter to verify that a selection function f on $\mathscr{E}_{\mathcal{L}}$ is complete just in case for every $S \in \mathscr{E}_{\mathcal{L}}$, there is $\Delta \subseteq \text{For}(\mathcal{L})$ such that $f(S) = [\![\Delta]\!]$.
15. Here we focus on some of Rott's results concerning belief revision. Rott also presents results concerning non-monotonic logic and belief contraction. For example, Rott shows that in the standard AGM framework condition α corresponds not only to posutlate $(K * 7)$, but also to postulate $(K \dot{-} 7)$ of belief contraction

[Rott, 2001, pp. 193–6] and to rule (Or) of non-monotonic reasoning (which demands observance of the following: From $\phi|\sim \chi$ and $\psi|\sim \chi$, infer $\phi \vee \psi|\sim \chi$) [Rott, 2001, pp. 201–4].

16. Rott claims in [Rott, 2001] that the formal results proved in the book offer a reduction of theoretical reason to practical reason. This claim goes beyond the formal results stated in the book and it has been questioned on philosophical grounds (see [Olsson, 2003]). There is a debate as well regarding whether the formal results offered by Rott offer decision-theoretic foundations for belief change. Isaac Levi has questioned this aspect of Rott's representation results in [Levi, 2004]. It is clear that Rott has proved very valuable formal results. It is perhaps more controversial how to interpret them.
17. See footnote 3 for a definition of maximal consistency.
18. Here we follow the the presentation (and in particular, the terminology) in [Hansson and Olsson, 1995]). The presentation in [Hansson and Olsson, 1995] might not capture all the subtleties of the philosophical ideas and arguments in [Levi, 1991]. For better or worse, the terminology used here is now more or less standard in the literature. Readers interested in Levi's ideas should consult [Levi, 1991].
19. This definition is more complex than the definition for partial meet contraction. The second clause in (17.1) is added to ensure that $\dot{-}$ satisfies postulate $(K \dot{-} 3)$. In contrast with remainder sets, when $\alpha \notin K$, it is possible for $\Gamma, \Gamma' \in S(K,\alpha)$ and $\Gamma \subset \Gamma'$. To take an example, consider a language with precisely two propositional variables p,q and a belief set $K := \mathrm{Cn}(\{p,q\})$. Then $\mathrm{Cn}(\{p\}), K \in S(K,\neg q)$ and $\mathrm{Cn}(\{p\}) \subseteq K$. We can construct a selection function δ for which $\delta(S(K,\neg q)) = \{\mathrm{Cn}(\{p\}), K\}$ and so $\bigcap \delta(S(K,\neg q)) = \mathrm{Cn}(\{p\})$. Thus, if we were to drop the second clause in (17.1), requiring that $K \dot{-} \alpha = \bigcap \delta(S(K,\alpha))$ for all α, the resulting contraction operation would violate $(K \dot{-} 3)$ (cf. [Hansson and Olsson, 1995, p. 108]). The qualification that (17.2) holds for all formulae $\alpha \in K \backslash \mathrm{Cn}(\emptyset)$ and not necessarily for formulae outside $K \backslash \mathrm{Cn}(\emptyset)$ is also needed. For example, consider again the language with two propositional variables, this time with a belief set K given by $K := \mathrm{Cn}(\{p\})$. Then $S(K, \neg p \wedge q) = \emptyset$. Now if (17.2) were required to hold for $\alpha \notin K$ as well, then since the definition of a selection function demands that $\delta(S(K,\neg p \wedge q)) = \{K\}$, it would follow that $K \in S(K, \neg p \wedge q)$, yielding a contradiction.
20. See also the introduction of [Levi, 2004].
21. Levi has defended Antitony by appealing to the use of partitions in the presentation of contraction. Many counterexamples to Antitony appeal to cases where the sentences α and β used in the postulate are mutually irrelevant. The use of partitions filters irrelevant cases, in the sense that the two sentences in question are potential answers to the same issue. One can certainly use a semantics where partitions of this sort are deployed. In [Arló Costa and Levi, 2006] such a semantics is used. But in [Arló Costa and Levi, 2006] a complete axiomatization is presented from which the postulates discussed here are derivable. In particular the postulates we are discussing here is derivable for any sentence α, β, without any further syntactic restrictions. Here we are considering the adequacy of postulates independently of the semantics utilized to validate them (the possible world semantics of Rott and Pagnucco, Levi's partitional semantics, etc.). But even if one only considers instances of this axiom where the two sentences are potential answers to the same issue, the requirement that any two representable arbitrary contractions obey this tidy entailment pattern seems too orderly to be correct.
22. A possible exception is the notion of *irrevocable revision* introduced in a completely different setting by Krister Segerberg.
23. The axioms are slightly stronger than stated below. See Definition 5.1 in [Hild and Spohn, 2008] for details.

18 Epistemic Logic

Paul Égré

Chapter Overview

1. Introduction: Knowledge, Belief, and Formal Epistemology	504
2. Basic Epistemic Logic	506
2.1 Syntax and Semantics	506
2.2 Main Axioms for Knowledge and Belief	508
3. Multi-Agent Systems and Interactive Epistemology	510
3.1 Group Knowledge	511
3.2 Common Knowledge and Games	513
4. Informational Dynamics	516
4.1 Belief Revision and Updates	516
4.2 Public Announcements	517
4.3 Belief Revision	519
4.4 Epistemic Actions	521
5. Logical Omniscience and Self-Knowledge	522
5.1 Logical Omniscience	523
5.2 Limitations on Self-Knowledge	525
6. Knowledge, Belief, and Justification	529
6.1 Combining Knowledge and Belief	529
6.2 Safety, Stability, Justification	530
7. Existence and Quantification	532
7.1 Intensionality and Belief Contexts	532
7.2 The *de re*/*de dicto* Distinction	533
7.3 Knowledge and Questions	536
8. Epistemic Paradoxes	538
8.1 Moore, Fitch, and the Surprise Examination	538
8.2 A Dynamic Perspective on the Paradoxes	539
9. Conclusion and Perspectives	541
Notes	541

1. Introduction: Knowledge, Belief, and Formal Epistemology

Epistemic logic is a branch of formal epistemology in which the notions of knowledge, belief, and information are described and investigated by means of formal logical methods. Contemporary research in epistemic logic was initiated by Jaakko Hintikka's seminal book *Knowledge and Belief: An Introduction to the Logic of the Two Notions*, which appeared in 1962. In his book, Hintikka proposed to apply the tools of formal semantics and model theory to analyse the truth conditions of sentences such as '*a* knows that *p*', '*a* believes that *p*', '*a* knows whether *p*', '*a* is uncertain as to whether *p*', '*a* knows who did so and so'.

As is true of much work done at the same period in analytic philosophy, Hintikka's original project was as much an attempt to clarify the meaning and logical form of sentences involving propositional attitude verbs such as 'believe' and 'know', as it was an attempt to formally represent the content of these two propositional attitudes and the general constraints governing them. Because of that, Hintikka's original project belongs both to the domain of natural language semantics and to the domain of epistemology. Part of Hintikka's epistemological project was to formally characterize the difference and the relation between the two attitudes of knowledge and belief, to clarify issues about iterated belief, iterated knowledge and introspection (such as 'does knowing imply knowing that one knows'?), and to cast light on Moore's paradox (why is it rationally inconsistent to say '*p* but I don't believe *p*'?). Part of his semantic project, on the other hand, was to make explicit the relation between belief, knowledge and existence, in particular to respond to the problem of quantification into belief and knowledge attributions (such as capturing the distinction between 'John knows that someone left' and 'there is someone of whom John knows that he left', a problem originally posed by Quine ([Quine, 1956]).

Epistemic logic started at about the same time intensional logics of various kinds were developed, including deontic logic (the logic of obligation, see [von Wright, 1951]), temporal logic (the logic of time, see [Prior, 1957]), and modal logic (see [Kripke, 1959], [Montague, 1960]). Like its siblings, epistemic logic first developed as a propositional modal logic of a particular kind, in which the modalities receive a doxastic or an epistemic interpretation (where $B_a p$ symbolizes '*a* believes that *p*', and $K_a p$ stands for '*a* knows that *p*'). While Hintikka's original perspective was mostly focused on the representation of the beliefs of a single agent, a second source of development in epistemic logic came a few years later from work done in game theory on the representation of group knowledge, in particular in the work of the economist Robert Aumann ([Aumann, 1976]). Decisions in game-theoretic situations are a function not only of the player's utilities, but also of the information each player can have about the information available to other players. Aumann in particular gave a set-theoretic formalization of the concept of common knowledge introduced before him by David

Lewis in his work on convention ([Lewis, 1969]). The interest of epistemic logic for the formal representation of information and uncertainty among groups of agents was fostered a bit later with work from the theoretical computer science community. Communication systems can be seen as networks of multiple agents exchanging information. As in game theory, the representation of the various information states of a multi-agent system can be modelled in a fruitful way using the framework of epistemic logic. This information-theoretic perspective has made room for a convergence of the modal perspective and of Aumann's set-theoretic perspective into a unified framework.

More recently, two further and complementary directions of research have emerged. The first of them, which Aumann has coined *interactive epistemology*, concerns the epistemic foundations of solution concepts in game theory (see [Aumann, 1999a], [Aumann, 1999b] [Brandenburger, 2007]). A general problem for game theorists concerns the dependence between profiles of strategies used by players in games, and the level of shared information (of common belief, of common knowledge) that they must have to sustain these strategies. In this area, epistemic logic is being used not only to formalize existing results, but also to give a precise account of the assumptions needed to secure specific outcomes in games. A second important source of development in epistemic logic has come from work done in *belief revision*. Hintikka's original epistemic logic is essentially static: formulas represent the state of information of a single agent at a given time, but they don't represent the effect of an agent learning new or contradictory information. Belief revision originally developed outside the framework of modal logic proper, in what is known as the AGM framework (see [Alchourrón et al., 1985], and Chapter 17 of this volume). Since the mid-1990s, however, the original framework of static epistemic logic has been extended into a variety of systems of dynamic epistemic logic. The resulting framework allows one to model information change and the effect of actions and announcements made by players at the successive stages of a game or of a communication process (see [Gerbrandy and Groeneveld, 1997], [van Benthem, 2002], [van Benthem et al., 2006], and [van Ditmarsch et al., 2007] for an overview).

In recent years, both the game-theoretic perspective and the dynamic perspective on information have found points of convergence. At the same time, further progress has been made on some of the epistemological and semantical issues Hintikka had put on the original agenda of epistemic logic. These concern the analysis of 'knowing-wh' constructions and the definition of systems of first-order epistemic logic ([Gerbrandy, 2000], [Aloni, 2005]), the problem of giving a fine-grained analysis of knowledge and justification (as opposed to mere true belief, see [Rott, 2004b], [Stalnaker, 2006], [Artemov, 2008]) , and the solution to various epistemic paradoxes (such as the Surprise Examination Paradox, and the Fitch Paradox, both of which relate to Moore's Paradox, see [van Benthem, 2004b] , [Gerbrandy, 2007]).

The present chapter is organized as follows. In Section 2, we present the basic syntax and semantics of propositional epistemic logic for a single agent. In Sections 3 and 4, we discuss two directions in which the basic framework has been generalized and applied: to the representation of group knowledge on the one hand, and to the treatment of informational dynamics on the other. Sections 5 and 6 deal with some classic issues in epistemology: Section 5 presents ways of relaxing some of the idealizations made in standard epistemic logic, in particular with the closure assumptions made on deduction and self-knowledge; Section 6 examines the articulation between knowledge, belief, and justification. In Section 7 we introduce first-order epistemic logic. In Section 8, finally, we conclude with a brief overview of some epistemic paradoxes and their treatment in dynamic epistemic logic.

2. Basic Epistemic Logic

2.1 Syntax and Semantics

Basic epistemic logic for a single agent, like basic modal logic (see [Blackburn et al., 2002] and Chapter 11 of this volume) can be seen as an extension of the language of standard propositional logic by means of an epistemic operator. Suppose given a set of propositional atoms $A = \{p, q, r, \ldots\}$. The language $\mathcal{L}_K$ of propositional epistemic logic for a single agent a is defined recursively as follows:

Definition 18.2.1 Syntax of basic epistemic logic:

$$\phi := p \mid \neg\phi \mid (\phi \wedge \phi) \mid K_a\phi$$

Let p stand for 'it is raining', then $K_a p$ represents the sentence: 'Ann knows that it is raining', and $\neg K_a \neg p$ represents the sentence: 'Ann does not know that it is not raining', or 'for all Ann knows, it is possible that it is raining'. Hintikka originally used P_a as shorthand for $\neg K_a \neg$; the notation $\widehat{K}_a$ is more commonly used today (see [van Ditmarsch et al., 2007]).

Intuitively, to say that a knows p means that p holds in every state of affairs compatible with the information available to a; dually, to say that a does not know that not p means that p holds in at least one state of affairs compatible with what a knows. To formalize those definitions, Hintikka originally proposed a semantics in terms of *model sets* rather than possible worlds. However, the fundamental intuition behind Hintikka's original semantics is essentially the same we find in possible world semantics. On Hintikka's approach, a model set μ is a collection of sentences satisfying some closure conditions and intended to represent 'the informal idea of a (partial) description of a possible state of affairs' ([Hintikka,

1962, p. 41]). Given a set Ω of model sets – which we call a model system – and a relation of alternativeness between them – which is intended to represent the notion of epistemic possibility for an agent a – Hintikka originally defined the truth of $P_a p$ relative to a model set μ and model system Ω as follows:

(C.P*) If $P_a p \in \mu$ and if μ belongs to a model system Ω, then there is in Ω at least one alternative μ^* (with respect to a) such that $p \in \mu^*$.

Today, it is more standard to evaluate knowledge sentences relative to Kripke models. A Kripke model $\mathcal{M} = (W, R_a, V)$ is a triple consisting of a non-empty set W of possible worlds, a binary relation R_a on $W \times W$ and a valuation function V mapping each atom in A to a subset of W. Thus, in a Kripke model, W is the counterpart of the model system Ω, each world $w \in W$ is the counterpart of a model set μ, and the relation R_a between worlds is the counterpart of Hintikka's alternativeness relation. The semantics works recursively as follows:

Definition 18.2.2 Relational semantics for propositional epistemic logic:

$\mathcal{M}, w \models p$	iff	$w \in V(p)$
$\mathcal{M}, w \models (\phi \wedge \psi)$	iff	$\mathcal{M}, w \models \phi$ and $\mathcal{M}, w \models \psi$
$\mathcal{M}, w \models \neg\phi$	iff	$\mathcal{M}, w \not\models \phi$
$\mathcal{M}, w \models K_a\phi$	iff	for every w' such that wR_aw', $\mathcal{M}, w' \models \phi$.

Basically, Kripke models serve to represent the notion of an agent's uncertainty. To appreciate the working of the semantics, consider the following very simple two-world model $\mathcal{M}$:

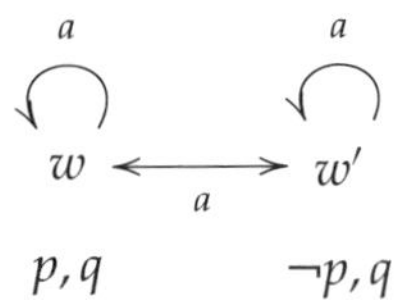

FIGURE 18.1 A model of Ann's uncertainty

Let q stand for 'it is raining' and p stand for 'the bank is open'. Let w represent the current world or state of affairs. We have that $\mathcal{M}, w \models K_a q$, but $\mathcal{M}, w \models \neg K_a p \wedge \neg K_a \neg p$. This describes a situation in which Ann knows that it is raining, but does not know whether the bank is open or not. Interestingly, the model makes predictions regarding iterations of K_a. For instance, we have that $\mathcal{M}, w \models$

$K_a(\neg K_a p \wedge \neg K_a \neg p)$, since in both w and w', Ann does not know whether the bank is open. This says that Ann knows that she does not know whether the bank is open.

2.2 Main Axioms for Knowledge and Belief

Everything we said so far could be used to handle belief rather than knowledge. To represent belief, introduce a belief operator B_a such that $B_a p$ represents that 'Ann believes that it is raining'. The same Kripke structures and truth definition can be used to handle belief, if we conceive of the relation R_a as representing doxastic rather than epistemic possibility. Relative to the operator B_a, the previous model could be used to represent that Ann believes that it is raining, is unsure whether the bank is open or not, and believes that she is unsure.

Hintikka, however, was interested in capturing the differences and commonalities between knowledge and belief depending on whether they satisfy certain general properties. The following table presents the axioms of central interest in epistemic logic:

K	$K_a(p \to q) \to (K_a p \to K_a q)$	Closure
T	$K_a p \to p$	Knowledge, Veridicality
D	$K_a \neg p \to \neg K_a p$	Consistency
4	$K_a p \to K_a K_a p$	Positive Introspection
5	$\neg K_a p \to K_a \neg K_a p$	Negative Introspection

The left column of the table indicates the standard name of the axioms in modal logic, and the right column their common appellation in the context of epistemic logic. Axiom **K**, or Kripke's axiom, corresponds to a property of closure of knowledge or belief under known implication. Axiom **T** is commonly referred to as the Knowledge axiom (see [Fagin et al., 1995]), or as the Veridicality or Factivity axiom, since it purports to characterize knowledge as opposed to belief: every known proposition must be true, whereas propositions merely believed can be false. Axiom **D** is weaker than **T** and merely rules out internal inconsistency, namely the possibility that an agent believes contradictory propositions. Axioms **4** and **5**, finally, are properties of self-knowledge: positive introspection means that one knows that one knows p whenever one knows p. Axiom **5** says that one knows that one does not know p whenever one does not know p.

As is known from correspondence results for relational semantics (see [Blackburn et al., 2002], [Fagin et al., 1995], Chapter 11 this volume), all of these axioms are valid exactly if certain frame properties are satisfied, namely if the relation of doxastic or epistemic possibility meets specific constraints, which are recalled below:

K	–	All frames
T	$\forall x(xR_ax)$	Reflexive frames
D	$\forall x\exists y(xR_ay)$	Serial frames
4	$\forall xyz(xR_ay \wedge xR_az \rightarrow xR_az)$	Transitive frames
5	$\forall xyz(xR_ay \wedge xR_az \rightarrow yR_az)$	Euclidean frames

A useful perspective on these axioms and on the frame properties to which they correspond from an epistemic point of view is given by the set-theoretical approach of belief and knowledge more familiar to economists, and originally used by Aumann in particular ([Aumann, 1976]). Instead of starting with a Kripke frame (W, R), the idea is to start from an information-theoretic structure (W, P_a) where P_a is a function that associates to each world w a set of possibilities (or epistemic alternatives to that world). The function P_a is standardly called an *information function* (for the agent a) (see [Osborne and Rubinstein, 1994]) or a *possibility correspondence* ([Bonanno and Battigalli, 1999]); $P_a(w)$ is called a's *belief state* in w. Given a valuation function V on W for the atoms, we can define $V(\phi)$ recursively as the set of worlds w such that $(W, P_a, V), w \models \phi$. The clauses for atoms and boolean compounds remain as before, and the clause for knowledge is:

Definition 18.2.3 Aumann-style semantics:

$$\mathcal{M}, w \models K_a\phi \qquad \text{iff} \qquad P_a(w) \subseteq V(\phi).$$

Intuitively, this says that a believes or knows ϕ iff the proposition expressed by ϕ is entailed by the information available to a in w, or by a's belief state.

The correspondence with Kripke's semantics is straightforward. From an information function, one can define an accessibility relation by letting wR_aw' iff $w' \in P_a(w)$. Conversely, given an accessibility relation, one can define an information function by letting $P_a(w) = \{w' \in W; wR_aw'\}$. From those definitions, the relational properties corresponding to axioms **T**, **D**, **4**, and **5** can be expressed more compactly as follows:

T	$w \in P_a(w)$
D	$P_a(w) \neq \emptyset$
4	$w' \in P_a(w) \Rightarrow P_a(w') \subseteq P_a(w)$
5	$w' \in P_a(w) \Rightarrow P_a(w) \subseteq P_a(w')$

Thus, reflexivity for **T** corresponds to the idea that the actual world should always be a possibility entertained by the agent. Seriality for **D** corresponds to the idea that one's belief state is not empty (does not entail the contradictory proposition). Transitivity for **4** corresponds to the idea that every epistemic alternative to an epistemic alternative is already an epistemic alternative to the current world. Finally, euclideanness for **5** implies that if w' and w'' are two possibilities relative to w, they should be possible relative to each other. Together **4** and **5** imply that if w' is a possibility relative to w, then both of them determine the same set of possibilities.

The previous axioms allow us to define various axioms systems for knowledge or belief, depending on which properties are considered relevant, and in combination with the axioms of propositional logic, the rule of necessitation: $\phi \therefore K\phi$, modus ponens ($\phi, \phi \rightarrow \psi \therefore \psi$) and uniform substitution ($\phi \therefore \phi[\psi/p]$), common to all systems based on Kripke semantics (for systems of normal modal logics, see [Blackburn et al., 2002] and Chapter 11 of this volume). Of those, the modal system **KD45** is a standard system for rational belief, since it includes consistency and the two axioms of self-knowledge, but fails veridicality. Adding **T** produces the system more commonly named **S5**, which is in fact equivalent to **KT5**. From a model-theoretic point of view, euclideanness and reflexivity imply symmetry and transitivity, and thus give rise to equivalence relations. **S5** models thus correspond to partition models of information: in such models, belief sets correspond to equivalence classes partitioning the universe W. A slightly weaker system than **S5** for knowledge is the system **KT4**, a.k.a. **S4**, of positively introspective knowledge. This system corresponds to the system originally favoured by Hintikka in his theory of knowledge.

The three axiom systems **KD45**, **S4**, and **S5** are among the most widely used systems to represent belief and knowledge in various areas, including computer science and game theory. As should be clear from the axioms, such systems purport to represent the beliefs of *idealized* and *rational* agents. The adequacy of each of the axioms we listed, and of their underlying semantics, has been questioned ever since Hintikka's seminal book, including by Hintikka himself, on epistemological grounds. Before addressing these epistemological issues in Section 5 below, in the next two sections we shall first highlight the fruitfulness of the general framework proposed by Hintikka for the treatment of group knowledge on the one hand, and informational dynamics on the other.

3. Multi-Agent Systems and Interactive Epistemology

Hintikka's original perspective was mainly the representation of the belief and knowledge of a single agent. Quickly, however, it became apparent that his framework can be extended to represent the beliefs of several agents. This representation is particularly useful to represent what an agent believes about what other agents believe, or what an agent knows about what others know. Belief about beliefs, like knowledge about knowledge, are central to strategic reasoning (in games), but also to represent the way information is distributed in complex communication networks (see [Fagin et al., 1995]).

3.1 Group Knowledge

Multi-agent epistemic logic is the extension of basic epistemic logic to deal with several agents. For each agent $i \in I$ (with I a finite set), an epistemic operator K_i is introduced:

Definition 18.3.1 Syntax of multi-agent epistemic logic:

$$\phi := p \mid \neg\phi \mid (\phi \wedge \phi) \mid K_i\phi$$

A multi-agent model is a Kripke model $(W, (R_i)_{i \in I}, V)$ with as many epistemic accessibility relations as there are agents to consider. The semantics is the same as in Section 2, namely each operator K_i is interpreted relative to R_i. For example, consider two scenarios. Consider the following models, where a denotes Ann and b denotes Bob:

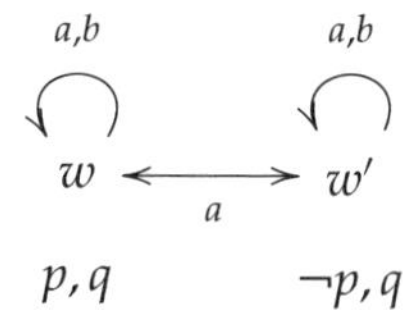

FIGURE 18.2 A model for the uncertainties of Ann and Bob

Suppose w is the actual world. $w \models K_b(p \wedge q)$ while $w \models K_a q \wedge \neg K_a p \wedge \neg K_a \neg p$. Moreover, we now have that $w \models K_b(\neg K_a p \wedge \neg K_a \neg p)$. This represents a situation in which Bob and Ann both know that it is raining, but only Bob knows that the bank is open. Moreover, Bob knows that Ann does not know that the bank is open. Furthermore, $w \models K_a K_b(\neg K_a p \wedge \neg K_a \neg p)$, that is Ann knows that Bob knows that she is ignorant.

Several notions of group knowledge can be defined in this framework. Given a group of agents $G \subseteq I$, it is useful first to introduce an operator E_G of *shared knowledge* to express that everyone in G knows ϕ, that is: $E_G\phi := \bigwedge_{i \in G} K_i\phi$. A weaker notion is the notion of *distributed knowledge*, to express that if the agents were to pool together their information, they would know ϕ. In the previous model, for instance, if a and b were to intersect their belief sets in w, they would both know p. Thus it is distributed knowledge between Ann and Bob that the bank is open, but Ann does not know it. Distributed knowledge within a group G is captured by means of the operator D_G.

A stronger notion than shared knowledge, originally due to Lewis [Lewis, 1969], Schiffer [Schiffer, 1972] and formalized in [Aumann, 1976], is the notion of *common knowledge*, intended to express that everyone knows ϕ, everyone knows that every known ϕ, and so forth *ad infinitum*.[1] Let $E^1_G\phi := E_G\phi$, and $E^{n+1}_G\phi = E_G E^n_G\phi$. The operator of common knowledge intuitively stands for the infinitary conjunction of all finite levels of shared knowledge: $C_G\phi = \bigwedge_{n\geq 1} E^n_G\phi$. For instance, in the previous model, it is in fact common knowledge between Ann and Bob that Ann does not know that the bank is open. Since we deal with only finitary conjunctions in the language, the operator C_G is standardly treated as a primitive symbol and we call $\mathcal{L}_{K,C}$ the extension of $\mathcal{L}_K$ with C_G, and $\mathcal{L}_{K,C,D}$ the full extension with distributed knowledge operators.

To capture the notions of shared knowledge, distributed knowledge, and common knowledge semantically, we define R_{E_G} as the union of the accessibility relations for all agents in group G, and R_{D_G} as their intersection, that is $R_{E_G} := \bigcup_{i\in G} R_i$, and $R_{D_G} := \bigcap_{i\in G} R_i$. Given a binary relation R, let R^* be the transitive closure of R (that is R^* is the smallest relation that contains R and such that aR^*c whenever aRb and bRc). Then R_{C_G} is defined as the transitive closure of R_{E_G}, namely $R_{C_G} := (R_{E_G})^*$.

Definition 18.3.2 Shared, Distributed, and Common Knowledge:

$$\mathcal{M}, w \models E_G\phi \quad \text{iff} \quad \text{for all } w' \text{ such that } wR_{E_G}w', \mathcal{M}, w' \models \phi.$$
$$\mathcal{M}, w \models D_G\phi \quad \text{iff} \quad \text{for all } w' \text{ such that } wR_{D_G}w', \mathcal{M}, w' \models \phi.$$
$$\mathcal{M}, w \models C_G\phi \quad \text{iff} \quad \text{for all } w' \text{ such that } wR_{C_G}w', \mathcal{M}, w' \models \phi.$$

The union and the intersection of a set of reflexive relations are reflexive, and similarly for the transitive closure of a reflexive relation. As a result, if for every $i \in G$, R_i is reflexive, then it follows that $E_G\phi \to \phi$, $D_G\phi \to \phi$, and $C_G\phi \to \phi$, namely the operators are veridical. If the R_i are not assumed to be reflexive, and purport to describe belief rather than knowledge, then E_G, D_G, and C_G are more adequately described as operators of *shared belief*, *distributed belief*, and *common belief*, respectively.

While shared knowledge can be defined in terms of the individual knowledge operators in the language, the operators of distributed knowledge and common knowledge each add expressive power to the basic language, as can be shown by means of standard techniques from modal logic (for proofs, see [Roelofsen, 2007] on distributed knowledge, and [van Ditmarsch et al., 2007, Chapter 8], on common knowledge). From an axiomatic point of view, the D operator inherits the common properties assumed of individual knowledge operators (i.e., **T**, **D**, **4** and **5**). Its distinguished properties are given by the following two axioms (see [Fagin et al., 1995]):

$$D_{\{a\}}\phi \leftrightarrow K_a\phi \qquad \text{for every } a \in I.$$

$$D_G\phi \rightarrow D_{G'}\phi \qquad \text{whenever } G \subseteq G'.$$

Likewise, the operator of common knowledge inherits the properties commonly assumed of individual operators (the same holds for common belief, except for negative introspection: when a proposition is not common belief, it needn't be common belief that it is not common belief). The characteristic properties of common knowledge are given by the following axiom and rule of inference:

$$C_G\phi \rightarrow E_G(\phi \wedge C_G\phi)$$

$$\text{from } \phi \rightarrow E_G(\phi \wedge \psi), \text{ infer } \phi \rightarrow C_G\psi.$$

The axiom is sometimes called the fixed point axiom, since when turned into an equivalence it actually provides an implicit definition of common belief: a sentence is common belief in a group exactly when everyone believes it and believes that it is common belief. The rule of inference is referred to as the induction rule: it says in particular that if ϕ is self-evident in the sense of being automatically believed by everyone, then it is thereby common belief. Note that from the fixed point axiom the infinitary definition of common knowledge could be recovered, by recursively rewriting $C_G\phi$ as $E_G(\phi \wedge C_G\phi)$ within $E_G(\phi \wedge C_G\phi)$.

While common knowledge and common belief have become central concepts in game theory in particular, there remains quite some discussion regarding the attainability of common knowledge, or the plausibility of the concept. Barwise [Barwise, 1988] presents a useful comparison of iterative, fixed point and 'shared event' pre-theoretic notions. From a logical point of view, the fixed point understanding of common knowledge bears a deep and mathematically non-trivial connection to the study of fixed point logics (see [Alberucci, 2002], [Lismont and Mongin, 2003], [van Benthem and Sarenac, 2004]; see also [Vanderschraaf and Sillari, 2009] for a very detailed overview on common knowledge).

3.2 Common Knowledge and Games

One of the areas in which notions of group knowledge are particularly useful is game theory. Lewis' original motivation for the definition of common knowledge was to deal with mutual expectations in situations in which agents have to coordinate. As pointed out in the literature, Lewis' original notion of common knowledge is in fact closer to common belief, and does not quite correspond to the iterative concept presented above (see [Cubitt and Sugden, 2003], [Sillari, 2005] for precise reconstructions of Lewis' definition). Starting with Aumann's [Aumann, 1976] work, however, the concepts of common belief and common knowledge presented above have come to play a central role when it comes to stating the precise conditions under which particular equilibria are attainable in games. To appreciate the centrality of the concept of common knowledge, we briefly review two examples from game theory, respectively intended to show the negative effect of lack of common knowledge in a game, or conversely the powerful effect of its presence.

The first example is useful to see the impact of lack of common knowledge in a game. The Email Game, defined by Rubinstein as a variant of Halpern's Coordinated Attack Problem (see [Fagin et al., 1995]), is a Bayesian Game in which two agents a and b have to choose between two actions A and B. The payoffs depend on whether the game is g_1 or g_2, which in turn depends on the state of nature, which only a can observe. Player a sends an email to b if the game is g_2, and no message otherwise, to inform b about the state of nature. Player b's machine sends an automatic response in case a message is received, and likewise for a. Both machines however have the same probability of transmission failure $\varepsilon > 0$. Thus, each agent sees on his screen the number of messages he sent at the end of the communication process, namely when the first transmission failure occurs, but not the other's number.

g_1	A	B
A	10,10	0, -5
B	-5,0	0,0

g_2	A	B
A	0,0	0, -5
B	-5,0	10,10

The informational structure of the game can be represented by coding each state as an ordered pair consisting of a's and b's respective numbers of messages sent after transmission failure occurs. Letting the atom g_1 (resp. g_2) represent the sentence 'the game is g_1' (resp. 'the game is g_2'), we see that g_1 holds only at the state (0,0):

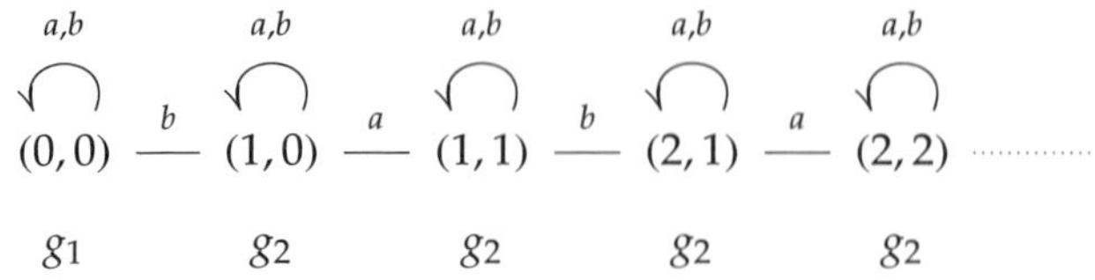

FIGURE 18.3 Epistemic structure of the Email Game

The striking result proved by Rubinstein ([Rubinstein, 1989]) is that *the Email Game has a unique Nash equilibrium in which both players always choose A*. (See Chapters 9 and 19 for a more detailed account of games and game theory.) This means that even when the game is g_2 and a and b have exchanged a possibly very large number of messages, as rational agents they will play the strategy profile (A,A) that is less profitable to both than (B,B). We shall not prove that result here (see [Osborne and Rubinstein, 1994]) but only highlight the intuitive reason why this may happen. Consider state (1,1) first: $(1,1) \models E_{a,b}g_2$, that is both a and b know the game played is g_2, but $(1,1) \models \hat{K}_a\hat{K}_b g_1$, that is a considers possible that b received 0 messages, and therefore that b thinks the game is g_1. More generally, since each state (n,n) or $(n,n-1)$ is connected to state $(0,0)$ by a path along the union of the accessibility relations R_a and R_b, it is never common knowledge between a and b that the game played is g_2. If it were common knowledge, then a and b could rationally play according to the Nash equilibrium (B,B) in g_2. Therefore, what the example suggests is that lack of common knowledge regarding the state of nature can have fairly dramatic consequences for the way ideal players should play.

We may now give an illustration of a positive result concerning the epistemic conditions for solution concepts. Paradigm cases of solution concepts include *Nash equilibrium* in strategic games, *iterated elimination of strictly dominated strategies* in strategic games and *backwards induction* in sequential games (a.k.a. subgame perfection). Aumann has been the main proponent of the programme consisting in characterizing the epistemic assumptions under which each of these solution concepts is forced in a game (see [Aumann, 1995], [Aumann and Brandenburger, 1995]). Each of those solution concepts has been extensively discussed in the literature. Our second example in this section thus concerns the connection between common belief and the iterated elimination of strictly dominated strategies in strategic games, following the presentation of [Stalnaker, 1994].

Formally, a *strategic game* can be defined as a structure $G = (N, (A_i, u_i)_{i \in N})$, where N is a set of players, A_i the set of actions or strategies available to each player, and u_i the utility attached by each player to action profiles (or outcomes). A *model for a game* G is a structure $M = (W, w, (R_i, P_i, a_i)_{i \in N})$, where each world $w \in W$ is the index of the action $a_i(w) \in A_i$ played by each player in w, $R_i(w)$ is the information state of i in w, and $P_i(w)$ represents the degree of i's belief about the actions played in w by the other players. Furthermore, each R_i is assumed to be serial, transitive, and euclidean, though not necessarily reflexive, meaning that players have consistent information and introspective access, but that the information is not necessarily veridical. Finally, whenever two worlds w and w' are such that wR_iw', then $a_i(w) = a_i(w')$, meaning that each agent knows her actions.

A player is rational in a state w if she maximizes her expected utility in w. Rationality can be defined in terms of the u_i, a_i, and P_i, namely of the utilities, actions, and partial beliefs of the player. Thus one can define the set of worlds in which each player is rational. An action a_i is strictly dominated if whatever actions taken by the other players, there is an alternative action (possibly a probability mix of alternative actions) that would yield i a better payoff.

The result we aim at, due to Berheim and Pearce, transposed into Stalnaker's framework, is that *in a game model $\mathcal{M}$ in which the players are all rational, if there* *is common belief between them that they are rational, then the set of actions played survives the iterated elimination of strictly dominated actions.* Conversely, given a strategic game G, for every strategy that survives iterated elimination of strictly dominated strategies, one can construct a canonical model for that game in which the strategy is played in the actual world and in which all players are rational at every world in the model, and so in which they have common belief in rationality.

The connection between common belief in rationality and iterated strict dominance is particularly telling because it shows how the information theoretic structure of game allows players to disregard particular strategies and thereby to act in an optimal way. A number of further connections between common belief, common knowledge, and equilibria could be mentioned. One of the particularly disputed issues concerns backwards induction in sequential games of perfect information, in particular due to a debate between Aumann and Stalnaker regarding the definition of what counts as rationality in sequential games. For lack of space, we refer the interested reader to the following papers on this issue: [Aumann, 1995], [Stalnaker, 1998], [Halpern, 2001], [Clausing, 2003], [de Bruin, 2004], and [Baltag et al., 2009]. Similarly, more detailed accounts of the epistemic foundations of game theory and on the incidence of common knowledge can be found in [Bonanno and Battigalli, 1999], [Vanderschraaf and Sillari, 2009], and [Roy, 2010].

4. Informational Dynamics

Everything we said so far concerns the representation of the information that is supposed to be available to agents at a given moment in time. The framework we described is *static* in that it does not describe the effect of agents *learning* new information. Since the 1980s, however, the basic framework of epistemic logic has been enlarged to deal with various notions of informational dynamics. Two distinct sources of development in this area can be distinguished. The first concerns belief revision, as originating from the AGM framework. The second concerns the effect of information updates through public announcements. Some fruitful connections and bridges between the two domains have been made, in particular in recent years (see [van Benthem, 2004a], [van Ditmarsch, 2005], [Aucher, 2008], [Baltag and Smets, 2008a]).

4.1 Belief Revision and Updates

Historically, notions of knowledge dynamics have come from the tradition of belief revision developed by Alchourrón, Gärdenfors, and Makinson in the 1980s. The AGM framework is different from the framework of epistemic logic in that AGM standardly represent belief states by so-called *knowledge bases*, namely by sets of formulae closed under logical consequence, rather than by means of Kripke structures. The AGM framework deals with the problem of how new information can be consistently accommodated into a corpus of knowledge.

Consider an agent like Ann who only knows q, namely that it is raining. If Ann comes to learn p, namely that the bank is open, then she need only *expand* her knowledge base with p. Now suppose Ann believed that it is raining and that the bank is closed, namely q and $\neg p$. If she learns that the bank is open, an expansion of the set $\{q, \neg p\}$ with p will produce the inconsistent set $\{q, \neg p, p\}$. To accommodate the information that p, Ann will need to *retract* the belief that $\neg p$ from her belief set, and then to expand it again with the information that p, so as to get to the consistent belief set $\{q, p\}$.[2]

Simple though it may seem, this very simple example contains the essential concepts of interest in the framework of belief revision. We shall not go here into the details of the AGM theory (see [Gärdenfors, 1988], [van Ditmarsch et al., 2007] for introductions). What we shall do, however, is to see how such processes of informational updates can be described semantically in the framework of epistemic logic. From a semantic point of view, our toy example allows us to distinguish two kinds of informational updates. When Ann learns information that is *compatible* with what she already believed, then the effect of expansion is to restrict her uncertainty, so to restrict the set of worlds compatible with her beliefs. On the other hand, when Ann learns information *incompatible* with what she believed, it should be apparent that more structure is needed to describe the effect of a contraction followed by an expansion.

4.2 Public Announcements

Let us consider the case of an update with information compatible with what Ann believes. Consider the model of Figure 18.1 again. The effect of Ann learning that p in w will be that the world w' is eliminated from her belief set.

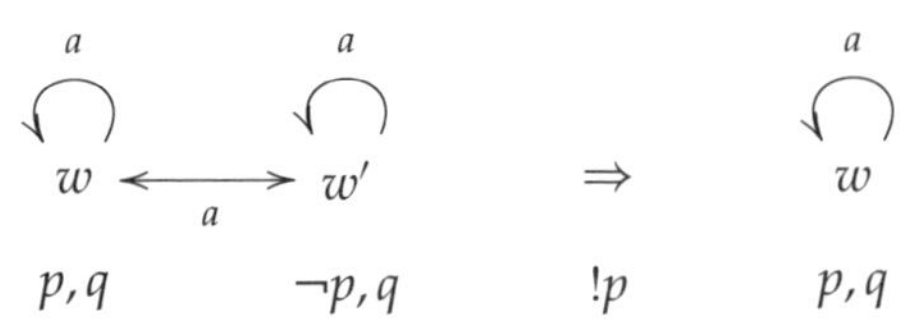

FIGURE 18.4 Updating with p

Thus, the effect is that Ann's original belief set is restricted. In the left model, $w \models \neg K_a \neg p$, in the right model, after the update with p (marked as $!p$), $w \models K_a p$, since now every world compatible with Ann's new informational state is a p-world. We may note that here we described the effect of Ann learning not only information compatible with her beliefs, but moreover true information. When dealing with several agents, we can examine in the same way the effect of all agents simultaneously learning information that is truthful. Such updates on the agents' information are called *public announcements*. The logic of updates by public announcements was developed independently by [Plaza, 1989] and [Gerbrandy and Groeneveld, 1997]. To describe the effect of updates by formulae on belief states, the language needs to be enriched with an update operator. We present the simplest language here, but the framework can be extended to accommodate common knowledge or distributed knowledge:

Definition 18.4.1 Syntax of basic public announcement logic (PAL):

$$\phi := p \mid \neg\phi \mid (\phi \wedge \phi) \mid K_i\phi \mid [!\phi]\phi$$

For example, a formula like $[!\phi]K_i\psi$ means that i knows ψ after learning that ϕ, or after it was publicly announced that ϕ. To model the effect of public announcements, we need to define the notion of model restriction. Given a model $\mathcal{M} = (W, (R_i)_{i\in I}, V)$, $\mathcal{M}|\phi$ is the model $\mathcal{M}' = (W', (R_i')_{i\in I}, V')$ where W' is the set of worlds in W that make ϕ true, R_i' is the intersection of R_i with $W' \times W'$, and V' is just like V, restricted to W'. The new clause for updates is the following:

Definition 18.4.2 Semantics for PAL

$$\mathcal{M}, w \models [!\phi]\psi \quad \text{iff} \quad \text{if } \mathcal{M}, w \models \phi, \text{ then } \mathcal{M}|\phi, w \models \psi.$$

The addition of update operators to the language allows one to represent the successive ways in which uncertainty is reduced in a game situation or in dialogues, under the assumption of truthfulness. A complete axiomatization of the logic is given by adding to standard axioms for epistemic logic the following reduction axioms:

$$\begin{array}{lcl} [!\phi]p & \leftrightarrow & (\phi \rightarrow p) \\ [!\phi]\neg\psi & \leftrightarrow & \neg[!\phi]\psi \\ [!\phi](\psi \wedge \chi) & \leftrightarrow & ([!\phi]\psi \wedge [!\phi]\chi) \\ [!\phi]K_i\psi & \leftrightarrow & (\phi \rightarrow K_i[!\phi]\psi) \\ [!\phi][!\psi]\chi & \leftrightarrow & [!\phi \wedge [!\phi]\psi]\chi \end{array}$$

What the above axioms show is that a sentence with update operators can be recursively transformed into a more complex sentence without update operators. A slightly more complex axiom system results when incorporating common knowledge with update operators (see [van Ditmarsch et al., 2007]). Likewise, it is possible to model the effect of public announcements that are not truthful, but that are *believed* to be true. Instead of eliminating states where ϕ is false, the announcement of ϕ simply removes epistemic accessibility to non-ϕ states for each agent. The reduction axiom $[!\phi]B_i\phi$ is modified accordingly in that case (see [van Ditmarsch et al., 2007, pp. 91–2]).

4.3 Belief Revision

As expressed by van Benthem [van Benthem, 2004a], public announcements describe a notion of update with *hard information*, namely true information that becomes later unrevisable. A different kind of update concerns revisions that might affect what an agent conceives as plausible or probable, and that may be revised again later. This includes, in particular, cases where the information is incompatible with what the agent believes.[3] Suppose Ann believes both that it is raining and that the bank is open, while in fact it is not raining and the bank is not open. If Ann is told that it is not raining, intuitively, Ann will accommodate that information so as to make minimal changes to her other beliefs.

One way to represent this, originally inspired from Lewis' similarity-based semantics for counterfactuals, consists in ordering belief worlds in terms of how plausible they are (see [Grove, 1988], [Spohn, 1988]). Several ways of implementing this are available (see [Board, 2004], [Baltag et al., 2009], [Pacuit, 2010] for definitions based on preorders). For instance, define a *doxastic epistemic model* as a structure of the form (W, d_i, V), where d is a function from $W \times W$ to natural numbers. Intuitively, $d_i(x, y)$ indicates the degree to which y is considered plausible relative to x for agent i. $d_i(x, y) \leq d_i(x, z)$ means that y is at least as plausible as z relative to x. Consider for instance:

w_1	w_2	w_3	w_4
p, q	$\neg p, q$	$p, \neg q$	$\neg p, \neg q$
0	1	1	2

FIGURE 18.5 A doxastic epistemic model

Here, the numbers 0, 1, and 2 represent the initial plausibility of each world relative to all the others (in this example we are assuming that each world is equally plausible relative to all others, but it need not be so in general): 0-degree worlds are most plausible worlds; 1-degree worlds are next most plausible worlds, and so on.

Plausibility allows us to define the semantics for belief. Let $\mathcal{M}, w \models B_i\phi$ be true iff for every w' such that $d_i(w, w')$ is minimal (namely such that $d_i(w, w') \leq$

$d_i(w, y)$ for every y), $\mathcal{M}, w' \models \phi$. This says that what an agent believes at a world are propositions true in the most plausible worlds relative to w. For instance, here at every world w in the model we have $\mathcal{M}, w \models B(p \wedge q)$.

Based on this, there are several ways of defining an appropriate notion of update corresponding to belief revision with ϕ. A standard way is to consider what is believed in the minimal worlds compatible with the information that ϕ. In our example, w_2 is the unique minimal world compatible with the information that $\neg p$, hence after revising her beliefs with $\neg p$, Ann will believe $\neg p \wedge q$.

To formally represent the effect of belief revision by ϕ, several possibilities exist. One is to use *conditional belief operators*, of the form $B^{\phi}\psi$ (see [van Benthem, 2004a], [Baltag et al., 2009]). Thus, $\mathcal{M}, w \models B^{\phi}\psi$ will be true if for every w' such that $\mathcal{M}, w' \models \phi$, and w' is least relative to w among ϕ-worlds, $\mathcal{M}, w' \models \psi$. For instance, in the above structure, $B^{\neg p}(\neg p \wedge q)$ holds at every world.

Another option is to use *revision operators* of the form $[*\phi]$, in order to compositionally derive truth conditions such that $[*\phi]B\psi$ will express that ψ is believed after a revision with ϕ ([Segerberg, 1995], [Aucher, 2008], van Ditmarsch [van Ditmarsch, 2005]). In this case, the update operator corresponds to an instruction to transform the initial model into a new model. Thus, one may view a revision by $\neg p$ as an operation that affects the ordering between worlds in the initial model. For instance, a revision by $\neg p$ may reassign plausibility as follows: all $\neg p$ worlds become *more* plausible than they were, all p worlds *less* plausible:

w_1	w_2	w_3	w_4
p, q	$\neg p, q$	$p, \neg q$	$\neg p, \neg q$
1	0	2	1

FIGURE 18.6 An update on plausibility

Note that the plausibility semantics introduced above for belief implies that in this new model $\mathcal{M}'$, $\mathcal{M}', w \models B(\neg p \wedge q)$. An interest of this perspective is that it makes room for the description of different belief revision policies. For instance, a different revision policy would say that a world retains the same degree of plausibility if it is $\neg p$, but decreases its plausibility if p, yielding:

w_1	w_2	w_3	w_4
p, q	$\neg p, q$	$p, \neg q$	$\neg p, \neg q$
1	1	2	2

FIGURE 18.7 A different revision policy

Here a revision with $\neg p$ would make Ann come to doubt whether p, though Ann would still believe q.[4] In contrast to public announcements, belief revisions therefore need not make an epistemic model shrink.

4.4 Epistemic Actions

Public announcements and belief revision policies may be viewed as particular cases of transformations of an epistemic structure into a new one. More transformations are conceivable. In the multi-agent case, for instance, an agent may learn some information privately, unbeknownst to others (by cheating in a game, or through outside informants). This raises the issue of whether the output model resulting from an input model can be described mathematically as the product of a particular *action* over the input model. This perspective, opened by Baltag, Moss and Solecki, suggests that one may differentiate private and public announcements, for instance, according to the model-theoretic structure of the actions or events to which they correspond.[5]

Consider for instance the model $\mathcal{M}$ to the left of the product sign in which Ann and Bob know that it is raining (q), but only Bob knows in w' that the Bank is open (p):

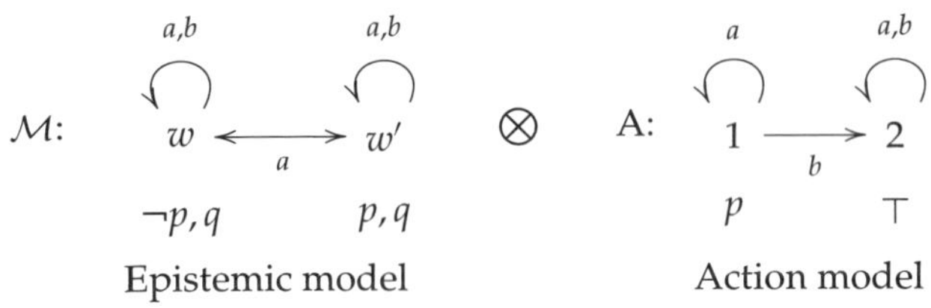

FIGURE 18.8 Epistemic model and Action model

In this model, $\mathcal{M}, w' \models K_b \neg K_a p$, i.e., Bob knows that Ann does not know p. If a public announcement that p were made, then the model would be reduced to the single world w', where Ann and Bob both know that p and q, and Bob knows that Ann knows p, and even where it would be common knowledge between Bob and Ann that Ann knows p.

Suppose however, that Ann privately learns that the bank is open (she looks up the information on the internet), and Bob is unaware of that. The idea of Baltag, Moss, and Solecki's approach is to represent the private action (or event) of learning as the model A to the right of the product sign. In this model, each formula at a world is taken to represent the precondition for each world. Here, 1 is a world where p holds (namely the bank is open, $Pre(1) = p$), and only Ann is aware of it (this explains why 2 is the only accessible world for Bob). 2 is a world where nothing happens ($Pre(2) = \top$), and both Ann and Bob have access to it. The following model represents the effect of the private announcement of p to Ann, and corresponds to the result of the product of the two above models:

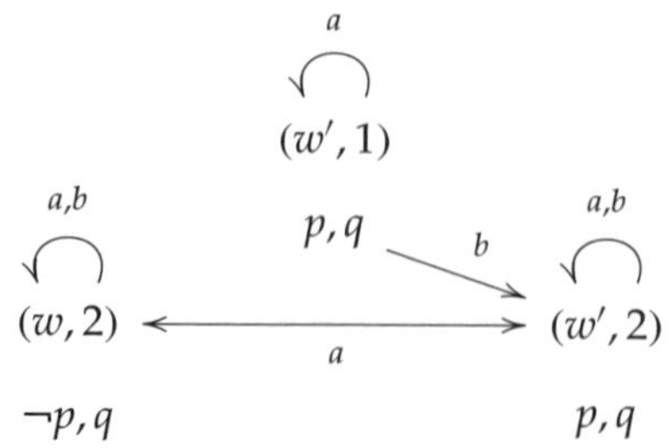

FIGURE 18.9 The effect of Ann privately learning p

This new model results from the previous one by requiring of each world (x, y) that $x \models Pre(y)$. This explains why the world $(w, 1)$ is not represented here. Furthermore, (x', y') is i-accessible to (x, y) provided xR_ix' and yR_iy': this explains why $(w', 1)R_b(w', 2)$, but not so for a. Finally, $(x, y) \models p$ provided $x \models p$ in the initial epistemic model. Usually, both epistemic models and action models are pointed models with a designated actual world. Here, if w' is the actual world in the epistemic model, and 1 the actual world in the action model, $(w', 1)$ is the new actual world. In this new model, it should be clear that it is not common knowledge between Ann and Bob in $(w', 1)$ that Ann knows p. Rather, Bob believes that Ann does not know p in $(w', 1)$, and in this case Ann knows that Bob believes it. As the model makes clear, accessibility relations are no longer reflexive as soon as agents can be unaware of the occurrence of particular actions.

The interest of the product approach is that the effect of a public announcement that p can be represented by the action model consisting of the single world 1 accessible to both a and b. Because of that, action models permit us to describe the structure of updates. The logic BMS, named after the authors, is a dynamic epistemic logic much like the logic of public announcements, with the main difference that updates now include the reference to the action models on which the updates happen. For instance, it is possible to write that $\mathcal{M}, w' \models [A, 1]B_aB_b\neg B_ap$, to mean that $M \bigotimes A, (w', 1) \models B_aB_b\neg B_ap$ provided $\mathcal{M}, w' \models Pre(1)$. Despite this very expressive syntax (which includes reference to models), the logic BMS is axiomatizable by means of reduction axioms analogous to the ones for.

5. Logical Omniscience and Self-Knowledge

In Sections 3 and 4, we presented applications of epistemic logic to the representation of group knowledge and of informational dynamics. We saw that dynamic epistemic logics of various sorts allow us to integrate these two perspectives. In particular, the effect of agents learning new information varies depending on the kind of information at stake (hard or soft information), but also on the procedure involved (such as public vs. private learning), both of which can be formally distinguished. In this section and the next, we consider epistemic logic in relation to the clarification of some more traditional issues in analytic epistemology. This section is particularly devoted to the idealizations encapsulated in Hintikka-Kripke's relational semantics and the resulting axioms for knowledge or belief. We discuss, in particular, various proposals that have been made to adapt Hintikka's semantics to the representation of logically bounded agents. The issues we are concerned with in this section essentially concern the representation of knowledge or belief from the perspective of a single agent, and we occasionally drop the subscript on K_a or B_a for ease of presentation.

5.1 Logical Omniscience

The standard Hintikka-Kripke semantics for static knowledge and belief implies that the corresponding operators obey the following closure properties:

K	$K(\phi \rightarrow \psi) \rightarrow (K\phi \rightarrow K\psi)$
N	$K\top$
M	$\phi \rightarrow \psi \therefore K\phi \rightarrow K\psi$
Re	$\phi \leftrightarrow \psi \therefore K\phi \leftrightarrow K\psi$
C	$K\phi \wedge K\psi \rightarrow K(\phi \wedge \psi)$
Nec	$\phi \therefore K\phi$

K implies that knowledge is closed under material implication. **N** implies that an agent knows all logical truths. **M** implies that knowledge is closed under valid implication. **Re** implies that knowledge is closed under logical equivalence, and **C** that it is closed under conjunction. **Nec** is the rule of generalization, or necessitation, which implies that every validity of the system is known automatically. These properties hold in all normal modal logics, and therefore in the standard systems of belief or knowledge **K45**, **S4**, and **S5** introduced in Section 2.

Because of that, it is widely admitted that such systems purport either to describe the beliefs of idealized agents, namely perfect reasoners capable of working out all the consequences of what they know; or otherwise that they describe the *implicit knowledge* available to ordinary agents. In order to model the knowledge explicitly available to agents who might not be perfect reasoners, a more fine-grained representation of the content of a belief state is needed. Thus, all available solutions to the problem of logical omniscience converge on the idea that some level of syntactic representation is needed to individuate belief states.

For instance, instead of using relational semantics for belief, one option is to use *neighbourhood semantics* (a.k.a. Montague-Scott semantics, see [Fagin et al., 1995]): in a state w, what an agent a believes is described by a set $\mathcal{B}_a(w)$ of possible worlds propositions that is not necessarily closed under logical entailment. In this case, $B_a\phi$ holds at w if the proposition expressed by ϕ belongs to $\mathcal{B}_a(w)$. Without special provisos, all of the above closure principles are blocked, except substitution of logically equivalent sentences (**Re**), since two logically equivalent sentences are true exactly in the same set of worlds.

Another option capable of blocking even (**Re**) is to preserve the standard relational semantics, but to add a level of syntactic representation. Two versions of this approach are the impossible worlds approach of [Rantala, 1982], and the awareness approach of [Fagin and Halpern, 1987]. An *impossible world structure* is a model $\mathcal{M} = (W, W^*, R_a, \sigma)$ such that W and W^* are sets of possible worlds, R_a is an accessibility relation between worlds, and σ is a syntactic assignment function that assigns sets of formulae to worlds in W and W^*.

On W, the set of 'logically possible worlds', σ works compositionally; on W^*, the set of 'logically impossible worlds', a formula can be assigned the value true at a world non-compositionally. For instance, a world $w \in W^*$ can satisfy $\phi \wedge \psi$ without satisfying either of the conjuncts. As usual, $\mathcal{M}, w \models K_a\phi$ iff for all w': if wR_aw', then $\mathcal{M}, w' \models \phi$. Consider for instance, the following structure $\mathcal{M}$ (Figure 18.10), in which w is a logically possible world, and w^* a logically impossible world.

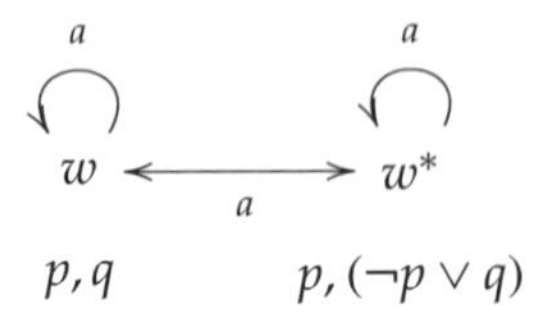

FIGURE 18.10 An impossible world structure

Below every world, we have indicated exactly which formulas are true: atoms for w, and arbitrary formulae in w^*. $\mathcal{M}, w \models K_a p$ and $\mathcal{M}, w \models K_a(p \rightarrow q)$, since every world satisfies p, and every world satisfies $\neg p \vee q$ (material implication). However, w^* does not make q true, hence $\mathcal{M}, w \not\models K_a q$. This is possible only because the truth of $(\neg p \vee q)$ in w^* does not require either $\neg p$ or q to be true there.

Essentially the same idea is in play in awareness structures, except that two operators are introduced in the language to mirror the difference between possible and impossible worlds: an operator K_a of implicit knowledge and an operator A_a of awareness. An awareness structure is a model $\mathcal{M} = (W, R_a, N_a, V)$: V

is now a standard valuation, R_a is as usual, but N_a assigns to each state w a set of formulae, the formulae that the agent is aware of. By definition,

$$\mathcal{M}, w \models K_a\phi \quad \text{iff} \quad \text{for every } w' \text{ such that } wR_aw', \mathcal{M}, w' \models \phi$$

$$\mathcal{M}, w \models A_a\phi \quad \text{iff} \quad \phi \in N_a(w)$$

$$\mathcal{M}, w \models X_a\phi \quad \text{iff} \quad \mathcal{M}, w \models K_a\phi \wedge A_a\phi.$$

This says that an agent knows ϕ explicitly iff ϕ is known implicitly and the agent is aware of ϕ.

A natural correspondence exists between awareness structures and impossible worlds structures (see [Fagin et al., 1995] or [Sillari, 2009]; try for instance, to turn the previous model into an awareness model). Moreover, both semantics can lead back to the standard Hintikka-Kripke semantics by imposing appropriate closure conditions on the syntactic functions σ or A_a.

Interestingly, all of these approaches are ways of blocking closure principles for knowledge statically. Some attempts have been made in the literature to resolve the logical omniscience problem in relation to informational dynamics. The idea here is that knowledge or belief should be conceived in relation to procedures. For instance, if I know p and I know that $(p \rightarrow q)$, I can know q if I perform an act of deduction, or relate the two sentences by applying the rule of modus ponens. Duc ([Duc, 1997]) gives the example of a system of dynamic epistemic logic in which the main idea is to assume that agents' knowledge is not closed statically, but such that one's knowledge can in principle be increased provided a particular rule is applied. In this system $K_ap \wedge K_a(p \rightarrow q)$ does not entail K_aq, but it holds that $K_ap \wedge K_a(p \rightarrow q) \rightarrow [\alpha]K_aq$, where $[\alpha]$ represents the effect of updating one's knowledge by the application of modus ponens. Parikh ([Parikh, 2008b]) similarly outlines several ways in which the folk concept belief can be analysed depending on which kind of update operation applies to it (update by a sentence, by witnessing an event, or by performing an inference). A more elaborate proposal along the lines of Duc's approach (but developed independently), finally, can be found in Artemov's justification based logic ([Artemov, 2008]), in which terms are used to mark the justification for a formula (see below).

5.2 Limitations on Self-Knowledge

The axioms **4** and **5** of positive and negative introspection also represent strong closure principles, since they guarantee that agents are automatically aware both of what they know and of what they are ignorant. Since Hintikka's book, there has been a consensus that negative introspection is an even stronger idealization on knowledge than positive introspection. As a result, the latter principle has been more vividly debated. Logically, **5** is a powerful axiom since in combination with **T** (in normal systems), it yields **4**. Axiom **4** is weaker in this regard, since **4** and **T** together with **K** do not imply **5**.

At least two related arguments can be given against the plausibility of **5**. The first concerns the *unawareness* of some propositions. Suppose I never heard of Lance Armstrong and the Tour de France. How could I then know that I fail to know that (or whether) Lance Armstrong won the Tour de France seven times? More generally, from **5** and **T** the *Brouwersche axiom* **B** follows, which says that $p \rightarrow K\neg K\neg p$, namely: every truth is such that I know I entertain it as possible.

A second argument concerns the occurrence of *false beliefs*, and the interaction between belief and knowledge. Suppose I have a false belief that p, and believe I know p (a case of misplaced self-confidence in p). If knowledge entails belief ($Kp \rightarrow Bp$), then since this is a case in which I don't know p, by **5** I know that I don't know p, and therefore I believe that I don't know p. So I believe that I know p, and believe that I don't know p. If belief is assumed to satisfy consistency (**D**), this is a contradiction. The upshot is that assuming $Kp \rightarrow Bp$ and consistency of belief, **5** rules out self-confidence in false propositions. Arguably, this argument is weaker than the former, since perfectly rational agents may sometimes be unaware of some true propositions, without ever having any false beliefs. On the other hand, both arguments make clear the sense in which **5** is an idealization of the ordinary notion of belief.

Hintikka's essay defends principle **4** (also called *the KK principle*), but Hintikka rejects the idea that **4** should hold due to the agent having special introspective powers. Rather, Hintikka's view is that $K\phi$ and $KK\phi$ come out 'virtually equivalent' on logical grounds (see [Stalnaker, 2006]). However the principle of positive introspection is usually seen as the expression of an internalist conception of knowledge and justification. On the internalist view, one's justification for one's beliefs or knowledge is accessible to oneself. This contrasts with the externalist view on which one's reasons to believe or know a proposition may not be fully open in this way. Williamson ([Williamson, 1994], [Williamson, 2000]) has argued forcefully against the plausibility of **4** in the context of a broader externalist conception of knowledge. Williamson's main argument against the plausibility of **4** involves what Williamson calls a *margin for error* principle for knowledge. The margin for error principle says that: 'in order to know p in context w, p should remain true in all contexts sufficiently similar to w'. Margins of error purport to account for the idea that knowledge is a form of *safe* or *reliable* belief, namely true belief that could not easily be false.[6] The principle extends the notion of factivity or veridicality of knowledge to neighbouring worlds, since $w \models Kp$ not only implies $w \models p$, but also that $w' \models p$ for any w' suitably related to w.

To formalize this notion, Williamson [Williamson, 1994] proposes a *margin for error semantics* for knowledge. Basically, a (fixed) margin for error model is a structure (W, d, α, V) where W and V are as usual, and α is a real valued parameter (representing the size of the margin), and d is a metric on $W \times W$ (a function from $W \times W$ satisfying $d(x,x) = 0$, $d(x,y) = d(y,x)$ and the triangular inequality. Williamson's semantics for knowledge then becomes:

Definition 18.5.1 Margin for error semantics (MS):

$$\mathcal{M}, w \models_{MS} K\phi \quad \text{iff} \quad \text{for every } w' \text{ such that } d(w,w') \leq \alpha, \mathcal{M}, w' \models_{MS} \phi.$$

This says that ϕ if known iff it is true in a neighbourhood of radius α around w. The induced logic for knowledge is the logic **KTB**.[7] In particular, the margin semantics validates neither positive nor negative introspection. For instance $w \models_{MS} K\phi$ means that ϕ holds throughout all worlds within distance α from w; but $w \models_{MS} KK\phi$ means that ϕ holds throughout all worlds within distance $2{\cdot}\alpha$ from w. Concretely, this means that knowing that one knows requires more safety than just knowing (a similar argument can be used to invalidate **5**).

Williamson has presented several arguments against the principle of positive introspection, all based on the observation that the assumption of margin of error and the principle of positive introspection are mutually inconsistent (see below the discussion of epistemic paradoxes). Arguably, however, the introspection principles can be maintained provided margin for error principles are restricted in the appropriate way. One of the problematic assumptions behind Williamson's semantics is the idea that each iteration of knowledge requires a new margin, of the same kind as the margin required for first-order knowledge (see [Dokic and Égré, 2009]). Based on this observation, Bonnay and Égré [Bonnay and Égré, 2009] put forward a two-dimensional semantics for epistemic logic, called *centred semantics*, in which a principled distinction is implemented between first-order knowledge (which requires a margin) and second-order knowledge (assumed to supervene only on first-order knowledge). The semantics, which can easily be adapted to margin models, is originally stated for standard Kripke models (W, R, V), and its two specific clauses are (boolean clauses are as expected):

Definition 18.5.2 Centred semantics (CS):

$$\mathcal{M}, (w, w') \models_{CS} p \quad \text{iff} \quad w' \in V(p) \qquad \text{(CS-at)}$$

$$\mathcal{M}, (w, w') \models_{CS} K\phi \quad \text{iff} \quad \text{for every } w'' \text{ such that } wRw'', \mathcal{M}, (w, w'') \models_{CS} \phi. \qquad \text{(CS-K)}$$

Finally, $\mathcal{M}, w \models_{CS} \phi$ is defined as $\mathcal{M}, (w, w) \models_{CS} \phi$. The second clause ensures, in particular, that all knowledge, including higher-order knowledge, is only relative to alternatives to the first-index, the second index fixing only the atomic information. The interest of the logic is that it makes **4** and **5** automatically valid over arbitrary one-dimensional structures (including non-transitive, non-euclidian structures). In contrast to standard Kripke semantics, iterations of knowledge operators thus permit to remain within worlds that are one step away from the world of evaluation. As shown in [Bonnay and Égré, 2009], Centred semantics can be generalized into a more complex multi-dimensional system, called *token semantics*, in which n iterations of operators involve making n steps along R to check for satisfaction, but such that iterations beyond n come for free. This gives rise to a family of logics intermediate in strength between **K** and **K45**, with weakened versions of the axioms **4** and **5**. In such systems, for instance, knowing need not automatically imply knowing that one knows, but knowing that one knows can guarantee that one will know that one knows that one knows.

Centred semantics follows a rather internalist inspiration. Halpern ([Halpern, 2008]) provides a middle-ground between this approach and Williamson's. Unlike Williamson or Bonnay and Égré, Halpern presents a standard two-dimensional epistemic logic based on two operators, an operator of subjective or internal knowledge, and an operator of objective or external definiteness. Both of these operators satisfy the usual introspection principles **4** and **5**. Their composition does not, however. Logically, this approach can be seen as a way of syntactically reflecting the truth conditions stated in (MS) for a single operator in terms of two operators: the standard knowledge operator, and a neighborhood operator. The same decomposition can be made of the truth conditions for (CS) in terms of a standard two-dimensional semantics for knowledge, and truth conditions for an actuality operator (see [Bonnay and Égré, 2009], [Bonnay and Égré, ta]).

A point worth emphasizing is that the choice between these various semantics ultimately depends on which view of higher-order knowledge is favoured, and on the problem of the relation between the first level and higher levels. From a logical point of view, the representation of self-knowledge happens to have interesting connections with the problem of logical omniscience. (CS), for instance, validates the rule of necessitation (**Nec**) over classes of models, but not within a model. If ϕ is true at every world of every model, so is $K\phi$. In contrast to standard Kripke semantics, however, a formula ϕ can be true everywhere in a model without $K\phi$ being true everywhere in the model. This fact can be used to represent the effect of agents *learning* validities (see below). Similarly, Bonnay and Égré ([Bonnay and Égré, 2009]) present a generalization of token semantics to several agents, to deal with well-known puzzles about common knowledge in which agents are intuitively in a position to attain a state practically comparable to common knowledge (better dubbed 'almost common knowledge', see [Rubinstein, 1989]) without computing all iterations of shared knowledge.

6. Knowledge, Belief, and Justification

One of the most debated issues in epistemology concerns the distinction between knowledge and belief. In most of what we covered so far, however, we handled belief and knowledge more or less interchangeably. In Hintikka's original semantics, in particular, the only difference between knowledge and belief lies in the assumption that knowledge is a veridical attitude, which implies that the associated accessibility relation is reflexive. This assumption, however, says little about the interplay between knowledge and belief. Several aspects of this question can be distinguished. The first concerns the definition of bimodal systems of knowledge and belief and the interaction between the corresponding modalities. The second concerns the possibility of either defining belief in terms of knowledge, or knowledge in terms of belief. The third, finally, concerns the incorporation into epistemic logic of some concept of justification, which is not represented in standard Kripke models.

6.1 Combining Knowledge and Belief

Hintikka's seminal work discusses some axioms concerning the relation between knowledge and belief. Among those are the following two principles:

$$K\phi \rightarrow B\phi \tag{KB}$$

$$B\phi \rightarrow KB\phi \tag{BKB}$$

KB says that everything that is known must be believed. BKB is a positive introspection axiom for belief, which says that one knows what one believes. In order to combine knowledge and belief, the most direct way thus is to define a bimodal language in which K and B are two primitive operators, each interpreted by distinct accessibility relations. A knowledge–belief model then is a structure (W, R^K, R^B, V), where R^K corresponds to epistemic accessibility, and R^B to doxastic accessibility.

Kraus and Lehmann [Kraus and Lehmann, 1988] give the details of such a system, in which they assume R^K to be an equivalence relation (so K is **S5**) and R^B to be serial (so B is **D**). From modal correspondence theory, the two bridge axioms KB and BKB can be seen to correspond to the following frame conditions:

$$R^B \subseteq R^K \tag{KB}$$

$$\text{if } xR^K y \text{ and } yR^B z \text{ then } xR^B z \tag{BKB}$$

From these conditions it follows that R^B is transitive and euclidean, and therefore that B satisfies positive and negative introspection, as well as $\neg B\phi \rightarrow K\neg B\phi$

(negative BKB). It follows moreover that $BK\phi \rightarrow K\phi$, a property sometimes named 'perfect belief' (see [Gochet and Gribomont, 2006] for a syntactic proof originally due to Voorbraak). A related property is the property called 'strong belief', which says that if I believe ϕ, then I believe I know ϕ:

$$B\phi \rightarrow BK\phi \tag{SB}$$

Perfect belief and strong belief together imply that $B\phi \leftrightarrow K\phi$, which makes the distinction between knowledge and belief collapse. Because of that Kraus and Lehmann do not include SB among their axioms.

Stalnaker ([Stalnaker, 2006]) shows that a more interesting interdefinabilily relation can be obtained from SB if knowledge is assumed to be **S4** rather than **S5**, belief is **KD45**, and all of KB, BKB, negative BKB and SB, are assumed as bridge axioms. Perfect belief does not follow then. However, $B\phi$ is then equivalent to $\neg K\neg K\phi$. This says that what is believed is that which one does not exclude to know. Lenzen ([Lenzen, 1978, p. 83]) proposes to see $\neg K\neg K\phi$ as a good equivalent of the operator 'being convinced that'. The resulting logic furthermore satisfies the commutation property **4.2**, which says that if I don't exclude knowing ϕ (if I am convinced that ϕ), I know I don't exclude ϕ:

$$\neg K\neg K\phi \rightarrow K\neg K\neg\phi \tag{4.2}$$

Lenzen ([Lenzen, 1978]) points out that one can then get an analysis of knowledge as true belief (or true strong belief) by assuming that $\phi \wedge \neg K\neg K\phi \rightarrow K\phi$. The latter axiom can be viewed as a particular case of axiom **4.4**:

$$\phi \wedge \neg K\neg K\psi \rightarrow K(\phi \vee \psi) \tag{4.4}$$

The addition of **4.2** or **4.4** to **S4** yields the logics **S4.2** and **S4.4** of increasing but intermediate strength between **S4** and **S5**.[8]

6.2 Safety, Stability, Justification

Admittedly, the definition of knowledge in terms of true strong belief is too crude to meet Gettier's celebrated puzzles showing that knowledge is more than justified true belief [Gettier, 1963]. Gettier's example shows that a belief can be true and can even rest on some internally valid justification, without that justification being adequate to make the belief into knowledge.

One of the possible responses to Gettier's puzzles is simply to abandon the idea that knowledge could be defined in terms of belief by means of supplementary conditions. [Williamson, 2000] thus contains several arguments for the idea that knowledge is a sui generis mental state, just like belief. Nevertheless,

Williamson considers that knowledge entails belief and is a form of safe belief. In Williamson's approach, safety is a condition directly imposed on knowledge by means of margin for error principles (see above), which require that what one believes be not only true, but furthermore true in all relevantly similar alternatives.

An approach partly related to the view of knowledge as safe belief is to be found in Lehrer and Paxson's [Lehrer and Paxson, 1969] analysis of knowledge as belief undefeated under revisions by new information. On this approach, knowledge is true belief that would remain true under any revision with a true proposition. The interest of this view is that it meshes quite nicely with ideas coming from belief revision and informational dynamics. In recent years, this particular analysis has been given attention from several formal epistemologists (see [Rott, 2004b], [Stalnaker, 2006], [Baltag and Smets, 2008b]).

Several ways of implementing that idea exist. To see this, consider the doxastic epistemic models introduced in Section 4, with plausibility orderings. Recall that a doxastic epistemic model is a structure (W, d, V), where $d(x, y)$ fixes the degree to which y is plausible relative to x. Baltag and Smets's rendering of the defeasibility analysis can be formulated in terms of the conditional belief operator introduced above in Section 4, that is, ϕ will be true in all the most plausible ψ-worlds for every true ψ:[9]

$$M, w \models K\phi \qquad \text{iff} \qquad M, w \models B^{\psi}\phi \text{ for any true } \psi.$$

A different proposal is made by [van Ditmarsch, 2005], who associates to each plausibility degree i a belief operator B^i in the language, such that $w \models B^i\phi$ iff for every w' such that $d(w', w) \leq i, M, w' \models \phi$. Intuitively, B^0 is an operator that selects the most plausible worlds, B^1 the same most plausible worlds and the next most plausible, and so on. Van Ditmarsch's suggestion is to view $K\phi$ as the (infinitary) conjunction of all $B^i\phi$: to say that ϕ is known, in this approach, means that ϕ is believed to any plausibility degree (or throughout all spheres around the evaluation world). Some care must be taken to ensure that K will have a reflexive accessibility relation, but a consequence of this will be that known propositions will be propositions that remain true under any new assignment of plausibility to worlds.

Several approaches finally deserve to be mentioned under the head of evidence-based logics of knowledge. Those approaches differ from standard epistemic logic or even from the previous analysis in that they do not relate knowledge merely to strength of belief, but to the methods used to acquire belief. They include in particular the epistemic logics developed by Kelly and Hendricks (see [Hendricks, 2005] for an exposition), and Artemov's work on justification based logics ([Artemov and Nogina, 2005], [Artemov, 2008]). Artemov's framework, inspired by his earlier work on provability logic with explicit proof terms, allows for formulae of the form $u : \phi$, to represent that u is a justification for ϕ (for a given agent). In particular, the framework makes it possible to represent that an agent believes a proposition under some justification and not under other justifications that might not be available to him or her. Because of that, it is possible to represent that an agent believes a true proposition on the basis of a wrong justification, if the justification he or she has is not factive (not such that $u : \phi \rightarrow \phi$). In this, Artemov's approach bears a relation to causal theories of knowledge (see [Goldman, 1967]; see also [Stalnaker, 2006] for insightful remarks on the comparison with defeasibility analyses).

7. Existence and Quantification

7.1 Intensionality and Belief Contexts

All of the systems of epistemic logic reviewed so far are built on propositional logic. One of Hintikka's aims, however, was to account for the interaction between epistemic operators, identity, and first-order quantifiers. The last chapter of [Hintikka, 1962] thus concerns the incorporation of epistemic operators to first-order logic and deals with the treatment of several classic puzzles in the philosophy of language originally put forward by Frege and Russell. These puzzles, following Quine's terminology, concern the intensionality or *referential opacity* of attitude contexts. Belief and knowledge operators can block the substitution *salva veritate* of coreferential singular terms in their scope. For instance, the truth of (18.1a) and (18.1b) is intuitively compatible with the truth of (18.1c):

Philipp knows that Cicero denounced Catiline. (18.1a)

Cicero is Tully. (18.1b)

Philipp does not know that Tully denounced Catiline. (18.1c)

A related problem concerns the rule of existential generalization, which classically permits to infer $\exists x P(x)$ from $P(a)$. From (18.1a) above, however, an application of this rule would allow us to infer that 'there is an x such that Philipp knows that x denounced Catiline'. One of the issues raised by Quine concerns the identity of this x: if this x is Cicero, then it appears that x is Tully too, and this seems to be in tension with the truth of (18.1c).

One of the achievements of Hintikka's work concerns the clarification of these issues. Hintikka's leading idea, in particular, is to handle referential opacity as what he calls *referential multiplicity*: on this approach, although two singular terms like 'Cicero' and 'Tully' have the same reference in the world of the speaker, they can have distinct denotations in the belief worlds of the agent. Concretely, this implies that each belief world comes equipped with a (possibly distinct) domain of individuals over which the same singular terms and predicates can take distinct denotations.

To illustrate the main idea, let c stand for 'Cicero', t for 'Tully', and a for 'Catiline', and $R(x,y)$ for 'x denounced y'. We get a logical translation of the previous sentences in the extension of first-order logic with the knowledge operator K:

$$K_p R(c,a) \tag{18.2a}$$

$$c = t \tag{18.2b}$$

$$\neg K_p R(t,a) \tag{18.2c}$$

Each of these sentences is interpretable over a pointed first-order Kripke model of the form (W, w, R_p, D, I) where W is a set of possible worlds, w is the actual world, R_p describes Philipp's epistemic accessibility relation over W, D is function associating to each world w a domain D_w of individuals, and I is an interpretation function that associates to each non-logical symbol and world w a denotation in D_w. To handle the example, assume that D_w is the same for every w, with $D_w = \{1,2,3\}$. Consider a two-world model with an equivalence relation for R_p such that $I(w,c) = I(w',c) = 1$, $I(w,a) = I(w',a) = 2$, and $I(w,t) = 1$ and $I(w',t) = 3$; suppose finally that $I(w,R) = I(w',R) = \{(1,2)\}$. In this model, 'Cicero' and 'Catiline' have a constant reference across worlds, but 'Tully' has a different reference in w and w'. (18.2b) is true in w, since c and t have the same denotation there, similarly (18.2a) is true, since every world satisfies $R(c,a)$, but $K_p R(t,a)$ is false, since in w the pair $(1,2)$ belongs to the interpretation of R, while in w' the pair $(3,2)$ does not. Intuitively, the model describes a case in which Philipp is confused about the reference of the singular terms 'Tully' and 'Cicero'.

Technically, the understanding of first-order epistemic logic would involve a more detailed presentation of quantified modal logic. We shall not go into all details here, but refer the reader to [Hughes and Cresswell, 1996], [Fitting and Mendelsohn, 1998], and [Aloni, 2005] for extended presentations. Historically and conceptually, however, it is fair to say that the epistemic interpretation of modalities has brought to light some particularly interesting issues in natural language semantics concerning the interplay of quantifiers with modal operators. In the rest of this section, we focus on two of these, which concern the *de re*/*de dicto* distinction on the one hand, and the interpretation of knowing-*wh* constructions on the other.

7.2 The *de re*/*de dicto* Distinction

In the previous section we mentioned the problem of existential generalization outside of the scope of a belief or knowledge operator. This problem can be seen as a particular instance of a broader distinction, which concerns the *de re* vs. *de dicto* interpretation of quantifiers in attitude sentences. Consider the following sentence concerning Ralph's beliefs about a lottery:

$$\text{Ralph believes that a ticket will win.} \tag{18.3}$$

The sentence is ambiguous, since it can mean either that there is a particular ticket about which Ralph believes that it will win, or rather that Ralph believes that some ticket or other will win, but no ticket in particular. Formally, the distinction can be captured as follows:

$$B_r \exists x(T(x) \wedge W(x)) \tag{18.4a}$$

$$\exists x(T(x) \wedge B_r W(x)) \tag{18.4b}$$

In (18.4a) the belief operator takes scope over the existential quantifier, which corresponds to the *de dicto* interpretation. In (18.4b) the existential quantifier takes scope over the belief operator, which corresponds to the *de re* reading. The interpretation of (18.4b) requires that the same individual in the actual world be a winner in all of Ralph's belief worlds; by contrast, (18.4a) is true provided every belief world contains a winning ticket, but that winning ticket can be a distinct individual in each world.

The *de re* vs. *de dicto* distinction makes it possible to understand why it is not in general possible to apply the rule of existential generalization in belief sentences. For instance, a sentence like 'Ralph believes that Santa Claus brought the presents' may be analysed as $B_r P(s)$. But from that sentence, it would be illegitimate to infer: $\exists x B_r P(x)$, if indeed no individual in the actual world can be such that Ralph has a *de re* belief about that individual. Intuitively, a *de dicto* belief does not imply the corresponding *de re* belief, but conversely, material that is scoped out of a belief operator cannot necessarily be scoped back in, and so similarly a *de re* belief need not imply the corresponding *de dicto* belief. In particular, none of the following principles is straightforward on epistemic grounds:

$$\exists x B\phi \rightarrow B\exists x\phi \qquad \text{(Importation)}$$

$$B\exists x\phi \rightarrow \exists x B\phi \qquad \text{(Exportation)}$$

$$\forall x B\phi \rightarrow B\forall x\phi \qquad \text{(Barcan formula)}$$

$$B\forall x\phi \rightarrow \forall x B\phi \qquad \text{(Converse Barcan formula)}$$

Logically, all of these equivalences will hold if domains of individuals are assumed to be identical across worlds.[10] They will not hold if domains are permitted to vary (see [Hughes and Cresswell, 1996], [Fitting and Mendelsohn, 1998]). The less obvious of these exceptions maybe concerns the Importation principle (Importation), which is generally assumed.[11] However, suppose Pierre believes that George W. Bush does not exist (but thinks he is a fictitious entity). In principle, one can say that there is someone of whom Pierre does not believe that he exists. It may be less obvious to infer that Pierre believes that there is someone who does not exist. One way to represent this is by having: $\exists x B \forall y(x \neq y)$. From this, one does not wish to infer that $B\exists x \forall y(x \neq y)$.

The interpretation of *de re* beliefs gives rise to further notorious problems, even assuming constant domain interpretations. These include in particular 'double vision puzzles' such as Quine's puzzle about Ralph, who believes of a certain man in a brown hat that he is a spy, and of a certain man seen at the beach that he is not a spy. As a matter of fact, the man in the brown hat and the man at the beach are one and the same person, namely Ortcutt ([Quine, 1956]). In this case we have:

$$\exists x(Hat(x) \wedge B_r Spy(x)) \tag{18.6a}$$

$$\exists x(Beach(x) \wedge B_r \neg Spy(x)) \tag{18.6b}$$

The difficulty here concerns the representation of these two *de re* beliefs, in particular under the assumption that Ralph cannot be blamed of inconsistency in this case. The problem has given rise to a large literature, including [Kaplan, 1968], [Gerbrandy, 2000] and [Aloni, 2005]. Following Kaplan, all of these authors have come to the conclusion that what is needed is a representation of methods of identification. A particularly elegant semantics of first-order epistemic logic with constant domains in which a family of such puzzles is solved is provided by Aloni's system of quantification under conceptual covers. A *conceptual cover* is defined as a set C of *individual concepts* (functions from W to D) such that in every world w, every individual d in D is picked out by exactly one individual concept in the cover ($d = c(w)$ for a unique c in C). Aloni's semantics can be described as Carnapian, since it assigns variables not to individuals in the domain but to individual concepts relative to a cover. In her system, the adequate logical representation of Quine's example becomes:

$$\exists x_n(Hat(x_n) \wedge B_r Spy(x_n)) \tag{18.7a}$$

$$\exists x_m(Beach(x_m) \wedge B_r \neg Spy(x_m)) \tag{18.7b}$$

Variables in Aloni's system are indexed, so that relative to an assignment g, $g(n)$ selects a conceptual cover, and $g(x_n)$ some concept in the cover $g(n)$. An open formula $\phi(x_n)$ is true in a model at a world w and relative to g iff the individual $g(x_n)(w)$ selected by $g(x_n)$ in w belongs to the interpretation of ϕ in w. Thus, the two sentences are jointly satisfiable if each of the variables is allowed to range over distinct cover. For example, the following model, taken from [Aloni, 2005], shows two conceptual covers $\{\mathbf{a}, \mathbf{b}\}$ and $\{\mathbf{c}, \mathbf{d}\}$ relative to a model with two worlds w, w' and common domain consisting of two individuals o (for Ortcutt) and p (for its epistemic counterpart), such that in w, Ralph's unique doxastic alternative is w' (and w' is self-related). In w and w', the only spy is p, and in the actual world w, o satisfies both the properties of having a brown hat, and of being seen at the beach. (18.7a) will be true relative to the first cover that maps x_n to **c**, and (18.7b) will be true relative to the second cover when x_m is mapped to **a**.

	a	**b**	**c**	**d**
w	o	p	o	p
w'	o	p	p	o

Thus the two sentences can be true together without contradictions, and covers provide a way of representing a notion of perspective or conceptualization of a domain (since **a** stands for the description 'the man seen at the beach' from Ralph's perspective, while **c** stands for the description 'the man in the brown hat' again from his perspective). Aloni ([Aloni, 2001], [Aloni, 2005]) shows that the semantics has a sound and complete axiomatization that differs from standard systems. A particularly interesting prediction of her system is that unlike standard systems of quantified modal logic with objectual quantification, it does not validate the necessity of identity $x_n = y_m \rightarrow \Box(x_n = y_m)$ (compare **a** and **c** in the above model), nor the converse $x_n \neq y_m \rightarrow \Box(x_n \neq y_m)$ (compare **a** and **d**), and yet it does not obliterate the distinction between *de re* and *de dicto* beliefs.

7.3 Knowledge and Questions

One application of quantifying into attitude contexts originally discussed by Hintikka concerns the analysis of *knowing wh-* constructions, in particular of *knowing who*. Hintikka ([Hintikka, 1962, p. 153]) suggested analysing a sentence like 'Watson knows who Dr Jekyll is' as $\exists x(K_w x = j)$. The argument he gave is that the *de re* occurrence of the variable x constrains x to denote the same individual in all of Watson's epistemic alternatives, suggesting that Watson can reliably identify the Dr. Jekyll as one and the same person.

By so doing, Hintikka furthermore observed that *knowing who* sentences can be analysed in terms of *knowing that*. Similarly, 'John knows whether p' can be analysed as 'John knows that p or John knows that not p'. Hintikka [Hintikka, 1975] thus lists a number of different constructions in terms of *know*, in particular all constructions involving embedded interrogative complements, such as *knowing which*, *knowing how*, *knowing where*, and so on, for which one can wonder whether it is possible to analyse them in quantified epistemic logic.

Following work done at the same time by Hamblin [Hamblin, 1973], Karttunen [Karttunen, 1977] and Groenendijk and Stokhof [Groenendijk and Stokhof, 1984], the semantic analysis of questions and their embedding under different verbs has gradually become a whole subfield of natural language semantics. While it would take us too far afield to enter into this subject, it is interesting to point out the existence of several connections between epistemic logic and the semantic analysis of embedded questions. At least three issues deserve particular mention.

The first is whether all constructions in terms of *know* can be analysed in terms of *know that*.[12] The second concerns the exact quantificational analysis of several of these constructions in relation to *knowing that*, and their derivation in an epistemic language with question forming operators.[13] Consider for instance the following two sentences:

John knows which students left. (18.8a)

John knows where one can buy an Italian newspaper. (18.8b)

One understanding of sentence (18.8a) in terms of 'knowing that' is:

(a) $\forall x(Student(x) \wedge Left(x) \rightarrow K_j(Student(x) \wedge Left(x)))$,

which says that John knows of every student who left that he is a student who left. Another is the conjunction of (a) with:

(b) $\forall x(Student(x) \wedge \neg Left(x) \rightarrow K_j(Student(x) \wedge \neg Left(x)))$,

namely John also knows that every student who did not leave is a student who did not leave. Groenendijk and Stokhof gave arguments for the second analysis as opposed to the first (defended by Karttunen). Contrast this with (18.8b). An intuitive paraphrase in this case is in terms of an existential quantifier:

(c) $\exists x(ItalianNews(x) \wedge K_j ItalianNews(x))$,

which says that there is a place where one can buy Italian newspapers such that John knows that one can buy Italian newspapers at that place. It is interesting to see that (c) puts a much weaker requirement on knowledge than even only (a).[14]

The third issue finally concerns the context-sensitivity of knowing-*wh* constructions. Hintikka [Hintikka, 1962] had pointed out that 'knowing who' can mean different things depending on the method of identification involved (see [Hintikka, 1962, p. 149]). Suppose for instance that you will win 10 euros if you can correctly guess which of two cards lying face down in front of you is the Ace of Hearts, the other card being the Ace of Spades. As pointed out by Aloni, 'knowing which card is the winning card' can mean different things in this case. Knowing that the Ace of Hearts is the winning card is in a sense sufficient to know which card is the Winning Card, but it does not gain you much. A more interesting sense in the context is knowing that it is the card on the left, or knowing that it is the card on the right, depending on the case. Because of that, such examples provide another fruitful application of Aloni's method of conceptual covers (see [Aloni, 2008]).

8. Epistemic Paradoxes

To complete our journey, we close this chapter with a discussion of some epistemic paradoxes. As in other areas of logic, the existence of paradoxes has been a continued source of stimulation and development for epistemic logic. Hintikka's original book contains a discussion of the Moore paradox. As it turns out, this particular paradox bears a deep connection to other epistemic paradoxes such as the Fitch Paradox, the Surprise Examination Paradox, and several variants thereof. In this section we focus our attention on those three paradoxes only. Our goal is to indicate, in particular, the way in which dynamic epistemic logic has changed the traditional, static perspective on those in recent years.

8.1 Moore, Fitch, and the Surprise Examination

Moore made the observation that while one can consistently utter sentences such as 'it is raining and yet John does not believe it', it is pragmatically inconsistent to utter: 'it is raining and I don't believe it'. The source of the inconsistency lies in the fact that one usually believes what one asserts. Hintikka put forward epistemic logic in particular to clarify the difference in status between the two sentences. Thus, a sentence such as $p \wedge \neg B_a p$ is satisfiable in a system as strong as **KD4**. However, in the same system one can show that the sentence $B_a(p \wedge \neg B_a p)$ leads to contradiction (see [Gochet and Gribomont, 2006]). The reason is that from $B_a(p \wedge \neg B_a p)$, one can infer $B_a p \wedge B_a \neg B_a p$, and so by **4**, $B_a B_a p \wedge B_a \neg B_a p$, hence $B_a(B_a p \wedge \neg B_a p)$, which contradicts **D**.

The epistemic Moore sentence $p \wedge \neg K_a p$ lies in turn at the bottom of the Fitch paradox. The Fitch paradox concerns the interaction of the knowledge operator K_a with the operator $\diamond$ of metaphysical possibility. The paradox originates in the principle of knowability, which says that every truth must be knowable:

$$\phi \rightarrow \diamond K\phi \qquad \text{(Ver)}$$

A paradox results from this principle if one assumes for $\Box$ a logic as weak as **K**, and for **K** a logic as weak as **T**. To get the paradox, it is enough to substitute the Moorean sentence $(p \wedge \neg Kp)$ for ϕ. From $K(p \wedge \neg Kp)$, in **KT** it follows that $Kp \wedge \neg Kp$, namely a contradiction. Hence (Ver) implies that $p \wedge \neg Kp \rightarrow \diamond\bot$. But since standardly $\diamond\bot \rightarrow \bot$, we have $\neg(p \wedge \neg Kp)$, namely $p \rightarrow Kp$. Since p is arbitrary, the latter implies that every truth is known, which thus precludes the intuitive possibility of unknown truths.

Before evaluating ways out of the Fitch paradox, let us consider the Surprise Examination Paradox. In one version of the story, a schoolmaster announces to his students that there will be an exam during the week, but that it will be a surprise (they will not know when it takes place). The students reason that it cannot be on Friday, since they would then expect it on Thursday evening, and would know it to happen the next day. By similar reasoning, they reason that it cannot happen on Thursday, nor on any of the previous days. Hence they conclude that there cannot be a surprise examination. On Wednesday the schoolmaster gives them a test, and sure enough they are surprised.

To see the connection with Moore's paradox, it is useful to envisage a limiting case, in which the week has only one day and the teacher announces on Sunday: 'you will have an exam tomorrow, and it will be a surprise'. If p stands for 'the exam is tomorrow', what the sentence then says is exactly: $p \wedge \neg Kp$, namely the Moorean sentence. The problem is that in order to believe the decree p, the students should believe both p to be true, and $\neg Kp$ to be true, which is self-contradictory in a system as weak as **KD4** in that case. Interestingly, this one day version of the paradox has led Kaplan and Montague [Kaplan and Montague, 1960] to the statement of a self-referential version of this paradox, called the Knower Paradox. Basically, the Knower Paradox is a statement p that says of itself that it is not known, namely a statement p such that $p \leftrightarrow \neg Kp$. If Kp, by contraposition $\neg p$. But if K is veridical, then p. Hence $\neg Kp$, namely p is not known. But if $\neg Kp$, then p. So p is true. But based on the proof, we come to know that p is true, which is inconsistent. As the reader can see, the Knower Paradox bears a close relationship to the Liar paradox, based on the sentence that says of itself that it is not true (see Chapter 13). In what follows, we focus only on the three paradoxes mentioned and set issues about self-reference aside.[15]

8.2 A Dynamic Perspective on the Paradoxes

Each of the aforementioned paradoxes has generated a very large literature.[16] In this section I will consider a family of approaches to these various puzzles that all recommend viewing them in the light of dynamic epistemic logic, rather than from the perspective of static epistemic logic.

In a short essay on the surprise paradox, Quine [Quine, 1953] points out that in the limiting case in which p means 'you will have an exam tomorrow' and $\neg Kp$ means 'you do not know it today', one should not take the truth of the decree $M := p \wedge \neg Kp$ for granted. As a matter of fact, what holds is that $K(M \rightarrow p)$, but if one does not know whether p, then what one should conclude is that one does not know whether M. Thus, although M is not knowable proper, the truth of M remains compatible with one's knowledge. For Quine, the source of the paradox thus lies in the wrong assumption that one knows the decree to be true. Quine's remark is insightful, but it raises a further issue, which is: what happens upon learning a true Moorean sentence in a state in which the sentence is initially true?

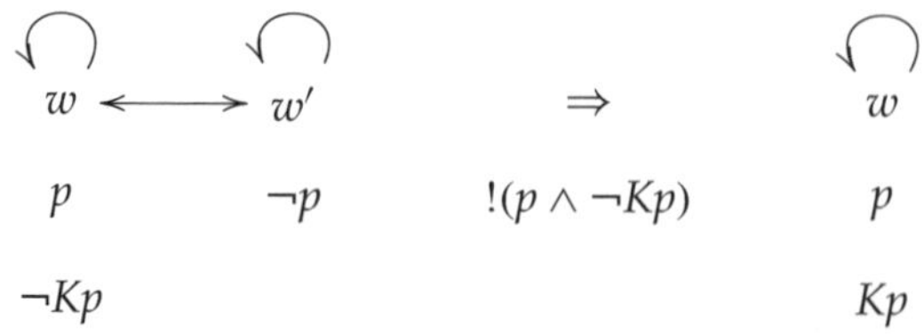

FIGURE 18.11 Moore's formula: a case of unsuccessful update

Consider the above model. As pointed out by Gillies [Gillies, 2001] and van Benthem [van Benthem, 2004b], if there is a public announcement that $p \wedge \neg Kp$, which initially holds in w, what happens is that the model is reduced to its left world. In the updated model, it then holds that $p \wedge Kp$. Thus, a crucial feature of Moorean sentences is that they do not satisfy a property called *success* in belief revision theory: upon learning that $p \wedge \neg Kp$, the fact $p \wedge \neg Kp$ does not hold any more, namely $M, w \nvDash [!(p \wedge \neg Kp)]p \wedge \neg Kp$. As a matter of fact, the Moorean sentence is even antisuccessful, since the update $!(p \wedge \neg Kp)$ in fact guarantees that $\neg(p \wedge \neg Kp)$. Based on this, van Benthem proposes an analysis of the Fitch paradox whose leading idea is to view the failure of the static verifiability principle as a reflection of the broader fact that not all formulae are successful. Viewed in this light, the lesson of the Fitch paradox is that one can *realize* that $p \wedge \neg Kp$, but one cannot not *know* this, precisely because the effect of realizing one's ignorance dissolves it dynamically.

Gerbrandy ([Gerbrandy, 2007]) proposes a similar analysis of the Surprise Paradox in terms of updates. Gerbrandy's idea is to view the teacher's announcement as another example of an unsuccessful update. Suppose that the pupils know that the exam will be Monday, Tuesday, or Wednesday, and represent the decree as follows: $S = ((m \wedge \neg Km) \vee (t \wedge [!\neg m]\neg Kt) \vee (w \wedge [!\neg m][!\neg t]\neg Kw))$. Let M be the structure in which the agent is initially uncertain between m, t, and w. Initially, $M, t \models S$, and $M, m \models S$, but $M, w \models \neg S$. Hence, $M, x \models [!S]\neg w$ for $x = m$ or t, namely learning the announcement rules out Wednesday as a possible exam day if the announcement is to be truthful. However, $M, t \models [!S]\neg S$, but $M, m \models [!S]S$. So if the exam is on Tuesday, it was true to say that it would be a surprise before the announcement, but it is false after that. However, it can still be a surprise if it takes place on Monday. By learning the teacher's initially true announcement, the pupils can therefore be led to belief states that no longer support the announcement being true. Interestingly, this suggests that an initially true principle can be used as a sound premise for reasoning, but may not adequately be iterated if it is not successful.[17] To be fair, we should point out that the dynamic approach to epistemic paradoxes does not entirely defuse their paradoxicality, since ways of strengthening the paradoxes are conceivable within the dynamic setting. Nevertheless, the dynamic setting highlights the special informational status of Moorean sentences and their kin.[18]

9. Conclusion and Perspectives

To conclude this chapter, it will be useful to highlight three aspects of epistemic logic which we did not explicitly cover in this chapter. For the most part, our effort has been to show the fruitfulness of Hintikka's framework to describe the basic attitudes of knowing information, believing some information, and of learning new information. In all we presented the basic concept is the notion of information compatible with one's available evidence. However, more work needs to be done in epistemic logic to represent and specify the very notion of evidence (see Section 6), as well as to specify to whom this evidence is available in ascribing individual or group attitudes (see e.g., [MacFarlane, ta] and [Yalcin, 2007] for recent work on the complexity and multi-dimensionality of epistemic and evidential constructions as 'might' and 'must'). A second issue which we did not go into here concerns the logic of belief, and the connection between models of plausibility such as the ones presented in Section 4 and the mathematical notion of probability. The epistemic and doxastic models we presented provide a qualitative description of the notion of uncertainty, while probability gives a quantitative measure of this notion (see Chapter 15). Several bridges exist between the two frameworks, including to represent probability operators in the object language of epistemic logic (see [Halpern, 2003] for a comprehensive textbook, see also [Aumann, 1999b] , [Kooi, 2003] , and [Meier, ta]). A third issue finally, which belongs in the general programme of modelling bounded rationality, concerns the interaction between agents with different logical or epistemological capabilities within the same group (see [Liu, 2008]). The logical omniscience problem is often viewed from the perspective of a single agent. When it comes to games and interaction, however, the problem becomes a broader issue, namely how to predict interesting outcomes in cases in which the agents have distinct observational, inferential, memory, or introspective capacities.

Notes

1. 'Common knowledge' is the term used by Lewis; Schiffer used 'mutual knowledge'. On the genealogy of the concept of common knowledge in Aumann's work, and its exact relation to Lewis' prior work, see Aumann's interesting testimony in [Hendricks and Roy, 2010].

2. This, in a nutshell, is the substance of the Levi identity, which characterizes belief revision with p as the composition of contraction with $\neg p$ and expansion with p.
3. Updates with incompatible information are possible in PAL, but they merely make the set of epistemically accessible worlds empty after the update. More structure is obviously needed to model the effect of revision towards consistent belief sets.
4. See [van Ditmarsch, 2005, p. 255], who calls this minimal belief revision.
5. Aucher [Aucher, 2008] favours talking of event models rather than action models. We stick to the terminology of action models, but the reader should indeed bear in mind that an action is an event of some kind, which may or may not change their informational state.
6. See [Égré, 2008] for a detailed discussion.
7. Williamson ([Williamson, 1994]) presents a variable semantics for knowledge on which **KT** is the resulting logic. See [Fara, 2002] for details and discussion.
8. See also [Halpern et al., 2009] for a recent survey of interdefinability results between knowledge and belief in bimodal systems. Another axiom intermediate between **4.2** and **4.4**, discussed in particular in [Stalnaker, 2006], is the axiom:

$$K(\phi \rightarrow \neg K \neg \psi) \vee K(\psi \rightarrow \neg K \neg \phi) \tag{4.3}$$

9. See [Pacuit, 2010] for a more detailed overview of various notions of belief definable in dynamic terms.
10. Note that this does not entail that *de re* beliefs are always equivalent to *de dicto*, or conversely, under the common domain assumption, due to restriction of quantifiers. For instance, suppose Pierre believes of Mary and Susan that they passed the test, without knowing that Mary and Susan are the only students. In principle, it is true to say that Pierre believes of every student that they passed the test ($\forall x(S(x) \rightarrow K_p P(x))$), but it does not imply that Pierre believes *de dicto* that every student passed the test ($K_p \forall x(S(x) \rightarrow P(x))$).
11. The Importation formula is also called the Ghilardi formula, and its converse the Converse Ghilardi formula (see [Corsi, 2002] and [Gochet and Gribomont, 2006]). The names Importation and Exportation are those used in [Aloni, 2005].
12. See [Lihoreau, 2008] for a recent volume with various contributions on this issue. See for instance Stanley and Williamson [Stanley and Williamson, 2001] on knowing how.
13. See for instance [Aloni et al., ta] for such an epistemic language.
14. See [Heim, 1994] and [Groenendijk and Stokhof, 1997] for classic expositions of these various readings.
15. See [Égré, 2005] for a survey on the Knower Paradox and its connection with provability logic, and [Dean and Kurokawa, 2009] for a recent contribution on the same topic.
16. See in particular [Broogard and Salerno, 2009] for a survey on the Fitch paradox.
17. See also [Bonnay and Égré, ta], which apply essentially the same strategy to a dynamic account of Williamson's margin for error paradox. Williamson's paradox, which we exposed in semantic terms in Section 5, can itself be seen as kindred to the Surprise paradox.
18. A partial syntactic characterization of successful and unsuccessful formulae appears in [van Ditmarsch et al., 2007]. A complete syntactic characterization has been found very recently by Holliday and Icard in [Holliday and Icard III, 2010]. A more detailed examination of Moorean sentences would also lead us into a discussion of epistemic modals such as 'might' and 'must' and their semantics in natural language. See in particular [Yalcin, 2007] for reasons to handle 'might' by means of a more complex semantics than Hintikka's relational semantics.

19 Logic of Decision

Paul Weirich

Chapter Overview

1. Introduction	544
2. Maximizing Utility	544
2.1 Decision Problems	544
2.2 Utility Maximization	545
2.3 Options	546
2.4 An Option's Utility	548
2.5 Utility Maximization's Assumptions	549
3. Analysing Utility	550
3.1 Multiattribute-Utility Analysis	550
3.2 Expected-Utility Analysis	553
4. Generalizations	558
4.1 Satisficing	559
4.2 Imprecision	559
4.3 Ratification	561
4.4 Infinite Utilities	562
5. Paradoxes	563
5.1 Newcomb's Problem	563
5.2 Allais's and Ellberg's Paradoxes	564
5.3 Paradoxes of Self-Location	566
5.4 The Two-Envelope Paradox	568
6. Extensions to Groups	569
6.1 Games	570
6.2 Social Choice	572
6.3 Trustee Decisions	573
7. Conclusion	573

1. Introduction

Decisions use practical reasoning. The reasoning resolves conflicts among goals and identifies means of reaching goals. Normative decision theory formulates principles of rationality that govern practical reasoning. It uses probability and utility as quantitative representations of beliefs and desires that form an agent's reasons for acts and assesses the strength of these reasons.

The phrase 'the logic of decision' is the title of Richard Jeffrey's textbook ([Jeffrey, 1990]) on decision theory. Jeffrey's attaching probability and utility to propositions (rather than, for example, dated commodity-bundles) highlights decision theory's roots in logic because it makes principles of practical reasoning resemble principles of theoretical reasoning.

Practical reasoning is dynamic in the sense that it moves from beliefs and desires to action. It is also dynamic in the sense that it directs formation and execution of multistep plans that respond to events occurring between the plan's steps. For example, a player who makes multiple moves in a game such as poker uses practical reasoning to formulate and execute a strategy for her sequence of moves. A good strategy responds to the moves other players make between her moves.

Normative decision theory divides into a branch that evaluates decisions and a branch that directs decisions. The evaluative branch advances requirements for decisions rather than directives for making decisions. Its principles evaluate a decision, even one already made, and do not offer decision procedures. This essay surveys evaluative decision theory. For systematicity, the survey takes stands on some controversial topics and, for balance, supplies references to rival points of view. The essay's sections treat utility maximization, utility analysis, generalization of utility maximization, difficult decision problems, and extension of decision theory to agents that are groups and to decisions made for others.

2. Maximizing Utility

This section explains the main principle of decision theory, the principle of utility maximization. It introduces the decision problems that the principle governs, the utilities of options that the principle assesses, and the assumptions that the principle makes.

2.1 Decision Problems

Suppose that a diner at a restaurant is ordering just one item from the menu. The diner faces a decision problem that she resolves by choosing an item. The dishes listed represent the diner's *options*, that is, decisions that she may make. She has preferences among her options. For example, she prefers pasta to fish. Because the menu is short and she has often visited the restaurant, she has a complete preference ranking of the menu's items. The ranking puts pasta at the top and fish lower. An assignment of numbers to items may represent the ranking. The higher the number assigned to an item, the higher it is in the ranking. Pasta may have the number 10, and fish the number 5. The numbers may also represent intensity of preference. If the diner likes pasta twice as much as fish, then the numbers for pasta and fish represent that as well as the diner's preference. The numbers representing preferences among options are the options' *utilities*. Some decision principles assume that options' utilities represent only the options' order in an agent's preference ranking, whereas other principles assume that options' utilities also represent intensities of preferences. To make utilities suitable for both types of principle, this essay assumes that they represent both order in the preference ranking and intensity of preference.

Choosing from a menu is a simple decision problem. A decision problem for an agent is any situation in which the agent has options and realizes one. The agent realizes an option even if she does nothing because doing nothing counts as an option. In complex decision problems options are hard to identify and comparing them is difficult. The agent may not have a preference ranking of her options.

2.2 Utility Maximization

Decision theory evaluates decisions for rationality and uses options' utilities to identify rational options in a decision problem. In textbook decision problems the agent's preferences rank all options, and a utility function represents those preferences. Decision theory's fundamental principle requires that an agent adopt an option at the top of her preference ranking of options. Realizing an option at the top of the preference ranking is equivalent to realizing an option with utility at least as great as any other option's utility, or *maximizing utility*.

Suppose that in a decision problem for an agent, O is the agent's set of options at a time and U is the agent's utility function, which goes from each option o in O to a real number. Then the agent maximizes utility if and only if she realizes an option $o \in O$ such that $U(o) \geq U(o')$ for all $o' \in O$.

Applying the principle of utility maximization requires identifying a set of options, assigning a utility to each, and comparing the utilities of options to discover which have maximum utility. Rationality in its ordinary sense, which the principle treats, is not by definition the same as utility maximization. Therefore, the principle makes the substantive claim that given certain circumstances rationality requires utility maximization.

The principle of utility maximization advances a necessary condition of rationality. Rationality may also require more than utility maximization, for example, having certain desires, such as a desire to satisfy other desires, and not having a *pure time preference*, that is, a preference for the lesser of two goods just because it will arrive earlier.

2.3 Options

The principle of utility maximization applies with respect to a decision problem's set of options. Making the principle precise requires describing the acts that form an agent's options in a decision problem.

An official may start a race by waving a flag. For the official, waving the flag is an option, but starting the race is not an option. The official fully controls waving the flag but not starting the race. Other agents contribute to starting the race. Rationality evaluates an agent's free acts that are in the agent's full control. Acts not free or not in the agent's full control may be evaluated for utility but are not evaluated for rationality. *Options* are possible free acts in an agent's full control. An option may be an act in the agent's *direct control*, that is, an act the agent performs at will, such as a decision, and may also be a sequence of acts that the agent directly controls at the times they are performed. Acts in an agent's full but not direct control, such as executing the steps of an extended plan, have components. If an act is simple and in the agent's full control, then it is in the agent's direct control. The principle of utility maximization evaluates an agent's realization of an option she directly controls at a time by comparing it with other options she directly controls at the time.

In many cases, an evaluation of an agent's realization of an option may, for simplicity, examine possible decisions only and ignore acts besides decisions that the agent also directly controls. The evaluation may substitute for the acts ignored decisions to perform them. Also, context affects the criteria for being an act in an agent's direct control. An evaluation may use relaxed criteria when convenient if using these criteria does not affect the evaluation's results. For example, an evaluation may treat opening the window, not just a decision to open the window, as an act in the agent's direct control. In typical cases, if the decision is rational, then so is the act.

The possible decisions that constitute an agent's options in a decision problem are the decisions that the agent might make at the time of the problem, for example, decisions to order an item from a menu. Individuating decisions by their content makes them exclusive, assuming that an agent makes only one decision at a time. A decision to order pasta and fish is not a decision to order pasta. If a diner makes only one decision at a time, then she does not make both of these decisions at the time. If her one decision at the time is to order pasta and fish, then at the time she does not make a second decision to order pasta. Her decision to order two items is incompatible with a decision to order one item, even if the acts forming the decisions' contents are compatible. An agent who decides to perform a combination of acts does not thereby decide to perform each component of the combination. The proposition that represents her decision's content is a conjunction of acts and does not entail a set of decisions each having one conjunct as its content. Letting D stand for decision and a_1 and a_2 stand for acts, $D(a_1\&a_2)$ is not equivalent to $D(a_1)$ & $D(a_2)$.

The principle of utility maximization, as noted, evaluates options directly controlled by comparison with rivals. If an act directly controlled is rational, then it maximizes utility among rival acts directly controlled. Rationality evaluates options fully but not directly controlled by evaluating their components. If an act is fully but not directly controlled, and all its components are rational, then the whole act is rational. Rationality does not require a replacement for the whole act while permitting each component to persist, for that requirement conflicts with the permissions. For example, rationality does not require a speaker to revise her comments and yet permit her to make each comment. She cannot revise her comments without changing some comment.

Decision theory treats solutions to decision problems, and game theory treats solutions to games. In games of strategy, the outcome of each agent's strategy depends on other agents' strategies. The strategy best for an agent typically depends on the strategies best for other agents. The agents' decision problems have interconnected solutions. This essay treats decision theory rather than game theory, but decision theory treats decisions that arise in games. Hence the essay treats some decision problems arising in games. For an introduction to game theory, see in this handbook Gabriel Sandu's chapter on game-theoretic semantics.

In a game of strategy, a *strategy profile* assigns exactly one strategy to each player. A strategy profile is a *Nash equilibrium* if and only if each strategy in the profile is a best response to the other strategies in the profile. A *sequential game* has multiple stages. At each stage in a sequential game, some player has a move to make. A strategy for a player specifies a move at each stage at which the player has a move to make.

In a sequential game, rationality evaluates a player's strategy stepwise. A strategy should be *dynamically consistent* in the sense that executing it does not require at any stage acting contrary to preferences at that stage. A stepwise evaluation of strategies discredits a Nash equilibrium whose realization requires some agent to be dynamically inconsistent. Players' strategies should together form a *rollback equilibrium*, that is, a Nash equilibrium assigning to each player a strategy that maximizes the player's utility whenever the player moves, and which may be discovered by proceeding from the end of the game back to the start. In compliance with rationality's general principle for evaluation of composite acts, evaluation of strategies works by applying utility maximization to their components rather than to the strategies themselves.

2.4 An Option's Utility

The principle of utility maximization presumes an assignment of utilities to options. The utility a rational, cognitively perfect agent assigns to a proposition is the agent's degree of desire that the proposition hold. The utility is a quantity that represents the agent's strength of desire. This interpretation of utility implicitly defines having a degree of desire towards a proposition with a theory of the attitude's causes and effects. Because propositions represent options, an option's utility for an ideal agent is a rational degree of desire to realize the option. A rational ideal agent's degrees of desire have the structure that utility theory describes. For example, they agree with preferences. Real agents, if rational, have degrees of desire that in simple cases approximate a rational ideal agent's degrees of desire.

This essay's traditional characterization of utility has rivals within decision theory. An alternative view, held by Binmore ([Binmore, 2009]), defines utility in terms of choices. Taking that definition strictly, utility does not explain choices. So the alternative view handicaps decision theory. The usefulness of a measure of utilities motivates the alternative view. However, the motivation is not compelling because utilities may be measured using choices without being defined by choices. In ideal conditions, a rational ideal agent's choices are consistent and reveal her preferences among her options. Assuming that her preferences extend to option types (that options in many decision problems may instantiate) and that her preferences among option types are constant, her choices furnish a means of discovering the utilities she assigns to her options.

An agent's degree of desire that a proposition hold depends on how she supposes the proposition's realization. An option's utility involves a particular form of supposition designed to make an option's utility comprehensive and yet accessible.

An option's utility evaluates *the option's world*. This is the possible world that would be realized if the option were realized. For simplicity, this essay assumes the existence of exactly one nearest world realizing an option, and takes that world to be the option's world. It is a *maximal proposition* specifying for everything the agent cares about whether it obtains. That an option's utility surveys the total outcome of the option's realization ensures that its evaluation of the option considers all relevant factors.

So that an agent has access to an option's utility, it evaluates the proposition that the option's world obtains. Unlike the option's world, this proposition is not maximal, although it is about a maximal proposition. An agent may not know which world would be realized if he were to realize a certain option and so may not know the utility he attributes to the option's world. He knows, however, the utility he attributes to the proposition that the option's world obtains. It is a probability-weighted average of the various worlds that might be the option's world. Hence, the option's utility equals the *expected utility* of its world. This characterization of an option's utility, which Weirich ([Weirich, 2010c], [Weirich, 2010b]) elaborates, follows Jeffrey [Jeffrey, 1990] in taking an option's utility to equal the expected utility of the option's outcome.

An ideal agent knows her own mental states and understands all propositions, including those that represent her options. She knows her beliefs and desires, and their quantitative representations. Nonetheless, an ideal agent may not be fully informed and may not know the utility she would assign to an option given full information. This may happen even in an ideal decision problem because standard idealizations do not remove uncertainty, a characteristic feature of typical decision problems. Given incomplete information about a lottery ticket's prospects, an ideal agent does not know what utility she would assign to owning the ticket if she had full information. She would assign a high utility if she were to know that the ticket will win and a low utility if she were to know that it will lose. However, she does not know whether it will win or lose.

So that an ideal agent has access to an option's utility despite incomplete empirical information, the principle of utility maximization takes an option's utility to equal the expected value of the option's informed utility rather than the option's informed utility. That is, the option's utility equals the expected utility of the option's world rather than the utility of the option's world. This makes an option o's utility equal to the option's expected utility $EU(o)$, taken as the expected utility of o's world. Consequently,

$$U(o) = EU(o) = \sum_i P(w_i \text{ given } o)U(w_i),$$

where w_i ranges over worlds that might be o's world. $U(o)$ is sensitive to information although $U(w_i)$ is not because w_i is a maximal proposition. Rationality requires that an ideal agent in an ideal decision problem realize an option that maximizes utility, expected utility, or the utility that an option's world obtains.

2.5 Utility Maximization's Assumptions

The principle of utility maximization is demanding but does not govern all agents in all decision problems. This section explains the cases it treats.

Some principles of rationality present standards to meet, and others present procedures to follow. The principle of utility maximization presents a standard of evaluation. It formulates a necessary condition of rational choice, not a procedure for choosing. Also, it evaluates only a choice and not also the choice's grounds. Because it takes an agent's utility assignment for granted, its evaluation is conditional and noncomprehensive. A nonconditional and comprehensive evaluation of an agent's decision asks not only whether the decision maximizes utility but also whether the agent's utility assignment is rational.

Rationality's demands are sensitive to an agent's circumstances and abilities. Nonideal agents and agents in nonideal decision problems may have excuses for failing to maximize utility. Utility maximization is a requirement of rationality for an ideal agent in an ideal decision problem. An ideal agent is cognitively unlimited and knows all logical and mathematical truths. A nonideal agent may not consider all his options because they overload his limited cognitive capacity, and may not make all relevant utility comparisons because some are too complex.

In a decision problem, most options are unrealized. They are possible but not actual acts. Utility maximization's comparison of options' utilities assumes that all options have utilities, not just options realized. Because utility attaches to propositional representations of possible acts, and not just to propositional representations of acts realized, all options may have utilities, and in an ideal decision problem they do because the agent precisely assesses each option. An ideal decision problem has an option of maximum utility and a stable basis for comparison of options' utilities. In a nonideal decision problem, an option of maximum utility may not exist. For example, options may have larger and larger utilities without end, as in a case allowing an employee to pick her own income. She has an infinite number of options, none of which has maximum utility.

For an ideal agent in an ideal decision problem, utility maximization is not just necessary but also sufficient for a rational decision if the agent is rational in all matters except perhaps the current decision problem. In that case, rationality in the decision problem completes the agent's full rationality.

3. Analysing Utility

An option's utility may be computed according to various principles for separating relevant considerations without omission or double counting. This section reviews two quantitative methods of separation: multiattribute-utility analysis and expected-utility analysis. Although a decision among options may rest on preferences that utility comparisons do not generate, if methods of separation generate utilities for options, then in ideal cases rational preferences agree with utility comparisons.

3.1 Multiattribute-Utility Analysis

Keeney and Raiffa [Keeney and Raiffa, 1993] present *multiattribute-utility* analysis. It divides an option's outcome into realizations of various objectives and computes the outcome's utility using the utilities of realizing the objectives.

Intrinsic-utility analysis, a general version of multiattribute-utility analysis that Weirich ([Weirich, 2001, Ch. 2]) introduces, takes an agent's objectives as realization of basic intrinsic desires and nonrealization of basic intrinsic aversions. It takes an option's outcome as the option's world and divides the world's utility into the intrinsic utilities of realizing the basic intrinsic desires and aversions that the world realizes. For simplicity, this section's formulation of intrinsic-utility analysis assumes certainty of the option's world.

Intrinsic-utility analysis distinguishes intrinsic and extrinsic desires, basic and derived preferences, and, in the terminology of economics, direct and indirect utility. It uses basic intrinsic desires and aversions to explain preferences among options and to explain utility assignments to options. An *intrinsic desire* is a desire for something for its own sake, and a *basic intrinsic desire* is an intrinsic desire for which no other intrinsic desires furnish reasons. Basic intrinsic aversions and attitudes of indifference have similar definitions. *Intrinsic utility* (*IU*) is a quantitative representation of basic intrinsic conative attitudes. It evaluates a proposition attending only to the logical consequences of the proposition's realization. Ordinary, or extrinsic, utility evaluates a proposition attending to the causal as well as the logical consequences of the proposition's realization. Because of its narrow scope, intrinsic utility is normally independent of information.

Let a possible world be a maximal consistent proposition that specifies for every basic intrinsic attitude (BIT) whether it is realized. In the cases intrinsic-utility analysis treats, the set of BITs is finite, and so the set of possible worlds is finite. A world, taken as a maximal consistent proposition, entails the objects of all BITs it realizes. All its relevant consequences are logical consequences. A world's utility therefore equals its intrinsic utility. A world's intrinsic utility, in turn, equals the sum of the intrinsic utilities of the objects of BITs that the world realizes. Therefore, the world's utility also equals that sum. This is the main principle of intrinsic-utility analysis.

A weak principle of *separation* for intrinsic utility takes the intrinsic utility of a whole as a function of the intrinsic utilities of its parts. A stronger principle of *additive separation*, that intrinsic-utility analysis adopts, takes the intrinsic utility of a whole as a sum of the intrinsic utilities of its parts. Two types of additive separation say that a BIT's realization contributes the same amount of Intrinsic utility to any world realizing the BIT. The types differ over whether the BIT's realization may affect realization of other BITs. The first type denies that changing a part of a whole ever changes the set of other parts. Given realization of a combination of BITs, it sums the intrinsic utilities of the objects of the BITs to obtain the intrinsic utility of the combination. According to the second type, realization of some BITs may entail realization of other BITs. To obtain the intrinsic utility of realizing a combination of BITs, it checks whether the combination entails realization of other BITs and then sums the intrinsic utilities of all objects of BITs whose realization the combination entails.

Some notation helps clarify the difference in types of additive separation. In statements of utility assignments, let a symbol for a BIT also stand for the attitude's object. Accordingly, if BIT stands for a basic intrinsic attitude, then *IU*(BIT) is the intrinsic utility of realizing that attitude. The first principle of separation asserts that $IU(BIT_1 \ \& \ BIT_2) = IU(BIT_1) + IU(BIT_2)$. This equality may fail if BIT_1 realized together with BIT_2 entails another BIT's realization. For example, BIT_1 and BIT_2 may be basic intrinsic desires for levels of pleasure during, for BIT_1, a certain temporal interval and, for BIT_2, an immediately succeeding temporal interval. Suppose that the levels of pleasure are the same so that joint realization of BIT_1 and BIT_2 entails satisfaction of BIT_3, a basic intrinsic desire for a constant level of pleasure during the combination of the two intervals. Given that the realization of BIT_1 and BIT_2 entails the realization of BIT_3, $IU(BIT_1 \ \& \ BIT_2) = IU(BIT_1 \ \& \ BIT_2 \ \& \ BIT_3) = IU(BIT_1) + IU(BIT_2) + IU(BIT_3)$. Although these equalities may not conform to the first principle of separation, they conform to the second principle of separation. To allow for such cases, this essay adopts the second principle: the intrinsic utility of realizing a combination of BITs is the sum of the intrinsic utilities of all objects of BITs whose realization the combination entails.

Realization of the combination of BITs characterizing a world does not entail any additional BIT's realization. The proposition characterizing a world explicitly specifies every BIT whose realization the proposition entails. Hence the formula for a world's intrinsic utility follows from both principles of additive separation. Both sum the intrinsic utilities of all the objects of BITs that the world realizes. The difference between the principles appears only when analysing the intrinsic utility of a nonmaximal combination of BITs, that is, a combination not characterizing a possible world. For an agent who has BITs towards health, pleasure, pain, and wisdom, it may be a combination of pleasure and wisdom.

Objections to the second principle of separation try to formulate counterexamples. However the objections do not establish that in their examples the objects of intrinsic utilities are objects of BITs and changing realization of one BIT does not entail changing realization of other BITs. A typical objection claims that the intrinsic utility of two pleasures differs from the sum of their intrinsic utilities. However, the objection does not establish that the desires for the pleasures are basic intrinsic desires, or does not establish that the two pleasures together do not entail realization of another BIT. For example, a person may like coffee and like tea but not like both at once. This case is not a counterexample if the person's basic intrinsic desires are for the taste of tea alone and for the taste of coffee alone. These desires are not realized when drinking coffee and tea together. Furthermore, even if a person has basic intrinsic desires for the taste of coffee and the taste of tea, she may also have a basic intrinsic aversion to their combination.

The intrinsic utility of their combination therefore sums the intrinsic utilities of realizing all three BITs. The sum may be negative.

Principles of separation may be restricted to worlds. Generalizing them to all propositions is controversial. A first attempt claims that the intrinsic utility of a proposition is the sum of the intrinsic utilities of the objects of the BITs whose realization the proposition entails. This principle works for conjunctions but not for disjunctions of BITs' objects. A better analysis takes the intrinsic utility of a proposition, represented as a disjunction of possible worlds, as the amount of intrinsic utility that the proposition entails, that is, the minimum of the intrinsic utilities of the worlds forming the disjuncts. Accordingly, $IU(\text{BIT}_1 \text{ or } \text{BIT}_2)$ is the smaller of $IU(\text{BIT}_1)$ and $IU(\text{BIT}_2)$.

3.2 Expected-Utility Analysis

Possible worlds yield another method of separating an option's utility into parts. The method computes a probability-utility product for each possible outcome and adds the products to obtain the option's *expected utility* (*EU*). The formula for an option *o* is

$$EU(o) = \sum_i P(w_i \text{ given } o)U(w_i),$$

where w_i ranges over the possible worlds that might be realized if *o* were realized, that is, the worlds that might be *o*'s world. According to the analysis, an option's utility equals its expected utility, as Section 2.4 states. The analysis governs a rational ideal agent.

An *expected-utility analysis* of an option's utility assumes that the utility of a chance for a possible world equals a probability-discounted utility of the world, namely, the world's probability-utility product. Then it assumes that an option's utility is the sum of the utilities of the chances for the possible worlds that might be the option's world.

A generalized form of expected-utility analysis allows using nonmaximal propositions called *states* to separate an option's utility into parts. States and outcomes of options in states have propositional representations and are individuated by the propositions that represent them. To obtain an option's utility, the analysis uses states that are exclusive and exhaustive, and so form a partition. It computes a probability-utility product for each possible outcome with respect to the partition of states, and adds the products to obtain the option's expected utility.

The formula for expected utility is simplest when options do not influence states. Then it asserts that

$$EU(o) = \sum_i P(s_i)U(o \text{ given } s_i).$$

$U(o \text{ given } s_i)$, a type of conditional utility, is implicitly defined by a theory of utility. It is not defined as the utility of the conjunction of the option and state $U(o \,\&\, s_i)$. A proposition's utility evaluates the proposition using a way of supposing the proposition's realization. A conjunction's evaluation may ask what if the conjunction *were* true, or it may ask what if the conjunction *is* true. To generate the nearest world with the proposition's realization, the proposition's subjunctive supposition attends to causal relations, whereas its indicative supposition attends to evidential relations. In the formula for expected utility, $U(o \text{ given } s_i)$ uses subjunctive supposition of o and indicative supposition of s_i. In contrast, $U(o \,\&\, s_i)$ uses a single type of supposition for the conjunction $o\&s_i$ and so the wrong type of supposition for either o or s_i. Using any single form of supposition for both the option and the state yields, as [Weirich, 1980] shows, an unreliable expected utility for the option.

When options influence states, the formula adjusts for their influence. One adjustment uses a type of conditional probability. It holds that

$$EU(o) = \sum_i P(s_i \text{ given } o)U(o \text{ given } (s_i \text{ if } o)).$$

$P(s_i \text{ given } o)$ is the probability s_i would have if o were realized. Use of the subjunctive mood signifies the supposition's attention to causal relations. $P(s_i \text{ given } o)$ is not defined as the ordinary conditional probability $P(s_i|o)$, that is, the ratio $P(s_i\&o)/P(o)$, because the ratio responds to evidential relations between o and s_i. $U(o \text{ given } (s_i \text{ if } o))$ is the utility o has if it is the case that s_i would obtain if o were realized. Use of the indicative mood to state the main supposition signifies its attention to evidential relations. In ordinary cases $U(o \text{ given } (s_i \text{ if } o))$ equals the simpler quantity $U(o \text{ given } s_i)$, and if states are also independent of options so that $P(s_i) = P(s_i \text{ given } o)$, then this paragraph's complex formula for expected utility yields the previous paragraph's simpler formula.

The general formula $EU(o) = \sum_i P(s_i \text{ given } o)U(o \text{ given } (s_i \text{ if } o))$ belongs to *causal decision theory*. Its conditional probabilities are causal. Some versions of causal decision theory define these probabilities as probabilities of subjunctive conditionals, or as probability images. However, a theory of their causes and effects may implicitly define them. *Evidential decision theory*, in contrast with causal decision theory, takes the conditional probabilities used to compute an option's expected utility with respect to a partition of states as ordinary conditional probabilities. Because ordinary conditional probabilities respond to evidential relations, they may award an act that is a sign, but not a cause, of good events an undeservedly high expected utility.

Jeffrey ([Jeffrey, 1990]) fully formulates evidential decision theory. Joyce ([Joyce, 1999]) fully formulates causal decision theory and also explains the reasons for favouring causal decision theory.

Causal decision theory's formula for expected utility assumes *partition invariance*, that is, that an option's expected utility is the same no matter which partition of states the formula employs. For example, imagine calculating the utility of a bet that George Washington and Abraham Lincoln were both presidents. One partition uses two states: (1) both men were presidents, and (2) not both men were presidents. Another partition uses four states: (1) Washington and Lincoln were presidents, (2) Washington was a president but Lincoln was not, (3) Washington was not a president but Lincoln was, and (4) neither Washington nor Lincoln was a president. According to causal decision theory's formula, the bet's expected utility has the same value using either partition of states.

Of course, some partitions more than others facilitate calculation of expected utilities. One partition has only the set of all states. Computing expected utility with respect to it is equivalent to asking directly for the option's utility. The computation does not facilitate discovery of the option's utility. It does not break down that utility. Wisely selecting a partition to calculate an option's expected utility is part of the art of decision making.

Some decision theorists, such as Savage ([Savage, 1954]), define probability and utility in terms of preferences and derive a weak form of the principle to maximize expected utility from axioms of preference. Savage's famous representation theorem establishes that if preferences satisfy certain axioms, then it is possible to construct probability and utility functions that represent the preferences as maximizing expected utility. The theorem is too complex to state and prove here. Kreps ([Kreps, 1988, pp. 115–36]) presents a compact version of the theorem's proof. Gilboa ([Gilboa, 2009, Chs. 10–12]) reviews and appraises the axioms of preference that the theorem assumes.

The weak form of the expected utility principle that the theorem supports claims that an agent should act 'as if' maximizing expected utility rather than claiming, as the traditional principle does, that an agent should maximize expected utility. Savage's axioms of preference are insufficient support for the traditional principle of expected utility maximization. The axioms take preferences among options for granted and do not give reasons for these preferences. Hence they lack the power to explain rational preferences among options, as Weirich ([Weirich, 2001, Ch. 1]) and Peterson ([Peterson, 2008]) argue. The traditional principle does not take probability and utility to be defined in terms of preferences among options. It uses probabilities and utilities of possible outcomes to explain rational preferences among options.

Some of Savage's axioms of preference are normative, and some are structural. The structural axioms ensure a set of preferences rich enough to constrain probability and utility functions representing the preferences so that the functions are unique (given a choice of scale for the utility function). The structural setup includes the assumption that functions from states to consequences may represent acts and that for every consequence some act yields the consequence in every state. This assumption excludes cases in which an agent cares about a consequence, such as risk, that only a chancy act generates.

Savage's representation theorem supports acting as if maximizing expected utility. Because of its structural assumptions, the theorem provides only restricted support for hypothetical expected-utility maximization. It does not cover cases that violate the structural assumptions. In contrast, the support for actual expected-utility maximization is general. It justifies the principle even when an agent calculates expected utilities for just a few salient options and does not have preferences among options rich enough to independently settle their expected utilities. It also justifies the principle when an agent is averse to risk, taken as a consequence of a risky option.

Binmore ([Binmore, 2009]) analyses Savage's framework for decisions. Savage's framework applies only to cases in which small worlds independent of acts represent every relevant possible state of the world. The framework does not apply to cases in which representation of relevant states requires large worlds that are not independent of acts. According to Binmore ([Binmore, 2009, Ch. 9]), independence restrictions limit applications of Savage's framework.

Binmore ([Binmore, 2009, Section 1.4]) raises questions about the type of independence that should obtain between an agent's preferences, his beliefs about his options, and his beliefs about the state of the world. Rationality requires one type of independence. An agent should not arrange to maximize utility by adjusting her preferences to fit her choice. She should rather adjust her choice to fit her preferences. However, rationality does not require other types of independence. An agent's adoption of an option may influence her beliefs about the state of the world. An agent's act is part of the world she inhabits. Similarly, an agent's beliefs about her set of options and her preferences among her options may influence her beliefs about the state of the world. Her set of options and her preferences are parts of the world, too. The independence conditions of Savage's framework simplify derivation of probabilities and utilities from preferences, but rationality does not impose those conditions. This section's general version of expected-utility analysis dispenses with them.

It is best to interpret Savage as showing how to use preferences among options to measure probabilities and utilities of outcomes, rather than as showing how to use these preferences to define the probabilities and utilities. This interpretation reconciles Savage's work with behavioural economics, which does not define probabilities and utilities in terms of preferences. Psychological studies of inconsistent preferences infer that if a subject is told that the chance of an event's occurrence is $x\%$, then the subject assigns a probability of $x\%$ to the event. This inference uses causes rather than effects of a subject's probability assignment to measure the assignment. Using effects such as preferences to infer probability assignments inaccurately attributes to subjects inconsistent probability assignments.

Jeffrey ([Jeffrey, 1990]) introduces probability and utility (desirability in Jeffrey's terminology) using preferences among propositions' realizations, including propositions representing possible acts. His text's centrepiece is a representation theorem showing that coherent preferences among options are as if the result of maximizing expected utility. The representation assigns an expected utility to each option, that is, a probability-weighted average of the utilities of the option's possible outcomes, which propositions represent. The representation theorem explicitly makes probabilities and utilities attach to propositions and incorporates conditional probabilities to accommodate the evidence that an option provides concerning states. It supports a weak form of the expected utility principle and also inferences from an agent's preferences among options to the agent's probability and utility assignments. Jeffrey's representation theorem, as Savage's, may be taken to ground probability's and utility's measurement rather than their meanings.

A classical decision theorist, such as Keynes ([Keynes, 1921]), instead of defining probabilities and utilities using preferences, takes them as rational degrees of belief and desire. They represent attitudes an agent has towards propositions. For example, an agent's probability that a proposition holds depends on only the agent's doxastic attitude towards that proposition, and not on a network of preferences among gambles involving the proposition and other propositions.

The standard axioms of probability constrain degrees of belief. These axioms, as formulated by Kolmogorov, require that an event have nonnegative probability, that the universal event have a probability equal to 1, and that the probability of a disjunction of incompatible events equal the sum of the events' probabilities. For ideal agents, who have no cognitive limits, the axioms form intuitively plausible constraints on degrees of belief. However, decision theorists advance various arguments to justify the constraints. Shimony ([Shimony, 1955]) advances a Dutch book argument showing that if an agent's degrees of belief violate the axioms, then he is open to a series of bets that guarantees a loss. Joyce ([Joyce, 1998]) advances a calibration argument showing that degrees of belief follow the axioms if they rationally estimate physical probabilities. Richard Pettigrew's chapter of this handbook contains a section on justifications of probabilism. The section analyses various arguments that rational degrees of belief obey the probability axioms.

Expected utilities depend on probabilities of states. Probabilities of states are subjective, but an agent's information, as well as the probability axioms, constrains them. For example, rationality may require assigning probability 1/2 to getting Heads on a toss of a symmetric coin, although the probability axioms do not impose this requirement.

The principle that an option's utility equals its expected utility constrains degrees of desire. For an ideal agent's degrees of desire, the constraint is intuitively plausible. An ideal agent's degree of desire that an option obtain should equal the agent's *expected degree of desire* that the option's outcome obtain, that is, a probability-weighted average of the agent's degrees of desire for the various possible outcomes that may be the option's outcome. For example, the agent's degree of desire to make a bet should be a probability-weighted average of the agent's degree of desire to win and the agent's (negative) degree of desire to lose.

An agent assigns a probability and a utility to a proposition representing an outcome using a way of understanding the proposition, as Weirich ([Weirich, 2010c]), ([Weirich, 2010b]) explains. A way of understanding a proposition is sometimes called a mode of presentation of, or means of grasping, the proposition. Although an agent's way of understanding a proposition influences the probability and utility she assigns to the proposition, decision principles may control for that influence by using only the assignment that the agent makes given a canonical way of understanding the proposition.

Options' utilities represent preferences. So the expected-utility principle, requiring an option's utility to equal its expected utility, has a companion requiring that an agent prefer one option to another if the first's expected utility is greater than the second's. The most common principle of preference among options, besides this companion principle, is *the principle of (strong) dominance*. This principle declares that one of two options is preferable if it is preferable in all the states of some partition. It assumes that the options do not influence the probabilities of the states. The principle of dominance may operate when options lack expected utilities, say, because possible outcomes do not have sharp utilities. However, the principle of dominance is compatible with the expected utility principle's companion. It yields the same preferences as expected-utility comparisons when expected utilities exist.

Intrinsic- and expected-utility analyses work together. Intrinsic-utility analyses yields utilities of worlds, and expected-utility analyses use these utilities to obtain utilities of options. Each type of utility analysis works within one dimension of utility analysis, and utility analysis is multidimensional, as Weirich ([Weirich, 2001]) elaborates.

4. Generalizations

The principle of utility maximization holds for ideal agents in ideal decision problems. Ideal agents are cognitively perfect, and, if utility maximization is advanced as both a necessary and a sufficient condition of rational choice, are fully rational except perhaps in the current decision problem. Ideal decision problems have an option of maximum utility, stable utility comparisons of options resting on their possible outcomes' probabilities and utilities, and only options with finite utilities. Generalizations of the principle of utility maximization govern cases with nonideal agents and nonideal decision problems. A typical generalization removes some idealizations but retains others. This section reviews four examples.

4.1 Satisficing

Simon ([Simon, 1982, pp. 250–1]) advances a generalization for humans, who have cognitive and practical limitations. He proposes *satisficing* as a decision procedure: pick the first satisfactory option you discover. For example, when selling a house, accept the first satisfactory offer. Transforming this procedure into a principle of evaluation yields a generalization of utility maximization: an option is rational if and only if it is satisfactory. An agent regards, and thereby *classifies*, options as satisfactory or as unsatisfactory. The agent's classification and utility assignment (if it exists) are coherent if and only if every satisfactory option's utility is higher than every unsatisfactory option's utility. In ideal cases an agent's classification of options agrees with her assignment of utilities to options.

For an ideal agent, an option is satisfactory if utility maximizing, but may be satisfactory without being utility maximizing. An ideal agent may classify some options as satisfactory without assigning utilities to any options. So the principle of satisficing applies to decision problems without a maximizing option, in particular, problems in which options do not have utility assignments. If a rational ideal agent identifies a utility-maximizing option, her aspiration level rises so that only maximizing options count as satisfactory. Therefore, in ideal cases satisficing yields utility maximization; it counts as a generalization of utility maximization that extends to nonideal cases.

The principle of satisficing relaxes some of utility maximization's idealizations and retains others. It assumes that the agent is rational in all matters except perhaps the current decision problem and that her decision problem is ideal except perhaps for the absence of utility assignments to options.

4.2 Imprecision

I. J. Good ([Good, 1952, p. 114]) addresses decision problems without sharp probabilities and utilities. He proposes maximizing expected utility with respect to a pair of probability and utility assignments compatible with the agent's doxastic and conative attitudes – for simplicity, her beliefs and desires. Such a pair of assignments is called a *quantization* of the agent's beliefs and desires. Expected-utility maximization with respect to a quantization is necessary for a rational decision if the agent and the decision problem are ideal except for the absence of sharp probabilities and utilities. It is sufficient as well if the agent is rational in all matters except perhaps the current decision problem. The principle generalizes expected-utility maximization because, when sharp probabilities and utilities exist, maximization with respect to a quantization is genuine maximization. Only the agent's actual probability and utility assignments are compatible with her beliefs and desires.

Assuming that choice works through preferences, the principle imposes a constraint on preferences among options. Rational preferences are compatible with expected-utility maximization under a quantization. Similarly, if a rational agent makes utility assignments to options, the assignments comply with this constraint.

An objection to Good's principle, along the lines of Elga's ([Elga, 2009]) objection, targets the principle's sufficiency for rational choice. Suppose that an agent has utilities for amounts of money that equal the amounts and has an unsharp probability for rain tomorrow that the interval [0.4, 0.6] represents. Applied case by case, Good's principle permits buying for $0.60 a gamble that pays $1 if it rains tomorrow and otherwise nothing. Then it permits selling the gamble for $0.40. However, the agent foresees a sure loss of $0.20 if he makes the pair of transactions.

A response to the objection shows how, in conditions where it is sufficient for rationality, Good's principle rejects the pair of transactions. After buying the gamble for $0.60, the consequences of selling it for $0.40 include a sure loss. Applying Good's principle circumspectly, the agent should not sell the gamble for that price. The sale does not maximize expected utility under a quantization.

A rational ideal agent following Good's principle and having a basic intrinsic desire only for money cares about avoiding sure losses and keeps track of decisions to prevent a series of transactions that ensures a loss. The cognitive demand is large. To simplify, a nonideal agent may pick one quantization of beliefs and desires and treat it as if it yielded his probability and utility assignments. A defence of Good's principle may acknowledge the benefit a nonideal agent gains by constraining the principle's application without revoking the licence the principle gives ideal agents. A rational ideal agent may maximize expected utility under any quantization, although a nonideal agent has pragmatic reasons for maximizing expected utility under a selected quantization.

The argument against Good's principle may contend that a rational agent focuses on the present and ignores the past. An agent who refuses to sell the gamble for $0.40 after purchasing it for $0.60 commits the fallacy of sunk costs, the argument holds. He refuses to sell only because of past decisions and not because of current beliefs and desires. According to the argument, a defence of Good's principle may not invoke past decisions.

The defence of Good's principle agrees that the principle may use only current beliefs and desires, but points out that past decisions may influence current beliefs and desires if they influence the foreseeable consequences of current options. The past decision to buy the gamble for $0.60 clearly affects the consequences of selling it for $0.40; the past decision makes selling yield a loss of $0.20. Taking account of all the consequences of current options before deciding is not fallacious. An agent who buys the gamble for $0.60 and sells it for $0.40, despite a foreseen loss, does not maximize utility under a quantization of beliefs and desires at each step. The second step, given its consequences, fails to maximize expected utility under a quantization of beliefs and desires at the time of the step.

4.3 Ratification

In some nonideal decision problems, comparison of options has an unstable basis. An option carries information that affects options' utilities. Although an option maximizes utility, it does not maximize utility given its adoption. Its adoption triggers regret. Such cases arise in games of strategy. Suppose that two agents are playing Matching Pennies with two pennies. The first wins if the pennies the agents display match, and the second wins if the pennies do not match. The second agent is good at predicting whether the first agent displays his penny with Heads up or Tails up. If the first agent displays Heads, he thereby has evidence that his opponent will display Tails to prevent a match. If he displays Tails, he thereby has evidence that his opponent will display Heads. Whatever the first agent does, he acquires evidence that the opposite choice would have been better. Heads maximizes utility for him if he thinks his opponent is likely to display Heads. Nonetheless, Heads does not maximize utility for him given its adoption because its adoption creates new evidence that his opponent displays Tails.

Jeffrey ([Jeffrey, 1990, Section 1.7]) presents a generalization of utility maximization that he calls the principle of ratification. It addresses cases in which an option's realization supplies evidence about its outcome. Suppose that the players in Matching Pennies may randomize their choices by flipping their pennies. Then the first agent may flip his penny, confident that his opponent cannot predict the result. Suppose he foresees that his opponent will respond by flipping also. Given that his opponent flips, the first agent's flipping maximizes utility, but so does his showing Heads and so does his showing Tails. Nevertheless, only his flipping is *self-ratifying*. Only it maximizes utility on the assumption that it is realized. *The principle of ratification* says that a rational choice is self-ratifying.

If both agents flip their pennies, their strategies (in this single-stage game, their choices) constitute a Nash equilibrium of their game. As Section 2.3 explains, a Nash equilibrium is a profile of strategies, consisting of one strategy for each agent, such that each strategy in the profile is a best response to the other. In an ideal version of Matching Pennies, the principle of ratification supports an agent's adopting his Nash strategy, that is, his part in the game's

Nash equilibrium. Only his Nash strategy is self-ratifying. Weirich ([Weirich, 2010a, Ch. 6]) provides details and generalizes the principle of ratification to suit all games of strategy.

Rational choices in games use the information a player's choice carries about other players' choices. Although a player does not possess the information until he makes his choice, he may anticipate having the information if he were to make the choice. Taking account of that information is compatible with causal decision theory. The formula for an option's expected utility uses causal conditional probabilities even when expected utility is calculated given a condition such as an option's realization. The condition just adds an assumption to the information used to calculate expected utilities.

4.4 Infinite Utilities

Suppose that in a nonideal decision problem, some options have infinite expected utilities. A problem arises immediately. Options with infinite utilities are not equally choiceworthy, contrary to utility comparisons. Suppose that an agent may choose between having eternal bliss with a 1% probability or with a 100% probability. The rational choice is the sure thing, even if both choices have infinite expected utility. A decision principle for such cases might use new mathematics to distinguish infinite amounts of utility.

The St. Petersburg gamble involves a fair coin tossed until Heads appears. The gamble pays \$2 if Heads first appears on the first toss, \$4 if Heads first appears on the second toss, \$8 if Heads first appears on the third toss, and so on ad infinitum. The expected monetary value of the gamble is $(1/2 \times 2) + (1/4 \times 4) + (1/8 \times 8) + \ldots$, or $1 + 1 + 1 + \ldots$. So its expected value is infinite, although it is not reasonable to pay much for the gamble. Daniel Bernoulli, the originator of the puzzle, used it to argue that money has diminishing marginal utility, and consequently the gamble's expected utility is less than its expected value. Switching from expected value to expected utility does not completely resolve the paradox, however. Suppose that for some possible agent in some possible world, the utility of money is linear and the supply of money is infinite. Then the gamble has infinite utility according to the expected-utility principle. Its utility seems to be less, however. Weirich ([Weirich, 1984]) explores the possibility that aversion to chance reduces the gamble's utility.

Nover and Hájek ([Nover and Hájek, 2004]) introduce a descendant of the St. Petersburg gamble that they call the Pasadena gamble. The probability-utility products for the Pasadena gamble form a conditionally convergent series. The terms of the series may be arranged so that it converges to any number, diverges to positive infinity, or diverges to negative infinity. Hence the gamble lacks an expected utility. Easwaran ([Easwaran, 2008]) proposes a way of generalizing the expected utility principle to handle such cases.

5. Paradoxes

Challenging decision problems, sometimes called paradoxes, motivate clarifications and refinements of decision theory. This section reviews a sample of paradoxes exercising contemporary decision theorists. It does not attempt to resolve these paradoxes; resolutions are too controversial to champion in a handbook. It just indicates promising paths to resolutions.

5.1 Newcomb's Problem

Section 3.2 presents a formula for an option's expected utility that uses causal conditional probabilities. Newcomb's problem, which Sobel ([Sobel, 1994, Ch. 2]) treats thoroughly, reveals a reason for using these special conditional probabilities.

In Newcomb's problem an agent may choose an opaque box or the opaque box together with a transparent box containing $1,000. The opaque box contains $1,000,000 if it has been predicted that the agent will take only the opaque box. Otherwise, that box is empty. The predictor is reliable. The agent knows these facts, and so if she takes just the opaque box has good reason to think that it contains $1,000,000. However, she is $1,000 ahead, whatever the opaque box contains, if she takes both boxes.

Evidential decision theory (EDT) uses the ordinary conditional probability $P(s_i|o)$ for a state s_i used to compute an option o's expected utility. Its formula for typical cases, as Section 3.2 explains, is $EU(o) = \sum_i P(s_i|o)U(o \text{ given } s_i)$. The conditional probability $P(s_i|o)$ is sensitive to correlation not just causation between o and s_i. To make the formula for expected utility sensitive to only an option's causal consequences, causal decision theory (CDT) replaces the ordinary conditional probability with the causal conditional probability $P(s_i \text{ given } o)$. Its formula for typical cases is $EU(o) = \sum_i P(s_i \text{ given } o)U(o \text{ given } s_i)$. CDT may interpret $P(s_i \text{ given } o)$ as the probability of the conditional that (if o were adopted, then s_i would obtain), or, for greater range, may implicitly define it using a theory of causal conditional probability.

In Newcomb's problem EDT supports one-boxing because it maximizes expected utility computed using ordinary conditional probabilities. In contrast, CDT supports two-boxing because it maximizes expected utility computed using causal conditional probabilities. Although one-boxing furnishes evidence that the opaque box contains $1,000,000, it does not cause the opaque box to contain $1,000,000.

Granting that two-boxing is rational given the agent's situation in Newcomb's problem, CDT's version of expected-utility maximization appears to be a correct principle of conditional rationality. Is two-boxing nonconditionally rational? This is controversial. It is rational for an agent to prepare for

Newcomb's problem by acquiring a one-boxing disposition—this disposition brings riches in Newcomb's problem. Does a two-boxer, who fails to acquire that disposition, act irrationally because her act stems from an irrational failure to acquire a one-boxing disposition? Her act is rational, Weirich ([Weirich, 2004, Section 7.3]) argues, because rationality's evaluation of her act given a one-boxing disposition is the same as its evaluation given a two-boxing disposition. Two-boxing, because dominant, is rational even if it springs from a disposition irrational to have. Failure to acquire a one-boxing disposition has no effect on rationality's conditional evaluation of two-boxing. Hence, the disposition's absence does not undermine the act's nonconditional rationality.

Binmore ([Binmore, 2009, p. 31]) holds that Savage's framework, requiring states that are independent of acts, suits Newcomb's problem, and he therefore rejects a representation of the problem that uses these states: (1) the prediction is correct and (2) the prediction is incorrect. The states are not independent of the agent's acts. CDT's partition invariant version of the expected-utility principle accepts the states. According to it, two-boxing has greater expected utility than one-boxing even using them. If the agent two-boxes and the prediction is correct, she does better by two-boxing than she would have done by one-boxing, because she gains the contents of the transparent box as well as the contents of the opaque box. If the agent two-boxes and the prediction is incorrect, she does better by two-boxing than she would have done by one-boxing, because she gains the contents of the transparent box as well as the contents of the opaque box. Because she gains from two-boxing in both cases, two-boxing has higher expected utility than one-boxing has.

5.2 Allais's and Ellberg's Paradoxes

As Section 3.2 mentions, a risk is a chance of a loss. An aversion to this chance is an aversion to the risk. Some versions of utility analysis define an agent's attitude to risk using the shape of her utility curve for a commodity. Aversion to risk has a technical sense whereby it is concavity of the utility function for the commodity, as Binmore ([Binmore, 2009, Section 3.7]) explains. Accordingly, a risk averse person prefers \$100 to a gamble that, given a toss of a fair coin, pays \$200 if Heads and \$0 if Tails, and so has an expected monetary value of \$100. However, the technical definition leaves risk unexplained, makes aversion to risk relative to a commodity, and does not distinguish aversion to risk from the commodity's diminishing marginal utility.

A richer, more accurate approach to risk in its ordinary sense takes an agent's attitude towards risk to be her attitude towards the risks that risky options involve. Because a risk is a probability that a bad event will occur, two types of risk exist. One depends on physical probabilities, and the other depends on subjective probabilities. Subjective probabilities equal objective probabilities when known, and subjective risks equal objective risks when known. Decision principles treat subjective risks, which are accessible to a decider. Also, for convenience, decision principles may count as a risk a subjective probability that a good event will occur. An agent is typically averse to having either a bad or a good event's occurrence depend on chance. An aversion to risk in the broad sense is an aversion to taking chances. It explains a desire for certainty that a bad event will not occur and that a good event will occur. Financial planners use the variance of the probability distribution of an investment's possible returns as a rough measure of risk and through questionnaires assess a client's aversion to risk.

Risk is a consequence of a risky act. Each possible outcome includes the risk the act entails. Recognizing this lowers evaluations of the act's possible outcomes in typical cases and thereby resolves Allais's and Ellsberg's paradoxes, Weirich ([Weirich, 1986]) argues. Aversion to risk explains typical preferences among options that the paradoxes construct. The paradoxes show that the principle of utility maximization should evaluate comprehensive outcomes including risk and not just monetary gains and losses.

In a version of Allais's paradox, an agent has a choice between \$3,000 and a 4/5 chance of \$4,000. He also has a choice between a 1/4 chance of \$3,000 and a 1/5 chance of \$4,000. The typical agent's preferences are for the sure thing in the first case and the chance of the larger prize in the second case. However, the inequalities $U(\$3{,}000) > (4/5)U(\$4{,}000)$ and $(1/4)U(\$3{,}000) < (1/5)U(\$4{,}000)$ are inconsistent. No utility function U represents the agent's preferences.

Treating comprehensive outcomes resolves the paradox. The chancy options have risk as a consequence, and aversion to risk explains preferences among the options. Suppose that R1, R2, and R3 stand for the risks involved in the three chancy options taken in order. Then the preferences imply these inequalities: $U(\$3{,}000) > (4/5)U(\$4{,}000 \text{ and R1})$ and $(1/4)U(\$3{,}000 \text{ and R2}) < (1/5)U(\$4{,}000 \text{ and R3})$. They are consistent.

A version of Ellsberg's paradox involves two urns. The first contains 50 white and 50 black balls. The second contains an unknown mixture of white and black balls. An agent has a choice between receiving \$100 if white is drawn from the first urn (W1) and receiving \$100 if white is drawn from the second urn (W2). She also has a choice between receiving \$100 if black is drawn from the first urn (B1) and receiving \$100 if black is drawn from the second urn (B2). A typical agent's preferences favour the chances involving the first urn in both cases. However, the inequalities $P(\text{W1})U(\$100) > P(\text{W2})U(\$100)$ and $P(\text{B1})U(\$100) > P(\text{B2})U(\$100)$ are inconsistent because probabilities obey the addition law. No probability assignment is compatible with these preferences.

Treating comprehensive outcomes that count risk as a consequence of risky acts also resolves this paradox. The risks arising from gambling with the first urn are less than those arising from gambling with the second urn because the agent knows more about the first urn than she does about the second urn. Aversion to risk therefore yields the typical preferences. Letting R1, R2, R3, and R4 stand for the risks in order, the preferences imply these inequalities: P(W1)U(\$100 and R1) > P(W2)U(\$100 and R2) and P(B1)U(\$100 and R3) > P(B2)U(\$100 and R4). They are consistent.

5.3 Paradoxes of Self-Location

A constellation of paradoxes involves propositions about an agent's location in space or time. The crucial propositions refer directly to the agent and locations using pronouns rather than descriptions. The paradoxes notice that an agent may know that she is here now without knowing who she is, which place is here, or which time is now. They ask whether standard decision principles accommodate such ignorance about her circumstances.

Piccione and Rubinstein ([Piccone and Rubinstein, 1997]) present the paradox of the absent-minded driver. At the end of an evening, a dinner guest plans to drive away from his host's house. He will take a highway that passes through two intersections. If he leaves the highway at the first intersection, he will get hopelessly lost. If he leaves the highway at the second intersection, he will reach his home. If he takes the highway past both intersections, he will reach a motel. His utility assignment for the outcomes of getting lost, reaching his home, and reaching the motel are respectively 0, 4, and 1. Because the driver is absent-minded, if he comes to the second intersection, he will not remember that he has already passed the first intersection. Therefore, he cannot distinguish arrivals at the first and second intersections. Given his absent-mindedness, his best plan is to stay on the highway past both intersections and reach the motel. Doing this has an expected utility of 1. The other implementable plan is to leave the highway at any intersection reached. This plan results in getting lost and has an expected utility of 0. However, when the driver reaches an intersection, the probability for him that it is the second intersection is 50%. So the expected utility of leaving the highway is $(0.5 \times 0) + (0.5 \times 4)$, or 2, whereas the expected utility of staying on the highway past all intersections is 1, as noted earlier. Consequently, the driver has an incentive to abandon the plan to stay on the highway past all intersections.

In this case the utility-maximizing plan seems to have steps that are not utility maximizing. The plan to stay on the highway past every intersection maximizes utility. However, at an intersection, given a 50% probability that it is the second intersection, leaving the highway maximizes utility. Does rational choice at an intersection conflict with the rational strategy for choice at each intersection? Aumann, Hart, and Perry ([Aumann et al., 1997]) and Rabinowicz ([Rabinowicz, 2003]) examine versions of the paradox that entertain mixed strategies and suggest resolutions of the paradox.

Elga ([Elga, 2000]) presents the problem of Sleeping Beauty. A subject in an experiment, Sleeping Beauty, learns that she will sleep from the start of Monday to the end of Tuesday except for a brief period Monday morning and possibly Tuesday morning. An amnestic drug will make her forget these periods of wakefulness. The experimenter tosses a fair coin to decide how often she will wake during the two-day period. If it lands Heads up, she will wake only Monday; and if it lands Tails up, she will wake both Monday and Tuesday. The subject knows all this before the experiment starts. When she wakes Monday (not knowing it is Monday rather than Tuesday), what is the probability given her information that the coin landed or will land Heads up? It seems that it is 1/2. That is what it was before the experiment, and it seems that she has not acquired new relevant information about the coin toss. However, she cannot distinguish three exclusive and exhaustive awakenings that she may experience during the experiment: (1) awaking Monday with Heads tossed or about to be tossed, (2) awaking Monday with Tails tossed or about to be tossed, and (3) awaking Tuesday with Tails tossed or about to be tossed. If each possible awakening has probability 1/3, then the probability of Heads is 1/3. This puzzle about probability generates a puzzle about decisions. When the subject awakens Monday, what probability should guide her decision about betting that the coin landed or will land Heads up?

The traditional Bayesian principle of *conditionalization* prescribes a method of updating probabilities as an agent gains, and does not lose, information. According to it, an agent moving from time t_1 to time t_2 should at t_2 assign to an event a probability equal to, according to the agent at t_1, the event's probability conditional on a proposition representing the information the agent gains from t_1 to t_2. If Sleeping Beauty assigns Heads probability 1/2 on Sunday and probability 1/3 on Monday, then, if her relevant information is the same on Sunday and on Monday, she violates the principle of conditionalization. Horgan ([Horgan, 2004]) claims that the subject both loses and gains relevant information concerning her location so that she does not violate the principle of conditionalization. Stalnaker ([Stalnaker, 2008, Section 3.4]) similarly argues that her revising the probability of Heads from 1/2 to 1/3 does not violate the principle of conditionalization because she gains new relevant information when she wakes. He proposes a new way of representing an agent's information about her location.

Bostrom ([Bostrom, 2002]) presents a problem for an assumption about probability assignments that he calls the Self-Sampling Assumption (SSA): observers should reason as if they were a random sample from the set of all observers in their reference class. The problem concerns Adam and Eve. Adam comes from a human population of two or from a human population of billions if he and Eve have offspring. Following SSA, he views himself as a random selection from the population of humans. According to Bayes's Theorem, if H is a hypothesis and E is evidence, then $P(H|E) = P(H)P(E|H)/P(E)$. Because of Bayes's Theorem,

Adam attributes greater probability to his coming from a population of two than to his coming from a population of billions. So he infers that his union with Eve is unlikely to yield offspring. His probability assignment in turn affects his decision about intercourse. Adam's deliberations seem misguided.

One response rejects SSA. Adam should not view himself as a random selection from the population of humans, but as the first male in that population. That he is the first male human does not give him information about the size of the human population or the consequences of intercourse. A less severe response proposes revising rather than rejecting SSA. Because the assumption has some initial plausibility, Bostrom suggests revising it to block Adam's counterintuitive reasoning.

5.4 The Two-Envelope Paradox

The two-envelope paradox comes in various versions. In the philosophical literature the problem arises for a single individual. See, for example, [Peterson, 2009, pp. 86–8]. The individual knows that two envelopes before her contain checks for amounts of money, and that one envelope contains twice the other's amount. A coin toss selects the envelope she receives. When she receives her envelope, she has an opportunity to trade it immediately for the other envelope. Should she exercise this option?

Suppose the amount in her envelope is x. The chance that the other envelope has $2x$ is $1/2$, and the chance that it has $(1/2)x$ is $1/2$. So the expected amount after switching is $(5/4)x$, and the expected gain is $(1/4)x$. It seems that she should switch. However, a similar argument, using y as the amount in the other envelope, concludes that the expected amount if she does not switch is $(5/4)y$, and the expected advantage from not switching is $(1/4)y$. It seems that she should not switch. Also, consider the difference between the amounts in the two envelopes. If she switches, she either gains or loses that difference, and the two outcomes have the same probability. So the expected gain from switching is 0. She apparently does just as well either switching or not switching. Because the three applications of the expected-utility principle yield conflicting advice, at least one has a flaw. Responses to the paradox often advance in some guise one of these applications of the expected-utility principle and put aside the others.

The literature in economics on the problem adds a twist by supposing that the two envelopes go to two individuals. The question, as Nalebuff ([Nalebuff, 1989]) presents it, is whether the two individuals should exercise their option to trade envelopes.

Some versions of the problem specify the possible pairs of amounts of money that may go into the envelopes. Each possible pair has one constituent twice as great as the other constituent. If there are a finite number of possible amounts that may go into the envelopes, then the argument for switching has a flaw. If x is the greatest possible amount, then switching generates a loss for sure. A similar problem arises if the number of possible amounts is infinite but bounded above. In any case, if the number of possible amounts is infinite, the probability distribution over possible amounts, if uniform, generates paradoxes by itself. So a nonparadoxical distribution is not uniform. Broome ([Broome, 1995]) presents some nonparadoxical distributions that for every possible value of x make the expected gain from switching greater than x. If, because the number of possible amounts is infinite and unbounded above, the expected gain from an envelope is infinite, problems concerning comparison of infinite expected gains arise. The expected difference between two options may not be partition invariant, for instance. The two-envelope paradox may therefore stem from familiar paradoxes concerning infinite quantities.

Some versions of the paradox suppose that the individual looks inside her envelope before deciding whether to switch. Looking seems to reveal no relevant information. However, if the number of possible amounts is finite and the envelope contains the greatest possible amount, the individual learns by looking that the other envelope contains less than her envelope does. So the information may be relevant.

Other versions of the paradox specify the mechanism that generates the amounts in the envelopes; the mechanism specified may alter the method used to give the individual an envelope. One mechanism randomly selects a pair of numbers from the set of permissible pairs. Another mechanism randomly selects an amount from the set of permissible amounts and places that amount in the individual's envelope. Then it randomly decides whether to put twice or half the amount in the other envelope. The second mechanism, but not the first, seems to furnish grounds for switching.

Some analyses of the paradox examine the role of the variable x in the argument for switching. A variable such as x under its assignment of value and a definite description such as 'the amount in the envelope' designate amounts of money in different ways. Does the argument for switching commit a fallacy of equivocation by sometimes treating the variable as a definite description? Horgan ([Horgan, 2000]) and Katz and Olin ([Katz and Olin, 2007]) address this question.

6. Extensions to Groups

Fundamental decision principles apply to individuals. Branches of decision theory extend the principles to groups. The extension is not straightforward because the fundamental principles use an agent's beliefs and desires to evaluate a decision, and a collective agent, lacking a mind, does not have beliefs and desires or decide, that is, form an intention. Some theorists hold that beliefs, desires, and intentions are functional states, and that a group, not just an individual, may be in these functional states. However a typical group of people lacks the structure of an individual's mind and so does not realize the functional states that are candidates for an individual's beliefs, desires, and intentions. This section assumes therefore that groups do not have mental states.

Despite lacking mental states, groups act. Rationality evaluates their acts. It evaluates a free act in a group's full control. A group's act constituted by its members' free and fully controlled acts qualifies for evaluation. Because a group acts through its members, and not directly, rationality evaluates a group's act by evaluating the act's components (just as it evaluates an individual's sequence of acts by evaluating the sequence's components). Suppose that rational acts of a group's members constitute a collective act. Then the collective act is rational. For rationality does not require a group to adopt an alternative act while permitting each member contributing to the collective act to perform her component.

Rationality may require a group to change its act while permitting each member's act to remain the same given that some other member changes her act. The group's requirement is consistent with the members' conditional permissions. However, unconditional permissions for the members' acts are incompatible with the requirement that the group's act change. The members' acts block a change in the group's act. Being consistent, rationality does not require a standing crowd to sit and yet permit each member of the crowd to stand. A standing crowd cannot sit unless some standing members sit. The crowd's requirement conflicts with the members' permissions, understood as nonconditional permissions that obtain for each member whatever other members do. Rationality issues consistent directives to individuals and the groups that they constitute.

6.1 Games

In a game of strategy, the players' strategies that together yield the game's outcome constitute a collective act. If all players select rational strategies, then their profile of strategies is rational. The players, if collectively rational, achieve a solution to the game. Also, according to a common objective characterization of a solution, the players achieve a solution only if their strategy profile is collectively rational under the assumption that the players are cognitively ideal, fully rational, and in possession of common knowledge of their game's features. Here *common knowledge* has its technical sense according to which the players' common knowledge of a proposition entails that each player knows the proposition, knows that each player knows the proposition, knows that each player knows that each player knows the proposition, and so on.

In a *noncooperative* game the players do not have opportunities to act jointly. They independently select strategies for playing the game. If the game has a single stage in which players act simultaneously, then no player's act causally influences another player's act. Their acts as well as their strategies for the whole game are independent. If the game is sequential, and has multiple stages, then one player's act at a stage may causally influence another player's act at a later stage. Nonetheless, their strategies for the whole game are independent.

A common standard for a solution to a game is the *joint rationality* of the players' strategies, that is, the rationality of each player's strategy given the entire profile of strategies. In typical circumstances, meeting the standard requires a *subjective Nash equilibrium*, that is, a strategy profile in which each player's strategy maximizes utility given the profile.

A strategy's rationality given the strategies of all differs from its rationality given knowledge of the strategies of all. Consequently, a subjective Nash equilibrium may differ from a game's Nash equilibrium, which, as Section 2.3 explains, is a strategy profile in which each player's strategy is a best response to the other players' strategies. A player may not know the other players' strategies or her best response to them. So rationality requires, rather than a strategy that is a best response to their strategies, a strategy that maximizes (expected) utility calculated with respect to the player's information. However, in ideal cases joint rationality yields a Nash equilibrium because each player uses strategic reasoning to anticipate others' strategies and knows her best response to them.

As Section 4.3 mentions, the principle of utility maximization generalized to take account of information an option's realization carries, forms the principle of ratification or self-support. In a game of strategy the generalized decision principle evaluates a strategy taking account of the information that the strategy's realization provides about other players' strategies and the strategy's outcome. It supports an agent's adoption of her Nash strategy in a game with a unique Nash equilibrium. Because of the principle, collective rationality yields joint rationality in games of strategy.

In a *cooperative* game the players have opportunities to act jointly. They may communicate and adopt binding contracts. Given these opportunities, the demands of collective rationality rise. Theorists commonly claim that the players, if collectively rational, achieve (weak) efficiency; that is, they realize a collective act such that no alternative is better for all.

Do cooperative games demonstrate the existence of principles of rationality besides utility maximization? Is efficiency a principle of rationality that governs the group, and so its members, independently of utility maximization? Given standard idealizations, including the players' full rationality, and hence their rational preparation for their game, efficiency is a requirement of collective rationality. However, as Weirich ([Weirich, 2009], [Weirich, 2010a, Ch. 11]) argues, it emerges from individuals' rationality, in particular, their compliance with a generalization of the principle of utility maximization.

A cooperative game has a cooperative representation showing how players may act jointly. It also has a noncooperative representation showing how individuals' acts may yield their joint acts. The players' utility maximization with respect to their strategies in the noncooperative representation generates a collective act that is efficient with respect to their strategies in the game's cooperative representation.

6.2 Social Choice

In a game of strategy, players' preferences among strategy profiles identify a solution. Assuming that the solution is unique, it is a strategy profile that the players in a technical sense collectively prefer to other strategy profiles. Methods of identifying solutions are methods of moving from individual preferences to technically defined collective preferences.

The literature on social choice treats aggregation of individual preferences to obtain technically defined social preferences. A function from individual preferences to social preferences represents an aggregation method. Social choice theory asks whether aggregation methods produce social preferences with certain desirable properties, such as transitivity. Popular aggregation methods fall short. For example, majoritarian methods fail to produce transitive social preferences. Indeed, the literature reveals many impossibility results, such as Arrow's theorem ([Arrow, 1963]), establishing that no aggregation method produces social preferences with various combinations of desirable features.

First principles of collective rationality derive a group's rationality from its members' rationality. Extending principles of individual rationality to groups using analogies between individuals and groups generates secondary principles of collective rationality that govern a group's acts only in special cases. Take, for example, the principle that an agent should select an option from the top of the agent's preference ranking of options. Suppose that collective preferences have a majoritarian definition. Condorcet's paradox of voting then shows that in some cases a group has intransitive collective preferences despite the rationality of its members. The principle to follow collective preferences does not govern such cases. Also, take the principle to maximize collective utility defined as a sum of members' utilities on an interpersonal scale. It does not yield a collectively rational act in all cases. For example, it is not rational for a pair of players to maximize collective utility in the Prisoner's Dilemma. Collective rationality requires collective-utility maximization only in special cases.

List and Pettit ([List and Pettit, 2011]) review the literature on methods of judgement aggregation. The methods the literature studies are generally analogical. A typical method seeks to technically define collective judgements so that they follow principles of rationality governing an individual's judgements, such as principles of consistency. Collective rationality requires a group's following analogical principles only if the rationality of all members entails the principles' satisfaction. If the rationality of all members does not entail the principles' satisfaction in certain cases, then the principles do not govern those cases.

Consider the principle of consistency for a committee's rulings. Suppose that the committee's members are unanimous and that unanimity suffices for a committee ruling. Then each member's consistency ensures that the committee's rulings are consistent. Collective rationality requires consistent rulings from a committee with unanimous ideal members in ideal conditions. However, given that in some cases a committee's rulings are inconsistent despite the rationality of its members, say, because of flaws in majoritarian methods, consistency is not a general requirement but rather a goal of collective rationality.

6.3 Trustee Decisions

A trustee may make a decision for a client. Although only one agent decides, a second agent's goals furnish the decider's objectives. A trustee decision involves a group of agents.

In trustee decisions, the trustee has the charge of selecting an option that serves the client's interests. The trustee's charge, taken broadly, is to decide as the client would if the client were rational and had the trustee's expert information. In some cases the trustee's charge is narrower. It may be to manage the client's business to maximize profits. Then instead of deciding as the client would if informed, the trustee's objective is to decide as the client would if informed and interested only in profits.

How should expected-utility maximization apply in trustee decisions? Its application combines a trustee's beliefs with a client's goals. The input for the decision principle comes from a pair of sources. The principle needs intrinsic-utility analysis to separate risk, typically an object of a basic intrinsic aversion, from elements of an option's outcome that, unlike risk, are independent of the probability distribution of possible outcomes. The trustee may use the analysis to construct for the client an informed attitude to risky options. Methods of separating risk from other consequences of risky options ground the risk-return school of financial planning.

7. Conclusion

Sections 1–5 survey standard evaluative decision theory and its generalization, refinement, and expansion. The survey is not exhaustive; [Bermúdez, 2009],

[Arló Costa and Helzner, 2010], and [Armendt, 2010], treat additional topics, for example.

This brief concluding section recommends two ways of enriching evaluative decision theory. One is to distinguish and explore various types of evaluation. For example, evaluating acts for comprehensive and nonconditional rationality supplements the noncomprehensive and conditional evaluations that the principle of utility maximization yields. The supplementary evaluations, in the case of a nonideal agent who has made mistakes, must consider the effects of the agent's mistakes on a current decision. Is a current decision irrational if it stems from an irrational probability assignment or an irrational goal? A second type of enrichment formulates principles of rationality for an agent's goals. They may, for example, prohibit pure time-preference and excessive aversion to risk. Although contemporary decision theory progresses well beyond the traditional principle of utility maximization, many more improvements are possible.

20 Further Reading

Leon Horsten and

Richard Pettigrew

Chapter Overview

1. Handbooks, Guides, Companions	575
2. Specialized Dictionaries	576
3. Electronic Sources	576
4. Sources for Specific Subjects	577
4.1 Classical First-Order Logic	577
4.2 Other Logics	577
4.2.1 Retaining Classical Logic	577
4.2.2 Extending Classical Logic	578
4.2.3 Changing Classical Logic	579
4.3 Modelling Rationality	580

In this chapter, we give a brief overview of the rich literature on the topics covered in this volume. We begin with handbooks, guides, and companions to the whole subject of philosophical logic – these are similar in format to the current volume. We also include references to online resources, such as encyclopedias and blogs. Then we turn to specific topics. Each contributor to the volume has provided us with a handful of the most important references in their area: typically, these include an historically important work, a seminal reference book, as well as central research articles or volumes in the area.

1. Handbooks, Guides, Companions

First, an overview of the subject by a single author:

1. *Philosophical Logic* Burgess, J. (Princeton : Princeton University Press), 2009

Then there are books with individual chapters written by different authors:

1. *Handbook of Philosophical Logic* (1st edition) Gabbay, D. M. and F. Guenthner (eds) (Dordrecht: Kluwer), 1983–1989
2. *Handbook of Philosophical Logic* (2nd edition) Gabbay, D. M. and F. Guenthner (eds) (Berlin: Springer), 2001–
3. *Oxford Handbook of Philosophy of Mathematics and Logic* Shapiro, S. (ed.) (New York: Oxford University Press), 2005
4. *Blackwell Guide to Philosophical Logic* Goble, L. (ed.) (Oxford: Blackwell), 2001
5. *A Companion to Philosophical Logic* Jacquette, D. (ed.) (Oxford: Blackwell), 2005

2. Specialized Dictionaries

1. *Key Terms in Logic* Russo, F. and J. Williamson (eds) (London: Continuum Press), 2010

3. Electronic Sources

1. *Stanford Encyclopedia of Philosophy* plato.stanford.edu
   An excellent online encyclopedia of philosophy with articles on a very wide variety of survey articles on topics in philosophical logic. The articles are written by leading philosophers in the area.
2. *Wikipedia* www.wikipedia.org
   This contains articles on most subjects in philosophical logic. The articles are written and revised by users. Inevitably, the quality is varied here, but it is often very good. More technical topics are treated best.
3. *FOM mailing list* cs.nyu.edu/mailman/listinfo/fom
   A mailing list to which anyone may subscribe and to which any subscriber may post. Discussions range from mathematical logic, foundations and philosophy of mathematics, to many central areas of philosophical logic. Many of the most important researchers in the subject contribute daily, as well as many young reseachers.
4. Blogs:
   (a) Brian Weatherson: tar.weatherson.org
   (b) Greg Restall: consequently.org
   (c) Peter Smith: www.logicmatters.net

Many books and articles on philosophical logic are electronically available. For instance, *Oxford Scholarship Online* (www.oxfordscholarship.com) contains electronic versions of books that have been published by Oxford University Press, while *Cambridge Companions Online* (cco.cambridge.org) contains electronic versions of the volumes in the *Cambridge Companions* series published by Cambridge University Press. If you are a student or researcher at an institution of higher education, you will probably have free access to at least some of these sources through your institution.

4. Sources for Specific Subjects

Full bibliographical information may be found in the bibliography to this volume.

4.1 Classical First-Order Logic

1. Logical Consequence
    - Seminal paper that both described the history of the subject and changed its direction: Kurt Gödel (1944). Russells Mathematical Logic. Reprinted in Paul Benacerraf and Hilary Putnam, eds, *Philosophy of Mathematics*, 2nd. ed. (Cambridge: Cambridge University Press, 1983), pp. 447–468, and in *Gödels Collected Works*, vol. 2 (Oxford: Oxford University Press, 1990), pp. 119–143.
    - Introductory survey: Samuel R. Buss (1998). An Introduction to Proof Theory. Available online at http://math.ucsd.edu/sbuss/ResearchWeb/handbookI/index.html
    - Introductory survey, not at all technical: Willard Van Orman Quine (1986). *Philosophy of Logic*.
    - (Seminal research article) Alfred Tarski (1983b). The Concept of Logical Consequence. English translation in *Tarskis Logic, Semantics, Metamathematics*, 2nd. ed. (Indianapolis, IN: Hackett, 1983), pp. 409–420.
    - Seminal article: Per Lindstrm (1969). On extensions of elementary logic.

4.2 Other Logics

4.2.1 Retaining Classical Logic

1. Quantification and Descriptions
    - (Seminal article) Russell (1905b). On denoting
    - (Seminal work) Neale (1990). *Descriptions*
    - (Seminal article) Smullyan (1948). Modality and descriptions
    - (Collected papers) Ostertag (1998). *Definite Descriptions: A Reader*
    - (Textbook) Kalish, Montague, and Mar (1992). *Logic: Techniques of Formal Reasoning*
2. Existence and Identity
    - (Seminal article) Russell (1905b). On denoting

- (Seminal article) Black (1962). The identity of indiscernibles
- (Seminal article) Quine (1948). On what there is
- (Reference work) Miller (2009). Existence
- (Reference work) Noonan (2009). Identity

4.2.2 Extending Classical Logic

1. Modal Logic
    - (Historically important work) Lewis and Langford (1932). *Symbolic Logic*
    - (Reference work) Hughes and Cresswell (1996). *A New Introduction to Modal Logic*
    - (Seminal article) Kripke (1963a). Semantical analysis of modal logic 1
    - (Seminal article) Kripke (1963b). Semantical analysis of modal logic 2
2. Tense Logic
    - (Seminal work) Prior (1967). *Past, Present, and Future*
    - (Reference work) Gabbay et al. (1994, 2000). *Temporal Logic*. Vols. 1 and 2
    - (Survey work) Burgess (2002). Basic tense logic
    - (Survey article) Hodkinson and Reynolds (2007). Temporal Logic
    - (Bibliographic source) Goldblatt (2005). Mathematical modal logic
3. Higher-Order Logic
    - (Seminal work) Frege (1879). *Begriffsschrift*
    - (Seminal article) Russell (1908). Mathematical logic as based of a theory of types
    - (Reference work) Shapiro (2000). *Foundations without Foundationalism*
    - (Survey article) Shapiro (2005). Higher-order ligic
    - (Survey article) Jané (2005). Higher-order logic reconsidered
    - (Seminal article) Boolos (1975). On second-order logic
    - (Seminal article) Boolos (1985). Nominalistic platonism
    - (Seminal work) Quine (1986). *Philosophy of Logic*
4. Mereology
    - (Seminal article) Leśniewski (1916). *Podstawy ogólnej teoryi mnogości. I*
    - (Seminal article) Leonard and Goodman (1940). The calculus of individuals and its uses
    - (Seminal work) Lewis (1991). *Parts of Classes*
    - (Seminal work) Simons (1987). *Parts*
    - (Survey article) Varzi (2011). Mereology

4.2.3 Changing Classical Logic

1. Negation
    - (Reference work) Horn (1989). *A Natural History of Negation*
    - (Introductory textbook) Priest (2008). *An Introduction to Non-Classical Logics*
    - (Survey article) Wansing (2001). Negation
    - (Seminal article) Priest (1979). The logic of paradox
    - (Seminal article) Dunn (1993). Star and Perp: Two Treatments of Negation
2. Vagueness
    - (Reference work) Williamson (1994). *Vagueness*
    - (Seminal article) Dummett (1975). Wang's paradox
    - (Seminal article) Sainsbury (1997). Concepts without boundaries
    - (Introductory text) Chapter 3 in Sainsbury (2009). Paradoxes
3. Indicative Conditionals
    - (Survey work) Bennett (2003). *A Philosophical Guide to Conditionals*
    - (Introductory article) Edgington (1995b). On conditionals
    - (Seminal article) Grice (1989a). Indicative conditionals
    - (Seminal article) Lewis (1976). Probabilities of conditionals and conditional probabilities
    - (Seminal article) Jackson (1979). On assertion and indicative conditionals
    - (Seminal article) Stalnaker (1968). A theory of conditionals
4. Truth and Paradox
    - (Seminal work) McGee (1991). *Truth, Vagueness, and Paradox*
    - (Reference work) Halbach (2010). *Axiomatic Theories of Truth*
    - (Survey article) Visser (1989). Semantics and the liar paradox
    - (Introductory text) Horsten (2010). The tarskian turn
    - (Seminal article) Tarski (1935). The concept of truth in formalized languages
    - (Seminal article) Kripke (1975b). An outline of a theory of truth
5. Game-Theoretic Semantics
    - (Seminal work) Hintikka (1996). *The Principles of Mathematics Revisisted*
    - (Reference work) Hintikka and Sandou (1997). Game-theoretic semantics
    - (Introductory text) Mann et al. (2010). *The Game of Logic*
    - (Seminal article) Hodges (1997). Compositional semantics for a language of imperfect information
    - (Seminal article) Sevenster and Sandu (2010). Equilibrium semantics of languages of imperfect information

4.3 Modelling Rationality

1. Probability
    - (Seminal work) Kolmogorov (1956). *Foundations of the Theory of Probability*
    - (Reference work) Howson and Urbach (1993). *Scientific Reasoning*
    - (Reference work) Gillies (2002). *Philosophical Theories of Probability*
    - (Seminal article) Ramsey (1931b). Truth and probability
    - (Seminal article) Lewis (1980). A subjectivist's guide to objective chance
2. Inductive Logic
    - (Seminal work) Carnap (1952). *The Continuum of Inductive Methods*
    - (Seminal article) Johnson (1932). Probability: The deductive and inductive problems
    - (Seminal article) de Finetti (1931). Sul significato soggettivo della probabilità
    - (Reference work) Carnap and Jeffrey (1971). *Studies in Inductive Logic and Probability*
    - (Reference work) Fitelson (2004). Inductive logic
    - (Seminal article) Gaifman (1964). Concerning measures on first-order calculi
    - (Seminal article) Landes et al. (t.a.) A survey of some recent results on spectrum exchangeability in inductive logic
3. Epistemic Logic
    - (Seminal work) Hintikka (1962). *Knowledge and Belief*
    - (Reference work) Fagin et al. (1995). *Reasoning about knowledge*
    - (Introductory text) van Ditmarsch et al. (2007). *Dynamic Epistemic Logic*
    - (Seminal article) van Benthem (2004b). What one may come to know
    - (Seminal research article) Stalnaker (2006). On the logics of knowledge and belief
4. Belief Revision
    - (Seminal work) Gärdenfors (1988). *Knowledge in Flux*
    - (Reference work) Hansson (1999). *A Textbook of Belief Dynamics*
    - (Seminal article) Alchourrón and Makinson (1985). On the logic of theory change: safe contraction
    - (Seminal article) Alchourrón et al. (1985). On the logic of theory change: Partial meet contraction and revision functions
    - (Seminal article) Gärdenfors and Makinson (1988). Revision of knowledge systems using epistemic entrenchment

5. Decision Theory
    - (Seminal work) Savage (1954). *The Foundations of Statistics*
    - (Seminal work) Jeffrey (1990). *The Logic of Decision (second edition)*
    - (Introductory textbook) Peterson (2009). *An Introduction to Decision Theory*
    - (Seminal work) Joyce (1999). *The Foundations of Causal Decision Theory*

Bibliography

[Abramsky and Väänänen, 2008] Abramsky, S. and Väänänen, J. (2008). From if to bi: A tale of dependence and separation. *Synthese*, 167:207–230.

[Adams, 1962] Adams, E. W. (1962). On rational betting systems. *Archive für Mathematische Logik und Grundlagenforschung*, 6:7–29, 112–128.

[Adams, 1965] Adams, E. W. (1965). The logic of conditionals. *Inquiry*, 8:166–197.

[Adams, 1975] Adams, E. W. (1975). *The Logic of Conditionals*. Reidel, Dordrecht, Holland.

[Adams, 1998] Adams, E. W. (1998). *A Primer of Probability Logic*. CSLI Publications, Stanford, CA.

[Adler, 2002] Adler, J. (2002). *Belief's Own Ethics*. MIT Press, Cambridge, MA.

[Alberucci, 2002] Alberucci, L. (2002). The modal mu-calculus and logics of common knowledge. PhD thesis, Universität Bern, Institut für Informatik und angewandte Mathematik.

[Alchourrón et al., 1985] Alchourrón, C. E., Gärdenfors, P., and Makinson, D. (1985). On the logic of theory change: Partial meet contraction and revision functions. *The Journal of Symbolic Logic*, 50(2):510–530.

[Alchourrón and Makinson, 1985] Alchourrón, C. E. and Makinson, D. (1985). On the Logic of Theory Change: Safe Contraction. *Studia Logica*, 44(4):405–422.

[Aloni, 2001] Aloni, M. (2001). Quantification under conceptual covers. PhD thesis, University of Amsterdam.

[Aloni, 2005] Aloni, M. (2005). Individual concepts in modal predicate logic. *Journal of Philosophical Logic*, 34(1):1–64.

[Aloni, 2008] Aloni, M. (2008). Concealed questions under cover. *Grazer Philosophische Studien*, 77(1):191–216.

[Aloni et al., ta] Aloni, M., Égré, P., and de Jager, T. (t.a.). Knowing whether A or B. *Synthese*.

[Alxatib and Pelletier, ta] Alxatib, S. and Pelletier, F. J. (t.a.). The psychology of vagueness: borderline cases and contradictions. *Mind and Language*.

[Anderson, 1959] Anderson, A. R. (1959). Church on ontological commitment. *The Journal of Philosophy*, 56:448–452.

[Anderson, 1974] Anderson, A. R. (1974). What do symbols symbolize?: Platonism. *Philosophia Mathematica*, s1–11(1–2):11–29.

[Anderson and Belnap, 1975] Anderson, A. R. and Belnap, N. D. (1975). *Entailment: Logic of Relevance and Necessity*, volume I. Princeton University Press, Princeton.

[Anderson et al., 1992] Anderson, A. R., Belnap, N. D., and Dunn, J. M. (1992). *Entailment: Logic of Relevance and Necessity*, volume II. Princeton University Press, Princeton.

[Anderson, 2001] Anderson, C. A. (2001). Alternative (1*): A criterion of identity for intensional entities. In Anderson, C. A. and Zelëny, M., editors, *Logic, Meaning and Computation: Essays in Memory of Alonzo Church*. Kluwer Academic Publishers, Dordrecht.

[Andjelković and Williamson, 2000] Andjelković, M. and Williamson, T. (2000). Truth, falsity, and borderline cases. *Philosophical Topics*, 28:211–242.

[Areces and ten Cate, 2007] Areces, C. and ten Cate, B. (2007). Hybrid logics. In [Blackburn and van Benthem, 2007, 821–868].

[Aristotle, 1933] Aristotle (1933). *Metaphysics*. Harvard University Press, Cambridge, MA. English translation by Hugh Tredennick.

[Arló Costa, 1999] Arló Costa, H. (1999). Qualitative and Probabilistic Models of Full Belief. In Buss, S., Hájek, P., and Pudlak, P., editors, *Proceedings of Logic Colloquim '98*, volume 13 of *Lecture Notes on Logic*. ASL, A. K. Peters.

[Arló Costa, 2001a] Arló Costa, H. (2001a). Bayesian epistemology and epistemic conditionals: On the status of the export-import laws. *The Journal of Philosophy*, 98(11):555–598.

[Arló Costa, 2001b] Arló Costa, H. (2001b). Hypothetical revision and matter-of-fact supposition. *Journal of Applied Non-Classical Logics*, 11(1–2):203–229.

[Arló Costa, 2006] Arló Costa, H. (2006). Decision-theoretic contraction and sequential change. In Olsson, E. J., editor, *Essays on the Pragmatism of Isaac Levi*. Cambridge University Press, Cambridge.

[Arló Costa, 2007] Arló Costa, H. (2007). The logic of conditionals. In Zalta, E. N., editor, *The Stanford Encyclopedia of Philosophy*. Stanford University, Stanford, CA.

[Arló Costa and Helzner, 2010] Arló Costa, H. and Helzner, J. (2010). Foundations of the decision sciences. Special issue of *Synthese*, 172(1).

[Arló Costa and Levi, 2006] Arló Costa, H. and Levi, I. (2006). Contraction: On the decision-theoretical origins of minimal change and entrenchment. *Synthese*, 152(1):129–154.

[Arló Costa and Liu, 2010] Arló Costa, H. and Liu, H. (2010). Saturatable contraction: A representation result. Manuscript, Carnegie Mellon University.

[Arló Costa and Pedersen, 2010a] Arló Costa, H. and Pedersen, A. P. (2010a). Belief and probability: A theory of high probability cores. Manuscript, Carnegie Mellon University.

[Arló Costa and Pedersen, 2010b] Arló Costa, H. and Pedersen, A. P. (2010b). Social norms, rational choice and belief change. In Olsson, E. J. and Enqvist, S., editors, *Belief Revision meets Philosophy of Science*, volume 21 of *Logic, Epistemology, and the Unity of Science*. Springer.

[Armendt, 2010] Armendt, B. (2010). Stakes and beliefs. *Philosophical Studies*, 147: 71–87.

[Arrow, 1951] Arrow, K. J. (1951). *Social Choice and Individual Values*. John Wiley & Sons, New York, 1st edition.

[Arrow, 1963] Arrow, K. J. ([1951] 1963). *Social Choice and Individual Values*. Yale University Press, New Haven, CT, 2nd edition.

[Artemov, 2008] Artemov, S. (2008). The logic of justification. *The Review of Symbolic Logic*, 1(04):477–513.

[Artemov and Nogina, 2005] Artemov, S. and Nogina, E. (2005). Introducing justification into epistemic logic. *Journal of Logic and Computation*, 15(6):1059–1073.

[Asher et al., 2010] Asher, N., Dever, J., and Pappas, C. (2010). Supervaluationism debugged. *Mind*, 118:901–933.

[Aucher, 2008] Aucher, G. (2008). *Perspectives on Belief and Change*. Dissertation. University of Otago and University of Toulouse.

[Aumann, 1976] Aumann, R. (1976). Agreeing to disagree. *Annals of Statistics*, 4:1236–1239.

[Aumann, 1995] Aumann, R. (1995). Backward induction and common knowledge of rationality. *Games and Economic Behavior*, 8:6–19.

[Aumann, 1999a] Aumann, R. (1999a). Interactive epistemology I: Knowledge. *International Journal of Game Theory*, 28:263–300.

[Aumann, 1999b] Aumann, R. (1999b). Interactive epistemology II: Probability. *International Journal of Game Theory*, 28:301–314.

[Aumann and Brandenburger, 1995] Aumann, R. and Brandenburger, A. (1995). Epistemic conditions for Nash equilibrium. *Econometrica*, 63(5):1161–1180.

[Aumann et al., 1997] Aumann, R., Hart, S., and Perry, M. (1997). The absent-minded driver. *Games and Economic Behavior*, 20:102–116.

[Avigad and Zach, 2009] Avigad, J. and Zach, R. (2009). The epsilon calculus. In Zalta, E. N., editor, *The Stanford Encyclopedia of Philosophy*. Stanford University, Stanford, CA, spring 2009 edition.

[Ayer, 1936] Ayer, A. J. (1936). *Language, Truth and Logic*. Victor Gollantz, London, 2nd, 1947 edition.

[Baltag and Smets, 2008a] Baltag, A. and Smets, S. (2008a). The logic of conditional doxastic actions. In van Rooij, R. and Apt, K., editors, *New Perspectives on Games and Interaction*, Texts in Logic and Games. Amsterdam University Press.

[Baltag and Smets, 2008b] Baltag, A. and Smets, S. (2008b). A qualitative theory of dynamic interactive belief revision. In Bonanno, G., van der Hoek, W., and Wooldridge, M., editors, *Logic and the Foundation of Game and Decision Theory (LOFT7)*, volume 3 of *Texts in Logic and Games*, pages 13–60. Amsterdam University Press.

[Baltag et al., 2009] Baltag, A., Smets, S., and Zvesper, J. (2009). Keep 'hoping' for rationality: a solution to the backwards induction paradox. *Synthese*, 169:301–333.

[Barwise, 1988] Barwise, J. (1988). Three views of common knowledge. In *TARK '88: Proceedings of the 2nd Conference on Theoretical Aspects of Reasoning about Knowledge*, pages 365–379, Morgan Kaufmann Publishers Inc., San Francisco, CA.

[Barwise and Cooper, 1981] Barwise, J. and Cooper, R. (1981). Generalized quantifiers and natural language. *Linguistics and Philosophy*, 4:159–219.

[Barwise and Etchemendy, 1987] Barwise, J. and Etchemendy, J. (1987). *The Liar: An Essay on Truth and Circularity*. CSLI Publications, Stanford.

[Barwise and Perry, 1981] Barwise, J. and Perry, J. (1981). Semantic innocence and uncompromising situations. *Midwest Studies in Philosophy*, 6:387–404.

[Barwise and Perry, 1983] Barwise, J. and Perry, J. (1983). *Situations and Attitudes*. MIT Press, Cambridge, MA.

[Barwise and Seligman, 1997] Barwise, J. and Seligman, J. (1997). *Information Flow: The Logic of Distributed Systems*. Cambridge University Press, Cambridge.

[Beall, 2003] Beall, J. C. (2003). *Liars and Heaps*. Oxford University Press, Oxford.

[Beall, 2007] Beall, J. C., editor (2007). *Revenge of the Liar: New Essays on the Paradox*. Oxford University Press, Oxford.

[Beall and van Fraassen, 2003] Beall, J. C. and van Fraassen, B. C. (2003). *Possibilities and Paradox: An Introduction to Modal and Many-valued Logic*. Oxford University Press, Oxford.

[Beaney, 2009] Beaney, M. (2009). Analysis. In Zalta, E. N., editor, *The Stanford Encyclopedia of Philosophy*. Stanford University, Stanford, CA, spring 2009 edition.

[Belnap, 1962] Belnap, N. D. (1962). Tonk, plonk, and plink. *Analysis*, 22: 130–134.

[Belnap, 1977a] Belnap, N. D. (1977a). How a computer should think. In Ryle, G., editor, *Contemporary Aspects of Philosophy*, pages 30–55. Oriel Press, Stocksfield.

[Belnap, 1977b] Belnap, N. D. (1977b). A useful 4-valued logic. In Dunn, J. M. and Epstein, G., editors, *Modern Uses of Many-Valued Logic*, pages 8–37. Reidel, Dordrecht.

[Belnap, 1992] Belnap, N. D. (1992). Branching space-time. *Synthese*, 92: 385–434.

[Belnap, 2001] Belnap, N. D. (2001). Double time references: Speech-act reports as modalities in an indeterminist setting. In Wolter, F., Wansing, H., de Rijke, M., and Zakharyaschev, M., editors, *Advances in Modal Logic*, volume 3, pages 1–22. CSLI Publications, Stanford, CA.

[Belnap, 2007] Belnap, N. D. (2007). An indeterminist view of the parameters of truth. In Müller, T., editor, *Philosophie der Zeit. Neue analytische Ansätze*, pages 87–113. Klostermann, Frankfurt a.M.

[Belnap, 2009] Belnap, N. D. (2009). Truth values, neither-true-nor-false, and supervaluations. *Studia Logica*, 91:305–334.

[Belnap et al., 2001] Belnap, N. D., Perloff, M., and Xu, M. (2001). *Facing the Future. Agents and Choices in Our Indeterminist World*. Oxford University Press, Oxford.

[Benacerraf and Putnam, 1983] Benacerraf, P. and Putnam, H., editors (1983). *Philosophy of Mathematics: Selected Readings*. Cambridge University Press, Cambridge, 2nd edition.

[Bencivenga, 1986] Bencivenga, E. (1986). Free logics. In Gabbay, D. M. and Guenthner, F., editors, *Handbook of Philosophical Logic*, volume III, pages 373–426. Reidel, Dordrecht.

[Bennett, 1998] Bennett, B. (1998). Modal semantics for knowledge bases dealing with vague concepts. In Cohn, A. G., Schubert, L. K., and Shapiro, S. C., editors, *Principles of Knowledge Representation and Reasoning*, pages 234–244, San Francisco, CA. Proceedings of the Sixth International Conference (KR'98), Morgan Kaufmann.

[Bennett, 2003] Bennett, J. (2003). *A Philosophical Guide to Conditionals*. Clarendon Press, Oxford.

[Bergman et al., 1990] Bergman, M., Moor, J., and Nelson, J. (1990). *The Logic Book*. McGraw-Hill Education, New York.

[Bermúdez, 2009] Bermúdez, J. L. (2009). *Decision Theory and Rationality*. Oxford University Press, Oxford.

[Beth, 1970] Beth, E. W. (1970). *Formal Methods*. Reidel, Dordrecht.

[Beziau, 2002] Beziau, J.-Y. (2002). S5 is a paraconsistent logic and so is first-order classical logic. *Logical Investigations*, 9:301–309.

[Biacino and Gerla, 1991] Biacino, L. and Gerla, G. (1991). Connection structures. *The Journal of Symbolic Logic*, 32:242–247.

[Binmore, 2009] Binmore, K. (2009). *Rational Decisions*. Princeton University Press, Princeton, NJ.

[Black, 1962] Black, M. (1962). The identity of indiscernibles. *Mind*, 61: 153–164.

[Blackburn, 2000] Blackburn, P. (2000). Representation, reasoning, and relational structures: A hybrid logic manifesto. *Logic Journal of the IGPL*, 8(3):339–365.

[Blackburn et al., 2002] Blackburn, P., de Rijke, M., and Venema, Y. (2002). *Modal Logic*. Cambridge University Press, Cambridge.

[Blackburn and van Benthem, 2007] Blackburn, P. and van Benthem, J. F. A. K., editors (2007). *Handbook of Modal Logic*. Elsevier, Amsterdam.

[Blackburn, 1986] Blackburn, S. (1986). How can we tell whether a commitment has a truth condition? In Travis, C., editor, *Meaning and Interpretation*, pages 201–232. Blackwell, Oxford.

[Blamey, 1986] Blamey, S. (1986). Partial logic. In Gabbay, D. M. and Guenthner, F., editors, *Handbook of Philosophical Logic*, volume III, pages 1–70. Reidel, Dordrecht.

[Blass and Gurevich, 1986] Blass, A. and Gurevich, Y. (1986). Henkin quantifiers and complete problems. *Annals of Pure and Applied Logic*, 32:1–16.

[Board, 2004] Board, O. (2004). Dynamic interactive epistemology. *Games and Economic Behavior*, 49:49–80.

[Bochman, 1990] Bochman, A. (1990). Mereology as a theory of part-whole. *Logique et Analyse*, 129:75–101.

[Bonanno and Battigalli, 1999] Bonanno, G. and Battigalli, P. (1999). Recent results on belief, knowledge and the epistemic foundations of game theory. *Research in Economics*, 53(2):149–225.

[Bonini et al., 1999] Bonini, N., Osherson, D., Viale, R., and Williamson, T. (1999). On the psychology of vague predicates. *Mind and Language*, 14:377–393.

[Bonnay and Égré, 2009] Bonnay, D. and Égré, P. (2009). Inexact knowledge with introspection. *Journal of Philosophical Logic*, 38(2):179–228.

[Bonnay and Égré, ta] Bonnay, D. and Égré, P. (t.a.). Knowing one's limits: an analysis in centered dynamic epistemic logic. In Girard, P., Marion, M., and Roy, O., editors, *Dynamic Epistemology: Contemporary Perspectives*, Synthese Library. Springer, Dordrecht.

[Boole, 1854a] Boole, G. (1854a). *An Investigation of the Laws of Thought on which are Founded the Mathematical Theories of Logic and Probabilities*. Macmillan.

[Boole, 1854b] Boole, G. (1854b). *The Laws of Thought*. Walton and Maberly, London.

[Boolos, 1975] Boolos, G. S. (1975). On second-order logic. *The Journal of Philosophy*, 72(16):509–527. Reprinted in [Boolos, 1998, 37–53].

[Boolos, 1984] Boolos, G. S. (1984). To be is to be a value of a variable (or to be some values of some variables). *The Journal of Philosophy*, 81:430–449. Reprinted in [Boolos, 1998, 54–72].

[Boolos, 1985] Boolos, G. S. (1985). Nominalistic platonism. *The Philosophical Review*, 94(3):327–344. Reprinted in [Boolos, 1998, 73-87].

[Boolos, 1993] Boolos, G. S. (1993). *The Logic of Provability*. Cambridge University Press, Cambridge.

[Boolos, 1998] Boolos, G. S. (1998). *Logic, Logic, and Logic*. Harvard University Press, Cambridge, MA.

[Boolos and Jeffrey, 1989] Boolos, G. S. and Jeffrey, R. C. (1989). *Computability and Logic*. Cambridge University press, Cambridge, 3rd edition.

[Bosch, 1983] Bosch, P. (1983). 'Vagueness' is context-dependence: A solution to the sorites paradox. In Ballmer, T. T. and Pinkal, M., editors, *Approaching Vagueness*, pages 189–210. North-Holland, Amsterdam.

[Bostrom, 2002] Bostrom, N. (2002). *Anthropic Bias : Observation Selection Effects in Science and Philosophy*. Routledge, New York.

[Boutilier, 1996] Boutilier, C. (1996). Iterated revision and minimal revision of conditional beliefs. *Journal of Philosophical Logic*, 25(3):262–305.

[Bradley, 2000] Bradley, R. (2000). A preservation condition for conditionals. *Analysis*, 60:219–222.

[Brandenburger, 2007] Brandenburger, A. (2007). The power of paradox: Some recent developments in interactive epistemology. *International Journal of Game Theory*, 35:465–492.

[Breitkopf, 1978] Breitkopf, A. (1978). Axiomatisierung einiger begriffe aus Nelson Goodmans *The Structure of Appearance*. *Erkenntnis*, 12:229–247.

[Bremer and Cohnitz, 2004] Bremer, M. and Cohnitz, D. (2004). *Information and Information Flow*. Ontos Verlag, Frankfurt.

[Broersen, 2009] Broersen, J. (2009). A complete stit logic for knowledge and action, and some of its applications. In Baldoni, M., Son, T. C., van Riemsdijk, M. B., and Winikoff, M., editors, *Declarative Agent Languages and Technologies VI, 6th International Workshop, DALT 2008, Estoril, Portugal, May 12, 2008, Revised Selected and Invited Papers*, volume 5397 of *Lecture Notes in Computer Science*, pages 47–59.

[Broogard and Salerno, 2009] Broogard, B. and Salerno, J. (2009). Fitch's paradox of knowability. In Zalta, E. N., editor, *The Stanford Encyclopedia of Philosophy*. Stanford University, Stanford, CA, spring 2009 edition.

[Broome, 1995] Broome, J. (1995). The two-envelope paradox. *Analysis*, 55: 6–11.

[Brouwer, 1927] Brouwer, L. E. J. (1927). Über definnitionberreiche von funktionen. *Mathematische Annalen*, 97:60–75. English translation by Stefan Bauer-Mengelberg in [van Heijenoort, 1967, 446–463].

[Bunt, 1979] Bunt, H. C. (1979). Ensembles and the formal semantic properties of mass terms. In Pelletier, F. J., editor, *Mass Terms: Some Philosophical Problems*, pages 249–277. Reidel, Dordrecht.

[Burge, 1979] Burge, T. (1979). Semantical paradox. In Martin, R. M., editor, *Recent Essays on Truth and the Liar Paradox*, pages 83–117. Clarendon Press, Oxford.

[Burgess, 1990] Burgess, J. A. (1990). The sorites paradox and higher-order vagueness. *Synthese*, 85:417–474.

[Burgess, 2001] Burgess, J. A. (2001). Vagueness, epistemicism, and response-dependence. *Australasian Journal of Philosophy*, 79:507–524.

[Burgess and Humberstone, 1987] Burgess, J. A. and Humberstone, I. L. (1987). Natural deduction rules for a logic of vagueness. *Erkenntnis*, 27:197–229.

[Burgess, 1984] Burgess, J. P. (1984). Basic tense logic. In [Gabbay and Guenthner, 1984, 89–133].

[Burgess, 1998] Burgess, J. P. (1998). Quinus Ab Omni Nævo Vindicatus. In Kazmi, A. A., editor, *Meaning and Reference*, volume 23, pages 25–66. University of Calgary Press, Calgary.

[Burgess, 1999] Burgess, J. P. (1999). Which modal logic is the right one? *Notre Dame Journal of Formal Logic*, 40:81–93.

[Burgess, 2002] Burgess, J. P. (2002). Basic tense logic. In [Gabbay and Guenthner, 2002, 1–42]. Almost identical to [Burgess, 1984].

[Burgess, 2003] Burgess, J. P. (2003). A remark on Henkin sentences and their contraries. *Notre Dame Journal of Formal Logic*, 44:185–188.

[Burgess, 2004] Burgess, J. P. (2004). *E Pluribus Unum*: Plural Logic and Set Theory. *Philosophia Mathematica*, 12(3):193–221.

[Burkhardt and Dufour, 1991] Burkhardt, H. and Dufour, C. A. (1991). Part/whole i: History. In Burkhardt, H. and Smith, B., editors, *Handbook of Metaphysics and Ontology*, pages 663–673. Philosophia, Munich.

[Burns, 1991] Burns, L. (1991). *Vagueness: An Investigation into Natural Languages amd the Sorites Paradox*. Kluwer, Dordrecht.

[Buss, 1998] Buss, S. (1998). An Introduction to Proof Theory. In Buss, S. editor, *Handbook of Proof Theory*. Elsevier, Amsterdam.

[Caicedo et al., 2009] Caicedo, X., Dechesne, F., and Janssen, T. M. V. (2009). Equivalence and quantifier rules for a logic with imperfect information. *Logic Journal of the IGPL*, 17:91–129.

[Caicedo and Krynicki, 1999] Caicedo, X. and Krynicki, M. (1999). Quantifiers for reasoning with imperfect information and σ_1^1-logic. In Carnielli, W. A. and Ottaviano, I. M. L., editors, *Contemporary Mathematics*, pages 17–31. American Mathematical Society.

[Campbell, 1974] Campbell, R. (1974). The sorites paradox. *Philosophical Studies*, 26:175–191.

[Cantini, 1996] Cantini, A. (1996). *Logical Frameworks for Truth and Abstraction*. North-Holland, Amsterdam.

[Cantor, 1895–7] Cantor, G. (1895–7). Beiträge zur begründung der transfiniten mengenlehre. *Mathematische Annalen*, 46, 49:481–512, 207–246.

[Cargile, 1969] Cargile, J. (1969). The sorites paradox. *British Journal for the Philosophy of Science*, 20:193–202.

[Carnap, 1928] Carnap, R. (1928). *Der logische Aufbau der Welt*. Weltkreis.

[Carnap, 1934] Carnap, R. (1934). *Logische Syntax der Sprache*. Springer, Wien. Translated as *The Logical Syntax of Language* (New York: Harcourt, Brace and Co, 1937).

[Carnap, 1935] Carnap, R. (1935). *Philosophy and Logical Syntax*. Kegan Paul, London.

[Carnap, 1946] Carnap, R. (1946). Modalities and quantification. *The Journal of Symbolic Logic*, 11:33–64.

[Carnap, 1947a] Carnap, R. (1947a). On the application of inductive logic. *Philosophy and Phenomenological Research*, 8:133–147.

[Carnap, 1947b] Carnap, R. (1947b). Reply to Nelson Goodman. *Philosophy and Phenomenological Research*, 8:461–462.

[Carnap, 1948] Carnap, R. (1948). *Naming and Necessity*. University of Chicago Press, Chicago, 2nd, 1956 edition.

[Carnap, 1950] Carnap, R. (1950). *Logical Foundations of Probability*. University of Chicago Press, Chicago.

[Carnap, 1952] Carnap, R. (1952). *The Continuum of Inductive Methods*. University of Chicago Press, Chicago.

[Carnap, 1980] Carnap, R. (1980). A basic system of inductive logic. In Jeffrey, R. C., editor, *Studies in Inductive Logic and Probability*, volume II, pages 7–155. University of California Press, Berkeley, CA.

[Carnap and Jeffrey, 1971] Carnap, R. and Jeffrey, R. C., editors (1971). *Studies in Inductive Logic and Probability*, volume I. University of California Press, Berkeley, CA.

[Cartwright, 1971] Cartwright, R. (1971). Identity and substitutivity. In Munitz, M. K., editor, *Identity and Individuation*. New York University Press, New York.

[Chambers, 1998] Chambers, T. (1998). On vagueness, sorites, and Putnam's intuitionistic strategy. *Monist*, 81:343–348.

[Chang and Keisler, 1973] Chang, C. C. and Keisler, J. (1973). *Model Theory*. Elsevier, Amsterdam.

[Chernoff, 1954] Chernoff, H. (1954). Rational selection of decision functions. *Econometrica*, 22(4):422–443.

[Chomsky, 1981] Chomsky, N. (1981). *Lectures on Government and Binding*. Foris, Dordrecht.

[Christensen, 1996] Christensen, D. (1996). Dutch-Book Arguments Depragmatized: Epistemic Consistency for Partial Believers. *The Journal of Philosophy*, 93(9): 450–479.

[Christensen, 2004] Christensen, D. (2004). *Putting Logic in its Place*. Oxford University Press, Oxford.

[Church, 1936] Church, A. (1936). A note on the entscheidungsproblem. *The Journal of Symbolic Logic*, 1:40–41.

[Church, 1951] Church, A. (1951). A formulation of the logic of sense and denotation. In Henle, P., Kallen, H. M., and Langer, S. K., editors, *Structure, Method, and Meaning. Essays in Honor of Henry M. Sheffer*. Liberal Arts Press, New York.

[Church, 1956] Church, A. (1956). Review of Hans Reichenbach, 'The Rise of Scientific Philosophy'. *The Journal of Symbolic Logic*, 21:396.

[Church, 1965] Church, A. (1965). Review of Karel Lambert, 'Existential Import Revisited'. *The Journal of Symbolic Logic*, 30:103–104.

[Church, 1973] Church, A. (1973). Outline of a revised formulation of the logic of sense and denotation (part I). *Noûs*, 7:24–33.

[Church, 1974] Church, A. (1974). Outline of a revised formulation of the logic of sense and denotation (part II). *Noûs*, 8:135–156.

[Clarke, 1981] Clarke, B. (1981). A calculus of individuals based on 'connection'. *Notre Dame Journal of Formal Logic*, 22:204–218.

[Clarke, 1985] Clarke, B. (1985). Individuals and points. *Notre Dame Journal of Formal Logic*, 26:61–75.

[Clausing, 2003] Clausing, T. (2003). Doxastic conditions for backward induction. *Theory and Decision*, 54(4):315–336.

[Cobreros, 2008] Cobreros, P. (2008). Supervaluationism and logical consequence: a third way. *Studia Logica*, 90:291–312.

[Cobreros, 2010] Cobreros, P. (2010). Supervaluationism and Fara's argument concerning higher-order vagueness. In Égré, P. and Klinedinst, N., editors, *Vagueness and Language Use*, pages 233–247. Palgrave Macmillan, Houndsmills.

[Cobreros, taa] Cobreros, P. (t.a.a). Paraconsistent vagueness: a positive argument. *Synthese*.

[Cobreros, tab] Cobreros, P. (t.a.b). Supervaluationism and classical logic. In Krifka, M., Nouwen, R., van Rooij, R., Sauerland, U., and Schmitz, H.-C., editors, *Proceedings of the Vagueness in Communication workshop (ESSLLI09)*.

[Cobreros et al., 2010] Cobreros, P., Egré, P., Ripley, D., and van Rooij, R. (2010). Tolerant, classical, strict. Unpublished manuscript.

[Cocchiarella, 1969] Cocchiarella, N. (1969). Existence entailing attributes, modes of copulation and modes of being of second order logic. *Noûs*, 3: 33–48.

[Copeland, 2002] Copeland, B. J. (2002). The genesis of possible worlds semantics. *Journal of Philosophical Logic*, 31(2):99–137.

[Corsi, 2002] Corsi, G. (2002). A unified completeness theorem for quantified modal logic. *The Journal of Symbolic Logic*, 67(4):1483–1510.

[Cox, 1979] Cox, R. T. (1979). On inference and enquiry, an essay in inductive logic. In Levine, R. D. and Tribus, M., editors, *The Maximum Entropy Formalism*, pages 119–167. MIT Press, Cambridge, MA.

[Cresswell, 1990] Cresswell, M. J. (1990). *Entities and Indices*. Kluwer, Dordrecht.

[Cross and Nute, 2001] Cross, C. and Nute, D. (2001). Conditional logic. In Gabbay, D. M. and Guenthner, F., editors, *Handbook of Philosophical Logic*, volume IV. Reidel, Dordrecht.

[Cubitt and Sugden, 2003] Cubitt, R. P. and Sugden, R. (2003). Common knowledge, salience and convention: A reconstruction of David Lewis' game theory. *Economics and Philosophy*, 19(2):175–210.

[Dancygier, 1998] Dancygier, B. (1998). *Conditionals and Predictions: Time, Knowledge and Causation in Conditional Constructions*. Cambridge University Press, Cambridge.

[Darwiche and Pearl, 1997] Darwiche, A. and Pearl, J. (1997). On the logic of iterated belief revision. *Artificial Intelligence*, 89(1–2):1–29.

[Davidson, 1971] Davidson, D. (1971). Reality without reference. *Dialectica*, 31:247–253. Reprinted in [Davidson, 1984, 215–225].

[Davidson, 1984] Davidson, D. (1984). *Inquiries into Truth and Interpretation*. Clarendon Press, Oxford.

[Davis, 1979] Davis, W. A. (1979). Indicative and subjunctive conditionals. *The Philosophical Review*, 88:544–564.

[de Bruin, 2004] de Bruin, B. (2004). *Explaining Games – On the Logic of Game Theoretic Explanations*. ILLC Dissertation Series.

[De Clercq and Horsten, 2005] De Clercq, R. and Horsten, L. (2005). Closer. *Synthese*, 146:371–393.

[de Finetti, 1931] de Finetti, B. (1931). Sul significato soggettivo della probabilità. Fundamenta Mathematicae.

[de Finetti, 1972] de Finetti, B. (1972). *Probability, Induction, and Statistics*. John Wiley & Sons, London.

[de Finetti, 1974] de Finetti, B. (1974). *Theory of Probability*, volume 1. John Wiley & Sons, New York.

[De Laguna, 1922] De Laguna, T. (1922). Point, line and surface as sets of solids. *The Journal of Philosophy*, 19:449–461.

[de Rouilhan and Bozon, 2006] de Rouilhan, P. and Bozon, S. (2006). The truth of if: Has Hintikka really exorcized Tarski's curse? In Auxier, R. E. and Hahn, L. E., editors, *The philosophy of Jaakko Hintikka*, Library of Living Philosophers, pages 683–705. Carus Publishing Company.

[Dean and Kurokawa, 2009] Dean, W. and Kurokawa, H. (2009). Knowledge, Proof and the Knower. In *Proceedings of the 12th Conference on Theoretical Aspects of Rationality and Knowledge*, pages 81–90. ACM.

[Declerck and Reed, 2001] Declerck, R. and Reed, S. (2001). *Conditionals: A Comprehensive Empirical Analysis*. Mouton de Gruyter, Berlin/New York.

[Dedekind, 1888] Dedekind, R. (1888). *Was sind und was sollen die Zahlen?* F. Vieweg, Braunschweig. English translation by Wooster W. Beman in [Dedekind, 1963, 29–115]. English translation in [Ewald, 1996, 790–833].

[Dedekind, 1963] Dedekind, R. (1963). *Essays on the Theory of Numbers*. Dover, New York. http://www.gutenberg.org/etext/21016.

[DeRose, 2002] DeRose, K. (2002). Assertion, knowledge, and context. *The Philosophical Review*, 111:167–203.

[DeRose, ta] DeRose, K. (t.a.). The conditionals of deliberation. *Mind*.

[di Maio, 1995] di Maio, M. C. (1995). Predictive probability and analogy by similarity. *Erkenntnis*, 43(3):369–394.

[Dietz, 2008] Dietz, R. (2008). Betting on borderline cases. *Philosophical Perspectives*, 22:47–88.

[Dietz, 2010] Dietz, R. (2010). On generalizing Kolmogorov. *Notre Dame Journal of Formal Logic*, 51:323–335.

[Dietz and Douven, 2010] Dietz, R. and Douven, I. (2010). Ramsey's test, Adams' thesis, and left-nested conditionals. *The Review of Symbolic Logic*, 3(3):467–484.

[Dietz and Moruzzi, 2010] Dietz, R. and Moruzzi, S. (2010). *Cuts and Clouds: Vagueness, Its Nature and Its Logic*. Oxford University Press, Oxford.

[Dimitracopoulos et al., 1999] Dimitracopoulos, C., Paris, J. B., Vencovská, A., and Wilmers, G. M. (1999). A multivariate natural prior probability distribution based on the propositional calculus. Technical Report 1999/6, Manchester Centre for Pure Mathematics. Available at www.maths.manchester.ac.uk/~jeff/.

[Dokic and Egré, 2009] Dokic, J. and Egré, P. (2009). Margin for Error and the Transparency of Knowledge. *Synthese*, 166(1):1–20.

[Döring, 1994] Döring, F. (1994). On the probabilities of conditionals. *The Philosophical Review*, 103:689–699.

[Dorr, 2010] Dorr, C. (2010). Iterating definiteness. In [Dietz and Moruzzi, 2010, 550–575].

[Douven, 2006] Douven, I. (2006). Assertion, knowledge, and rational credibility. *The Philosophical Review*, 115:449–485.

[Douven, 2007] Douven, I. (2007). On Bradley's preservation condition for conditionals. *Erkenntnis*, 67:111–118.

[Douven, 2008] Douven, I. (2008). The evidential support theory of conditionals. *Synthese*, 164:19–44.

[Douven, 2009] Douven, I. (2009). Assertion, Moore, and Bayes. *Philosophical Studies*, 144:361–375.

[Douven, 2010] Douven, I. (2010). The pragmatics of belief. *Journal of Pragmatics*, 42:35–47.

[Douven et al., 2009] Douven, I., Decock, L., Dietz, R., and Égré, P. (2009). Vagueness: a conceptual spaces approach. Unpublished manuscript.

[Douven and Dietz, ta] Douven, I. and Dietz, R. (t.a.). A puzzle about Stalnaker's hypothesis. *Topoi*.

[Douven and Verbrugge, 2010] Douven, I. and Verbrugge, S. (2010). The Adams Family. *Cognition*, 117:302–318.

[Drake, 1974] Drake, F. (1974). *Set Theory: An Introduction to Large Cardinals*. North-Holland, Amsterdam.

[Dubois et al., 2007] Dubois, D., Esteva, F., Godo, L., and Prade, H. (2007). Fuzzy-set based logics – an history-oriented presented of their main developments. In Gabbay, D. M. and Woods, J., editors, *The Handbook of the History of Logic*, volume 8, *The Many Valued and Nonmonotonic Turn in Logic*, pages 325–449. Elsevier, Amsterdam.

[Duc, 1997] Duc, H. N. (1997). Reasoning about rational, but not logically omniscient, agents. *Journal of Logic and Computation*, 7(5):633–648.

[Dummett, 1959] Dummett, M. A. E. (1959). Wittgenstein's philosophy of mathematics. *The Philosophical Review*, 58:324–348. Reprinted in [Dummett, 1978, 166–185]; page references to reprint.

[Dummett, 1975] Dummett, M. A. E. (1975). Wang's paradox. *Synthese*, 30:301–324. Reprinted in [Keefe and Smith, 1997,99–118]; page references to reprint.

[Dummett, 1978] Dummett, M. A. E. (1978). *Truth and Other Enigmas*. Duckworth, London.

[Dummett, 1981] Dummett, M. A. E. (1981). *Frege: Philosophy of Language*. Harvard University Press, Cambridge, MA, 2nd edition.

[Dummett, 1991] Dummett, M. A. E. (1991). *The Logical Basis of Metaphysics*. Harvard University Press, Cambridge, MA.

[Dummett, 2000] Dummett, M. A. E. (2000). *Elements of Intuitionism*. Oxford University Press, Oxford, 2nd edition.

[Dunn, 1976] Dunn, J. M. (1976). Intuitive semantics for first-degree entailments and 'coupled trees'. *Philosophical Studies*, 29:149–168.

[Dunn, 1993] Dunn, J. M. (1993). Star and perp. *Philosophical Perspectives*, 7: 331–357.

[Eagle, 2004] Eagle, A. (2004). Twenty-One Arguments Against Propensity Analyses of Probability. *Erkenntnis*, 60:371–416.

[Earman, 1985] Earman, J. (1985). Concepts of projectibility and the problems of induction. *Noûs*, XIX:521–535.

[Earman, 1992] Earman, J. (1992). *Bayes or Bust?* MIT Press.

[Easwaran, 2008] Easwaran, K. (2008). Strong and weak expectations. *Mind*, 117: 633–641.

[Eberle, 1967] Eberle, R. (1967). Some complete calculi of individuals. *Notre Dame Journal of Formal Logic*, 8:267–278.

[Eberle, 1968] Eberle, R. (1968). Yoes on non-atomic systems of individuals. *Noûs*, 2:399–403.

[Eberle, 1969] Eberle, R. (1969). Non-atomic systems of individuals revisited. *Noûs*, 3:431–434.

[Eberle, 1970] Eberle, R. (1970). *Nominalistic Systems*. Reidel, Dordrecht.

[Edgington, 1993] Edgington, D. (1993). Wright and Sainsbury on higher-order vagueness. *Analysis*, 53:193–200.

[Edgington, 1995a] Edgington, D. (1995a). Conditionals and the Ramsey test. *Proceedings of the Aristotelian Society*, 69:67–86.

[Edgington, 1995b] Edgington, D. (1995b). On conditionals. *Mind*, 104:235–329.

[Edgington, 1997] Edgington, D. (1997). Vagueness by degrees. In [Keefe and Smith, 1997, 294–316].

[Edgington, 2001] Edgington, D. (2001). Conditionals. In Goble, L., editor, *The Blackwell Guide to Philosophical Logic*, pages 385–414. Blackwell, Oxford.

[Égré, 2005] Égré, P. (2005). The knower paradox in the light of provability interpretations of modal logic. *Journal of Logic, Language and Information*, 14(1): 13–48.

[Égré, 2008] Égré, P. (2008). Reliability, margin for error and self-knowledge. In Pritchard, D. and Hendricks, V. F., editors, *New Waves in Epistemology*, pages 215–250. Palgrave Macmillan.

[Égré and Bonnay, 2010] Égré, P. and Bonnay, D. (2010). Vagueness, uncertainty and degrees of clarity. *Synthese*, 174:47–78.

[Eklund, 2005] Eklund, M. (2005). What vagueness consists in. *Philosophical Studies*, 125:27–60.

[Eklund, 2010] Eklund, M. (2010). Vagueness and second-level indeterminacy. In [Dietz and Moruzzi, 2010, 63–76].

[Elga, 2000] Elga, A. (2000). Self-locating belief and the sleeping beauty problem. *Analysis*, 60:143–147.

[Elga, 2009] Elga, A. (2009). Subjective probabilities should be sharp. *Philosophers' Imprint*, 10(5).

[Enderton, 1972] Enderton, H. B. (1972). *A Mathematical Introduction to Logic*. Academic Press, San Diego.

[Enderton, 2001] Enderton, H. B. (2001). *A Mathematical Introduction to Logic*. Academic Press, San Diego, 2nd edition.

[Engel, ta] Engel, P. (t.a.). Formal methods in philosophy: shooting right without collateral damage. In Czarnecki, T., Kijania-Placek, K., and Wolenski, J., editors, *The Analytical Way*. 6th European Congress of Analytic Philosophy, College Publications.

[Eschenbach, 1994] Eschenbach, C. (1994). A mereotopological definition of 'point'. In Eschenbach, C., Habel, C., and Smith, B., editors, *Topological Foundations of Cognitive Sciences*. Graduiertenkolleg Kognitionswissenschaft, Hamburg. Bereicht Nr. 37.

[Etchemendy, 1999] Etchemendy, J. (1999). *The Concept of Logical Consequence*. CSLI Publications, Stanford, CA, 2nd edition.

[Etlin, 2009] Etlin, D. (2009). The problem of noncounterfactual conditionals. *Philosophy of Science*, 76:676–688.

[Evans, 1977] Evans, G. (1977). Pronouns, quantifiers and relative clauses (i). *Canadian Journal of Philosophy*, 7:187–208.

[Evans and Over, 2004] Evans, J. S. B. T. and Over, D. E. (2004). *If*. Oxford University Press, Oxford.

[Ewald, 1996] Ewald, W. B. (1996). *From Kant to Hilbert: A Source Book in the Foundations of Mathematics*, volume 2. Oxford University Press, Oxford.

[Fagin and Halpern, 1987] Fagin, R. and Halpern, J. Y. (1987). Belief, awareness, and limited reasoning. *Artificial Intelligence*, 34(1):39–76.

[Fagin et al., 1995] Fagin, R., Halpern, J. Y., Moses, Y., and Vardi, M. (1995). *Reasoning about Knowledge*. The MIT Press, Cambridge, MA.

[Fara, 2000] Fara, D. G. (2000). Shifting sands: An interest-relative theory of vagueness. *Philosophical Topics*, 28:45–81.

[Fara, 2001] Fara, D. G. (2001). Phenomenal continua and the sorites. *Mind*, 110: 905–935.

[Fara, 2002] Fara, D. G. (2002). An anti-epistemicist consequence of margin for error semantics for knowledge. *Philosophy and Phenomenological Research*, 64(1): 127–142.

[Fara, 2003] Fara, D. G. (2003). Gap principles, penumbral consequence, and infinitely higher-order vagueness. In [Beall, 2003, 195–222].

[Feferman, 1960] Feferman, S. (1960). Arithmetization of metamathematics in a general setting. *Fundamenta Mathematicae*, 49:35–92.

[Feferman, 1991] Feferman, S. (1991). Reflecting on incompleteness. *The Journal of Symbolic Logic*, 56:1–49.

[Feferman, 2006] Feferman, S. (2006). What kind of logic is 'independence friendly' logic? In Auxier, R. E. and Hahn, L. E., editors, *The philosophy of Jaakko Hintikka*, Library of Living Philosophers, pages 453–469. Carus Publishing Company.

[Ferme and Rott, 2004] Ferme, E. and Rott, H. (2004). Revision by comparison. *Artificial Intelligence*, 157(1–2):5–47.

[Festa, 1996] Festa, R. (1996). Analogy and exchangeability in predictive inferences. *Erkenntnis*, 45:229–252.

[Fetzer, 1981] Fetzer, J. H. (1981). *Scientific Knowledge: Causation, Explanation, and Corroboration*. Boston Studies in the Philosophy of Science. Reidel, Dordrecht.

[Fetzer, 1982] Fetzer, J. H. (1982). Probabilistic Explanations. *PSA*, 2:194–207.

[Field, 1980] Field, H. (1980). *Science without Numbers: A Defence of Nominalism*. Blackwell, Oxford.

[Field, 1994] Field, H. (1994). Disquotational truth and factually defective discourse. *The Philosophical Review*, 103:405–452.

[Field, 2000] Field, H. (2000). Indeterminacy, degree of belief, and excluded middle. *Noûs*, 34:1–30.

[Field, 2003] Field, H. (2003). No fact of the matter. *Australasian Journal of Philosophy*, 81:457–480.

[Field, 2008] Field, H. (2008). *Saving Truth from Paradox*. Oxford University Press, Oxford.

[Field, 2010] Field, H. (2010). The magic moment: Horwich on the boundaries of vague terms. In [Dietz and Moruzzi, 2010, 200–208].

[Fine, 1975] Fine, K. (1975). Language, truth and logic. *Synthese*, 30:265–300.

[Fine, 1994] Fine, K. (1994). Compounds and aggregates. *Noûs*, 28:137–158.

[Fine, 1995] Fine, K. (1995). Part-whole. In Smith, B. and Smith, D. W., editors, *The Cambridge Companion to Husserl*, pages 463–485. Cambridge.

[Fine, 1999] Fine, K. (1999). Things and their parts. *Midwest Studies in Philosophy*, 23:61–74.

[Finger et al., 2002] Finger, M., Gabbay, D. M., and Reynolds, M. A. (2002). Advanced tense logic. In [Gabbay and Guenthner, 2002, 43–203].

[Fitch, 1950] Fitch, F. (1950). Actuality, possibility, and being. *The Review of Metaphysics*, 3:367–384.

[Fitelson, 2004] Fitelson, B. (2004). Inductive logic. In Zalta, E. N., editor, *The Stanford Encyclopedia of Philosophy*. Stanford University, Stanford, CA.

[Fitelson, 2006] Fitelson, B. (2006). Inductive logic. In Sarkar, S. and Pfeifer, J., editors, *The Philosophy of Science*, volume I, pages 384–394. Routledge, New York and Abingdon.

[Fitting and Mendelsohn, 1998] Fitting, M. and Mendelsohn, R. L. (1998). *First-order modal logic*. Kluwer Academic Publishers, Dordrecht.

[Føllesdal, 1961] Føllesdal, D. (1961). *Referential Opacity and Modal Logic*. Routledge, New York and London, 2004 edition.

[Forbes, 1983] Forbes, G. (1983). Thisness and vagueness. *Synthese*, 54:235–259.

[Forrest, 2010] Forrest, P. (2010). Mereotopology without mereology. *Journal of Philosophical Logic*, 39:229–254.

[Frege, 1879] Frege, G. (1879). Begriffsschrift: Eine der arithmetischen nachgebildete Formelsprache des reinen Denkens. In [van Heijenoort, 1967]. Translated and reprinted in [van Heijenoort, 1967].

[Frege, 1892a] Frege, G. (1892a). Über begriff und gegenstand. *Vierteljahrschrift für wissenschraftliche Philosophie*, 16:192–205. English translation by Geach, Peter T. in [Frege, 1960, 42–55].

[Frege, 1892b] Frege, G. (1892b). Über sinn und bedeutung. *Zeitschrift für Philosophie und philosophische Kritik*, 100:25–50.

[Frege, 1893] Frege, G. (1893). *Grundgesetze der Arithmetik*, volume 1. Pohle, Jena.

[Frege, 1960] Frege, G. (1960). *Translations from the Philosophical Writings*. Basil Blackwell, Oxford, 2nd edition.

[Frege, 1979] Frege, G. (1979). Dialogue with pünzer on existence. In Hermes, H., Kambartel, F., and Kaulbach, F., editors, *Posthumous Writings*. The University of Chicago Press, Chicago.

[Frege, 1980] Frege, G. (1980). *The Foundations of Arithmetic*. Northwestern University Press, Evanston. Translated by J. L. Austin.

[Friedman, 1999] Friedman, H. (1999). A complete theory of everything. http://www.math.ohio-state.edu/~friedman/manuscripts.htm.

[Friedman and Sheard, 1987] Friedman, H. and Sheard, M. (1987). Axiomatic theories of self-referential truth. *Annals of Pure and Applied Logic*, 33:1–21.

[Gabbay and Guenthner, 1989] Gabbay, D. M. and Guenthner, F., editors (1983–1989). *Handbook of Philosophical Logic*. Kluwer, Dordrecht, First edition. 4 volumes.

[Gabbay and Guenthner, 1984] Gabbay, D. M. and Guenthner, F., editors (1984). *Handbook of Philosophical Logic*, volume II. Reidel, Dordrecht.

[Gabbay and Guenthner, 2002] Gabbay, D. M. and Guenthner, F., editors (2002). *Handbook of Philosophical Logic*, volume VII. Kluwer, Dordrecht, 2nd edition.

[Gabbay et al., 1994] Gabbay, D. M., Hodkinson, I., and Reynolds, M. A. (1994). *Temporal Logic. Mathematical Foundations and Computational Aspects*, volume 1. Oxford University Press, Oxford.

[Gabbay et al., 2000] Gabbay, D. M., Reynolds, M. A., and Finger, M. (2000). *Temporal Logic. Mathematical Foundations and Computational Aspects*, volume 2. Oxford University Press, Oxford.

[Gaifman, 1964] Gaifman, H. (1964). Concerning measures on first order calculi. *Israel Journal of Mathematics*, 2:1–18.

[Gaifman, 1971] Gaifman, H. (1971). Applications of de Finetti's theorem to inductive logic. In Carnap, R. and Jeffrey, R. C., editors, *Studies in Inductive Logic and Probability*, volume I, pages 235–251. University of California Press, Berkeley and Los Angeles.

[Gaifman, 1992] Gaifman, H. (1992). Pointers to truth. *The Journal of Philosophy*, 89:223–261.

[Gaifman, 2010] Gaifman, H. (2010). Vagueness, tolerance and contextual logic. *Synthese*, 174:5–46.

[Galison, 1997] Galison, P. (1997). *Image & Logic. A material culture of microphysics.* University of Chicago Press, Chicago.

[Galliani, 2009] Galliani, P. (2009). *Game values and equilibria for undetermined sentences of Dependence Logic.* Master of Logic Series 2008-08. Universiteit van Amsterdam, ILLC.

[Galton, 1984] Galton, A. (1984). *The Logic of Aspect.* Oxford University Press, Oxford.

[Gärdenfors, 1982] Gärdenfors, P. (1982). Rules for Rational Changes of Belief. In Pauli, T., editor, *Philosophical Essays Dedicated to Lennart Åqvist on His Fiftieth Birthday*, volume 34. Philosophical Society and Department of Philosophy, University of Uppsala.

[Gärdenfors, 1986] Gärdenfors, P. (1986). Belief revisions and the Ramsey test for conditionals. *The Philosophical Review*, 95:81–93.

[Gärdenfors, 1988] Gärdenfors, P. (1988). *Knowledge in Flux. Modeling the Dynamics of Epistemic States.* The MIT Press, Cambridge, MA.

[Gärdenfors and Makinson, 1988] Gärdenfors, P. and Makinson, D. (1988). Revisions of Knowledge Systems Using Epistemic Entrenchment. In *TARK '88: Proceedings of the 2nd Conference on Theoretical Aspects of Reasoning about Knowledge*, pages 83–95, San Francisco, CA. Morgan Kaufmann Publishers Inc.

[Garson, 2006] Garson, J. W. (2006). *Modal Logic for Philosophers.* Cambridge University Press, Cambridge.

[Geach, 1962] Geach, P. T. (1962). *Reference and Generality.* Cornell University Press, Ithaca.

[Gentzen, 1934] Gentzen, G. (1934). Untersuchungen über das logische schliessen. *Mathematische Zeitschrift*, 39:176–210. English translation by M. E. Szabo in [Gentzen, 1969, 68-131].

[Gentzen, 1969] Gentzen, G. (1969). *Collected Papers.* North-Holland, Amsterdam.

[Gerbrandy, 2000] Gerbrandy, J. (2000). Identity in epistemic semantics. *Logic, Language and Computation*, 3:147–159.

[Gerbrandy, 2007] Gerbrandy, J. (2007). The surprise examination in dynamic epistemic logic. *Synthese*, 155(1):21–33.

[Gerbrandy and Groeneveld, 1997] Gerbrandy, J. and Groeneveld, W. (1997). Reasoning about information change. *Journal of Logic, Language and Information*, 6:147–169.

[Gettier, 1963] Gettier, E. (1963). Is justified true belief knowledge? *Analysis*, 23:121–123.

[Gibbard, 1981] Gibbard, A. (1981). Two recent theories of conditionals. In Harper, W. L., Stalnaker, R., and Pearce, G., editors, *Ifs*, pages 211–247. Reidel, Dordrecht.

[Gilboa, 2009] Gilboa, I. (2009). *Theory of Decision under Uncertainty.* Cambridge University Press, Cambridge.

[Gillies, 2001] Gillies, A. S. (2001). A new solution to Moore's paradox. *Philosophical Studies*, 105:237–250.

[Gillies, 2000] Gillies, D. (2000). Varieties of Propensity. *British Journal for the Philosophy of Science*, 51:807–835.

[Gillies, 2002] Gillies, D. (2002). *Philosophical Theories of Probability*. Cambridge University Press, Cambridge.

[Glanzberg, 2003] Glanzberg, M. (2003). Against truth-value gaps. In [Beall, 2003, 151–194].

[Glibowski, 1969] Glibowski, E. (1969). The application of mereology to grounding of elementary geometry. *Studia Logica*, 24:109–125.

[Gochet and Gribomont, 2006] Gochet, P. and Gribomont, P. (2006). Epistemic logic. In Gabbay, D. M. and Woods, J., editors, *The Handbook of the History of Logic*, volume 7, *Logic and the Modalities in the Twentieth Century*. Elsevier, Amsterdam.

[Gödel, 1930] Gödel, K. (1930). Die vollständigkeit der axiome des logischen funktionenkalküs. *Monatshefte für Mathematik und Physik*, 37:349–360. English translation by Stefan Bauer-Mengelberg in [van Heijenoort, 1967, 582–591].

[Gödel, 1931] Gödel, K. (1931). Über formal unentscheidbare sätze der principia mathematica und verwandter systeme i. *Monatshefte für Mathematik und Physik*, 38:173–198. English translation by Stefan Bauer-Mengelberg in [van Heijenoort, 1967, 596–616].

[Gödel, 1933a] Gödel, K. (1933a). Eine interpretation des intuitionistischen aussagenkalkuls. In *Ergebnisse eines mathematisches Kolloquiums*, volume 4, pages 39–40. Springer, Vienna.

[Gödel, 1933b] Gödel, K. (1933b). The present situation in the foundations of mathematics. In [Gödel, 1995], pages 45–53.

[Gödel, 1944] Gödel, K. (1944). Russell's mathematical logic. In Schilpp, P. A., editor, *The Philosophy of Bertrand Russell*. Tudor Publishing Company, New York.

[Gödel, 1944b] Gödel, K. (1944b). Russell's mathematical philosophy. In Schilpp, P. A., editor, *The Philosophy of Bertrand Russell*, pages 125–153. Northwestern University Press, Evanston and Chicago. Reprinted in [Benacerraf and Putnam, 1983, 447–469].

[Gödel, 1995] Gödel, K. (1995). *Collected Works*, volume III. Oxford University Press, Oxford.

[Goguen, 1969] Goguen, J. (1969). The logic of inexact concepts. *Synthese*, 19:325–373.

[Goldblatt, 1974] Goldblatt, R. (1974). Semantic analysis of orthologic. *Journal of Philosophical Logic*, 3:19–35.

[Goldblatt, 2005] Goldblatt, R. (2005). Mathematical modal logic: A view of its evolution. In Gabbay, D. M. and Woods, J., editors, *Handbook of the History of Logic*, volume 5, pages 1–98. Elsevier, Amsterdam.

[Goldblatt, 2006] Goldblatt, R. (2006). Mathematical modal logic: A view of its evolution. *Journal of Applied Logic*, 1:309–392.

[Goldblatt, 2007] Goldblatt, R. (2007). Mathematical modal logic: A view of its evolution. In Gabbay, D. M. and Woods, J., editors, *Handbook of the History of Logic*, volume 7, pages 1–98. Elsevier, Amsterdam.

[Goldman, 1967] Goldman, A. I. (1967). A causal theory of knowing. *The Journal of Philosophy*, 64(12):357–372.

[Gómez-Torrente, 1997] Gómez-Torrente, M. (1997). Two problems for an epistemicist view of vagueness. *Philosophical Issues*, 8:237–245.
[Gómez-Torrente, 2002] Gómez-Torrente, M. (2002). Vagueness and margin for error principles. *Philosophy and Phenomenological Research*, 64:107–125.
[Gómez-Torrente, 2010] Gómez-Torrente, M. (2010). The sorites, linguistic preconceptions, and the dual picture of vagueness. In [Dietz and Moruzzi, 2010, 228–253].
[Good, 1952] Good, I. J. (1952). Rational decisions. *Journal of the Royal Statistical Society, Ser. B*, 14:107–114.
[Goodman, 1946] Goodman, N. (1946). A query on confirmation. *The Journal of Philosophy*, 43:383–385.
[Goodman, 1947] Goodman, N. (1947). On infirmities in confirmation-theory. *Philosophy and Phenomenological Research*, 8:149–151.
[Goodman, 1951] Goodman, N. (1951). *The Structure of Appearance*. Harvard University Press, Cambridge, MA.
[Goodman, 1954] Goodman, N. (1954). *Fact, Fiction and Forecast*. The Athlone Press.
[Goodman, 1956] Goodman, N. (1956). A world of individuals. In *The Problem of Universals. A Symposium*, pages 13–31. Notre Dame University Press, Notre Dame. reprinted in [Goodman, 1972, 155–171].
[Goodman, 1958] Goodman, N. (1958). On relations that generate. *Philosophical Studies*, 9:65–66. Reprinted in [Goodman, 1972, 171–172].
[Goodman, 1966] Goodman, N. (1966). *The Structure of Appearance*. Bobbs-Merrill, New York.
[Goodman, 1972] Goodman, N. (1972). *Problems and Projects*. Bobbs-Merril, Indianapolis.
[Goodman and Quine, 1947] Goodman, N. and Quine, W. V. O. (1947). Steps toward a constructive nominalism. *The Journal of Symbolic Logic*, 12:105–122.
[Greaves and Wallace, 2006] Greaves, H. and Wallace, D. (2006). Justifying conditionalization: Conditionalization maximizes expected epistemic utility. *Mind*, 115:607–632.
[Grice, 1989a] Grice, H. P. (1989a). Indicative conditionals. In *Studies in the Way of Words*, pages 58–85. Harvard University Press, Cambridge MA.
[Grice, 1989b] Grice, H. P. (1989b). Logic and conversation. In *Studies in the Way of Words*, pages 22–40. Harvard University Press, Cambridge MA.
[Groenendijk and Stokhof, 1984] Groenendijk, J. and Stokhof, M. (1984). Studies in the semantics of questions and the pragmatics of answers. PhD thesis, University of Amsterdam.
[Groenendijk and Stokhof, 1997] Groenendijk, J. and Stokhof, M. (1997). Questions. In van Benthem, J. F. A. K. and ter Meulen, A., editors, *Handbook of Logic and Language*. Elsevier Science Publishers, Amsterdam.
[Grove, 1988] Grove, A. (1988). Two Modellings for Theory Change. *Journal of Philosophical Logic*, 17(2):157–170.
[Grove et al., 1994] Grove, A. J., Halpern, J. Y., and Koller, D. (1994). Random worlds and maximum entropy. *Journal of Artificial Intelligence Research*, 2:33–88.
[Grzegorczyk, 1951] Grzegorczyk, A. (1951). Undecidability of some topological theories. *Fundamenta Mathematicae*, 38:137–152.
[Grzegorczyk, 1955] Grzegorczyk, A. (1955). The system of Leśniewski in relation to contemporary logical research. *Studia Logica*, 3:77–95.

[Gupta and Belnap, 1993] Gupta, A. and Belnap, N. D. (1993). *The Revision Theory of Truth*. MIT Press.

[Haegeman, 2005] Haegeman, L. (2005). *The Syntax of Negation*. Cambridge University Press, Cambridge.

[Hájek, 1989] Hájek, A. (1989). Probabilities of conditionals—revisited. *Journal of Philosophical Logic*, 18:423–428.

[Hájek, 1994] Hájek, A. (1994). Triviality on the cheap? In Eells, E. and Skyrms, B., editors, *Probability and Conditionals*, pages 113–140. Cambridge University Press, Cambridge.

[Hájek, 1997] Hájek, A. (1997). 'Mises Redux'—Redux: Fifteen Arguments against Finite Frequentism. *Erkenntnis*, 45:209–227.

[Hájek, 2008] Hájek, A. (2008). Dutch Book Arguments. In Anand, P., Pattanaik, P., and Puppe, C., editors, *The Oxford Handbook of Corporate Social Responsibility*, pages 173–195. Oxford University Press, Oxford.

[Hájek, 2009] Hájek, A. (2009). Fifteen Arguments against Hypothetical Frequentism. *Erkenntnis*, 70:211–235.

[Hájek and Hall, 2002] Hájek, A. and Hall, N. (2002). Induction and probability. In Machamer, P. and Silberstein, R., editors, *The Blackwell Guide to the Philosophy of Science*, pages 149–172. Blackwell, Oxford.

[Hájek and Pudlák, 1993] Hájek, P. and Pudlák, P. (1993). *Metamathematics of First-Order Arithmetic*. Springer, Berlin.

[Halbach, 2009] Halbach, V. (2009). Reducing compositional to disquotational truth. *The Review of Symbolic Logic*, 2:786–798.

[Halbach, 2010] Halbach, V. (2010). *The Logic Manual*. Oxford University Press, Oxford.

[Halbach, ta] Halbach, V. (t.a.). *Axiomatic Theories of Truth*. Cambridge University Press, Cambridge.

[Halbach and Horsten, 2006] Halbach, V. and Horsten, L. (2006). Axiomatizing Kripke's theory of truth. *The Journal of Symbolic Logic*, 71: 677–712.

[Halbach et al., 2003] Halbach, V., Leitgeb, H., and Welch, P. (2003). Possible worlds semantics for modal notions conceived as predicates. *Journal of Philosophical Logic*, 32:179–223.

[Hall, 1994] Hall, N. (1994). Back in the CCCP. In Eells, E. and Skyrms, B., editors, *Probability and Conditionals*, pages 141–160. Cambridge University Press, Cambridge.

[Halldén, 1963] Halldén, S. (1963). A pragmatic approach to modal theory. *Acta Philosophica Fennica*, 16:53–64.

[Halpern, 2001] Halpern, J. Y. (2001). Substantive rationality and backward induction. *Games and Economic Behavior*, 37:425–435.

[Halpern, 2003] Halpern, J. Y. (2003). *Reasoning about Uncertainty*. The MIT Press, Cambridge, MA.

[Halpern, 2008] Halpern, J. Y. (2008). Intransitivity and vagueness. *The Review of Symbolic Logic*, 1(04):530–547.

[Halpern et al., 2009] Halpern, J. Y., Samet, D., and Segev, E. (2009). Defining knowledge in terms of belief: The modal logic perspective. *The Review of Symbolic Logic*, 2:469–487.

[Hamblin, 1973] Hamblin, C. L. (1973). Questions in Montague English. *Foundations of Language*, 10(1):41–53.

[Hansson, 1991] Hansson, S. O. (1991). Belief Contraction without Recovery. *Studia Logica*, 50(2):251–260.

[Hansson, 1996] Hansson, S. O. (1996). Hidden Structures of Belief. In Fuhrmann, A. and Rott, H., editors, *Logic, Action, and Information: Essays on Logic in Philosophy and Artificial Intelligence*, pages 79–100. Walter de Gruyter, Berlin and New York.

[Hansson, 1999] Hansson, S. O. (1999). *A Textbook of Belief Dynamics: Theory Change and Database Updating*. Kluwer Academic Publishers, Dordrecht.

[Hansson, 2000] Hansson, S. O. (2000). Formalization in philosophy. *Bulletin of Symbolic Logic*, 2:162–175.

[Hansson, 2009] Hansson, S. O. (2009). Logic of Belief Revision. In Zalta, E. N., editor, *The Stanford Encyclopedia of Philosophy*. Stanford University, Stanford, CA, spring 2009 edition.

[Hansson and Olsson, 1995] Hansson, S. O. and Olsson, E. J. (1995). Levi Contractions and AGM Contractions: A Comparison. *Notre Dame Journal of Formal Logic*, 36(1):103–119.

[Harper, 1975] Harper, W. L. (1975). Rational belief change, Popper functions and counterfactuals. *Synthese*, 30(1–2):221–262.

[Harper, 1977] Harper, W. L. (1977). Rational Conceptual Change. In Suppe, F. and Asquith, P. D., editors, *PSA 1976*, volume 2, pages 462–494, East Lansing, MI. Philosophy of Science Association.

[Harper et al., 1981] Harper, W. L., Stalnaker, R., and Pearce, G., editors (1981). *Ifs*. Reidel, Dordrecht.

[Harris, 1982] Harris, J. H. (1982). What's so logical about 'logical' axioms? *Studia Logica*, 41:159–171.

[Heck, 1993] Heck, R. G. (1993). A note on the logic of (higher-order) vagueness. *Analysis*, 53/4:201–208.

[Heck, 2003] Heck, R. G. (2003). Semantic accounts of vagueness. In [Beall, 2003, 106–127].

[Hegselmann and Krause, 2006] Hegselmann, R. and Krause, U. (2006). Truth and cognitive division of labour: first steps towards a computer aided social epistemology. *Journal of Artificial Societies and Social Simulation*, 9. http://jasss.soc.surrey.ac.uk/9/3/10.html.

[Heim, 1994] Heim, I. (1994). Interrogative semantics and Karttunen's semantics for know. In Buchalla, R. and Mittwoch, A., editors, *IATL 1, Akademon, Jerusalem*, pages 128–144.

[Hellman, 1969] Hellman, G. (1969). Finitude, infinitude, and isomorphism of interpretations in some nominalistic calculi. *Noûs*, 3:413–425.

[Hellman, 1989] Hellman, G. (1989). *Mathematics without Numbers*. Clarendon, Oxford.

[Hendricks, 2005] Hendricks, V. F. (2005). *Mainstream and Formal Epistemology*. Cambridge University Press, Cambridge.

[Hendricks and Roy, 2010] Hendricks, V. F. and Roy, O., editors (2010). *Epistemic Logic: 5 Questions*. Automatic Press, New York.

[Hendry, 1980] Hendry, H. E. (1980). Two remarks on the atomistic calculus of individuals. *Noûs*, 14:235–237.

[Hendry, 1982] Hendry, H. E. (1982). Complete extensions of the calculus of individuals. *Noûs*, 16:453–460.

[Henkin, 1949] Henkin, L. (1949). The completeness of the first-order functional calculus. *The Journal of Symbolic Logic*, 14:159–166.

[Henkin, 1961] Henkin, L. (1961). Some remarks on infinitely long formulas. In Bernays, P., editor, *Infinitistic Methods. Proceedings of the Symposium on Foundations of Mathematics*, pages 167–183. Pergamon Press and PWN, New York.

[Henkin et al., 1971] Henkin, L., Monk, J. D., and Tarski, A. (1971). *Cylindric Algebras, part 1*. North-Holland, Amsterdam.

[Henry, 1991] Henry, D. (1991). *Medieval Mereology*. Grüner, Amsterdam.

[Heyting, 1971] Heyting, A. (1971). *Intuitionism: An Introduction*. North-Holland, Amsterdam, 3rd edition.

[Heyting, 1972] Heyting, A. (1972). *Intuitionism: An Introduction*. North-Holland, Amsterdam.

[Hilbert, 1899] Hilbert, D. (1899). *Grundlagen der Geometrie*. Teubner, Leipzig.

[Hilbert, 1903] Hilbert, D. (1903). *Grundlagen der Geometrie*. B. G. Tuebner, Leipzig, 2nd edition.

[Hilbert, 1926] Hilbert, D. (1926). Über das Unendliche. *Mathematische Annalen*, 95:161–190. Translated as 'On the Infinite' in [van Heijenoort, 1967].

[Hilbert, 1927] Hilbert, D. (1927). Die grundlagen der mathematik. *Abhandlungen aus dem mathematischen Seminar der Hamburgischen Universität*, 6:65–85. English translation by Stefan Bauer-Mengelberg and Dagfinn Føllesdal in [van Heijenoort, 1967, 464–479].

[Hilbert and Bernays, 1939] Hilbert, D. and Bernays, P. (1939). *Grundlagen der Mathematik*, volume 2. Julius Springer, Berlin.

[Hild and Spohn, 2008] Hild, M. and Spohn, W. (2008). The measurement of ranks and the laws of iterated contraction. *Artificial Intelligence*, 172(10):1195–1218.

[Hill and Paris, unp.] Hill, A. and Paris, J. B. (unpublished). A note on support by analogy. In preparation.

[Hill et al., 2002] Hill, M. J., Paris, J. B., and Wilmers, G. M. (2002). Some observations on induction in predicate probabilistic reasoning. *Journal of Philosophical Logic*, 31:43–75.

[Hintikka, 1962] Hintikka, J. (1962). *Knowledge and Belief: An Introduction to the Logic of the Two Notions*. Cornell University Press, Ithaca.

[Hintikka, 1965] Hintikka, J. (1965). Towards a theory of inductive generalization. In Bar-Hillel, Y., editor, *Logic, Methodology and Philosophy of Science, Proceedings of the 1964 International Congress*, pages 274–288, North-Holland, Amsterdam. Studies in Logic and the Foundations of Mathematics.

[Hintikka, 1966] Hintikka, J. (1966). A two dimensional continuum of inductive methods. In Hintikka, J. and Suppes, P., editors, *Aspects of Inductive Logic*, pages 113–132. North-Holland, Amsterdam.

[Hintikka, 1974] Hintikka, J. (1974). Quantifiers vs. quantification theory. *Linguistic Inquiry*, 5:153–177.

[Hintikka, 1975] Hintikka, J. (1975). Different constructions in terms of the basic epistemological verbs: A survey of some problems and proposals. In *The Intensions of Intentionality and Other New Models for Modalities*, pages 1–25. Reidel, Dordrecht.

[Hintikka, 1983] Hintikka, J. (1983). *The Game of Language*. Reidel, Dordrecht.

[Hintikka, 1996] Hintikka, J. (1996). *The Principles of mathematics revisited*. Cambridge University Press, Cambridge.

[Hintikka and Sandu, 1989] Hintikka, J. and Sandu, G. (1989). Informational independence as a semantic phenomenon. In Fenstad, J. E., Frolov, I. T., and Hilpinen, R., editors, *Logic, Methodology and Philosophy of Science*, volume VIII, pages 571–589. Elsevier Science, Amsterdam.

[Hintikka and Sandu, 1997] Hintikka, J. and Sandu, G. (1997). Game-theoretical semantics. In van Benthem, J. and ter Meulen, A. editors., *Handbook of Logic and Language*. Elsevier, Amsterdam.

[Hodges, 1997] Hodges, W. (1997). Compositional semantics for a language of imperfect information. *Logic Journal of the IGPL*, 5:539–563.

[Hodges and Lewis, 1968] Hodges, W. and Lewis, D. K. (1968). Finitude and infinitude in the atomic calculus of individuals. *Noûs*, 2:405–410.

[Hodkinson and Reynolds, 2007] Hodkinson, I. and Reynolds, M. (2007). Temporal logic. In [Blackburn and van Benthem, 2007, 655–720].

[Holliday and Icard III, 2010] Holliday, W. H. and Icard III, T. F. (2010). Moorean phenomena in epistemic logic. In Beklemishev, L., Goranko, V., and Shehtman, V., editors, *Advances in Modal Logic*, volume 8, pages 167–187. College Publications, London.

[Hoover, 1979] Hoover, D. N. (1979). Relations on probability spaces and arrays of random variables. Technical report, Institute of Advanced Study, Princeton.

[Horgan, 2000] Horgan, T. (2000). The two-envelope paradox, nonstandard expected utility, and the intentionality of probability. *Noûs*, 34:578–603.

[Horgan, 2004] Horgan, T. (2004). Sleeping beauty awakened: New odds at the dawn of the new day. *Analysis*, 64:10–21.

[Horn, 1989] Horn, L. (1989). *A Natural History of Negation*. University of Chicago Press, Chicago.

[Horsten, 2004] Horsten, L. (2004). A note concerning the notion of satisfiability. *Logique et Analyse*, 185–188:463–468.

[Horsten, 2010] Horsten, L. (2010). Perceptual indiscriminability and the concept of a color shade. In [Dietz and Moruzzi, 2010, 209–227].

[Horsten, ta] Horsten, L. (t.a.). *The Tarskian Turn. Deflationism and axiomatic truth*. Cambridge University Press, Cambridge.

[Horsten and Douven, 2008] Horsten, L. and Douven, I. (2008). Formal methods in the philosophy of science. *Studia Logica*, 89:151–162.

[Horsten and Leitgeb, 2001] Horsten, L. and Leitgeb, H. (2001). No future. *Journal of Philosophical Logic*, 30:259–265.

[Horty, 2001] Horty, J. F. (2001). *Agency and Deontic Logic*. Oxford University Press, Oxford.

[Horwich, 2000] Horwich, P. (2000). The sharpness of vague terms. *Philosophical Topics*, 28:83–92.

[Hottinger, 1988] Hottinger, S. (1988). *Nelson Goodman's Nominalismus und Methodologie*. Berner Reihe philosophische Schriften, Bd. 7, Bern, Stuttgart; Haupt.

[Hovda, 2009] Hovda, P. (2009). What is classical mereology? *Journal of Philosophical Logic*, 38:55–82.

[Howson and Urbach, 1993] Howson, C. and Urbach, P. (1993). *Scientific Reasoning: The Bayesian approach*. Open Court, La Salle, 2nd edition.

[Hughes and Cresswell, 1996] Hughes, G. E. and Cresswell, M. J. (1996). *A New Introduction to Modal Logic*. Routledge, London.

[Husserl, 1913] Husserl, E. (1913). *Logische Untersuchungen*. Niemeyer, Halle, 2nd edition. 2 Volumes; originally published by Niemeyer 1901.

[Hyde, 1994] Hyde, D. (1994). Why higher-order vagueness is a pseudo-problem. *Mind*, 103:35–41.

[Hyde, 1997] Hyde, D. (1997). From heaps and gaps to heaps and gluts. *Mind*, 106:641–660.

[Hyde, 2003] Hyde, D. (2003). Higher-orders of vagueness reinstated. *Mind*, 112:301–305.

[Hyde, 2007] Hyde, D. (2007). Logics of vagueness. In Gabbay, D. M. and Woods, J., editors, *The Handbook of the History of Logic*, volume 8, *The Many Valued and Nonmonotonic Turn in Logic*, pages 285–324. Elsevier, Amsterdam.

[Hyde, 2008] Hyde, D. (2008). *Vagueness, Logic and Ontology*. Ashgate, Aldershot.

[Hyde and Colyvan, 2008] Hyde, D. and Colyvan, M. (2008). Paraconsistent vagueness: Why not? *Australasian Journal of Logic*, 6:107–121.

[Iamhoff, 2008] Iamhoff, R. (2008). Intuitionism in the philosophy of mathematics. In Zalta, E. N., editor, *The Stanford Encyclopedia of Philosophy*. Stanford University, Stanford, CA.

[Jackson, 1979] Jackson, F. (1979). On assertion and indicative conditionals. *The Philosophical Review*, 88:565–589. Reprinted, with postscript, in [Jackson, 1991, 111–135]; page references to reprint.

[Jackson, 1987] Jackson, F. (1987). *Conditionals*. Blackwell, Oxford.

[Jackson, 1991] Jackson, F., editor (1991). *Conditionals*. Oxford University Press, Oxford.

[Jackson, 2002] Jackson, F. (2002). Language, thought and the epistemic theory of vagueness. *Language and Communication*, 22:269–279.

[Jané, 2005] Jané, I. (2005). Higher-order logic reconsidered. In Shapiro, S., editor, *Oxford Handbook of Philosophy of Mathematics and Logic*, pages 781–810. Oxford University Press, Oxford.

[Janicki, 2005] Janicki, R. (2005). Basic mereology with equivalence relations. In Jedrzejowicz, J. and Szepietowski, A., editors, *Mathematical Foundations of Computer Science 2005*, volume 3618 of *Lecture Notes in Computer Science*, pages 507–519. Springer, Berlin Heidelberg.

[Janssen and Dechesne, 2006] Janssen, T. M. V. and Dechesne, F. (2006). Signalling: a tricky business. In van Benthem, J. F. A. K., Heinzmann, G., Rebuschi, M., and Visser, H., editors, *The Age of Alternative Logics: Assessing the Philosophy of Logic and Mathematics Today*, pages 223–242. Kluwer Academic Publishers, Dordrecht.

[Jaśkowki, 1969] Jaśkowki, S. (1969). Propositional calculus for contradictory deductive systems. *Studia Logica*, 24:143–260. Originally published in Polish in 1948.

[Jaynes, 1957a] Jaynes, E. T. (1957a). Information theory and statistical mechanics I. *Physical Review*, 106:620–630.

[Jaynes, 1957b] Jaynes, E. T. (1957b). Information theory and statistical mechanics II. *Physical Review*, 108:171–190.

[Jeffrey, 1990] Jeffrey, R. C. ([1965] 1990). *The Logic of Decision*. University of Chicago Press, Chicago, 2nd edition. Paperback.

[Jeffrey, 1977] Jeffrey, R. C. (1977). Mises *Redux*. In Butts, R. E. and Hintikka, J., editors, *Basic Problems in Methodology and Linguistics*, University of Western Ontario Series in Philosophy of Science. Springer, London.

[Jeffrey, 1983] Jeffrey, R. C. (1983). *The Logic of Decision*. University of Chicago Press, Chicago, 2nd edition.

[Jeffrey, 2004] Jeffrey, R. C. (2004). *Subjective Probability: The Real Thing*. Cambridge University Press, Cambridge.

[Jeffrey, 2006] Jeffrey, R. C. (2006). *Formal Logic*. Hackett, Indianapolis, 4th edition.

[Jennings, 1994] Jennings, R. E. (1994). *The Genealogy of Disjunction*. Oxford University Press, Oxford.

[Johansson, 1937] Johansson, I. (1937). Der Minimalkalkuel, ein reduzierter intuitionistischer Formalismus. *Compositio Mathematica*, 4:119–136.

[Johnson, 1932] Johnson, W. E. (1932). Probability: The deductive and inductive problems. *Mind*, 41:409–423.

[Joosten and Visser, 2000] Joosten, J. and Visser, A. (2000). The interpretability logic of all reasonable arithmetical theories. *Erkenntnis*, 53:3–26.

[Joyce, 1998] Joyce, J. M. (1998). A nonpragmatic vindication of probabilism. *Philosophy of Science*, 65:575–603.

[Joyce, 1999] Joyce, J. M. (1999). *The Foundations of Causal Decision Theory*. Cambridge University Press, Cambridge.

[Joyce, 2009] Joyce, J. M. (2009). Accuracy and Coherence: Prospects for an Alethic Epistemology of Partial Belief. In Huber, F. and Schmidt-Petri, C., editors, *Degrees of Belief*, volume 342 of *Synthese Library*, pages 263–297. Springer, Dordrecht.

[Kalish et al. 1992] Kalish, D., Montague, R., and Mar, G. (1992). *Logic: Techniques of Formal Reasoning*, 2nd edition. Thompson Learning, London.

[Kallenberg, 2005] Kallenberg, O. (2005). *Probabilistic Symmetries and Invariance Principles*. Springer, Dordrecht. ISBN-10:0-387-25115-4.

[Kamp, 1968] Kamp, H. (1968). Tense logic and the theory of linear order. PhD thesis, University of California at Los Angeles.

[Kamp, 1971] Kamp, H. (1971). Formal properties of 'now'. *Theoria*, 37: 227–273.

[Kamp, 1975] Kamp, H. (1975). Two theories about adjectives. In Keenan, E. L., editor, *Formal Semantics of Natural Language*. Cambridge University Press, Cambridge.

[Kamp, 1981] Kamp, H. (1981). The paradox of the heap. In Mönnich, U., editor, *Aspects of Philosophical Logic: Some Logical Forays into Central Notions of Linguistics and Philosophy*, pages 225–277. Reidel, Dordrecht.

[Kant, 1787] Kant, I. (1787). *Critik der reinen Vernunft*. J. F. Hartknoch, Riga, 2nd edition.

[Kaplan, 1968] Kaplan, D. (1968). Quantifying in. *Synthese*, 19(1):178–214.

[Kaplan, 1970] Kaplan, D. (1970). What is Russell's theory of descriptions? In Yourgrau, W. and Breck, A., editors, *Physics, Logic and History*, pages 277–288. Plenum Press, New York.

[Kaplan and Montague, 1960] Kaplan, D. and Montague, R. (1960). A paradox regained. *Notre Dame Journal of Formal Logic*, 1(3):79–90.

[Karttunen, 1977] Karttunen, L. (1977). Syntax and semantics of questions. *Linguistics and Philosophy*, 1(1):3–44.

[Katsuno and Mendelzon, 1989] Katsuno, H. and Mendelzon, A. O. (1989). A Unified View of Propositional Knowledge Base Updates. In *Proceedings of the 11th International Joint Conference on Artifical Intelligence*, volume 2, pages 1413–1419. Morgan Kaufmann Publishers Inc, San Francisco.

[Katsuno and Mendelzon, 1991a] Katsuno, H. and Mendelzon, A. O. (1991a). On the Difference between Updating a Knowledge Base and Revising It. In Allen, J. A., Fikes, R., and Sandewell, E., editors, *Principles of Knowledge Representation and Reasoning: Proceeding of the Second International Conference*, pages 387–394, Morgan Kaufmann, San Mateo, CA.

[Katsuno and Mendelzon, 1991b] Katsuno, H. and Mendelzon, A. O. (1991b). Propositional knowledge base revision revision and minimal change. *Artificial Intelligence*, 52(3):263–294.

[Katz and Olin, 2007] Katz, B. and Olin, D. (2007). A tale of two envelopes. *Mind*, 116:903–926.

[Keefe, 1998] Keefe, R. (1998). Vagueness by numbers. *Mind*, 107:565–579.

[Keefe, 2000] Keefe, R. (2000). *Theories of Vagueness*. Cambridge University Press, Cambridge.

[Keefe, 2003] Keefe, R. (2003). Unsolved problems with numbers: Reply to Smith. *Mind*, 112:291–293.

[Keefe and Smith, 1997] Keefe, R. and Smith, P., editors (1997). *Vagueness: A Reader*. MIT Press, Cambridge, MA.

[Keeney and Raiffa, 1993] Keeney, R. and Raiffa, H. ([1976] 1993). *Decisions with Multiple Objectives: Preferences and Value Tradeoffs*. Cambridge University Press, Cambridge.

[Keisler, 1970] Keisler, H. J. (1970). Logic with the quantifier 'there exist uncountably many'. *Annals of Mathematical Logic*, 1:1–93.

[Kelly, 1998] Kelly, K. (1998). Iterated belief revision, reliability, and inductive amnesia. *Erkenntnis*, 50(1):57–112.

[Kemeny, 1955] Kemeny, J. G. (1955). Fair bets and inductive probabilities. *The Journal of Symbolic Logic*, 20(3):263–273.

[Kemeny, 1963] Kemeny, J. G. (1963). Carnap's theory of probability and induction. In Schilpp, P. A., editor, *The Philosophy of Rudolf Carnap*, pages 711–738. Open Court, La Salle, IL.

[Keynes, 1921] Keynes, J. M. (1921). *A Treatise on Probability*. Macmillan, London.

[Kirk and Raven, 1957] Kirk, G. S. and Raven, J. E. (1957). *The Presocratic Philosophers: A Critical History with a Selection of Texts*. Cambridge University Press, Cambridge.

[Kleene, 1952] Kleene, S. C. (1952). *Introduction to Metamathematics*. North-Holland, Amsterdam.

[Klein, 1893] Klein, F. (1893). Vergleichende betrachtungen über neuere geometrische forschungen. *Mathematische Annalen*, 43:63–100.

[Kleinknecht, 1992] Kleinknecht, R. (1992). Mereologische strukturen der welt. *Wissenschaftliche Zeitschrift der Humboldt-Universität zu Berlin, Reihe Geistes- und Sozialwissenschaften*, 41:40–53.

[Koellner, 2010] Koellner, P. (2010). Strong logics of first and second order. *Bulletin of Symbolic Logic*, 16(1):1–36.

[Kolmogorov, 1956] Kolmogorov, A. N. (1956). *Foundations of the Theory of Probability*. Chelsea Publishing Company, New York.

[Kooi, 2003] Kooi, B. (2003). Knowledge, chance and change. PhD thesis, University of Groningen.

[Koons, 1994] Koons, R. C. (1994). A new solution to the sorites problem. *Mind*, 103:439–449.

[Körner, 1966] Körner, S. (1966). *Experience and Theory*. Routledge and Kegan Paul, London.

[Korzybski, 1933] Korzybski, A. (1933). *Science and Sanity*. International Non-Aristotelian Publishing Company, New York.

[Koslow, 1992] Koslow, A. (1992). *A Structuralist Theory of Logic*. Cambridge University Press, Cambridge.

[Kourousias and Makinson, 2007] Kourousias, G. and Makinson, D. (2007). Parallel interpolation, splitting, and relevance in belief change. *The Journal of Symbolic Logic*, 72(3):994–1002.

[Kraus and Lehmann, 1988] Kraus, S. and Lehmann, D. (1988). Knowledge, belief and time. *Theoretical Computer Science*, 58(1-3):155–174.

[Krauss, 1969] Krauss, P. H. (1969). Representation of symmetric probability models. *The Journal of Symbolic Logic*, 34(2):183–193.

[Kreisel, 1967] Kreisel, G. (1967). Informal rigor and completeness proofs. In Lakatos, I., editor, *Problems in the Philosophy of Mathematics*, pages 138–186. North-Holland, Amsterdam.

[Kreisel, 1969] Kreisel, G. (1969). Informal rigour and completeness proofs. In Hintikka, J., editor, *The Philosophy of Mathematics*, pages 78–94. Oxford University Press, London.

[Kremer and Kremer, 2003] Kremer, P. and Kremer, M. (2003). Some supervaluation-based consequence relations. *Journal of Philosophical Logic*, 32:225–244.

[Kreps, 1988] Kreps, D. (1988). *Notes on the Theory of Choice*. Westview Press, Boulder, CO.

[Kripke, 1959] Kripke, S. A. (1959). A completeness theorem in modal logic. *The Journal of Symbolic Logic*, 24(1):1–14.

[Kripke, 1963a] Kripke, S. A. (1963a). Semantical analysis of modal logic 1, normal propositional calculi. *Zeitschrift für Mathematische Logik und Grundlagen der Mathematik*, 9:113–116.

[Kripke, 1963b] Kripke, S. A. (1963b). Semantical considerations on modal logic. *Acta Philosophica Fennica*, 16:83–94.

[Kripke, 1965] Kripke, S. A. (1965). Semantical analysis of intuitionist logic I. In Crossley, J. N. and Dummett, M. A. E., editors, *Formal systems and Recursive Functions*, pages 92–129. North-Holland, Amsterdam.

[Kripke, 1972a] Kripke, S. A. (1972a). *Naming and Necessity*. Harvard University Press, Cambridge, MA.

[Kripke, 1972b] Kripke, S. A. (1972b). Naming and necessity. In Davidson, D. and Harman, G., editors, *Semantics of Natural Language*, pages 253–355, 763–769. Reidel, Dordrecht.

[Kripke, 1975a] Kripke, S. A. (1975a). An outline of a theory of truth. *The Journal of Philosophy*, 72:690–716.

[Kripke, 1975b] Kripke, S. A. (1975b). Outline of a theory of truth. In Martin, R. M., editor, *Recent Essays on Truth and the Liar Paradox*, pages 53–81. Clarendon Press, Oxford.

[Kripke, 1979] Kripke, S. A. (1979). Speaker's reference and semantic reference. In French, P. A., Uehling, Jr., T. E., and Wettstein, H. K., editors, *Contemporary Perspectives in the Philosophy of Language*, pages 6–27. University of Minnesota Press, Minnesota.

[Kunen, 1980] Kunen, K. (1980). *Set Theory, An Introduction to Independence Proofs*. North-Holland, Amsterdam.

[Kyburg Jr., 1961] Kyburg Jr., H. E. (1961). *Probability and the Logic of Rational Belief*. Wesleyan University Press, Middletown, CT.

[Lackey, 2007] Lackey, J. (2007). Norms of assertion. *Noûs*, 41:594–626.

[Lakoff, 1973] Lakoff, G. (1973). Hedges: A study in meaning criteria and the logic of fuzzy concepts. *Journal of Philosophical Logic*, 2:458–508.

[Lambert, 2001] Lambert, K. (2001). Free logics. In Goble, L., editor, *The Blackwell Guide to Philosophical Logic*. Blackwell, Oxford.

[Landes, 2009] Landes, J. (2009). The principle of spectrum exchangeability with inductive logic. PhD thesis, University of Manchester. Available at www.maths.manchester.ac.uk/~jeff/.

[Landes et al., 2008] Landes, J., Paris, J. B., and Vencovská, A. (2008). Some aspects of polyadic inductive logic. *Studia Logica*, 90:3–16.

[Landes et al., 2009a] Landes, J., Paris, J. B., and Vencovská, A. (2009a). Instantial relevance in polyadic inductive logic. In Ramanujam, R. and Sarukkai, S., editors, *Proceedings of the 3nd India Logic Conference, ICLA 2009, Chennai, India*, pages 162–169. Springer LNAI 5378.

[Landes et al., 2009b] Landes, J., Paris, J. B., and Vencovská, A. (2009b). Representation theorems for probability functions satisfying spectrum exchangeability in inductive logic. *International Journal of Approximate Reasoning*, 51(1):35–55.

[Landes et al., ta] Landes, J., Paris, J. B., and Vencovská, A. (t.a.). A survey of some recent results on spectrum exchangeability in polyadic inductive logic. *Synthese*. DOI:10.1007/s11229-009-9711-9.

[Landman, 1991] Landman, F. (1991). *Structures in Semantics*. Kluwer, Dordrecht.

[Lavine, 1998] Lavine, S. (1998). *Understanding the Infinite*. Harvard University Press, Cambridge, MA.

[Lawry, 2006] Lawry, J. (2006). *Modelling and Reasoning with Vague Concepts*. Springer, Berlin.

[Lehrer and Paxson, 1969] Lehrer, K. and Paxson, T. (1969). Knowledge: Undefeated justified true belief. *The Journal of Philosophy*, 66:225–237.

[Leibniz, 1966] Leibniz, G. W. (1966). *Logical Papers*. Clarendon Press, Oxford. Translated by G. H. R. Parkinson.

[Leitgeb, 2005] Leitgeb, H. (2005). What truth depends on. *Journal of Philosophical Logic*, 34:155–192.

[Leitgeb, 2007] Leitgeb, H. (2007). A new analysis of quasi-analysis. *Journal of Philosophical Logic*, 36:181–226.

[Leitgeb, 2010] Leitgeb, H. (2010). Reducing belief simpliciter to degrees of belief. Manuscript, Bristol.

[Leitgeb, ta] Leitgeb, H. (t.a.). Logic in general philosophy of science: old things and new things. *Synthese*.

[Leitgeb and Pettigrew, 2010a] Leitgeb, H. and Pettigrew, R. (2010a). An Objective Justification of Bayesianism I: Measuring Inaccuracy. *Philosophy of Science*,

77:201–235.

[Leitgeb and Pettigrew, 2010b] Leitgeb, H. and Pettigrew, R. (2010b). An Objective Justification of Bayesianism II: The Consequences of Minimizing Inaccuracy. *Philosophy of Science*, 77:236–272.

[Lemmon, 1965] Lemmon, E. J. (1965). *Beginning Logic*. Thomas Nelson and Sons, London.

[Lemmon et al., 1977] Lemmon, E. J., Scott, D., and Segerberg, K. (1977). *The Lemmon Notes: An Introduction to Modal Logic*, volume 11 of *American Philosophical Quarterly Series*. Blackwell, Oxford.

[Lenzen, 1978] Lenzen, W. (1978). *Recent Work in Epistemic Logic*. North-Holland, Amsterdam.

[Leonard and Goodman, 1940] Leonard, H. and Goodman, N. (1940). The calculus of individuals and its uses. *The Journal of Symbolic Logic*, 5:45–55.

[Leśniewski, 1916] Leśniewski, S. (1916). *Podstawy ogólnej teoryi mnogości. I [On the foundation of mathematics]*. Prace Polskiego Kola Naukowego w Moskwie, Moskow.

[Levi, 1980] Levi, I. (1980). *The Enterprise of Knowledge: An Essay on Knowledge, Credal Probability, and Chance*. MIT Press, Cambridge, MA.

[Levi, 1991] Levi, I. (1991). *The Fixation of Belief and its Undoing: Changing Beliefs Through Enquiry*. Cambridge University Press, Cambridge.

[Levi, 1996] Levi, I. (1996). *For the Sake of the Argument*. Cambridge University Press, Cambridge.

[Levi, 2003] Levi, I. (2003). Counterexamples to Recovery and the Filtering Condition. *Studia Logica*, 73(2):209–218.

[Levi, 2004] Levi, I. (2004). *Mild Contraction*. Oxford University Press, Oxford.

[Lewis, 1917] Lewis, C. I. (1917). The issues concerning material implication. *Journal of Philosophy, Psychology, and Scientific Methods*, 14:350–356.

[Lewis, 1918] Lewis, C. I. (1918). *A Survey of Symbolic Logic*. University of California Press, Berkeley, CA.

[Lewis and Langford, 1932] Lewis, C. I. and Langford, H. (1932). *Symbolic Logic*. Century, New York.

[Lewis, 1969] Lewis, D. K. (1969). *Convention*. Harvard University Press, Cambridge, MA.

[Lewis, 1970a] Lewis, D. K. (1970a). General semantics. *Synthese*, 22:18–67.

[Lewis, 1970b] Lewis, D. K. (1970b). Nominalistic set theory. *Noûs*, 4:225–240.

[Lewis, 1973] Lewis, D. K. (1973). *Counterfactuals*. Blackwell, Oxford.

[Lewis, 1975] Lewis, D. K. (1975). Languages and language. *Minnesota Studies in the Philosophy of Science*, 7:3–35.

[Lewis, 1976] Lewis, D. K. (1976). Probabilities of conditionals and conditional probabilities. *The Philosophical Review*, 85(3):297–315. Reprinted, with postscript, in [Jackson, 1991, 76–101].

[Lewis, 1979] Lewis, D. K. (1979). Scorekeeping in a language game. *Journal of Philosophical Logic*, 8:339–359.

[Lewis, 1980] Lewis, D. K. (1980). A subjectivist's guide to objective chance. In Jeffrey, R. C., editor, *Studies in Inductive Logic and Probability*, volume II, pages 263–293. University of California Press, Berkeley, CA.

[Lewis, 1986a] Lewis, D. K. (1986a). *On the Plurality of Worlds*. Basil Blackwell, Oxford.

[Lewis, 1986b] Lewis, D. K. (1986b). Probabilities of conditionals and conditional probabilities ii. *The Philosophical Review*, 95(4):581–589.

[Lewis, 1991] Lewis, D. K. (1991). *Parts of Classes*. Blackwell, Oxford.

[Lewis, 1994] Lewis, D. K. (1994). Humean Supervenience Debugged. *Mind*, 103:473–490.

[Lewis, 1999] Lewis, D. K. (1999). Why Conditionalize? In *Essays in Metaphysics and Epistemology*, Cambridge Studies in Philosophy, pages 403–407. Cambridge University Press, Cambridge.

[Libardi, 1994] Libardi, M. (1994). Applications and limits of mereology. from the theory of parts to the theory of wholes. *Axiomathes*, 1:13–54.

[Lihoreau, 2008] Lihoreau, F., editor (2008). *Knowledge and Questions*. Grazer Philosophische Studien 77.

[Lindström, 1969] Lindström, P. (1969). On extensions of elementary logic. *Theoria*, 35:1–11.

[Lindström, 1991] Lindström, S. (1991). A Semantic Approach to Nonmonotonic Reasoning: Inference Operations and Choice. Uppsala Prints and Preprints in Philosophy 6, Department of Philosophy, University of Uppsala.

[Linnebo, 2003] Linnebo, Ø. (2003). Plural quantification exposed. *Noûs*, 37(1):71–92.

[Linnebo, ta] Linnebo, Ø. (t.a.). Pluralities and sets. Forthcoming in *The Journal of Philosophy*.

[Linnebo and Nicolas, 2008] Linnebo, Ø. and Nicolas, D. (2008). Superplurals in English. *Analysis*, 68(3):186–197.

[Linnebo and Rayo, 2011] Linnebo, Ø. and Rayo, A. (2011). Hierarchies ontological and ideological. *Mind*.

[Linsky, 1971] Linsky, L., editor (1971). *Reference and Modality*. Oxford University Press, Oxford.

[Lismont and Mongin, 2003] Lismont, L. and Mongin, P. (2003). Strong completeness theorems for weak logics of common belief. *Journal of Philosophical Logic*, 32(2):115–137.

[List and Pettit, 2011] List, Christian and Philip Pettit, 2011. *Group Agency: The Possibility, Design, and Status of Corporate Agents*. Oxford: Oxford University Press.

[Liu, 2008] Liu, F. (2008). Changing for the better: Preference dynamics and agent diversity. PhD thesis, Institute for logic, language and computation (ILLC).

[Löwe and Müller, ta] Löwe, B. and Müller, T. (t.a.). Data and phenomena in conceptual modelling. *Synthese*.

[Lowe, 1996] Lowe, E. J. (1996). Conditional probability and conditional beliefs. *Mind*, 105:603–615.

[Luce, 1956] Luce, R. D. (1956). Semi-orders and a theory of utility discrimination. *Econometrica*, 24:178–191.

[Łukasiewicz, 1970] Łukasiewicz, J. (1970). On three-valued logic. In Borkowski, L., editor, *Jan Łukasiewicz: Selected Works*, pages 87–88. North-Holland, Amsterdam. Originally published in Polish in 1920.

[Łukasiewicz and Tarski, 1930] Łukasiewicz, J. and Tarski, A. (1930). Untersuchungen über den aussagenkalkül. *Comptes rendus des séances de la Société des Sciences et des Lettres de Varsovie*, cl. 3, 23:1–21, 30–50. Reprinted in [Tarski, 1983a, 38–59].

[MacColl, 1906] MacColl, H. (1906). *Symbol Logic and Its Applications*. Logmans, Green and Co., London.

[MacFarlane, 2010] MacFarlane, J. (2010). Fuzzy epistemicism. In [Dietz and Moruzzi, 2010, 438–463].

[MacFarlane, ta] MacFarlane, J. (t.a.). Epistemic modals are assessment-sensitive. In Weatherson, B. and Egan, A., editors, *Epistemic Modality*. Oxford University Press, Oxford.

[Machina, 1976] Machina, K. F. (1976). Truth, belief, and vagueness. *Journal of Philosophical Logic*, 5:47–78.

[Maher, 1993] Maher, P. (1993). *Betting on Theories*. Cambridge Studies in Probability, Induction, and Decision Theory. Cambridge University Press, Cambridge.

[Maher, 2001] Maher, P. (2001). Probabilities for multiple properties: The models of hesse, carnap and kemeny. *Erkenntnis*, 55:183–216.

[Maher, 2006] Maher, P. (2006). A conception of inductive logic. *Philosophy of Science*, 73:513–520.

[Makinson, 1987] Makinson, D. (1987). On the status of the Postulate of Recovery in the logic of theory change. *Journal of Philosophical Logic*, 16(4): 383–394.

[Makinson, 1997] Makinson, D. (1997). On the force of some apparent counter-examples to Recovery. In Valdés, E. G., editor, *Normative Systems in Legal and Moral Theory*, pages 475–481. Duncker and Humblot, Berlin. Festschrift for Carlos Alchourrón and Eugenio Bulygin.

[Mann et al., 2010] Mann, A., Sandu, G., and Sevenster, M. (2010). *The Game of Logic: A New Approach to Independence-Friendly Logic*. Cambridge University Press, Cambridge.

[Marcus, 1946] Marcus, R. B. (1946). A functional calculus of first order based on strict implication. *The Journal of Symbolic Logic*, 11:1–16.

[Marcus, 1947] Marcus, R. B. (1947). Identity of individuals in a strict functional calculus of second order. *The Journal of Symbolic Logic*, 12:12–15.

[Mares, 2004] Mares, E. D. (2004). *Relevant Logic: A Philosophical Interpretation*. Cambridge University Press, Cambrdge.

[Mares, taa] Mares, E. D. (t.a.a). Conjunction and relevance. *Journal of Logic and Computation*.

[Mares, tab] Mares, E. D. (t.a.b). The nature of information: a relevant approach. *Synthese*.

[Mares et al., ta] Mares, E. D., Seligman, J., and Restall, G. (t.a.). Situation theory 2: Constraints and channels. In van Benthem, J. F. A. K. and ter Meulen, A., editors, *Handbook of Logic and Language*. Elsivier, Amsterdam, 2nd edition.

[Martin, 1943] Martin, R. M. (1943). A homogeneous system of formal logic. *The Journal of Symbolic Logic*, 8:1–23.

[Martin, 1958] Martin, R. M. (1958). *Truth and Denotation*. Routledge and Kegan Paul, London.

[Martin, 1965] Martin, R. M. (1965). Of time and the null individual. *The Journal of Philosophy*, 62:723–736.

[Massey, 1969] Massey, G. J. (1969). Tense logic! Why bother? *Noûs*, 3:17–32.

[Mates, 1972] Mates, B. (1972). *Elementary Logic*. Oxford University Press, Oxford and New York, 2nd edition.

[Mautner, 1946] Mautner, F. I. (1946). An extension of Klein's Erlanger program. *American Journal of Mathematics*, 68:345–384.

[McArthur, 1976] McArthur, R. P. (1976). *Tense Logic*, volume 111 of *Synthese library*. Reidel, Dordrecht.

[McCarty, 2008] McCarty, D. C. (2008). Completeness and incompleteness for intuitionistic logic. *The Journal of Symbolic Logic*, 73:1315–1327.

[McGee, 1985a] McGee, V. (1985a). A counterexample to modus ponens. *The Journal of Philosophy*, 82:462–471.

[McGee, 1985b] McGee, V. (1985b). How truth-like can a predicate be? A negative result. *Journal of Philosophical Logic*, 14:399–410.

[McGee, 1989] McGee, V. (1989). Conditional probabilities and compounds of conditionals. *The Philosophical Review*, 98:485–541.

[McGee, 1991] McGee, V. (1991). *Truth, Vagueness and Paradox. An essay on the logic of truth*. Hackett, Indianapolis.

[McGee, 1992] McGee, V. (1992). Maximal consistent sets of instances of Tarski's schema (T). *Journal of Philosophical Logic*, 21:235–241.

[McGee, 1997] McGee, V. (1997). How we learn mathematical language. *The Philosophical Review*, 106(1):35–68.

[McGee and McLaughlin, 1995] McGee, V. and McLaughlin, B. (1995). Distinctions without a difference. *Southern Journal of Philosophy*, (suppl.) 33: 203–251.

[McKinsey, 1941] McKinsey, J. C. C. (1941). A solution to the decision problem for the Lewis systems S2 and S4, with an application to topology. *The Journal of Symbolic Logic*, 6:117–134.

[Meier, ta] Meier, M. (t.a.). An infinitary probability logic for type spaces. *Israel Journal of Mathematics*.

[Meinong, 1960] Meinong, A. (1960). The theory of objects. In Chisholm, R., editor, *Realism and the Background of Phenomenology*. Free Press, Glencoe, IL.

[Mellor, 1998] Mellor, D. H. (1998). *Real Time II*. Routledge, London.

[Meyer et al., 2002] Meyer, T., Heidema, J., Labuschagne, W., and Leenen, L. (2002). Systematic Withdrawal. *Journal of Philosophical Logic*, 31(5):415–443.

[Miller, 2009] Miller, B. (2009). Existence. In Zalta, E. editor, *Stanford Encyclopedia of Philosophy*. http://plato.stanford.edu/archives/fall2009/entries/existence/.

[Miller, 1996] Miller, D. W. (1996). Propensities and Indeterminism. In O'Hear, A., editor, *Karl Popper: Philosophy and Problems*, pages 121–147. Cambridge University Press, Cambridge.

[Milne, 2008] Milne, P. (2008). Betting on fuzzy and many-valued propositions. In Pelis, M., editor, *The Logica Yearbook 2008*, pages 137–146. College Publications, London.

[Monk, 1976] Monk, J. D. (1976). *Mathematical Logic*. Springer, Berlin.

[Montagna and Mancini, 1994] Montagna, F. and Mancini, A. (1994). A minimal predictive set theory. *Notre Dame Journal of Formal Logic*, 35:186–203.

[Montague, 1960] Montague, R. (1960). Pragmatics. In *Formal Philosophy: Selected Papers of Richard Montague*. Yale University Press, New Haven, NJ.

[Montague, 1963] Montague, R. (1963). Syntactical treatments of modality, with corollaries on reflexion principles and finite axiomatizability. *Acta Philosophica Fennica*, 16:153–167. Reprinted in [Montague, 1974, 286–302].

[Montague, 1970] Montague, R. (1970). English as a formal language. In Thomason, R. H., editor, *Formal Philosophy: Selected Papers of Richard Montague*, pages 188–221. Yale University Press, New Haven and London.

[Montague, 1974] Montague, R. (1974). *Formal Philosophy*. Yale University Press, New Haven and London.

[Morreau, 1992] Morreau, M. (1992). Epistemic semantics for counterfactuals. *Journal of Philosophical Logic*, 21(1):33–62.

[Mortensen and Nerlich, 1978] Mortensen, C. and Nerlich, G. (1978). Physical topology. *Journal of Philosophical Logic*, 7:209–223.

[Mostowski, 1957] Mostowski, A. (1957). On a generalization of quantifiers. *Fundamenta Mathematicae*, 44:12–36.

[Müller, taa] Müller, T. (t.a.a). Formal methods in the philosophy of natural science. In Stadler, F., editor, *The Present Situation in the Philosophy of Science*. Springer, Dordrecht.

[Müller, tab] Müller, T. (t.a.b). Towards a theory of limited indeterminism in branching space-times. *Journal of Philosophical Logic*. DOI = 10.1007/s10992-010-9138-2.

[Nalebuff, 1989] Nalebuff, B. (1989). The other person's envelope is always greener. *Journal of Economic Perspectives*, 3:171–181.

[Nayak, 1994] Nayak, A. (1994). Iterated belief change based on epistemic entrenchment. *Erkenntnis*, 41(3):353–390.

[Neale, 1990] Neale, S. (1990). *Descriptions*. MIT Press, Cambridge, MA.

[Niebergall, 2000] Niebergall, K.-G. (2000). On the logic of reducibility: axioms and examples. *Erkenntnis*, 53:27–61.

[Niebergall, 2005] Niebergall, K.-G. (2005). Zur nominalistischen behandlung der mathematik. In Steinbrenner, J., Scholz, O., and Ernst, G., editors, *Symbole, Systeme, Welten: Studien zur Philosophie Nelson Goodmans*, pages 235–260. Synchron Wissenschaftsverlag der Autoren, Heidelberg.

[Niebergall, 2007] Niebergall, K.-G. (2007). Zur logischen stärke von individuenkalkülen. In Bohse, H. and Walter, S., editors, *Ausgewählte Sektionsbeiträge der GAP. 6*. Sechster Internationaler Kongress der Gesellschaft für Analytische Philosophie, Berlin, 11–14 September 2006. (CD-ROM) Paderborn: mentis 2007.

[Niebergall, 2009a] Niebergall, K.-G. (2009a). Calculi of individuals and some extensions: an overview. In Hieke, A. and Leitgeb, H., editors, *Reduction – Abstraction – Analysis*, pages 335–354, Frankfurt, Paris, Lancaster, New Brunswick. Proceedings of the 31th International Ludwig Wittgenstein-Symposium in Kirchberg, 2008, Ontos Verlag.

[Niebergall, 2009b] Niebergall, K.-G. (2009b). On 2nd order calculi of individuals. *Theoria*, 24(2):169–202.

[Nix and Paris, 2006] Nix, C. J. and Paris, J. B. (2006). A continuum of inductive methods arising from a generalized principle of instantial relevance. *Journal of Philosophical Logic*, 35(1):83–115.

[Nix and Paris, 2007] Nix, C. J. and Paris, J. B. (2007). A note on binary inductive logic. *Journal of Philosophical Logic*, 36(6):735–771.

[Nolan, 2003] Nolan, D. (2003). Defending a possible-worlds account of indicative conditionals. *Philosophical Studies*, 116:215–269.

[Noonan, 2009] Noonan, H. (2009). Identity. In Zalta, E. editor, *Stanford Encyclopedia of Philosophy*. http://plato.stanford.edu/archives/win2009/entries/identity/.

[Nover and Hájek, 2004] Nover, H. and Hájek, A. (2004). Vexing expectations. *Mind*, 113:237–249.

[Oaklander and Smith, 1994] Oaklander, N. and Smith, Q., editors (1994). *The New Theory of Time*. Yale University Press, New Haven, CT.

[Øhrstrøm and Hasle, 1995] Øhrstrøm, P. and Hasle, P. F. V. (1995). *Temporal Logic—from Ancient Ideas to Artificial Intelligence*, volume 57 of *Studies in Linguistics and Philosophy*. Kluwer, Dordrecht.

[Oliver and Smiley, 2005] Oliver, A. and Smiley, T. J. (2005). Plural descriptions and many-valued functions. *Mind*, 114:1039–1068.

[Olsson, 2003] Olsson, E. J. (2003). Belief Revision, Rational Choice and the Unity of Reason. *Studia Logica*, 73(2):219–240.

[Orłowska, 1985] Orłowska, E. (1985). Semantics of vague concepts. In Dorn, G. and Weingartner, P., editors, *Foundations of Logic and Linguistics: Problems and Their Solutions*, pages 465–482. Plenum Press, New York.

[Osborne, 2004] Osborne, M. J. (2004). *An introduction to game theory*. Oxford University Press, Oxford.

[Osborne and Rubinstein, 1994] Osborne, M. J. and Rubinstein, A. (1994). *A Course in Game Theory*. MIT Press, Cambridge, MA.

[Ostertag, 1998] Ostertag, G. (1998). *Definite Descriptions: A Reader*. MIT Press, Cambridge, MA.

[Pacuit, t.a.] Pacuit, E. (t.a.). Logics of informational attitudes and informative actions. *Journal of the Indian Council of Philosophical Research*.

[Pagin, 2010] Pagin, P. (2010). Vagueness and central gaps. In [Dietz and Moruzzi, 2010, 254–272].

[Parikh, 1999] Parikh, R. (1999). Belief revision and language splitting. In *Proc. Logic, Language and Computation*, pages 266–278. CSLI.

[Parikh, 2008a] Parikh, R. (2008a). Beth definability, interpolation and language splitting. *Synthese*, 179(2):211–221.

[Parikh, 2008b] Parikh, R. (2008b). Sentences belief and logical omniscience or what does deduction tell us? *The Review of Symbolic Logic*, 1(4):514–529.

[Paris, 1994] Paris, J. B. (1994). *The Uncertain Reasoner's Companion*. Cambridge University Press, Cambridge.

[Paris, 1999] Paris, J. B. (1999). Common sense and maximum entropy. *Synthese*, 117:75–93.

[Paris, 2001] Paris, J. B. (2001). On the distribution of probability functions in the natural world. In Hendricks, V. F., Pedersen, S. A., and Jørgensen, K. F., editors, *Probability Theory: Philosophy, Recent History and Relations to Science*, pages 125–145. Synthese Library 297.

[Paris and Vencovská, 1989] Paris, J. B. and Vencovská, A. (1989). On the applicability of maximum entropy to inexact reasoning. *International Journal of Approximate Reasoning*, 3(1):1–34.

[Paris and Vencovská, 1990] Paris, J. B. and Vencovská, A. (1990). A note on the inevitability of maximum entropy. *International Journal of Approximate Reasoning*, 4(3):183–224.

[Paris and Vencovská, 2001] Paris, J. B. and Vencovská, A. (2001). Common sense and stochastic independence. In Corfield, D. and Williamson, J., editors, *Foundations of Bayesianism*, pages 203–240. Kluwer Academic Press, Amsterdam.

[Paris and Vencovská, 2009] Paris, J. B. and Vencovská, A. (2009). A general representation theorem for probability functions satisfying spectrum exchangeability. In Ambros-Spies, K., Löwe, B., and Merkle, W., editors, *CiE 2009, Springer LNCS 5635*, pages 379–388.

[Paris and Vencovská, ta] Paris, J. B. and Vencovská, A. (t.a.). Symmetry's end?, *Erkenntnis*.

[Paris and Vencovská, unp.] Paris, J. B. and Vencovská, A. (unpublished). Symmetry principles in polyadic inductive logic. To be submitted to the *Journal of Logic, Language and Information*.

[Parsons, 1977] Parsons, C. (1977). What Is the Iterative Conception of Set? In Butts, R. E. and Hintikka, J., editors, *Logic, Foundations of Mathematics, and Computability Theory*, pages 335–367. Reidel, Dordrecht. Reprinted in [Benacerraf and Putnam, 1983] and [Parsons, 1983a].

[Parsons, 1983a] Parsons, C. (1983a). *Mathematics in Philosophy*. Cornell University Press, Ithaca, NY.

[Parsons, 1983b] Parsons, C. (1983b). Sets and modality. In *Mathematics in Philosophy*, pages 298–341. Cornell University Press, Cornell, NY.

[Parsons, 1990] Parsons, C. (1990). The structuralist view of mathematical objects. *Synthese*, 84:303–346.

[Parsons, 2008] Parsons, C. (2008). *Mathematical Thought and Its Objects*. Cambridge University Press, Cambridge.

[Parsons, 1980] Parsons, T. (1980). *Nonexistent Objects*. Yale University Press, New Haven, CT.

[Parsons, 2000] Parsons, T. (2000). *Indeterminate Identity: Metaphysics and Semantics*. Clarendon Press, Oxford.

[Pawlak, 1991] Pawlak, Z. (1991). *Rough Sets: Theoretical Aspects of Reasoning about Data*. Kluwer, Dordrecht.

[Peano, 1891] Peano, G. (1891). Sul concetto de numero. *Revista di Matematica*, 1:87–102, 256–267.

[Pearl and Goldszmidt, 1996] Pearl, J. and Goldszmidt, M. (1996). Qualitative probabilities for default reasoning, Belief Revision, and causal modeling. *Artificial Intelligence*, 84(1–2):57–112.

[Pedersen, 2008] Pedersen, A. P. (2008). Rational Choice and Formal Epistemology. Master's thesis, Carnegie Mellon University, Department of Philosophy.

[Pelletier, 1979] Pelletier, F. J., editor (1979). *Mass Terms: Some Philosophical Problems*. Reidel, Dordrecht.

[Peréz-Montoro, 2007] Peréz-Montoro, M. (2007). *The Phenomenon of Information: A Conceptual Approach to Information Flow*. Rowman and Littlefield, Lanham, MD.

[Perry, 1970] Perry, J. (1970). The same F. *The Philosophical Review*, 79: 191–200.

[Perry, 1977] Perry, J. (1977). Frege on demonstratives. *The Philosophical Review*, 86:474–497.

[Peterson, 2008] Peterson, M. (2008). *Non-Bayesian Decision Theory: Beliefs and Desires as Reasons for Action*. Springer, New York.

[Peterson, 2009] Peterson, M. (2009). *An Introduction to Decision Theory*. Cambridge University Press, Cambridge.

[Piccone and Rubinstein, 1997] Piccone, M. and Rubinstein, A. (1997). The absent-minded driver paradox: synthesis and responses. *Games and Economic Behavior*, 20:121–130.

[Pinkal, 1983] Pinkal, M. (1983). On the limits of lexical meaning. In Bäuerle, R., Schwarze, C., and von Stechow, A., editors, *Meaning, Use, and Interpretation of Language*. de Gruyther, Berlin.

[Pinkal, 1995] Pinkal, M. (1995). *Logic and Lexicon: The Semantics of the Indefinite*. Kluwer, Dordrecht.

[Plaza, 1989] Plaza, J. (1989). Logics of public communications. In Emrich, M. L., Pfeifer, M. S., Hadzikadic, M., and Ras, Z. W., editors, *Proceedings, 4th International Symposium on Methodologies for Intelligent Systems*, pages 201–216.

[Pnueli, 1977] Pnueli, A. (1977). The temporal logic of programs. In *18th Annual Symposium on Foundations of Computer Science*, pages 46–57.

[Pontow, 2004] Pontow, C. (2004). A note on the axiomatics of theories in parthood. *Data & Knowledge Engineering*, 50:195–213.

[Pontow and Schubert, 2006] Pontow, C. and Schubert, R. (2006). A mathematical analysis of theories of parthood. *Data & Knowledge Engineering*, 59:107–138.

[Popper, 1957] Popper, K. (1957). The propensity interpretation of the calculus of probability, and the Quantum Theory. In Körner, S., editor, *Observation and Interpretation*, Proceedings of the Ninth Symposium of the Colston Research Society. Butterworth, London.

[Popper, 1959] Popper, K. (1959). The propensity interpretation of probability. *British Journal for the Philosophy of Science*, 10:25–42.

[Popper, 1990] Popper, K. (1990). *A World of Propensities*. Thoemmes Press, Bristol.

[Post, 1921] Post, E. (1921). Introduction to a general theory of propositions. *American Journal of Mathematics*, 43:163–185.

[Pour-El and Kripke, 1967] Pour-El, M. B. and Kripke, S. A. (1967). Deduction-preserving 'recursive isomorphisms' between theories. *Bulletin of the American Mathematical Society*, 73:145–148.

[Pratt and Lemon, 1997] Pratt, I. and Lemon, O. (1997). Ontologies for plane, polygonal mereotopology. *Notre Dame Journal of Formal Logic*, 38:225–245.

[Pratt and Schoop, 1998] Pratt, I. and Schoop, D. (1998). A complete axiom system for polygonal mereotopology of the real plane. *Journal of Philosophical Logic*, 27:621–658.

[Pratt and Schoop, 2000] Pratt, I. and Schoop, D. (2000). Expressivity in polygonal, plane mereotopology. *The Journal of Symbolic Logic*, 65:822–838.

[Pratt-Hartmann and Schoop, 2002] Pratt-Hartmann, I. and Schoop, D. (2002). Elementary polyhedral mereotopology. *Journal of Philosophical Logic*, 31:469–498.

[Prawitz, 2006] Prawitz, D. (2006). *Natural Deduction: A Proof-Theoretic Study*. Dover, Mineola, NY.

[Priest, 1979] Priest, G. (1979). The logic of paradox. *Journal of Philosophical Logic*, 8:219–241.

[Priest, 1987] Priest, G. (1987). *In Contradiction*. Kluwer, Amsterdam.

[Priest, 1991] Priest, G. (1991). Sorites and identity. *Logique et Analyse*, 135–6: 293–296.

[Priest, 2005] Priest, G. (2005). *Towards Non-Being. The Logic and Metaphysics of Intentionality*. Clarendon Press, Oxford.

[Priest, 2006] Priest, G. (2006). *In Contradiction*. Oxford University Press, Oxford, 2nd edition.

[Priest, 2008] Priest, G. (2008). *An Introduction to Non-Classical Logic: From If to Is*. Cambridge University Press, Cambridge, 2nd edition.

[Prior, 1957] Prior, A. N. (1957). *Time and Modality*. Oxford University Press, Oxford.

[Prior, 1959] Prior, A. N. (1959). Thank goodness that's over. *Philosophy*, 34:12–17.

[Prior, 1960] Prior, A. N. (1960). The runabout inference ticket. *Analysis*, 21:38–39.

[Prior, 1963] Prior, A. N. (1963). Is the concept of referential opacity really necessary? *Acta Philosophica Fennica*, XVI:189–199. Proceedings of a Colloquium on Modal and Many-Valued Logics.

[Prior, 1967] Prior, A. N. (1967). *Past, Present and Future*. Oxford University Press, Oxford.

[Prior, 1976] Prior, A. N. (1976). *Papers in Logic and Ethics*. Duckworth, London.

[Prior and Fine, 1977] Prior, A. N. and Fine, K. (1977). *Worlds, Times and Selves*. Duckworth, London.

[Przełecki, 1976] Przełecki, M. (1976). Fuzziness and multiplicity. *Erkenntnis*, 10:371–380.

[Psillos, 1999] Psillos, S. (1999). *Scientific Realism. How science tracks truth*. Routledge, London and New York.

[Putnam, 1962] Putnam, H. (1962). The analytic and the synthetic. In Feigl, H. and Maxwell, G., editors, *Scientific Explanation, Space, and Time, Minnesota Studies in the Philosophy of Science, volume 3*, pages 358–397. University of Minnesota Press, Minneapolis. Reprinted in [Putnam, 1975, 33–69].

[Putnam, 1975] Putnam, H. (1975). *Mind, Language, and Reality. Philosophical Papers*, volume 2. Cambridge University Press, Cambridge.

[Putnam, 1980] Putnam, H. (1980). Models and reality. *The Journal of Symbolic Logic*, 45:464–483. Reprinted in [Benacerraf and Putnam, 1983, 421–444].

[Putnam, 1983] Putnam, H. (1983). Vagueness and alternative logic. *Erkenntnis*, 19:297–314.

[Putnam, 1985] Putnam, H. (1985). A quick Read is a wrong Wright. *Analysis*, 45:203.

[Quine, 1946] Quine, W. V. (1946). Concatenation as a basis for arithmetic. *The Journal of Symbolic Logic*, 10:105–114.

[Quine, 1953] Quine, W. V. (1953). On a supposed antinomy. In *The ways of paradox, and other essays*. Harvard University Press, Cambridge, MA.

[Quine, 1936] Quine, W. V. O. (1936). Truth by convention. In Lee, O. H., editor, *Philosophical Essays for A. N. Whitehead*, pages 90–124. Longmans, New York. Reprinted in [Quine, 1976, 77–106].

[Quine, 1940] Quine, W. V. O. (1940). *Mathematical Logic*. Harvard University Press, Cambridge, MA.

[Quine, 1943] Quine, W. v. O. (1943). Notes on existence and necessity. *The Journal of Philosophy*, XL:113–127.

[Quine, 1948] Quine, W. V. O. (1948). On what there is. In *From A Logical Point of View: Logico-Philosophical Essays*. Harper and Row, New York and Evanston, 2nd edition.

[Quine, 1951a] Quine, W. V. O. (1951a). *Mathematical Logic*. Harper and Row, New York, revised edition.

[Quine, 1951b] Quine, W. V. O. (1951b). Two dogmas of empiricism. *The Philosophical Review*, 60:20–43. Reprinted in [Quine, 1980, 20–46].

[Quine, 1956] Quine, W. V. O. (1956). Quantifiers and propositional attitudes. *The Journal of Philosophy*, 8(5):177–187.

[Quine, 1976] Quine, W. V. O. (1976). *The Ways of Paradox*. Harvard University Press, Cambridge, MA, 2nd edition.

[Quine, 1980] Quine, W. V. O. (1980). *From a Logical Point of View*. Cambridge University Press, Cambridge, MA, 2nd edition.

[Quine, 1982] Quine, W. V. O. (1982). *Methods of Logic*. Harvard University Press, Cambridge, MA, 4th edition.

[Quine, 1985] Quine, W. V. O. (1985). Events and reification. In LePore, E. and McLaughlin, B., editors, *Actions and Events*, pages 162–171. Blackwell, Oxford.

[Quine, 1986] Quine, W. V. O. (1986). *Philosophy of Logic*. Harvard University Press, Cambridge, MA, 2nd edition.

[Rabinowicz, 2003] Rabinowicz, W. (2003). Remarks on the absentminded driver. *Studia Logica*, 73:241–256.

[Rabinowicz and Lindström, 1994] Rabinowicz, W. and Lindström, S. (1994). How to model relational belief revision. In Prawitz, D. and Westerstahl, D., editors, *Logic and Philosophy of Science in Uppsala*. Kluwer, Amsterdam.

[Raffman, 1994] Raffman, D. (1994). Vagueness without paradox. *The Philosophical Review*, 103:43–74.

[Raffman, 1996] Raffman, D. (1996). Vagueness and context-sensitivity. *Philosophical Studies*, 81:175–192.

[Rakić, 1997] Rakić, N. (1997). Past, present, future, and special relativity. *British Journal for the Philosophy of Science*, 48:257–280.

[Ramsey, 1931a] Ramsey, F. P. (1931a). Philosophy. In Braithwaite, R. B., editor, *The Foundations of Mathematics and Other Logical Essays*, pages 263–269. Routledge and Kegan Paul, London.

[Ramsey, 1931b] Ramsey, F. P. (1931b). Truth and probability. In Braithwaite, R. B., editor, *The Foundations of Mathematics and Other Logical Essays*, pages 156–198. Routledge and Kegan Paul, London.

[Ramsey, 1990] Ramsey, F. P. (1990). General propositions and causality. In Mellor, D. H., editor, *Philosophical Papers*, pages 145–163. Cambridge University Press, Cambridge. Originally published 1929.

[Rantala, 1982] Rantala, V. (1982). Impossible worlds semantics and logical omniscience. *Intensional Logic: Theory and Applications*.

[Ray, 1973] Ray, P. (1973). Independence of Irrelevant Alternatives. *Econometrica*, 41(5):987–991.

[Rayo, 2006] Rayo, A. (2006). Beyond Plurals. In Rayo, A. and Uzquiano, G., editors, *Unrestricted Quantification: New Essays*. Oxford.

[Rayo, 2008] Rayo, A. (2008). Vague representation. *Mind*, 117:329–373.

[Rayo, 2010] Rayo, A. (2010). A metasemantic account of vagueness. In [Dietz and Moruzzi, 2010, 23–45].

[Rayo and Williamson, 2003] Rayo, A. and Williamson, T. (2003). A completeness theorem for unrestricted first-order languages. In [Beall, 2003, 331–356].

[Rayo and Yablo, 2001] Rayo, A. and Yablo, S. (2001). Nominalism through De-Nominalization. *Noûs*, 35(1):74–92.

[Read, 1988] Read, S. (1988). *Relevant Logic: The Philosophical Interpretation of Inference*. Blackwell, Oxford.

[Reichenbach, 1947] Reichenbach, H. (1947). *Elements of Symbolic Logic*. Macmillan, London.

[Reichenbach, 1949] Reichenbach, H. (1949). *The Theory of Probability*. University of California Press, Berkeley, CA.

[Rescher, 1969] Rescher, N. (1969). *Many-Valued Logic*. McGraw-Hill, New York.

[Rescher and Urquhart, 1971] Rescher, N. and Urquhart, A. (1971). *Temporal Logic*. Springer, Wien.

[Resnik, 1986] Resnik, M. (1986). Frege's Proof of Referentiality. In Haaparanta, L. and Hintikka, J., editors, *Frege Synthesized*. Reidel, Dordrecht.

[Richard, 2010] Richard, M. (2010). Indeterminacy and truth-value gaps. In [Dietz and Moruzzi, 2010, 464–481].

[Ridder, 2002] Ridder, L. (2002). *Mereologie. Ein Beitrag zur Ontologie und Erkenntnistheorie*. Klostermann, Frankfurt a. M.

[Rieger, 2006] Rieger, A. (2006). A simple theory of conditionals. *Analysis*, 66:233–240.

[Roelofsen, 2007] Roelofsen, F. (2007). Distributed knowledge. *Journal of Applied Non-Classical Logics*, 17(2):255–273.

[Roeper, 1997] Roeper, P. (1997). Region-based topology. *Journal of Philosophical Logic*, 26:251–309.

[Rolf, 1981] Rolf, B. (1981). Topics on vagueness. PhD thesis, Lunds Universitet.

[Romeijn, 2006] Romeijn, J. W. (2006). Analogical predictions for explicit similarity. *Erkenntnis*, 64(2):253–280.

[Rosenberg, 1970] Rosenberg, J. (1970). Notes on Goodman's nominalism. *Philosophical Studies*, 21:19–24.

[Rott, 1991] Rott, H. (1991). Two Methods of Constructing Contractions and Revisions of Knowledge Systems. *Journal of Philosophical Logic*, 20(2): 149–173.

[Rott, 1993] Rott, H. (1993). Belief Contraction in the Context of the General Theory of Rational Choice. *The Journal of Symbolic Logic*, 58(4):1426–1450.

[Rott, 2001] Rott, H. (2001). *Change, Choice and Inference: A Study of Belief Revision and Nonmonotonic Reasoning*. Oxford University Press, Oxford.

[Rott, 2003] Rott, H. (2003). Coherence and conservatism in the dynamics of belief ii: Iterated belief change without dispositional coherence. *Journal of Logic and Computation*, 1(13):111–145.

[Rott, 2004a] Rott, H. (2004a). A counterexample to six fundamental principles of belief formation. *Synthese*, 139(2):225–240.

[Rott, 2004b] Rott, H. (2004b). Stability, strength and sensitivity: Converting belief into knowledge. *Erkenntnis*, 61(2):469–493.

[Rott and Pagnucco, 1999] Rott, H. and Pagnucco, M. (1999). Severe Withdrawal (and Recovery). *Journal of Philosophical Logic*, 28(5):501–547.

[Routley and Meyer, 1972a] Routley, R. and Meyer, R. K. (1972a). Semantics for entailment II. *Journal of Philosophical Logic*, 1:53–73.

[Routley and Meyer, 1972b] Routley, R. and Meyer, R. K. (1972b). Semantics for entailment III. *Journal of Philosophical Logic*, 1:192–208.

[Routley and Meyer, 1973] Routley, R. and Meyer, R. K. (1973). Semantics for entailment. In Leblanc, H., editor, *Truth, Syntax, and Modality*. North-Holland, Amsterdam.

[Routley and Routley, 1972] Routley, R. and Routley, V. (1972). The semantics of first-degree entailment. *Noûs*, 6:335–395.

[Roy, t.a.] Roy, O. (t.a.). Epistemic logic and the foundations of decision and game theory. *Journal of the Indian Council of Philosophical Research*.

[Roy, 2006] Roy, T. (2006). Natural derivations for Priest, An Introduction to Non-classical Logic. *Australasian Journal of Logic*, 5:47–192.

[Rubinstein, 1989] Rubinstein, A. (1989). The electronic mail game: Strategic behavior under 'almost common knowledge'. *The American Economic Review*, 79(3):385–391.

[Russell, 1902] Russell, B. (1902). Letter to Frege. Printed in [van Heijenoort, 1967, 124–125].

[Russell, 1903] Russell, B. (1903). *The Principles of Mathematics*. Cambridge University Press, Cambridge.

[Russell, 1905a] Russell, B. (1905a). The existential import of propositions. *Mind*, 14:398–401.

[Russell, 1905b] Russell, B. (1905b). On denoting. *Mind*, 14:479–493.

[Russell, 1908] Russell, B. (1908). Mathematical logic as based on a theory of types. *American Journal of Mathematics*, 30:222–262.

[Russell, 1914] Russell, B. (1914). *On Our Knowledge of the External World*. Allen and Unwin, London.

[Russell, 1923] Russell, B. (1923). Vagueness. *Australasian Journal of Philosophy and Psychology*, 1:84–92. Reprinted in [Keefe and Smith, 1997, 61–8].

[Russell, 1926] Russell, B. (1926). *Our Knowledge of the External World*. Allen and Unwin, London, 2nd edition.

[Russell, 1956] Russell, B. (1956). Logical atomism. In Smith, R. C., editor, *Bertrand Russell. Logic and Knowledge. Essays 1901–1950*. Allen and Unwin, London.

[Russell, 1994] Russell, B. (1994). On meaning and denotation. In Urquhart, A. and Lewis, A. C., editors, *The Collected Papers of Bertrand Russell, Volume 4: Foundations of Logic 1903-1905*, pages 314–358. Routledge, London and New York.

[Ryle, 1979] Ryle, G. (1979). Bertrand Russell: 1872–1970. In Roberts, G. W., editor, *Bertrand Russell Memorial Volume*, pages 15–21. George Allen and Unwin, London.

[Sainsbury, 1986] Sainsbury, R. M. (1986). Degrees of belief and degrees of truth. *Philosophical Papers*, 15:97–106.

[Sainsbury, 1990] Sainsbury, R. M. (1990). Concepts without boundaries. Inaugural lecture, Kings College London. Reprinted in [Keefe and Smith, 1997, 251–264].

[Sainsbury, 1991] Sainsbury, R. M. (1991). Is there higher-order vagueness? *Philosophical Quarterly*, 41:167–182.

[Sainsbury, 2009] Sainsbury, M. (2009). *Paradoxes*, Third edition. Cambridge University Press, Cambridge, UK.

[Salmon, 2001] Salmon, N. (2001). The very possibility of language. a sermon on the consequences of missing Church. In Anderson, C. A. and Zelëny, M., editors, *Logic, Meaning and Computation: Essays in Memory of Alonzo Church*. Kluwer Academic Publishers, Dordrecht.

[Sandu, 1998] Sandu, G. (1998). If-logic and truth-definition. *Journal of Philosophical Logic*, 27:143–164.

[Sandu and Pietarinen, 2003] Sandu, G. and Pietarinen, A. (2003). Informationally independent connectives. In Mints, G. and Muskens, R., editors, *Logic, Language and Computation*, pages 23–41. CSLI Publications.

[Savage, 1954] Savage, L. J. (1954). *The Foundations of Statistics*. John Wiley & Sons, New York.

[Scheffler, 1979] Scheffler, I. (1979). *Beyond the Letter*. Routledge and Kegan Paul, London.

[Schiffer, 1972] Schiffer, S. R. (1972). *Meaning*. Oxford University Press, Oxford.

[Schiffer, 1999] Schiffer, S. R. (1999). The epistemic theory of vagueness. *Philosophical Perspectives*, 13:481–503.

[Schiffer, 2003] Schiffer, S. R. (2003). *The Things We Mean*. Clarendon Press, Oxford.

[Schuldenfrei, 1969] Schuldenfrei, R. (1969). Eberle on nominalism in non-atomic systems. *Noûs*, 3:427–430.

[Schwabhaüser et al., 1983] Schwabhaüser, W., Szmielew, W., and Tarski, A. (1983). *Metamathematische Methoden in der Geometrie*. Springer, Berlin.

[Schwartz, 1987] Schwartz, S. P. (1987). Intuitionism and sorites. *Analysis*, 47:179–183.

[Schwartz and Throop, 1991] Schwartz, S. P. and Throop, W. (1991). Intuitionism and vagueness. *Erkenntnis*, 34:347–356.

[Segerberg, 1971] Segerberg, K. (1971). *An Essay in Classical Modal Logic*. Filosofiska Institutionen vid Uppsala Universitet, Uppsala.

[Segerberg, 1995] Segerberg, K. (1995). Belief revision from the point of view of doxastic logic. *Logic Journal of the IGPL*, 3(4):535–553.

[Sen, 1969] Sen, A. (1969). Quasi-transitivity, rational choice and collective decisions. *The Review of Economic Studies*, 36(3):381–393.

[Sen, 1971] Sen, A. (1971). Choice Functions and Revealed Preference. *The Review of Economic Studies*, 38:307–317.

[Sen, 1977] Sen, A. (1977). Social Choice Theory: A Re-Examination. *Econometrica*, 45(1):53–89.

[Serchuk et al., ta] Serchuk, P., Hargreaves, I., and Zach, R. (t.a.). Vagueness, logic and use: four experimental studies on vagueness. *Mind and Language*.

[Sevenster, 2006] Sevenster, M. (2006). *Branches of Imperfect Information: Logic, Games, and Computation*. Universiteit van Amsterdam, ILLC.

[Sevenster and Sandu, 2010] Sevenster, M. and Sandu, G. (2010). Equilibrium semantics of languages of imperfect information. *Annals of Pure and Applied Logic*, 161(5):618–631.

[Shapiro, 1987] Shapiro, S. (1987). Principles of reflection and second-order logic. *Journal of Philosophical Logic*, 16:309–333.

[Shapiro, 1999] Shapiro, S. (1999). Do not claim too much: Second-order logic and first-order logic. *Philosophia Mathematica*, 7:42–64.

[Shapiro, 2000] Shapiro, S. (2000). *Foundations without Foundationalism: A Case for Second-Order Logic*. Oxford University Press, Oxford.

[Shapiro, 2005] Shapiro, S. (2005). Higher-order logic. In Shapiro, S., editor, *Oxford Handbook of Philosophy of Mathematics and Logic*, pages 751–780. Oxford University Press, Oxford.

[Shapiro, 2006] Shapiro, S. (2006). *Vagueness in Context*. Clarendon Press, Oxford.

[Sharvy, 1969] Sharvy, R. (1969). Things. *Monist*, 53:488–504.

[Shepard, 1973] Shepard, P. (1973). A finite arithmetic. *The Journal of Symbolic Logic*, 38:232–248.

[Shimony, 1955] Shimony, A. (1955). Coherence and the axioms of confirmation. *The Journal of Symbolic Logic*, 20:1–28.

[Shoesmith and Smiley, 1978] Shoesmith, D. J. and Smiley, T. J. (1978). *Multiple-Conclusion Logic*. Cambridge University Press, Cambridge.

[Shore and Johnson, 1980] Shore, J. E. and Johnson, R. W. (1980). Axiomatic derivation of the principle of maximum entropy and the principle of minimum cross-entropy. *IEEE Transactions on Information Theory*, IT-26:26–37.

[Sillari, 2005] Sillari, G. (2005). A logical framework for convention. *Synthese*, 147(2):379–400.

[Sillari, 2009] Sillari, G. (2009). Quantified logic of awareness and impossible possible worlds. *The Review of Symbolic Logic*, 1(04):514–529.

[Simon, 1982] Simon, H. (1982). *Models of Bounded Rationality*, volume 2. MIT Press, Cambridge, MA.

[Simons, 1982] Simons, P. (1982). Class, mass and mereology. *History and Philosophy of Logic*, 4:157–180.

[Simons, 1987] Simons, P. (1987). *Parts: A Study in Ontology*. Clarendon Press, Oxford.

[Simons, 1991] Simons, P. (1991). Free part-whole theory. In Lambert, K., editor, *Philosophical Applications of Free Logic*, pages 285–306. Oxford University Press, Oxford.

[Simons, 1992] Simons, P. (1992). Vagueness and ignorance. *Aristotelian Society*, (suppl.) 66:163–177.

[Skolem, 1920] Skolem, T. (1920). Logisch-kombinatorische untersuchungen über die erfüllbarkeit oder beweisbarkeit mathematischer sätze nebst einem theoreme über dichte mengen. Videnskapsselskapets skrifter I. Matematisk-naturvidenskabelig klasse 3.

[Skolem, 1923] Skolem, T. (1923). Einige bemerkungen zur axiomatischen begründung der mengenlehre. In *Matematikerkongressen i Helsingfors den 4-7 Juli 1922. Den femte skandinaviska matematikerkongressen, Redogörelse*, pages 217–232, Helsinki. Akademiska Bokhandeln. English translation by Stefan Bauer-Mengelberg in [van Heijenoort, 1967, 254–263].

[Skyrms, 1993] Skyrms, B. (1993). Analogy by similarity in hyper-Carnapian inductive logic. In Earman, J., Janis, A. I., Massey, G. J., and Rescher, N., editors, *Philosophical Problems of the Internal and External Worlds*, pages 273–282. University of Pittsburgh Press.

[Slote, 1966] Slote, M. (1966). The theory of important criteria. *The Journal of Philosophy*, 63:211–224.

[Smith, 2009] Smith, A. (2009). Kernel, cumulative, and safe contractions. Master's thesis, Department of Philosophy, Carnegie Mellon University.

[Smith, 1996] Smith, B. (1996). Mereotopology: A theory of parts and boundaries. *Data & Knowledge Engineering*, 20:287–303.

[Smith and Varzi, 2000] Smith, B. and Varzi, A. C. (2000). Fiat and bona fide boundaries. *Philosophy and Phenomenological Research*, 60:401–420.

[Smith, 2003] Smith, N. J. J. (2003). Vagueness by numbers? No worries. *Mind*, 112:283–290.

[Smith, 2008] Smith, N. J. J. (2008). *Vagueness and Degrees of Truth*. Oxford University Press, Oxford.

[Smith, 2010] Smith, N. J. J. (2010). Degree of belief is expected truth value. In [Dietz and Moruzzi, 2010, 491–506].

[Smullyan, 1948] Smullyan, A. F. (1948). Modality and descriptions. *The Journal of Symbolic Logic*, 13:31–37. Reprinted in [Linsky, 1971, 35–43].

[Smullyan, 1957] Smullyan, R. (1957). Languages in which self-reference is possible. *The Journal of Symbolic Logic*, 22:55–67.

[Soames, 1999] Soames, S. (1999). *Understanding Truth*. Oxford University Press, New York.

[Sobel, 1994] Sobel, J. H. (1994). *Taking Chances: Essays on Rational Choice*. Cambridge University Press, Cambridge.

[Sorensen, 1985] Sorensen, R. (1985). An argument for the vagueness of 'vague'. *Analysis*, 45:134–137.

[Sorensen, 1988] Sorensen, R. (1988). *Blindspots*. Clarendon Press, Oxford.

[Sorensen, 2001] Sorensen, R. (2001). *Vagueness and Contradiction*. Clarendon Press, Oxford.

[Spohn, 1988] Spohn, W. (1988). Ordinal conditional functions: A dynamic theory of epistemic states. In Harper, W. L. and Skyrms, B., editors, *Causation in Decision, Belief Change, and Statistics*, volume II, pages 105–134. Kluwer Academic Publishers, Amsterdam.

[Spohn, 1990] Spohn, W. (1990). A General Non-Probabilistic Theory of Inductive Reasoning. In Schachter, R. D., Levitt, T. S., Kanal, L. N., and Lemmer, J. F., editors, *Uncertainty in Artificial Intelligence*, volume 4. North-Holland, Amsterdam.

[Spohn, 1998] Spohn, W. (1998). A general non-probabilistic theory of inductive inference. In Harper, W. L. and Skyrms, B., editors, *Causation in Decision, Belief Change and Statistics*, pages 105–134. Reidel, Dordrecht.

[Spohn, 2010] Spohn, W. (2010). *Ranking Theory: A tool for epistemology*. Oxford University Press, Oxford.

[Stalker, 1994] Stalker, D., editor (1994). *Grue! The New Riddle of Induction*. Open Court, Chicago.

[Stalnaker, 1968] Stalnaker, R. (1968). A theory of conditionals. In Rescher, N., editor, *Studies in Logical Theory*, pages 98–112. Blackwell, Oxford.

[Stalnaker, 1970] Stalnaker, R. (1970). Probability and conditionals. *Philosophy of Science*, 37:64–80.

[Stalnaker, 1975] Stalnaker, R. (1975). Indicative conditionals. *Philosophia*, 5: 269–286.

[Stalnaker, 1994] Stalnaker, R. (1994). On the evaluation of solution concepts. *Theory and Decision*, 37(42).

[Stalnaker, 1998] Stalnaker, R. (1998). Belief revision in games: forward and backward induction. *Mathematical Social Sciences*, 36:31–56.

[Stalnaker, 2006] Stalnaker, R. (2006). On logics of knowledge and belief. *Philosophical Studies*, 128:169–199.

[Stalnaker, 2008] Stalnaker, R. (2008). *Our Knowledge of the Internal World*. Clarendon Press, Oxford.

[Stalnaker, 2009] Stalnaker, R. (2009). Iterated belief revision. *Erkenntnis*, 70: 189–209.

[Stanley and Williamson, 2001] Stanley, J. and Williamson, T. (2001). Knowing how. *The Journal of Philosophy*, pages 411–444.

[Strawson, 1950] Strawson, P. F. (1950). On referring. *Mind*, 59:320–344.

[Suppes, 1968] Suppes, P. (1968). The desirability of formalization in science. *The Journal of Philosophy*, 65:651–664.

[Szpilrajn, 1930] Szpilrajn, E. (1930). Sur l'extension de l'ordre partiel. *Fundamenta Mathematicae*, 16:386–389.

[Tappenden, 1993] Tappenden, J. (1993). The liar and the sorites paradoxes: towards a unified treatment. *The Journal of Philosophy*, 90:551–577.

[Tarski, 1929] Tarski, A. (1929). Foundations of the geometry of solids (les fondements de la geometrie de corps). *Annales de la Societé Polonaise de Mathématique, Krakow*, pages 29–33.

[Tarski, 1935] Tarski, A. (1935). Der Wahrheitsbegriff in den formalisierten Sprachen. *Studia Philosophica*, 1:261–405. English translation by J. H. Woodger as 'The Concept of Truth in Formalized Languages' in [Tarski, 1983a, 152–278].

[Tarski, 1936] Tarski, A. (1936). Über den begriff der logischen folgerung. *Actes du Congrès International de Philosophie Scientifique*, 7:1–11. English translation by J. H. Woodger in [Tarski, 1983a, 409-420].

[Tarski, 1949] Tarski, A. (1949). Arithmetical classes and types of boolean algebras. *Bulletin of the American Mathematical Society*, 55:63.

[Tarski, 1983a] Tarski, A. (1983a). *Logic, Semantics, Metamathematics*. Hackett, Indianapolis, 2nd edition. Translated by J. H. Woodger.

[Tarski, 1983b] Tarski, A. (1983b). On the concept of logical consequence. In Tarski, A. (1983). *Logic, Semantics, Meta-mathematics*, pages 409–420. Hackett, Indianapolis.

[Tarski, 1986] Tarski, A. (1986). What are logical notions? *History and Philosophy of Logic*, 7:143–154.

[Tarski and Lindenbaum, 1934] Tarski, A. and Lindenbaum, A. (1934). Über die beschränktheit der ausdrucksmittel deduktiver theorien. *Ergebnisse eines mathematischen Kolloquiums*, 7:15–22. English translation by J. H. Woodger in [Tarski, 1983a, 384–392].

[Tarski et al., 1953] Tarski, A., Mostowski, A., and Robinson, R. M. (1953). *Undecidable theories*. North-Holland, Amsterdam.

[Teller, 1976] Teller, P. (1976). Conditionalization, observation, and change of preference. In Harper, W. L. and Hooker, C. A., editors, *Foundations of Probability Theory, Statistical Inference, and Statistical Theories of Science*. Reidel, Dordrecht.

[Tennant, 2006] Tennant, N. (2006). New foundations for a relational theory of theory revision. *Journal of Philosophical Logic*, 35(5):489–528.

[Tennant, forthcoming] Tennant, Neil (forthcoming). Parts, Classes, and *Parts of Classes*: An Anti-Realist Reading of Lewisian Mereology *Synthese*.

[Thomason, 1970] Thomason, R. H. (1970). Indeterminist time and truth value gaps. *Theoria*, 36:264–281.

[Thomason, 1984] Thomason, R. H. (1984). Combinations of tense and modality. In [Gabbay and Guenthner, 1984, 135–165].

[Thomason, 2002] Thomason, R. H. (2002). Combinations of tense and modality. In [Gabbay and Guenthner, 2002, 205–234]. Reprint of [Thomason, 1984].

[Thomason, 1972] Thomason, S. K. (1972). Semantic analysis of tense logic. *The Journal of Symbolic Logic*, 37:150–158.

[Tichý, 1988] Tichý, P. (1988). *The Foundations of Frege's Logic*. Walter de Gruyter, Berlin.

[Trotsky, 1973] Trotsky, L. (1973). The ABC of dialectical materalism. In *Problems of Everyday Life & Other Writings on Culture and Science*. Monad Press, New York.

[Tye, 1990] Tye, M. (1990). Vague objects. *Mind*, 99:535–557.

[Tye, 1994] Tye, M. (1994). Sorites paradoxes and the semantics of vagueness. *Philosophical Perspectives*, 8:189–206.

[Tye, 1997] Tye, M. (1997). On the epistemic theory of vagueness. *Philosophical Issues*, 8:247–251.

[Uckelman and Uckelman, 2007] Uckelman, S. L. and Uckelman, J. (2007). Modal and temporal logics for abstract space–time structures. *Studies in History and Philosophy of Modern Physics*, 38(3):673–681.

[Unger, 1979] Unger, P. K. (1979). There are no ordinary things. *Synthese*, 41: 117–154.

[Unger, 1980] Unger, P. K. (1980). The problem of the many. *Midwest Studies in Philosophy*, 5:411–467.

[Unger, 1990] Unger, P. K. (1990). *Identity, Consciousness and Value*. Oxford University Press, Oxford.

[Urquhart, 1986] Urquhart, A. (1986). Many-valued logic. In Gabbay, D. M. and Guenthner, F., editors, *Handbook of Philosophical Logic*, volume III, pages 71–116. Kluwer, Dordrecht.

[Uzquiano, 2003] Uzquiano, G. (2003). Plural quantification and classes. *Philosophia Mathematica*, 11(1):67–81.

[van Benthem, 1982] van Benthem, J. F. A. K. (1982). The logical study of science. *Synthese*, 51:431–472.

[van Benthem, 1983] van Benthem, J. F. A. K. (1983). *The Logic of Time*. Reidel, Dordrecht.

[van Benthem, 1991] van Benthem, J. F. A. K. (1991). *The Logic of Time*. Kluwer, Dordrecht, 2nd edition.

[van Benthem, 2002] van Benthem, J. F. A. K. (2002). 'One is a lonely number': on the logic of communication. In Chatzidakis, Z., Koepke, P., and Pohlers, W., editors, *Logic Colloquium '02*, pages 96–129. ASL and A. K. Peters. Available at http://staff.science.uva.nl/~johan/Muenster.pdf.

[van Benthem, 2004a] van Benthem, J. F. A. K. (2004a). Dynamic logic for belief revision. *Journal of Applied Non-Classical Logics*, 14(2):129–155.

[van Benthem, 2004b] van Benthem, J. F. A. K. (2004b). What one may come to know. *Analysis*, 64(2):95–105.

[van Benthem, 2006] van Benthem, J. F. A. K. (2006). The epistemic logic of if games. In Auxier, R. E. and Hahn, L. E., editors, *The philosophy of Jaakko Hintikka*, Library of Living Philosophers, pages 481–513. Carus Publishing Company.

[van Benthem and Sarenac, 2004] van Benthem, J. F. A. K. and Sarenac, D. (2004). The geometry of knowledge. In J-Y Béziau, A. Costa Leite and A. Facchini, editors, *Aspects of Universal Logic*, Centre de Recherches Sémiologiques, Université de Neuchatel, pages 1–31.

[van Benthem et al., 2006] van Benthem, J. F. A. K., van Eijck, J., and Kooi, B. (2006). Logics of communication and change. *Information and Computation*, 204(11):1620–1662.

[van Deemter, 1996] van Deemter, K. (1996). The sorites fallacy and the context-dependence of vague predicates. In Makoto, M., Piñón, C., and de Swart, H., editors, *Quantifiers, Deduction, and Context*, pages 59–86. CSLI Publications, Stanford, CA.

[van Ditmarsch, 2005] van Ditmarsch, H. P. (2005). Prolegomena to dynamic logic for belief revision. *Synthese*, 147(2):229–275.

[van Ditmarsch et al., 2007] van Ditmarsch, H. P., van der Hoek, W., and Kooi, B. (2007). *Dynamic Epistemic Logic*. Springer, Dordrecht.

[van Fraassen, 1976] van Fraassen, B. C. (1976). Probabilities of conditionals. In Harper, W. L. and Hooker, C. A., editors, *Foundations of Probability Theory, Statistical Inference, and Statistical Theories of Science*, volume I, pages 261–301. Reidel, Dordrecht.

[van Fraassen, 1980] van Fraassen, B. C. (1980). *The Scientific Image*. Clarendon Library of Logic and Philosophy. Clarendon Press, Oxford.

[van Fraassen, 1984] van Fraassen, B. C. (1984). Belief and the will. *The Journal of Philosophy*, 81:235–256.

[van Fraassen, 1995] van Fraassen, B. C. (1995). Fine-grained opinion, probability, and the logic of full belief. *Journal of Philosophical Logic*, 24(4):349–377.

[van Heijenoort, 1967] van Heijenoort, J., editor (1967). *From Frege to Gödel*. Harvard University Press, Cambridge, MA.

[van Inwagen, 1994] van Inwagen, P. (1994). Composition as identity. *Philosophical Perspectives*, 8:207–220.

[van Lambalgen and Hamm, 2005] van Lambalgen, M. and Hamm, F. (2005). *The Proper Treatment of Events*. Blackwell, Oxford.

[van Rooij, 2009] van Rooij, R. (2009). Vagueness and linguistics. In Ronzitti, G., editor, *The Vagueness Handbook*. Springer, Berlin.

[van Rooij, 2010] van Rooij, R. (2010). Vagueness, tolerance, and non-transitive entailment. Unpublished manuscript.

[Vanderschraaf and Sillari, 2009] Vanderschraaf, P. and Sillari, G. (2009). Common knowledge. In Zalta, E. N., editor, *The Stanford Encyclopedia of Philosophy*. Stanford University, Stanford, CA, spring 2009 edition.

[Varzi, 1996] Varzi, A. C. (1996). Parts, wholes, and part-whole relations: the prospects of mereotopology. *Data & Knowledge Engineering*, 20: 259–286.

[Varzi, 2003] Varzi, A. C. (2003). Higher-order vagueness and the vagueness of 'vague'. *Mind*, 112:295–299.

[Varzi, 2005] Varzi, A. C. (2005). The vagueness of 'vague': Rejoinder to Hull. *Mind*, 114:695–702.

[Varzi, 2007] Varzi, A. C. (2007). Supervaluationism and its logics. *Mind*, 116: 633–676.

[Varzi, 2011] Varzi, A. C. (2011). Mereology. In Zalta, E. editor, *Stanford Encyclopedia of Philosophy*. http://plato.stanford.edu/archives/spr2011/entries/mereology/.

[Vaught, 1964] Vaught, R. L. (1964). The completeness of logic with the added quantifier 'there are uncountably many.'. *Fundamenta Mathematicae*, 54: 303–304.

[Veblen, 1904] Veblen, O. (1904). A system of axioms for geometry. *Transactions of the American Mathematica Society*, 5:343–384.

[Vencovská, 2006] Vencovská, A. (2006). Binary induction and Carnap's continuum. In *Proceedings of the 7th Workshop on Uncertainty Processing (WUPES), Mikulov, Czech Republic*. Available at www.utia.cas.cz/files/mtr/articles/data/vencovska.pdf.

[Venn, 1876] Venn, J. (1876). *The Logic of Chance*. Macmillan and Co., London, 2nd edition.

[Visser, 1989] Visser, A. (1989). Semantics and the liar paradox. In Gabby, D. M. and Guenthner, F., editors, *Handbook of Philosophical Logic*, volume IV, pages 617–706.

[von Mises, 1957] von Mises, R. (1957). *Probability, Statistics, and Truth*. George Allen and Unwin Ltd., 2nd edition.

[von Neumann, 1928] von Neumann, J. (1928). Zur theorie der gesellschaftsspiele. *Mathematische Annalen*, 100:295–320.

[von Neumann and Morgenstern, 1944] von Neumann, J. and Morgenstern, O. (1944). *Theory of Games and Economic Behavior*. Princeton University Press, Princeton, NJ. 2nd edition published 1947.

[von Wright, 1951] von Wright, G. H. (1951). *An Essay in Modal Logic*. North-Holland, Amsterdam.

[von Wright, 1957] von Wright, G. H. (1957). *Logical Studies*. Routledge and Kegan Paul, London.

[Waismann, 1951] Waismann, F. (1951). Verifiability. In Flew, A., editor, *Logic and Language*, pages 117–144. Basil Blackwell, Oxford. 1st series.

[Walton, 1992] Walton, D. (1992). *Slippery Slope Arguments*. Clarendon Press, Oxford.

[Wang, 1955] Wang, H. (1955). On formalization. *Mind*, 64:226–238.

[Wansing, 2011] Wansing, H. Negation (2011). In Goble, L. editor, *Blackwell Guide to Philosophical Logic*. Blackwell, Oxford.

[Weatherson, 2005] Weatherson, B. (2005). True, truer, truest. *Philosophical Studies*, 123:47–70.

[Weatherson, 2010] Weatherson, B. (2010). Vagueness as indeterminacy. In [Dietz and Moruzzi, 2010, 77–90].

[Weintraub, 2004] Weintraub, R. (2004). On sharp boundaries for vague terms. *Synthese*, 138:233–245.

[Weirich, 1980] Weirich, P. (1980). Conditional utility and its place in decision theory. *The Journal of Philosophy*, 77:702–715.

[Weirich, 1984] Weirich, P. (1984). The St. Petersburg gamble and risk. *Theory and Decision*, 17:193–202.

[Weirich, 1986] Weirich, P. (1986). Expected utility and risk. *British Journal for the Philosophy of Science*, 37:419–442.

[Weirich, 2001] Weirich, P. (2001). *Decision Space: Multidimensional Utility Analysis*. Cambridge University Press, Cambridge.

[Weirich, 2004] Weirich, P. (2004). *Realistic Decision Theory: Rules for Nonideal Agents in Nonideal Circumstances*. Oxford University Press, New York.

[Weirich, 2009] Weirich, P. (2009). Does collective rationality entail efficiency. *Logic Journal of the IGPL*. DOI: 10.1093/jigpal/jzp064.

[Weirich, 2010a] Weirich, P. (2010a). *Collective Rationality: Equilibrium in Cooperative Games*. Oxford University Press, New York.

[Weirich, 2010b] Weirich, P. (2010b). Probabilities in decision rules. In Eells, E. and Fetzer, J. H., editors, *The Place of Probability in Science*. Springer, New York.

[Weirich, 2010c] Weirich, P. (2010c). Utility and framing. *Synthese. Realistic Standards for Decisions*, Special Issue edited by Paul Weirich.

[Wheeler, 1979] Wheeler, S. S. (1979). On that which is not. *Synthese*, 41:155–194.

[Whitehead, 1929] Whitehead, A. N. (1929). *Process and Reality*. Macmillan, New York.

[Whitehead and Russell, 1910] Whitehead, A. N. and Russell, B. (1910). *Principia Mathematica*, volume I. Cambridge University Press, Cambridge, 2nd, 1925 edition.

[Whitehead and Russell, 1925] Whitehead, A. N. and Russell, B. (1925). *Principia Mathematica*. Cambridge University Press, Cambridge, 2nd edition. 3 volumes.

[Williamson, 2010] Williamson, J. (2010). *In Defense of Objective Bayesianism*. Oxford University Press, Oxford.

[Williamson, 1986] Williamson, T. (1986). Criteria of identity and the axiom of choice. *The Journal of Philosophy*, 861:380–394.

[Williamson, 1994] Williamson, T. (1994). *Vagueness*. Routledge, London.

[Williamson, 1995] Williamson, T. (1995). Definiteness and knowability. *Southern Journal of Philosophy*, (suppl.) 33:171–192.

[Williamson, 1996a] Williamson, T. (1996a). Knowing and asserting. *The Philosophical Review*, 105:489–523.

[Williamson, 1996b] Williamson, T. (1996b). Putnam on the sorites paradox. *Philosophical Papers*, 25:47–56.

[Williamson, 1997a] Williamson, T. (1997a). Imagination, stipulation and vagueness. *Philosophical Issues*, 8:215–228.

[Williamson, 1997b] Williamson, T. (1997b). Replies to commentators. *Philosophical Issues*, 8:255–265.

[Williamson, 1999] Williamson, T. (1999). On the structure of higher-order vagueness. *Mind*, 108:127–144.

[Williamson, 2000] Williamson, T. (2000). *Knowledge and Its Limits*. Oxford University Press, Oxford.

[Williamson, 2002] Williamson, T. (2002). Epistemicist models: Comments on Gómez-Torrente and Graff. *Philosophy and Phenomenological Research*, 64:143–150.

[Williamson, 2003a] Williamson, T. (2003a). Everything. In Hawthorne, J. and Zimmerman, D. W., editors, *Philosophical Perspectives 17: Language and Philosophical Linguistics*. Blackwell, Boston and Oxford.

[Williamson, 2003b] Williamson, T. (2003b). Vagueness in reality. In Loux, M. J., and Zimmermann, D. W., editors, *The Oxford Handbook of Metaphysics*, pages 690–715. Oxford University Press, Oxford.

[Williamson, 2007a] Williamson, T. (2007a). Evidence in philosophy. In [Williamson, 2007c, 208–246].

[Williamson, 2007b] Williamson, T. (2007b). Must do better. In [Williamson, 2007c, 278–292].

[Williamson, 2007c] Williamson, T. (2007c). *The Philosophy of Philosophy*. Blackwell, Oxford.

[Wittgenstein, 1953] Wittgenstein, L. (1953). *Logical Investigations*. Basil Blackwell, Oxford.

[Woodruff, 1970] Woodruff, P. (1970). Logic and truth-value gaps. In Lambert, K., editor, *Philosophical Problems in Logic*. Reidel, Dordrecht.

[Woods, 1997] Woods, M. (1997). *Conditionals*. Oxford University, Oxford.

[Wright, 1976] Wright, C. (1976). Language mastery and the sorites paradox. In Evans, G. and McDowell, J., editors, *Truth and Meaning: Essays in Semantics*, pages 223–247. Oxford University Press, Oxford.

[Wright, 1987] Wright, C. (1987). Further reflections on the sorites paradox. *Philosophical Topics*, 15:227–290.

[Wright, 1992] Wright, C. (1992). Is higher-order vagueness coherent? *Analysis*, 52:129–139.

[Wright, 2001] Wright, C. (2001). On being in a quandary: relativism, vagueness, logical revisionism. *Mind*, 60:45–98.

[Wright, 2007] Wright, C. (2007). On quantifying into predicate position. In Leng, M., Paseau, A., and Potter, M., editors, *Mathematical Knowledge*, pages 150–174. Oxford University Press, Oxford.

[Wright, 2010] Wright, C. (2010). The illusion of higher-order vagueness. In [Dietz and Moruzzi, 2010, 523–549].

[Yablo, 1982] Yablo, S. (1982). Grounding, dependence, and paradox. *Journal of Philosophical Logic*, 11:117–137.

[Yalcin, 2007] Yalcin, S. (2007). Epistemic modals. *Mind*, 116(464):983–1026.

[Yoes Jr., 1967] Yoes Jr., M. G. (1967). Nominalism and non-atomic systems. *Noûs*, 1:193–200.

[Zalta, 1983] Zalta, E. N. (1983). *Abstract Objects: An Introduction to Axiomatic Metaphysics*. Reidel, Dordrecht.

[Zardini, 2008] Zardini, E. (2008). A model of tolerance. *Studia Logica*, 90:337–368.

[Zeman, 1973] Zeman, J. J. (1973). *Modal Logic: The Lewis-Modal Systems*. Clarendon, Oxford.

[Zermelo, 1930] Zermelo, E. (1930). Über Grenzzahlen und Mengenbereiche. *Fundamenta Mathematicae*, 16:29–47. Translated in [Ewald, 1996].

[Zynda, 2000] Zynda, L. (2000). Representation Theorems and Realism about Degrees of Belief. *Philosophy of Science*, 67(1):45–69.

General Index

σ-algebra 408, 452
σ-additive probability measure 453
stit logic 348
$Ł_3$ 187
de re/de dicto distinction 533
sui generis epistemic probability 424
$\mathbf{K}_t$ 331
T 313

abstract situations 204
accessibility relation 326
actualist frequentism 409
Adams' Thesis (about conditionals) 395
additive separation 551
admissible interpretations 151
AGM model 456
AGM postulates 456, 460
alethic modal logic 300
algebra 408, 452
Allais's paradox 565
analyticity 301
antitony 486
Aristotelian modality 342
Arrow's theorem 572
aspect 346
Atom Exchangeability Principle 436
atom of an algebra 408
atomic algebra 408
Axiom of Choice 109
axiomatization stage 5
axioms of modal logic 508

backwards linear 328
Barcan formula 318
basic intrinsic attitude 551
Bayesian conditionalization 420
Becker's rule 304
being 64, 65
belief base contraction 486
belief basis 455
belief revision 516
belief set 454, 457
best-system analysis of probability 417
BHK interpretation 196
BIT 551
boolean algebra 184, 284
borderline vague 134
branching space-time 347
branching time 339
Brouwer's axiom (for modal logic) 526
Brouwer–Heyting–Kolmogorov interpretation 197

calculus of individuals 272
canonical model 312
canonical terms 320
categoricity (of a theory) 111
causal decision theory 554
chance 418
circular time 332
classical negation 183
coherence constraint 473
common knowledge 512, 570
compactness (for first-order logic) 110
Compactness (of first-order logic) 41
complete propositional selection function 475
completed infinity 196
completeness 6
completeness (for first-order logic) 110
completeness stage 6
Completeness Theorem (for first-order logic) 38
completeness theorem for first-order logic 7
completeness theorem for modal logic 309

Compositional theory of truth 363
comprehension axioms (for second-order logic) 108
computer simulations 21
concept 71
conceptual analysis 16
conceptual cover 535
conceptual modelling 21
conditional probability 453
conditional sorites 131
conditionalization 567
conditionals 383
conditioning 453
Condorcet's paradox 572
connectedness 131, 151
Constant Exchangeability Principle 436
constraint 204
construction system 3
constructive proof 196
contextual definition 85
contextualism 151
Continuum of inductive methods 438
contraction 458
converse Barcan formula 318
converse Dutch book theorem 421
cooperative game 571
Coordinated Attack Problem (Halpern) 514
core 493
correspondence 337
correspondence theorem 313
countably additive 409
countably additive probability space 409
course of values 92
criterion of identity 60
Curry's paradox 191
cyclic negation 188

D4 215
de dicto attitude 89
de Finetti Representation Theorem 441
De Morgan negation 213
De Morgan's laws 183
de re attitude 89
decidability 6, 315
decision 544
decision procedure 315
Dedekind continuity 335
Dedekind–Peano arithmetic 290
deductively closed 310
deep structure 85
definite description 16, 57, 79
definite truth 137
degree of belief 419
demonstrability 301
denotation 70, 78
deontic necessity 300
designated values 190
dialetheism 190
dialethism 49
Diodorean modality 342
direct Barcan formula 318
direct control 546
direct discriminability 154
direct reference theory 58
discreteness 335
disquotational theory of truth 357
doxastic commitments 454
Dutch book argument 421
Dutch book theorem 421
dynamically consistent 547

EFQ 189
Ellsberg's paradox 565
Email Game (Rubinstein) 514
empty name 80
entrenchment-based severe withdrawal 484
epistemic entrenchment 461
epistemic necessity 300
epistemic paradoxes 538
epistemic probability 407
epistemic state 496
epistemic utility function 424
epistemicism 144
equilibrium (in a strategic game) 255
equilibrium semantics 219
equivalence relation 55
evidential decision theory 554
evidential situation 197

ex falso quodlibet 181
existence 61, 65
existential commitment 74
existential quantifier 34, 65
expected degree of desire 558
expected utility 549, 553
expected utility interpretation 423
expected-utility analysis 553
explicit definition 85
expressiveness 322
expulsiveness 485
extensive game 219

fair betting odds interpretation 420
falsum 199
field 452
finite model property 315
first moment 334
first-order logic 33
Fitch paradox 538
fixed margin model 145
fixed point 193
FOL 33
formal language stage 4
formal necessity 301
formalization 4
frame 328
free logic 67
Frege–*Grundgesetze* theory of descriptions 92
Frege–Carnap theory of descriptions 93
Frege–Hilbert theory of descriptions 90
full control 546
full meet contraction 459

Gödel coding 352
Gödel Incompleteness Theorem (for Peano Arithmetic) 41
Gödel numbering 352
Gödel's diagonal lemma 356
Gale–Stewart theorem 221
game of strategy 547
game theory 547
game-theoretical semantics 221
generalized quantifier 100
global constraint 204
grammatical mood 300
Grove connection 468
grue paradox 19, 429

Henkin model 109
Henkin semantics (for second-order logic) 109
Henkin, Leon 109
hereditariness property 198
higher-order logic 44
higher-order vagueness 140
historical modality 341
history 339
Humphrey's paradox 416
hypothetical frequentism 412

identity 55
identity of indiscernibles 60
IF logic 218
imaging 495
implosion principle 157
Import–Export Principle (for nested conditionals) 387
impossible situation 204
impossible worlds 524
improper description 90
incomplete symbol 84
indefinite description 79
Independence-Friendly logic 218
indicative conditional 384
indifference relation 131
indirect discriminability 154
indiscernibility of identicals 55
indiscernibility of identicals with respect to properties 56
inductive logic 428
infinite utility 562
infinite valued logics 187
information 203
information preservation 207
informational value 480
intensional context 56, 82
intensity of preference 545
interactive epistemology 510
intrinsic desire 551
intrinsic utility 551

intrinsic-utility analysis 551
intuitionism 49
intuitionist logic 196
intuitionist mathematics 195
is-ness 65

Jeffrey conditioning 454
Johnson's Sufficientness Principle 437
judgement aggregation 572
justification stage 5

K_3 186
kernel contraction 488
KK principle 526
Kleene's strong three-valued logic 185
knowing-wh 536
Kreisel counterexample 232
Kreisel form 232
Kreisel's squeezing argument 115
Kripke's theory of truth 370

Löwenheim–Skolem Theorem (for FOL) 41, 110
last moment 334
law of the excluded middle 48, 181
left linearity 333
Leibniz' law 87
Levi contraction 480
Levi identity 462
Lewis' triviality results (for Stalnaker's hypothesis) 393
liar paradox 352
liar sentence 352
limiting relative frequency 412
Lindenbaum's lemma 311
linear 328
linearity 333
linguistic necessity 301
local constraint 205
logic of agency 348
logical analysis 103
logical consequence 29
logical form 85
logical necessity 301
logical probability 407
logical terms 42
logically perfect language 90
long-run propensity interpretation 413
lottery paradox 150
LP 190
Łukasiewicz's three-valued logic 187

main lemma 311
majoritarian methods 572
many-valued logic 185
margin of error 146
Margin or error semantics (for epistemic logic) 527
Matching Pennies 561
material conditional 385
mathematical modelling 1
mathematical probability 408
maxichoice contraction 459
maximal consistent 310
maximal proposition 548
Meinongian object 81
mereological algebra 284
mereological atoms 278
mereological complement 276
mereological fusions 272
mereologlical sums 272
mereology 271
mereotopology 294
metaphysical necessity 300
metric operator 343
minimal logic 202
minimal modal logic 304
Minimax theorem 219
mixed strategy (in a game) 258
model 18
model (of first-order / predicate logic) 36
model theory 182
modes of being 74
moment of context 343
moment of evaluation 343
Moore's paradox 504, 538
multi-agent systems 510
multiattribute-utility analysis 550

narrow scope 87
Nash equilibrium 514, 547
necessitation rule 304
necessity of distinctness 319
necessity of identity 318
negation 181
neo-Meinongian theory 81
Newcomb's problem 563
nihilism 133
non-existent entities 67
noncooperative game 570
normal bimodal logic 331
now 342

Ockhamist temporal logic 340
ontological commitment 74
operator 183
option's world 548

package principle 422
paracomplete logic 156
paraconsistent logic 157, 189
paradox of the absent-minded driver 566
paradoxes of material implication 208, 390
Parmenides' paradox 62
part–whole relation 271
partial definition 139
partial information 203
partial meet contraction 458
partial meet revision 463
partition invariance 555
Pasadena gamble 562
Peano arithmetic 290
Peano arithmetic (second-order) 111
Peircean temporal logic 339
persistent revision 471
Pettit, Philip 572
physical probability 407
platonism 195
plural logic 112
plural quantification 47
Popper function 409
possible world 315
possible worlds models 18
potential infinity 196
potential relative to 311
practical reasoning 544
predicate calculus 33
predicative 109
preference ranking 545
presupposition 185
Principal Principle 417
principle of bivalence 181
principle of consistency 184
principle of dominance 558
principle of double negation 185
Principle of Instantial Relevance 437
principle of non-contradiction 184
principle of ratification 561
Priorean tense logic 328
Prisoners' dilemma 252
priviliged terms 320
probabilism 420
probabilistic models 19
probability measure 452
probability space 408
Proof of Completeness Theorem for First-order Logic 39
proof theory 182
propensity 415
proper description 85
proper name 78
property 60
propositional calculus 31
propositional choice-based revision function 475
propositional function 70
propositional selection function 465
provability logic 316
public announcement logic 518
pure time preference 546

quality 60
quantified modal logic 317
quantifier 77
quantization 559
quasi-categoricity (of ZF_2) 114
quasi-order 501

R 208
Ramsey, F. P. 398
rational selection function 473
reference failure 80

reference sequence 413
reference type 410
reflection principle (in set theory) 115
Reichenbach's Axiom 439
relational theory of meaning 203
relative frequency 410
relative identity 61
relative interpretation 288
relative possibility 308, 326
relevant logic 207
remainder set 458
replacement rule 305
representation theorem 423
restricted conditional probability space 409
Revision theory of truth 378
rigidity 320
risk aversion 564
Robinson arithmetic 290
rollback equilibrium 547
Routley Meyer semantics 215
Routley star operator 213
Russell's paradox 45
Russell's theory of descriptions 69, 83

S4 313
S5 313
safe contraction 488
satisficing 559
saturable set 479
Savage's representation theorem 555
scope 86
scope indicator 86
scoring rule 424
second-order logic 107
selection function 459
self-ratifying 561
Self-Sampling Assumption 567
semantic stage 5
semantical game 222
sense 70
sentential calculus 31
separation 551
sequential game 547
serial frame 334
set theory 115
severe withdrawals 460, 483
sharpening of a language 166
signalling 241
since 344
single-case propensity interpretation 415
singular term 62, 77
situation semantics 203
Skolem form 229
Skolem functions 218
Skolemization 229
Sleeping Beauty problem 567
slingshot argument 96
social choice theory 572
sorites condition 131
sorites paradox 130
soundness 5
soundness theorem for modal logic 309
sphere semantics 466
sphere-based revision 466, 470
St. Petersburg gamble 562
Stalnaker's Hypothesis 393
standard semantics (for second-order logic) 109
standard translation 337
state 553
state description 433
statement 91
strategic game 252
strategy profile 547
strengthened liar sentence 190
strict implication 302
strict partial ordering 328
strong completeness 331
Strong Kleene logic 160
Strong Kleene scheme 371
strong modal operator 326
strongly paracomplete 158
strongly paraconsistent 158
subjective Nash equilibrium 571
subjunctive conditional 384
Substitution Lemma (for first-order logic) 230
substitutivity of identity 55
supertruth 166
supervaluation 341

supervaluation scheme 375
supervaluationism 165
Surprise Examination paradox 539
syllogism 6, 29

Tarski-biconditionals 352
temporal Barcan formula 348
temporal predicate logic 347
Tennant, Neil 471
tense 346
textbfK 304
The Analogy Principle 446
tolerance 131
trace 101
tractability 322
transformative analysis 103
trichotomy 328, 333
trump semantics (for IF logic) 243
truth tables (for the sentential calculus) 32
truth-functional 82
two-envelope paradox 568

universal quantifier 34
universally free logic 67
until 344
utility 421, 545, 548
utility function 545
utility maximization 545

vagueness 48, 129
validity 301
valuation 31
variable assignment (for first-order logic) 37
variable margin model 145
variable sharing property 208

weak completeness 331
Weak Irrelevance Principle 436
weak modal operator 326
weakly paracomplete 158
weakly paraconsistent 158
wide scope 86

Zermelo–Fraenkel set theory 114

Author Index

Alchourrón, C.E. 456
Anderson, Alan Ross 73
Anselm of Canterbury 348
Aristotle 2, 4, 6, 29, 184, 325, 342
Arló-Costa, Horacio 493, 494
Aumann, R. 504
Ayer, Alfred 103

Barwise, Jon 96, 100
Belnap, N. 50, 192, 347, 348
Bernoulli, Daniel 562
Beziau, Jean-Yves 194
Binmore, Kenneth 548, 556, 564
Black, Max 60
Boole, G. 3, 4, 31
Boolos, G. 47, 113
Bostrom, Nick 567, 568
Boutilier, Craig 498
Broome, John 569
Brouwer, L. E. J. 49, 215

Carnap, R. 3, 90, 316, 429
Chomsky, Noam 101
Church, A. 41, 70
Cooper, Robin 100

Darwiche, Adnan 496
Diodorus Cronus 342
Dummett, Michael 196
Dunn, Michael 192

Easwaran, Kenneth 562
Edgington, Dorothy 164
Elga, Adam 560, 567

Føllesdal, Dagfinn 96
Fara, Delia 134, 143, 155, 169
Fetzer, James 416
Fine, Kit 165
Frege, G. 3, 4, 7, 33, 59, 70, 78, 79, 90, 91, 103, 326

Gärdenfors, Peter 456, 478, 495
Gödel, K. 4, 7, 39, 96, 99, 303
Gaifman, Haim 135
Geach, Peter 61
Gentzen, G. 50
Goldblatt, Robert 206
Goldszmidt, Moises 500
Good, Irving 559
Goodman, Nelson 3, 272
Greaves, Hilary 426
Grove, Adam 465

Hájek, Alan 562
Hamblin, Charles 335
Hansson, Sven 478, 486, 488
Harper, William 456
Henkin, L. 217
Heyting, Arend 196
Hintikka, J. 217, 504
Hyde, Dominic 141

Jaśkowski, Stanisław 194
Jeffrey, Richard 454, 544, 554, 557
Johansson, Ingebringt 202
Joyce, James 426, 554, 557

Kalish, Donald 104
Kant, Immanuel 319
Kaplan, David 100
Katsuno, Hirofumi 495
Keefe, Rosanna 165
Keeney, Ralph 550
Kelly, Kevin 500
Keynes, John 557
Kleene, Stephen 160, 185
Kolmogorov, Andrey 197, 453
Koons, Robert 142
Kreisel, Georg 115
Kripke, S. 18, 58, 59, 78–80, 197, 300, 313, 320, 331, 370

Leibniz, G. W. 31, 59, 60, 315
Leitgeb, Hannes 426, 493
Lemmon, John 331
Leśniewski, Stanisław 272
Levi, I. 456, 479, 480
Lewis, Clarence 301, 302, 326
Lewis, D. 241, 276, 417–19, 426, 495
Lindström, Sten 472
List, Christian 572
Łukasiewicz, Jan 163, 187

Makinson, David 456, 478
Meinong, Alexius 76
Mendelzon, Alberto 495
Meyer, Robert 215
Miller, David 416
Montague, Richard 100, 104
Moore, G. E. 538

Neale, Stephen 101
Nover, Harris 562

Pagnucco, M. 484
Parikh, Rohit 470
Parmenides 62
Pearl, Judea 496, 500
Pedersen, Arthur 494
Perry, John 61, 96, 203
Pettigrew, Richard 426
Piccione, Michele 566
Plato 73
Popper, Karl 414, 416
Post, Emil 188
Prawitz, Dag 201
Priest, Graham 157, 162, 190, 195
Prior, Arthur 99, 325, 326, 328, 337, 339, 342, 350

Quine, W. V. O. 3, 119

Raiffa, Howard 550
Rott, Hans 462, 465, 472, 475, 476, 478, 484, 489, 491, 498, 501
Routley, Richard 213, 215
Routley, Val 213
Rubinstein, Ariel 566
Russell, Bertrand 3, 4, 16, 57, 64, 72, 78, 81–4, 86, 87, 103
Ryle, Gilbert 85

Sainsbury, Mark 139
Savage, Leonard 555
Sharvy, Richard 99
Shimony, Abner 557
Simon, Herbert 559
Slote, Michael 73
Smullyan, Arthur 98
Sorensen, Roy 141
Spinoza, Baruch 326
Spohn, Wolfgang 456
Stalnaker, Robert 492
Strawson, Peter 91

Tarski, A. 4, 5, 7, 18, 37, 352
Thomason, Richmond 340

Unger, Peter 133

Van Deemter, Kees 153
van Fraassen, Bas 492
Von Wright, Henrik 300

Wallace, David 426
Weatherson, Brian 139
Weirich, Paul 551, 558
Williamson, Timothy 140, 144, 148, 168
Wright, Crispin 142

Philosophy